Handbook of

TRANSFUSION MEDICINE

Handbook of
TRANSFUSION
MEDICINE

Edited by

Christopher D. Hillyer, M.D.

Professor, Department of Pathology and Laboratory Medicine,
Emory University School of Medicine; and Director, Transfusion
Medicine Program, Emory University, Atlanta, Georgia

Frank J. Strobl, M.D., PhD

Assistant Professor, Department of Pathology
and Laboratory Medicine, University of Pennsylvania
School of Medicine, Philadelphia, Pennsylvania

Krista L. Hillyer, M.D.

Medical Director, American Red Cross Blood Services, Southern Region;
and Assistant Professor, Department of Pathology and Laboratory
Medicine, Emory University School of Medicine, Atlanta, Georgia

Leigh C. Jefferies, M.D.

Associate Professor, Pathology and Laboratory Medicine;
and Associate Director, Blood Bank / Transfusion Medicine Section,
University of Pennsylvania Medical Center, Philadelphia, Pennsylvania

Leslie E. Silberstein, M.D.

Professor, Department of Pathology, Harvard Medical School;
and Director, Joint Program in Transfusion Medicine Program at Children's Hospital Boston,
Brigham and Women's Hospital, Dana-Farber Cancer Institute, Boston, Massachusetts

ACADEMIC PRESS

A Harcourt Science and Technology Company

San Diego   San Francisco   Boston   New York   London   Sydney   Tokyo

Academic Press
A Harcourt Science and Technology Company
525 B Street, Suite 1900, San Diego, California 92101-4495, USA
http://www.academicpress.com

Academic Press
Harcourt Place, 32 Jamestown Road, London NW1 7BY, UK
http://www.academicpress.com

Library of Congress Catalog Card Number: 2001090828

International Standard Book Number: 0-12-348775-7

PRINTED IN THE UNITED STATES OF AMERICA
01 02 03 04 05 06 QW 9 8 7 6 5 4 3 2 1

Contents

SECTION

IV

SPECIALIZED COMPONENT PROCESSING OR TESTING

SECTION

V

HEMATOPOIETIC STEM CELLS AND RELATED CELLULAR PRODUCTS

SECTION

VI

SPECIALIZED TRANSFUSION SITUATIONS

Books are to be returned on or before
the last date below.

Contributors

Numbers in parentheses indicate the pages on which the authors' contributions begin.

Kirsten Alcorn (333), Blood Bank, Washington Hospital Center, Washington, DC 20010

Mary Beth Allen (19), Department of Pathology, Emory Hospital, Atlanta, Georgia 30322

Chester Andrzejewski, Jr. (115, 323), Department of Pathology, Baystate Medical Center, Springfield, Massachusetts 01199

Sheilagh Barclay (91), Transfusion Medicine Program, Emory University School of Medicine, Atlanta, Georgia 30307

Joanne L. Becker (333), Roswell Park Cancer Institute, Buffalo, New York 14263

Douglas P. Blackall (189), Department of Pathology & Medicine, UCLA Division Transfusion Medicine, UCLA Medical Center, Los Angeles, California 90095

Scott R. Burger (171), Research & Development, MERIX Bioscience, Inc., Durham, North Carolina 27704

Silvana Z. Bucur (39, 47, 79), Emory University School of Medicine, Atlanta, Georgia 30322

Raymond L. Comenzo (137, 143), Blood Bank Boston Medical Center, Boston University School of Medicine, Boston, Massachusetts 02118

Joan Debelak (179), Department of Laboratory Medicine, School of Medicine, Yale University, New Haven, Connecticut 06520-8035

Walter Dzik (125), Blood Transfusion Service, Massachusetts General Hospital, Harvard Medical School, Boston, Massachusetts 02114

Anne F. Eder (253), Children's Hospital of Philadelphia, Philadelphia, Pennsylvania 19104-4399

Cherie S. Evans (349), American Red Cross-BBACCC, Oakland, California 94618

David F. Friedman (275), Children's Hospital of Philadelphia, Philadelphia, Pennsylvania 19104

Theresa W. Gillespie (63), Winship Cancer Institute, Emory University School of Medicine, Atlanta, Georgia 30322

Christopher D. Hillyer (29, 39, 47, 53, 63), Department of Pathology & Laboratory Medicine, Transfusion Medicine Program, Emory University, Atlanta, Georgia 30322

Krista L. Hillyer (29), Department of Pathology & Laboratory Medicine, Emory University, Atlanta, Georgia 30322

Diane S. Krause (179), Department of Laboratory Medicine, School of Medicine, Yale University, New Haven, Connecticut 06520-8035

Jonathan D. Kurtis (301), Department of Pathology & Laboratory Medicine, University of Pennsylvania Medical Center, Philadelphia, Pennsylvania 19104-6523

Peter J. Larson (263), Abramson Research Center, Children's Hospital of Philadelphia, Philadelphia, Pennsylvania 19104

Cathy Litty (3, 11), Transfusion Medicine, Department of Pathology and Laboratory Medicine, St. Christopher's Hospital for Children, Philadelphia, Pennsylvania 19134

Catherine S. Manno (231), Transfusion Medicine, Children's Hospital of Philadelphia, Philadelphia, Pennsylvania 19104

Una O'Doherty (293), Department of Pathology & Laboratory Medicine, University of Pennsylvania Medical Center, Philadelphia, Pennsylvania 19104

Eleanor S. Pollak (197), Abramson Research Center, Children's Hospital of Philadelphia, Philadelphia, Pennsylvania 19104

Eline T. Luning Prak (209), Department of Pathology & Laboratory Medicine, University of Pennsylvania Medical Center, Philadelphia, Pennsylvania 19104

Maria Jose A. Ribeiro (69), Division of Hematology & Oncology, Emory University, Atlanta, Georgia 30322

John D. Roback (53, 129, 285), Department of Pathology & Laboratory Medicine, Emory University, Atlanta, Georgia 30322

Nancy C. Rose (221), Obstetrics & Gynecology, Maternal/Fetal Medicine Division, University of Pennsylvania Medical Center, Philadelphia, Pennsylvania 19104

Karen S. Roush (197), Department of Pathology & Laboratory Medicine, Emory University School of Medicine, Atlanta, Georgia 30322

Bruce Sachais (307), Department of Pathology, University of Pennsylvania Medical Center, Philadelphia, Pennsylvania 19104

Kathleen Sazama (355, 359), Pathology & Laboratory Medicine, MCP Hahneman University, Philadelphia, Pennsylvania 19102-1192

Deborah A. Sesok-Pizzini (247, 259), University of Pennsylvania Medical Center, Philadelphia, Pennsylvania 19104

Hua Shan (237), Department of Pathology and Laboratory Medicine, MUSC/Children's Hospital, Charleston, South Carolina 29425-0707

Don Siegel (107), Pathology and Laboratory Medicine, University of Pennsylvania Medical Center, Philadelphia, Pennsylvania 19104-4283

Steve Sloan (149, 161), Blood Bank, Children's Hospital Boston, Boston, Massachusetts 02115

Christopher P. Stowell (345), Blood Transfusion Service, Massachusetts General Hospital, Boston, Massachusetts 02114-2696

Frank J. Strobl (301, 315), Pathology and Laboratory Medicine, University of Pennsylvania Medical Center, Philadelphia, Pennsylvania 19104-4283

William J. Swiggard (293), Department of Pathology & Laboratory Medicine, University of Pennsylvania Medical Center, Philadelphia, Pennsylvania 19104

Mark A. Valleca (179), Department of Laboratory Medicine, School of Medicine, Yale University, New Haven, Connecticut 06520-8035

David Wuest (137, 143), Hematology Service, Memorial Sloan-Kettering Cancer Center, New York, New York 10021

Preface

The *Handbook of Transfusion Medicine* is a valuable resource for anyone who transfuses blood products and derivatives. It is an up-to-date, practical, easy-to-read, and comprehensive but cutting-edge manual, textbook, and reference. The *Handbook's* clinical approach allows it to be effective for any practitioner, including those in surgery, anesthesia, pediatrics, hematology, oncology, and others, in addition to the transfusion medicine specialist. Its breadth of information, concise, clear writing style, and attention to relevant technical detail allow the *Handbook* to be an excellent starting point and an ongoing resource for medical and technical students and medical, pediatric, and pathology residents and fellows. We have constructed the book's style and design to accentuate its utility as a comprehensive textbook, as well as a quick reference. Chapters are organized into 10 sections that lead, focus, and aid the reader and allow him or her to acquire specific information as well as to grasp broad concepts in transfusion medicine. These sections include blood collection and testing, blood component description, preparation and usage, red blood cell antigens and antibodies, specialized component processing, hematopoietic and other stem cells, specialized transfusion situations, transfusion reactions, infectious complications of transfusion, therapeutic apheresis, and quality and other related issues. Specifically, acute bleeding and massive transfusion, the transfusion of the patient with a coagulopathy, and the transfusion of obstetrics, pediatric, immunocompromised, and platelet refractory patients are addressed; these sections allow the transfusion practitioner to become knowledgeable and comfortable with these special transfusion situations. Great attention has been given to making this book easy to read, focused, and timely, and current references are provided to allow the reader to access additional information as desired. We hope you will use this *Handbook* frequently in your practice. Any comments or input will always be welcomed and appreciated.

C.D.H.
Atlanta, GA

Acknowledgments

The editors acknowledge the aid and support given to us by a number of key individuals. These include Sue Rollins and Sherry Ballenger for their outstanding technical and professional support; Ms. Hilary Rowe, Academic Press, for her continued patience, expertise, and guidance; and our families—Krista, Whitney, Peter, Margot, Pete, and Sally Hillyer; Chris Hillyer and Jim and Berthene Lankford; Patricia, Hannah, Jakob, Anna, and Jakob Strobl; and Henry, Alex, and Isadora Weinfeld and Donald and Carmela Jefferies—for their love and personal support. The editors dedicate this *Handbook* to all of our friends, trainees, colleagues, and mentors who share in our goal of advancing the field of transfusion medicine.

List of Abbreviations

AHTR	Acute Hemolytic Transfusion Reaction
AIHA	Autoimmune Hemolytic Anemia
ALL	Acute Lymphocytic Leukemia
AML	Acute Myelocytic Leukemia
ANH	Acute Normovolemic Hemodilution
ARDS	Adult Respiratory Distress Syndrome
ATL	Adult T-cell Leukemia/Lymphoma
BM-HPC	Bone-marrow-derived Hematopoietic Stem Cells
BMT	Bone Marrow Transplant
CJD	Creutzfeldt-Jakob Disease
CLL	Chronic Lymphocytic Leukemia
CML	Chronic Myelocytic Leukemia
CMV	Cytomegalovirus
CTL	Cytotoxic T-lymphocyte
DAT	Direct Antiglobulin Test
DDAVP	Desmopressin
DHTR	Delayed Hemolytic Transfusion Reaction
DIC	Disseminated Intravascular Coagulopathy
DLI	Donor Leukocyte Infusion
DMSO	Dimethyl Sulfoxide
EBV	Epstein-Barr Virus
ECP	Extracorporeal Photochemotherapy
EIA	Enzyme Immuno Assay
EPO	Erythropoietin
FFP	Fresh Frozen Plasma
FNHTR	Febrile Non-hemolytic Transfusion Reaction
FSP	Fibrin-split Products
G-CSF	Granulocyte Colony Stimulating Factor
GM-CSF	Granulocyte-Monocyte Colony Stimulating Factor
GTX	Granulocyte Tranfusion
GVHD	Graft-versus-host Disease
HBIg	Hepatitis B Immunoglobulin
HBV	Hepatitis B Virus
HCV	Hepatitis C Virus
HDN	Hemolytic Disease of the Newborn
HIT	Heparin-induced Thrombocytopenia
HLA	Human Leukocyte Antigen
HPA	Human Platelet Antigen
HPC	Hematopoietic Progenitor Cell
IAT	Indirect Antiglobulin Test
ISIg	Immune Serum Globulin
ITP	Idiopathic Thrombocytopenic Purpura
IVIg	Intravenous Immunoglobulin
MHC	Major Histocompatibility Complex
MNC	Mononuclear Cells
NAIT	Neonatal Alloimmune Thrombocytopenia
NAT	Nucleic Acid Testing
NDMP	National Donor Marrow Program
nvCJD	New-variant Creutzfeldt-Jakob Disease
PB-HPC	Peripheral-blood-derived Hematopoietic Progenitor Cells
PCH	Paroxysmal Cold Hemoglobinuria
PCR	Polymerase Chain Reaction
PNH	Paroxysmal Nocturnal Hemoglobinuria
PPF	Plasma Protein Fraction
PRBC	Packed Red Blood Cells
RhIg	Rh Immunoglobulin
RIBA	Recombinant Immunoblot Assay
SCD	Sickle Cell Disease
SD-plasma	Solvent-detergent Plasma
TA-GVHD	Transfusion-associated Graft-versus-host Disease
TPE	Therapeutic Plasma Exchange
TPO	Thrombopoietin
TRALI	Transfusion-related Acute Lung Injury
TRIM	Transfusion-related Immunomodulation
TTP	Thrombotic Thrombocytopenic Purpura
UCB	Umbilical Cord Blood
VZIg	Varicella-Zoster Immunoglobulin

SECTION I

BLOOD COLLECTION, TESTING, AND PROCESSING

Introduction to Blood Donors and Donation

CATHY LITTY

St. Christopher's Hospital for Children
Philadelphia, Pennsylvania

Most of the blood used in the United States is collected from volunteer donors at specialized blood centers. The American Red Cross (ARC) is the largest supplier of blood components in the United States and is represented by 38 individual regions that collect almost 50% of the total U.S. blood supply. Approximately 47% of the nation's blood supply is drawn by independent community blood centers, and the remaining 3% is drawn at hospital collection sites and U.S. military facilities. In 1997, 12.5 million units of blood were collected in the United States, of which 11.9 million were allogeneic donations. The remainder was autologous blood collected from patients prior to surgery for their own individual uses. During that same year, 11.4 million units were transfused (Goodnough *et al.*, 1999). Because of the different blood groups and types that are needed by patients at certain times, this nearly 1-million-unit difference (12.5 − 11.4 million = 0.9 million) does not represent a surplus of blood in the United States for the year 1997. In fact, a nationwide blood shortage has been predicted to occur by 2002.

Substantial advances in the safety of the blood supply have been made in recent years, owing to expanded health screening of donors and an increased number and quality of infectious disease tests performed on donated blood. Infectious disease testing requirements for donated blood products are discussed in detail in Chapter 2. This chapter focuses on blood donor medical history and physical requirements, the donation process, and special categories of donors.

I. REGULATORY AGENCIES

In the United States, the Food and Drug Administration (FDA) regulates the blood industry. Blood centers must be licensed by the FDA in order to collect and distribute blood products. The FDA is concerned with all aspects of blood center operations that affect the safety, quality, identity, potency, or purity of blood components. The blood industry is responsible for upholding regulations for biologic products provided in the Code of Federal Regulations (CFR), as well as Good Manufacturing Practice (GMP) regulations, originally devised for the drug industry. The American Association of Blood Banks (AABB) is a professional association and accrediting organization with a voluntary membership. *Standards for Blood Banks and Transfusion Services* (Menitove, 2000), published by the AABB, details specific requirements for donor screening, health history exams, and infectious disease testing.

II. VOLUNTEER DONORS

A. General Recruitment

Blood centers depend on volunteer donors to provide blood and blood components to meet their communities' needs. Money or substantial gifts are not offered in return for donation, as such payment could be an incentive to be dishonest regarding the health history, thus defeating the purpose of this portion of the screening process. Successful recruitment of blood donors is dependent upon many factors, including community education, donor motivation, and convenience. It is estimated that less than 5% of the population of the United States donates blood; yet this represents the donor source for the entire country. Low blood donation rates may be influenced by fear of needles, the unfounded fear of AIDS transmission by blood donation, crowded donor sites, lengthiness of the donation process, or history of an adverse donor reaction. Blood center personnel work to make donation a pleasant experience for donors to ensure that they return. Regular, repeat blood donors are desirable, since rates for positive infectious disease tests have been reported to be lower in this group as compared to first-time blood donors (Starky *et al.*, 1989).

B. Recruitment of Specific Donor Populations

1. Group O Donors

Recruitment specifically for group O donors is often conducted in addition to general recruitment. Group O blood is used in excess of its representation among recipient blood groups, as group O packed red blood cells (PRBCs) may be used safely in recipients who are either group A, B, or AB.

Donors who are O, Rh-negative are in even greater demand. Group O, Rh-negative PRBCs are frequently used for trauma patients with unknown blood types who need uncrossmatched blood emergently in order to avoid Rh immunization of Rh-negative recipients. For women of childbearing age who are Rh-negative, the transfusion of Rh-negative PRBCs is desirable to avoid exposure to the Rh(D) antigen in order to prevent the formation of an anti-Rh(D) antibody; this antibody may cause hemolytic disease of the newborn (HDN) in an Rh(D)-positive fetus of a subsequent pregnancy. Group O, Rh-negative PRBCs are also often used for neonatal transfusions, as ABO antigens may not be fully formed at birth (e.g., the blood typing of the neonate may be inaccurate), and also in cases of HDN due to ABO and/or Rh incompatibility.

2. Other Special Donor Categories

Although targeted recruitment is most common for O, Rh-negative donors, other special needs exist for certain populations. Cytomegalovirus (CMV)-negative blood is indicated for patients at risk for CMV disease

(see Chapter 15). IgA-negative plasma components are indicated for patients with IgA deficiency who have antibodies against IgA, placing them at risk for serious anaphylactic reactions (see Chapter 30). Recruitment of donors known to have rare blood types that lack certain RBC antigens (other than ABO) may be necessary in specific cases. Also, group AB patients can only receive AB fresh frozen plasma (FFP) because all other types of plasma will contain anti-A, anti-B, or both. When group AB patients have high-volume plasma needs (as in the case of burns, thrombotic thrombocytopenic purpura [TTP] requiring plasma exchange, etc.), special AB donor recruitment may be necessary because AB plasma is often not available in large quantities, as only 4% of the general population is group AB.

C. Screening

Donor screening is vital to ensure the safety of both the blood supply and the blood donors. Donors are selected based on an abbreviated physical examination, a health history questionnaire, and a hemoglobin measurement. Historical records of donor deferrals and previously positive infectious disease test results (see Chapter 2 for a detailed discussion of infectious disease testing requirements for donated blood) are maintained so that relevant information for each potential donor may be checked against past records. As a result, either past high-risk behavior or test results previously positive for infectious disease markers will prevent a particular donor's blood from being used, even if the same information is not elicited on a current donation.

1. Physical Examination

The physical exam includes weight, vital signs, and inspection of the upper extremities of each potential donor. Five hundred milliliters of whole blood is drawn for the donation and an additional 25 ml is collected for infectious disease testing. The donor must weigh at least 110 pounds so that the amount collected does not exceed 15% of his or her blood volume. Donor temperature must not exceed 99.5°F (36.5°C), as an elevation in temperature above 99.5°F may indicate systemic infection that could lead to bacterial contamination of the blood product. Donor heart rate must be regular and between 50 and 100 beats per minute. The highest acceptable systolic blood pressure is 180 mm Hg, while the highest acceptable diastolic pressure is 100 mm Hg. Heart rate or blood pressure values outside of these ranges may indicate hemodynamic instability or illness that could lead to a donor reaction upon phlebotomy. The donor's upper extremities are examined for skin lesions that may signify infection or raise suspicion for a history of drug use.

2. Health History

The health history questions serve to protect the donor and/or the future recipient. Donor safety questions inquire about illnesses or preexisting conditions that would put the donor at risk of adverse reactions due to the phlebotomy of 525 ml of whole blood over approximately 15 minutes (see Table 1). For recipient safety, specific information about high-risk behavior is elicited to identify donors who may carry transfusion-transmissible diseases. For certain viral diseases, such history is relied upon to eliminate donors who may be in the window period—that time period during the acute stage of infection before laboratory tests can detect any serologic markers for disease.

The donor is also queried about recent and current medications. Ingestion of medication known to have teratogenic or mutagenic effect is cause for deferral (see Table 2). If the donor has recently ingested medications known to inhibit platelet function, the donor should not be used as the sole source of platelets for the recipient. A history of other medications may reveal underlying diseases that would put the donor at risk during donation. Also, questions targeted at recipient safety identify those donors at risk for illness or disease that current methods of infectious disease testing will not address, such as Creutzfeldt-Jakob disease and malaria.

Donors are provided with educational materials explaining the signs and symptoms of AIDS and descriptions of high-risk behavior, and they are urged not to donate blood if any of the information supplied applies to them. Blood centers offer an opportunity for confidential unit exclusion (CUE). This allows the donation to be marked for disqualification in an indistinguishable way, such as barcoded stickers, to facilitate anonymous self-deferral. This is particularly relevant for blood drives in the workplace, at schools, or at other organizations where potential donors may have close

TABLE 1 Elements of Health History Investigated for Protection of the Blood Donor

- History of diseases of the heart, liver, or lungs
- History of cancer
- History of abnormal bleeding
- Pregnancy (or within 6 weeks of having been pregnant)
- Time interval from last donation (at least 56 days between whole blood donations are required)

TABLE 2 Elements of Health History Investigated for Protection of the Recipient

- History of viral disease: hepatitis after age 11, HIV, HTLV
- Behavior that puts individuals at risk for above viruses: IV drug abuse, male sex with another male, female prostitutes, sex with anyone with risk factors in the last 12 months
- Exposure to HIV group O (20% missed on current tests): born in or resided in African country of concern, received a transfusion while visiting country, had sexual contact with anyone born/resided in such country
- Other exposure to blood-borne pathogens: tattoo, mucous membrane exposure to blood, accidental needle stick with needle contaminated with blood, sexual contact with person with or at risk for HIV or hepatitis in the last 12 months
- Received blood transfusion or tissue transplant in past 12 months
- History of syphilis or gonorrhea in past 12 months
- Malaria: had malaria in last 3 years, immigrant from endemic area in last 3 years, travel to endemic area in last 12 months
- Creutzfeldt-Jakob disease (CJD) risk: family history, received tissue graft known to be a source of CJD (dura mater), received human pituitary-derived hormones such as growth hormone or gonadotropins
- New variant CJD: lived in United Kingdom for 6 months or greater between 1980 and through 1996, exposure to non-U.S.-licensed drug product from cattle, such as bovine insulin
- Recent immunizations and vaccinations: rubella, MMR, polio
- History of babesiosis or Chagas' disease
- History of ingestion of medications known to be teratogenic: isotretinoin, finistride, etretinate, acitretin

association with one another outside the donation event.

3. Hemoglobin and Hematocrit

In order to donate blood, the donor's hemoglobin concentration (hgb) must be at least 12.5 g/dl, or the hematocrit must be 38% or higher. This helps to ensure that the donor has adequate iron stores and that the final product will have the appropriate RBC content to provide a therapeutic dose of RBCs to the recipient. The required minimum of 8 weeks (56 days) between donations serves to protect donors from iron depletion. In one study, an average of 253 mg of iron was removed upon 1 donation of 500 ml (Milman and Sondergaasrd, 1984). Most blood centers do not recommend that all donors take iron supplements. A balanced diet should adequately replace this deficit. Iron deficiency in blood donors is primarily a concern in menstruating women; it has been shown that 30 to 40 mg of daily iron supplements prevents iron deficiency in regular, repeat female blood donors (Simon et al., 1984).

D. Phlebotomy for Whole Blood Donation

1. Procedure

Phlebotomy is performed with the donor in either a supine or a reclining position. The collection set consists of a 16 gauge needle, tubing, sterile collection bag containing anticoagulant, and attached secondary bags for component preparation. Proper preparation of the skin (antecubital fossa) prior to phlebotomy using an iodophore compound helps to both ensure sterility of the collection and prevent cellulitis in the donor. The collection bag is positioned below the level of the donor's arm to facilitate blood flow. The donor is observed throughout the collection for any possible adverse reactions, and the collection bag is agitated periodically to mix the anticoagulant with the whole blood. If the donation is not completed within 15 minutes, the collection is stopped, since blood flow will not have been of adequate rapidity to prevent coagulation. Collection bags commonly used may be filled to 450–550 ml.

2. Physiologic Effects on Donor

The amount of whole blood removed will decrease the donor's hemoglobin by approximately 1 g/dl. The amount of plasma removed from the donor is not sufficient to cause significant changes in his or her blood chemistry values. Of importance, donors are instructed to drink four 8-oz. glasses of nonalcoholic liquids following donation in order to replace lost volume. Donors are observed postdonation for adverse reactions and signs of hypovolemia before leaving the donation site. During the immediate postdonation period it is recommended that donors refrain from strenuous activities.

E. Apheresis Donation

Apheresis is the collection of a specific blood component (or components) by a specialized cell separator instrument that centrifuges the donor's whole blood, separates and removes the desired component (e.g., plasma, platelets, RBCs, or granulocytes), and returns all other blood components to the donor. The automated collection system uses a sterile, disposable kit containing collection bags and tubing. As with whole blood donation, any materials that come in contact with donor blood are discarded. The anticoagulant used in the apheresis procedure is citrate (ACD) that is rapidly metabolized by the donor.

1. Donor Requirements

In general, the same physical examination, hematocrit values, and infectious disease testing required of whole blood donors apply to apheresis donors. Additional requirements may apply, depending upon the specific component to be collected.

a. Plateletpheresis

Plateletpheresis donors are allowed to donate more frequently than are whole blood donors. If a platelet donation occurs less than 4 weeks following the last platelet donation, the donor platelet count must be at least 150,000/ml for the donor to proceed. Procedures should not be more frequent than twice a week or 24 times a year, with at least 48 hours between each plateletpheresis donation. In addition to the routine blood donor requirements, platelet donors cannot ingest aspirin-containing medications within 3 days of donation, as this medication would interfere with platelet function. The same 56-day interval that is required between whole blood donations is required between whole blood donation and any subsequent apheresis donation.

b. Plasmapheresis

For plasmapheresis donors, if plasma is collected more often than every 4 weeks, donor protein levels and quantitative IgG and IgM measurements must be determined to be within normal limits prior to the next donation.

c. Granulocyte Donors

For granulocyte donors, there are special concerns regarding the use of agents administered to enhance granulocyte yields. Steroids are often given to granulocyte donors prior to donation in order to temporarily increase granulocyte levels in the circulation. Before use of corticosteroids, donors should be questioned about history of and risk for hypertension, diabetes, and peptic ulcer. Granulocyte colony-stimulating factor (G-CSF), a hematopoietic growth factor, has been administered to donors prior to granulocyte collection to increase donor granulocyte production. G-CSF is generally well tolerated by donors but can cause bone pain, headache, insomnia, and fatigue. Hydroxyethyl starch (HES) is added to the whole blood as it is drawn into the cell separator to enhance separation of granulocytes from RBCs. HES acts as a volume expander and may cause headache or peripheral edema. Small amounts of HES are retained in the body, and the long-term effects of this are unknown. Due to the side effects of steroids, HES, and G-CSF, it is important to exclude donors who have any preexisting conditions that may be exacerbated by these agents.

2. Apheresis Collection Procedure

Procedures for aseptic preparation of the phlebotomy site are the same as for whole blood donation. Some procedures can be performed utilizing discontinuous flow (i.e., draw and return) via 1 venipuncture site. Alternatively, machines with continuous flow necessitate 2 venipuncture/phlebotomy lines, generally using both antecubital fossae. Apheresis procedures generally require 1–2 hours, during which donors are monitored for signs of adverse reactions. Donors are instructed to hydrate themselves and to avoid overexertion in the postdonation period.

F. Adverse Reactions

The majority of blood donation attempts are accomplished without adverse effects. Adverse reactions of some type occur in 1–5% of donations (Popovsky et al., 1995).

1. Vasovagal

For both whole blood donation and apheresis donation, vasovagal reactions with or without syncope are the most common adverse reactions. Watching the actual venipuncture, fear of needles, or a neurophysiologic response to donation itself may trigger a vasovagal reaction. Symptoms include pallor, sweating, hyperventilation, and bradycardia, and these may progress to syncope. Bradycardia is used to distinguish a vasovagal reaction from hypovolemia; with hypovolemia, the heart rate would be expected to rise. The donation process is discontinued at the first sign of a vasovagal reaction. Treatment may include application of cold compresses, elevation of the lower extremities, and administration of oral fluids.

2. Hematomas and Infections

Hematomas may result from phlebotomy due to extravasation of blood into tissues at the venipuncture site. If noted during collection, the entire procedure is discontinued, and firm pressure with or without a cold compress is applied to the area while elevating the arm. Accidental arterial sticks represent a rare complication, indicated by rapid filling of the collection bag with pulsatile flow. Treatment requires firm pressure applied to the venipucture site for at least 10 minutes and subsequent confirmation of a radial pulse. Donors may occasionally experience phlebitis or cellulitis following the donation process, sometimes necessitating treatment with antibiotics. Infrequently the needle may come in contact with the median nerve or one of its branches, causing pain and possible nerve injury.

3. Cardiac and Pulmonary

Serious cardiac reactions such as myocardial infarction are extremely rare. When they occur, the donor generally has undiagnosed coronary vascular disease. Clots at the venipuncture site leading to pulmonary embolus are equally rare and are typically secondary to a preexisting hypercoagulable condition.

4. Reactions Specific to Apheresis Donation

While all of the above reactions may occur with whole blood or apheresis donation, some additional reactions are specific to the apheresis procedure. Apheresis donors may experience a reaction from the effects of citrate, a component of the anticoagulant used to keep the blood from clotting in the extracorporeal circuit. Some of the citrate is returned to the donor along with his or her own blood in the procedure. Sensitivity to citrate varies from individual to individual. Citrate binds ionized calcium and may cause symptoms of hypocalcemia, such as perioral or digital paresthesias. Slowing the apheresis procedure and/or administering oral calcium-containing tablets to the donor will often ameliorate mild citrate toxicity. The liver rapidly metabolizes citrate, so that when the procedure is completed, symptoms of citrate toxicity quickly resolve.

G. Special Donor Categories

1. Directed Donors

Directed donors provide a blood component or components for a specific, intended recipient. Directed donors must meet all criteria for allogeneic blood donation (including satisfactory medical history and infectious disease testing). Sometimes directed donors are needed in order to adequately transfuse a patient who needs a rare type of blood, such as patients with antibodies against common or high-frequency blood group antigens needing PRBCs lacking those particular antigens. In this case, siblings or close relatives are the most likely compatible PRBC donors. Directed donors may also be selected on the basis of human leukocyte antigen (HLA) type to provide platelets for a patient requiring HLA-matched platelets. These directed, single-donor platelets are collected by apheresis.

In some cases, an individual may feel more secure if the blood he or she receives is donated by a relative or friend rather than a stranger. Thus, the most common directed donations are whole blood donations from friends or relatives. *However, directed donations are no safer than other volunteer, allogeneic donations as regards the transfusion transmission of infectious diseases* (Williams *et al.*, 1992; Litty *et al.*, 1998). The general blood supply is provided primarily by repeat donors, while directed donors are most often first-time donors. First-time donors have a higher rate of positive testing for infectious diseases, probably because they have not been previously screened (Starky *et al.*, 1989). Most transfusion services and blood centers facilitate, but do not necessarily promote, directed donation unless rare units are required.

2. Dedicated Donors

A dedicated donor is a directed donor who donates for the same individual repeatedly. This reduces the number of blood donors to which the recipient is exposed. In fact, for neonates and small children, one blood donor can often supply one child's entire transfusion requirement. This "minimal donor exposure" concept grew out of concerns over blood safety and risks of transfusion-transmissible diseases more than a decade ago—concerns that are less applicable in today's environment (Menitove, 1991).

A real advantage to dedicated platelet donation is the mitigation of platelet refractoriness that may result from repeated exposure to random donor platelet donations. Dedicated donors are often highly motivated individuals such as the patient's family members. Healthy dedicated donors may donate whole blood for the same patient more than once within a 56-day interval, provided they satisfy all medical history and health requirements, including a hgb > 12.5 g/dl and approval from their physicians.

3. Autologous Donors

Blood collection for subsequent reinfusion to the donor is referred to as autologous donation. Methods of autologous blood collection include preoperative blood donation, intraoperative blood salvage, acute normovolemic hemodilution, and postoperative blood collection. The motivation for the use of autologous blood is to avoid allogeneic exposure and the attendant risk of infectious disease transmission. These methods of autologous blood collection are described in detail in Chapter 4.

References

Aubuchon, J. P., and Popovsky, M. A. (1991). The safety of preoperative blood donation in the nonhospital setting. *Transfusion* **31**, 513–517.

Clements, D. H., Sculco, T. P., Burke, S. W., Mayer, K., and Levine, D. B. (1992). Salvage and reinfusion of postoperative sanguineous wound drainage. *J Bone Joint Surg* **74A**, 646–651.

De Andrade, J. R., Jove, M. J., Landon, G., Frei, D., *et al.* (1996). Baseline hemoglobin as a predictor of risk of transfusion and response to Epoetin alfa in orthopedic surgery patients. *Am J Orthoped* **25**, 533–542.

Ereth, M. H., Oliver, W. C., Beynan, F. M., Mullany, C. J., and Orszulak, T. A. (1993). Autologous platelet rich plasma does not reduce transfusion of homologous blood products in patients undergoing repeat valvular surgery. *Anesthesiology* **79**, 540–547.

Etchason, J., Petz, L., Keeler, E., Calhoun, L., *et al.* (1995). The cost effectiveness of preoperative autologous blood donations. *N Engl J Med* **332**, 719–724.

Goldberg, M. A., McCutchen, J. W., Jove, M., Di Cesare, P., *et al.* (1996). A safety and efficacy comparison study of two dosing regimens of Epoetin alpha in patients undergoing major orthopedic surgery. *Am J Orthopedics* **25**, 544–552.

Goodnough, L. T., Rudnick, S., Price, T., *et al.* (1989). Increased preoperative collection of autologous blood with recombinant human erythropoietin therapy. *N Engl J Med* **321**, 1163–1168.

Goodnough, L. T., and Monk, T. G. (1996). Evolving concepts in autologous blood procurement and transfusion: Case reports of perisurgical anemia complicated by myocardial infaction. *Am J Med* **101**, 2A–33S.

Goodnough, L. T., Brecher, M. E., Kanter, M. H., *et al.* (1999). Transfusion medicine. First of two parts. Blood transfusion. *N Engl J Med* **340**, 438–447.

Goodnough, L. T., Brecher, M. E., Kanter, M. H., *et al.* (1999). Transfusion medicine. Second of two parts. Blood conservation. *N Engl J Med* **340**, 525–533.

Kanter, M. H., van Maanen, D., Anders, K. H., *et al.* (1996). Preoperative blood donations before elective hysterectomy. *JAMA* **276**, 798–801.

Kickler, T. S., and Spivak, J. L. (1988). Effect of repeated whole blood donations on serum immunoreactive erythropoietin levels in autologous donors. *JAMA* **260**, 65–67.

Leitman, S. F., Oblitas, J. M., and Emmons, R., *et al.* (1996). Clinical efficacy of daily G-CSF recruited granulocyte transfusions in patients with severe neutropenia and life-threatening infections. *Blood* **88**, 331a.

Litty, C. A., Lugo, J., and McDevitt, C. (1998). Comparison of donors who confirm positive for HIV and HCV in the directed and allogeneic populations. *Transfusion* **38**, 51S.

Menitove, J. E. (1991). Minimal exposure transfusion concept. *J Clin Apher* **6**, 194–195.

Menitove, J. E., Ed. (2000). "Standards for Blood Banks and Transfusion Services," 20th ed. AABB, Bethesda, MD.

Milman, N., and Sondergaasrd, M. (1984). Iron stores in male blood donors evaluated by serum ferritin. *Transfusion* **24**, 464–468.

Popovsky, M. A., Whitaker, B., and Arnold, N. L. (1995). Severe outcomes of allogeneic and autologous blood donation: Frequency and characterization. *Transfusion* **35**, 734–737.

Shore-Lesserson, L., Reich, D. L., DePerio, M., *et al.* (1995). Autologous platelet rich plasmapheresis: Risk versus benefit in repeat cardiac operations. *Anesth Analg* **81**, 229–235.

Simon, T. L., Hunt, W. C., and Garry, P. J. (1984). Iron supplementation for menstruating female blood donors. *Transfusion* **24**, 469–472.

Spiess, B. D., Sasetti, R., McCarthy, F. J., *et al.* (1992). Autologous blood donation: Hemodynamics in a high-risk patient population. *Transfusion* **32**, 17–22.

Starky, J. M., MacPherson, J. L., Bolgiano, D. C., *et al.* (1989). Markers for transfusion-transmitted disease in different groups of blood donors. *JAMA* **262**, 3452–3454.

Stehling, L., and Zauder H. L. (1991). Acute normovolemic hemodilution. *Transfusion* **31**, 857–868.

Williams, A. E., Klienman, S., Gilcher, R. O., *et al.* (1992). The prevalence of infectious disease markers in directed vs. homologous blood donations. *Transfusion* **32**, 45S.

Williamson, K. R., and Taswell, H. F. (1991). Intraoperative blood salvage: A review. *Transfusion* **31**, 662–675.

2

Infectious Disease Testing

CATHY LITTY

St. Christopher's Hospital for Children
Philadelphia, Pennsylvania

Infectious disease testing of donated blood is performed to enhance the safety of the blood supply. Infectious disease testing has improved markedly in the past decade, is very sensitive and specific, and is extremely effective in limiting transfusion-transmitted disease. The Food and Drug Administration (FDA) regulates this process, and both the FDA and the American Association of Blood Banks (AABB) require testing of all donated blood for infectious agents, utilizing a variety of test assays. Testing performed to limit transfusion-transmissible infectious diseases in the blood supply are listed in Table 1.

I. TEST ASSAYS

A. Screening Assays

The majority of donor blood testing is performed utilizing the enzyme immunoassay (EIA) method, in which an enzyme is used as an immunochemical label for detection of serum antibodies directed against a variety of infectious agent, or specific viral, antigens. Reagent antigen (or antibody) coated on beads or microtiter wells is incubated with patient serum to bind the respective antibody (or antigen) to be detected. An enzyme-labeled second antibody is added to form an immune complex. Bound enzyme reacts upon substrate added to create a color change which is quantitated spectrophotometrically and expressed as optical density (OD).

Blood samples to be tested for infectious diseases are collected at the time of blood donation from the donated unit. A 7-ml blood sample (serum or plasma)

is an adequate volume with which to perform all testing requirements in most situations.

Tests are run in batches according to each manufacturer's specifications, usually requiring 2.5–4.5 hours. Initially negative tests are termed "nonreactive," and no further testing is necessary for nonreactive specimens. Repeat testing is required to resolve initially "reactive" or positive specimens. Each initially reactive test is repeated in duplicate to rule out the possibility of a false-positive test. If two of three, or three of three, repeat tests are reactive, the test is considered a "repeat reactive."

False-positive tests may occur due to technical errors or to biologic factors, such as the presence of anti-human leukocyte antigen (HLA) antibodies or nonspecific serum antibodies cross reacting with test agents. Repeat reactive screening assays proceed to confirmation assays. Regardless of the result of the confirmation test, the donation is not used. Therefore, all repeat reactive screening results lead to the destruction of the donated blood, and the donor is subsequently deferred. Turnaround time for infectious disease testing in most laboratories is not less than 12 hours. This time restraint has an impact on directed donations, because directed donor units cannot be made available quickly enough if the intended recipient has an urgent or emergent need for transfusion.

B. Confirmatory Assays

The Western blot is the most commonly used methodology for confirmation of the EIAs that initially detect serum antibodies. Detection of anti-HIV-1/2

TABLE 1 Testing Required for Donated Blood Products in Order to Limit Infectious Disease Transmission

Infectious disease	Testing performed
Human immunodeficiency virus (HIV)	HIV-1 and HIV-2 antibodies (anti-HIV-1/2)
	HIV p24 antigen (HIV p24)
	HIV RNA[a]
Hepatitis B	Hepatitis B surface antigen (HBsAg)
	Hepatitis B core antibody (anti-HBc)
Hepatitis C	Hepatitis C antibody (anti-HCV)
	HCV RNA[a]
Adult T-cell leukemia/lymphoma, human T-cell lymphotropic virus (HTLV)-associated myelopathy, and tropical spastic paraparesis	HTLV antibodies, types I and II (anti-HTLV-I/II)
Syphilis	Treponemal antibody

[a] Nucleic acid testing (NAT) for HIV and HCV is widely performed, but has not yet been mandated.

and anti-HTLV-I/II by EIA are confirmed by Western blot. Anti-HCV (hepatitis C virus) is confirmed with a recombinant immunoblot assay (RIBA), a variation of the Western blot. In the RIBA, component proteins of purified inactive virus, or recombinant antigens, are electrophoretically separated into bands based on molecular weight. If antibodies are present in the donor serum, they react with the separated proteins. According to the band pattern that forms, the blot is interpreted as either positive, indeterminate, or negative. Indeterminate Western blots may be seen during sero-conversion or may represent false-positive results. Studies following donors with inderminate HIV Western blot tests without risk factors for HIV or evidence of HIV disease have shown that if results do not become positive after 6 months, the donor can be assured that he or she is very unlikely to be infected with HIV (Henrard et al., 1994).

The two EIAs that detect the viral antigens HIV p24 antigen and hepatitis B virus (HBV) surface antigen (HBsAg) are confirmed using a neutralization assay. Viral-specific antibodies are incubated with the donor's serum or plasma specimen. If the offending antigen is present, an immune complex will form, preventing a reaction from occurring. The EIA is then repeated, but the result, measured by absorbance via spectrophotometry, is reduced or "neutralized." The original reaction must be reduced by $\geq 50\%$ for the repeat reactive EIA to be confirmed as positive. In the absence of a positive confirmatory test result, the EIA is interpreted as a false positive. Follow-up evaluation is recommended, but often fails to reveal evidence of a disease process. In certain cases, when repeat reactive tests are positive but confirmatory tests are negative, donors may be considered for "reentry" (reinstatement as blood donors).

C. Nucleic Acid Testing (NAT)

After infection with either HIV or HCV, viral RNA is the first marker that is detectable in an individual's blood. Thus, circulating RNA may be detected by nucleic acid amplification (NAT) procedures, such as the polymerase chain reaction (PCR) or transcription-mediated amplification. Some blood collection facilities in the United States initiated NAT testing for HIV and HCV using pooled donor samples in 1999. Studies suggest this testing resulted in reduction of the window period for HIV from 16 to 10 days and for HCV from 66 to 10–30 days (see Table 2; Wilkenson et al., 1999). The definition of a "positive" NAT result is reserved for results obtained on an individual donation that is "reactive" by NAT and "confirmed" by a different NAT procedure.

The low level of HBV present during the window period (prior to detection of hepatits B surface antigen [HBsAg] by licensed, serologic tests) does not, as of this writing, allow sensitive detection of HBV DNA by NAT when applied to pooled plasma specimens. Therefore, NAT for HBV is not used for pooled testing at this time. HBV NAT will most likely require testing of single donated blood products to significantly reduce residual HBV risk.

TABLE 2 Window Period Risk Estimates for Voluntary Blood Donations in the United States (Schrieber et al., 1996a, b)

	Window period (days)	Risk per unit	NAT window period reduction (days)	Risk with NAT (undiluted) per unit
HIV	22	1:493,000	11	1:986,000
HIV plus p24	16	1:676,000	5	1:986,000
HTLV	51	1:641,000	N/A	N/A
HCV	66	1:127,000	43	1:368,000
HBV	59	1:63,000	25	1:110,000

It is anticipated that NAT applied to single donation samples will be used to test for HIV, HCV, and HBV in the near future. Its widespread use will significantly improve the safety of the blood supply and may lead to a review and possible revision of the required tests for blood donations.

D. Sensitivity and Specificity

Sensitivity is a measure of how accurately a test detects evidence of infection, or exposure to infection (e.g., a positive test result), in an individual who is truly infected or has been exposed to the infectious agent. A test with a high sensitivity will have a low false-negative rate. Specificity is a measure of how accurately a test identifies as test-negative those individuals who are not truly infected nor previously exposed to the infectious agent. If a test has a high specificity, it will have a low false-positive rate.

Sensitivity and specificity are not the only measures of how well tests perform. The underlying probability of disease in the population being tested impacts on the positive predictive value of tests. In populations with a low prevalence of infectious disease, like volunteer blood donors, there will be a high false-positive rate, and thus the positive predictive value of the tests is low. In contrast, the positive predictive value of the same test will be higher when applied to populations with risk factors and/or signs and symptoms of infectious disease.

The screening tests used to test donors are very sensitive. It is exceedingly unlikely that positive donors will be missed unless they are in the window period after infection before an antibody response, an antigen, or a nucleic acid can be detected. The confirmatory assays used for donor counseling and reentry have greater specificity than screening assays.

II. INDIVIDUAL TEST ASSAYS

A. Human Immunodeficiency Virus (HIV) Types 1 and 2

1. Background

The first recognized transfusion-transmitted case of HIV was in 1982. Deferral of donors with self-identified behavioral risk factors for HIV was then initiated and became the primary method of protecting the blood supply from HIV until 1985. In May 1985, the first screening test for anti-HIV was implemented in blood collection facilities nationwide. Since that time,

transfusion-transmitted HIV has increasingly become exceedingly rare. In 1992, new versions of EIA assays that test for HIV types 1 and 2 were introduced. The addition of HIV p24 antigen testing occurred in March 1996, and has identified only a few ($\sim$ 5) additional HIV-contaminated (antigen positive, antibody-negative) blood products out of 40 million donations since that time.

2. Current Testing

The currently required screening tests for HIV include tests for antibody against HIV types 1 and 2, and a test for HIV p24 antigen. Most centers use a combined EIA, which detects antibodies to HIV types 1 and 2. The p24 antigen, a marker of early infection or very advanced disease, becomes detectable in a donor approximately 16 days after exposure, while the antibody detected by EIA requires about 22 postexposure days to develop.

The licensed Western blot confirmatory assay is for HIV-1 only. A positive Western blot has at least two of the following three diagnostic bands present: p24, gp41, and gp120/160. The requirement for the presence of the p31 band was dropped in 1993 in order to reduce the reporting of indeterminate readings in seroconverting individuals in high-risk groups (Kleinman *et al.*, 1998). Confirmatory rates are low; only 6/100,000 blood donations are positive for anti-HIV antibodies (PHS, 1996). Confirmatory tests for HIV-2 are not licensed by the FDA, and thus are considered to be for research use only. If the confirmatory test for HIV-1 is negative or indeterminate, a second EIA test specific for HIV-2 is performed that is different from the screening EIA. When false-positive results are suspected, research tests may be used, if available, for donor counseling purposes (Sullivan *et al.*, 1998). As described above, NAT is also currently in use to detect HIV RNA.

3. Current Risk of Window Period Transmission

The risk of transfusion-transmitted viral infections may be calculated by multiplying incidence rates of seroconversion among blood donors (measured in over 500,000 repeat donors) by the window period (see Table 2). The risk of acquiring HIV from a window period donor based on testing for both antibody and p24 antigen was calculated to be 1 in 676,000 units transfused. The risk has been diminished further due to implementation of NAT for HIV RNA, which reduces the window period. The residual risk (following NAT implementation) per unit for HIV transmission is estimated to be 1/986,000 (Schreiber *et al.*, 1996a). The residual risks (following NAT implementation) of transfusion-transmitted HCV and HBV are estimated at 1/368,000 and 1/110,000 units, respectively (Schreiber *et al.*, 1996a).

B. Hepatitis

1. Hepatitis B

a. Background

Hepatitis B was the first transfusion-transmissible viral disease identified. Testing for hepatitis B surface antigen (HBsAg) has been conducted since the early 1970s. Prior to 1996 the risk of transfusion transmission of HBV was believed to be 1 in 200,000 units transfused (Dodd, 1992). It was known that rare transmission from seronegative donors was possible due to low serum levels of viral particles that were below the level of detection of available assays.

b. Current Tests

Current screening tests for HBsAg detect levels of viral particles ≥ 0.1 to 0.2 ng/ml. After exposure to hepatitis B virus, HBsAg is detectable in serum in about 2–6 weeks and becomes undetectable at about 12 weeks. Donor samples that are repeat reactive for HBsAg are confirmed using a neutralization assay.

Antibody to surface antigen does not begin to appear until 16 to 20 weeks after exposure. Thus, there is a time period in hepatitis B infection when HBsAg has declined, anti-HBs is not detectable, the individual is still infectious, and antibody to hepatitis B core antigen (anti-HBc) is the only infectious disease marker present. Vaccines for hepatitis B do not contain the core antigen and vaccinated individuals should not in general develop antibody to this antigen. Because of the high number of false positives for anti-HBc (Busch, 1998), donors are not deferred until the second positive donation or until testing is confirmed by another manufacturer's test or method. *Still, components from a donation with any positive testing for HBV will not be used.* In some instances, the plasma from donations with positive testing for anti-HBc may be used for fractionation into manufactured derivatives, such as intravenous immune globulin (IVIg).

c. Current Risk of Transmission

The risk of acquiring HBV infection has been calculated at 1 in 63,000 units transfused (Schreiber *et al.*, 1996a). The calculation was adjusted to account for transient antigenemia and long-term carriers, and this rate may be higher than previously appreciated. The historically lower rate actually observed may be at-

tributable to absence of, or low-level symptoms in, the recipient population.

2. Hepatitis C

a. Background

Hepatitis C, prior to its identification by molecular biologic techniques in 1989, was known as non-A, non-B hepatitis. Before that time, there was no means by which to test specifically for this virus, and as a result, surrogate tests for HCV were introduced. The AABB began requiring alanine aminotransferase (ALT) testing in 1986 and anti-HBc testing in 1987 for the purpose of attempting to identify donors with HCV.

The FDA licensed a first-generation test for hepatitis C virus in May 1990. This EIA utilized a recombinant antigen from the HCV genome. In March 1992, second-generation testing was licensed. This EIA 2.0 used three recombinant proteins to detect serum HCV antibody and substantially increased the detection rate of HCV in volunteer blood donors. The licensed confirmatory assay for second-generation testing was a recombinant immunoblot assay (RIBA 2.0). Third-generation HCV EIA testing began in 1996, adding yet another HCV-specific antigen to the test and increasing sensitivity and specificity. A third-generation RIBA confirmatory assay for EIA 3.0 screening was licensed by the FDA in February 1999.

Due to this increased ability to detect hepatitis C in donated blood, the use of surrogate testing was reconsidered. The AABB eliminated the requirement for ALT testing in June of 1995, by which time second-generation HCV testing was in place. The risk of acquiring HCV decreased to 1/3300 units transfused in 1990 with first-generation testing (Donahue *et al.*, 1992), and then to 1/103,000 units transfused with second-generation testing (Schreiber *et al.*, 1996a). Many blood centers continue to use the ALT test, however, because it is required in the European Pharmaceutical industry. Results need to be twice the upper limit of normal before blood is discarded and donors are deferred. Anti-HBc testing has continued, as it is thought to be useful for identification of donors infectious for the hepatitis B virus.

b. Current Testing and Risk of Window Period Transmission

Currently, HCV testing of the blood supply is performed with third-generation EIAs and RIBAs. The reported risk of HCV transfusion transmission at 1/103,000 units transfused was calculated using the second-generation HCV test window period of 82 days

(Schreiber *et al.*, 1996a). In 1996, however, donor screening progressed to third-generation assays having a window period of 66 days. The resulting projected HCV risk declined another 20%, to roughly 1/127,000 units transfused (Schreiber *et al.*, 1996b). Addition of NAT is projected to decrease the risk for HCV transfusion transmission to 1:368,000 (Schreiber *et al.*, 1996a).

C. Human T-Cell Lymphotropic Virus, Types I/II (HTLV-I/II)

1. Background

HTLV-I is associated with adult T-cell lymphoma-leukemia (ATL) and HTLV-associated myelopathy (HAM), although only a small percentage of individuals harboring the virus manifest these diseases. HTLV-II is thought to only be able to cause HAM, and with less frequency than type I. Blood donors have been tested for HTLV since 1988 when testing was initially aimed at HTLV-I. There was test cross-reactivity to HTLV-II, but it is likely that 43% of HTLV-II cases were missed (Hjelle *et al.*, 1993). In 1997, assays that were able to reliably detect both types of HTLV were licensed, and within 6 months, most blood centers were testing with these combined assays.

2. Current Tests and Residual Risk

Although the licensed EIA tests are for both HTLV types, most tests do not reliably distinguish between the two. There is no FDA requirement to perform confirmatory testing for HTLV, and the donor may donate again, but is deferred after a second repeat reactive EIA for HTLV. Most centers do, however, perform confirmatory testing for donor counseling purposes. These Western blot tests for HTLV (capable of distinguishing between HTLV types I and II) are not FDA-licensed at this time. Some blood centers may choose to confirm with a second EIA, but a false positive may still manifest itself when testing again by the same methodology. The risk of a window period transmission of HTLV-I/II is reported to be 1 in 641,000 units transfused (Schreiber *et al.*, 1996a).

The flu vaccine has been linked with false positive results of the HTLV EIA. This nonspecific reaction is present in less than 5% of individuals who receive the flu vaccine and is usually no longer present after 6 months (Simonsen *et al.*, 1995). Other false-positive test results do occur and the etiology of these results has not been determined.

D. Syphilis

1. Background

In 1938, screening of blood donations for syphilis was instituted. Since 1969, there have been no transfusion-transmitted syphilis cases reported in the United States. In Europe, 2 cases have been reported, and in both of these instances the donors tested negative for syphilis (Herrera *et al.*, 1997).

Testing donors for syphilis does not appear to be useful in preventing infection with the *Treponema pallidum* spirochete. If the donor is spirochetemic, the serologic tests will not yet be positive, and donors with antibodies are no longer infectious. In addition, spirochetes cannot survive at temperatures of 1–6°C for more than 3 days. Therefore, PRBCs, which are stored at 4°C, and fresh frozen plasma (FFP), which is stored at −18°C, are very unlikely to transmit the agent. In contrast, platelets are stored at room temperature and therefore have potential risk.

Interestingly, screening tests for syphilis may identify some donors who in fact have Lyme disease. It is still believed, however, that syphilis testing serves as a surrogate marker for donors who may have early stage HIV infection.

2. Current Tests and Residual Risks

Screening tests often utilize nontreponemal material, since syphilis antibodies agglutinate lipoidal substances collectively known as reagin. The Venereal Disease Research Laboratory (VDRL) test or Rapid Plasma Reagin (RPR) tests are examples. Specific treponemal tests, such as fluorescent treponemal antibody absorption tests (FTA-ABS) and *T. pallidum* hemagglutination assays (TPHA), are used as confirmatory tests. Blood centers may use automated, specific treponemal assays; however, these are FDA-licensed only as screening tests.

As described above, the period of spirochetemia in syphilis is brief, and the organisms survive for only a few days at 1–6°C. Therefore, in combination with current testing and other screening measures, the residual risk of transmission of syphilis by blood transfusion is exceedingly rare.

E. Cytomegalovirus (CMV)

Unlike the screening assays discussed above, tests for CMV are not required by the FDA. Not all donors need to be tested for CMV, as only a certain subset of patients are at risk to develop disease due to this ubiquitous virus (see Chapters 15 and 34). Donors who test positive for CMV are not deferred and the components they donate may be used for most recipients. Blood center screening tests for CMV may use EIA or latex agglutination technology, or may be performed by automated testing using a hemagglutination method.

There is a 1–4% rate of CMV transmission from transfusion of CMV-seronegative cellular blood components. This is due to a combination of factors, including window period transmission with donor viremia, declining antibody titers over time, intrinsic false-negative rates of the tests, or perhaps reactivation of virus with neutralization of antibody (Hillyer *et al.*, 1994). While some studies suggest that a proportion of CMV-seronegative donors have CMV DNA in peripheral blood mononuclear cells, these findings cannot be confirmed.

III. DONOR ISSUES

A. Donor Reentry

When screening tests are repeat reactive for viral diseases, confirmatory testing results are performed. Donors who confirm positive need to be notified confidentially, counseled, and encouraged to seek medical follow-up. When the repeat reactive EIA is believed to be a false positive, there are some reentry protocols approved by the FDA. Of note, in cases where there is not a confirmatory assay licensed by the FDA, there is not an approved reentry, such as for HTLV-I/II.

Reentry protocols exist for donors who have had a false-positive screening test result for anti-HIV and HBsAg, and for HIV p24 antigen. Reentry is also possible after false-positive tests or successful treatment for syphilis. Not all blood centers elect to consider reentry, and thus some donors with false-positive screening results will be indefinitely deferred. All protocols require that the donation with a false-positive screening test result must be followed by a negative confirmatory assay. If the confirmatory result is indeterminate or positive there can be no reentry, even if the donor is believed to be negative for the infectious disease.

For reentry, a repeat sample must be drawn at a later time period. This follow-up sample must be negative by both screening and confirmatory testing. If the donor has the same false-positive results on the screening test, but does not confirm as positive, he or she may not be reentered.

For HBsAg reentry, the donor must also be nonreactive for anti-HBc on the initial donation and at the time of subsequent testing. For HIV reentry, if the original positive test is the combined HIV-1/2, then the second, different HIV-2 EIA must have been negative on the original donation and at the time of subsequent testing. For reentry after false-positive HIV p24, all tests for HIV antibody have to be nonreactive.

B. Donor Counseling

Donors with confirmatory positive tests may be counseled and referred for further medical follow-up when appropriate. Donors with false-positive test results may also be counseled to be reassured that the results are not indicative of exposure to a specific infectious agent. In 1997, it was estimated that 36,000 donors per year are indefinitely deferred annually because of laboratory results that cannot be explained (Ownby *et al.*, 1997). Due to the presence of highly sensitive infectious disease tests (responsible for the low risk of infectious disease transmission in the blood supply today), comparatively high numbers of false positives may occur in the blood donor population.

References

Busch, M. P. (1997). To thy (reactive) donors be true! *Transfusion* **37**, 117–119.

Busch, M. P. (1998). Prevention of transmission of hepatitis B, hepatitis C and human immunodeficiency virus infections through blood transfusion by anti-HBc testing. *Vox Sang* **74**, 147–154.

Busch, M. P., Lee, L. L. L., Satten, G. A., *et al.* (1995). Time course of detection of viral and serologic markers preceding human immunodeficiency virus type 1 seroconversion: Implications for screening of blood and tissue donors. *Transfusion* **35**, 91–97.

Dodd, R. Y. (1992). The risk of transfusion-transmitted infection. *N Engl J Med* **327**, 419–420.

Donahue, J. G., Munoz, A., Ness, P., *et al.* (1992). The declining risk of post-transfusion hepatitis C virus infection. *N Engl J Med* **327**, 369–373.

Henrard, D. R., Phillips, J., Windsor, I., *et al.* (1994). Detection of human immunodeficiency virus type 1 p24 and plasma RNA: Relevance to indeterminate serologic tests. *Transfusion* **34**, 376–380.

Herrera, G. A., Lacritz, E. M., Janssen, R. S., *et al.* (1997). Serologic test for syphilis as a surrogate marker for human immunodeficiency virus infection among United States blood donors. *Transfusion* **37**, 836–840.

Hillyer, C. D., Emmens, R. K., Zago-Novaretti, M., and Berkman, E. M. (1994). Methods for the reduction of transfusion-transmitted cytomegalovirus infection: Filtration versus the use of seronegative donor units. *Transfusion* **34**, 929–934.

Hjelle, B., Wilson, C., Cyrus, S., *et al.* (1993). Human T-cell leukemia virus type II infection frequently goes undetected in contemporary US blood donors. *Blood* **81**, 1641–1644.

Kleinman, S., Busch, M. P., Hall, L., *et al.* (1998). False-positive HIV-1 test results in a low-risk screening setting of voluntary blood donation. *JAMA* **280**, 1080–1085.

Lee, H. H., and Allain, J. P. (1998). Genomic screening for blood-borne viruses in transfusion settings. *Vox Sang* **74**, 119–123.

Ownby, H. E., Korelitz, J. J., Busch, M. P., *et al.*, and the Retrovirus Epidemiology Donor Study (1997). Loss of volunteer blood donors because of unconfirmed enzyme immunoassay screening results. *Transfusion* **37**, 199–205.

PHS (1996). U.S. Public Health Service guidelines for testing and counseling blood and plasma donors for human immunodeficiency virus type 1 antigen. *MMWR* **45**, 1.

Schreiber, G. B., Busch, M. P., Kleinman, S. H., *et al.*, and the Retrovirus Epidemiology Donor Study (1996a). The risk of transfusion transmitted viral infections. *N Engl J Med* **334**, 1685–1690.

Schreiber, G. B., Busch, M. P., and Kleinman, S. H. (1996b). Authors reply to letter to the editor. *N Engl J Med* **335**, 1610.

Simonsen, L., Buffington, J., Shapiro, C. N., *et al.* (1995). Multiple false reactions in viral antibody screening assays after influenza vaccination. *Am J Epidem* **141**, 1089–1096.

Sullivan, M. T., Guido, E. A., Metier, R. P., *et al.* (1998). Identification and characterization of an HIV-2 antibody-positive blood donor in the United States. *Transfusion* **38**, 189–193.

Wilkinson, S. L., and Shoos Lipton, K. (1999). NAT implementation. American Association of Blood Banks Association Bulletin 99-3.

3

Component Preparation and Storage

MARY BETH ALLEN

Transfusion Medicine Program
Department of Pathology and Laboratory Medicine
Emory University School of Medicine
Atlanta, Georgia

I. INTRODUCTION

The process of transfusion therapy begins with the collection of a volume of blood by either whole blood donation or by an apheresis collection procedure. This blood may then be divided into its separate constituent parts of packed red blood cells, plasma, white cells, and platelets by a process known as "component preparation." Separation of whole blood into specific components allows selective transfusion therapy based on specific patient need, and also maximizes the number of recipients who may benefit from any single donation. In this chapter, we will explore the fundamental technical principles of the component preparation process.

II. ANTICOAGULANT-PRESERVATIVE SOLUTIONS

Anticoagulant-preservative solutions serve to (1) maintain blood products in an anticoagulated state, (2) support the metabolic activity of red cells while in storage, and (3) minimize the effects of cellular degradation during storage. Several anticoagulant-preservative solutions, including ACD-A, CPD, CP2D, and CPDA-1, are currently approved by the Food and Drug Administration (FDA) for storage of blood and blood components. Most anticoagulant-preservative solutions contain *C*itrate, which maintains the anticoagulated state through chelation of calcium; *D*extrose and/or *A*denine, which support ATP synthesis in the stored cells; and Sodium *P*hosphate (hence CPDA), which provides phosphate for ATP synthesis and serves as a

buffer to minimize the effects of decreasing pH in the stored products.

Additive systems, such as AS-1, AS-3, and AS-5, extend the shelf life of red cell products via an adenine additive solution. Additive collection systems consist of a primary collection bag containing anticoagulant-preservative solution, and multiple attached satellite bags. One satellite bag is empty, and will eventually contain separated plasma. Another satellite bag contains the additive solution, which must be added to the anticoagulated red cell product within 72 hours of collection. The resulting red cell product has a final hematocrit of approximately 60%, which improves the product's flow rate and adds to ease of administration. Table 1 summarizes the essential constituents in the various approved anticoagulant-preservative solutions.

III. PROCESS OF COMPONENT PREPARATION

Donor blood is collected into a sterile container which contains an appropriate anticoagulant-preservative solution. Collection should begin with a single venipuncture, with minimal tissue trauma. Collections should be completed within 15 minutes to ensure optimal component quality. Sterility should be maintained by use of sterile solutions and disposables, and through proper aseptic technique. The collection container should be mixed frequently to ensure thorough mixing of whole blood with the anticoagulant.

Commercially available collection sets contain 63 ml of anticoagulant-preservative solution, which results in

TABLE 1

System and solutions	Storage period (days)	Final hematocrit	Solution constituents						
			Trisodium citrate	Citric acid	Dextrose	Monobasic sodium phosphate	Adenine	Mannitol	Sodium chloride
Anticoagulant preservatives									
ACD-A	21	80%	X	X	X				
CPD	21	80%	X	X	X	X			
CP2D	21	80%	X	X	X	X			
CPDA-1	35	80%	X	X	X	X	X		
Additive solutions									
AS-1	42	60%			X	X	X	X	X
AS-3	42	60%			X	X	X		X
AS-5	42	60%			X		X	X	X

a solution to blood ratio of 1.4 to 10. The standard collection volume is 10.5 ml per kilogram of donor body weight. The collection volume should be proportionate for the anticoagulant-preservative solution in the blood collection container. Low-volume collections may be used for red cell transfusions if appropriately labeled, but may not be used for the manufacture of components.

Products collected in integral or "closed" systems may be maintained for the recommended storage periods for that specific component. However, once a blood component has been entered or "opened," shelf life is shortened to either 24 hours for components which are stored at 1–6°C, or 4 hours for components which are stored at 20–24°C.

Whole blood donations may be separated into multiple component parts of "red blood cells," "platelets," and a plasma portion (see Figure 1). The red cells may be further modified by irradiation, washing. leukoreduction, freezing, deglycerolization, or rejuvenation, as described elsewhere. The plasma portion may be frozen within 6–8 hours of collection to create fresh frozen plasma. Plasma frozen after 8 hours but within 24 hours of collection may be labeled as "Plasma Frozen within 24 Hours." Like fresh frozen plasma, it has a 1-year shelf life when stored at < 18°C, but contains approximately 90% of the labile coagulation factors found in fresh frozen plasma. Alternately, plasma may be separated within 5 days of the red cell expiration date to create "liquid plasma," if stored at 1–6°C, or "plasma," if stored at −18°C or below. Fresh frozen plasma may be further processed to create "cryoprecipitated antihemophilic factor" (cryoprecipitate) and "plasma cryoprecipitate reduced."

IV. LABELING

The labeling process should conform with the most recent version of the *United States Industry Consensus Standard for the Uniform Labeling of Blood and Blood Components Using ISBT 128* or the *1985 Food and Drug Administration Uniform Labeling Guidelines.* Labels must be firmly affixed to the container, and contain clear, eye-readable information. Information may also be barcoded or machine readable, and any manual handwritten corrections should be in legible, indelible, moisture-proof ink. Products must be uniquely labeled using a numeric or alphanumeric system, so that any product can be identified and traced from its origin to its final disposition. Original unique identifying numbers affixed by the collecting facility may not be removed or obscured. Any intermediate facility which manipulates the product may add additional identifying labels to the product; however, a maximum of two unique identifiers may be visible on a product at any given time.

Labeling at the time of collection or component preparation must include the following information:

- Name of the product
- A unique numeric or alphanumeric identifier
- Type of anticoagulant (except for washed, frozen, deglycerolized, or rejuvenated red cells)
- Approximate volume collected from the donor
- Name of any sedimenting agent
- Identification of facility performing the collection or modification

The final product label prior to distribution to the transfusing facility should contain the following addi-

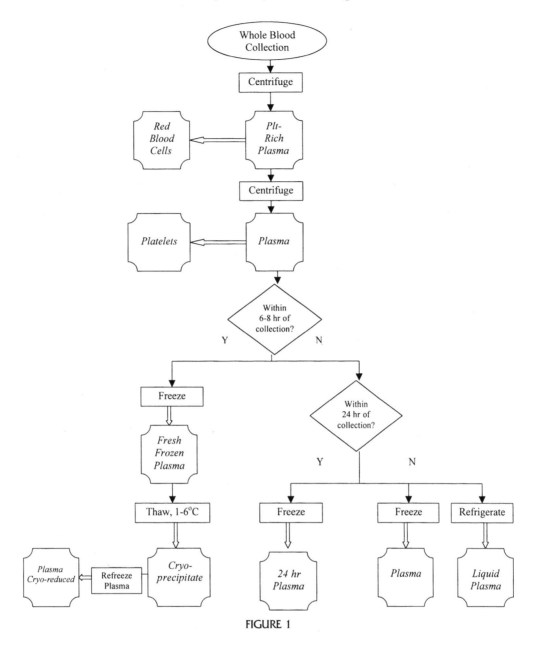

FIGURE 1

tional information:

- Temperature of storage
- Expiration date and (when appropriate) time
- Identification of facility
- ABO group and Rh type
- Specificity of unexpected red cell antibodies (except for cryoprecipitate, or washed, frozen, deglycerolized, or rejuvenated red cells)
- Indication of autologous, volunteer, or paid donor (where applicable)
- Instructions to the transfusionist, which include:
  —"See Circular of Information for the Use of Human Blood and Blood Components"

—"Properly identify intended recipient"
—"This product may transmit infectious agents"
—"Caution: Federal law prohibits dispensing without a prescription"
—"Rx Only"

V. CURRENT GOOD MANUFACTURING PRACTICE AND REGULATORY ISSUES

Blood banks and transfusion services are subject to an array of regulatory and inspection agencies, depending on their scope of practice and organizational structure. Facilities in the United States which collect blood from donors or perform more than minimal component preparation are subject to regulation by the Food and Drug Administration. Other U.S. inspection and accreditation agencies may include, but are not limited to, the American Association of Blood Banks (AABB), the Joint Commission on Accreditation of Healthcare Organizations (JCAHO), the Clinical Laboratory Improvement Act of 1988 (CLIA), the College of American Pathologists (CAP), the Occupational Safety and Health Administration (OSHA), the Foundation for the Accreditation of Hematopoietic Cell Therapy (FAHCT), and the International Organization for Standardization (ISO), as well as state and local governments. While each regulatory agency has its own specific requirements, there is much commonality among them.

In general terms, organizations must have a quality system which ensures appropriate performance in at least the following essential areas: (1) organization, (2) personnel, (3) equipment, (4) supplier issues, (5) process control, (6) documents and records, (7) incidents, errors, and accidents, (8) assessments, both internal and external, (9) process improvement, and (10) facilities and safety. The quality system must be structured so that responsibility for the organization's performance is clearly defined. Personnel must be appropriately trained, and their job performance and competency assessed annually. Equipment must comply with a regular schedule of preventive maintenance, and repair and quality control records must be maintained. Suppliers of goods or services must meet the customer's written expectations. Process control must be exercised to ensure the consistency and accuracy of all work processes. Documents and records must be maintained, stored, and archived in a manner which ensures retrievability and confidentiality. There must be a system to document, investigate, and prevent incidents, errors, and accidents. Internal and external assessments of the facility should be performed and documented. Process improvement must be implemented through systematic review of organizational performance for opportunities for improvement. Finally, the facility must ensure safe environmental conditions in the workplace for employees, visitors, and products.

VI. PREPARATION OF INDIVIDUAL COMPONENTS

A. Red Blood Cells

Red blood cells (RBCs) are obtained by centrifugation or sedimentation of donor whole blood, followed by removal of the plasma portion. Red blood cells retain the original expiration date of the parent product (35–42 days) if plasma removal is performed in an integral or "closed" system. If plasma removal is performed in an entered or "open" system, the red blood cell expiration date is shortened to 24 hours from the time of entry to protect against bacterial contamination. Red blood cells may be further modified by washing (the removal of plasma proteins to prevent allergic transfusion reactions), irradiation (exposure to radioactive sources to prevent graft-versus-host disease, GVHD), and leukoreduction (the removal of white blood cells to prevent alloimmunization and viral disease transmission, and to reduce the occurrence of febrile transfusion reactions).

Biochemical changes which result during storage of red cell products are known as "storage lesion." These changes include (1) a decreasing percentage of viable cells, (2) decreasing pH, (3) decreasing levels of ATP and 2,3-DPG, and (4) increasing plasma potassium and free hemoglobin levels. The cumulative effect of the storage lesion determines the length of the red cell product shelf life for each respective anticoagulant-preservative solution. The maximum shelf life is generally defined as the longest storage period which will still result in viability of 70% of the red cells 24 hours following infusion. Typically, the storage lesion does not have clinically significant consequences in transfused adult recipients. However, special care may be warranted for pediatric and neonatal recipients, and recipients with renal failure, to avoid potassium overload.

Red blood cells may also be frozen to provide extended storage for units with rare phenotypes, or for autologous units intended for future use. Frozen red cells may be stored for 10 years at $-65°C$ or below. The freezing process must be carefully managed to maintain post-thaw viability of the red cells. Strictly controlled rates of freezing and the use of cryoprotective agents such as glycerol, dimethyl sulfide (DMSO), or hydroxyethyl starch (HES) help maintain cell viability.

Frozen units must then be thawed and washed (deglycerolized) prior to infusion. Removal of the cryoprotective agents is normally accomplished by a series of wash procedures, using solutions of decreasing osmolarity. This gradual removal of cryoprotectant is essential to prevent hemolysis of the red cell product, and to return the product to a transfusable state of osmotic equilibrium.

B. Platelets

Platelet concentrates are platelet products separated from centrifuged whole blood within 8 hours of collection. The whole blood may not be stored below 20–24°C prior to platelet production. The platelet concentrates should be suspended in a volume of plasma sufficient to maintain a minimum pH of 6.2 by the end of their allowable 5-day shelf life. Platelet concentrates must contain a minimum of 5.5×10^{10} platelets in at least 75% of the products prepared by the facility. Platelet concentrates are stored at room temperature (20–24°C) with gentle agitation to prevent platelet aggregation, and should be visibly inspected prior to administration.

Platelet viability is markedly affected by storage conditions, and suboptimal storage conditions may contribute to a degradation of platelet morphology known as "platelet storage lesion." Optimal storage parameters include (1) maintenance at temperatures between 20 and 24°C, (2) constant, gentle agitation to prevent aggregation, (3) an adequate volume of plasma to maintain the pH at 6.2 or above, (4) appropriate anticoagulant-preservative solution to maintain platelet metabolism, and (5) storage in a plastic container with sufficient surface area to support adequate oxygen exchange.

C. Fresh Frozen Plasma

Fresh frozen plasma components are prepared by separating plasma from red blood cells of a single donor, and placing the plasma at −18°C. The separation must occur within 8 hours of collection with CPD, CP2D, or CPDA-1 anticoagulant, and within 6 hours of collection with ACD anticoagulant. Plasma stored at −18°C has a 1-year shelf life. Thawed fresh frozen

TABLE 2   Derivatives and Their Functions

Product name	Purpose or indication
Coagulation factor concentrates	
Antithrombin III	Treatment of ATIII deficiency
Factor VIII	Treatment of hemophilia A
Factor IX	Treatment of hemophilia B
Factor XIII	Treatment of factor XIII deficiency
Protein C concentrate	Treatment of protein C deficiency
Gamma globulins	
Albumin, 5 or 25%	Plasma expander
Plasma protein fraction, 5%	Plasma expander
Immune serum globulins	
Cytomegalovirus immune globulin	Prophylaxis following CMV exposure
Immune serum globulin	Treatment of hypogammaglobulinemia
Immune serum globulin intravenous	Treatment for immune globulin deficiency
IgM-enriched immune globulin	Treatment of septicemia and septic shock
Hepatitis B immune globulin	Prophylaxis following hepatitis B exposure
Rabies immune globulin	Prophylaxis following rabies exposure
Rh immune globulin	Prevention of Rh hemolytic disease of newborn
Rubella immune globulin	Prophylaxis following rubella exposure
Tetanus immune globulin	Prophylaxis following tetanus exposure
Vaccinia immune globulin	Prophylaxis following smallpox exposure
Varicella–Zoster immune globulin	Prophylaxis following chicken pox or herpes exposure
Miscellaneous	
Alpha$_1$ proteinase inhibitor	Treatment of genetic-acquired emphysema
Fibrinolysin	Treatment for intravascular clotting
Haptoglobin	Treatment of viral hepatitis and pernicious anemia
Serum cholinesterase	Treatment of apnea following succinyl-choline chloride therapy

plasma may be stored at 1–6°C, and outdates 24 hours from the time of thawing.

D. Cryoprecipitate

Cryoprecipitated antihemophilic factor (Cryo) is the cold-insoluble portion of plasma processed from fresh frozen plasma by thawing the plasma unit at 1–6°C. The cryoprecipitate is then promptly centrifuged, separated from the plasma portion, and refrozen within 1 hour. Cryoprecipitates contain a minimum of 150 mg of fibrinogen, and 80 IU of coagulation factor VIII. Cryoprecipitates may be stored up to 1 year at a temperature of −18°C or lower. Thawed cryoprecipitates must be used within 4 hours after thawing.

E. Derivative Products

Derivative products used to treat specific factor deficiencies or coagulopathies may also be manufactured

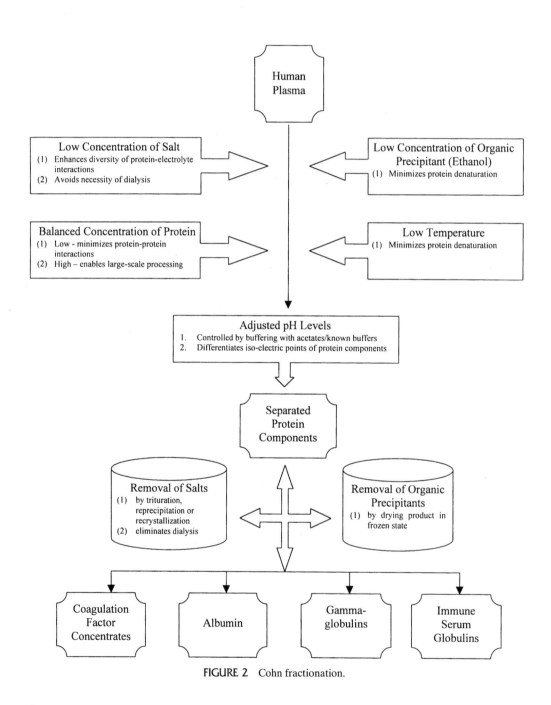

FIGURE 2 Cohn fractionation.

from plasma. The most common derivatives include albumin, gamma globulins, immune serum globulins, and coagulation factor concentrates. Table 2 summarizes currently available derivatives and their clinical function. The manufacturing process for derivatives has changed through the years as new technologies emerge for viral inactivation. Cohn fractionation (Fig. 2), a cold ethanol fractionation process, was the first large-scale method of isolating plasma proteins or derivatives. Methods of viral inactivation now include heat treatment, pasteurization, and solvent/detergent treatment. The recent development of recombinant products, manufactured from sources other than human plasma, appears to have minimized or eliminated the risk of viral transmission by transfusion of these products.

VII. STORAGE AND SHIPPING REQUIREMENTS

A. Storage Requirements

Storage of blood and blood products should be designed to maintain maximum viability in a strictly controlled and secure environment. These designated storage areas (refrigerators, freezers, rotators, environmental chambers, etc.) may contain only blood, blood components, derivatives, donor samples, patient samples, tissue for transplantation, and/or related reagent or laboratory supplies. These storage areas must have adequate capacity and air circulation to ensure optimal storage conditions for the particular blood products. Storage devices should have a system for continuous temperature monitoring, and for recording of temperatures at least every 4 hours. If components are stored at room temperature, outside a controlled storage device, the ambient room temperature should be recorded every 4 hours.

Audible alarm systems should be present on all refrigerators and freezers, and be set to activate at a temperature which will allow rescue of the blood or blood products before they reach unacceptable temperatures. The audible alarms must be present in an area with sufficient personnel coverage to ensure an immediate response. Backup power sources must be available for all critical storage devices, in the event of power failure. Written instructions must specify corrective action to be taken in the event of a storage device failure, in order to prevent loss of valuable blood and blood components.

B. Shipping Requirements

Like storage, shipping conditions should be defined which maintain maximum viability of blood and blood components. All blood and blood products should be inspected upon packing and again upon receipt to ensure the container is intact and the product is normal in appearance. Any product which appears abnormal should not be shipped by the vendor, and should be quarantined and handled according to the facility's standard operating procedure for nonconforming products. Similarly, products which appear abnormal upon receipt should be placed in a segregated storage area, and the nonconformity should be investigated and documented. Upon receipt, all products should be inspected to ensure the container is undamaged, and that labeling of the product is appropriate. If platelet products are shipped, they may not be out of a state of gentle agitation for longer than 24 hours.

References

Code of Federal Regulations, 21 CFR 610.53© (1998, rev. annually). GPO, Washington, DC.

Cohn, E. J., Strong, L. E., Hughes, W. L., et al. (1946). Preparation and properties of serum and plasma proteins. IV. A system for the separation into fractions of the protein and lipoprotein components of biological tissues and fluids. J Am Chem Soc 68, 459.

Goodnough, L. T., Brecher, M. E., Kanter, M. H., AuBushcon, J. P. (1999). Medical progress: Transfusion medicine—Blood transfusion. N Engl J Med 340(7), 525–533.

Huh, Y. O., Lichtiger, B., Giacco, G. G., et al. (1989). Effect of donation time on platelet concentrates and fresh frozen plasma. Vox Sang 56, 21–24.

Kistler P., and Nitschmann, H. (1962). Large scale production of human plasma fractions. Eight years experience with the alcohol fractionation procedure of Nitschmann, Kistler and Lergies. Vox Sang 7, 414–424.

Menitove, J., Ed. (1999). "Standards for Blood Banks and Transfusion Services," 19th ed. AABB, Bethesda.

Meryman, H. T., and Homblower, M. (1972). A method for freezing and washing RBCs using a high glycerol concentration. Transfusion 12, 145–146.

Simon, T. L., Marcus, C. S., Myhre, B. A., and Nelson, E. J. (1987). Effects of AS-3 nutrient-additive solution of 42 and 49 day storage of red cells. Transfusion 27, 178–182.

Valeri, C. R., Pivacek, L. E., Gray, A. D., et al. (1989). The safety and therapeutic effectiveness of human red cells stored at −8°C for as long as 21 years. Transfusion 29, 429–437.

Vengelen-Tyler, V., Ed. (1996). "Technical Manual," 12th ed. AABB, Bethesda.

Wagstaff, W. (1998). GMP in blood collection and processing. Vox Sang 74 (Suppl. 2), 513–521.

BLOOD COMPONENTS

C H A P T E R

4

Packed Red Blood Cells and Related Products

KRISTA L. HILLYER
CHRISTOPHER D. HILLYER
Department of Pathology and Laboratory Medicine
Emory University School of Medicine
Atlanta, Georgia

Packed red blood cells (PRBCs) are the most commonly transfused blood component; approximately 12 million units of PRBCs are transfused each year in the United States. Symptomatic anemia, the need for increased oxygen delivery, is the major indication for the transfusion of PRBCs. This need usually arises from acute blood loss or chronic anemia. Each patient's clinical circumstance is unique and should be thoroughly evaluated prior to, during, and following transfusion.

PRBCs are made from volunteer donations of whole blood, and a number of different preparations of PRBCs are available to meet the needs of the individual patient in the context of his or her current clinical situation. These PRBC preparations are reviewed below. The *Technical Manual* (Vengelen-Tyler *et al.*, 1996) and the *Standards for Blood Banks and Transfusion Services* (Menitove, 1999), both published by the American Association of Blood Banks, are excellent references that provide the details regarding all types of PRBC products described throughout this chapter.

I. PRODUCT DESCRIPTION

A. Whole Blood (WB)

Whole blood contains red blood cells (RBCs), plasma proteins, clotting factors, $\sim 10^9$ white blood cells (WBCs), and ~ 60 ml of anticoagulant-preservative solution. The total volume of a unit of WB is 500 ml $\pm$ 10%. The final hematocrit of a WB product ranges from 35 to 40%. The storage period is 21–35 days, depending on the type of anticoagulant-preservative used (see also Chapter 3).

B. Packed Red Blood Cells

The final volume and shelf-life of the different PRBC products depend on the type of anticoagulant-preservative or additive solution added. These solutions are described below.

1. CPDA-1

To create this PRBC product, 200–250 ml of plasma is removed from WB; some residual plasma is present, as well as approximately 10^8 leukocytes. The CPDA-1 anticoagulant-preservative solution consists of *C*itrate for anticoagulation, *P*hosphate buffers, *D*extrose for cell maintenance, and *A*denine to improve red cell viability; this allows a shelf-life of 35 days. The hematocrit of CPDA-1 PRBC units is < 80%. The volume is ~ 250 ml.

2. AS-1 (Additive Solution)

Most of the plasma has been removed from this PRBC product, and 100 ml of the AS-1 saline–nutrient solution has been added. This nutrient solution consists of dextrose, adenine, mannitol, and sodium chloride, which allows a shelf-life of 42 days. AS-1-preserved PRBCs also contain approximately 10^8 leukocytes. The hematocrit is ~ 50–60%. The final volume is ~ 350 mL.

C. Leukocyte-Reduced Packed Red Blood Cells (LR-PRBCs)

In LR-PRBCs, the leukocytes in each PRBC unit are significantly reduced (>99.9%) via leukocyte-reduction filters. Filtration is currently the most effective leukocyte-reduction method (see Chapter 14). These leukocyte-reduction filters may be utilized immediately following blood collection (prestorage leukoreduction), in the laboratory prior to issue of the component, or at the time of transfusion (bedside filtration). LR-PRBCs should contain ≥ 85% of the original PRBC mass. The volume and shelf-life of LR-PRBCs are identical to those of PRBCs, depending on the anticoagulant-preservative solution used. If indicated for the prevention of febrile nonhemolytic transfusion reactions (see Section VI.C and Chapter 30), final leukocyte count per RBC unit should be $< 5 \times 10^8$. Likewise, the final WBC count should be $< 5 \times 10^6$ if this product is intended to reduce the rate of alloimmunization or CMV transmission (see Section VI.C).

D. Frozen, Deglycerolized Packed Red Blood Cells (F / D-PRBCs)

PRBCs may be frozen for up to 10 years in the presence of a cryopreservative such as glycerol. PRBCs are frozen in this manner primarily to maintain a reserve of rare, antigen-negative PRBC units for use by those patients alloimmunized to high-frequency RBC antigens. After thawing, the glycerol is removed by washing before transfusion of the product. The end product is resuspended in normal saline and contains > 80% of the original PRBC mass, approximately 10^7 leukocytes (i.e., approximately 90% leukoreduced), minimal residual plasma, < 1% residual glycerol (420 mOsm), and < 300 mg of residual free hemoglobin. The final volume is ~ 250 ml. Once thawed and deglycerolized the F/D-PRBCs must be transfused within 24 hr.

E. Washed Packed Red Blood Cells (W-PRBCs)

PRBCs are washed with 1–2 liters of normal saline to reduce the amount of non-RBC elements (primarily plasma proteins, but also WBCs) in the component, in order to reduce the recurrence of severe allergic or anaphylactic reactions to plasma proteins. The RBC count is reduced by ~ 20%, the WBC count by ~ 85% (~ 10^7 residual WBC count), and plasma by ~ 99%. The final volume is ~ 300 ml; the washed PRBCs are suspended in saline rather than plasma. Following the washing procedure, W-PRBCs must be transfused within 24 hours.

F. Irradiated Red Blood Cells

PRBCs (and other blood components) may be irradiated for the prevention of transfusion-associated graft-versus-host disease (TA-GVHD). The minimum recommended dose of gamma radiation is 25 Gy. Because of the increase in supernatant potassium levels following irradiation, irradiated PRBCs must be transfused within 28 days of irradiation or before the original expiration date, whichever comes first.

II. COLLECTION, STORAGE, AND HANDLING

A WB unit is collected from a volunteer donor into the primary blood bag of a multiple-bag system, where it is mixed with anticoagulant-preservative solution. The WB unit is stored at room temperature if platelets are to be made from the unit; otherwise it is refrigerated. The WB unit is then centrifuged, usually within 8 hours of collection, to separate the WB into its two major components: PRBCs and platelet-rich plasma. The PRBCs remain in the primary blood bag following removal of the plasma. The duration of storage of the PRBC unit (at 1–6°C in a monitored, blood storage refrigerator) depends on the type of additive preservative solution with which it is mixed (35–42 days). The collection, storage, and handling processes for PRBCs are described in more detail in Chapter 3.

III. SPECIAL PREPARATION AND PROCESSING

A. Leukocyte Reduction

Leukocyte-reduction filtration (see Chapter 14) is currently the most effective means of reducing WBC count to $< 5 \times 10^6$ per PRBC unit (to reduce the risk of febrile nonhemolytic transfusion reactions, HLA alloimmunization, CMV transmission, and transfusion-related immunomodulation (Anonymous, 1998; Bowden et al., 1995; Sniecinski et al., 1988). Presently, most LR-PRBC units have $< 10^6$ residual WBCs. For quality control testing, the few remaining WBCs in the PRBC unit may be manually counted in a Nageotte chamber or measured by flow cytometry.

Filtration for leukocyte reduction may be performed at the blood center (prestorage leukoreduction) in the blood bank, or at the bedside. Special bedside filter sets are made for use with PRBC components and cannot be used interchangeably with filters made for platelets. In order to prevent the potentially serious and costly adverse effects caused by leukocytes in blood components, it has been advocated that all cellular blood products in the United States should be leukocyte-reduced ("universal prestorage leukoreduction") prior to being dispensed from the blood center. This policy has been recently adopted in Canada and Great Britain.

B. Washing

PRBC units are washed with 1–2 liters of normal saline in a semiautomated machine for the removal of ~99% of plasma proteins, electrolytes, and antibodies in order to prevent recurrent allergic and anaphylactic transfusion reactions. Another indication for which washed PRBCs are utilized is post-transfusion purpura. This procedure requires 20–30 minutes of additional preparation time.

C. Thawing and Deglycerolization

See Chapter 3 for a detailed description of the cryopreservation process, utilized primarily for the maintenance of a bank of rare, phenotype-specific, antigen-negative PRBCs. Thawing of the frozen PRBC unit takes place in a protective canister within a dry warmer or waterbath, and requires 20–40 minutes. The deglycerolization process that follows takes an additional 20–40 minutes. The physician should be aware of these added issues and time constraints when ordering frozen PRBCs, and the selection of such units for transfusion should be made in conjunction with the transfusion medicine physician and/or blood bank director.

D. Irradiation

PRBC units are irradiated for the purpose of preventing TA-GVHD (Anderson and Weinstein, 1990) using either self-contained irradiators or radiotherapy units (see Chapter 16). Depending on the size, location, and availability of the irradiation equipment, the process may require a few minutes to several hours.

IV. ABO AND Rh COMPATIBILITY

ABO and Rh compatibility are the two most important factors in the selection of appropriate blood for transfusion. *The infusion of ABO-incompatible blood into a recipient may cause severe acute hemolytic transfusion reactions, and anti-D alloantibody formed in a female recipient of childbearing age may lead to severe hemolytic disease of the newborn, both of which may have fatal consequences.* The routine compatibility tests

performed prior to transfusion are described below (see also Chapter 12 for a more detailed description of pretransfusion compatibility testing methodologies).

A. Compatibility Tests Performed

1. ABO Group

The ABO group of the recipient is determined for both red cells and serum. ABO typing may be performed using a number of different systems, but the basic concepts are identical. The recipient's RBCs are tested with known anti-A, anti-B, and anti-A,B (optional) sera to determine which antigens are present ("front type"). For instance, if the recipient's RBCs agglutinate in the presence of anti-A serum, but not anti-B, this would indicate the presence of A antigen on the RBCs and that the recipient is type A. In addition to front typing, the recipient's serum is also mixed with reagent RBCs of known A and B groups to identify naturally occurring ABO antibodies, if present ("back type"). For example, the same recipient's serum should agglutinate B reagent RBCs, showing the presence of anti-B in the recipient's serum, which should be present in a group A individual. ABO typing is performed on both the recipient's RBCs and his or her serum to best validate each test result, after which the appropriate ABO group may be chosen for transfusion to the recipient. ABO and Rh compatibility guidelines are listed in Table 1.

2. Rh Typing

There are more than 40 antigens in the Rh blood group system, but the D antigen is the most immunogenic and the most clinically relevant. Thus, it is the

TABLE I ABO and Rh Compatibility Guidelines

Patient's blood type	Compatible PRBCs	Compatible plasma components
A	A, O	A, AB
B	B, O	B, AB
AB	A, B, AB, O	AB
O	O	A, B, AB, O
Rh-positive	Rh-positive, Rh-negative	N/A
Rh-negative	Rh-negative[a]	N/A

[a] In emergency situations in which Rh-negative blood is not available, Rh-positive blood may be administered to a Rh-negative individual; however, this practice is best confined to male patients and postmenopausal women.

Rh(D) antigen that is referred to when describing a patient's Rh type: patients with the D antigen on their red cells are described as Rh-positive; patients without the D antigen are Rh-negative. In contrast to the ABO system, individuals who are Rh-negative and have not been exposed to Rh(D)-positive RBCs do not have anti-Rh(D) circulating in their serum.

Donors who initially test Rh-negative by standard methods are further tested for the presence of the weak D antigen with more specific antisera to ensure that the RBCs are truly Rh-negative. Although a donor may initially test Rh-negative, if the weak D antigen is present, he or she is considered to be Rh-positive (since weak D may be immunogenic). The donor Rh type chosen for transfusion is based on the Rh type of the recipient. ABO and Rh compatibility guidelines are listed in Table 1.

3. Antibody Screen

An antibody screen (indirect antibody [indirect Coombs'] test, or IAT) is performed on the patient's serum to determine the presence of unexpected RBC alloantibodies. Unexpected RBC alloantibodies are those alloantibodies formed to blood group antigens (e.g., K, Fy^a, Jk^a) following exposure to foreign RBCs, as might occur with prior transfusions, or in a prior pregnancy in which there was fetomaternal hemorrhage. If the recipient's antibody screen is positive, further testing is performed to identify the specificity of the antibody (if it is of clinical significance) so that antigen-negative PRBC products may be made available.

4. Crossmatch

The major crossmatch procedure is performed prior to the administration of any WB or PRBC component, except in emergency situations (see 5, below). In the major crossmatch, donor PRBCs and recipient serum are tested for compatibility using a 2% suspension of donor PRBCs in saline, adding recipient serum, centrifuging the sample, and observing for agglutination. If the antibody screen is positive or the patient has a history of an alloantibody, an additional test (antihuman globulin at 37°C) is performed.

The "computer crossmatch" is a method by which the laboratory testing for ABO compatibility is replaced by a computer check of the donor and recipient records, following antibody screening. Where used, this method has allowed for the vast majority of RBC units to be released without a laboratory crossmatch, and no hemolytic transfusion reactions have been reported (Safwenberg *et al.*, 1997).

5. The Emergency Transfusion

Minimum time requirements are necessary to provide compatible PRBCs; these are listed in Chapter 20. However, in an emergency situation where there is no time for ABO/Rh typing of the recipient and PRBC transfusion is needed immediately, *Group O, Rh-negative PRBC products ("universal donor") may be administered to all recipients.* Most blood banks require the signature of a physician for "emergency release" units to ensure that it is clear that the units are not crossmatched.

In a similar emergency situation, if Group O, Rh-negative products are unavailable, but Group O, Rh-positive products are available, Group O, Rh-positive RBCs may be administered to untyped recipients. However, it is preferable to limit this practice to male recipients and postmenopausal female recipients, to avoid the risk of immunizing a Rh-negative woman who may later become pregnant with a Rh-positive fetus.

For a semiemergent situation in which a limited amount of time is available before transfusion is necessary, type-specific (i.e., group A) but non-crossmatched blood may be administered. If a patient sample can be quickly delivered to the blood bank immediately upon the patient's arrival, or a prior sample is available, an emergency crossmatch can often be performed within 15–20 minutes. A complete test series of ABO and Rh type, antibody screen, and crossmatch requires at least 45–60 minutes if the antibody screen is negative. More time is needed if the antibody screen is positive, in order to identify the antibody's specificity, and to locate and prepare antigen-negative PRBC units, if indicated.

6. Incompatible RBC Administration

Occasionally, donor PRBCs that are incompatible with recipient serum will have been, or must be, administered. This occurs most often in the setting of an emergency, in which the first few "emergency release" units are later crossmatched and found to be incompatible with the recipient sample. Likewise, for a patient with autoimmune hemolytic anemia or an alloantibody to a high-incidence RBC antigen who is symptomatic and needs PRBC transfusion immediately, it is often impossible to locate crossmatch-compatible PRBCs. (If time permits, national rare donor registries of the American Red Cross or the American Association of Blood Banks may have rare, antigen-negative, frozen PRBCs for use by a recipient with a clinically significant alloantibody, or local rare donor files may locate a donor who could be called to emergently donate for a specific patient.)

If incompatible PRBCs must be issued, the donor PRBC blood bag should be labeled "incompatible" and the patient's physician should be notified as to the circumstances. If an alloantibody is present in the recipient's serum, the physician should be told whether or not it has been identified, and if so, if it is expected to be clinically significant. He or she should also be counseled to proceed cautiously with the PRBC transfusion and to be alert for the signs and symptoms of an acute hemolytic transfusion reaction.

V. ORDERING AND ADMINISTRATION

A. Informed Consent

Prior to initiating transfusion, except in emergency situations, the physician or his or her designee should obtain informed consent from the intended recipient of the PRBC transfusion (Sazama, 1997, and Chapter 41). Informed consent implies that the physician has explained to the patient the treatment plan, its risks and benefits, and available alternatives. Consent should be documented in the patient record, either by entering a detailed written note or by completing a standardized form signed by the patient or his or her guardian. If the patient is unable to consent to the transfusion and no responsible family member is present, or the situation requires emergent action, this information should also be reflected in the patient record.

B. The Physician's Order

A written order from the physician including the patient's first and last name, hospital identification number, the specific PRBC product requested, and the anticipated date of transfusion should be sent to the blood bank. If any of this information is incorrect or illegible, the blood bank must refuse to accept the request, in order to prevent misidentification and the potentially fatal consequences of a hemolytic transfusion reaction. The following orders are typically used:

1. Type and Screen

If the anticipated likelihood of transfusion of PRBCs is low (less than 10%), or if the need is not urgent, the "type and screen" is the appropriate order. In a type and screen procedure, the ABO group and Rh type of the patient are determined, and an antibody screen is performed on the patient's serum (requiring approximately 30–45 minutes). If no unexpected alloantibodies are identified in the screening procedure, crossmatch-

compatible units may usually be provided within 15–30 minutes of the subsequent order to crossmatch.

2. Type and Crossmatch

This is the appropriate order when the likelihood of PRBC transfusion is 10% or greater, and/or this need is likely to occur within 2 to 4 hours of the request. The patient's ABO group and Rh type are determined, an antibody screen is performed on the patient's serum, and the appropriate number of PRBC units are crossmatched.

Most hospitals set their own guidelines as to which surgical procedures require either no blood order, a type and screen, or a type and crossmatch, and the maximum, standard number of PRBC units that will be crossmatched for each procedure. These guidelines are known as maximum surgical blood order schedules (MSBOSs; Boral et al., 1979). Valuable blood resources are unnecessarily sequestered when crossmatches are ordered for procedures in which PRBCs are not likely to be utilized; thus, the MSBOS is valuable, should be based on local blood usage, and should be revised periodically to conserve blood resources.

C. Component and Recipient Identification

The blood sample sent to the blood bank must be labeled with the recipient's name and hospital identification number and must exactly match the information on the recipient's hospital identification arm band. A blood bank staff member labels the donor PRBC unit and matches it with the recipient's identification number prior to issue of the PRBC unit from the blood bank. The recipient and the labeled donor PRBC unit must again be identified by the person who will administer the transfusion immediately prior to its initiation. These steps are crucial, as clerical errors (e.g., patient misidentification) are the leading cause of fatal hemolytic transfusion reactions (Linden et al., 1992).

D. Transfusion Equipment

1. Needles and Catheters

An 18-gauge or larger needle or catheter is typically recommended for infusion of RBC components; smaller gauge catheters (23 gauge or larger) may be used if veins are small, but slower flow rates (and, potentially, hemolysis) will result.

2. Filters

Standard 170-micron in-line filters are used in the administration of all PRBC products for the purpose of trapping cellular debris and coagulated proteins formed during storage. One 170-micron filter may be used for the infusion of up to 4 units of PRBCs. Leukocyte reduction filters may be added to the infusion system to reduce the WBC count per PRBC unit to $< 5 \times 10^6$; these are discussed in Chapter 14.

3. Blood Warmers

Blood warming decreases the risk of cardiac arrhythmias, cardiac arrest, and cold-induced coagulopathies, and thus is helpful to: (1) patients receiving multiple transfusions of PRBCs at fast rates, including those receiving exchange transfusions, and those receiving plasma exchange, and (2) patients with cold agglutinin disease. Routine blood warming is not recommended. Appropriate blood warming devices are those having a visible thermometer and an audible alarm. The warmed temperature of the blood should not exceed 42°C. If blood is warmed and not used, it must be discarded. PRBCs should not be heated in a microwave oven, under hot tap water, or in a hot-water bath, without appropriate monitoring devices. Excessive warming may cause extensive RBC hemolysis (Iserson and Huestis, 1991).

4. Infusion Control Pumps and Pressure Devices

One unit of PRBCs must be infused within 4 hours of the time the transfusion begins. If gravity alone does not allow for an infusion rate sufficient to complete the transfusion within this time period, a pump may be used to exert positive pressure on the PRBC product and increase flow rate (Ciavarella and Snyder, 1988). A pressure device similar to a blood pressure cuff may also be used to increase the flow rate.

5. Intravenous Solutions

Medications and other intravenous fluids should *never* be added to PRBC products, with the exceptions of normal saline, plasma, 5% albumin, and plasma protein fraction. Five percent dextrose in water hemolyzes RBCs. Lactated Ringer's solution contains calcium that can create clots in the tubing. If blood is to be infused through an intravenous line that has been previously used for administration of other fluids, the tubing should be flushed with normal saline prior to beginning the PRBC infusion.

E. Infusion Rate

PRBC products may be administered as quickly as the patient is able to tolerate, up to 100 ml/min, but 1 unit of PRBCs must be infused within a 4-hour time

period to reduce the risk of bacterial contamination. If very slow rates are known to be required for a particular patient, the blood bank may be asked to split the PRBC product into aliquots, each of which may be infused over a 4-hour period.

VI. USES AND CONTRAINDICATIONS

PRBC products should be administered only for the purpose of increasing oxygen-carrying capacity in anemic patients who are at risk for ischemic events (see reports from the American Society of Anesthesiologists Task Force on Blood Component Therapy [Anonymous, 1996] and the American College of Physicians [Anonymous, 1992]). The decision to transfuse should be made based on the individual patient's vital signs, symptoms of decreased tissue oxygenation, cardiopulmonary reserve, rate and magnitude of blood loss, and oxygen consumption rate, and on the presence of significant atherosclerotic disease. The use of rigid "transfusion triggers" (i.e., the concept of transfusing PRBCs solely based on a hemoglobin value = 10 g/dl or hematocrit = 30%) has little scientific support and is outdated. *As a general concept only, transfusion of PRBCs is rarely indicated when the hemoglobin concentration is greater than 10 g/dl, and is almost always indicated when hemoglobin concentration is less than 6 g/dl.*

A. Whole Blood

Few, if any, clinical situations require the use of WB. After WB is stored for 24 hours, platelet function is essentially lost. The heat-labile coagulation factors V and VIII decrease to 5–30% by 21 days of storage, and thus WB is not considered adequate to prevent bleeding caused by deficiencies of these factors. Some leukocytes continue to maintain function throughout storage and cytokines accumulate, contributing to WBC-related transfusion complications (see Chapter 14). Donor WB is not routinely provided by blood centers or stored in blood banks. WB is, however, often used for autologous transfusion (see Section VIII.A).

B. Packed Red Blood Cells

Transfusion of PRBCs is indicated for anemic patients needing increased oxygen-carrying capacity, as explained in the beginning of Section VI, above.

C. Leukoreduced Packed Red Blood Cells

LR-PRBCs are indicated in the following circumstances:

1. To prevent febrile nonhemolytic transfusion reac-

tions (FNHTRs). WBCs synthesize and release various cytokines during storage that can cause the fever and chills associated with FNHTRs. If the WBC count in cellular blood products is reduced to $<5 \times 10^8$, the incidence of FNHTRs may be decreased; however, this is most effective when the WBCs are removed at the time of collection of the whole blood from the donor (prestorage leukoreduction).

2. To prevent the transmission of cylomegalovirus (CMV) by PRBCs to patients at risk, including neonates, CMV-negative pregnant women, and immunocompromised or immunoincompetent patients (i.e., bone marrow transplant, HIV-infected). CMV is believed to be latent within WBCs, and the reduction in WBC count of the PRBC product to $<5 \times 10^6$ reduces the risk of CMV transmission to that of CMV-seronegative PRBC transfusion.

3. To prevent or delay the onset of alloimmunization in patients receiving multiple transfusions (e.g., patients with sickle cell disease, thalassemia, acute leukemia, or aplastic anemia). These patients are continually exposed to human leukocyte antigen (HLA) proteins on residual WBCs within the PRBC units, and thus are more likely than a singly transfused patient to become alloimmunized to these HLA epitopes. Anti-HLA antibodies in the alloimmunized patient can cause refractoriness to platelet and granulocyte transfusion. Therefore, reduction of the WBC count of the PRBC product to $<5 \times 10^6$ to reduce the risk of HLA alloimmunization is prudent for these patient populations.

4. To reduce the immunomodulatory effect of blood transfusions, thereby reducing postoperative infections (Tartter *et al.*, 1998), viral reactivation (Hillyer *et al.*, 1999), and (potentially) tumor recurrence (Vamvakis, 1996). To avoid all of these potentially serious complications, it has been advocated that all cellular blood products should be leukoreduced prior to release from the blood center ("universal prestorage leukoreduction"). For these reasons, the United States will likely move to 100% leukoreduction within the next few years. Still, some authorities debate the magnitude of this immunomodulatory effect.

D. Frozen, Deglycerolized Packed Red Blood Cells

F/D-PRBCs are indicated in the following circumstances:

1. For use by those patients with alloantibodies to high-frequency red blood cell antigens, rare, antigen-negative PRBC units are stored in repositories. The American Association of Blood Banks (AABB) main-

tains a Rare Donor File to locate blood for patients needing rare or uncommon blood which can be accessed via the clinical institution's nearest AABB-accredited Immunohematology Reference Laboratory.

2. Autologous donations may be frozen if the time to the surgical procedure is extended past the storage date for the collected WB.

E. Washed Packed Red Blood Cells

Washing red cells with saline removes plasma from the PRBC product. W-PRBCs are indicated to reduce the recurrence of severe allergic or anaphylactic transfusion reactions thought to be caused by plasma proteins, including patients with IgA deficiency and anti-IgA antibodies (if IgA-deficient PRBCs are unavailable). Post-transfusion purpura is another indication for the use of washed PRBCs.

F. Irradiated Red Blood Cells

Gamma irradiation prevents the occurrence of TA-GVHD by altering the DNA in the WBCs such that they cannot divide and therefore cannot engraft in the recipient. Thus, irradiated PRBCs should be administered in the following circumstances:

1. Low-birth-weight neonates
2. Intrauterine transfusions
3. Bone marrow transplant patients
4. Patients with congenital immunodeficiency syndromes
5. Patients with Hodgkin's or non-Hodgkin's lymphoma undergoing intensive treatment
6. Patients being treated for acute leukemia
7. Units donated by a blood relative

VII. EXPECTED RESPONSE AND POTENTIAL ADVERSE EFFECTS

A. Expected Clinical Response

The expected increase, after receiving 1 unit of PRBCs or WB, in a nonbleeding patient's hematocrit is 3%, and in hemoglobin, 1 g/dl.

B. Potential Adverse Effects

Adverse effects of RBC transfusion include transfusion reactions of all types, including transfusion-transmitted infections. These are discussed extensively in Chapters 28–35, 37.

VIII. ALTERNATIVES TO ALLOGENEIC PRBC TRANSFUSION

A. Autologous Donation

Autologous donation of whole blood by preoperative collection is an alternative to allogeneic RBC transfusion (Goodnough and Brecher, 1998). In stable patients scheduled to undergo surgical procedures for which blood transfusion is likely, such as orthopedic procedures, vascular, cardiac, or thoracic surgery, and radical prostatectomy, preoperative autologous blood collection can significantly reduce exposure to allogeneic blood (AABB Technical Manual, 1999).

1. Donor Screening

Criteria are less rigid for donor selection in the case of autologous donation. However, the patient's hemoglobin should be no less than 11 g/dl at the time of donation.

Contraindications to autologous donation include:

a. Evidence of infection and risk of bacteremia
b. Aortic stenosis
c. Unstable angina
d. Active seizure disorder
e. Myocardial infarction or cerebrovascular accident within 6 months of donation
f. High-grade left main coronary artery disease
g. Cyanotic heart disease
h. Uncontrolled hypertension
i. Patients with significant pulmonary or cardiac disease not yet cleared for surgery by their physicians

2. Supplemental Iron

Patients should take supplemental iron prior to collection to maintain iron stores.

3. Volume Collected

For donors weighing > 50 kg, 450–500 ml of whole blood is typically collected. For patients weighing < 50 kg, there is a proportional reduction in the volume collected. In either case, the volume of whole blood collected should not exceed 10.5 ml/kg of the donor's body weight.

4. Collection Schedule and Storage

The collection of whole blood units should be scheduled as far in advance of surgery as possible, to allow adequate erythropoiesis for the prevention of anemia by the time surgery takes place. A weekly schedule of autologous whole blood collection is the most common practice, with the last collection occurring no less than 72 hours prior to surgery.

Autologous whole blood units can be safely stored refrigerated for up to 35 days. If the time period between collection and surgery is such that the unit may expire prior to its use, the unit may be frozen until the time of surgery.

5. Infectious Disease Testing

Autologous donations shipped outside the collection facility must be tested for syphilis, HIV, hepatitis C, and hepatitis B. If the unit tests positive for HIV or for hepatitis B surface antigen, the physician must submit a written request for use of the unit prior to its release. A "Biohazard" label must be added to those units that are positive for infectious disease markers.

6. Labeling and Release of Units

In addition to labeling with the patient's name and an identification number, each autologous unit must be labeled "Autologous Donor" and "For Autologous Use Only."

The AABB *Standards* do *not* permit the "crossing over" of autologous units for allogeneic use, because autologous donors often do not meet the strict requirements for allogeneic donors.

B. Intraoperative Blood Collection (Cell Salvage)

Intraoperative blood collection is typically performed by machines that collect the shed blood and then wash and concentrate the RBCs, creating 225-ml units of RBCs suspended in normal saline with a hematocrit of 55%. These washed, concentrated RBCs may then be further processed and stored, or they may be directly reinfused into the patient (Goodnough *et al.*, 1996). If collected aseptically and properly labeled, these RBCs may be stored for 6 hours at room temperature or for up to 24 hours at 1–6° C (if storage at 1–6°C is initiated within 6 hours of the beginning of the collection). *Contraindications to intraoperative blood collection include patients with malignant neoplasms or infections and the presence of a contaminated operative field.*

C. Acute Normovolemic Hemodilution (ANH)

ANH is the collection of whole blood from a patient immediately prior to anticipated blood loss and the subsequent replacement of this blood volume with an acellular fluid, such as crystalloid or colloid. The collected whole blood is later reinfused to the patient, typically as soon as major blood loss has ended (Goodnough *et al.*, 1992). The reported benefit of ANH is the reduction of RBC loss, due to the fact that whole blood shed during the surgical procedure is at lower hematocrit levels following ANH. This technique has been primarily utilized in Europe. For more information regarding ANH, the reader is referred to the AABB *Technical Manual* and Goodnough *et al.* (1992).

D. Postoperative Blood Collection

Postoperative blood collection is the recovery of blood from surgical drains for reinfusion to the patient. This sanguineous drainage is usually filtered and is not uniformly washed. The salvaged blood is dilute, may be hemolyzed, and may contain high levels of cytokines. Most programs set 1400 ml as an upper limit on the volume that may be reinfused. Transfusion must begin within 6 hours following collection, or the blood must be discarded. This procedure is used primarily in cardiac and orthopedic surgeries.

E. Pharmacologic Methods

1. DDAVP

Desmopressin (DDAVP) causes release of stored von Willebrand factor (vWF) and thereby increases levels of circulating factor VIII. Side effects include headache, tachycardia, hypertension, and tachyphylaxis (Manucci, 1998).

a. Indications

DDAVP is used as adjunctive therapy to decrease surgical blood loss intraoperatively and for bleeding in patients with hemophilia A, von Willebrand syndromes, and other platelet function disorders.

b. Contraindications

The use of DDAVP is contraindicated in patients with Type IIb von Willebrand disease.

2. Aprotinin

Aprotinin is a serine protease inhibitor that is used to decrease surgical blood loss, primarily in cardiac surgery patients, by an unknown mechanism thought to affect platelet adhesiveness and aggregation (Peters and Noble, 1999). It appears to also decrease postoperative blood loss, especially in those patients receiving aspirin prior to surgery. Side effects include anaphylaxis to bovine protein and renal toxicity.

F. Topical Agents (Fibrin Sealants, Fibrin Glue)

Topical agents serve as both physical and chemical barriers to blood loss (Radosevich *et al.*, 1997). They have a wide variety of uses, including: (1) decreasing blood loss in neurologic, ocular, gynecologic, obstetric, hepatic, and pulmonary surgical procedures, (2) skin graft fixation, and (3) decreasing all types of surgical bleeding in hemophiliacs. Rare side effects include anaphylaxis to bovine protein, if contained in the sealant.

G. Blood Substitutes

A number of blood substitutes, primarily human, bovine, or recombinant hemoglobin-based products, are currently undergoing Phase III trials after passing preliminary FDA safety studies. None of these hemoglobin-based products is available for patient use at this time in the United States. The only oxygen-carrying volume expander approved for use in the United States is a first-generation perfluorocarbon emulsion, and its use is restricted to perfusion of the coronary arteries following angioplasty. However, in the future, blood substitutes may become acceptable alternatives to PRBCs in certain situations, particularly in cases of trauma and massive transfusion (Dietz *et al.*, 1996).

H. Erythropoietin

Recombinant human erythropoietin (EPO) stimulates the bone marrow to produce new red cells. By day 7 of EPO treatment, the equivalent of 1 unit of blood is produced; by day 28, the equivalent of 5 units is produced. EPO is approved for use in the correction of anemia caused by: (1) end-stage renal disease, (2) AZT treatment in HIV-positive patients, and (3) chemotherapy in cancer patients, as well as in (4) anemic patients undergoing surgery. Other potential indications include the anemia of chronic disease, bone marrow transplantation, myelodysplastic syndromes, prematurity, and sickle cell disease (Goodnough *et al.*, 1997).

EPO is administered to anemic patients prior to surgery in order to decrease the need for intra- or postoperative allogeneic transfusion by: (1) increasing the hemoglobin concentration, (2) allowing for autologous predonation, or (3) enhancing the benefits of acute normovolemic hemodilution. In general, clinical trials suggest that EPO therapy is most beneficial in patients scheduled for surgery with initial hemoglobin values of 11–13 g/dl and anticipated blood losses of 1000 to 3000 ml.

Dosing regimens vary widely, with general recommendations as follows:

1. Medical patients: 50–300 U/kg IV or SQ 3×/week, with decreases in dosage once the hematocrit reaches desired levels.
2. Surgical patients (presurgical therapy):
   a. 100–600 U/kg SQ weekly × 4, or
   b. 600 U/kg SQ weekly × 2, then 300 U/kg SQ on the day of surgery + acute normovolemic hemodilution.

Contraindications to the use of EPO include those patients in whom therapy will result in polycythemia and in patients with uncontrolled hypertension. Side effects of EPO therapy include hypertension and polycythemia.

I. Jehovah's Witnesses

The religious beliefs of a Jehovah's Witness prevent him or her from receiving blood or blood product transfusions (Bennett and Shulman, 1997). As most blood substitutes are hemoglobin-based, these are not likely to become a viable alternative to PRBCs for these patients. Many Jehovah's Witnesses also refuse autologous predonation, as the blood is separated from the body for a period of time. Potential alternatives for reducing blood loss in Jehovah's Witnesses during surgery include hemodilution, autotransfusion devices, and intraoperative blood salvage (see Section B above). Recombinant human erythropoietin may be acceptable to some of these patients (see Section H above), although the product does contain a small amount of human albumin.

References

Anonymous (1992). Practice strategies for elective red blood cell transfusion. American College of Physicians. *Ann Intern Med* **116**, 403–406.

Anonymous (1996). Practice guidelines for blood component therapy: A report from the American Society of Anesthesiologists Task Force on Blood Component Therapy. *Anesthesiology* **84**, 732–747.

Anonymous (1998). Guidelines on the clinical use of leucocyte-depleted blood components. Blood Transfusion Task Force. *Transfusion Med* **8**, 59–71.

Anderson, K. C., and Weinstein, H. J. (1990). Transfusion-associated graft-versus-host disease. *N Engl J Med* **323**, 315–321.

Bennett, D. R., and Shulman, I. A. (1997). Practical issues when confronting the patient who refuses blood transfusion therapy. *Am J Clin Pathol* **107**, S23–S27.

Boral, L. I., Dannemiller, F. J., Stanford, W., *et al.* (1979). A guideline for anticipated blood usage during elective surgical procedures. *Am J Clin Pathol* **71**, 680–684.

Bowden, R. A., Slichter, S. J., Sayers, M., *et al.* (1995). A comparison of filtered leukocyte-reduced and cytomegalovirus (CMV) seronegative blood products for the prevention of transfusion-associated CMV infection after marrow transplant. *Blood* **86**, 3598–3603.

Ciavarella, D., and Snyder, E. (1988). Clinical use of blood transfusion devices. *Transfusion Med Rev* **2**, 95–111.

Dietz, N. M., Joyner, M. J., and Warner, M. A. (1996). Blood substitutes: Fluids, drugs, or miracle solutions? *Anesth Analg* **82**, 390–405.

Goodnough, L. T., and Brecher, M. E. (1998). Autologous blood transfusion. *Intern Med* **37**, 238–245.

Goodnough, L. T. Brecher, M. E., and Monk, T. G. (1992). Acute normovolemic hemodilution in surgery. *Hematology* **2**, 413–420.

Goodnough, L. T., Monk, T. G., Sicard, G., *et al.* (1996). Intraoperative salvage in patients undergoing elective abdominal aortic aneurysm repair: An analysis of cost and benefit. *J Vasc Surg* **24**, 213–218.

Goodnough, L. T., Monk, T. G., and Andriole, G. L. (1997). Current concepts: Erythropoietin therapy. *N Engl J Med* **336**, 933–938.

Hillyer, C. D., Lankford, K. V., Roback, J. D., *et al.* (1999). Transfusion of the HIV seropositive patient: Immunomodulation, viral reactivation, and limiting exposure to EBV (HHV-4), CMV (HHV-5), and HHV-6, 7, and 8. *Transfusion Med Rev* **13**, 1–17.

Iserson, K. V., and Huestis, D. W. (1991). Blood warming: Current applications and techniques. *Transfusion* **31**, 558–571.

Linden, J. V., Paul, B., and Dressler, K. P. (1992). A report of 104 transfusion errors in New York state. *Transfusion* **32**, 601–606.

Manucci, P. M. (1998). Drug therapy: Hemostatic drugs. *N Engl J Med* **339**, 245–253.

Menitove, J. E., Ed. (1999). "Standards for Blood Banks and Transfusion Services," 19th ed. AABB, Bethesda, MD.

Peters, D. C., and Noble, S. (1999). Aprotinin: An update of its pharmacology and therapeutic use in open heart surgery and coronary artery bypass surgery. *Drugs* **57**, 233–260.

Radosevich, M., Goubran, H. I., and Burnouf, T. (1997). Fibrin sealant: Scientific rationale, production methods, properties, and current clinical use. *Vox Sang* **72**, 133–143.

Ryden, S. E., and Oberman, H. A. (1975). Compatibility of common intravenous solutions with CPD blood. *Transfusion* **15**, 250–255.

Safwenberg, J., Hogman, C. F., and Cassemar, B. (1997). Computerized delivery control—A useful and safe complement to the type and screen compatibility testing. *Vox Sang* **72**, 162–168.

Sazama, K. (1997). Practical issues in informed consent for transfusion. *Am J Clin Pathol* **107**, S72–S74.

Sniecinski, I., O'Donnell, M. R., Nowicki, B., and Hill, L. R. (1988). Prevention of refractoriness and HLA-alloimmunization using filtered blood products. *Blood* **71**, 1402–1407.

Strautz, R. L., Nelson, J. M., Meyer, E. A., and Shulman, I. A. (1989). Compatibility of ADSOL-stored red cells with intravenous solutions. *Am J Emerg Med* **7**, 162–164.

Tartter, P. I., Mohandas, K., Azar P., *et al.* (1998). Randomized trial comparing packed red blood cell transfusion with and without leukocyte depletion for gastrointestinal surgery. *Am J Surg* **176**, 462–466.

Vamvakis, E. C. (1996). Transfusion-associated cancer recurrence and postoperative infection: Meta-analysis of randomized, controlled clinical trials. *Transfusion* **36**, 175–186.

Vengelen-Tyler, V., Ed. (1999). "Technical Manual," 13th ed. AABB, Bethesda, MD.

5

Fresh Frozen Plasma and Related Products

SILVANA Z. BUCUR
CHRISTOPHER D. HILLYER
Emory University School of Medicine
Atlanta, Georgia

Over the past several years, reviews of transfusion practices have demonstrated increased use of fresh frozen plasma (FFP). Since its administration is not free of untoward effects, vigilance against inappropriate use is of utmost importance. To this end, knowledge of plasma constituents and appreciation of the effectiveness of plasma transfusion in specific clinical situations are essential. Physicians should also be aware of alternative products and treatments and reserve the use of plasma products for those clinical situations in which FFP has been proven to be effective as primary therapy or in circumstances in which the more specific therapy is not available.

I. PRODUCT DESCRIPTION

Plasma, the aqueous component of blood, consists of approximately 85–90% water, representing 6–8% of total body water (40–50 ml/kg body weight), and has a density of 1.055–1.063 g/ml attributed to its various constituents, including protein and colloids (7%), and nutrients, crystalloids, hormones, and vitamins (2–3%; see Table 1). The concentration of anions, cations, and proteins in plasma confers its ionic properties. The bicarbonate content is the primary element responsible for the pH of normal plasma, which varies from 7.33 to 7.43 with temperature ranges from 37 to 4°C, respectively.

Human plasma contains more than 120 proteins with various physical and functional properties. Of these plasma proteins, albumin is the most abundant (3500–5000 mg/dl), and among its numerous functions, it also has an important role in maintaining normal plasma colloid oncotic pressure and blood pressure. Other plasma proteins include immunoglobulins and over 20 complement components, of which C3 predominates at a concentration of 1200 μg/dl. The albumin and immunoglobulin fractions can be isolated from plasma through the cold ethanol fractionation method developed by Cohn.

Variable concentrations of proteins involved in hemostasis, transport, and enzymatic processes are also present in plasma (see Table 1). One milliliter of fresh frozen plasma contains 2–3 mg fibrinogen, 60 μg factor XIII, 5–10 μg von Willebrand factor (vWF), and approximately 1 unit of activity of each of the stable coagulation factors (II, VII, IX, X). The concentration of factor VIII in plasma is approximately 100 ng/ml and is found primarily bound to vWF. Plasma also contains variable amounts of vitamins, hormones, and trace elements, although it is the protein fraction containing the coagulation factors for which plasma is most commonly prescribed.

One unit of plasma product is the quantity of plasma that has been obtained from 1 unit of whole blood and usually contains between 180 and 300 ml of anticoagulated plasma. Plasmapheresis of a single donor can yield as much as 500–800 ml of plasma per unit, known as "jumbo-plasma units." Plasma products contain minimal numbers of red blood cells (RBCs) and platelets, but may contain a small number of viable white blood cells (WBCs).

Several different plasma products can be retrieved from whole blood and through plasmapheresis. "Source plasma" or "single-donor plasma" is obtained through

TABLE 1 Selected Plasma Constituents

1. Solutes	4. Proteins (*continued*)
Ionic solutes	(b) Albumins
Na^+ (142 mEq/liter) Cl^- (103 mEq/liter)	Albumin (3500–5000 mg/dl—60% of plasma colloid)
K^+ (4 mEq/liter) HCO_3^- (27 mEq/liter)	Vitamin-D-binding protein (40 mg/dl)
Mg^2 (2 mEq/liter) HPO_3^- (2 mEq/liter)	Other (trace amounts)
Ca^{2+} (5 mEq/liter) SO_4^{2-} (1 mEq/liter)	(c) Coagulation and fibrinolytic proteins
Nonionic solutes	Factor V (10 μg/ml) Fibrinogen (2–3 mg/ml)
Glucose (100 mg/dl)	Factor VII (very low) Prothrombin (100 μg/ml)
Cholesterol (200 mg/dl)	Factor VIII (100 ng/ml) Antithrombin III (0.15 mg/dl)
Triglycerides (200 mg/dl)	Factor IX (5 μg/ml) Heparin cofactor II (0.06 mg/dl)
Urea (15 mg/dl)	Factor X (10 μg/ml) alpha$_2$-antiplasmin (7 mg/dl)
Other ionic solutes (21 mEq/liter)	Factor XI (5 μg/ml) von Willebrand factor (5–10 μg/ml)
2. Trace elements	Factor XII (30 μg/ml) Protein C inhibitor (0.004 mg/dl)
copper (70–155 μg/dl)	Factor XIII (60 μg/ml) Plasminogen (20 mg/dl)
Zinc (70–150 μg/dl)	Other: prekallikrein, kininogens, protein S, protein C,
Iron (40–160 μg/dl)	tissue factor pathway inhibitor
Silicon (40–1000 μg/dl)	(d) Complement components
Other: selenium, manganese, nickel, chromium	C1–9 (25–1200 μg/dl, with C3 the most abundant at 1200 μg/dl)
(trace amounts)	Properdin (25 μg/dl)
3. Vitamins	Factors D, H, I, P (35–500 μg/dl)
Ascorbic acid (600–1000 μg/dl)	S protein or vitronectin (500 μg/dl)
Cobalamin (0.02–0.08 μg/dl)	C4 bp and C8 bp (500 μg/dl)
Folate (0.2–0.9 μg/dl)	(e) Other proteins
Retinol (30–65 μg/dl)	Haptoglobin (variable)
Tocopherol (500–2000 μg/dl)	C-reactive protein (variable)
4. Proteins	alpha$_1$-antitrypsin (25 mg/dl)
(a) Immunoglobulins	C1 inhibitor (200 μg/dl)
IgG (5–14 mg/dl)	Other transport proteins and enzymes (variable)
IgM (0.3–2.5 mg/dl)	
IgA (0.5–5.0 mg/dl)	
IgD (0.01 mg/dl)	
IgE (trace amounts)	

plasmapheresis of a single donor, and is stored at −20°C after its collection. "Recovered plasma" refers to plasma obtained from whole blood of a regular donor. "Fresh-frozen plasma" is separated from whole blood within 6 hours of collection and is stored frozen at −18°C or colder for up to 1 year. "Plasma frozen within 24 hours" (F24) is separated from whole blood within 24 hours of collection and is stored frozen at −18°C or colder for up to 1 year. "Cryosupernatant" is plasma depleted of its cryoprecipitate fraction and then refrozen. "Solvent/detergent plasma" (SD plasma) is a pooled product that has undergone viral inactivation

through treatment with a combination of an organic solvent and a nonionic detergent.

II. COLLECTION, STORAGE, AND HANDLING

There are two principal methods by which plasma may be obtained. One method is through centrifugation or sedimentation of whole blood that separates the cellular blood elements from plasma. Alternatively, plasma may be collected through plasmapheresis as a single product or as a by-product of platelet or RBC apheresis. For the purpose of plasma product preparation, the whole blood may be collected in any anticoagulant preservative solution except heparin, and may be processed either fresh (within 6–24 hours of collection) or outdated (within 26 days of collection). While the intrinsic quality of plasma is donor dependent, the conditions during collection, processing, and storage will affect the final concentration of its various components, specifically the concentration of coagulation proteins. Although obtained through different methods, both source plasma from plasmapheresis and recovered plasma retrieved from whole blood may be fractionated into various plasma derivatives.

Because the processing and storage conditions for recovered plasma are not as stringent as those for source plasma, over the last several years its use has been restricted to clinical situations in which coagulation factor levels are not critical (e.g., trauma, intraoperatively). Although no longer in use in the United States, "liquid plasma" was obtained from whole blood within 5 days of its expiration date, refrigerated, and fractionated into other plasma components. Processing and storage requirements for FFP, F24, cryosupernatant, and SD plasma are described in the section above. Only FFP, F24, SD plasma, source plasma, and cryosupernatant are FDA-approved for use in the United States.

III. SPECIAL PREPARATION AND PROCESSING

A. SD Plasma

Solvent detergent is a newly approved method of viral inactivation applied to plasma. This method uses an organic solvent, tri(n-butyl)phosphate (TNBP), in combination with a nonionic detergent, Triton X-100. Solvent/detergent treatment of plasma results in highly efficient inactivation of lipid-enveloped viruses, including human immunodeficiency virus (HIV) 1/2, hepatitis B virus (HBV), hepatitis C virus (HCV), human T-lymphocytotropic virus (HTLV) I/II, hepatitis G virus (HGV), vesicular stomatitis virus, Sindbis virus, and Sendai virus. SD plasma is made from pools of 2500 donor units and has similar concentrations of most coagulation factors that FFP does.

Although solvent/detergent treatment inactivates lipid-enveloped viruses, which represent the vast majority of pathogenic viruses, it has no effect on nonenveloped viruses. Therefore, there were initial concerns that administration of such a product may lead to the transmission of some non-lipid-enveloped viruses, especially parvovirus B19 and hepatitis A virus (HAV).

However, SD plasma is now tested for parvovirus B19 and HAV by PCR, and high-titer pools are destroyed. Also, since SD plasma is a pooled product, it contains the respective antiviral antibodies; thus, transmission of such viruses is likely not a clinically significant concern (Smak Gregoor *et al.*, 1993; Prowse *et al.*, 1997; Horowitz *et al.*, 1992).

B. Cryoprecipitate and Cryosupernatant

These are prepared by thawing FFP at 4°C. During this procedure, a white precipitate (cryoprecipitate) forms that is rich in factor VIII, von Willebrand factor, fibrinogen, and factor XIII, and is separated from the plasma supernatant (cryosupernatant). Each product is refrozen and stored frozen separately for future use.

IV. ABO AND Rh COMPATIBILITY

As FFP contains isohemagglutinins it must be ABO compatible with the recipient. However, specific compatibility tests are not performed prior to infusion of FFP (see Table 1 of Chapter 4). If the recipient's blood group is not known prior to FFP administration, AB plasma should be administered. While Rh type may be recorded on the label of FFP units, alloimmunization is rarely a consideration, as there are so few RBCs in FFP.

V. ORDERING AND ADMINISTRATION

A. Thawing and Storage

FFP is thawed over 20–30 minutes at a temperature between 30 and 37°C. The activities of the labile coagulation factors, specifically factors V and VIII, decrease gradually after thawing, and thus it is recommended that FFP is best used soon after thawing. FFP is typically outdated 24 hours after thawing when it is stored at 1–6°C. For certain indications, however, some blood banks use thawed plasma for up to 5 days, if it is appropriately labeled as "thawed plasma." The process of freeze–thaw compromises the integrity, and thus the activity, of coagulation proteins. Thus, FFP should not be refrozen.

B. Dosing

The dose or volume of FFP to be administered is primarily dependent on the clinical situation, the patient's plasma volume, and the desired incremental increase in coagulation factor activity. In general, a starting dose of 8–15 ml/kg is recommended and can be ordered as number of milliliters to be infused, though in clinical practice the volume is approximated to the nearest whole unit of FFP. For example, if a 70-kg patient is given a starting dose of 15 ml/kg, this results in a total dose of 1050 ml, or approximately 4–6 units, of FFP.

An infusion rate of 4–10 ml/min has been suggested, although FFP may be infused as rapidly as tolerated. The frequency and duration of infusion of FFP depend on the half-life of the factor being repleted and the clinical response. Although normalization of laboratory parameters is an oft-desired endpoint, it may not always be possible to assess the true clinical response to FFP infusion by this criterion.

VI. USES AND CONTRAINDICATIONS FOR PLASMA ADMINISTRATION

A. Uses

Plasma transfusion is indicated for patients who either are actively bleeding or are at risk for bleeding due to an underlying coagulopathy presumed to be responsive to plasma infusion (Table 2). An increased risk of bleeding is assumed if the prothrombin time (PT) is >16–18 seconds or the partial thromboplastin time (PTT) is >55–60 seconds, or exceeds 1.5–1.8 times the control. It should be noted that for patients with no previous history of a bleeding disorder, abnormal PT and/or PTT values are not good predictors for intraoperative bleeding. As FFP contains normal (not concentrated) plasma levels of all coagulation factors, its administration in usual quantities will not increase coagulation factor levels by more than 20–30% without high-volume dosing, causing a concurrent risk of volume overload (Cohen, 1993). Specific disorders for which plasma is administered are listed below. At present, the indications for FFP, F24, and solvent/detergent plasma are equivalent.

1. Multiple Coagulation Factor Deficiencies

a. Liver Disease

The liver is the site of synthesis for most coagulation proteins. Patients with severe liver dysfunction manifest a coagulopathy of complex pathophysiology involving decreased and/or abnormal synthesis of coagulation proteins, increased consumption of clotting factors and platelets, activation of fibrinolysis, and thrombocy-

TABLE 2 Indications and Contraindications for Plasma Administration

A. Indications

Bleeding diathesis associated with multiple acquired coagulation factor deficiency:

End-stage liver disease

Massive transfusions

Disseminated intravascular coagulation

Rapid reversal of warfarin effect

Plasma infusion or exchange for TTP or Refsum's disease

Congenital coagulation defects (except when specific therapy is available)

C1-esterase inhibitor deficiency

B. Relative indications and investigational uses

Burns

Meningococcal sepsis

Acute renal failure in the context of multiorgan failure

C. Absolute contraindications

Reconstitution of "packed" red cells, except in neonatal exchange transfusion

Volume expansion

Source of nutrients

Immunodeficiency

Wound healing

topenia due to hypersplenism. Laboratory assessment of the coagulopathy reveals initially increased PT and/or PTT values, and as the disease worsens, the thrombin time (TT) may become prolonged, fibrin split products may be increased, and decreased fibrinogen levels may be present. Thrombocytopenia may be present at any stage of the disease (Kemkes-Matthes et al., 1991; Garcia-Avello et al., 1998; Blanchard et al., 1981). The coagulopathy associated with severe liver disease places the patient at increased risk of bleeding, especially during invasive procedures (liver biopsy, paracentesis).

To minimize the risk of bleeding prior to invasive procedures or to attain hemostatic control during episodes of active bleeding, infusion of FFP (and possibly platelets, prothrombin complex, or antithrombin III concentrate) may be indicated. Although laboratory parameters assessing hemostasis are abnormal in patients with liver disease, there is poor correlation between laboratory abnormalities and risk of bleeding in these patients. Therefore, the goal of therapy with FFP infusion is to achieve clinical hemostasis rather than to normalize laboratory parameters (PT, PTT), as these may be impossible to correct even with the administration of large quantities of FFP. The administration of

FFP at a dose of 15–20 ml/kg, the equivalent of approximately 6 units every 4–6 hours, has been shown to be effective therapeutically (Humphries, 1994; Gerlach *et al.*, 1993).

b. Familial Coagulation Factor Deficiencies

Most congenital, isolated coagulation factor deficiencies are treated with specific factor concentrates (see Chapter 9). However, rare isolated and multiple congenital factor deficiencies are amenable to treatment with FFP. In these patients, the level of the deficient factor is measured, and a dose of FFP is administered which has been calculated to contain an adequate amount of the deficient factor (CAP, 1994; Am. Soc. Anesth., 1996).

2. Rapid Reversal of Warfarin Effect

For certain coagulation factors (II, VII, IX, X) and regulatory proteins (C and S), a vitamin-K-dependent, post-translational gamma-carboxylation of the amino terminus is necessary to render them active. Impaired gamma-carboxylation, either due to warfarin therapy or as a result of liver disease, impedes the coagulation cascade and can lead to bleeding.

This warfarin-induced coagulopathy is completely reversed within 48 hours after the discontinuation of warfarin, provided that the patient's vitamin K level is normal. Although, administration of vitamin K corrects the coagulopathy within 12–18 hours, FFP infusion as a single dose of 15–20 ml/kg is appropriate when rapid reversal of warfarin-induced coagulopathy is necessary, as in active bleeding, trauma, or emergent surgical procedures. It is noteworthy that in order to attempt to simply correct the laboratory abnormalities (e.g., PT), infusion of a very large volume of FFP may be required, creating potential for volume overload in susceptible patients. *Thus, therapy with FFP should be guided by clinical response and not by normalization of laboratory parameters, since, as previously stated, these laboratory values may not correlate with risk of bleeding* (CAP, 1994; Am. Soc. Anesth., 1996; Cohen, 1993).

3. Massive Transfusion

Dilutional coagulopathy is a state of relative deficit of coagulation factors and platelets induced by massive transfusion—the infusion of large volumes (greater than 1 blood volume replaced within 24 hours) of replacement fluids that lack coagulation factors, such as PRBC or crystalloid solutions. Studies evaluating the effectiveness of FFP in massive transfusions have demonstrated that adequate hemostasis is usually achieved through correction of the dilutional throm-

bocytopenia by platelet transfusion. However, many physicians transfuse FFP in addition to platelets. It is unlikely that FFP transfusion is beneficial in stable patients without evidence of acute hemorrhage who have received < 10 units of PRBCs within a 24-hour period (Hewson *et al.*, 1985; CAP, 1994).

4. Disseminated Intravascular Coagulation (DIC)

DIC is characterized by increased consumption of coagulation factors and platelets combined with concomitant deposition of thrombin in the microvasculature. It is a syndrome caused by a variety of clinical conditions, including obstetric complications, gram-negative sepsis, liver disease, trauma, promyelocytic leukemia, and other malignancies. Both bleeding and thrombosis are clinical manifestations of DIC. The laboratory abnormalities reflect activation of the coagulation system with subsequent depletion of clotting factors, activation of the fibrinolytic system, and activation of inhibitor proteins (see Chapter 23).

Treatment of the underlying cause for the DIC is essential. Additional therapy is dictated by the predominant clinical feature present, thrombosis versus bleeding. In general, treatment with FFP is directed at repleting the clotting factors consumed in DIC. FFP infusions are useful in ameliorating the bleeding diathesis or decreasing the bleeding risk in the face of an elevated PT (> 1.5 times control). Hyperfibrinolysis with decreased fibrinogen level (< 100 mg/dl) may also be treated with FFP if volume overload is not a concern, although cryoprecipitate provides a greater concentration of fibrinogen per unit volume than does FFP. It is especially important to replete coagulation factors with FFP in the setting of DIC associated with liver disease, as the coagulopathy is aggravated by the inadequate hepatic synthetic capacity (Bick, 1992; Rubin and Colman, 1992; Tsuzuki *et al.*, 1990).

Bleeding unresponsive to FFP and cryoprecipitate replacement therapy may be treated with antifibrinolytic agents (ϵ-aminocaproic acid or tranexamic acid), with or without heparin, or other agents proven to be effective in experimental DIC, including antithrombin III (AT III) and protein C concentrates. When thrombosis is the predominant clinical feature, low doses of heparin alone or in combination with antifibrinolytic agents (ϵ-aminocaproic acid) may be beneficial (Rubin and Colman, 1992).

5. Thrombotic Thrombocytopenic Purpura (TTP)

Four related clinical entities with common clinical features have been described: TTP, childhood hemolytic-uremic syndrome (1-LUS), adult HUS, and HELLP (a pregnancy-related syndrome of "hemolysis, elevated liver enzymes, and low platelets"). While these entities have commonalities in clinical presentation and manifestations, they differ in pathogenesis.

In acquired, nonfamilial TTP, the pathophysiologic mechanism involves a serum metalloproteinase that is needed to reduce ULvWF multimers to active vWF, which is required for normal platelet adhesion. This metalloproteinase can be affected either quantitatively or qualitatively by an IgG inhibitor that impedes its normal function. Acquired, nonfamilial TTP is treated with therapeutic plasma exchange (TPE) using FFP or cryosupernatant as replacement fluid. Thus, TPE allows for: (1) the removal of the IgG inhibitor, and (2) replacement of normal serum metalloproteinase. This is the standard and recommended therapy for TTP. In recurrent or refractory TTP, cryosupernatant is often used as replacement fluid instead of FFP, as cryosupernatant contains fewer of the ULvWF multimers that enhance platelet aggregability and lead to vascular endothelial damage (Obrador et al., 1993; Rock et al., 1991, 1992).

Although the standard therapeutic approach for patients with TTP is TPE with either FFP or cryosupernatant, simple infusions of FFP have induced remission in a few patients with these disorders, primarily in those patients with familial TTP who have a congenital deficiency of the metalloproteinase and no IgG inhibitor. The other syndromes, HUS and HELLP, are not due to deficiencies of the metalloproteinase and are not responsive to either FFP infusion or TPE (Martin et al., 1995).

6. Contraindications

The use of FFP is not without risk. Because of the risks of FFP administration and the availability of better alternative therapies, *there is no justification for the use of FFP to expand plasma volume, augment albumin concentration, or improve the nutritional status of debilitated patients.* Although FFP contains immunoglobulins, fractionated immunoglobulin preparations are preferred over FFP infusion as replacement therapy for immunodeficiency states. Burns, meningococcal sepsis, and acute renal failure with multiorgan failure may be associated with either low-grade or overt DIC, and in this context, coagulation factor replacement with FFP may be beneficial, although other alternative therapies might best be considered first (AT III concentrate, protein C concentrate) (Churchwell et al., 1995; Cohen, 1993).

VII. EXPECTED RESPONSE AND POTENTIAL ADVERSE EFFECTS

A. Response

The effectiveness of FFP depends on the severity of the bleeding diathesis, the underlying etiology, and the tolerance of the patient to large volumes of FFP. Patients with impaired cardiac or renal function are susceptible to volume overload and thus particular attention should be paid to both total volume infused and the rate of infusion. It is of paramount importance for the physician to understand that effectiveness of FFP should be assessed more often on clinical grounds, as the normalization of laboratory parameters may not always be possible.

B. Adverse Effects

Administration of FFP can be associated with fever, chills, and allergic reactions, ranging from mild urticaria to anaphylaxis and pulmonary complications. These rare but potentially life-threatening allergic reactions are presumably due to donor antibodies that react with the recipient's WBCs. The presence of isohemagglutinins in FFP may cause mild hemolytic reactions or result in a positive direct antiglobulin test if "out-of-group" or incompatible plasma is administered. The IgA-deficient recipient possessing antibodies to IgA needs to receive plasma that does not contain IgA in order to avoid life-threatening anaphylaxis. In such cases, IgA-deficient plasma must usually be procured from a national rare donor registry. Most mild to moderate allergic reactions, however, may be treated symptomatically (Isbister, 1993, Hanson et al., 1988; Pineda and Taswell, 1975).

FFP transfusions pose little or no risk of transmission of cell-associated infections such as cytomegalovirus or HTLV-I and -II. Although the transmission of other viral agents (HBV, HCV, and HIV) is greatly reduced through the use of thorough donor questionnaires and screening tests, it is not completely eliminated. Thus, other methods of inactivating viruses have been recently instituted, including solvent/detergent treatment (Section III). The solvent/detergent process inactivates lipid-encapsulated viruses and preserves the structural and functional integrity and concentration of most coagulation proteins (Horowitz et al., 1992; Schreiber et al., 1996). There are no specific adverse effects of solvent/detergent plasma or contraindications to its use.

VIII. ALTERNATIVES

The use of FFP in the treatment of coagulation factor deficiencies should be restricted to those situations in which specific factor concentrates are not available or immediate control of bleeding is essential. Thus, replacement with factor concentrates, including factor VIII, prothrombin complex concentrates (PCCs), activated factor IX complex, activated factor VII, AT III, C1-esterase inhibitor concentrate, and alpha$_1$-antitrypsin concentrate, should be considered when clinically indicated (see Chapter 9). *In clinical situations in which the major factor to be repleted is fibrinogen, cryoprecipitate is the best alternative.* In addition, the use of DDAVP and cryoprecipitate in the treatment of von Willebrand's disease may be more appropriate than the use of FFP. Replacement of vitamin-K itself should be considered in vitamin-K-deficient states or in nonemergent situations requiring the reversal of warfarin effect.

References

Am. Soc. Anesth. (1996). Practice guidelines for blood component therapy. A report by the American Society of Anesthesiologists Task Force on Blood Component Therapy. *Anesthesiology* **84**, 732–736.

Bick, R. L. (1992). Disseminated intravascular coagulation. *Hematol Oncol Clin N Am* **6**, 1259–1285.

Blanchard, R. A., Furie, B. C., Jorgensen, M., *et al.* (1981). Acquired vitamin K dependent carboxylation deficiency in liver disease. *N Engl J Med* **305**, 242–248.

Churchwell, K. B., McManus, M. L., Kent, P., *et al.* (1995). Intensive blood and plasma exchange for treatment of coagulopathy in Meningococcemia. *J Clin Apher* **10**, 171–177.

Cohen, H. (1993). Avoiding misuse of fresh frozen plasma. *Br Med J* **307**, 395–396.

College of American Pathologists (CAP) (1994). Practice parameters for the use of fresh-frozen plasma, cryoprecipitate, and platelets. *JAMA* **271**, 777–810.

Garcia-Avello, A., Lorente, J. A., Cesar-Perez, J., *et al.* (1998). Degree of hypercoagulability and hyperfibrinolysis is related to organ failure and prognosis after burn trauma. *Thromb Res* **89**, 59–64.

Gerlach, H., Rossaint, R., Bechstein, W. O., *et al.* (1993). "Goal-directed" transfusion management leads to distinct reduction of fluid requirement in liver transplantation. *Semin Thromb Hemost* **19**, 282–285.

Hanson, L. A., Bjorkander, J., Carisson, B., *et al.* (1988). The heterogeneity of IgA deficiency. *J Clin Immunol* **8**, 159–162.

Hewson, J. R., Neame, P. B., Kumar, N., *et al.* (1985). Coagulopathy related to dilution and hypotension during massive transfusion. *Crit Care Med* **13**, 387–391.

Horowitz, B., Bonomo, R., Prince, A. M., *et al.* (1992). Solvent/detergent-treated plasma: A virus-inactivated substitute for fresh frozen plasma. *Blood* **79**, 826–831.

Humphries, J. E. (1994). Transfusion therapy in acquired coagulopathies. *Hematol Oncol Clin N Am* **8**, 1181–1201.

Isbister, J. P. (1993). Adverse reactions to plasma and plasma components. *Anaesth Intens Care* **21**, 31–38.

Kemkes-Matthes, B., Bleyl, H., and Matthes, K. S. (1991). Coagulation activation in liver disease. *Throm Res* **64**, 253–261.

Martin, J. N., Jr., Files, J. C., Blake, P. G., *et al.* (1995). Postpartum plasma exchange for atypical preeclampsia-eclampsia as HELLP (hemolysis, elevated liver enzymes, and low platelets) syndrome. *Am J Obstet Gynecol* **172**, 1107–1125.

Obrador, G. T., Ziegler, Z. R., Shadduck, R. K., *et al.* (1993). Effectiveness of cryosupernatant therapy in refractory and chronic relapsing thrombotic thrombocytopenic purpura. *Am J Hematol* **42**, 217–220.

Pineda, A. A., and Taswell, H. F. (1975). Transfusion reactions associated with anti IgA antibodies: Report of four cases and review of the literature. *Transfusion* **15**, 10–15.

Prowse, C., Ludlam, C. A., and Yap, P. L. (1997). Human parvovirus B19 in blood products. *Vox Sang* **72**, 1–10.

Rock, G. A., Shumak, K. H., Buskard, N. A., *et al.* (1991). Comparison of plasma exchange with plasma infusion in the treatment of thrombotic thrombocytopenic purpura. *N Engl J Med* **325**, 393–397.

Rock, G., Shumak, K., Kelton, J., *et al.* (1992). Thrombotic thrombocytopenic purpura: Outcome in 24 patients with renal impairment treated with plasma exchange. *Transfusion* **32**, 710–714.

Rubin, R. N., and Colman, R. W. (1992). Disseminated intravascular coagulation. Approach to treatment. *Drugs* **44**, 963–971.

Schreiber, G. B., Busch, M. P., Kleinman, S. H., *et al.* (1996). The risk of transfusion-transmitted viral infections. *N Engl J Med* **334**, 1685–1690.

Smak Gregoor, P. H. J., Harvey, M. S., Briaet, E., *et al.* (1993). Coagulation parameters of COP fresh-frozen plasma and CDP cryoprecipitate-poor plasma after storage at 4°C for 28 days. *Transfusion* **33**, 735–738.

Tsuzuki, T., Toyama, K., Nakaysu, K., *et al.* (1990). Disseminated intravascular coagulation after hepatic resection. *Surgery* **107**, 172–176.

6

Cryoprecipitate and Related Products

SILVANA Z. BUCUR
CHRISTOPHER D. HILLYER
Emory University School of Medicine
Atlanta, Georgia

Cryoprecipitate, also known as cryoprecipitate antihemophilic factor (cryoprecipitate AHF), is the blood product containing the highest concentrations of factor VIII (FVIII:C), von Willebrand's factor (vWF), factor XIII (FXIII), fibrinogen, and fibronectin. Until the 1980s, cryoprecipitate was the preferred product for the treatment of von Willebrand's disease and mild hemophilia A (Brettler and Levine, 1989). With the development of methods for extracting factor concentrates from plasma, improved procedures for viral inactivation of these factor concentrates, and recombinant protein manufacture, the role of cryoprecipitate in the management of these entities has largely become secondary (see Chapter 9). Cryoprecipitate is currently used for the treatment of: (1) congenital and acquired states of hypofibrinogenemia, (2) factor XIII deficiency, (3) certain types of von Willebrand's disease, and (4) rarely for mild hemophilia A.

I. PRODUCT DESCRIPTION

Cryoprecipitate is the insoluble cold precipitate formed when fresh frozen plasma (FFP) is thawed between 1 and 6°C. The recovered precipitate is then refrozen in 10–15 ml of plasma within 1 hour of separation and stored at −18°C or colder for a maximum of 1 year. Each bag of cryoprecipitate contains between 150 and 250 mg of fibrinogen, 30–60 mg of fibronectin, and 80–120 units of factor VIII. In addition, cryoprecipitate also contains approximately 40–70% (80 units) of original von Willebrand factor and approximately 30% (40–60 units) of the initial concentration of factor XIII.

II. COLLECTION, STORAGE, AND HANDLING

A. Collection and Storage

Cryoprecipitate is the protein fraction extracted from FFP through cold precipitation at 1–6°C. Once obtained, cryoprecipitate is resuspended in 5–10 ml of residual plasma and refrozen at −18°C or colder for a maximum of 1 year. Rarely, the cryoprecipitate is frozen in 15–20 ml of plasma, and this product is known as "wet cryo." Wet cryo has the advantage of not requiring diluent when pooled.

B. Preparation for Transfusion

Prior to administration, the frozen cryoprecipitate is thawed for 5–10 minutes in a water bath at 30–37°C. The final product consists of a pool of several units collected from different donors, as indicated on the label. Depending on the initial method of preparation, specifically the amount of plasma the cryoprecipitate was packaged in, pooling to a single container may require that each unit of cryoprecipitate be mixed with 5–10 ml of diluent (i.e., normal saline) so as to ensure complete removal of the product from its container. As multiple freeze–thaw procedures may denature the protein content of cryoprecipitate, it is recommended that cryoprecipitate not be refrozen. Once thawed, cryoprecipitate should be kept at room temperature (20–24°C) and transfused within 6 hours of thawing or 4 hours after the unit has been entered and pooled.

III. SPECIAL PREPARATION AND PROCESSING

A. Quality Assurance

To ensure quality, assays to determine FVIII:C activity and fibrinogen content are performed on selected cryoprecipitate units by the collection facility. While cryoprecipitate prepared from solvent/detergent (SD)-treated plasma is not commercially available at the time this writing, this product is expected to contain adequate concentrations of fibrinogen and FVIII:C when it becomes available (Pehta, 1996; Smak Gregoor *et al.*, 1993).

B. Fibrin Glue and Fibrin Sealant

Cryoprecipitate has been used to make fibrin glue, the extemporaneous, non-FDA-approved mixture of fibrinogen and thrombin prepared in the operating room and applied immediately to the surgical site to stop bleeding. However, a newly FDA-approved, commercially available product called fibrin sealant is now available, consisting of a protein concentrate–fibrinolytic inhibitor solution and a thrombin solution that when mixed using a warming device or water bath (at 37°C) and applied to the bleeding site forms a solid, adherent sealant. The fibrinogen source in fibrin sealant is a lyophilized, heat-treated, sealer protein concentrate made from pooled human cryoprecipitate. This concentrate is reconstituted in a fibrinolysis inhibitor solution (bovine source), containing both aprotinin and plasmin. The hemostatic mechanism of both fibrin glue and fibrin sealant relies upon the action of thrombin on fibrinogen, resulting in fibrin clot formation at the site of bleeding.

IV. ABO AND Rh COMPATIBILITY

Since cryoprecipitate is packaged in small volumes of plasma, it may contain small amounts of anti-A and anti-B antibodies. While an ABO-compatible product is preferred, especially for pediatric patients with small blood volumes, compatibility testing is not required prior to its use. Since cryoprecipitate does not contain RBCs, Rh testing is unnecessary for this product.

V. ORDERING AND ADMINISTRATION

As cryoprecipitate is stored frozen and several units must be pooled prior to each use to obtain an adequate single dose, approximately 20–30 minutes must be allowed for thawing and pooling of the product. The dose of cryoprecipitate to be administered is ideally based on the level of the specific factor that needs to be repleted (fibrinogen, vWF, FVIII:C, or FXIII) and the desired factor incremental increase. In fibrinogen deficiency, the number of units of cryoprecipitate necessary to correct the fibriniogen deficit can be calculated with the following formula:

$$\text{Desired fibrinogen increment (g/liter)} = (0.2 \times \text{number of units of cryo})/ \text{plasma volume in liters.}$$

Alternatively, an empirically estimated dose of 1 unit of cryoprecipitate for every 5 kg of body weight can be used. In general, the initial infusion rate of cryoprecipitate is 5–10 ml/min, and may be increased as the patient tolerates. Although fibrinogen has a half-life of 3–5 days, in order to estimate the frequency of transfusion of cryoprecipitate, its diffusion phase half-life (12 hours) and its recovery (50%) need to be taken into account. Therefore, in congenital hypofibrinogenemia, an every-other-day schedule of cryoprecipitate infusion is recommended during episodes of active bleeding.

There is no standardized assay to determine the vWF content and/or activity in each unit of cryoprecipitate, and thus, *an approximate dose of 1 unit of cryoprecipitate per 10 kg of body weight is recommended for the replacement of vWF.* Because of the variability in procoagulant activity and half-life of vWF in the cryoprecipitate product, replacement by daily infusions is recommended.

Cryoprecipitate administration in factor XIII deficiency is based on the long half-life of factor XIII (9 days) and the very low plasma levels of factor XIII required for hemostasis (less than 5%). Thus, the general recommendation for cryoprecipitate administration for factor XIII deficiency is 1 unit of cryoprecipitate per 10–20 kg of body weight administered every 2–3 weeks.

VI. USES AND CONTRAINDICATIONS

Cryoprecipitate AHF is the blood product richest in FVIII:C, vWF, FXIII, fibrinogen, and fibronectin. In general, this blood product is used to treat bleeding

TABLE 1   Clinical Uses for Cryoprecipitate

Uses

   Fibrinogen deficiency (congenital and acquired)

   Von Willebrand's disease

   Factor XIII deficiency

   Renal stone removal

   Orthotopic liver transplantation (OLT)

   Poststreptokinase therapy (hyperfibrinogenolysis)

Possible uses

   Uremic bleeding

   Fibrin glue/sealant

   Hemophilia A (when FVIII:C concentrates are not available)

   Wound healing (as a source of fibronectin)[a]

Contraindications

   Sepsis (postoperative)

[a] Investigational.

conditions associated with deficiencies in fibrinogen or factor XIII. Occasionally, it may be used to treat bleeding related to von Willebrand's disease (vWD) and mild forms of hemophilia A, though factor VIII concentrates are recommended. No clear indications exist for the use of cryoprecipitate as a source of fibronectin. Indications and contraindications to cryoprecipitate administrations are listed in Table 1. (For dose and frequency of administration, refer to Section V.) Indications for the use of cryoprecipitate are as follows:

A. Fibrinogen Deficiency States

The prothrombin time (PT) and partial thromboplastin time (PTT) are usually prolonged when plasma fibrinogen level falls below 100 mg/dl, even in the face of normal levels of all other coagulation factors. This situation occurs in acquired or congenital hypofibrinogenemia and in L-asparaginase therapy. Although isolated hypofibrinogenemia is rarely associated with bleeding, correction of the fibrinogen deficit is necessary during active bleeding and prior to invasive procedures. Humphries (1994) reviewed the use of cryoprecipitate and found that its major indication is as a fibrinogen source for replacement in:

1. Acquired conditions associated with hypofibrinogenemia, dysfibrinogenemia, liver disease, orthotopic liver transplant (OLT), disseminated intravascular coagulopathy (DIC), and L-asparaginase therapy.

2. Congenital afibrinogenemia. In this consumptive coagulopathy associated with a variety of other disorders, depletion of other coagulation factors in addition to fibrinogen occurs. Therefore, FFP may also be needed as source of replacement for these factors. The rate of degradation of the specific coagulation factors needed dictates the infusion rate of cryoprecipitate and/or FFP in any individual patient (Francis and Armstrong, 1982; Humphries, 1994; Am. Soc. Anesth., 1996).

B. Von Willebrand's Disease

vWD is an autosomally inherited hemorrhagic disorder, in which vWF, a complex adhesive glycoprotein required for normal platelet interaction with the endothelium, is either deficient or dysfunctional. In the circulation, vWF associates with FVIII and stabilizes it; thus, FVIII levels may also be decreased in this disorder. In general, DDAVP (desmopressin acetate, [1-deamino, 8-D-arginine]vasopressin), which potentiates the release of vWF and FVIII:C from vascular endothelium, has been used as the primary therapy for most types of vWD. Cryoprecipitate provides sufficient quantities of both vWF and FVIII:C in a relatively small volume. Several reviews of the efficacy of cryoprecipitate in vWD (Ewenstein, 1997; Menache and Aronson, 1997; Rinder et al., 1997) confirm that cryoprecipitate (or FVIII:C concentrate) is considered the primary therapy in approximately 10–20% of patients with vWD who are unresponsive to DDAVP (vWD type 2B and type 3). Cryoprecipitate may also be used as replacement therapy in cases of DDAVP tachyphylaxis following repeated use. The effectiveness of cryoprecipitate is assessed both clinically and through monitoring of bleeding times (Ewenstein, 1997; de la Fuente et al., 1985; Menache and Aronson, 1997; Rinder et al., 1997; Brettler and Levine, 1989).

C. Factor XIII Deficiency

Factor XIII, also known as fibrin-stabilizing factor, is essential to the formation of an insoluble fibrin clot. Factor XIII deficiency is inherited in an autosomal recessive fashion and manifests initially in the neonatal period as umbilical stump bleeding. The most feared complication of factor XIII deficiency is spontaneous intracranial hemorrhage, and thus prophylactic therapy is required but relatively simple to institute, in view of factor XIII's long biologic half-life and the low level of plasma factor XIII activity necessary to prevent bleeding (Girolami et al., 1991; Am. Soc. Anesth., 1996; CAP, 1994).

D. Fibrin Glue and Fibrin Sealant

Fibrin glue causes a hemostatic fibrin plug, resulting from the action of thrombin on fibrinogen. Commer-

TABLE 2 Clinical Uses for Fibrin Glue and Fibrin Sealant

Approved uses

    Cardiopulmonary bypass

    Heparinized patients undergoing coronary artery bypass
      graft (CABG)

    Splenic injury

    Colostomy closure

Selected investigational uses

    Orthopedic surgery: meniscal tear repair

    Urology: establishing patency of vasovasostomy

    Neurosurgery: wound closure; securing prosthetic devices; nerve
      anastomoses; dura repair; fascial repair

    Pediatric surgery: postoperative neonatal chylothorax repair

cially prepared fibrin sealants have been recently licensed in the United States and are indicated as adjuncts to conventional surgical hemostatic techniques (e.g., suture, ligature, and cautery) in heparinized cardiopulmonary bypass surgical patients, in the treatment of splenic injuries, and in the closure of colostomies. In addition, fibrin sealant has been used successfully as an investigational tool in many other types of surgical procedures that require tight hemostatic control to preserve tissue integrity (see Table 2; Donovan, 1995; Kollias and Fox, 1996; Martinowitz et al., 1996; McCarthy, 1993; Rousou et al., 1989; Schlag, 1994).

E. Wound Healing

Several small studies have demonstrated that the application of cryoprecipitate directly to wounds promotes healing, with this beneficial effect attributed to its fibronectin content. Definitive data are anticipated (Powell and Doran, 1991).

F. Uremic Bleeding

The pathogenetic mechanism of uremic bleeding is thought to be related primarily to impaired platelet function and defective interaction between platelets and vascular entothelium. It seems reasonable, then, that a source of active vWF (e.g., cryoprecipitate) might correct the platelet dysfunction associated with uremia. Although cryoprecipitate administration does not affect platelet aggregability, it does shorten the bleeding time. A few small studies and case reports have reported an improvement in bleeding time and prevention of postoperative bleeding following cryoprecipitate administration. Other studies have shown that as many as 50% of uremic patients fail to respond to cryoprecipitate. Based on these small studies, there is some evidence to suggest a beneficial role of cryoprecipitate in the treatment of uremic bleeding, although more definitive data are needed. Other therapeutic modalities for the treatment of uremic bleeding include dialysis, PRBC and platelet transfusions, and the administration of erythropoietin (EPO), DDAVP, and conjugated estrogens (Weigert and Schafer, 1998; CAP, 1994; Am. Soc. Anesth., 1996).

G. Other Uses

Cryoprecipitate infused into the renal pelvis along with thrombin and calcium forms an easily removable gel that entraps renal stones and small particulates, and thus facilitates the complete removal of renal stones (Fischer et al., 1980). Cryoprecipitate administration as a source of fibrinogen decreases bleeding during orthotopic liver transplantation and poststreptokinase therapy.

In mild forms of hemophilia A (plasma factor VIII levels of 5–50% of normal), bleeding may occur following trauma or surgical intervention. DDAVP, the preferred treatment, stimulates the release of factor VIII from storage sites in sufficient quantities to provide adequate hemostatic levels of factor VIII. With repeated use of DDAVP, however, tachyphylaxis often develops; in addition, rare patients may initially demonstrate inadequate response to DDAVP. In these circumstances of inadequate response to DDAVP, cryoprecipitate may be used successfully to stop acute bleeding. However, with the availability of virally inactivated and recombinant factor VIII products, the use of cryoprecipitate for this indication has become infrequent.

VII. EXPECTED RESPONSE AND POTENTIAL ADVERSE EFFECTS

A. Clinical Response

Response to cryoprecipitate administration should be assessed by clinical improvement, as well as by laboratory assays dictated by the clinical situation. These may include fibrinogen level in cases of hypofibrinogenemia, and bleeding time, PT, and PTT in patients with vWD.

B. Adverse Effects

1. Cryoprecipitate

As with the administration of any blood product, the recipient is at increased risk of developing fever, chills, allergic reactions, and transmission of infectious disease following cryoprecipitate administration. As cryoprecipitate contains isohemagglutinins, a positive direct antiglobulin test (DAT) may develop following transfusion; rarely, hemolysis can occur following its administration (Isbister, 1993; Schreiber *et al.*, 1996).

2. Fibrin Sealant

The only major adverse reaction and therefore *a contraindication to fibrin sealant use is prior allergic or anaphylactoid reactions to bovine protein.* Occasionally, the development of factor V antibodies after exposure to bovine thrombin has resulted in abnormal thrombin time (TT) assays and has been incriminated in postoperative bleeding.

VIII. ALTERNATIVES

A. Hemophilia A

Currently, cryoprecipitate is rarely used in the treatment of hemophilia A. Either pooled or recombinant human factor VIII concentrate is administered to patients with hemophilia A during episodes of active bleeding. These products have the advantage of relatively little or no risk of viral transmission, either because of the virucidal procedure used (as in solvent/detergent pooled plasma-derived concentrates) or by virtue of manufacturing techniques (recombinant-derived factors; Manucci, 1993; Schreiber *et al.*, 1996).

B. vWD

Until recently, the mainstay of therapy in vWD was cryoprecipitate, which carries a small but real risk of transmission of blood-borne viruses. Plasma-derived, virucidally treated products containing vWF have become available and have proved effective in the treatment of vWD, though they are not yet licensed for this indication. Currently, products such as intermediate- and high-purity factor VIII/vWF concentrates are clinically preferred over cryoprecipitate for the treatment of vWD nonresponsive to DDAVP (primary therapy). With the development of a variety of other factor concentrates, including vWF concentrate, factor XIII concentrate, and fibrinogen, the only indication for cryoprecipitate administration in the near future may be hypofibrinogenemia.

References

Am. Soc. Anesth. (1996). Practice guidelines for blood component therapy. A report by the American Society of Anesthesiologists Task Force on Blood Component Therapy. *Anesthesiology* **84**, 732–738.

Brettler, D. B., and Levine, P. H. (1989). Factor concentrates for treatment of hemophilia. Which one to choose? *Blood* **73**, 2067–2073.

College of American Pathologists (1994). Practice parameter for the use of fresh-frozen plasma, cryoprecipitate, and platelets. *JAMA* **271**, 777–781.

De la Fuente, B., Kasper, C. K., Rickles, F. R., *et al.* (1985). Response of patients with mild and moderate hemophilia A and von Willebrand's disease to treatment with desmopressin. *Ann Intern Med* **103**, 6–14.

Donovan, J. F., Jr. (1995). Microscopic vasovasostomy: Current practice and future trends. *Microsurgery* **16**, 325–332.

Ewenstein, B. M. (1997). Von Willebrand's disease. *Annu Rev Med* **48**, 525–542.

Fischer, C. P., Sonda, L. P., and Dionko, A. C. (1980). Further experience with cryoprecipitate coagulum in renal calculus surgery: A review of 60 cases. *J Urol* **126**, 432–436.

Francis, J. L., and Armstrong, D. J. (1982). Acquired dysfibrinogenemia in liver disease. *J Clin Pathol* **35**, 667–675.

Girolami, A., Sartori, M. T., and Simioni, P. (1991). An updated classification of factor XIII defect. *Br J Haematol* **77**, 565–566.

Humphries, J. E. (1994). Transfusion therapy in acquired coagulopathies. *Transfusion Med* **8**, 1181–1201.

Isbister, J. P. (1993). Adverse reactions to plasma and plasma components. *Anaesth Intens Care* **21**, 31–38.

Kemkes-Matthes, B., Bleyl, H., and Matthes, K. S. (1991). Coagulation activation in liver disease. *Thromb Res* **64**, 253–261.

Kollias, S. L., and Fox, J. M. (1996). Meniscal repair. Where do we go from here? *Clin Sports Med* **15**, 621–630.

Mannucci, P. M. (1993). Modern treatment of hemophilia: From the shadows toward the light. *Thromb Haemost* **70**, 17–23.

Martinowitz, U., Schulman, S., Horoszowski, H., *et al.* (1996). Role of fibrin sealants in surgical procedures on patients with hemostatic disorders. *Clin Ortho Related Res* **328**, 65–75.

McCarthy, P. M. (1993). Fibrin glue in cardiothoracic surgery. *Transfusion Med Rev* **7**, 173–229.

Menache, D., and Aronson, D. L. (1997). New treatments of von Willebrand disease: Plasma derived von Willebrand factor concentrates. *Thromb Hemost* **78**, 566–570.

Pehta, J. C. (1996). Clinical studies with solvent detergent-treated products. *Transfusion Med Rev* **10**, 303–311.

Powell, F. S., and Doran, J. E. (1991). Current status of fibronectin in transfusion medicine: Focus on clinical studies. *Vox Sang* **60**, 193–202.

Rinder, M. R., Richard, R. E., and Rinder, H. M. (1997). Acquired von Willebrand's disease: A concise review. *Am J Hematol* **54**, 139–145.

Rousou, J., Levitsky, S., Gonzalez-Lavin, L., *et al.* (1989). Randomized clinical trials of fibrin sealant in patients undergoing restemotomy or reoperation after cardiac operations. A multicenter study. *J Thorac Cardiovasc Surg* **97**, 194–203.

Schlag, G. (1994). Immuno's fibrin sealant: The European experience. Abstract from Symposium on Fibrin Sealant: Characteristics and Clinical Uses. Uniformed Services University of the Health Sciences, Bethesda, MD, Dec. 8–9.

Schreiber, G. B., Busch, M. P., Kleinman, S. H., *et al.* (1996). The risk of transfusion-transmitted viral infections. *N Engl J Med* **334**, 1685–1690.

Smak Gregoor, P. H. J., Harvey, M. S., Briaet, E., *et al.* (1993). Coagulation parameters of CDP fresh-frozen plasma and CDP cryoprecipitate-poor plasma after storage at 4C for 28 days. *Transfusion* **33**, 735–738.

Weigert, A. L., and Schafer, Al. (1998). Uremic bleeding: Pathogenesis and therapy. *Am J Med Sci* **316**, 94–104.

Platelets and Related Products

JOHN D. ROBACK
CHRISTOPHER D. HILLYER
Department of Pathology and Laboratory Medicine
Emory University School of Medicine
Atlanta, Georgia

Platelet transfusions are a critical facet of supportive therapy for thrombocytopenic patients. In fact, estimates indicate that up to 20% of hematology/oncology inpatient costs are associated with platelet transfusions (Gordon *et al.*, 1996). Furthermore, a subset of thrombocytopenic patients will require specialized components, including filtered, gamma-irradiated, and cross-matched or human leukocyte antigen (HLA)-matched platelets. Provision of these components can add significantly to the baseline costs. This chapter discusses standard and specialized platelet preparations with an emphasis on indications, contraindications, expected results, and potential adverse effects associated with transfusion of these components. A thorough understanding of the proper use of platelet transfusions will lead to the most clinically beneficial and cost-effective use of these blood components.

I. PLATELET PRODUCTS

A. Product Description

Two types of platelet components are available in most hospital settings (Figure 1): platelet concentrates (also known as "random-donor platelets") and apheresis-derived platelets (also known as "single-donor platelets" or "platelets, pheresis"). The hemostatic activity of platelets is similar in either component. However, platelet concentrate and apheresis-derived platelet units differ in methods of preparation, platelet con-

tents, and potential for adverse effects in the transfusion recipient.

1. Platelet Concentrates

Each platelet concentrate unit is prepared from a single donated unit of whole blood by sequential low and high *g*-force centrifugation. A single platelet concentrate has a volume of approximately 50 ml. Although the mandated minimum platelet content is 5.5×10^{10} platelets in 75% of platelet concentrates randomly selected for testing, platelet concentrates usually contain approximately 7×10^{10} platelets. The remaining volume of the platelet concentrate is composed of donor plasma and preservative–anticoagulant solution. Prior to transfusion, the blood bank staff pools 5–8 platelet concentrates (each from a different donor) into a single product containing approximately $3-6 \times 10^{11}$ platelets. The resulting pooled platelet product is designated as a single dose. Each pooled platelet product exposes the transfusion recipient to the risk of infectious disease transmission from 5–8 donors. In addition, each pooled platelet product also contains 10^8 or more leukocytes, which may produce adverse transfusion reactions as described below.

2. Apheresis Platelets (APs)

Each apheresis-derived platelet component is prepared from a single volunteer platelet donor by a semiautomated apheresis process. Donors are con-

Platelet components	Special processing	Uses as:
Platelet concentrates Apheresis-derived platelets	CMV-seronegative Filtered Washed Gamma-irradiated Volume-reduced ABO-matched Crossmatched HLA-matched UVB-irradiated	I. Prophylactic transfusions, patients with: • Plts < 10,000/µl • Plts < 20,000/µl with fever, bleeding, coagulation disorders, neonates II. Prophylactic transfusions, prior to invasive procedures in patients with: • Plts < 50,000/µl for minor surgery • Plts < 100,000/µl for major surgery III. Therapeutic transfusions • Plts < 50,000/µl with active hemorrhage

FIGURE 1 Summary of platelet components and their processing and clinical uses.

nected via a peripheral line (or lines) to a blood cell separator that withdraws and fractionates whole blood from the donor, removes platelets to a separate container, and then returns the remaining platelet-depleted blood to the donor. ACD-A or similar citrate-containing anticoagulants are usually employed. The process takes about 1–2 hours and results in an apheresis-derived platelets unit with approximately 3–8 $\times 10^{11}$ platelets in 200 ml final volume (mandated minimum platelet content is 3×10^{11} platelets in 75% of units tested).

Apheresis-derived platelets usually contain 10^4 to 10^6 leukocytes ($\leq 1\%$ of the leukocytes in a platelet concentrate pool; as such, they are labeled "leukoreduced") and less than 0.5 ml of contaminating RBCs, demonstrating the selectivity of the apheresis process. As compared to platelet concentrates, the low leukocyte content of apheresis-derived platelets reduces the risk of leukocyte-mediated adverse transfusion effects, such as febrile transfusion reactions. In addition, each apheresis-derived platelet unit exposes the recipient to only the infectious disease risk of a single donor. Finally, apheresis-derived platelets can be efficiently crossmatched and HLA-matched for transfusion to refractory recipients.

B. Collection, Storage, and Handling

Platelet concentrates and apheresis-derived platelets, prepared from volunteer donors as described above, are stored at 20–24°C with continuous gentle agitation. Under these conditions, platelets remain functional for up to 7 days. However, the maximum platelet storage period is limited to 5 days, because platelet units stored beyond this point carry an unacceptably high risk of bacterial contamination.

C. Special Preparation, Processing, and Selection

A number of options are available for selecting or preparing platelet components to meet the requirements of specific transfusion situations. Each utilizes special devices and/or methodologies that increase the cost of the transfused unit, require a significant amount of time to perform, and can alter the component expiration date. (The reader is also referred to Chapters 4, 14, and 17 for detailed descriptions of leukoreduced, gamma-irradiated, washed, and volume-reduced blood components, as well as a description of the selection process of special platelet components for platelet-refractory patients.)

1. CMV-Seronegative Components

Cytomegalovirus (CMV), a member of the human herpes virus family (HHV-5), infects and remains latent in peripheral blood leukocytes, including those of the monocyte/macrophage lineage (Bolovan-Fritts et al., 1999). CMV can be transmitted by blood transfusions from an infected donor, presumably with latently infected leukocytes acting as the vector. Prior to the institution of preventative measures, 13–37% of immunocompromised transfusion recipients contracted CMV infections from unscreened blood components (Miller et al., 1991; Yeager et al., 1981). In many of these patients, transfusion-transmitted CMV (TT-CMV) infections caused significant morbidity and mortality.

The "gold standard" for the prevention of TT-CMV is the use of blood components from donors without detectable anti-CMV antibodies (CMV-seronegative), since these donors have likely not been infected with CMV. The incidence of TT-CMV following transfusion of CMV-seronegative blood components is 0–3% (Bowden et al., 1995). However, less than half of the blood donor population is CMV-seronegative, and thus seronegative components are often in limited supply. As discussed below, the use of leukoreduced components has recently emerged as an alternative to the use of CMV-seronegative components for the prevention of TT-CMV.

2. Leukoreduced Units

Platelet components can be passed through polyester-based filtration devices to remove the leukocytes that can produce adverse transfusion reactions. Most commercially available leukoreduction filtration devices remove over 99.9% of leukocytes (3-$\log_{10}$ leukoreduction), resulting in $1–5 \times 10^6$ or fewer residual leukocytes per filtered platelet concentrate pool. Since the apheresis process effectively depletes leukocytes from apheresis-derived platelets (process leukoreduction), these components are considered leukoreduced. If further leukoreduction is desired, apheresis-derived platelets can be filtered prior to transfusion. Although currently the subject of investigation, leukoreduction to 10^6 or fewer leukocytes per platelet dose is generally considered sufficient to decrease the risk of CMV transmission, febrile nonhemolytic reactions, and HLA alloimmunization (these and other adverse effects of platelet transfusions are discussed in more detail in Chapters 15, 24, and 29). In some clinical settings, such as in adult patients with acute myelogenous leukemia, the use of filtered blood components for these indications has proven cost-effective (Balducci et al., 1993).

Filtration in the blood center at the time of component processing (prestorage leukoreduction) is generally considered superior to bedside filtration during transfusion. Prestorage leukoreduction prevents increases in levels of white blood cell (WBC)-synthesized proinflammatory cytokines during platelet storage, which can result in febrile transfusion reactions. In addition, prestorage leukoreduction is performed by trained technical personnel and is subject to tight quality control. Given current trends in transfusion practices, it appears likely that within 2–5 years leukocyte reduction filtration will be uniformly applied to all platelet and packed red blood cell (PRBC) components prior to transfusion.

3. Washing

For patients who experience repetitive episodes of allergic transfusion reactions, even with diphenhydramine premedication, platelet components washed with saline may be beneficial. While platelet washing removes plasma proteins that can initiate allergic reactions, washing can also lead to loss and functional impairment of platelets.

In practice, washed platelets are rarely used. However, if IgA-deficient components are not available, the use of washed components may be necessary to prevent potentially life-threatening anaphylactic reactions in IgA-deficient transfusion recipients who express anti-IgA antibodies (Toth et al., 1998). Patients with paroxysmal nocturnal hemoglobinuria have historically been transfused with washed components, although the clinical benefits of this approach have been questioned. A more detailed discussion of the uses of washed blood components can be found in Chapter 17.

4. Gamma Irradiation

Transfusion recipients with acquired or congenital immunodeficiencies, immature immune systems, or those receiving closely HLA-matched platelet transfusions (especially from first-degree family members) are at risk of developing transfusion-associated graft-versus-host disease (TA-GVHD). There is no effective treatment for TA-GVHD, which has an associated mortality rate of > 90%. However, TA-GVHD can be effectively prevented by exposing blood components to a minimum of 2500 cGy of gamma irradiation prior to transfusion to at-risk patients (see Chapter 4).

5. Volume Reduction

In some transfusion situations, for example, in the pediatric setting, it may be necessary to transfuse a given dose of platelets in a reduced volume of plasma. Alternatively, it may be desirable to volume-reduce group O units with high-titer isohemagglutinins prior to transfusion of non-group O recipients. The preparation of volume-reduced platelets by centrifugation can be performed in the blood bank. However, it should be noted that up to half of the platelets in the original unit may be lost during volume reduction (Simon and Sierra, 1984), and thus the utility of this procedure has been questioned. Please see Chapter 17 for further discussion of volume-reduced components.

6. Fresh, ABO-Matched Platelets

Post-transfusion platelet increments decrease with increasing storage time of platelet units. Thus, in clinically refractory patients, a trial of fresh (< 24 hours old), ABO-matched platelets is warranted prior to the use of either crossmatched or HLA-matched components. The role of ABO compatibility in the efficacy of platelet transfusion is discussed in more detail below.

7. Crossmatched Platelets

Thrombocytopenic patients can display alloimmune refractoriness to transfused platelet units that are selected at random from the blood bank inventory. These patients can often be supported with specific platelet units selected by crossmatching or HLA matching. Conceptually, platelet crossmatching is analogous to PRBC crossmatching. Using the recipient's serum, platelet components are screened to identify donor platelets lacking surface epitopes targeted by the recipient's alloantibodies. Because of differences in patient selection criteria and crossmatching methodologies, the fraction of crossmatched transfusions that result in acceptable increases in platelet counts has varied in different studies, ranging from 50 to 90% (Freedman *et al.*, 1984; Heal *et al.*, 1987). The use of crossmatch-compatible platelets is discussed in more detail in Chapter 24.

8. HLA-Matched Platelets

Since most antibodies causing platelet refractoriness are directed against allogeneic HLA class I epitopes, transfusion of platelets from related or unrelated donors whose HLA type closely matches the recipient's can often produce good platelet increments. The use of HLA-matched platelet transfusions in the refractory patient is discussed in more detail in Chapter 23.

D. ABO and Rh Compatibility

For many patients requiring platelet transfusions, the ABO and Rh phenotypes of the blood donor and recipient are not critical factors. However, these issues can become important for patients who require numerous platelet transfusions and/or become refractory to platelet transfusions.

1. Effects on Platelet Responses

The small amounts of A and B antigen found on platelets can lead to poor platelet increments in recipients with high titers of the corresponding anti-A and/or anti-B isoagglutinins (Carr *et al.*, 1990; Lee and Schiffer, 1989). While ABO-mismatched platelet transfusions are not specifically contraindicated, and indeed are often performed, thrombocytopenic patients that develop refractoriness to nonmatched platelet components may display improved responses to ABO-matched platelets. In addition, transfusion of ABO-mismatched platelets has also been found to accelerate the development of platelet refractoriness (Carr *et al.*, 1990). Thus, for patients likely to require long-term platelet support, it may be prudent to supply leukocyte-filtered, ABO-compatible platelets, in order to delay the development of platelet alloimmunization.

2. Potential for RBC Hemolysis

Platelet components usually contain less than 1 ml of RBCs, an amount insufficient to produce significant clinical effects if hemolyzed by recipient antibodies after transfusion. However, this level of RBC contamination can produce recipient sensitization to RBC alloantigens, including the development of anti-Rh antibodies, which in turn can complicate future PRBC transfusions, or future pregnancies in the case of an Rh-negative woman. It Rh-positive platelets must be given to patients in whom Rh immunization could pose a future problem, an appropriate dose of RhIg should be concomitantly administered based on an estimate of 1 ml of RBCs per platelet concentrate (see Chapter 10, Table 5, for indications and dosages of RhIg following exposure to blood products).

The concentrations of anti-A, anti-B, and/or anti-Rh immunoglobulins in the small volume of plasma in a single platelet dose is usually insufficient, when passively transfused, to produce clinically evident hemolysis of recipient PRBCs. However, there have been case reports of RBC hemolysis in patients who received either large volumes of ABO-incompatible plasma or plasma with high-titer isoagglutinins (Pierce *et al.*, 1985). For this reason, the volume of mismatched plasma transfused to adult recipients should be limited to approximately 1 liter per week. Alternatively, the volume of mismatched plasma could be decreased by volume-reducing platelets prior to transfusion.

E. Ordering and Administration

A single therapeutic dose, even for severely thrombocytopenic patients, is approximately 4×10^{11} platelets. This dose is present in a single pool of 5–8 platelet concentrates (approximately 1 platelet concentrate per 10 kg body mass) or in a single apheresis-derived platelet unit. However, the appropriate number of units can also be calculated based on the required increase in platelet count. Special processing of platelet components (e.g., leukocyte-filtered or CMV-seronegative) should be ordered in consultation with blood bank staff. All transfused units should have a 170-μm filter interposed between the unit and the recipient to prevent infusion of microaggregates and fibrin clots.

The transfusion should initially proceed at a rate not greater than 5 ml per minute for the first 15 minutes. The transfusion should be completed within 4 hours of starting, and prior to component expiration. As with the transfusion of all blood components, the recipient should be monitored closely during infusion for signs of adverse transfusion sequelae, including febrile, allergic, and anaphylactic reactions. Mild allergic reactions can be treated by slowing or stopping the infusion and administering diphenhydramine; the transfusion can then be restarted when the symptoms have resolved. Patients who experience recurrent allergic reactions can be premedicated with diphenhydramine before initiating transfusion. For all other reactions, the transfusion must be discontinued, and the unused fraction of the component returned to the blood bank for further investigation. After stopping the transfusion, febrile reactions can be managed with acetaminophen.

F. Uses and Contraindications

Platelet transfusions are administered either to prevent future bleeding (prophylactic transfusions) or to stop ongoing bleeding (therapeutic transfusions). While the peripheral blood of a healthy adult usually contains 150,000–450,000 platelets/μl, most clinically stable nonbleeding patients tolerate platelet counts as low as 5000–10,000/μl without experiencing major hemorrhages (Gaydos et al., 1962; Gmur et al., 1991). In patients with coexisting clinical conditions, such as fever, bleeding may occur at higher platelet counts than would otherwise be expected.

1. Prophylaxis against Spontaneous Hemorrhage

Platelet transfusions are administered more frequently for prophylaxis against spontaneous hemorrhage than for treatment of ongoing hemorrhage (Pisciotto et al., 1995). Studies suggest that about 7000 platelets/μl/day are required to maintain endothelial integrity in normal individuals (Hanson and Slichter, 1985). Thus, thrombocytopenic patients are likely to experience significant hemorrhage when they have insufficient platelets (< 5000–$10,000/\mu$l) to meet baseline hemostatic requirements. For years, most practitioners utilized a 20,000 platelet/μl transfusion trigger, administering a single platelet dose (1 platelet concentrate pool or an apheresis-derived platelet unit) each day in which the platelet count was below 20,000/μl. However, recent prospective clinical trials (Rebulla et al., 1997; Wandt et al., 1998) have demonstrated that stable thrombocytopenic patients without coexisting

conditions (equivalent to 80% of all thrombocytopenic patients in one study [Rebulla et al., 1997]) can be safely managed with a more restrictive 10,000 platelet/μl trigger. A threshold of 20,000/μl was used for patients with fever, active bleeding, and coexisting coagulation disorders. A higher threshold is also acceptable in neonates. These thresholds were recently endorsed in a statement from the Royal College of Physicians of Edinburgh consensus conference on platelet transfusion (Contreras, 1998). Use of these more restrictive transfusion triggers significantly decreased platelet usage 20–40% when compared to patients managed with a 20,000/μl transfusion trigger. Decreased platelet transfusion, in turn, reduces the probability of adverse transfusion effects, including infectious disease transmission and alloimmunization, and also provides healthcare cost savings. Platelet usage might be further decreased by replacing prophylactic transfusions in thrombocytopenic patients with an alternative approach of transfusing platelets only for specific clinical indications, such as grossly evident hemorrhage. Studies have shown that platelet transfusions for specific indications can be as effective as prophylactic transfusions (Patten, 1992). Nonetheless, the majority of clinicians continue to practice prophylactic transfusion of platelets.

2. Prophylaxis Prior to Invasive Procedures

The standard of practice in many institutions is to utilize prophylactic platelet transfusions to achieve preoperative platelet counts of 50,000 and 100,000 platelets/μl prior to minor and major surgical procedures, respectively. Unfortunately, there are few good studies to support these guidelines. In one clinical series, 95 patients with acute leukemia who underwent a total of 167 surgical procedures were transfused preoperatively in order to reach intraoperative platelet counts of 50,000/μl (Bishop et al., 1987). Bleeding requiring PRBC transfusion was significantly associated with the extent of the operation, and presence of fever, or coexisting coagulation abnormalities. For example, while 48% of major operations (including laparotomy, craniotomy, and thoracotomy) were associated with intraoperative blood loss of > 500 ml and/or perioperative transfusions of > 4 units, only 21% of minor procedures (including catheter and chest tube placement, thoracentesis, and biopsies) reached these levels. None of the patients died of bleeding postoperatively, and all bleeding episodes were controlled with few complications. Thus, while these results suggest that all surgical patients may be safely managed with a 50,000 platelet/μl transfusion trigger, at present there is insufficient evidence to recommend a change in the standard practices described above.

3. Therapeutic Platelet Transfusions for Ongoing Bleeding

In patients with active hemorrhage and platelet counts of less than approximately 50,000–80,000 platelets/μl, therapeutic transfusions are indicated. Platelet transfusions may also be useful in bleeding patients with platelet function defects. Continuous clinical assessment is required to determine the appropriate duration of platelet support.

4. Contraindications

Microvascular hemorrhage from mucous membranes, along with petechiae and ecchymoses not associated with trauma, are hallmarks of thrombocytopenic bleeding that may be responsive to platelet transfusions. In contrast, bleeding localized to a single anatomic site is usually not responsive to platelet transfusions, unless the patient is severely thrombocytopenic, and is best treated with surgery or other localized interventions. Likewise, brisk hemorrhage (> 5 ml blood loss/kg body weight/h) is likely the result of an anatomic lesion that requires surgical intervention. Platelet transfusions are also usually contraindicated for patients with thrombotic thrombocytopenic purpura (TTP) and heparin-induced thrombocytopenia (HIT). In TTP and HIT, platelet transfusions may be harmful, as they can exacerbate thrombotic complications. However, in the bleeding patient with an extremely low platelet count, transfusion of platelets is occasionally unavoidable, no matter what the underlying illness.

In the setting of immune thrombocytopenic purpura (ITP), platelet transfusions are unlikely to be of clinical benefit without additional interventions. Platelet transfusions alone are rarely effective in treating uremic bleeding, although they may be useful in conjunction with dialysis. Uremic bleeding may also respond to desmopressin (DDAVP) administration or to PRBC transfusions to maintain the hematocrit above 30%.

G. Expected Response and Potential Adverse Effects

1. Expected Platelet Increment

For an average-sized adult, the transfusion of 1 apheresis-derived platelet unit or 1 platelet concentrate pool should increase the platelet count by 30,000–50,000 platelets/μl at 10–60 minutes post-transfusion. While this guideline is sufficient for patients requiring minimal platelet support, marrow transplant recipients and other thrombocytopenic patients receiving multiple transfusions should have more accurate evaluations performed to compare actual and expected platelet increments. Patients who do not demonstrate a sufficient increase in platelet count post-transfusion may have developed immune refractoriness to platelet transfusions. The corrected count increment (CCI) is often used to identify and manage these patients (see Chapter 24).

2. Adverse Effects

Many of the adverse reactions that occur with transfusion of blood·components are described in detail elsewhere in this volume, and thus are not discussed below. These include infectious disease transmission, transfusion-related acute lung injury (TRALI), TA-GVHD, and febrile and allergic transfusion reactions. However, the adverse reactions that are of specific concern to platelet transfusions are described below.

a. Hypotensive Reactions

These reactions were initially described in patients medicated with angiotensin-converting enzyme (ACE) inhibitors who were receiving platelet transfusions leukoreduced at bedside with negatively charged filters (Hume et al., 1996). This reaction manifests primarily with marked hypotension, sometimes accompanied by respiratory distress. Unlike TRALI, new infiltrates are not observed on the chest X-ray. Fever, flushing, and gastrointestinal discomfort are not usually reported. Usually these reactions resolve rapidly after the transfusion is discontinued.

The pathogenesis of these reactions is still under investigation. It is thought that bradykinin is generated from donor plasma during leukocyte filtration (Takahashi et al., 1995). ACE inhibitors are proposed to potentiate hypotensive reactions by inhibiting bradykinin breakdown. Although these reactions are not yet thoroughly characterized, it is reasonable to use prestorage leukoreduced platelet components for patients taking ACE inhibitors, rather than performing leukoreduction filtration at the bedside.

b. HLA and HPA Alloimmunization

Individuals exposed to foreign HLA proteins, also known as major histocompatibility (MHC) proteins, through either previous transfusion or pregnancy, can generate antibodies against the allogeneic HLA epitopes. They may also produce antibodies against human platelet antigens (HPA) expressed on platelets. In some studies, up to 85% of heavily transfused patients generated antibodies against allogeneic HLA and/or HPA proteins (Friedman *et al.*, 1996). However, a 40–50% rate of HLA and/or HPA alloimmunization in the absence of specific interventions is more commonly seen (TRAP, 1997). Because platelets express HLA class I A and B proteins, anti-HLA alloantibodies can cause poor responses to platelet transfusions. Anti-HPA antibodies can also result in platelet refractoriness.

In hematology/oncology patients that are expected to require substantial platelet support, and thus be exposed to numerous transfusions of allogeneic blood, it may be more prudent to attempt to prevent alloimmunization than to identify compatible platelets after alloimmunization has occurred. The recently completed Trial to Reduce Alloimmunization to Platelets (TRAP) demonstrated that leukoreduction by filtration can significantly decrease the incidence of alloimmune platelet refractoriness to ≤ 3–4% of patients (TRAP, 1997). Irradiation with intermediate-wavelength ultraviolet (UV-B) light also decreased the incidence of alloimmunization, although this modality is not available in many blood banks. Neither filtration nor UV-B irradiation prevented alloimmune refractoriness in patients with previously developed HLA alloantibodies, nor did they influence the development of anti-HPA antibodies. Prevention of HLA alloimmunization and management of platelet refractory patients are discussed in more detail in Chapter 24.

c. Transfusion-Related Immunomodulation (TRIM)

In the 1970s, improved renal allograft survival was reported for patients that received allogeneic blood transfusions prior to transplantation (Opelz *et al.*, 1973). Subsequently, additional examples of beneficial effects of TRIM have been described. However, the possibility of detrimental immunosuppressive effects of blood transfusion has also come under investigation. For example, an increased rate of postoperative infection in transfused surgical patients has been reported. Subsequent comparisons between surgical patients receiving either nonfiltered or filtered blood components demonstrated that those transfused with filtered components had significantly fewer postoperative infections, shorter hospital stays, and reduced medical costs (Jensen *et al.*, 1996; Vamvakas and Carven, 1998). These results indicate a role for leukocytes in transfusion-mediated immunosuppression. Transfusions may also lead to reactivation of latent viruses in the recipient (Hillyer *et al.*, 1999). TRIM and the use of leukoreduction filtration to mitigate the effects of TRIM are discussed in more detail in Chapter 14.

II. ALTERNATIVES

There is not likely to be a significant decline in the clinical use of platelet transfusions in the foreseeable future. Nonetheless, several alternatives to platelet transfusions are currently undergoing basic laboratory investigations and/or early clinical trials. With the exception of rHuIL-11 and frozen autologous platelets, none of these alternatives is approved for routine clinical use at this time.

A. Thrombopoietin (TPO)

Molecular cloning, sequencing, and expression of the growth factor TPO, the physiologic regulator of thrombopoiesis, closely followed the discovery of Mpl, the TPO receptor (Kaushansky, 1995). Subsequently, recombinant human TPO (rHuTPO) has been studied as a potential adjunct to standard platelet transfusion therapy.

rHuTPO stimulates the generation of megakaryocyte progenitor cells. In clinical trials, rHuTPO effectively increased circulating platelet counts in oncology patients receiving non-myelosuppressive chemotherapy, but unfortunately it has been ineffective in myeloablated patients, the group that requires the majority of platelet transfusions (reviewed in Kuter *et al.*, 1999). rHuTPO has not yet been approved for routine clinical use. Nonetheless, there is optimism that rHuTPO may eventually prove useful in selected clinical settings, for example, in increasing platelet production in apheresis platelet donors.

B. Other Thrombopoietic Growth Factors

In clinical trials, 30% of oncology patients that received 50 μg/kg recombinant human interleukin 11 (rHuIL-11) while undergoing chemotherapy did not require platelet transfusions, compared to 4% of placebo-control patients (Tepler *et al.*, 1996). Treated patients experienced some side effects, including fatigue and cardiovascular symptoms. rHuIL-11 is currently approved for use in patients undergoing myelosuppressive, but not myeloablative chemotherapy. Early clinical trials are also investigating the use of other hematopoietic growth factors, including IL-3, IL-6, stem cell factor, and progenipoietin G (a Flt-3 ligand/G-CSF fusion protein) for thrombocytopenic patients.

C. Frozen Autologous Platelets

Prior to myeloablation and resulting thrombocytopenia, patients may donate autologous platelets which can be cryopreserved with DMSO for subsequent use. These platelets can be thawed and transfused years after donation to produce acceptable platelet increments in alloimmunized thrombocytopenic patients (Daly et al., 1979). However, because of losses of up to 50% of platelets upon thawing, the use of frozen autologous platelets has not become routine.

D. HLA Class-I-Eluted Platelets

Brief exposure of platelets to an acidic solution of pH 3.0 can selectively strip up to 90% of HLA class I molecules from platelets, while leaving thrombogenic platelet glycoproteins such as GPIIIa intact. The resulting HLA class-I-eluted platelets should not be targeted by anti-HLA antibodies in alloimmunized patients. These platelet preparations have been effective in the treatment of alloimmunized thrombocytopenic patients in limited clinical trials (Novotny et al., 1996; Shanwell et al., 1991). However, severe adverse reactions have also been observed.

E. Infusible Platelet Membrane Microvesicles (IPMs)

IPMs are generated from outdated platelets by a combination of centrifugation, freeze-thaw lysis, heating to inactivate viruses, and lyophilization. The resulting preparation is composed of spherical microvesicles about one-third the size of normal platelets which retain the GPIb receptor and platelet factor 3 procoagulant proteins, but are deficient in HLA antigens. In studies with thrombocytopenic rabbits, IPMs decreased bleeding times from 15 minutes to 6 minutes (Chao et al., 1996). No clinical trials have been reported.

F. Other Alternatives

Other platelet substitutes have also been designed to sidestep the presence of anti-HLA alloantibodies in refractory thrombocytopenic patients. For example, thrombogenic platelet membrane proteins have been inserted into red blood cell membranes and liposomes in the absence of HLA proteins (Coller et al., 1992). In a recent study, fibrinogen-coated albumin microcapsules (Synthocytes) were shown to correct the bleeding time in thrombocytopenic rabbits (Levi et al., 1999). Clinical trials have not been reported with any of these substitutes.

References

Balducci, L., Benson, K., Lyman, G. H., et al. (1993). Cost-effectiveness of white cell-reduction filters in treatment of adult acute myelogenous leukemia. Transfusion 33, 665–670.

Bishop, J. F., Schiffer, C. A., Aisner, J., et al. (1987). Surgery in acute leukemia: A review of 167 operations in thrombocytopenic patients. Am J Hematol 26, 147–155.

Bolovan-Fritts, C. A., Mocarski, E. S., and Wiedeman, J. A. (1999). Peripheral blood CD14(+) cells from healthy subjects carry a circular conformation of latent cytomegalovirus genome. Blood 93, 394–398.

Bowden, R. A., Slichter, S. J., Sayers, M., et al. (1995). A comparison of filtered leukocyte-reduced and cytomegalovirus (CMV) seronegative blood products for the prevention of transfusion-associated CMV infection after marrow transplant. Blood 86, 3598–3603.

Carr, R., Hutton, J. L., Jenkins, J. A., et al. (1990). Transfusion of ABO mismatched platelets leads to early platelet refractoriness. Br J Haematol 75, 408–413.

Chao, F. C., Kim, B. K., Houranieh, A. M., et al. (1996). Infusible platelet membrane microvesicles: A potential transfusion substitute for platelets. Transfusion 36, 536–542.

Coller, B. S., Springer, K. T., Beer, J. H., et al. (1992). Thromboerythrocytes. In vitro studies of a potential autologous, semi-artificial alternative to platelet transfusions. J Clin Invest 89, 546–555.

Contreras, M. (1998). Final statement from the consensus conference on platelet transfusion. Transfusion 38, 796–797.

Daly, P. A., Schiffer, C. A., Aisner, J., et al. (1979). Successful transfusion of platelets cryopreserved for more than 3 years. Blood 54, 1023–1027.

Freedman, J., Hooi, C., and Garvey, M. B. (1984). Prospective platelet crossmatching for selection of compatible random donors. Br J Haematol 56, 9–18.

Friedman, D. F., Lukas, M. B., Jawad, A., et al. (1996). Alloimmunization to platelets in heavily transfused patients with sickle cell disease. Blood 88, 3216–3222.

Gaydos, L. A., Freireich, E. J., and Mantel, N. (1962). The quantitative relation between platelet count and hemorrhage in patients with acute leukemia. N Engl J Med 266, 905–909.

Gmur, J., Burger, J., Schanz, U., et al. (1991). Safety of stringent prophylactic platelet transfusion policy for patients with acute leukaemia. Lancet 338, 1223–1226.

Gordon, G., Kallich, J., Erder, M., et al. (1996). Inpatient platelet use and costs for patients with hematological diseases: A comparison between refractory and nonrefractory patients. Blood 88, 334a.

Hanson, S. R., and Slichter, S. J. (1985). Platelet kinetics in patients with bone marrow hypoplasia: Evidence for a fixed platelet requirement. Blood 66, 1105–1109.

Heal, J. M., Blumberg, N., and Masel, D. (1987). An evaluation of crossmatching, HLA, and ABO matching for platelet transfusions to refractory patients. Blood 70, 23–30.

Hillyer, C. D., Lankford, K. V., Roback, J. D., et al. (1999). Transfusion of the HIV seropositive patient: Immunomodulation, viral reactivation, and limiting exposure to EBV (HHV-4), CMV (HHV-5) and HHV-6, 7, and 8. Transfusion Med Rev 13, 1–17.

Hume, H. A., Popovsky, M. A., Benson, K., et al. (1996). Hypotensive reactions: A previously uncharacterized complication of platelet transfusion? Transfusion 36, 904–909.

Jensen, L. S., Kissmeyer-Nielsen, P., Wolff, B., et al. (1996). Randomised comparison of leucocyte-depleted versus buffy-coat-poor blood transfusion and complications after colorectal surgery. Lancet 348, 841–845.

Kaushansky, K. (1995). Thrombopoietin: The primary regulator of platelet production. Blood 86, 419–431.

Kuter, D. J., Cebon, J., Harker, L. A., et al. (1999). Platelet growth factors: Potential impact on transfusion medicine. Transfusion 39, 321–332.

Lee, E. J., and Schiffer, C. A. (1989). ABO compatibility can influence the results of platelet transfusion. Results of a randomized trial. *Transfusion* **29**, 384–389.

Levi, M., Friederich, P. W., Middleton, S., *et al.* (1999). Fibrinogen-coated albumin microcapsules reduce bleeding in severely thrombocytopenic rabbits. *Nat Med* **5**, 107–111.

Miller, W. J., McCullough, J., Balfour, H. H., Jr., *et al.* (1991). Prevention of cytomegalovirus infection following bone marrow transplantation: A randomized trial of blood product screening. *Bone Marrow Transplant* **7**, 227–234.

Novotny, V. M., Huizinga, T. W., van Doorn, R., *et al.* (1996). HLA class I-eluted platelets as an alternative to HLA-matched platelets. *Transfusion* **36**, 438–444.

Opelz, G., Sengar, D. P., Mickey, M. R., *et al.* (1973). Effect of blood transfusions on subsequent kidney transplants. *Transplant Proc* **5**, 253–259.

Patten, E. (1992). Controversies in transfusion medicine. Prophylactic platelet transfusion revisited after 25 years: con. *Transfusion* **32**, 381–385.

Pierce, R. N., Reich, L. M., and Mayer, K. (1985). Hemolysis following platelet transfusions from ABO-incompatible donors. *Transfusion* **25**, 60–62.

Pisciotto, P. T., Benson, K., Hume, H., *et al.* (1995). Prophylactic versus therapeutic platelet transfusion practices in hematology and/or oncology patients. *Transfusion* **35**, 498–502.

Rebulla, P., Finazzi, G., Marangoni, F., *et al.* (1997). The threshold for prophylactic platelet transfusions in adults with acute myeloid leukemia. Gruppo Italiano Malattie Ematologiche Maligne dell'Adulto. *N Engl J Med* **337**, 1870–1875.

Shanwell, A., Sallander, S., Olsson, I., *et al.* (1991). An alloimmunized, thrombocytopenic patient successfully transfused with acid-treated, random donor platelets. *Br J Haematol* **79**, 462–465.

Simon, T. L., and Sierra, E. R. (1984). Concentration of platelet units into small volumes. *Transfusion* **24**, 173–175.

Takahashi, T., Abe, H., Nakai, K., and Sekiguchi, S. (1995). Bradykinin generation during filtration of platelet concentrates with a white cell-reduction filter [letter]. *Transfusion* **35**, 967.

Tepler, I., Elias, L., Smith, J. W., 2nd, *et al.* (1996). A randomized placebo controlled trial of recombinant human interleukin-11 in cancer patients with severe thrombocytopenia due to chemotherapy. *Blood* **87**, 3607–3614.

Toth, C. B., Kramer, J., Pinter, J., *et al.* (1998). IgA content of washed red blood cell concentrates. *Vox Sang* **74**, 13–14.

TRAP (1997). Leukocyte reduction and ultraviolet B irradiation of platelets to prevent alloimmunization and refractoriness to platelet transfusions. The Trial to Reduce Alloimmunization to Platelets Study Group. *N Engl J Med* **337**, 1861–1869.

Vamvakas, E. C., and Carven, J. H. (1998). Allogeneic blood transfusion, hospital charges, and length of hospitalization: A study of 487 consecutive patients undergoing colorectal cancer resection. *Arch Pathol Lab Med* **122**, 145–151.

Wandt, H., Frank, M., Ehninger, G., *et al.* (1998). Safety and cost effectiveness of a $10 \times 10(9)/L$ trigger for prophylactic platelet transfusions compared with the traditional $20 \times 10(9)/L$ trigger: A prospective comparative trial in 105 patients with acute myeloid leukemia. *Blood* **91**, 3601–3606.

Yeager, A. S., Grumet, F. C., Hafleigh, E. B., *et al.* (1981). Prevention of transfusion-acquired cytomegalovirus infections in newborn infants. *J Pediatr* **98**, 281–287.

8

Granulocytes

THERESA W. GILLESPIE
CHRISTOPHER D. HILLYER
Emory University School of Medicine
Atlanta, Georgia

Granulocyte transfusions (GTX) have been utilized in the setting of severe neutropenia with progressive infections or neutrophil dysfunction, with varying results and controversy as to their therapeutic efficacy. The historical incidence of significant toxicity associated with GTX, combined with extensive improvements in antimicrobial therapy and a shortened period of bone marrow aplasia due to use of hematopoietic growth factors, has diminished the enthusiasm for this intervention. However, more recent advances in granulocyte collection, with administration of growth factors to enhance granulocyte yield, have produced renewed interest in this therapy.

I. PRODUCT DESCRIPTION

Granulocyte collections are apheresis-derived white blood cell (WBC) components, usually about 300 ml in volume, composed of a mixture of granulocytes and other elements that are considered contaminants, such as red blood cells (RBCs; 6–7 g/dL of hemoglobin per granulocyte product), platelets, and citrated plasma (Klein *et al.*, 1996).

II. COLLECTION, STORAGE, AND HANDLING

A. Collection

Granulocytes are collected by apheresis of volunteer donors who are primed with: (1) corticosteroids, or (2) hematopoietic growth factors (investigational), using centrifugal separation with a commercial blood cell separator. In general, granulocyte collections are performed employing standard leukapheresis tubing sets via peripheral venous access. However, inadequate peripheral access or the need for multiple daily collections may justify the use of central venous catheters. Final granulocyte yield depends on the total volume of blood processed as well as the initial peripheral blood neutrophil count of the donor. Routinely, 7–12 liters of blood are processed through a continuous flow blood cell separator over a period of 2–4 hours, and this may be performed daily or every other day, for 4–5 days (Price, 1995).

With a median density of 1.080 g/ml, granulocytes have a higher density than either young RBCs or reticulocytes (Chanock and Gorlin, 1996). Apheresis techniques that utilize a density gradient frequently result in collection of large numbers of red cells as well as granulocytes. The donor's erythrocyte sedimentation rate (ESR) may predict the collection efficiency of granulocytes, and increasing the donor's ESR can heighten the granulocyte yield. Interventions to increase collection efficiency have included administration of agents to the donor that promote density gradient formation, for example, hydroxyethyl starch (HES).

Both pentastarch and hetastarch have been reported to enhance granulocyte collection yield, but also are associated with side effects experienced by the donor, including edema and weight gain (Bensinger *et al.*, 1993). Hetastarch is recommended due to its improved collection efficiency as well as to its more tolerable toxicity profile compared to pentastarch (Adkins *et al.*, 1998b; Price, 1995). A combination of sodium citrate, at a lower dose than for platelet collection (30 ml of 56.5% trisodium citrate), in HES (500 ml of 6% HES) added to the donor line (1 part anticoagulant to 10–12 parts blood processed) may be used as the anticoagulant (Adkins *et al.*, 1997).

B. Donor Stimulation

Historically, donors for granulocyte collection were stimulated with corticosteroids, primarily dexamethasone, to increase granulocyte production in the donor and thus increase product yield. More recently, however, the use of hematopoietic growth factors, such as recombinant human granulocyte colony-stimulating factor (rHuG-CSF), is being investigated in clinical trials. Normal donors are usually given a subcutaneous injection of G-CSF at a dose of 5 μg/kg, starting 12–16 hours prior to collection, or they may receive a combination of G-CSF and corticosteroids. Stimulation of donors with G-CSF appears to not only significantly expand the granulocyte yield by 2- to 15-fold to allow collection of more granulocytes as compared to collections using dexamethasone stimulation alone, but also functions to increase the number of circulating leukocytes for several days, and apparently neutrophil progenitors as well (Liles *et al.*, 1997). Although granulocyte-macrophage colony-stimulating factor (GM-CSF) has also been investigated in donor stimulation, G-CSF is the preferred growth factor, as it appears to have greater efficacy and reduced toxicity. Nevertheless, donors who receive stimulation with G-CSF may experience mild bone pain, headache, malaise, and occasionally a flu-like syndrome, including fever, myalgia, and fatigue. Most donor symptoms are reduced with use of antipyretics and analgesics.

The optimal recommended donor stimulation regimen has not yet been established; however, good results have been reported following a G-CSF regimen using 5 μg/kg SQ qd, given 12 hours prior to the initial collection. Using this regimen, granulocyte counts can be increased to 20,000–40,000 neutrophils/μL (Price, 1995) and are collected every other day, to a maximum of 5 collections for men and 4 collections for women (Adkins *et al.*, 1998a). In order to have sufficient doses of donor granulocytes for daily transfusions, recipients often require at least two compatible donors, each of whom undergoes apheresis and granulocyte collection every other day (Jendiroba *et al.*, 1998).

C. Cell Yield

Administration of oral dexamethasone can lead to two- to threefold increases from baseline in peripheral blood granulocyte counts in the donor, with subsequent granulocyte collections in the range of 17×10^9 (Liles *et al.*, 1997). In comparison, use of daily stimulation of donors with G-CSF can result in 7- to 10-fold increases in granulocyte counts from baseline, with cell yields of 4–5×10^{10} or more granulocytes (Adkins *et al.*, 1998a, b; Strauss, 1995). The combination of G-CSF with dexamethasone has been reported as superior in inducing neutrophilia as compared to either agent alone, with significant increases in circulating neutrophil counts (9- to 12-fold increase).

D. Dosage

As an unlicensed product, granulocyte collections have no official product specifications established by the FDA. Standards set by the American Association of Blood Banks (AABB) have required the granulocyte apheresis unit to include a minimum of 1×10^{10} granulocytes in at least 75% of units tested (AABB, 1994). The minimal dosage (about 1.5×10^8 cells/kg) should in fact be at least twice this amount in order to detect a significant granulocyte increment in adults (Dutcher, 1989). Pediatric dosage of GTX is considered an average of 1×10^9/kg/day, with neonatal dosage about 1–2×10^9/kg (Vamvakas and Pineda, 1996). Daily administration of granulocytes exceeding the minimum standard, for a period of 4–7 days, is recommended in order to achieve positive outcomes in the setting of progressive, non-treatment-responsive infection in severely neutropenic patients.

The number of mature granulocytes in the product is the primary measure of product efficacy, although control of infection in the neutropenic recipient is the ultimate indicator of the product's effectiveness. Routine testing of granulocyte quality is not performed; however, *in vitro* assays for granulocyte function should be done whenever the collection technique is modified significantly.

E. Storage

No specific standards have been set for granulocyte storage, although it is known that the product's viability and optimal functioning diminish quickly over time. Thus, granulocytes should be given to the designated recipient as soon as possible after collection, within 6 hours if possible, and absolutely within 24 hours after collection, at which time the product is considered expired. No preservative solution has been licensed that is specific for granulocytes. Collected granulocytes should be handled minimally, with no agitation, and stored at room temperature (20 to 24°C; Klein *et al.*, 1996).

Granulocytes are capable of transmitting the infectious agents for which other blood component transfusions are considered at risk, such as HIV, HTLV, CMV, and hepatitis viruses (Klein *et al.*, 1996). As such, GTX must undergo the same standard screening and testing for these transfusion-transmitted infectious agents. Because of the brief period of time before the product expires, every effort to facilitate the various steps is justified. Emergency release of the product prior to obtaining results of viral testing is often authorized by the physician ordering the transfusion. Some centers will elect to have viral testing performed on the donor prior to collection, rather than on the product itself, in order to eliminate time delays. Results are usually considered valid for the rest of the collection interval, or up to a period of 2 weeks from the time of initial testing.

III. SPECIAL PREPARATION AND PROCESSING

A. HLA Matching

If the patient is not alloimmunized, histocompatible donors for granulocytes are not considered necessary, in general. A useful rule may be to evaluate the patient's response to platelet transfusions: if the recipient has been experiencing reasonable increments postplatelet transfusion, the recipient is less likely to require HLA-matched GTX.

For alloimmunized patients, the efficacy of granulocyte transfusions is dubious. In these cases, HLA-compatible family members or donors, particularly if they have given platelet donations that have resulted in appropriate increments, should be used for granulocyte collection. A donor who meets the set criteria for alloimmunized patients, that is, HLA-matched, should be collected from for several consecutive days to produce sufficient quantities of granulocytes for clinical efficacy (Stroncek *et al.*, 1996). Although granulocyte-specific antibody assays exist, many institutions are unable to perform these tests, or are unable to run these assays on a routine basis when needed, and their predictive value is not known. Thus, the initial HLA antibody screen may be the most useful tool for HLA compatibility assessment.

B. Irradiation

Granulocyte products contain significant numbers of T-lymphocytes, with an accompanying risk of transfusion-associated graft-versus-host disease (TA-GVHD) in immunocompromised patients. A small risk for immunocompetent recipients also exists, but is not considered clinically significant. To prevent the occurrence of TA-GVHD, blood components should be gamma irradiated prior to administration to immunosuppressed individuals. Although granulocytes collected by centrifugation may be damaged by irradiation, *it is recommended that GTX be irradiated prior to administration to at-risk patients*. The dose of 2500 cGy, administered in close proximity to the time the product will be infused, is considered adequate. Earlier established doses of 1500 cGy are no longer considered sufficient for the prevention of TA-GVHD (Klein *et al.*, 1996).

IV. ABO AND Rh COMPATIBILITY AND CROSSMATCHING

Because routinely there are large numbers of RBCs present in the collected GTX product, granulocyte transfusions should be ABO compatible with the recipient and undergo RBC crossmatching. In some centers, RBCs in granulocyte units are maximally depleted by sedimentation prior to transfusion.

V. ORDERING AND ADMINISTRATION

When the clinical decision has been made to proceed with granulocyte transfusions, GTX should be given each day at a minimum dose of 1×10^{10} neutrophils for adults, but ideally at a higher dose of $2-3 \times 10^{10}$. The granulocyte infusions should be ordered daily until either the infection resolves, the peripheral granulocyte count is sustained at more than 0.5×10^9/liter, or the infusions are deemed to be ineffective or poorly tolerated by the recipient (Strauss, 1994).

The product is given at the bedside, infused without a filter, or with only a 170-μ filter. A leukocyte depletion filter cannot be used (Chanock and Gorlin, 1996). The recipient should be monitored for immediate occurrence of adverse reactions, with modification of the infusion rate or administration of medication for symptom management as appropriate (Schiffer *et al.*, 1983). The need for premedication with antihistamines, antipyretics, and/or corticosteroids is not universally recognized, particularly in recipients who are not alloimmunized. To reduce the incidence of transfusion reactions, the GTX should be given slowly, at a rate of about $1-2 \times 10^{10}$ cells per hour. Should a significant transfusion reaction occur, the infusion should be slowed or stopped and then restarted when the reaction has diminished (Klein *et al.*, 1996). Corticosteroids may be given to help alleviate toxicity, and IV meperidine may reduce severe rigors. Serologic testing for recipient leukocyte antibodies may be performed if a serious transfusion reaction occurs.

VI. USES AND CONTRAINDICATIONS

A. Indications

Although the trend to administer granulocyte transfusions has decreased over the past several decades, when analyzing the data from earlier trials, most of these studies were limited by insufficient dosage of the product, absence of growth factor mobilization of donors to enhance the collection yield, and obsolete collection technology, such as use of filtration leukapheresis. More recent studies have supported the use of granulocytes in specific clinical settings, although institutional standards may not reflect this view.

In general, the clinical indications for granulocyte transfusions include severe neutropenia ($< 0.5 \times 10^9$ polymorphonuclear leukocytes (PMNs)/liter blood), and: (1) progressive, nonresponsive, documented bacterial, yeast, or fungal infections that have failed to improve after 48 hours of antimicrobial therapy; (2) a prolonged period of neutropenia in bone marrow transplant (BMT) recipients, by which time the marrow is expected to have been recovered; (3) congenital granulocyte dysfunction; or (4) bacterial infections in neonates (Klein *et al.*, 1996). The role of GTX in fungal infections in BMT recipients, particularly when the marrow graft has been T-cell depleted, is unclear and requires further study (Chanock and Gorlin, 1996).

In the past, granulocyte transfusions were used as prophylaxis for infections in severely neutropenic patients. However, in view of the controversial nature of GTX efficacy, the risk of transfusion transmission of infectious agents, and the toxicities associated with the product, *the prophylactic use of GTX is not recommended* (Dutcher, 1989; Vamvakas and Pineda, 1997).

B. Contraindications

Because of several reports of severe pulmonary toxicities occurring with concurrent administration of amphotericin B in patients also receiving GTX, current guidelines advise separating administration of amphotericin B from granulocyte transfusions by at least 4 hours or more (Chanock and Gorlin, 1996).

VII. EXPECTED RESPONSES AND POTENTIAL ADVERSE EFFECTS

A. Efficacy

Previous studies (1975–1982) of granulocyte transfusions, when reviewed carefully, have been unable to definitively determine the efficacy of this intervention (Higby *et al.*, 1975; Herzig *et al.*, 1977; Winston *et al.*, 1982). Important variables have been identified that predict less positive outcomes in these earlier studies, including: (1) a low dose of granulocyte transfusion (including insufficient number of transfusions and a low dose of granulocytes per dose), (2) sustained bone marrow dysfunction after prolonged aplasia or neutropenia, (3) collection of granulocytes by filtration leukapheresis, and (4) leukocyte incompatibility (Vamvakas and Pineda, 1996). Current controlled trials using high-dose GTX in appropriate patients are needed.

B. Expected Response

The post-transfusion increment after granulocyte administration is unpredictable. A mean 1-hour increment in recipient absolute neutrophil count (ANC) of near 1000 per μL on day 1 has been reported, with post-transfusion increments, as well as mean peak ANC, observed to decline on subsequent days for each granulocyte transfusion (Adkins *et al.*, 1997). The granulocyte increment post-transfusion does not appear to correlate with granulocyte dose given; thus, the clinical significance of the post-transfusion increment is unclear. Ideally, recipients will experience a sustained increase of their granulocyte count to above 500 PMN/μL as a result of the transfusions.

C. Potential Adverse Effects

Adverse events experienced by recipients of GTX, even when compatible transfusions are administered to nonalloimmunized patients, can be significant (Schiffer *et al.*, 1983). Common risks include fever and shaking chills, both of which can result from too-rapid infusions. If the recipient is alloimmunized, the side effects can be exacerbated and include pulmonary toxicities such as hypoxia, dyspnea, and pulmonary infiltrates.

VIII. ALTERNATIVES

The decision to proceed with granulocyte transfusion is dependent on many parameters, including clinical criteria for predicted efficacy, availability of compatible donors, and patient tolerance. If no benefit is seen or predicted to be derived from this approach, the decision may be to stop or not initiate granulocyte transfusions (Vamvakas and Pineda, 1996). Because of the controversy surrounding this approach to severe neutropenia, an acceptable alternative may be to pursue aggressive or investigational antimicrobial therapy, and to continue administration of hematopoietic growth factors to reduce duration of the neutropenic period, rather than proceed with granulocyte transfusions.

References

AABB (1994). "Standards for Blood Banks and Transfusion Services," 16th ed. AABB, Bethesda, MD.

Adkins, D., Spitzer, G., Johnston, M., et al. (1997). Transfusions of granulocyte-colony-stimulating factor-mobilized granulocyte components to allogeneic transplant recipients: Analysis of kinetics and factors determining posttransfusion neutrophil and platelet counts. Transfusion 37, 737–748.

Adkins, D., Ali , S., Despotis, G., Dynis, M., and Goodnough, L. T. (1998a). Granulocyte collection efficiency and yield are enhanced by the use of a higher interface offset during apheresis of donors given granulocyte-colony-stimulating factor. Transfusion 38, 557–564.

Adkins, D., Johnston, M., Walsh, J., et al. (1998b). Hydroxyethyl-starch sedimentation by gravity ex vivo for red cell reduction of granulocyte apheresis components. J Clin Apher 13, 56–61.

Bensinger, W. I., Price, T. H., Dale, D. C., et al. (1993). The effects of daily recombinant human granulocyte stimulating factor administration on normal granulocyte donors undergoing leukapheresis. Blood 81, 1883–1888.

Chanock, S. J., and Gorlin, J. B. (1996). Granulocyte transfusions. Time for a second look. Infect Dis Clin N Am 10, 327–343.

Dutcher, J. P. (1989). The potential benefit of granulocyte transfusion therapy. Cancer Invest 7, 457–462.

Herzig, R. H., Herzig, G. P., Graw, R. G., Bull, M. I., and Ray, K. K. (1977). Successful granulocyte transfusion therapy for gram-negative septicemia. A prospectively randomized controlled study. N Engl J Med 296, 701–705.

Higby, D. J., Yates, J. W., Henderson, E. S., and Holland, J. F. (1975). Filtration leukapheresis for granulocyte transfusion therapy. Clinical laboratory studies. N Engl J Med 292, 761–766.

Jendiroba, D. B., Lichtiger, B., Anaissie, E., et al. (1998). Evaluation and comparison of three mobilization methods for the collection of granulocytes. Transfusion 38, 722–728.

Klein, H. G., Strauss, R. G., and Schiffer, C. A. (1996). Granulocyte transfusion therapy. Semin Hematol 33, 359–368.

Liles, W. C., Huang, J. E., Llewellyn, C., et al. (1997). A comparative trial of granulocyte-colony-stimulating factor and dexamethasone, separately and in combination, for the mobilization of neutrophils in the peripheral blood of normal volunteers. Transfusion 37, 182–187.

Price, T. H. (1995). Blood center perspective of granulocyte transfusions: Future applications. J Clin Apher 10, 119–123.

Schiffer, C. A., Aisner, J., Dutcher, J. P., and Wiernik, P. H. (1983). Sustained post-transfusion granulocyte count increments following transfusion of leukocytes obtained from donors with chronic myelogenous leukemia. Am J of Hematol 15, 65–74.

Strauss, R. G. (1994). Granulocyte transfusion therapy. Hematol Oncol Clin N Am 6, 1159–1166.

Strauss, R. G. (1995). Clinical perspectives of granulocyte transfusions: Efficacy to date. J Clin Apher 10, 114–118.

Stroncek, D. F., Leonard, K., Eiber, G., et al. (1996). Alloimmunization after granulocyte transfusions. Transfusion 36, 1009–1015.

Vamvakas, E. C., and Pineda, A. A. (1996). Meta-analysis of clinical studies of the efficacy of granulocyte transfusion in the treatment of bacterial sepsis. J Clin Apher 11, 1–9.

Vamvakas, E. C., and Pineda, A. A. (1997). Determinants of the efficacy of prophylactic granulocyte transfusions: A meta-analysis. J Clin Apher 12, 74–81.

Winston, D. J., Ho, W. G., and Gale, R. P. (1982). Therapeutic granulocyte transfusions for documented infections. A controlled trial in ninety-five infectious granulocytopenic episodes. Ann Intern Med 97, 509–512.

9

Coagulation Factor Concentrates

MARIA JOSE A. RIBEIRO

Emory University Comprehensive Hemophilia Program
Atlanta, Georgia

The evolution of the treatment of bleeding disorders over the last 40 years has been significant. Prior to 1960, the only treatment available for bleeding episodes in hemophilia, and other factor deficiency patients, was the infusion of frozen plasma. In 1964, Pool discovered that plasma cryoprecipitate contained coagulation factor VIII (FVIII) in high concentrations. Single units, and eventually pooled cryoprecipitate, became widely available for the treatment of hemophilia A and von Willebrand disease (vWD; Pool *et al.*, 1964). Investigators then started to purify FVIII from cryoprecipitate using chromatography. These products were referred to as "intermediate-purity FVIII concentrates." Further purification steps, including monoclonal antibody immunoaffinity chromatography, led to the production of "high-purity FVIII concentrates." Even with these high-purity preparations, the transmission of viral infections remained substantial. By the mid-1980s, all plasma-derived factor concentrates were treated with viral attenuation steps (Rosendaal *et al.*, 1991). Still, there was great concern about the transmission of hepatitis B, hepatitis C, and HIV, along with non-lipid-enveloped viruses and as yet unknown human pathogens. This concern led to the production of recombinant antihemophilic factor concentrates. The production of recombinant coagulation factor products was made possible through the cloning of the desired factor gene, optimization of the expression system, and characterization of a unique cell line through recombinant DNA technology.

Collectively, these new coagulation factor products have allowed remarkable advances in the care of patients with bleeding disorders. These products permit rapid reversal of bleeding episodes through self-infu-sion, and therefore limit irreversible joint damage and other bleeding complications. It is now possible to perform long-term prophylaxis at home to prevent bleeding episodes, minimizing costly hospitalizations and the effects of chronic joint bleeding and its attendant morbidity (Hoyer, 1994).

The discussion below categorizes factor concentrates by production method. Information regarding product description and storage was obtained from product monographs. These monographs should be consulted for more detailed information.

I. GENERAL PRINCIPLES OF THE USAGE OF FACTOR CONCENTRATES

While current plasma-derived concentrates for the treatment of bleeding disorders undergo rigorous virucidal treatment, there is still a great deal of concern about the potential for transmission of known, and unknown, human pathogens via blood products. In April 1998, the National Hemophilia Foundation Public Health Service's Advisory Committee on Blood Safety and Availability recommended the following: "Every effort should be made to make recombinant clotting factors available to all who will benefit from them, and all barriers to conversion from human plasma-derived concentrates to recombinant clotting factors should be removed" (National Hemophilia Foundation, MASAC recommendation 89).

The principles listed in Table 1 should be used to prescribe factor concentrates; these will apply for the treatment of all disorders discussed in this chapter.

TABLE 1   Principles for the Administration of Coagulation
Factor Concentrates

- The calculated and infused dose of factor concentrate should be verified by measuring the plasma factor level pre- and postinfusion

- The entire contents of each vial of factor concentrate should be administered, even if it exceeds the calculated dose

- FVIII and FIX should be infused by slow IV push at a rate not to exceed 3 ml/min in adults and 100 U/min in young children

- Continuous infusion of FVIII and purified FIX concentrates is permissible, as these factor concentrates are stable for at least 12 hours at room temperature, and there is no evidence of significant risk of proteolytic inactivation, degradation, or bacterial contamination when the product is infused within this time frame (Bona *et al.*, 1989)

II. TREATMENT OF HEMOPHILIA A

Hemophilia A, or FVIII deficiency, is a congenital bleeding disorder characterized by an increased tendency to bleed, either spontaneously or following trauma. For many years, intracranial hemorrhage was the leading cause of death in hemophilia A patients, and hemarthrosis was, and still is, a cause of severe pain and permanent disability. The severity of hemophilia A is classified according to the patient's plasma FVIII level. Patients with FVIII levels of less than 1% are considered to have severe disease; those with FVIII levels between 1 and 5% are considered moderate; and those with FVIII levels > 5% are considered to have mild disease. Most patients with mild disease can be treated with desmopressin acetate (DDAVP) and/or inhibitors of fibrinolysis; these products will be discussed further in the section on von Willebrand disease. Patients with moderate or severe hemophilia require FVIII infusions for bleeding episodes and for the prevention of bleeding during surgery; some of these patients will also require prophylactic FVIII infusion.

In general, the dose of FVIII to be infused should be calculated based on the patient's body weight in kilograms and the desired FVIII level to be achieved. The desired FVIII level will vary: a level of 40% is desirable for the treatment of early hemarthrosis without large effusions, and a level of 100% is desirable for life-threatening bleeding episodes, such as intracranial hemorrhage.

Each FVIII unit per kilogram of body weight will increase the plasma FVIII level by approximately 2%. The half-life is approximately 8–12 hours. The interval of administration by IV bolus varies from 8 to 24 hours, depending on the initial biodistribution and the desired FVIII level to be maintained. The dosage calculations for FVIII are described in Figure 1.

- **Bolus dose (U)** = weight (kg) X (% desired FVIII level) X 0.5

- **Continuous infusion dose (U):** (expected level of 100%) = 4-5 U/kg/hour. Dose should be individualized according to post-infusion FVIII levels.

FIGURE 1   Dosage calculations for FVIII preparations. Bolus dose (U) = weight (kg) × (% desired level) × 0.5. Continuous infusion dose (U): the expected FVIII level of 100% is 4–5 U/kg/h, but the dose should be individualized according to postinfusion FVIII levels. From Bona *et al.* (1989).

FVIII concentrates fall into three categories, as described below: (1) recombinant products; (2) monoclonal antibody purified products; and (3) intermediate and "high-purity" FVIII products.

A. Recombinant FVIII (rFVIII) Concentrates (Schwartz *et al.*, 1990)

1. First-Generation rFVIII Products: Kogenate, Bioclate, Helixate, and Recombinate

a. Product Description

Recombinant FVIII is a glycoprotein synthesized by a genetically engineered Chinese hamster ovary (CHO) cell line or by a baby hamster kidney (BHK) cell line. The cells secrete rFVIII into the cell culture medium, which is then purified utilizing a series of chromatography columns. The final product has a combination of heterogeneous heavy and light chains similar to those found in plasma-derived FVIII. The rFVIII is then lyophilized into a powder that contains human albumin as a stabilizer. The final product is available in single-dose bottles that contain 250, 500, and 1000 IU per bottle. Von Willebrand factor (vWF) is coexpressed in some of the rFVIII preparations, but these preparations contain no more vWF than 2 ng/IU of rFVIII, and thus do not have any clinically beneficial effects if administered to patients with vWD.

b. Dose and Interval of Administration

The dosing of rFVIII concentrates follows the general recommendation for FVIII dosing as discussed above for plasma-derived products. The pulse dose can be administered at a rate of up to 10 ml/min. Continuous infusion therapy can also be performed with these recombinant products.

c. Uses

Recombinant products are the generally accepted and recommended treatment of choice for patients with hemophilia A whose inhibitor titers are less than 10 Bethesda units (BU). Recombinant products cannot be used for the treatment of vWD because they do no contain significant levels of vWF.

Some concern still exists about the inclusion of human albumin as a stabilizer for the final product. Recombinant products that do not contain human albumin in any of the manufacturing steps are undergoing clinical trials at this time.

d. Adverse Reactions

Infusion-related minor events include flushing, nausea, and fatigue, and occur in < 0.1% of infusions. The development of anti-FVIII antibodies (FVIII inhibitors) represents a significant adverse effect in a number of patients (see Section III below). However, the rate of development of clinically relevant FVIII inhibitors with recombinant products is similar to the rate associated with plasma-derived products.

2. Second-Generation Recombinant FVIII Products: B-Domain-Deleted Recombinant FVIII (BDDrFVIII)

The B domain of the FVIII molecule is not necessary for procoagulant activity, but is the epitope to which most FVIII inhibitors are directed. Thus, it has been eliminated in BDDrFVIII, a novel form of rFVIII which is being developed for the treatment of hemophilia A for the purpose of minimizing subsequent FVIII inhibitor formation. This product is produced in CHO cell lines, purified from the culture medium in 5 chromatographic steps, and modified to achieve the final B-domain-deleted product, which undergoes several steps for viral inactivation. The final product contains no serum albumin (Pollmann and Alerdof, 1999). These products will soon become available commercially and should have minimal immunogenicity and little, if any, stimulation of FVIII inhibitor formation.

B. Monoclonal Antibody-Purified FVIII Products: Monarc-M, Hemophil M, and Monoclate P (Hrinda et al., 1990; Kasper et al., 1991)

1. Product Description

The monoclonal purified FVIII concentrates are sterile, lyophilized preparations of FVIII in concentrated form with a specific FVIII activity range of 2 to 15 IU/mg of protein. They are prepared from pooled human plasma by immunoaffinity chromatography utilizing a murine monoclonal antibody to FVIII, followed by ion exchange chromatography.

2. Dose and Interval of Administration

See the general instructions for FVIII administration described above. These monoclonal products can be administered at a rate up to 10 ml/min.

3. Uses

Monoclonal products are indicated for the prevention and control of hemorrhagic episodes in patients with hemophilia A. Most authorities recommend that monoclonal products not be used in patients who have never been exposed to plasma-derived products; that is, *patients who have never been exposed to plasma-derived products should receive only recombinant products* (unless recombinant products are not available). Monoclonal FVIII products are not effective in controlling bleeding episodes in patients with vWD.

4. Adverse Reactions

Allergic reactions may occur, especially in individuals allergic to mouse protein. Other hypersensitivity reactions manifested by hives, generalized urticaria, wheezing, hypotension, and anaphylaxis may occur. Despite a long history of viral safety, as with all plasma-derived products, the risk of transmission of infectious agents with monoclonal products cannot be eliminated. The reported risk of development of FVIII inhibitors in severe patients with hemophilia A treated with these products is 2%. This incidence is similar to the risk associated with other plasma-derived and rFVIII concentrates.

C. Other Plasma-Derived FVIII Concentrates: Humate P, Koate DVI, Alphanate, and Profilate (Allain et al., 1980; Mannucci et al., 1992)

1. Product Description

These products are derived from human plasma and manufactured via specific steps into intermediate or low-grade concentrates. They are heat treated, solvent/detergent treated, and/or pasteurized to minimize the risk of viral transmission.

2. Dose and Interval of Administration

Dose and interval are as described above.

3. Uses

Although these products are FDA-approved for the treatment of hemophilia A, most experts recommend that they should not be used for this indication when a recombinant product is available and its use is feasible, due to the remote possibility of transmission of human pathogens. The use of this class of FVIII products may be preferable to non-virucidally treated products (e.g., cryoprecipitate) and for the treatment of diseases for which a specific recombinant product is not available (e.g., vWD). Please see Section IV regarding the treatment of vWD for further discussion of these products.

III. TREATMENT OF HEMOPHILIA A PATIENTS WITH INHIBITORS TO FVIII

Patients with hemophilia A may develop neutralizing antibodies to exogenous FVIII. Those antibodies are called FVIII inhibitors. Overall, 6 to 7% of all patients with hemophilia A will develop anti-FVIII antibodies. Some patients without hemophilia A may also develop spontaneous inhibitors to FVIII; this condition is an autoimmune disorder sometimes called "acquired hemophilia" or "acquired FVIII deficiency."

The level of FVIII inhibitor is measured by the Bethesda assay and is reported in Bethesda units. If the inhibitor titer is less than 10 BU, a clinical response can be obtained by increasing the dose of FVIII units infused to 100 or 200 U/kg. Inhibitor titers higher than 10 BU correlate with an inability to respond to even massive doses of FVIII infusion. In those cases, xenogenic (porcine) FVIII or other factor concentrate products that bypass FVIII in the coagulation cascade are available and are described below. Some authorities currently consider recombinant factor VIIa (rFVIIa; see Section B below) as primary therapy for patients with FVIII inhibitors. Historically, as there is often little or no cross-reactivity between human FVIII inhibitors and porcine FVIII, porcine FVIII concentrate has been considered primary therapy. Consultation is recommended.

A. Porcine FVIII: Hyate:C (Travis *et al.*, 1997; Brettler *et al.*, 1989; Rubinger *et al.*, 1997)

1. Product Description

Porcine FVIII is a freeze-dried concentrate highly purified by ion exchange chromatography by means of polyelectrolytes. Each vial contains 400 to 700 U of porcine FVIII. The specific activity of porcine FVIII is greater than 50 U/mg of protein. Trace amounts of porcine vWF are present in a ratio of FVIII/vWF of 10:1.

2. Dose and Interval of Administration

An initial dose of 100 to 150 U/kg is recommended for patients with anti-human FVIII inhibitor titers of 10–50 BU/ml. Based on the reported half-life of 10 to 11 hours in patients without inhibitors, it is recommended that porcine FVIII be administered every 6–8 hours. The intravenous infusion rate should not exceed 2–5 ml per minute. Porcine FVIII has also been administered successfully by continuous infusion over 12 hours. Consultation is recommended in the dosage and administration of this product.

3. Uses

Porcine FVIII is used in the treatment and prevention of bleeding in the following circumstances:

1. Congenital hemophilia A patients with antibodies (inhibitors) to human FVIII
2. Previously nonhemophiliac patients with spontaneously acquired inhibitors to human FVIII (acquired FVIII deficiency)

All patients treated with porcine FVIII should have antibody titers less than 20 BU/ml against porcine FVIII, and preferably less than 10 BU, for this therapy to be effective. The treatment of hemophilia A patients with both anti-human and anti-porcine FVIII inhibitors is beyond the scope of this text.

4. Adverse Reactions

Mild infusion reactions characterized by fever, headache, nausea, vomiting and skin rash can be treated with steroids and or antihistamines. Acute thrombocytopenia of unknown etiology has been reported; therefore, monitoring of the platelet count is recommended. Anaphylaxis is a contraindication for any further use of the product.

B. Recombinant Factor VIIa (rFVIIa): NovoSeven (Hedner, 1996; Lusher, 1996; Hay *et al.*, 1997)

1. Product Description

rFVIIa is molecularly similar to plasma-derived VIIa. The gene for human FVII is cloned and expressed in BHK cells, which secrete the rFVIIa into a medium containing newborn calf serum. The single-chain secreted product is converted to the active two-chain form of rFVIIa during a chromatographic purification process. This process also removes viruses. The single-use vials contain either 1.2 mg (60 KIU) or 4.8 mg (240 KIU). After reconstitution with the appropriate volume of sterile water, the final product contains 0.6 mg/ml rFVIIa.

2. Dose and Interval of Administration

It is recommended that a dose of 90 μg/kg be given over 2 to 5 minutes every 2 hours until hemostasis is achieved.

3. Uses

rFVIIa is indicated for the treatment of bleeding episodes in patients with hemophilia A or B who have inhibitors to factors VIII or IX. Up to 3 doses have been used for treatment of early bleeding complications in home therapy. Because rFVIIa is not plasma-derived, *it should be considered the product of choice for patients who have never been exposed to plasma*, that is, patients who have only been treated with rFVIII and have developed inhibitors to the rFVIII. rFVIIa can also be used to treat persons with FVII deficiency (NHF, 1999, MASAC no. 356). Simultaneous use of activated prothrombin complex concentrates should be avoided.

4. Adverse Reactions

Fever, allergic reactions, headache, pruritus, and rash have been reported. Presence of clinical thrombosis or intravascular coagulation should lead to discontinuation or dose reduction depending on the patient's symptoms.

C. Prothrombin Complex Concentrates (PCCs): Konyne 80, Bebulin, Profilnine SD, and Proplex T (Leissinger, 1999)

1. Product Description

The nonactivated PCCs are plasma-derived concentrates that contain coagulation factors II, VII, IX, and X. Each one of these products undergoes some form of viral inactivation processing. There is a marked difference in factor contents in different preparations and between lots. The reader is referred to each product's package insert for more information regarding factor content.

2. Dose and Interval of Administration

The recommended dose for factor IX (FIX) is 75–80 U/kg as a single dose or at 8-hour intervals. For Bebulin the dose is 80–100 bypassing U/kg at 4- to 12-hour intervals as needed.

3. Uses

PCCs used for hemophilia A patients with FVIII inhibitors have the advantage of a longer half-life and lower cost than rFVIIa. They have been used safely for the treatment of bleeding episodes in the home infusion setting. The main disadvantage of those products when compared to rFVIIa is that the prothrombin complexes are plasma-derived, and as such have some risk of viral transmission. PCCs also can be used to treat patients with deficiencies of factor II, VII, and X, and some patients with inhibitors to FIX. Profilnine has nontherapeutic levels of factor VII; therefore, Profilnine is not indicated for use in the treatment of factor VII deficiency.

4. Adverse Reactions

Serious thrombotic complications such as DIC and coronary artery thrombosis have been reported with administration of high doses of PCCs.

D. Anti-inhibitor Coagulant Complexes (Activated Prothrombin Complex Concentrates, or APCCs): Autoplex T and Feiba VH (Barthels, 1999; Teitel, 1999)

1. Product Description

APCCs are heat-treated, plasma-derived anti-inhibitor concentrates that have a high content of factors VIIa, IXa, and Xa.

2. Dose and Interval of Administration

For spontaneous mild to moderate bleeding episodes, Autoplex T is administered at a dose of 30–50 U/kg every 8 hours. For surgical or severe bleeding, the recommended dose is 75–100 U/kg every 4–8 hours. The recommended dosage for Feiba VH is 50 to 100 U/kg at 12-hour intervals, with careful monitoring for the development of DIC.

3. Uses

APCCs are used to treat bleeding in patients with FVIII inhibitors. It has also been used in some nonhemophiliacs with acquired inhibitors to factors VIII, XI, and XII, and vWF.

4. Adverse Reactions

Life-threatening thromboembolic events may occur in the course of treatment with preparations containing prothrombin complexes. Fever, chills, headache, flushing, and change in pulse may occur if the infusion is given too fast. Hypotension may occur if the infusion is accomplished more than 1 hour after reconstitution because of the accumulation of prekallikrein activator. Allergic reactions reported range from mild urticaria to severe anaphylactic reactions. APCCs carry the risk for the transmission of infectious agents, such as parvovirus.

IV. TREATMENT OF VON WILLEBRAND DISEASE (vWD)

vWF plays an important role in platelet aggregation and in FVIII activity. Patients with vWF deficiency present with a history of bleeding, predominantly from mucosal sites. Although they may have prolonged bleeding times and PTT, sometimes the diagnosis can only be made by repeated measurements of ristocetin cofactor (which measures plasma levels of vWF) and FVIII levels. vWD is classified according to the type of defect in the vWF multimer protein and its functional activity. Treatment varies according to disease type.

• Type 1 vWD is characterized by decreased functional activity of vWF but normal multimer morphology. It should be treated with DDAVP and with agents that inhibit fibrinolysis. It is considered the mildest form of vWD, and accounts for 70% of cases.
• In Type 2 vWD there is both an abnormal multimeric vWF protein and decreased functional activity of vWF. Type 2a vWD, the most common subtype of Type 2 vWD (25% of all cases of vWD), is characterized by decreased platelet aggregation in response to ristocetin and can sometimes be treated with DDAVP. Type 2b vWD must be treated with FVIII concentrates containing high ristocetin cofactor activity (i.e., high levels of vWF). Type 2b vWD is rare. Because the abnormal molecule in Type 2b can cause thrombocytopenia in response to DDAVP administration, *DDAVP is contraindicated for use in patients with Type 2b vWD.*
• In Type 3 vWD the patient has a total absence of both ristocetin cofactor activity and vWF antigen; thus, DDAVP has no beneficial effect on Type 3 vWD. Patients with Type 3 vWD therefore require FVIII infusion for the treatment and prevention of bleeding episodes. Type 3 vWD is the most severe form of vWD and accounts for only 3–5% of all cases.

There is no recombinant vWF concentrate commercially available at the time of this writing. Some of the FVIII preparations have a higher ristocetin cofactor activity (i.e., vWF content) than others, and these high-vWF-content FVIII preparations may be used to treat patients with vWD.

A. Humate-P

1. Product Description

This is a lyophilized, pasteurized, concentrate of human plasma-derived FVIII and vWF. It has a low concentration of nonfactor proteins. The amount of FVIII activity in Humate-P is listed on the label of each vial, and its ristocetin cofactor (RCof) to FVIII ratio approximates 2 to 1.

2. Dose and Administration

For vWD patients, treatment typically consists of the infusion of Humate-P with 40–80 U of RCof activity (14–32 U of FVIII) per kilogram every 8 to 12 hours for major bleeding episodes and prior to invasive procedures. One infusion per day is usually sufficient for minor bleeding episodes or invasive procedures. The dose calculation is based on expected *in vivo* recovery of 1.5 IU/dl per IU Rcof administered. The administration of 1 IU of FVIII is expected to lead to a rise in RCof of approximately 3.3 to 4 IU/dl. The maximum rate of IV bolus administration is 4 ml/min.

3. Uses

Humate-P is the only plasma-derived product with an FDA-approved indication for adults and children with vWD when DDAVP is known, or suspected, to be inadequate for treatment. Because the RCof is reported on each vial, Humate-P is recommended as the product of choice for the treatment of vWD until the time a recombinant vWF product becomes available. Although Humate-P also carries an indication for adult patients with hemophilia A, it is not recommended for the treatment of hemophilia A whenever a recombinant FVIII or higher-purity FVIII concentrate is available.

4. Adverse Reactions

As with all plasma-derived products, with Humate-P there is a risk of transfusion-transmitted infection. Allergic reactions may occur in up to 6% of patients receiving Humate-P. Other adverse events include phlebitis, chills, vasodilation, and paresthesia.

B. Koate DVI (Double Viral Inactivated)

1. Product Description

Koate DVI is a FVIII concentrate purified from pooled human plasma which has been treated with the solvent/detergent tri-n-butyl phosphate and polysorbate 80 and heated in lyophilized form in the final container at 80°C for 72 hours. The specific activity after the addition of human albumin is in the range of 9–22 IU/mg protein.

2. Dosing and Administration

Dosage and administration are the same as for the general FVIII concentrates (see Section II).

3. Uses

Koate DVI is indicated for the treatment of hemophilia A. Although is it is not specifically indicated for the treatment of vWD, it may be used as long as levels are monitored and the ratio of FVIII/RCof in each vial is known.

4. Adverse Reactions

These are the same as described above for Humate-P.

C. Alphanate

1. Product Description

Alphanate is a lyophilized FVIII concentrate extracted from pooled human plasma. There is a step in the purification process that employs heparin-coupled, cross-linked agarose that has an affinity to the heparin-binding domain of the vWF–FVIII complex. Alphanate undergoes several viral inactivation processes, including solvent detergent and heat treatment. FVIII IU content is listed on the label of each vial. It has been reported that Alphanate has approximately 2 IU of FVIII for each IU of RCof.

2. Dosing and Administration

These are similar to the above guidelines for Humate-P, but higher doses may be required to achieve the same effect.

3. Uses

Although Alphanate is indicated for the prevention or treatment of bleeding in patients with hemophilia A, it is not recommended for hemophilia A whenever treatment with recombinant products is a viable option. Alphanate can be used in the treatment of vWD, but higher doses than those of Humate-P may be required to achieve the same effectiveness.

4. Adverse Reactions

These are the same as described above for Humate-P.

V. TREATMENT OF HEMOPHILIA B

Hemophilia B is a congenital bleeding disorder caused by deficiency of factor IX (FIX). Clinically, it is indistinguishable from hemophilia A. Although the PCCs contain FIX, they should not be used in the treatment of hemophilia B because there are now purified FIX concentrates with an extremely low thrombogenic potential. Commercially available FIX concentrates include recombinant FIX (rFIX; BeneFix) and two plasma-derived FIX products, Alphanine and Mononine. *rFIX is recommended as the treatment of choice for patients with hemophilia B*, especially for those patients who have not been previously exposed to plasma-derived products.

A. rFIX:BeneFix (Abshire *et al.*, 1998; Thompson *et al.*, 1998)

1. Product Description

BeneFix is the only rFIX concentrate available at present. It contains a FIX molecule produced by recombinant DNA technology that has structural and functional characteristics similar to those of endogenous FIX. The specific activity of BeneFix is greater or equal to 200 IU/mg of protein. The mean increase in circulating FIX activity is 0.8 IU/dl (range, 0.4 to 1.4) per IU/kg infused, and the mean biologic half-life is 19.4 (range, 11 to 36) hours. The *in vivo* recovery of BeneFix is significantly lower (28%) than the recovery using a highly purified plasma-derived FIX product. There is no significant difference in biological half-life between the two products.

2. Dose and Interval of Administration

BeneFix dosages are calculated as follows:

$$(\text{FIX units required}) = \text{body weight (kg)} \\ \times \text{desired FIX increase (\%)} \\ \times 1.2 \, \text{U/kg}.$$

Higher doses may be necessary and the actual doses should be based on follow-up FIX plasma levels.

3. Uses

BeneFix is approved for the treatment and prevention of bleeding for patients with hemophilia B. It is *not* effective in the treatment of other factor deficiencies or bleeding due to anticoagulation with warfarin or liver failure.

4. Adverse Reactions

Patients have reported nausea, altered taste, discomfort at the IV site, burning sensation in jaw and skull, allergic rhinitis, lightheadedness, and headache with use of FIX-containing products. Patients may experience allergic reactions. The use of FIX-containing products may be potentially hazardous in patients with signs of fibrinolysis and in patients with DIC.

A. Plasma-Derived Purified FIX Concentrates: Mononine and Alphanine SD (Shapiro *et al.*, 1996; Mannucci *et al.*, 1990; Limentani and Gowell, 1993; Kim *et al.*, 1992)

1. Product Description

Mononine is a lyophilized concentrate of FIX extracted from pooled human plasma by immunoaffinity chromatography. Alphanine SD is a purified, solvent/detergent-treated, virus-filtered preparation of not less than 150 U FIX per milligram of total protein.

2. Dose and Interval of Administration

The dosage of Alphanine can be calculated as follows:

$$(IU \text{ required}) = \text{body weight (kg)} \\ \times \text{desired FIX increase (\%)} \\ \times 1 \text{ IU/kg.}$$

The desired postinfusion plasma FIX level varies from 20 to 50%, according to the severity of the bleeding or the type of invasive procedure. Infusions are generally required daily because of the FIX plasma half-life of 22 to 25 hours.

3. Uses

Alphanine and Mononine are indicated for the treatment of hemophilia B.

4. Adverse Reactions

Mild allergic reactions can be prevented by slowing the rate of infusion. The risk of thrombotic events is felt to be small, but caution has been advised when administering to patients with known liver disease. There is a low rate of anti-FIX (inhibitor) formation.

VI. TREATMENT OF RARE CONGENITAL BLEEDING DISORDERS

There are several congenital bleeding disorders (e.g., factors V and X) for which no specific factor concentrates are available. In these cases, PCCs may be effective treatment options. For patients with FVII deficiency, rFVIIa is the most specific FVII product available on the market. Concentrates of FXI and FXIII are in development but are not currently FDA-approved in the United States. In most cases in which a specific concentrate is not available, FFP must be used for treatment, although FXIII deficiency can be treated with cryoprecipitate. For patients with congenital afibrinogenemia, cryoprecipitate is the treatment of choice.

References

Abshire, T., Shapiro, A., Gill, J., *et al.* (1998). Recombinant FIX (rFIX) in the treatment of previously untreated patients with severe or moderately severe hemophilia B. Presented at XXIII International Congress of the World Federation of Haemophilia, May 17–22, The Hague, The Netherlands.

Allain, J. P., Verrous, F., and Soulier, J. P. (1980). *In vitro* and *in vivo* characterization of FVIII preparations. *Vox Sang* **38**, 68–80.

Barthels, M. (1999). Clinical efficacy of prothrombin complex concentrates and 7 recombinant factor VIIa in the treatment of bleeding episodes in patients with FVIII and IX inhibitors. *Thromb Res* **95**, S31–S38.

Bona, R. D., Weinstein, R. A., Weisman, S. J., *et al.* (1989). The use of continuous infusion factor concentrates in the treatment of hemophilia. *Am J Hematol* **32**, 8–13.

Brettler, D. B., Forsberg, A. D., Levine, P. H., *et al.* (1989). The use of porcine FVIII concentrate (Hyate C) in the treatment of patients with inhibitor antibodies to FVIII. *Arch Intern Med* **149**, 1381–1385.

Hay, C. R. M., Negrier, C., and Ludlam, C. A. (1997). The treatment of bleeding in acquired hemophilia with recombinant factor VIIa: A multicenter study. *Thromb Haemost* **78** (Suppl. 1), 1463–1467.

Hedner, U. (1996). Dosing and monitoring NovoSeven treatment. *Haemostasis* **26** (Suppl. 1), 102–108.

Hoyer, L. W. (1994). Hemophilia A. *N Engl J Med* **330**, 38–47.

Hrinda, M. E., Feldman, F., and Schreiber, A. B. (1990). Preclinical characterization of a new pasteurized monoclonal antibody purified FVIIIC. *Semin Hematol* **27** (Suppl. 2), 19–24.

Kasper, C. K., Kim, H. C., Gomperts, E. D., *et al.* (1991). In vivo recovery and survival of monoclonal-antibody-purified FVIII concentrates. *Thromb Haemost* **66**, 730–733.

Kim, H. C., McMillan, C. W., White, G. C., *et al.* (1992). Purified FIX using monoclonal immunoaffinity technique: Clinical trials in hemophilia B and the comparison to prothrombin complex concentrates. *Blood* **79**, 568–575.

Leissinger, C. A. (1999). Use of prothrombin complex concentrates and activated prothrombin complex concentrates as prophylatic therapy in haemophilia patients with inhibitors. *Haemophilia* **5** (Suppl. 3), 25–32.

Limentani, S. A., and Gowell, K. P. (1993). Evaluation of the purity and specific activity of two preparations of ultra-pure FIX. *Blood* **82**, 596a.

Lusher, J. M. (1996). Recombinant VIIa (NovoSeven) in the treatment of internal bleeding in patients with FVIII and FIX inhibitors. *Haemostasis* **26** (Suppl. 1), 124–130.

Mannucci, P. M., Bauer, K. A., Gringeri, A., *et al.* (1990). Thrombin generation is not increased in the blood of patients with hemophilia B patients after the infusion of a purified FIX concentrate. *Blood* **76**, 2540–2545.

Mannucci, P. M., Tenconi, P. M., Castamanan, G., and Rodeghiero, F. (1992). Comparison of four virus-inactivated plasma concentrates for the treatment of severe von Willebrand disease: A cross-over randomized trial. *Blood* **79**, 3130–3137.

Pollman, H., and Alerdof, L. (1999). Albumin-free formulated recombinant FVIII preparations—How big a step forward? *Thromb Haemost* **82**, 1370–1371.

Pool, J. G., Hershgold, E. J., and Pappenhagen, A. R. (1964). High-potency antihaemophilic factor concentrate prepared from cryoglobulin in precipitate. *Nature* **203**, 312.

Rosendaal, F. R., Smith, C., and Briet, E. (1991). Hemophilia treatment in historical perspective: A review of medical and social developments. *Ann Hematol* **62**, 5–15.

Rubinger, M., Houston, D.S., Schwetz, N., *et al.* (1997). Continuous infusion of porcine FVIII in the management of patients with FVIII inhibitors. *Am J Hematol* **56**, 112–118.

Schwartz, R. S., Abildgaard, C. F., Aledorf, L. M., *et al.* (1990). Human recombinant DNA-derived antihemophilic factor (FVIII) in the treatment of hemophilia. *N Engl J Med* **323**, 1800–1805.

Shapiro, A. D., Ragni, M.V., Lusher, J. M., *et al.* (1996). Safety and efficacy of monoclonal antibody purified FIX concentrate in previously untreated patients with hemophilia B. *Thromb Haemost* **75**, 30–35.

Teitel, J. M. (1999). Recombinant FVIIa versus aPCCs in haemophiliacs with inhibitors: Treatment and cost considerations. *Haemophilia* **5** (Suppl. 3), 43–49.

Thompson, A., Brackman, H., Negrier, C., *et al.* (1998). Recombinant FIX is an effective and safe replacement therapy for previously treated patients with hemophilia B. Presentation at XXIII International Congress of the World Federation of Haemophilia, May 17–22, The Hague, The Netherlands.

Travis, S. F., Burns, N., and Hofmann, P. (1997). Porcine FVIII (Hyate:C): In vitro stability, microbiological safety and the effect of heparin. *Hemophilia* **3**, 254–258.

10

Albumin, Gamma Globulin, and Related Derivatives

SILVANA Z. BUCUR

Emory University School of Medicine
Atlanta, Georgia

Plasma contains more than 700 proteins of various structures and functions, of which only albumin, immunoglobulin, and clotting factor concentrates are available for administration. Efficient use of plasma mandates that these components be isolated, purified, and concentrated for individual use. In turn, efficient and appropriate use of the plasma derivatives requires that the clinician be thoroughly familiar with their indications and therapeutic benefits, as well as with possible complications associated with their administration. Plasma components may be extracted from various plasma products, including from plasma obtained through centrifugation of whole blood and plasmapheresis, and from fresh frozen plasma (FFP). Extraction of these components is accomplished through the cold ethanol fractionation method described by Cohn, and results in the separation of various protein fractions, including albumin, immunoglobulin, and factor concentrates VII, VIII, and IX.

I. PRODUCT DESCRIPTION

A. Albumin

Human albumin, a highly soluble, negatively charged, symmetric protein, weighing approximately (MW) 66,500 kDa, is synthesized in the liver. It has a relatively small total body store (300–350 g in a 70-kg person) and a rapid turnover rate of approximately 15 g (4–5% of total albumin) per day. Although the majority (60–65%) of albumin is found in the extravascular compartment, in the circulation albumin accounts for 70–80% of plasma colloid oncotic pressure, with a life span of approximately 15–20 days. The highly negative

charge makes albumin a versatile transport protein, efficiently binding a variety of compounds, including therapeutic and toxic agents.

Commercially available albumin preparations include albumin (human) 5% solution, albumin (human) 25% solution, and plasma protein fraction (human) 5% solution (PPF). All albumin solutions are prepared from pooled plasma, are balanced to physiological pH, contain 145 mEq of sodium and less than 2 mEq potassium per liter, and do not contain any preservatives or coagulation factors. Albumin (human) 5% and 25% solutions contain 1 and 25 g of human albumin per 100 ml of buffered diluent, respectively. Plasma protein fraction (human) 5% solution is a sterile, nonpyrogenic protein solution containing 12.5 g of protein per 250 ml, with at least 83% albumin, less than 17% alpha and beta globulins, and less than 1% gamma globulins.

B. Immunoglobulin

Serum immunoglobulins are synthesized by plasma cells and compose approximately one-third of the total plasma protein content. Immunoglobulins are composed of two heavy (H) and two light (L) polypeptide chains, each consisting of structurally variable and constant regions. Immunoglobulins recognize and bind antigens, form antigen–antibody complexes, facilitate antigen recognition by the accessory lymphoid tissue, and enhance antigen elimination by monocyte macrophages in the reticuloendothelial system. Immunoglobulins can also activate complement and bind Fc receptors. Complex mechanisms regulate the synthesis of various classes, subclasses, and isotypes of immunoglobulins, each with specific corporeal distribution and functions.

Several types of immunoglobulin preparations are available for clinical use (see Table 1). The immunoglobulin solutions used clinically are prepared from pooled human plasma and consist predominantly of IgG (all subclasses and allotypes) with only minute amounts of IgM and IgA.

1. Multispecific immunoglobulin preparations for intravenous administration (IVIg) are commercially available in either liquid or lyophilized form, ready to be reconstituted at a 3–12% concentration in various carbohydrate solutions. Few are available in ready-to-use formulations.

2. Immunoglobulin preparations for intramuscular or subcutaneous (immune serum globulin, ISg) administration are usually 16% solutions and are intended for intramuscular (IM) or subcutaneous (SQ) administration only, as they aggregate *in vitro* to form complexes with anticomplementary activity.

3. Hyperimmune or high-titer, disease-specific immunoglobulin preparations contain high concentrations of disease-specific antibodies, including antibodies against hepatitis B virus (HBIg), varicella-zoster (VZIg), rabies (RIg), and tetanus (TIg), and can be administered either intramuscularly or intravenously (IV).

4. Rh immunoglobulin (RhIg) is a concentrated solution of IgG anti-D available for IM or IV administration.

5. Animal antisera and antitoxins are products collected from animals (equine or bovine) immunized to various venins (spiders, snakes), toxins (tetanus, diphtheria), and antithymocyte (ATG) globulin, and are used for immunosuppression.

6. Monoclonal antibodies, including murine anti-CD3 (OKT3) antibodies against T lymphocytes, are used for immunosuppression after bone marrow transplantation, while murine anti-CD20 is a targeted therapy for B-cell lymphoma.

7. Antibody fragments, including the Fab portion of bovine IgG, bind digoxin and digitoxin, and are used clinically to treat digitalis toxicity.

TABLE 1   Commercially Available Immunoglobulin Preparations

1. Multispecific: Immunoglobulin (human)
   Intravenous immune globulin (IVIg)
   Intramuscular/subcutaneous immune globulin (ISg)
2. Specific: Hyperimmune globulin (human)
   Hepatitis B immune globulin (HBIg)
   Rabies immune globulin (RIg)
   Cytomegalovirus immune globulin (CMV-Ig)
   Tetanus immune globulin (TIg)
   Varicella-zoster immune globulin (VZIg)
   Rh(D) immune globulin (RhIg)
   Vaccinia immune globulin (VIg)
   Western equine encephalitis immune globulin
   Respiratory syncytial virus immune globulin
   Pertussis immune globulin
   Pseudomonas immune globulin
3. Animal immune products
   a. Antitoxins (equine)
      Tetanus antitoxin
      Diphtheria antitoxin
   b. Antivenins (equine)
      Latrodectus mactans antivenin
      Crotalidas polyvalent antivenin
      Micrurus fulvius antivenin
   c. Antiserum (equine)
      Botulism antiserum
      Rabies antiserum
      Antithymocyte globulin (ATg)
   d. Monoclonal antibodies (murine)
      Anti-CD3
      Anti-CD20
   e. Antibody fragments (ovine)
      Digoxin immune Fab fragments

II. COLLECTION, STORAGE, AND HANDLING

A. Albumin and PPF

Human albumin and PPF are prepared from either pooled fresh plasma or pooled fresh frozen plasma. Albumin preparations are stored in ready-to-use bottles, at room temperature (not to exceed 30°C). Freezing should be avoided.

B. Immunoglobulin Fractions

The human immunoglobulin fractions can be isolated from either pooled normal plasma or pooled hyperimmune plasma. Some immunoglobulin products may be of animal origin, including equine, bovine, and ovine. Both liquid and lyophilized IVIg products may be stored at room temperature (not to exceed 25°C) or refrigerated. Reconstitution with sterile water immediately prior to administration is necessary for most immunoglobulin preparations. Intramuscular immunoglobulin preparations may be stored at 2–8°C for up to 3 years. Freezing may increase IgG aggregability, and therefore should be avoided.

III. SPECIAL PREPARATION AND PROCESSING

Plasma derivatives are prepared from plasma using the cold ethanol fractionation process described by Edwin Cohn. This process involves the differential precipitation of albumin and immunoglobulins under conditions of varying protein and ethanol concentrations, pH, ionic strength, and temperature. Once plasma is obtained, it may be frozen at −18°C within 6 hours of separation to result in FFP or within 24 hours of separation to create "plasma frozen within 24 hours" (F24), or it may be immediately fractionated into its derivatives. These products, FFP, F24, and fresh plasma, undergo an initial step of cryoprecipitation to separate the Factor VIII (FVIII) and von Willebrand Factor (vWF) fraction and to remove other coagulation factors and protease inhibitors from the cryosupernatant plasma before the cold ethanol fractionation method is applied. The protein fractions are collected through centrifugation or filtration, while the ethanol is extracted from the precipitated protein fraction either through freeze drying or ultrafiltration. This processing method removes all isohemagglutinins and other antibodies, thus permitting the administration of albumin preparations without compatibility testing.

Unlike immunoglobulin preparations intended for intramuscular or subcutaneous administration, the immunoglobulin fraction intended for intravenous use must be treated by various methods, including enzymatic and chemical processes, in order to remove the aggregated immunoglobulins. Commercially available liquid or lyophilized IVIg preparations are relatively free of high-molecular-weight aggregates.

Although Cohn's fractionation method inactivates human immunodeficiency virus (HIV), hepatitis C virus may still persist after this process; thus other methods are required for its inactivation. The albumin fraction undergoes pasteurization (heated for 10 hours at 60°C), which is an effective method of inactivating hepatitis viruses. The immunoglobulin preparations are subjected to viral inactivation by the solvent/detergent method, followed by the removal of the reactants through filtration. In short, the solvent/detergent method involves the addition of a solvent, 0.3% tri-n-butyl phosphate (TNBP), and a detergent, 0.2% sodium cholate. Immunoglobulin preparations are also exposed to heat (30°C) for at least 6 hours. This sequence of processes effectively inactivates all enveloped and nonenveloped viruses, including HIV, hepatitis B and C viruses, and the nonenveloped virus prototype Reovirus type 3.

IV. ABO AND Rh COMPATIBILITY

Plasma derivatives, including albumin, PPF, and immunoglobulin, are acellular products. Blood group isohemagglutinins have been removed from albumin products; thus, neither serological testing nor ABO and Rh compatibility is necessary prior to their administration. Immunoglobulin preparations contain low concentrations of blood group isohemagglutinins; thus, administration of high doses may result in a positive direct antiglobulin test and rarely hemolysis.

V. ORDERING AND ADMINISTRATION

A. Albumin and PPF

Human albumin 5% and 25% and PPF 5% are supplied in ready-to-use bottles. No special preparation is required when administered without dilution. All albumin solutions must be administered intravenously. Albumin preparations may be used as diluents or in conjunction with blood, plasma, glucose, or saline solutions. The volume and rate of administration for the various human albumin preparations should be adjusted to the individual's specific clinical indication and subsequent clinical response. In general, for adults, 5% albumin solutions should not be infused at rates exceeding 100 ml/h. Following the administration of 25% albumin solution, there is a prompt increase in plasma volume and blood pressure coupled with a fall in hemoglobin level. Thus, this preparation should be administered slowly, at 1–2 ml/min, in individuals with normal blood volume. The rate of PPF 5% infusion must not exceed 10 ml/min or hypotension may ensue. Although the protein dose and volume of albumin preparations are dictated clinically, in certain situations predictable responses may be obtained with specific doses. Thus, in adults with hypovolemia with or without shock, the initial dose of albumin is approximately 25–50 g, while in adults with hypoproteinemia with or without edema, the dose of albumin should not exceed 2 g per day.

B. IVIg and ISg

In general, the dose of IVIg and ISg is 400–1000 mg/kg body weight/day, administered at various schedules conforming to clinical indications. To prevent systemic reactions, IVIg must be administered slowly. For a 10% immunoglobulin solution, an initial infusion rate of 0.01–0.02 ml/kg body weight/min for 30 minutes is recommended, with close observation of the patient. If well tolerated, the infusion rate may then be gradually increased to a maximum of 0.08 ml/kg body weight/min.

VI. USES AND CONTRAINDICATIONS

A. Albumin Preparations

Hypoproteinemia may be due to decreased hepatic synthesis (liver disease, poor nutritional status, infection, malignancy, congenital analbuminemia), increased catabolism (burns, thyrotoxicosis, pancreatitis), excessive extracorporeal loss (nephrotic syndrome, protein-losing enteropathy, exfoliative dermatosis), or redistribution from the intravascular compartment to the interstitium (cirrhosis with ascites, inflammatory processes, surgery). In general, a mild decrease in serum albumin level is not associated with significant clinical consequences and does not require parenteral albumin therapy. In certain situations, albumin infusion has a definite therapeutic benefit (see Table 2), while in others, the role of albumin is still being investigated (Gines et al., 1988; Luca et al., 1995).

1. Paracentesis

Administration of albumin immediately following large-volume paracentesis prevents massive intra/extravascular volume shifts, and thus mitigates the subsequent complications of hyponatremia and renal insufficiency.

2. Nephrotic Syndrome

Massive ascites or anasarca associated with nephrotic syndrome unresponsive to steroid and diuretic therapy may improve with albumin infusions. In this setting, administration of 25% albumin will increase the plasma colloid oncotic pressure, which may subsequently promote diuresis and mobilization of the interstitial fluid into the intravascular compartment (Weiss et al., 1984).

3. Ovarian Hyperstimulation Syndrome (OHSS)

In women undergoing fertility enhancement procedures, the administration of exogenous gonadotrophins may be associated with a potentially lethal syndrome characterized by fluid and electrolyte shifts, ascites, and

TABLE 2   Uses of Albumin

Indications
    Following large-volume paracenthesis
    Nephrotic syndrome resistant to potent diuretics
    Ovarian hyperstimulation syndrome
    Volume/fluid replacement in plasmapheresis
Relative indications
    Adult respiratory distress syndrome
    Cardiopulmonary bypass pump priming
    Fluid resuscitation in shock, sepsis, and burns
    Neonatal kernicterus
    To reduce enteral feeding intolerance
Investigational
    Cadaveric renal transplantation
    Cerebral ischemia
    Stroke
Contraindications
    Correction of hypoalbunemia or hypoproteinemia
    Nutritional deficiency—total parenteral nutrition
    Preeclampsia
    Red blood cell suspension
    Simple volume expansion (e.g., in surgical or burn patients)
    Wound healing

renal dysfunction. Prophylactic albumin infusions at the time of oocyte retrieval decreases the incidence and severity of OHSS (Shaker et al., 1996; Shoham et al., 1994).

4. Hypovolemia

Eighty percent of the plasma colloid oncotic pressure is delivered by serum albumin. Minutes after intravenous administration, 90% of the albumin is found intravascularly, and equilibration with the extravascular compartment occurs at 4 hours postinfusion, with further gradual loss to the interstitium over several days. Although crystalloid and artificial colloid solutions may also effectively expand the plasma volume, they are redistributed into the extravascular space at a much faster rate than human albumin. Several studies evaluating albumin infusions in a variety of clinical situations requiring simple volume expansion have not shown a significant benefit over the use of crystalloid or colloid solutions. In postsurgical patients with intrapulmonary shunting and in patients with acute respiratory distress syndrome (ARDS), infusion of 25% albumin solution followed by diuretic administration may be of some therapeutic benefit (Subcommittee of the Victorian Drug Use Advisory Committee, 1992; Erstad *et al.*, 1991).

5. Hypoproteinemia

The transient benefit of parenteral albumin solution in patients with chronic hypoalbuninemia, including cirrhosis, protein-losing enteropathy, malabsorption, and chronic pancreatitis, is only rarely a valid reason for its use.

6. Burns

Initial fluid resuscitation in burn patients is primarily accomplished with crystalloid solutions. In severe burns, continual protein loss may be repleted by albumin infusions (Subcommittee, 1992; Erstad *et al.*, 1991).

B. Immunoglobulin Preparations

The administration of immunoglobulin preparations: (1) provides prophylactic passive immunity to susceptible individuals exposed to various infectious diseases, (2) enhances bacterial opsonization in patients with deficient humoral immunity, and (3) through Fc receptor blockade reduces the clearance of antibody-coated platelets in immune thrombocytopenic purpura. Usages for IVIg are listed in Table 3. In addition to the therapeutic uses of hyperimmune globulin briefly described in Table 4, prophylactic cytomegalovirus immune globulin (CMVIg) has been associated with increased first-year survival in orthotopic liver transplant recipients.

1. Primary Immunodeficiency Syndromes

Periodic prophylactic ISg or IVIg administration has been shown to decrease the number of febrile episodes, infectious complications, and hospitalizations, and improve survival in patients with primary immunodeficiency syndromes (Cunningham-Rundles *et al.*, 1984; Skull and Kemp, 1996).

2. Hypogammaglobulinemia

Patients with conditions associated with deficient humoral immunity, including chronic lymphocytic leukemia and multiple myeloma, are predisposed to frequent bacterial infections. Prophylactic therapy with IVIg has been demonstrated to decrease the incidence and severity of infections in these patients (Skull and Kemp, 1996; Boughton *et al.*, 1995; Cooperative Group for the Study of Immunoglobulin in Chronic Lymphocytic Leukemia, 1988).

TABLE 3   Common Uses for Intravenous Immunoglobulin

Primary immunodeficiency syndromes
  Common variable immunodeficiency
  X-linked agammaglobulinemia
  Severe combined immunodeficiency
  Ataxia-telangiectasia
  Wiskott–Aldrich syndrome
  IgG subclass deficiency
Neurological disorders
  Chronic inflammatory demyelinating polyneropathy
  Guillain–Barre syndrome
Inflammatory disorders
  Refractory dermatomyositis
  Refractory polymyositis
Disorders with high risk for infections due
  to hypogammaglobulinemia
  Chronic lymphocytic leukemia (CLL)
  Multiple myeloma (MM)
Infectious disorders
  Cytomegalovirus interstitial pneumonitis, post-bone
    marrow transplantation
  Pediatric HIV infection
  Mucocutaneous lymph node syndrome
Immune-mediated disorders
  Idiopathic thrombocytopenic purpura (refractory)
  Thrombocytopenia refractory to platelet transfusion[a]
  Autoimmune hemolytic anemia (refractory, warm-type AIHA)[a]
  Immune neutropenia[a]
  FVIII inhibitor (refractory)[a]
  Graves ophthalmopathy[a]
  Vasculitis (refractory)[a]
  Antiphospholipid syndrome in pregnancy[a]
  Myasthenia gravis (refractory)[a]
  Systemic lupus erythematosus (refractory)[a]
  Post-transfusion purpura[a]

[a] Relative indications.

3. Pediatric Human Immunodeficiency Virus (HIV) Infections

Children infected with HIV have defects in both humoral and cellular immunity, predisposing them to recurrent, life-threatening infections. IVIg therapy in pediatric HIV-infected patients enhances bacterial opsonization and neutralization. Prophylactic IVIg has reduced the number of serious infections and the number of hospitalizations, and has improved the survival rate and quality of life, of children with HIV (Mofenson *et al.*, 1992).

TABLE 4   Uses of and Doses for Hyperimmune Globulin Therapy

Indication	Clinical situations	Immunoglobulin preparation	Dose and route of administration
Hepatitis B: Prophylaxis	Sexual, oral, ophthalmic exposure to hepatitis B or hepatitis B carrier Perinatal exposure Exposure to blood products	Hepatitis B immunoglobulin (HBIg)	Adults: Single dose of HBIg, 0.06 ml/kg IM (max, 5 ml), within 24 hours and up to 14 days of exposure, and HBV vaccine repeated at 1 and 6 months Infants: Single dose of HBIg, 0.5 ml IM, and HBV vaccine repeated at 1 and 6 months Blood transfusion: Larger doses of HBIg or ISg, 0.12 ml/kg (max, 10 ml) IM
Hepatitis A: Prophylaxis	Susceptible individuals exposed to hepatitis A (household contact) Institutional outbreaks Needle exposure Newborns of infected mothers Foreign travel in endemic areas Primate (chimpanzee) exposure	Intramuscular immunoglobulin (ISg)	Individual and institutional exposure: Single dose of 0.02 mL/kg IM during the incubation period (up to 6 days before onset of disease) confers protection for 6–8 weeks Foreign travel: 0.02 ml/kg IM, if intended stay is longer than 3 months, 0.06 ml/kg IM repeated every 3–4 months Animal handlers: Single dose of 0.06 ml/kg IM every 5 months
Measles: Prophylaxis and treatment	After exposure of nonvaccinated and high-risk children (malignancy)	Intramuscular immunoglobulin (ISg)	Prophylaxis: 0.05 ml/kg IM immediately after exposure Lessens severity: 0.01 ml/kg IM immediately after exposure Lessens severity: 0.05 ml/kg IM early after onset of illness
Rabies: Prophylaxis	All animal bites in which rabies cannot be ruled out, in nonimmunized individuals	Rabies immunoglobulin (human and equine) (RIg)	Half the dose of RIg should be injected locally (wound site) and the other IM; rabies vaccine should be given at different site
Varicella-zoster: prophylaxis	After exposure to chicken pox in susceptible and/or immunocompromised individuals, and newborns	Varicella-zoster immunoglobuline (VZIg)	Single dose of VZIg, 125 U/10 km IM, within 5 days of exposure (max, 625 U); confers immunity up to 3 weeks
Tetanus: Prophylaxis and treatment	In nonimmunized individuals following a severe injury or bite	Tetanus immunoglobulin (human TIg or bovine or equine tetanus antitoxin [TAT])	Prophylaxis: TIg in doses of 250–500 U IM with tetanus toxoid; alternatively, TAT, 3000–5000 U IM Treatment: TIg in doses of 500–3000 U (half of the dose at the wound site, half of the dose at a different site IM); alternatively, TAT at a dose of 50,000–100,000 U, with 20,000 U administered IV, and the rest IM—primary immunization with tetanus toxoid upon recovery
Diphtheria: Treatment	Suspected and documented cases of diphtheria	Diphtheria antitoxin (equine)	Dose dependent on systemic severity, location, and extensiveness of membranes: severe cases, IV administration; mild cases, IM administration
Hemolytic disease of the newborn (Rh(D)): Prophylaxis	Rh(D)-negative individuals exposed to Rh-positive blood Rh(D)-negative women after delivery of Rh-positive infant, miscarriage or abortion Immune thrombocytopenic purpura (ITP) in Rh(D)-positive individuals	Anti-D immunoglobulin (RhIg)	Prevention of Rh(D) immunization: Single dose of RhIg, 300 μg IV or IM per 15 ml of fetal blood or 30 ml of whole blood exposure ITP: Initial dose of 50 μg/kg IV if hemoglobin greater than 10 mg/dl, or 25–40 μg/kg IV if hemoglobin is less than 10 mg/dl; additional doses of 25–50 μg/dl IV may be administered if clinically indicated

4. Immune Thrombocytopenic Purpura (ITP)

ITP is an autoimmune disorder, characterized by production of antiplatelet antibodies with subsequent removal and destruction of platelets by the reticuloendothelial system (RES). Studies have demonstrated that IVIg increases platelet counts acutely, is at least as effective as corticosteroid therapy in pediatric ITP, and is useful in chronic ITP refractory to corticosteroids or splenectomy. Several mechanisms of action of IVIg have been proposed, including Fc receptor blockade with subsequent decline in the platelet clearance rate by RES, suppression of antibody synthesis, and/or enhanced antiviral immunity. Although IVIg is at least as effective in ITP as other therapeutic modalities, increased cost limits its widespread use. Thus, IVIg is indicated in refractory ITP unresponsive to corticosteroids or splenectomy or in those for whom corticosteroids are contraindicated, in the treatment of acute bleeding episodes, prophylactically in patients at high risk for intracranial hemorrhage, and prior to surgery (Imholz *et al.*, 1988; Rosthoj *et al.*, 1996; Ratko *et al.*, 1995; Barrios *et al.*, 1993; Blanchette *et al.*, 1994).

C. Rh Immune Globulin

Uses of RhIg and relative contraindications are listed in Table 5. The major indications for RhIg are for: (1) the prophylaxis of Rh(D) immunization resulting from pregnancy or blood transfusion, and (2) the treatment of ITP.

1. Prophylaxis of Rh(D) Immunization Resulting from Pregnancy

RhIg attaches to the Rh(D) antigen on fetal red blood cells (RBCs) and prevents Rh(D) immunization by interfering with the mother's primary immune response to the Rh(D) antigen. RhIg as an antepartum and postpartum prophylaxis is recommended for those Rh(D)-negative pregnant women who have not yet made antibodies to the Rh(D) antigen and who have a D-positive (or potentially D-positive) fetus.

1. For routine antepartum prophylaxis, a single dose of 300 g RhIg (IM or IV) at 28 weeks of gestation is recommended.
2. For postpartum prophylaxis, a dose of 300 g IM of RhIg should be administered within 72 hours of delivery. If excessive fetomaternal hemorrhage is suspected to have occurred at the time of delivery, an additional 300 g RhIg for every 15 ml of fetal RBCs entering the maternal circulation should be administered (Bowman *et al.*, 1978).

TABLE 5   Rh Immune Globulin Usage

Clinical indication	Dose and route of administration
Routine antepartum prophylaxis (28 weeks gestation)	300 g RhIg IM or IV
Routine postpartum prophylaxis (Rh-positive infant)	300 g RhIg IM, additional 300-g doses for every additional 15 ml of fetomaternal hemorrhage
Abortion, miscarriage, ectopic pregnancy	
< 12 weeks gestation	50 g RhIg IM
> 12 weeks gestation	300 g RhIg IM or IV
Amniocentesis or other invasive procedures	
< 34 weeks gestation	300 g RhIg IM or IV, every 12 weeks until delivery
< 34 weeks gestation	300 g RhIg IM or 120 g RhIg IV
Obstetrical complications and accidents	300 g RhIg IM or IV
Prophylaxis following transfusion of Rh(D)-positive blood products	20 g RhIg/ml RBC IM 18 g RhIg/ml RBC IV
Immune thrombocytopenic purpura (ITP)	
Hemoglobin > 10 g/dl	50 g/kg RhIg IV
Hemoglobin < 10 g/dl	25–40 g/kg RhIg IV

Relative contraindications

Rh(D)-negative mothers carrying an Rh(D)-negative fetus (i.e., no risk of exposure to Rh(D) antigen)

Rh(D)-positive or weak D-positive women

Rh(D)-negative women already immunized to the Rh(D) antigen (anti-D titer greater than 1:4 typically represents active immunization)

Rh(D)-negative individuals without childbearing potential

RhIg should also be administered to an Rh(D)-negative pregnant woman following any procedure or incident during which fetomaternal hemorrhage is likely to have occured, for example, amniocentesis or abortion. More details regarding indications for RhIg prophylaxis in pregnancy may be found in Table 5 and in Chapter 25.

2. Prophylaxis of Rh(D) Immunization Resulting from Blood Transfusion

Prophylaxis of Rh-negative individuals after exposure to blood products, including PRBCs, platelets, and granulocytes, depends on the volume of transfused

PRBCs, the clinical situation, and the childbearing potential of the recipient. In these situations, the dose of RhIg needed to ensure suppression of Rh immunization should be calculated using the formula 20 g RhIg per milliliter of RBCs transfused. RhIg should be administered within 72 hours of the exposure (see Table 5; Berkman *et al.*, 1988; NIH Consensus Conference, 1990; Ratko *et al.*, 1995; Seligmann *et al.*, 1983).

3. Treatment of ITP

In patients with ITP who are Rh(D)-positive and not yet splenectomized, RhIg promotes the destruction of the patient's own Rh(D)-positive RBCs within the spleen, packing the spleen with antibody-coated RBCs and thus effectively avoiding RES-associated platelet destruction. Administration of intravenous RhIg to Rh(D)-positive, nonsplenectomized ITP patients has been demonstrated to be as effective as IVIg in treating ITP (Blanchette *et al.*, 1994; Scaradavou *et al.*, 1997; Hartwell, 1998).

VII. EXPECTED RESPONSE AND ADVERSE EFFECTS

A. Clinical Response

Therapeutic responses to infusions of albumin and immunoglobulin preparations are assessed clinically. Monitoring of laboratory parameters is not possible or necessary.

B. Adverse Effects

1. Albumin and PPF

Albumin preparations are associated with a wide variety of side effects. Some side effects are related to rapid infusion rates and include circulatory overload in individuals with compromised cardiac function, pulmonary edema, hypotension, and rarely hypocalcemia. Allergic-type reactions including erythema, urticaria, and anaphylaxis are due to sensitization to plasma proteins. In patients with renal impairment, PPF infusions have occasionally been associated with metabolic acidosis. Albumin infusions may lead to RBC dilution with a subsequent decline in hemoglobin and hematocrit. Rarely, hypocoagulability has been reported following PPF infusion and is related to the variable concentrations of platelet factor 4 and beta-thromboglobulin in the preparation.

2. IVIg and ISg

Intramuscular administration of ISg may be associated with pain at the site of injection as well as systemic reactions including urticaria, flushing, fever, headache, and angioedema. These allergic reactions may be due to the preservatives, the solution stabilizers, or the protein itself. Many of the untoward effects of IVIg preparations are rate-dependent and include flushing, anxiety, headache, nausea, vomiting, and malaise. Allergic-type reactions range from mild, including pruritus, urticaria, and joint pain, to serious, including dyspnea and bronchospasm. Individuals with a previous history of allergic reactions are more susceptible to anaphylactoid reactions to IVIg. *Rarely, immunoglobulin preparations may contain IgG anti-A and/or anti-B antibodies and, when administered in large quantities, may cause hemolysis.* Uncommon side effects of high-dose IVIg therapy are reversible acute renal failure, aseptic meningitis, and cerebrovascular accidents (CVAs). Immunoglobulin products of animal origin (equine, bovine, or ovine) require sensitivity testing prior to administration, as individuals may be allergic to heterologous proteins (Berkman *et al.*, 1988; NIH, 1990; Ratko *et al.*, 1995; Seligmann *et al.*, 1983).

VIII. ALTERNATIVES

Natural and artificial plasma substitutes (see Table 6) administered intravenously increase and/or maintain plasma colloid oncotic pressure and subsequently expand the intravascular plasma volume. Although several artificial plasma colloid preparations are available for clinical use, they have not supplanted the need for albumin. Dextran is a solution of polymerized glucose molecules in 0.9% sodium chloride. Dextran 70 and dextran 40 are preparations in which the molecular weight of the glucose polymers vary (70,000 or 40,000 kDa). Hydroxyethyl starch is a highly branched polymer of amylopectin with hydroxyethyl ether attached to the glucose moiety. Gelatins are derived from animal collagen and available preparations have variable molecular weights and structures, conferring an unpredictable pharmacokinetic profile as well as colloid oncotic effect (Alexander, 1978).

Monoclonal antibodies directed at various cellular antigens, cytokines, and receptors may be used in the future as specific therapies for certain autoimmune disorders and malignancies. Monoclonal antibodies or fragments may also be used for the efficient delivery of radioisotopes to various organs and tumor beds, as well

TABLE 6 Plasma Volume Expander Solutions

Solution	Usage	Distribution
I. Crystalloids		80% of crystalloid solution is found in the interstitial space. Only 20% remains intravascularly after infusion
A. Saline solution		
1. 0.9% NaCl	Volume expansion	
2. 7.5% NaCl	Volume expansion	
3. 3% NaCl	Hyponatremic states, hypochloremic states	
B. Lactated Ringer's solution	Volume expansion	
C. Balanced electrolyte solution	Volume expansion	
II. Colloid solutions		
A. Natural		90% of albumin solutions and PPF are found intravascularly within minutes of administration Loss into the interstitium occurs slowly, over 7–10 days
1. Albumin		
a. 5% solution	Volume expansion, plasma exchange	
b. 25% solution	Peripheral edema	
2. Plasma protein fraction (5%PPF)	Plasma exchange, volume expansion	
3. Fresh frozen plasma (FFP)	Plasma exchange, coagulation factor deficiencies	
B. Artificial		
1. Dextrans		
a. Dextran 40	Antithrombotic prophylaxis, promoting capillary blood flow	
b. Dextran 70	Volume expansion	50% is found intravascularly, 8 hours after infusion
2. Hydroxyethyl starch		
a. Hetastarch	Volume expansion	Comparable to albumin solutions
b. Pentastarch	Leukocyte collection	
3. Gelatins[a]		50% is found intravascularly, 4–5 hours postinfusion
a. Oxypolygelatin	Volume expansion	
b. Modified fluid gelatin	Volume expansion	
c. Urea-linked gelatin	Volume expansion	

[a] Not commercially available in the United States.

as employed in the treatment of drug overdose and intoxication.

References

Alexander, B. (1978). Effects of plasma expanders on coagulation and hemostasis: Dextran, hydroxyethyl starch, and other macromolecules revisited. *Prog Clin Biol Res* **19**, 293–330.

Barrios, N. J., Humbert, J. R., and McNeil, J. (1993). Treatment of acute idiopathic thrombocytopenic purpura with high-dose methylprednisolone and immunoglobulin. *Acta Haematol* **89**, 6–9.

Berkman, S. A., Lee, M. L., and Gale, R. P. (1988). Clinical uses of intravenous immunoglobulins. *Semin Hematol* **25**, 140–158.

Blanchette, V., Imbach, P., Andrew, M., *et al.* (1994). Randomized trial of intravenous immunoglobulin G, intravenous anti-D, and oral prednisone in childhood acute immune thrombocytopenic purpura. *Lancet* **344**, 703–707.

Boughton, B. J., Jackson, N., Lim, S., *et al.* (1995). Randomized trial of intravenous immunoglobulin prophylaxis for patients with chronic lymphocytic leukaemia and secondary hypogammaglobulinemia. *Clin Lab Haem* **17**, 75–80.

Bowman, J. M., Chown, B., Lewis, M., *et al.* (1978). Rh immunization during pregnancy: Antenatal prophylaxis. *Can Med Assoc J* **118**, 623–627.

Cooperative Group for the Study of Immunoglobulin in Chronic Lymphocytic Leukemia (1988). Intravenous immunoglobulin for the prevention of infection in chronic lymphocytic leukemia: A randomized, controlled clinical trial. *N Engl J Med* **319**, 902–910.

Cunningham-Rundles, C., Siegal, F. P., and Smithwick, E. M., *et al.* (1984). Efficacy of intravenous immunoglobulin in primary humoral immunodeficiency disease. *Ann Intern Med* **101**, 435–439.

Demling, R. H. (1987). Fluid replacement in burned patients. *Surg Clin N Am* **67**, 15–30.

Erstad, B. L., Gales, B. J., and Rappaport, W. D. (1991). The use of albumin in clinical practice. *Arch Intern Med* **51**, 901–911.

Gines, P., Tito, L., Arroyo, V., *et al.* (1988). Randomized comparative study of therapeutic paracentesis with and without intravenous albumin in cirrhosis. *Gastroenterology* **94**, 1493–1502.

Hartwell, E. A. (1998). Use of Rh immune globulin. *Am J Clin Pathol* **110**, 281–292.

Imholz, B., Imbach, P., Baumgartner, C., *et al.* (1988). Intravenous immunoglobulin (i.v. IgG) for previously treated acute or for chronic idiopathic thrombocytopenic purpura (ITP) in childhood: A prospective multicenter study. *Blut* **56**, 63–68.

Luca, A., Garcia-Pagan, J. C., Bosch, J., *et al.* (1995). Beneficial effects of intravenous albumin infusion on the hemodynamic and humoral changes after total paracentesis. *Hepatology* **22**, 753–758.

Mofenson, L. M., Moye, J., Jr., Bethel, J., *et al.* (1992). Prophylactic intravenous immunoglobulin in HIV-infected children with CD4 + counts of 0.20×10^9/L or more. Effect on viral, opportunistic and bacterial infections. *JAMA* **268**, 483–488.

NIH Consensus Conference (1990). Intravenous immune globulin. Prevention and treatment of disease. *JAMA* **264**, 3189–3193.

Ratko, T. A., Burnett, D. A., Foulke, G. E., *et al.* (1995). Consensus statement—Recommendations for off-label use intravenously administered immunoglobulin preparations. *JAMA* **273**, 1865–1870.

Rosthoj, S., Nielsen, S., Karup, F., *et al.* (1996). Randomized trial comparing intravenous immunoglobulin with methylprednisolone pulse therapy in acute idiopathic thrombocytopenic purpura. *Acta Paediatr* **85**, 910–915.

Scaradavou, A., Woo, B., Woloski, B. M. R., *et al.* (1997). Intravenous anti-D treatment of immune thrombocytopenic purpura: Experience in 272 patients. *Blood* **89**, 2689–2700.

Seligmann, M., Cunningham-Rundles, C., Hanson, L. A., *et al.* (1983). Appropriate uses of human immunoglobulin in clinical practice: IUIS WHO Notice 1983. *Clin Exp Immunol* **52**, 417–420.

Shaker, A. G., Zosmer, A., Dean, N., *et al.* (1996). Comparison of intravenous albumin and transfer of fresh embryos with cryopreservation of all embryos for subsequent transfer in prevention of ovarian hyperstimulation syndrome. *Fertil Steril* **65**, 992–996.

Shoham, Z., Weissman, A., Barash, A., *et al.* (1994). Intravenous albumin for the prevention of severe ovarian hyperstimulation syndrome in an in vitro fertilization program: A prospective, randomized, placebo-controlled study. *Fertil Steril* **62**, 137–142.

Skull, S., and Kemp, A. (1996). Treatment of hypogammaglobulinemia with intravenous immunoglobulin, 1973–93. *Arch Dis Child* **74**, 527–530.

Subcommittee of the Victorian Drug Usage Advisory Committee (1992). Human albumin solutions: Consensus statements for use in selected clinical situations. *Med J Aust* **157**, 340–345.

Weiss, R. A., Schoeneman, M., and Greifer, I. (1984). Treatment of severe nephrotic edema with albumin and furosemide. *NY State J Med* **84**, 384–386.

RED BLOOD CELL ANTIGENS AND ANTIBODIES

Red Blood Cell Antigens and Human Blood Groups

SHEILAGH BARCLAY

Transfusion Medicine Program
Emory University School of Medicine
Atlanta, Georgia

Since Landsteiner's discovery of the ABO system in 1900, there has been tremendous growth in the understanding of human blood groups. More than 250 red blood cell (RBC) antigens have been described and categorized by the Working Party of the International Society of Blood Transfusion into 23 major systems (Daniels, 1995, 1996). Those RBC antigens not assigned to major systems have been grouped into five collections, a series of low-prevalence antigens and a series of high-prevalence antigens. This chapter describes the RBC antigens that are most commonly encountered in the clinical practice of transfusion medicine (see Table 1). A detailed description of all known blood group systems can be found in the text, *Applied Blood Group Serology* (Issitt, 1998).

Blood group antigens are determined by either carbohydrate moieties linked to proteins or lipids, or by amino acid (protein) sequences. Specificity of the carbohydrate-defined RBC antigens is determined by terminal sugars; genes code for the production of enzymes that transfer these sugar molecules onto a protein or lipid. Specificity of the protein-defined RBC antigens is determined by amino acid sequences that are directly determined by genes. The proteins that carry blood group antigens are inserted into the RBC membrane in one of three ways: single-pass, multipass, or linked to phosphotidylinositol.

Many factors influence the clinical significance of alloantibodies formed against RBC antigens. The prevalence of different RBC antibodies depends on both the prevalence of the corresponding RBC antigen in the population and the relative immunogenicity of the

antigen. The clinical importance of a RBC antibody depends on both its prevalence in a population and whether it is likely to cause RBC destruction (hemolytic transfusion reactions) or hemolytic disease of the newborn (HDN). The type and degree of transfusion reactions and the degree of clinical HDN caused by antibodies to each blood group antigen system will be reviewed in this chapter. The overall clinical significance of antibodies to each of the major blood group antigens is summarized in Table 2.

I. ABO BLOOD GROUP SYSTEM

A. Antigens

Three genes control the expression of the ABO antigens: *ABO*, *Hh*, and *Se* (see Figure 1). The *H* gene codes for the production of an enzyme transferase that attaches L-fucose to the RBC membrane-anchored polypeptide or lipid chain. In the presence of the *A* gene-encoded transferase, N-acetyl-galactosamine is attached, which confers "A" specificity. In the presence of the *B* gene-encoded transferase, galactose is added and confers "B" specificity. If no *A* or *B* gene/enzyme is present, the H specificity remains, and the individual is of group O. If *A* and *B* genes/transferases are both present, "AB" specificity is defined. If the *H* gene is absent, L-fucose is not added to the precursor substance, and even if the *A* and/or *B* genes and their respective enzymes are present, the A and B antigens

TABLE 1   Summary of the Major Blood Group Antigens

Antigens	Systems	Antibody IgM	IgG	Transfusion reactions	HDN[a]	% compatible/prevalence in U.S. population Caucasians	African-Americans
A	ABO	X			None–mod	40%	27%
B	ABO	X			None–mod	11%	20%
D	Rh	X	X	Mild–severe	Mild–severe	15%	8%
C	Rh		X	Mild–severe	Mild	32%	73%
E	Rh	X	X	Mild–mod	Mild	71%	78%
c	Rh		X	Mild–severe	Mild–severe	20%	4%
e	Rh		X	Mild–mod	Rare	2%	2%
K	Kell	X	X	Mild–severe	Mild–severe	91%	98%
k	Kell		X	Mild–mod	Mild–severe	0.2%	< 1%
Kp[a]	Kell		X	Mild–mod	Mild–mod	98%	> 99%
Kp[b]	Kell		X	None–mod	Mild–mod	< 1%	< 1%
Js[a]	Kell		X	None–mod	Mild–mod	> 99%	80%
Js[b]	Kell		X	Mild–mod	Mild–mod	< 0.1%	1%
Fy[a]	Duffy		X	Mild–severe	Mild–severe	34%	90%
Fy[b]	Duffy		X	Mild–severe	Mild	17%	77%
Jk[a]	Kidd		X	None–severe	Mild–mod	23%	8%
Jk[b]	Kidd		X	None–severe	None–mild	26%	51%
M	MNS	X		None	None	22%	26%
N	MNS	X		None	None	28%	25%
S	MNS		X	None–mod	None–severe	48%	69%
s	MNS		X	None–mild	None–severe	11%	6%
U	MNS		X	Mild–severe	Mild–severe	0%	< 1%
Le[a]	Lewis	X		Few	None	78%	77%
Le[b]	Lewis	X		None	None	28%	45%
Lu[a]	Lutheran	X	X	None	None–mild	92%	95%
Lu[b]	Lutheran	X	X	Mild–mod	Mild	< 1%	< 1%
Do[a]	Dombrock		X	Rare	+DAT/No HDN	33%	45%
Do[b]	Dombrock		X	Rare	None	18%	11%
Co[a]	Colton		X	None–mod	Mild–severe	< 0.1%	< 0.1%
Co[b]	Colton		X	None–mod	Mild	90%	90%
P1	P	X		Rare	None	21%	6%

[a] HDN, hemolytic disease of the newborn.

TABLE 2   Clinical Significance of Antibodies to the Major Blood Group Antigens

Usually clinically significant	Sometimes clinically significant	Clinically insignificant if not reactive at 37°C	Generally clinically insignificant
A and B	At[a]	A₁	Bg
Diego	Colton	H	Chido/Rogers
Duffy	Cromer	Le[a]	Cost
H in O_h	Dombrock	Lutheran	JMH
Kell	Gerbich	M, N	Knops
Kidd	Indian	P1	Le[b]
P, PP1P[k]	Jr[a]	Sd[a]	Xg[a]
Rh	Lan		
S, s, U	LW		
Vel	Scianna		
	Yt		

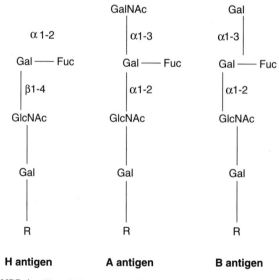

FIGURE 1   Chemical structure of A, B, and H RBC antigens.

TABLE 3

Phenotypes	Prevalence	
	Caucasians	African-Americans
A	40%	27%
B	11%	20%
AB	4%	4%
O	45%	49%

cannot be constructed. The secretor gene (*Se*) controls the individual's ability to secrete soluble A, B, and H antigens into body fluids and secretions. There are about 800,000–1,000,000 copies of the A antigen per group A RBC; 600,000–800,000 copies of the B antigen per group B RBC; and 800,000 copies of the AB antigen per group AB RBC.

B. Phenotypes

The prevalences in the United States of the four phenotypes associated with the ABO blood group system are listed in Table 3.

C. Antibodies

The antibodies of the ABO system are "naturally occurring" in that they are formed as a result of exposure to ABH-like substances from the gastrointestinal tract, occurring *in utero* or immediately postpartum. Thus, there is development of antibodies against whichever ABH antigens are absent on the person's own RBCs. ABO antibodies are usually a mixture of IgM and IgG, and they efficiently fix complement (see Chapter 20 regarding the immune response to blood group antigens).

D. Clinical Significance

Clinically, ABO is the most important RBC antigen system, as circulating A and B antibodies are complement-fixing and thus can cause intravascular hemolysis. Also, ABO incompatibility is the most common cause of HDN in the United States; however, the clinical significance of HDN caused by ABO incompatibility is typically none to moderate, and only rarely severe.

II. Rh BLOOD GROUP SYSTEM

A. Antigens

The Rh system contains at least 45 antigens, of which the major antigens are D, C, E, c, and e. The Rh system is a complex system, and controversy over its genetics has resulted in the development over time of multiple nomenclature systems. In 1943 Wiener proposed the idea of a single Rh locus with multiple alleles. Fisher and Race later inferred the existence of reciprocal alleles on three different but closely linked Rh loci (1946). Later, Rosenfield proposed a numerical system for the Rh blood group system based on serological data (1962).

The isolation of the Rh antigen-containing components of the RBC membrane led to the definitive identification of several nonglycosylated fatty acid acylated Rh polypeptides. Genomic studies have identified two distinct Rh genes, *RHD* and *RHCE* (Cherif-Zahar *et al.*, 1991; Le Van *et al.*, 1992). The presence of *RHD* determines Rh(D) antigen activity. Rh(D)-negative individuals have no Rh(D) gene, and thus have no Rh(D) antigen. The *RHCE* gene codes for both Cc and Ee polypeptides. There are four possible alleles: RHCE, RHCe, RhcE, and Rhce.

With the exception of the A and B antigens, Rh(D) is the most important RBC antigen in transfusion practice. The Rh(D) antigen has greater immunogenicity than virtually all other RBC antigens. Expression of Rh(D) antigen varies quantitatively and qualitatively among individuals. Weakened D reactivity can be caused by three mechanisms: (1) If a *C* gene is on the chromosome opposite the D gene (trans position), the D antigen may be weakened. (2) There can be a qualitative difference in the D antigen in which an individual lacks a portion of the D antigen molecule (and if exposed to the D antigen may produce an antibody to the portion of the D antigen that they lack). This condition is called "partial D" and is defined in terms of the specific D epitopes possessed (Lomas *et al.*, 1993; Cartron, 1994). (3) The Rh(D) gene in some individuals (primarily African-Americans) codes for a Rh(D) antigen that reacts more weakly.

B. Phenotypes

The prevalences of the various phenotypes associated with the Rh blood group system are listed in Table 4.

C. Antibodies

As was stated previously, the Rh(D) antigen has greater immunogenicity than virtually any other RBC antigen, followed by Rh(c) and RH(E). Most Rh antibodies result from exposure to human RBCs through pregnancy or transfusion. Rh antibodies are almost always IgG and do not bind complement: thus, they

TABLE 4

Antigens	Phenotypes[a]	Prevalence		
		Caucasians	African-Americans	Asians
CcDe	R_1r	34.9%	21.0%	8.5%
CDe	R_1R_1	18.5%	2.0%	51.8%
CcDEe	R_1R_2	13.3%	4.0%	30.0%
cDe	R_0r	2.1%	45.8%	0.3%
cDEe	R_2r	11.8%	18.6%	2.5%
cDE	R_2R_2	2.3%	0.2%	4.4%
CDEe	R_1R_z	0.2%	Rare	1.4%
CcDe	R_2R_z	0.1%	Rare	0.4%
CDE	R_zR_z	0.01%	Rare	Rare
cde	rr	15.1%	6.8%	0.1%
Cce	r'r	0.8%	Rare	0.1%
cEe	r''r	0.9%	Rare	Rare
CcEe	r'r''	0.05%	Rare	Rare

[a] Shorthand terminology based on Wiener's Rh-Hr notation (Weiner, 1942).

lead to extravascular rather than intravascular RBC destruction.

D. Clinical Significance

The most common Rh antibody is anti-D; 15% of the U.S. Caucasian population lacks the Rh(D) antigen. Since it is a potent immunogen, the likelihood of an Rh(D)-negative person becoming immunized to Rh(D) following exposure to Rh(D)-positive RBCs is great. It is standard practice to type all donors and recipients for the Rh(D) antigen, and to give Rh(D)-negative PRBCs to Rh(D)-negative recipients. The use of Rh(D)-positive blood for Rh(D)-negative recipients should be restricted to acute emergencies when Rh(D)-negative PRBCs are not available. Once formed, anti-D can cause severe and even fatal HDN. Anti-D is capable of causing mild to severe delayed transfusion reactions. Antibodies to other Rh antigens have also been implicated in both hemolytic transfusion reactions and HDN.

E. Percentage of Compatible Donors

The prevalence in the population of the antigen-negative phenotype determines the ease of, or difficulty in, providing compatible PRBCs for transfusion (see Table 5).

TABLE 5

Phenotypes	Prevalence		
	Caucasians	African-Americans	Asians
D-negative	15%	8%	1%
C-negative	32%	73%	7%
E-negative	71%	78%	61%
c-negative	20%	4%	53%
e-negative	2%	2%	4%

III. KELL BLOOD GROUP SYSTEM

A. Antigens

The primary antigens of the Kell system are K and k. Other antithetical antigens are $Kp^a/Kp^b/Kp^c$, as well as Js^a/Js^b, K11/K17, and K14/K24. Three unpaired low-prevalence antigens and seven high-prevalence antigens complete the system. The antigens are carried on a single-pass membrane glycoprotein (type II).

B. Phenotypes

The prevalences in the U.S. population of the various phenotypes associated with the Kell blood group system are listed in Table 6.

C. Antibodies

Kell system antibodies are generally of the IgG type, react best at body temperature, and rarely bind complement. Anti-K is strongly immunogenic and is frequently found in the serum of transfused K-negative patients. Anti-k, -Kp^a, -Kp^b, -Js^a, and -Js^b are less

TABLE 6

Phenotypes	Prevalence	
	Caucasians	African-Americans
K− k+	91%	98%
K+ k−	0.2%	Rare
K+ k+	8.8%	2%
Kp(a+ b−)	Rare	0%
Kp(a− b+)	97.7%	100%
Kp(a+ b+)	2.3%	Rare
Js(a+ b−)	0%	1%
Js(a− b+)	100%	80%
Js(a+ b+)	Rare	19%

commonly observed in the United States. Anti-Ku is sometimes seen in immunized K_o persons.

D. Clinical Significance

Anti-K may cause both severe HDN and immediate and delayed hemolytic transfusion reactions. Anti-k, -Kp[a], -Kp[b], -Js[a], and -Js[b] occur less often than anti-K, but when present may cause HDN and hemolytic transfusion reactions. The other Kell system antibodies have the potential to cause HDN and hemolytic transfusion reactions, but due to their high (or low) frequencies, these reactions seldom occur.

E. Percentage of Compatible Donors

The prevalence in the population of the antigen-negative phenotype determines the ease of, or difficulty in, providing compatible PRBCs for transfusion, as shown in Table 7.

IV. DUFFY BLOOD GROUP SYSTEM

A. Antigens

The Duffy system comprises five antigens (Fy[a], Fy[b], Fy3, Fy5, and Fy6). The molecule containing the Duffy antigens is a multipass membrane glycoprotein, and there are approximately 13,000 Duffy antigen sites per RBC in persons homozygous for Fy[a] or Fy[b]. RBCs

from heterozygotes have about 6000 antigen sites per RBC. These heterozygous RBCs show weaker agglutination than homozygous cells in serologic tests, a phenomenon called the "dosage" effect. Duffy antigen frequency varies significantly in different racial groups. Interestingly, the Fy3 glycoprotein is the receptor for the malarial parasites *Plasmodium vivax* and *P. knowlesi*; thus, Fy(a − b −) RBCs resist infection by certain malarial organisms.

B. Phenotypes

The U.S. prevalences of the four phenotypes associated with the Duffy blood group system are listed in Table 8.

C. Antibodies

Duffy antibodies are almost always IgG, though rarely IgM, and only rarely bind complement. In spite of the high percentage of the Fy:-3 phenotype in African-Americans, anti-Fy3 is rare.

D. Clinical Significance

Anti-Fy[a] may cause mild to severe transfusion reactions and HDN. Fy[b] is a poor immunogen, and anti-Fy[b] antibodies are only infrequently implicated as the cause of transfusion reactions and HDN.

E. Percentage of Compatible Donors

The prevalence of the antigen-negative phenotype determines the ease of, or difficulty in, providing compatible PRBCs for transfusion, as listed in Table 9.

V. KIDD BLOOD GROUP SYSTEM

A. Antigens

Three antigens make up the Kidd system: Jk[a], Jk[b], and Jk3. The carrier molecule is a multipass membrane

TABLE 7

Phenotypes	Prevalence	
	Caucasians	African-Americans
K-negative	91%	98%
k-negative	0.2%	< 1.0%
Kp[a]-negative	98%	> 99%
Kp[b]-negative	< 1.0%	< 1.0%
Js[a]-negative	> 99%	80%
Js[b]-negative	< 0.1%	1%

TABLE 8

Phenotypes	Prevalence				
	Caucasians	African-Americans	Asians/Chinese	Asians/Japanese	Asians/Thai
Fy(a + b −)	17%	9%	90.8%	81.5%	69%
Fy(a − b +)	34%	22%	0.3%	0.9%	3%
Fy(a + b +)	49%	1%	8.9%	17.6%	28%
Fy(a − b −)	Very rare	68%	0%	0%	

TABLE 9

| Phenotypes | Prevalence | |
	Caucasians	African-Americans
Fy^a-negative	37%	90%
Fy^b-negative	17%	77%
Fy3-negative	0%	68%
Fy5-negative	0%	68%
Fy6-negative	0%	68%

protein, and there are 11,000–14,000 Kidd antigens per RBC.

B. Phenotypes

The prevalences in the United States of the four phenotypes associated with the Kidd blood group system are listed in Table 10.

C. Antibodies

Kidd antibodies are usually IgG, but may be a mixture of IgG and IgM. *They often bind complement and may cause intravascular hemolysis.* It is not uncommon for anti-Jk^a antibody titers to fall rapidly following initial elevation and become undetectable in future antibody screening procedures. If the patient is then exposed to Jk^a antigen-positive PRBCs, a rapid rise in anti-Jk^a titer (anamnestic or "rebound" phenomenon) is often observed, leading to hemolysis.

D. Clinical Significance

Because Kidd antibodies often bind complement, severe hemolytic transfusion reactions are possible. However, only mild HDN is generally seen. Because of the above-described characteristic rapid decline in antibody levels, a delayed transfusion reaction, with marked hemolysis of transfused PRBCs within a few hours, can

TABLE 10

| Phenotypes | Prevalence | | |
	Caucasians	African-Americans	Asians
Jk(a+ b−)	26.3%	51.1%	23.2%
Jk(a− b+)	23.4%	8.1%	26.8%
Jk(a+ b+)	50.3%	40.8%	49.1%
Jk(a− b−)	Rare	Rare	0.9% (Polynesians)

TABLE 11

| Phenotypes | Prevalence | |
	Caucasians	African-Americans
Jk^a-negative	23%	8%
Jk^b-negative	26%	51%

be seen in subsequent exposures to anti-Jk^a-positive PRBCs.

E. Percentage of Compatible Donors

The prevalence of the antigen-negative phenotype determines the ease of, or difficulty in, providing compatible PRBCs for transfusion, as listed in Table 11.

VI. MNS BLOOD GROUP SYSTEM

A. Antigens

Forty antigens make up the MNS system. The major antigens are M, N, S, s, and U. MNS antigens are carried on single-pass membrane sialoglycoproteins. The M and N antigens are located on glycophorin A, while the S and s antigens are located on glycophorin B. Also included are a number of low-prevalence antigens whose reactivity is attributed to either one or more amino acid substitutions, a variation in the extent or type of glycosylation, or the existence of a hybrid sialoglycoprotein.

B. Phenotypes

The prevalences of the numerous phenotypes associated with the MNS blood group system are listed in Table 12.

C. Antibodies

Anti-M antibodies are typically IgM, but may have an IgG component. Anti-N is almost always IgM. Both may be present as seemingly "naturally occurring" antibodies. Anti-S, anti-s, and anti-U are usually IgG and occur following RBC stimulation. Antibodies to M and N may frequently show "dosage" effects, reacting more strongly with RBCs with homozygous expression of these antigens. Anti-U is rare, but should be considered when serum from a previously transfused or preg-

TABLE 12

Phenotypes	Prevalence	
	Caucasians	African-Americans
M+ N− S+ s−	6%	2%
M+ N− S+ s+	14%	7%
M+ N− S− s+	10%	16%
M+ N+ S+ s−	4%	2%
M+ N+ S+ s+	22%	13%
M+ N+ S− s+	23%	33%
M− N+ S+ s−	1%	2%
M− N+ S+ s+	6%	5%
M− N+ S− s+	15%	19%
M+ N− S− s−	0%	0.4%
M+ N+ S− s−	0%	0.4%
M− N+ S− s−	0%	0.7%

TABLE 14

Phenotypes	Prevalence	
	Caucasians	African-Americans
P1-negative	21%	6%

nant African-American person contains antibody to an unidentified high-prevalence antigen.

D. Clinical Significance

Anti-M has only been implicated in transfusion reactions or HDN in rare cases. Anti-N has no known clinical significance. Antibodies to S and s are capable of causing hemolytic transfusion reactions and HDN. Anti-U has been implicated in mild to severe hemolytic transfusion reactions and in HDN.

E. Percentage of Compatible Donors

The prevalence of the antigen-negative phenotype determines the ease of, or difficulty in, providing compatible PRBCs for transfusion, as listed in Table 13.

TABLE 13

Phenotypes	Prevalence	
	Caucasians	African-Americans
M-negative	22%	30%
N-negative	28%	25%
S-negative	48%	69%
s-negative	11%	3%
U-negative	0%	< 1%

VII. P BLOOD GROUP SYSTEM

A. Antigens

The sequential transcription of multiple genes is required for the expression of the P1 antigen; this occurs upon the addition of a galactosyl residue to paragloboside. The P, Pk, and LKE antigens were previously included in the P system, but because a different locus and biochemical pathway has been found to be involved in their production, they have been moved to the Globoside (Glob) collection.

B. Antibodies

Anti-P1 is IgM and is naturally occurring in many P1-negative individuals. Complement binding by anti-P1 is rare.

C. Clinical Significance

Antibodies to P1 rarely cause transfusion reactions and have not been implicated in HDN.

D. Percentage of Compatible Donors

The prevalence of the antigen-negative phenotype determines the ease of, or difficulty in, providing compatible PRBCs for transfusion, as listed in Table 14.

VIII. LUTHERAN BLOOD GROUP SYSTEM

A. Antigens

The Lutheran system comprises four antithetical pairs, Lu^a and Lu^b, Lu6 and Lu9, Lu8 and Lu 14, and Au^a and Au^b, as well as 10 high-prevalence antigens. The antigens are carried on a single-pass membrane glycoprotein with 1500−4000 copies per RBC, depending on zygosity.

TABLE 15

Phenotypes	Prevalence in most populations
Lu(a+ b−)	0.2%
Lu(a− b+)	92.4%
Lu(a+ b+)	7.4%
Lu(a− b−)	Rare

B. Phenotypes

The prevalences of the four phenotypes associated with the Lutheran blood group system are listed in Table 15.

C. Antibodies

The most common Lutheran antibodies are anti-Lu^a and anti-Lu^b, but they are not often encountered. Lutheran antibodies may be IgG or IgM, are generally not reactive at body temperature, and rarely bind complement.

D. Clinical Significance

The Lutheran antigens are not well developed at birth, and anti-Lu^a has not been reported as a cause of HDN. Anti-Lu^a has not been associated with hemolytic transfusion reactions. Anti-Lu^b may cause accelerated destruction of transfused PRBCs, but only causes (if at all) mild HDN.

E. Percentage of Compatible Donors

The prevalence of the antigen-negative phenotype determines the ease or difficulty in providing compatible PRBCs for transfusion, as listed in Table 16.

IX. LEWIS BLOOD GROUP SYSTEM

A. Antigens

Le^a and Le^b make up the antigens of the Lewis system. The Lewis antigens are not intrinsic to RBCs, but are located on type 1 glycosphingolipids that are adsorbed onto RBCs from the plasma. The biosynthesis of Lewis antigens results from the interaction of two independent genetic loci, *Le* and *Se*. The transferase that is the product of the *Le* gene attaches fucose to the subterminal GlcNAc of type 1 oligosaccharides. This confers Le^a activity. The *Se* gene determines a transferase that attaches a fucose to the terminal Gal, but only if the adjacent GlcNAc is already fucosylated. This configuration has Le^b activity. Le^b is adsorbed onto the RBC preferentially over Le^a. Individuals possessing both *Le* and *Se* genes will have RBCs that express Le^b, but not Le^a. RBCs from a person with *Le* but not *Se* genes will express Le^a.

B. Phenotypes

The prevalence of phenotypes associated with this blood group system is listed in Table 17.

C. Antibodies

Anti-Le^a and anti-Le^b are frequently occurring antibodies made by Le(a− b−) individuals, usually in the absence of a foreign RBC stimulus. Lewis antibodies are predominantly IgM, and some bind complement.

D. Clinical Significance

Most Lewis antibodies react at colder temperatures than body temperature and are not clinically significant. Since Lewis antigens are poorly developed at birth, and IgM antibodies cannot cross the placenta,

TABLE 16

	Prevalence	
Phenotypes	Caucasians	African-Americans
Lu^a-negative	92%	95%
Lu^b-negative	<1%	<1%

TABLE 17

	Prevalence	
Phenotypes	Caucasians	African-Americans
Le(a+ b−)	22%	23%
Le(a− b+)	72%	55%
Le(a− b−)	6%	22%
Le(a+ b+)	Rare	Rare

TABLE 18

	Prevalence	
Phenotypes	Caucasians	African-Americans
Le^a-negative	78%	77%
Le^b-negative	28%	45%

Lewis antibodies do not cause HDN. Anti-Le^a has caused hemolytic transfusion reactions on rare occasions.

E. Percentage of Compatible Donors

The prevalence of the antigen-negative phenotype determines the ease of, or difficulty in, providing compatible PRBCs for transfusion, as listed in Table 18.

X. DIEGO BLOOD GROUP SYSTEM

A. Antigens

The Diego system includes two pairs of antithetical antigens, Di^a/Di^b and Wr^a/Wr^b, and three low-prevalence antigens, Wd^a, Rb^a, and WARR. The antigens are carried on a multipass membrane protein, having 1,000,000 antigen copies per RBC.

B. Phenotypes

The prevalences of phenotypes associated with the Diego blood group system are listed in Table 19.

C. Antibodies

Anti-Di^a and anti-Di^b antibodies are of the IgG class and do not bind complement. Anti-Wr^a and anti-Wr^b may be IgM or IgG and do not bind complement.

TABLE 19

	Prevalence		
Phenotypes	Caucasians	African-Americans	Asians
Di(a+ b−)	< 0.01%	< 0.01%	< 0.01%
Di(a− b+)	> 99.9%	> 99.9%	90%
Di(a+ b+)	< 0.1%	< 0.1%	10%

D. Clinical Significance

Anti-Di^a is not common because of the low prevalence of the antigen in the United States, but *anti-Di^a can cause RBC destruction, and should be considered clinically significant when present.* Anti-Wr^a has been implicated in mild to severe transfusion reactions and in HDN.

XI. CARTWRIGHT BLOOD GROUP SYSTEM

A. Antigens

The high-prevalence antigen Yt^a and the low-prevalence antigen Yt^b make up the Cartwright system. The antigens are carried on the PI-linked glycoprotein, acetylcholinesterase. There are 10,000 antigen copies per RBC.

B. Antibodies

The Cartwright antibodies are IgG and do not bind complement. Because of the high prevalence in the United States of the Yt^a antigen, anti-Yt^a is not common. Because Yt^b is a poor immunogen, anti-Yt^b is also uncommon.

C. Clinical Significance

Many examples of anti-Yt^a have been shown to be clinically benign *in vivo*, while other examples have shown increased RBC destruction in *in vivo* survival studies.

XII. Xg BLOOD GROUP SYSTEM

A. Antigens

Xg was the first blood group system to be assigned to the X chromosome. The sole antigen, Xg^a, is located on a single-pass glycoprotein (type 1).

B. Phenotypes

The prevalence of phenotypes associated with the blood group system is listed in Table 20.

C. Antibodies

Xg^a is a poor immunogen. IgG-type antibodies to Xg^a are more common than IgM. Some examples of anti-Xg^a are naturally occurring.

TABLE 20

	Prevalence	
Phenotypes	**Males**	**Females**
Xg(a+)	65.6%	88.7%
Xg(a−)	34.4%	11.3%

TABLE 21

Phenotype	**Prevalence**
Xg^a-negative	11% (females)
Xg^a-negative	34% (males)

D. Clinical Significance

Antibodies to Xg^a have not been shown to cause transfusion reactions or HDN.

E. Percentage of Compatible Donors

The prevalence of the antigen-negative phenotype determines the ease of, or difficulty in, providing compatible RBCs for transfusion, as listed in Table 21.

XIII. SCIANNA BLOOD GROUP SYSTEM

A. Antigens

There are three antigens associated with the Scianna system: the high-prevalence Sc:1, the low-prevalence Sc:2, and Sc:3 (which is present if either Sc:1 or Sc:2 is present). These antigens are carried on a membrane glycoprotein.

B. Phenotypes

The prevalences of phenotypes associated with the blood group system are listed in Table 22.

C. Antibodies

Antibodies to Scianna RBC antigens are rare, but if present are generally of the IgG type.

D. Clinical Significance

Antibodies directed against Scianna antigens have not been reported to cause transfusion reactions. Anti-

TABLE 22

Phenotypes	**Caucasians**
Sc:1, -2	99.7%
Sc:1, 2	0.3%
Sc:-1, 2	Very rare
Sc:-1, -2	Very rare

Sc1 may cause a positive direct antiglobulin test (DAT), but it does not cause clinically significant HDN.

E. Percentage of Compatible Donors

The prevalence of the antigen-negative phenotype determines the ease of, or difficulty in, providing compatible PRBCs for transfusion, as listed in Table 23.

XIV. DOMBROCK BLOOD GROUP SYSTEM

A. Antigens

The five antigens that make up the Dombrock system are Do^a Do^b, and the high-prevalence antigens Hy, Gy^a, and Jo^a. The antigens are carried on a GPI-linked glycoprotein.

B. Phenotypes

The prevalences of phenotypes associated with the blood group system are listed in Table 24.

C. Antibodies

Dombrock system antibodies are IgG and do not bind complement. Do^a and Do^b are poor immunogens, and thus antibodies to Do^a and Do^b are not common. Gy^a is highly immunogenic, but because of its high prevalence in the United States, anti-Gy^a is rarely observed.

TABLE 23

Phenotypes	**Prevalence in Caucasians**
Sc1-negative	Very rare
Sc2-negative	99.7%

TABLE 24

Phenotypes	Do^a	Do^b	Gy^a	Hy	Jo^a	Prevalence	
						Caucasians	African-Americans
Do(a+ b−)	+	0	+	+	+	18%	11%
Do(a+ b+)	+	+	+	+	+	49%	44%
Do(a− b+)	0	+	+	+	+	33%	45%
Gy(a−)	0	0	0	0	0%	Rare	0%
Hy−	0	wk	wk	0	0	0%	Rare
Jo(a−)	wk	0/wk	+	wk	0	0%	Rare

wk, weak.

TABLE 25

Phenotypes	Prevalence	
	Caucasians	African-Americans
Do^a-negative	33%	45%
Do^b-negative	18%	11%
Hy-negative	0%	Rare
Gy^a-negative	Rare	0%
Jo^a-negative	0%	Rare

TABLE 26

Phenotypes	Prevalence in most populations
Co(a+ b−)	90%
Co(a− b+)	0.5%
Co(a+ b+)	9.5%
Co(a− b−)	< 0.01%

D. Clinical Significance

Anti-Do^a may cause increased RBC destruction, but anti-Do^b is considered to be clinically insignificant. Antibodies to Do^a, Do^b, Gy^a, Hy, and Jo^a have not been observed to cause HDN.

E. Percentage of Compatible Donors

The prevalence of the antigen-negative phenotype determines the ease of, or difficulty in, providing compatible PRBCs for transfusion, as listed in Table 25.

XV. COLTON BLOOD GROUP SYSTEM

A. Antigens

Three antigens have been assigned to the Colton system: Co^a, Co^b, and Co3. The antigens are carried on a multipass membrane glycoprotein that is part of the water transport protein CHIP-1.

B. Phenotypes

The prevalences of the four phenotypes associated with the Colton blood group system are listed in Table 26.

C. Antibodies

Antibodies to Colton system antigens are IgG and rarely bind complement (with the exception of Co3).

D. Clinical Significance

Colton antibodies are rare, but anti-Co^a has been implicated in both HDN and RBC destruction. Anti-Co3 has been known to cause severe HDN.

E. Percentage of Compatible Donors

The prevalence of the antigen-negative phenotype determines the ease of, or difficulty in, providing compatible PRBCs for transfusion, as listed in Table 27.

TABLE 27

Phenotypes	Prevalence in all populations
Co^a-negative	0.1%
Co^b-negative	90%
Co3-negative	< 0.01%

TABLE 28

TABLE 28

| Phenotypes | Prevalence | |
	Most populations	Finnish
LW(a+ b−)	97%	93.9%
LW(a+ b+)	3%	6%
LW(a− b+)	Rare	0.1%

TABLE 29

Phenotypes	Prevalence in all populations
Ch-negative	4%
Rg-negative	2%
Wh-negative	85%

XVI. LW BLOOD GROUP SYSTEM

A. Antigens

The LW system is composed of the antigens LW^a, LW^{ab}, and LW^b. The antigens are located on a single-pass membrane glycoprotein (type 1). The original anti-Rh described by Landsteiner and Wiener was in fact directed against the LW antigen. There is a phenotypic relationship between LW and the Rh(D) antigen. In adults, Rh(D)-negative RBCs have less expression of LW than do Rh(D)-positive RBCs.

B. Phenotypes

The prevalences of phenotypes associated with the LW blood group system are listed in Table 28.

C. Antibodies

LW antibodies may be IgM or IgG and do not bind complement. Autoanti-LW antibodies have been reported upon immunosuppression of LW antigens, observed primarily during pregnancy and in patients with certain malignancies.

D. Clinical Significance

While Rh(D)-negative, LW+ PRBCs have successfully been transfused to persons with anti-LW in their sera, LW antibodies may cause accelerated destruction of LW+ RBCs. Significant HDN has not been reported.

XVII. CHIDO—ROGERS BLOOD GROUP SYSTEM

A. Antigens

The Chido-Rogers (Ch/Rg) system is made up of nine antigens (Ch1, Ch2, Ch3, Ch4, Ch5, Ch6, Rg1, Rg2, and WH) that reside on the complement component C4. Chido antigens are located in the C4d region of C4B. Rogers antigens are located in the C4d region of C4A. C4A and C4B are glycoproteins that are adsorbed onto the RBC membrane from the serum.

B. Antibodies

Ch/Rg antibodies are IgG, do not bind complement, and may be neutralized (in serological tests) with serum or plasma from Ch/Rg antigen-positive individuals.

C. Clinical Significance

Antibodies to Ch/Rg antigens are usually clinically insignificant, but they may create difficulties in interpretations of serological investigations. None have been known to cause HDN or increased RBC destruction.

D. Percentage of Compatible Donors

The prevalence of the antigen-negative phenotype determines the ease of or difficulty in providing compatible PRBCs for transfusion, as listed in Table 29.

XVIII. Hh BLOOD GROUP SYSTEM

A. Antigens

The H antigen is the precursor molecule on which the A and B antigens are built. On group O RBCs, there are no A or B antigens, and the RBC membrane expresses abundant H antigen. The amount of H antigen present on RBCs by blood group is, in order of diminishing quantity, $O > A_2 > B > A_2B > A_1 > A_1B$. In the absence of the *H* gene, no L-fucose is added to the protein or lipid, and thus no H antigen is expressed (the Bombay phenotype). In the Bombay phenotype, even in the presence of functional *A* and *B* genes, no A or B antigen will be formed; the RBCs will type as group O, and hence the notation O_h.

B. Antibodies

A few group A_1, A_1B, and B individuals have so little unconverted H antigen that they produce an anti-H that is a weak IgM antibody reactive at colder than room temperatures. The anti-H formed by an O_h individual is potent, reacts over a thermal range of 4–37°C, and rapidly destroys RBCs with A, B, and/or H antigens.

C. Clinical Significance

The benign anti-H formed by group A_1, A_1B, and B individuals is not considered clinically significant. However, *the anti-H formed by O_h individuals can cause hemolysis* in vivo *and subsequent severe hemolytic transfusion reactions*. HDN may occur in O_h mothers.

D. Percentage of Compatible Donors

The only suitable PRBCs for transfusion to a Bombay patient are those from another Bombay (O_h) individual.

XIX. XK BLOOD GROUP SYSTEM

A. Antigens

Only one antigen is included in the XK system, Kx. It is carried on a multipass membrane protein. The antigen is present on virtually all RBCs, but its expression is weak on RBCs of common Kell phenotypes. Elevated levels of Kx are found on the null (K_o) RBC. Absence of the Kx protein is associated with McLeod syndrome, a spectrum of abnormalities including weakened Kell system antigens, bizarre RBC shapes, chronic anemia, and abnormalities of the nervous system, of the cardiac system, and of granulocytes. The McLeod phenotype has been identified in patients with chronic granulomatous disease (CGD) and appears to result from deletion of a portion of the X chromosome that includes the *XK* locus (as well as *X-CGD*).

B. Antibodies

Anti-Kx is IgG and does not bind complement. Males with both McLeod syndrome and CGD may be immunized by PRBC transfusion and produce anti-KL (anti-Kx + anti-Km). Only Kx-negative PRBCs are compatible. Non-CGD males may make anti-Km; both Kx-negative and K_o PRBCs are compatible.

C. Clinical Significance

Mild delayed transfusion reactions have been reported with anti-Kx. HDN caused by anti-Kx has not been described.

XX. GERBICH BLOOD GROUP SYSTEM

A. Antigens

The Gerbich system contains seven antigens: Ge2, Ge3, and Ge4 (all high prevalence), and Wb, Ls^a, An^a, and Dh^a (all low prevalence). The antigens are carried on single-pass membrane sialoglycoproteins (type 1). Ge3 and Ge4 are located on glycophorin C (GPC), while Ge2, Ge3, and An^a are located on glycophorin D (GPD). Dh^a and Wb are located on altered forms of GPC. Ls^a is found on altered forms of GPC and GPD. Rare phenotypes that lack one or more of the high-prevalence antigens are Leach (Ge:-2, -3, -4), Gerbich (Ge:-2, -3, 4), and Yus (Ge:-2, 3, 4).

B. Antibodies

Antibodies to Gerbich antigens are usually IgG, but may also have an IgM component. Some examples of anti-Ge2 and anti-Ge3 are hemolytic *in vitro*.

C. Clinical Significance

Gerbich system antibodies are of variable clinical significance. Some have been known to cause mild HDN and slightly accelerated RBC clearance.

XXI. CROMER BLOOD GROUP SYSTEM

A. Antigens

Ten antigens have been assigned to the Cromer system: Cr^a, Tc^a, Dr^a, Es^a, IFC, UMC, and Wes^b (all high prevalence), and Tc^b, Tc^c, and Wes^a (all low prevalence). These antigens are located on the complement regulating protein decay accelerating factor (DAF).

B. Antibodies

Antibodies to Cromer system antigens are extremely rare. Most examples of anti-Cr^a, anti-Wes^b, and anti-Tc^a have been found in serum from African-American individuals.

C. Clinical Significance

The clinical significance of Cromer antibodies is variable. Some have reportedly caused decreased survival of transfused PRBCs *in vivo*. Antibodies to Cromer system antigens have not been known to cause HDN.

XXII. KNOPS BLOOD GROUP SYSTEM

A. Antigens

Kn^a, Kn^b, McC^a, Sl^a, and Yk^a make up the Knops system. These antigens are located on the single-pass membrane glycoprotein (type 1) CR1, the primary complement receptor on RBCs.

B. Antibodies

The antibodies directed against Knops antigens are IgG and do not bind complement. They were previously grouped among the "high-titer, low-avidity (HTLA) antibodies." These antibodies commonly show weak, variable reactivity in the antiglobulin phase of testing, but may continue to react *in vitro*, even at high dilutions. This variable reactivity has been shown to be a direct reflection of the number of CR1 sites on RBCs.

C. Clinical Significance

The Knops-directed antibodies are of no known clinical significance, since they do not cause transfusion reactions, increased RBC destruction, or HDN.

XXIII. INDIAN BLOOD GROUP SYSTEM

A. Antigens

The Indian system is comprised of two antigens, the low-prevalence In^a and the high-prevalence In^b. The antigens are located on a single-pass membrane glycoprotein (CD44). CD44 has wide tissue distribution and is a cellular adhesion molecule involved in immune stimulation and signaling between cells. These antigens were discovered in persons from the Indian subcontinent (hence the name "Indian").

B. Antibodies

Indian antibodies are of the IgG class and do not bind complement.

TABLE 30

Phenotypes	Prevalence			
	Caucasians	African-Americans and Asians	Indians	Iranians
In^a-negative	99.9%	99.9%	96%	89.6%
In^b-negative	1%		4%	

C. Clinical Significance

Both anti-In^a and anti-In^b have been known to cause transfusion reactions and decreased RBC survival *in vivo*. Both antibodies may cause a positive DAT, but neither has been known to cause HDN.

D. Percentage of Compatible Donors

The prevalence of the antigen-negative phenotype determines the ease of, or difficulty in, providing compatible PRBCs for transfusion, as listed in Table 30.

XXIV. COST BLOOD GROUP COLLECTION

A. Antigens

The antigens in this collection are Cs^a and Cs^b. They demonstrate variable expression on RBCs.

B. Antibodies

Antibodies to Cost antigens are IgG and do not bind complement. These antibodies were formerly grouped with the HTLA antibodies.

C. Clinical Significance

The Cost antibodies are not considered clinically significant.

XXV. Er BLOOD GROUP COLLECTION

A. Antigens

The Er collection consists of the high-prevalence Er^a and the low-prevalence Er^b.

B. Antibodies

The Er antibodies are IgG and do not bind complement.

C. Clinical Significance

Anti-Er^a has not been implicated in transfusion reactions, but has been reported to cause a positive DAT without evidence of clinical HDN.

XXVI. Ii BLOOD GROUP COLLECTION

A. Antigens

The antigens included in this collection, I and i, are not the products of alleles, but instead arise from the sequential actions of multiple glycosyltransferases. The antigens are located on precursor A, B, and H active oligosaccharide chains (I on branched type 2 chains and i on linear type 2 chains). Fetal RBCs carry few branched oligosaccharide chains and therefore are rich in i and poor in I. With age, the linear chains are modified by the addition of branched structures. This branching configuration confers I specificity. So-called compound antigens have been described: IA, IB, IAB, IH, IP, ILe^bH, iH, iP1, and iHLe^b.

B. Phenotypes

The prevalences of phenotypes associated with the blood group system are listed in Table 31.

C. Antibodies

Anti-I is an IgM antibody that reacts preferentially at room temperature or colder and binds complement. Only rare examples of I-negative adult RBCs exist; thus, alloanti-I is rare. Since all RBC have trace amounts of i, anti-i is considered an autoantibody.

D. Clinical Significance

Anti-I is a low-titer, cold-reactive, common autoantibody that reacts within a narrow thermal range. *Anti-I assumes pathological significance in cold agglutinin disease and mixed-type autoimmune hemolytic anemia,* in which it behaves as a complement-binding hemolytic antibody with a high titer and wide thermal range. Although rare, accelerated RBC destruction has been

TABLE 31

	Prevalence		
Phenotypes	Adult	Cord	i adult
I	Strong	Weak	Trace
i	Weak	Strong	Strong

TABLE 32

Phenotypes	Antigens	Prevalence	Antibodies
P$_1$	P, P1, P^k	79%	None
P$_2$	P, P^k	21%	Anti-P1
P$_1^k$	P1P^k	Rare	Anti-P
P$_2^k$	P^k	Rare	Anti-P
p	None	Rare	Anti-P, P1, P^k

observed in the adult I phenotype in the presence of alloanti-I. HDN has not been associated with Ii antibodies.

XXVII. GLOBOSIDE COLLECTION

A. Antigens

Three antigens are included in this collection: P, P^k, and LKE. The sequential action of multiple gene products is required for expression of these antigens. Lactosyl ceramide is the precursor substance.

B. Phenotypes

The prevalences of phenotypes associated with the blood group system are listed in Table 32.

C. Antibodies

The antibodies directed against the Globoside antigens may be IgG or IgM. They may bind complement, and some are hemolytic *in vitro*.

D. Clinical Significance

Anti-P has been implicated in rare cases of severe transfusion reactions, but has been known to cause only mild HDN. The biphasic autohemolysin in paroxysmal cold hemoglobinuria (PCH) has anti-P specificity. Anti-P, P1, P^k (formerly Tj^a) is a potent IgM antibody that has caused hemolytic transfusion reactions and occasional cases of HDN.

XXVIII. LOW-PREVALENCE ANTIGENS

A. Antigens

Low-prevalence antigens occur in <1% of most populations, have no known alleles, and cannot be placed in a particular blood group system or collection. They include By, Bi^a, Bp^a, BOW, Bx^a, Chr^a, ELO, Fr^a,

TABLE 33

	Antibody			
Antigen	IgM/IgG	Complement-binding	Hemolytic transfusion reaction	HDN
Vel	IgM or IgG	Yes	None–severe	+ DAT, no HDN
Lan	IgG	Some	None–severe	None–mild
At^a	IgG	No	None–mod	+ DAT, no HDN
Jr^a	IgG	Some	↓ RBC survival	+ DAT, no HDN
Ox^a	IgG	No	↓ RBC survival	None
JMH	IgG	No	None	None
EMM	IgG	Some	Unknown	Unknown
AnWj	IgG	Rare	Severe 1 case	None
MER2	IgG	Some	None	Unknown
Sd^a	IgG	No	Rare	None
Duclos	IgG	Unknown	Unknown	Unknown
PEL	IgG	Unknown	↓ RBC survival	None

↓, decreased.

HJK, HOFM, Hey, Hg^a, JFV, JONES, Je^a, Kg, Li^a, Milne, Mo^a, NFLD, Ol^a, Os^a, Pt^a, Rb^a, Re^a, Sw^a, SW1, To^a, Tr^a, Vg^a, Wd^a, and Wu.

B. Antibodies

These antibodies may be IgM or IgG. Antibodies to low-prevalence antigens may cause HDN. The relative prevalence of antigens and antibodies in the population make incompatibilities rare.

XXIX. HIGH-PREVALENCE ANTIGENS

A. Antigens

High-prevalence antigens occur in > 90% of the U.S. population, but have no known alleles and cannot be placed in a specific blood group system or collection.

B. Antibodies

Since the individual who makes high-prevalence antibody lacks the respective antigen, antibodies directed against high-prevalence antigens are rarely encountered. When these antibodies do occur, it may be exceedingly difficult to find compatible blood for the patient (as most of the population is positive for the antigen; see Table 33).

References

Cartron, J. P. (1994). Defining the Rh blood group antigens. Biochemistry and molecular genetics. *Blood Rev* **8**, 199–212.

Cherif-Zahar, B., Mattei, M. G., Le Van Kim, C., Bailly, P., Cartron, J. P., and Colin, Y. (1991). Localization of the human Rh blood group gene structure to chromosome 1p34.3–1p36.1 region by in situ hybridization. *Hum Genet* **86**, 98.

Daniels, G. L., Anstee, D. J., Cartron, J. P., *et al.* (1995). Blood group terminology 1995: From the Working Party on Terminology for Red Cell Surface Antigens. *Vox Sang* **69**, 265–279.

Daniels, G. L., *et al.* (1996). Report of ISBT Working Party. Makuhari, Japan.

Issitt, P. D., Anstee, D. J. (1998). "Applied Blood Group Serology," 4th ed. Montgomery Scientific Publication, Durham, NC.

Le Van, K. C., Mouro, I., Cherif-Zahar, B., *et al.* (1992). Molecular cloning and primary structure of the human blood group Rh D polypeptide. *Proc Natl Acad Sci USA* **89**, 10925.

Lomas, C., Tippett, P., Thompson, K. M., *et al.* (1984). Demonstration of seven epitopes of the Rh antigen D using monoclonal anti-D antibodies and red cells from D categories. *Vox Sang* **57**, 261–264.

Lomas, C., McColl, K., and Tippett, P. (1993). Further complexities of the Rh antigen D disclosed by testing category D^{II} cells with monoclonal anti-D. *Transfusion Med* **3**, 67–69.

Mollison, P. L., Englefriet, C. P., and Contreras, M. (1987). "Blood Transfusion in Clinical Medicine," 8th ed. Blackwell Science, Cambridge, MA.

Rosenfield, R. E., Allen, F. H., Jr., Swisher, S. N., and Kochwa, S. (1962). A review of Rh serology and presentation of a new terminology. *Transfusion* **2**, 187.

Wiener, A. S. (1943). Genetic theory of the Rh blood types. *Proc Soc Exp Biol Med* **54**, 316–319.

CHAPTER

12

Pretransfusion Compatibility Testing

DON L. SIEGEL

Department of Pathology and Laboratory Medicine
University of Pennsylvania Medical Center
Philadelphia, Pennsylvania

In order for blood components to be safely issued and transfused to recipients, a series of laboratory tests are performed to ensure immunological compatibility. The process, summarized in Figure 1 for the preparation of packed red blood cells (PRBCs), involves typing patient RBCs, screening patient sera for the presence of unexpected RBC-reactive antibodies, and crossmatching as a final check to verify compatibility between donor RBCs and patient sera. The overall goal of this testing is to help prevent the occurrence of potentially fatal hemolytic transfusion reactions. From the perspective of the requesting physician, it should be appreciated that such reactions, although rare, are almost always due to clerical errors that result in the transfusion of ABO-mismatched blood. Most commonly, these errors occur on patient wards by the inadvertent mislabeling of blood specimens or by the transfusion of a patient with blood intended for someone else. Therefore, the importance of proper patient identification during blood specimen collection and blood component administration cannot be overemphasized.

The following discussion focuses on those aspects of pretransfusion compatibility testing most relevant to the practicing physician. For more in-depth reading, clinicians are encouraged to consult several excellent transfusion medicine and blood banking texts (Mollison *et al.*, 1997; Issitt and Anstee, 1998; Vengelen-Tyler, 1999).

I. ABO ANTIGENS AND TYPING

A. Medical Aspects

1. ABO Blood Group

The ABO antigens comprise a set of complex carbohydrates present on RBC glycoproteins and glycolipids. They are also expressed on many other tissues and secreted in a soluble form in a number of body fluids. They are produced by the action of enzymes called glycosyl transferases encoded by the *H*, *A*, and *B* genes (see Figure 1 of Chapter 18). The H transferase attaches fucose (Fuc) to terminal galactose residues, forming the H antigen. The A transferase couples a N-acetylgalactosamine (GalNAc) to the H antigen, forming the A antigen. The B antigen is formed when the B transferase attaches a D-galactose (Gal) to H precursor. Thus, blood group A individuals have functioning H and A transferases, blood group B individuals have functioning H and B transferases, blood group AB individuals have three functioning transferases, and blood group O individuals have only a functioning H transferase.

2. Immune Response to ABO Antigens

The immune response to the ABO antigens is predominantly T-cell independent and results in the pro-

---

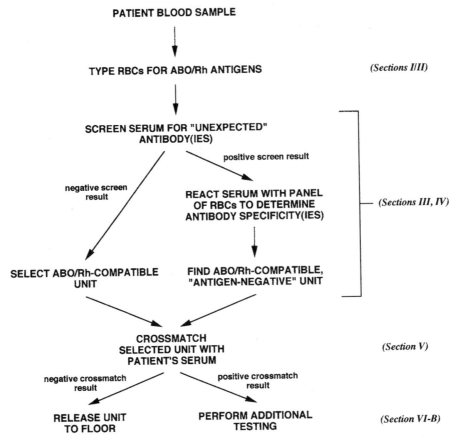

FIGURE 1 Flow chart outlining pretransfusing compatibility testing. Approximate times required to complete each procedure are discussed in Section VI.

duction of high-titer IgM antibodies referred to as "isohemagglutinins." Due to the presence of cross-reacting carbohydrate structures on environmental agents such as gut flora, anti-A and anti-B antibodies occur "naturally" (i.e., in the absence of previous transfusion or pregnancy), depending on an individual's ABO blood group. For example, plasma from an A blood group individual contains anti-B, blood group B plasma contains anti-A, and blood group O plasma contains both anti-A and anti-B (see Table 1). Therefore, all patients except those of blood group AB are capable of experiencing an acute hemolytic transfusion reaction if transfused with the incorrect ABO group even if they have never been exposed to foreign blood in the past.

Such transfusion reactions can be particularly severe because IgM antibodies efficiently activate complement upon RBC binding, causing brisk intravascular hemolysis. The resulting circulating antigen–antibody immune complexes may lead to renal failure, shock, disseminated intravascular coagulation, and death. This form

TABLE 1 ABO Blood Groups, Isohemagglutinins, and Blood Component Compatibility

ABO blood group	Frequency (%)	RBC antigen	Serum antibodies	Compatible	
				RBCs	FFP
O	45	None ("H")	Anti-A, anti-B	O only	O, A, B, AB
A	41	A	Anti-B	A, O	A, AB
B	10	B	Anti-A	B, O	B, AB
AB	4	A, B	Neither	AB, A, B, O	AB only

of immune hemolysis contrasts sharply with IgG-mediated reactions (e.g., reactions due to alloantibodies to Rh and many other RBC protein antigens—see below), in which antibody-coated RBCs are phagocytosed in the reticular endothelial system (extravascular hemolysis) over a more extended period of time.

B. Transfusion Practice

1. Selection of PRBCs and Plasma

The presence or absence of anti-A and anti-B antibodies in a patient's plasma and A and B antigens on donor PRBCs form the basis for choosing compatible RBCs or plasma, respectively (see Table 1). This leads to the concept of universal donors and recipients. For example, blood group O PRBCs can be transfused to anyone, as can plasma from a blood group AB individual. Conversely, AB recipients can receive PRBCs of any ABO group, whereas O recipients can receive any type of plasma. In practice, however, efforts are made to provide "type-specific" (vs. "type-compatible") products so as to reserve the "universal" products for those patients who can only be given O PRBCs or AB plasma. An additional reason for routinely providing type-specific units to patients is to avoid ABO blood typing discrepancies caused by the presence of more than one RBC group in subsequent patient blood samples.

2. Selection of Platelets

Platelets also express ABO antigens and the results of some studies have shown a somewhat decreased survival of transfused platelets that are incompatible with the recipient's plasma. Furthermore, the incompatible plasma introduced by unmatched platelet units may react with the recipient's RBCs. However, neither of these effects is generally clinically significant. Therefore, although it may be of some benefit to provide "type-specific" platelets for transfusion, *it is not usually considered necessary to delay a needed transfusion if ABO-matched platelets are unavailable* (Vengelen-Tyler, 1999). If large numbers of incompatible platelets are transfused, some institutions may "volume reduce" platelet units to minimize the amount of incompatible plasma transfused.

C. ABO Testing

Determining a patient's ABO group comprises two laboratory tests that provide complementary results (see Figure 2). In performing the "forward typing" reaction, the blood bank technologist determines the

Forward Typing for ABO

drop of patient's RBCs + drop of anti-A typing serum → look for agglutination
drop of patient's RBCs + drop of anti-B typing serum → look for agglutination

Four possible results:

| reaction with: | | |
anti-A	anti-B	Patient's ABO Group
O	O	O
+	O	A
O	+	B
+	+	AB

Reverse Typing for ABO

drop of patient's serum + drop of reagent A RBCs → look for agglutination
drop of patient's serum + drop of reagent B RBCs → look for agglutination

Four possible results:

| reaction with: | | |
A RBCs	B RBCs	Patient's ABO Group
+	+	O
O	+	A
+	O	B
O	O	AB

FIGURE 2 Steps involved in forward and reverse typing for ABO antigens.

presence or absence of A and B antigens on patient RBCs by mixing them with either anti-A or anti-B typing serum. Since the typing sera comprise decavalent IgM isoagglutinins, cross-linking of RBCs which bear the appropriate A and/or B antigens results in agglutination visualized by the unaided eye. As a confirmatory test, a "reverse typing" is performed in which drops of patient sera are mixed with RBCs that are known to be either blood group A or B, and the formation of RBC agglutinates is determined. This second test takes advantage of the phenomenon described above in which individuals constituitively produce isohemagglutinins to the ABO antigens they lack. It should be noted, however, that newborns, the elderly, and certain seriously ill patients may not necessarily produce the expected isohemagglutinins (based on forward typing results). For this reason, only forward types are performed on newborns, and forward–reverse typing discrepancies are not uncommon when testing the latter two patient populations.

II. Rh ANTIGENS AND TYPING

A. Medical Aspects

The antigens of the ABO blood group system represent only a minute fraction of the more than 400 total blood group antigens that may be present on patient RBCs. Typing patient and donor cells for every known antigen with the intention of providing perfectly matched blood would be a practical and fiscal impossibility. Fortunately, such extensive testing is not required for a number of reasons, the most important of which is that exposure to the majority of foreign RBC antigens through transfusion does not lead to the production of clinically significant alloantibodies.

The Rh(D) antigen of the Rh blood group system is one notable exception. As many as 80% of Rh(D)-negative patients exposed to Rh(D)-positive RBCs may develop high-titer, high-affinity, anti-Rh(D) IgG antibodies that may persist for the rest of their lives, even if they are never exposed to the antigen again. The resulting antibodies can cause hemolytic transfusion reactions and can cross the placenta, causing hemolytic disease of the newborn, when present in a Rh(D)-negative pregnant woman carrying an Rh(D)-positive fetus. Therefore, because of its extraordinary immunogenicity, the Rh(D) antigen is the only blood group antigen (other than ABO) for which it is routine to prophylactically match blood prior to transfusion. Although there are over 40 different members of the Rh blood group system, the Rh(D) antigen is the most immunogenically important, so it is customary to refer to Rh(D)-positive individuals as "Rh-positive" and those that lack Rh(D) as "Rh-negative," even though both groups of patients may be positive or negative for any of the other Rh antigens.

B. Transfusion Practice

1. Selection of PRBCs

Individuals are routinely typed for the presence or absence of the Rh(D) antigen and provided with antigen-matched PRBCs. Note that Rh-positive patients can be transfused with either Rh-positive or -negative PRBCs—the absence of Rh(D) will cause no harm—but it is deemed prudent to reserve the rarer Rh-negative blood ($\sim$15% of donor units) for Rh-negative individuals. In cases of emergent trauma and/or massive transfusion in which the patient's Rh(D) status is unknown, efforts are made to provide Rh-negative PRBCs until the appropriate testing can be completed. When Rh-negative blood is in short supply, it may be necessary to transfuse Rh-negative patients with Rh-positive units. In such scenarios, Rh-negative units are reserved for women of childbearing age and for patients whose serum contains anti-Rh(D) from a previous sensitization.

2. Selection of Platelets

Unlike the A and B antigens that are expressed on both platelets and PRBCs, the Rh(D) antigen is only present on RBCs. Theoretically, the selection of platelet units for transfusion should be independent of the Rh status of the donor. However, a transfusion of platelets may introduce as much as 5 ml of donor RBCs, which may be sufficient to alloimmunize an Rh-negative patient to the Rh(D) antigen. Therefore, *the standard of care is to avoid transfusing Rh-negative patients, particularly women of childbearing age, with platelet units derived from Rh-positive donors.* If such units are unavailable and platelet transfusion must be undertaken, the administration of Rh immune globulin (RhIg) should be considered. A standard 300-μg dose of RhIg, which may inhibit the immunizing potential of up to 15 ml of Rh-positive RBCs, would therefore neutralize the effects of Rh(D)-positive cells from several mismatched platelet transfusions.

3. Selection of Plasma Products

With respect to the transfusion of plasma products, the Rh(D) status of the donor is not an issue, since plasma does not contain cellular or soluble material capable of inducing Rh(D) immune responses.

C. Rh(D) Testing

Determining the presence or absence of the Rh(D) antigen on RBCs is a one-step or a multistep process, depending on the particular typing reagent used at a given institution. Classically, anti-Rh-containing typing serum comprised only anti-Rh(D) molecules of the IgG isotype which, although bivalent, were not capable of directly agglutinating antigen-positive RBCs. As a result, Rh typing consisted of incubating RBCs with sera, washing away unbound antibody, and adding rabbit antihuman IgG (also called "Coombs reagent"). If the cells were Rh(D)-positive, the Coombs reagent would induce agglutination by cross-linking anti-Rh(D)-coated cells (a positive "indirect Coombs test").

More recently, however, many blood banks have begun to use anti-Rh(D) typing reagents that include an IgM anti-Rh(D) component that can directly agglutinate Rh(D)-positive cells. With these newer types of reagents, Rh(D) typing is performed in a manner similar to that for the ABO forward typing reaction (see above). Furthermore, in some institutions, Rh(D) typing reagents consist of blends of IgM and IgG anti-Rh(D); if the IgM-mediated direct agglutination is negative, the reaction is carried to the "Coombs phase" to allow the higher-affinity IgG molecules to detect patient cells that express weaker forms of the Rh(D) antigen. Regardless of reagent, however, Rh(D) typing does not include a reverse typing, since anti-Rh(D) is not "naturally" present in the sera of Rh-negative individuals.

III. TYPING FOR OTHER RBC ANTIGENS

As noted above, routine typing of donor and/or patient RBCs for the presence of antigens other than A, B, and Rh(D) is generally not required. However, when unexpected RBC alloantibodies are detected in a patient's serum (see below), prospective PRBC units will be typed for the alloantigen(s) so that antigen-negative PRBCs can be selected for transfusion. In addition, the patient's own RBCs will undergo extended phenotyping to confirm their negativity for the respective alloantigen(s). For certain patient populations, for example, chronically transfused individuals with sickle cell disease, prophylactically matching for additional RBC antigens, such as Rh(C), Rh(E), Kell, and Duffy, will decrease the likelihood of alloantibody formation and may result in better long-term management and decreased morbidity (Tahhan *et al.*, 1994).

IV. ANTIBODY SCREEN AND IDENTIFICATION

A. Medical Aspects

Unlike the anti-A and anti-B isohemagglutinins, the production of "naturally occurring" antibodies to other RBC antigens does not generally occur. Therefore, unlike in the ABO antigen–antibody system in which patients (except those of blood group AB) are already predisposed toward having potentially fatal hemolytic transfusion reactions from ABO-mismatched transfusions, transfusion reactions due to non-ABO antibodies result from "immune" antibodies, that is, antibodies formed in response to previous immunization by foreign RBCs from transfusion or pregnancy. The purpose of the antibody screen and identification is to detect the possible presence of such serum alloantibodies and determine their antigen specificity (or specificities) so that PRBCs can be provided that lack those antigens.

In order to accurately interpret the clinical significance of antibody screening and identification results, it is important for physicians to provide the blood bank staff with relevant recent historical information. For example, an antibody screen on an obstetrical patient who had recently been given a dose of Rh(D) immune globulin following an amniocentesis would reveal the presence of anti-Rh(D) alloantibodies. Without knowing the appropriate clinical history, these serological results due to the passive administration of an IgG could be misinterpreted as the *de novo* formation of clinically significant alloantibodies that would put the patient's fetus at risk for hemolytic disease of the newborn.

B. Transfusion Practice

If the antibody screen reveals the presence of unexpected serum antibodies, it is then necessary to consider their clinical significance (i.e., their potential for causing hemolysis *in vivo*—see Chapter 18) before initiating a search for antigen-negative donor units. As a general rule, antibodies reactive at 37°C and/or by the indirect antiglobulin (Coombs) test are potentially clinically significant (Issitt and Anstee, 1998). Conversely, serum antibodies that react with reagent cells at room temperature or below (but not at 37°C) are not routinely hemolytic. However, there are a number of exceptions to this rule and an institution's blood bank staff is experienced in deciding which alloantibody specificities necessitate the selection of antigen-negative units. If clinically significant alloantibodies are determined to be present in the patient's serum, ABO/Rh-compatible PRBC units are selected that lack the additional antigen(s). PRBC units lacking appropriate alloantigens are also selected for patients who have a history of alloantibody production, even though the antibodies may not be currently detectable.

In emergent situations in which the patient's ABO/Rh type is known but time does not permit completion of the antibody screen, ABO-compatible, Rh(D) type-specific units can be issued as described below for uncrossmatched blood products.

C. Laboratory Procedure

Alloantibody identification consists of performing agglutination tests with patient sera and sets of reagent RBCs that have been extensively characterized with respect to their antigenic composition. The first step consists of screening a patient's serum against two such reagent cells as illustrated in Figure 3. This initial pair of cells is selected so that between them, all common, clinically significant antigens will be represented. If the patient's serum does not react with either cell, then the patient is said to have a "negative screen" and random ABO/Rh-specific compatible units are selected from inventory and crossmatched (see Figure 1). If one or both cells are reactive, then the patient has a "positive screen," and further testing is required to determine the specificity of any serum alloantibodies.

For example, in the hypothetical case illustrated in Figure 3, both screening cells are reactive. The patient's serum is then run against a larger set of reagent cells (typically 8 to 12 cells referred to as a "panel") as illustrated in Figure 4. This particular set of cells is chosen so that the resulting pattern of reactivity (last column of Figure 4) will identify the antibody specificities to an acceptable degree of certainty, in this case an

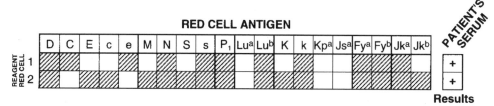

FIGURE 3 A sample anagram for a pair of reagent screening cells. Columns represent different RBC antigens. Each line depicts a cell's antigenic composition as indicated by hatch marks. Hypothetical results for the screening of a patient's serum reactive with both screening cells are shown on the right.

alloantibody with anti-Fy^a specificity. Based on these results, a group of ABO/Rh-compatible units are assembled and typed with the appropriate antisera to identify antigen-negative units. In this case, ABO/Rh-compatible, Fy^a-negative units will be sought and testing will proceed to the crossmatching step.

V. CROSSMATCHING

A. Medical Aspects

The crossmatch consists of testing the patient's serum against an aliquot of PRBCs from each of the PRBC units selected as a final check for immunologic compatibility. One of the crossmatch's main functions is to verify the correct selection of ABO blood group, since acute hemolytic transfusion reactions caused by anti-A and/or anti-B isohemagglutinins can result in

significant morbidity or even death. If the patient's serum is reactive during the crossmatch, previously performed testing will be reviewed and/or repeated (patient typing and screening results, typing and labeling of PRBC unit, etc.) to determine the reason for incompatibility.

B. Transfusion Practice

In most routine situations, crossmatch-compatible, ABO/Rh(D)-specific PRBC units can be provided for transfusion. However, in certain circumstances, patients may need to receive uncrossmatched blood or blood that is crossmatch-incompatible.

1. Uncrossmatched Blood

In emergency situations in which the patient's blood group is not known (and time does not permit its

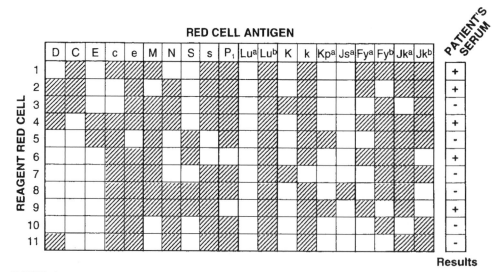

FIGURE 4 A sample anagram for a panel of reagent RBCs. Columns represent different RBC antigens. Each line depicts a cell's antigenic composition as indicated by hatch marks. Hypothetical patient serum results are consistent with the presence of an Fy^a alloantibody.

TABLE 2 Compatibility Testing of RBC Products in Emergency Situations

Time (minutes) until RBC transfusion is needed	RBC product provided
5	Universal donor—group O (no type, no crossmatch)
10	1. ABO type-specific (no crossmatch) or
	2. ABO type-specific and emergency crossmatch (if a previous sample is available for this patient)
15	ABO type-specific and emergency crossmatch (no previous sample available)
60	ABO/Rh type-specific, antibody screen,[a] and crossmatch

[a] If antibody screen is positive, time may be 2 hours or longer.

determination), uncrossmatched, blood group O PRBCs that are either Rh(D)-positive or -negative can be administered. The minimum time requirements for compatibility testing and provision of PRBCs in emergency situations are listed in Table 2. In an emergency situation, it is preferable to give Rh(D)-negative blood if available, particularly if the patient is a female of childbearing age.

Once the units are released from the blood bank and a patient's pretransfusion blood sample is received, blood typing will immediately commence so that additional urgently needed units may be type-specific. As time permits, crossmatched blood will follow and compatibility testing be performed retrospectively on residual samples of RBCs from the transfused uncrossmatched units.

In deciding whether or not to assume the risks of transfusing uncrossmatched, potentially incompatible blood, the risk versus benefit of delaying transfusion until compatibility testing is completed must be carefully considered (see Table 2). It should be appreciated that the odds are greatly in favor of an uneventful transfusion, particularly in younger patient populations where the incidence of unexpected clinically significant serum alloantibodies may be as low as 1–2% (Giblett, 1977; Walker et al., 1989; Mollison et al., 1997).

2. Crossmatch-Incompatible Blood

One of the more complex scenarios is the case in which the patient's antibody screen is positive against both screening cells as well as with all reagent red cells in the panel. Such a serology may result from the

presence of multiple alloantibodies to a number of alloantigens or from a single alloantibody against a "high-frequency" (i.e., nearly 100% prevalent) antigen the patient does not possess. In practice, compatible, antigen-negative units may not be found within the time frame in which the patient requires transfusion, and blood that is crossmatch-incompatible may need to utilized.

Anti-RBC autoantibodies are made by patients who have the diagnosis of autoimmune hemolytic anemia. Such antibodies react not only with the patient's own RBCs, but with all allogeneic cells as well, thus making crossmatch compatibility an impossibility. A direct agglutination test, or "direct Coombs test," performed on the patient's RBCs may help differentiate autoantibodies from alloantibodies as the cause of panreactive serum screening results. In either case, it should be appreciated that crossmatch-incompatible RBCs, when transfused, do not necessarily have shortened survival or cause harm to the patient. In every case, a clinical judgment has to be made (ideally in consultation with an institution's blood bank physician), weighing the need for transfusion versus the potential for RBC hemolysis. To aid in decision-making, an "*in vivo* crossmatch" can be performed in which a small amount (5–10 ml) of a PRBC unit is transfused slowly over a 10- to 15-minute period of time, during which the patient is closely monitored for signs and symptoms of a hemolytic transfusion reaction. Transfusion is temporarily halted while a blood sample is drawn and sent to the blood bank for analysis. The intravascular hemolysis of even such a small amount of transfused RBCs will result in pink-tinged serum visible by eye. Lack of visible hemolysis suggests that the remainder of the unit can be transfused without posing a serious threat to the patient. More sophisticated red cell survival studies employing ^{51}Cr-labeled cells can be utilized if desired (Mollison et al., 1997).

C. Laboratory Procedures

There are two basic types of crossmatch procedures, and the one chosen to be performed depends on the results of the patient's antibody screen. If the screen is negative and the patient's records do not show a history of RBC alloantibodies, then an "immediate spin" crossmatch may be performed in which serum and cells are centrifuged at room temperature and immediately assessed for direct agglutination. This type of crossmatch is valuable as it serves as a final check for ABO compatibility. For patient specimens with historical or currently positive antibody screening results, a "full crossmatch" ("Coombs crossmatch") is performed in which the serum–cell incubation mix is incubated for

~ 15 minutes at 37°C, washed, and resuspended in Coombs' reagent. This more extended type of crossmatch verifies that the ABO/Rh-compatible, antigen-negative, PRBC units are indeed compatible with the patient's alloantibody-containing serum.

VI. TURNAROUND TIMES

A. Common Clinical Scenarios

1. Type and Crossmatch

When the need for transfusion is likely, then a "type and crossmatch" for an appropriate number of PRBC units is requested. This type of blood request would necessarily include an antibody screen and, if positive, an antibody identification panel. An appreciation of the time involved in performing these tests is valuable. In normal circumstances, ABO/Rh typing should require ~ 10–20 minutes and the antibody screen an additional ~ 30 minutes. If the screen is negative, then an "immediate spin crossmatch" can be performed which only requires ~ 5 minutes. If the screen is positive, the antibody identification panel would require ~ 45 minutes to perform. Once the alloantibody specificities are identified in the patient's serum, additional time is required to find ABO/Rh-compatible units which lack the appropriate alloantigens. In the hypothetical case of a single alloantibody described above, a request for 1 unit of PRBCs would require the blood bank technologist to randomly select 3 or 4 ABO/Rh-compatible units off the shelf and then type each of them for the Fy^a antigen, so as to find at least one unit that lacks Fy^a (approximate frequency of negativity for the Fy^a antigen in the donor population is ~ 34%). Typing the units for Fy^a would require ~ 30 minutes. Since "full crossmatches" would be required, crossmatching would take ~ 30 minutes before the units could be released to the floor or operating room. Therefore, even under ideal conditions, a request to type and crossmatch 1 unit could take as long as 3 hours to complete.

2. Type and Screen

When the likelihood of transfusion is too low to justify the reservation of crossmatched units, a "type and screen" can be ordered. If the screen is positive, any alloantibody specificities are determined so as to identify complex serologies prior to a request for blood. A type and screen would also be the type of test requested on all pregnant women early in pregnancy. The goal is to identify Rh-negative patients who will need to receive RhIg at the appropriate times during gestation, as well as to identify the presence of RBC alloantibodies that may put her fetus at risk for hemolytic disease of the newborn.

B. Special Situations

For most of the blood components transfused per year, pretransfusion compatibility testing is a straightforward process, as described in the previous sections. However, each step can lead to a number of unexpected or unusual outcomes which in turn can lead to significant delays in providing blood for patients. In the case of a patient with multiple alloantibodies or a single alloantibody to a "high-frequency" antigen (see above), time will be required to react the patient's serum against additional panels of reagent RBCs as well as to type donor units for the absence of any such antigens. In some cases, compatible RBCs may have to be acquired through national rare donor programs, which may result in delays on the order of days. For patients with RBC autoantibodies, the panreactive nature of their sera may camouflage the presence of serum alloantibodies. In such situations, specialized testing (described in Chapter 20) is required to separate the two sets of reactivity, which may require 12–24 hours to complete.

References

Giblett, E. (1977). Blood group alloantibodies: An assessment of some laboratory practices. *Transfusion* **17**, 299–308.

Issitt, P. D., and Anstee, D. J. (1998). "Applied Blood Group Serology." Montgomery Scientific, Miami.

Mollison, P. L., Engelfreit, C. P., and Contreras, M. (1997). "Blood Transfusion in Clinical Medicine." Blackwell Scientific, Oxford.

Tahhan, H. R., Holbrook, C. T., Braddy, L. R., *et al.* (1994). Antigen-matched donor blood in the transfusion management of patients with sickle cell disease. *Transfusion* **34**, 562–569.

Vengelen-Tyler, V. (1999). "Technical Manual," 13th ed. American Association of Blood Banks, Bethesda.

Walker, R., Lin, D., and Hatrick, M. (1989). Alloimmunization following transfusion. *Arch Pathol Lab Med* **113**, 254–261.

Special Investigations in the Workup of Unexpected Antibodies and Blood Group Incompatibilities

CHESTER ANDRZEJEWSKI

Department of Pathology
Baystate Health System
Springfield, Massachusetts
and
Tufts University School of Medicine
Boston, Massachusetts

Fundamental immunologic principles underlie all of the various serologic techniques applied by blood banks today. The interaction of antibodies (from the patient or as special reagents) with antigen (from donor RBC) either in a tube, on a slide, or in a matrix setting (e.g., dextran acrylamide gel beads) serves as the basic process through which clinically important antibodies are identified.

I. UNDERLYING IMMUNOLOGIC PRINCIPLES

A. Direct Agglutination

First introduced by early immunologists in the 19th century, direct agglutination was used by Karl Landsteiner and colleagues in the discovery of the ABO blood group system, as well as in the recognition and identification of other RBC antigens (Widmann, 1997). This technique, whereby RBCs simply mixed with serum or plasma resulted in visible agglutination, could be enhanced by various maneuvers (e.g., centrifugation). Sufficient for the identification of certain types of antigens (e.g., polysaccharides) and classes of antibodies (e.g., IgM), technical limitations of direct agglutination became apparent when more complex protein antigen structures and IgG antibodies were evaluated. Not until the introduction of antiglobulin reagents for use in

indirect agglutination (Coombs *et al.*, 1945) could investigation of those systems easily proceed and be applied to routine clinical practice.

B. Indirect Agglutination

The indirect antiglobulin test (IAT), also known as the indirect Coombs test, is an immunologic technique used to detect antibody (typically IgG) in the patient's serum or plasma. This is accomplished through the use of another antibody produced in a nonhuman species that cross-links patients' antibodies that have bound to RBCs used as substrate in the reaction. Unlike IgM antibodies, which because of their large pentameric configuration can easily cross-link antibody-coated RBCs, bivalent IgG molecules cannot easily overcome the steric hindrances between individual RBCs necessary to result in agglutination. The use of an antiglobulin reagent directed to the constant region of the bound IgG facilitates the interaction, resulting in agglutination. Theoretically antiglobulin reagents can be prepared against a variety of antibody classes and/or immune mediators (e.g., complement components); however, in practice only antiglobulin reagents against IgG and/or C3 fragments are used routinely due to their clinical significance. The use of the IAT allows for antibody screening procedures to detect unexpected

TABLE I   Diagnostic Studies Performed by Blood Banks in Antibody Identification and Monitoring

- Antibody screens
- Standard/select RBC panel testing
- Reactivity enhancement methods
- RBC phenotyping
- Titration analysis
- Direct antiglobulin test (DAT) and elution studies
- Cold antibody-related testing
  —Prewarm screens
  —Mini cold panels
  —Thermal amplitude titrations
- Warm autoantibody-related testing
  —Autoabsorptions
  —Differential adsorption techniques
- Miscellaneous methodologies

TABLE 2   Factors of Importance in Assessment of a Patient's Medical History

- Past/concurrent medical problems
- Pregnancy history if female
- Transfusion history
  —Time
  —Types of products
- Past/concurrent medications
- Immunoserologic workup: Initial, repeat, referral
- Laboratory studies performed elsewhere
  —Data
  —Availability of pretransfusion specimens for testing

antibodies (Boral and Henry, 1977), as well as for crossmatching of donor blood with patient samples so as to obtain a compatible blood transfusion (Oberman *et al.*, 1982). Variations of this theme are used in the further workup of complex patients by blood banks in their special investigation sections or by donor center reference labs. A listing of some of the more commonly performed studies is found in Table 1. Individual services may differ slightly in the application of these methods, but the overall goals of identifying a clinically significant alloantibody or autoantibody remain paramount.

II. STRATEGIES IN THE EVALUATION OF CLINICALLY SIGNIFICANT ALLOANTIBODIES

A. Preliminary Considerations

All blood bank investigations should begin the same way, that is, by an analysis of the patient's immunoserologic problem in the context of clinical history and other available laboratory data. Factors critical in the assessment of a patient's medical history that may help direct subsequent investigative actions are shown in Table 2. An exhaustive examination is unnecessary. However, knowledge of recent or past transfusions, pregnancies, and medications, as well as results from previous immunoserologic investigations performed in the laboratory or elsewhere, will increase the speed and efficiency of the investigation.

B. Patients with Positive Antibody Screens with or without History of Clinically Significant Alloantibodies

1. General Approaches

Patients with positive antibody screens and a history of clinically significant alloantibodies can be evaluated by using abbreviated cell panels lacking the appropriate antigen (see Figure 1). Nonreactivity in such a setting is consistent with no new clinically significant antibodies, and crossmatch-compatible, antigen-negative blood can be quickly provided. For individuals with positive antibody screens and no history of clinically significant antibodies, identification of the antibody requires testing the specimen against a panel of selected RBC specimens arrayed such that all of the major clinical antigen systems are accounted for.

2. Panel Testing

Cell panels prepared by commercial suppliers are readily available for the above purpose. Generally, a sample of the patient's own cells is also included, that is, the autocontrol. All phases of reactivity should be assessed for agglutination and/or hemolysis. If no pattern emerges, antibody enhancement methods such as additive solutions, pH adjustments, temperature adjustments, lengthening the incubation time, and altering the antigen/antibody ratio should be considered. If a new clinically significant antibody is identified, the patient's RBCs should be phenotyped for the lack of that antigen, and crossmatch-compatible antigen-negative blood selected for transfusion. When no clear pattern can be discerned, reassessment of panel reactivity

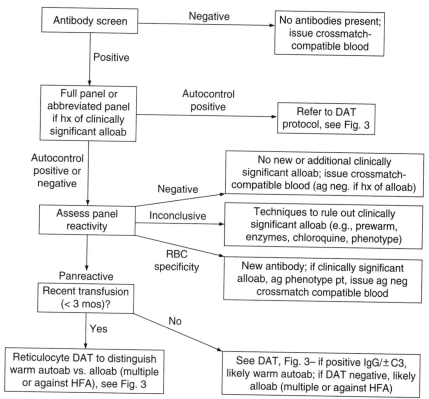

FIGURE 1 Investigation for presence of anti-red blood cell antibodies. DAT, direct antiglobulin test; alloab, alloantibody; autoab, autoantibody; ag, antigen; hx, history; pt, patient; and HFA, high-frequency antigen.

should occur. In individuals with inconclusive findings, a workup for interference from cold-reactive antibodies should be performed. For panels exhibiting agglutination of all cells (panagglutinins), careful examination of the possibility of a mixture of clinically significant alloantibodies should be considered. Similarly, antibodies to high-frequency antigens or to paternal antigens, as in cases of suspected hemolytic disease of the newborn (HDN), may need to be evaluated.

C. Evaluation of Cold-Reactive Autoantibodies

1. Mini Cold Panels and Prewarmed Techniques

Interference by cold-reactive autoantibodies is common. Their reactivity may hamper serologic investigations, especially if they possess heightened thermal amplitudes. The use of a mini cold panel and prewarmed antibody screening techniques may alleviate some of the difficulties. In the former method, the patient's serum is incubated with group O adult RBCs that have a high density of the I blood group antigen and with umbilical cord blood cells rich in the i blood group antigen. Incubations are typically conducted at 18°C for 15 minutes, with agglutination judged after centrifugation. Agglutination with any cell type is positive for the presence of these antibodies. A prewarmed screen can be performed to determine if the antibody binds at 37°C and hence may be of clinical significance. In this technique both serum and substrate RBCs are warmed to 37° for a sufficient period of time prior to their juxtaposition, after which they are incubated for a short interval, also at 37°C. Washes are conducted with prewarmed saline, and after the addition of anti-IgG reagent, agglutination is assessed. No reactivity implies cold-reactive antibodies.

Depending on the clinical history, further evaluation of those antibodies may be desired and is described below. Generally, these antibodies are of little clinical concern. If the prewarmed screen continues to show reactivity, potential clinically significant alloantibodies have not been ruled out. Further workup at this point might be directed at the identification of such other antibodies as high-titer, low-avidity antibodies, for example, anti-Bg or other antibodies.

2. Thermal Amplitude Titrations and Other Methods

In individuals in whom cold-reactive autoantibodies may be pathologic, additional studies may be desired. Pathologic cold agglutinins typically exist in higher titers, with an elevated thermal reactivity range well beyond 4°C. Evidence for complement activation may be found on the RBC membrane. The use of thermal amplitude titrations may be of benefit in the assessment of these antibodies' pathogenicity. In this technique, serum dilutions from the patient are incubated at various temperatures (e.g., 37, 30, 23, and 4°C) and the degree of agglutination, if any, at each dilution is recorded. The antibody's titer (i.e., the reciprocal of the highest dilution showing reactivity) is recorded for each temperature range. If specificity analysis is also desired, reagent screening cells (I +) and cord blood cells (i +) can be run in parallel. Most nonproblematic cold-reactive antibodies will demonstrate thermal range optima around 4°C with titers less than 1:64. Antibody reactivity occurring above 23°C (room temperature), especially if found with elevated titers, may signify a clinically significant antibody. Thermal amplitude titrations may be used to follow selected therapeutic interventions to assess their efficacy. Occasionally, cold autoadsorption studies to detect clinically significant alloantibodies masked by the cold-reactive autoantibody may need to be performed. The use of warmed saline washes and/or ZZAP (a mixture of proteolytic enzyme and sulfhydryl reducing reagent) to clear antibody off the RBC surface enhances the process (Branch and Petz, 1982). In patients recently transfused, serum adsorption with rabbit red cells may remove antibodies directed to I and IH antigens; however, clinically significant alloantibodies may also be depleted by the technique (Marks *et al.*, 1980; Dzik *et al.*, 1986).

D. Problems with Warm-Reactive Autoantibodies

1. Pathogenicity Potentials and Chief Concerns

Also commonly encountered in blood bank workups are problems with warm-reactive autoantibodies. These antibodies generate much concern because of their IgG nature and optimal reactivity at body temperatures, and the fact that crossmatch-incompatible blood may need to be issued quickly. Just as with their cold

counterparts, warm-reactive autoantibodies may vary in the pathogenicity spectrum—from benign variants to highly pathological mediators of immune RBC destruction. Discernment of their character in this regard is not always easily achieved.

Also similar to cold-reactive autoantibodies and other naturally occurring autoantibodies, many warm-reactive autoantibodies probably function in normal immune-homeostatic mechanisms to help remove aged RBCs from the circulation. Enhancement of this process may lead to accelerated immune clearance, resulting in anemia in some patients. When the blood bank confronts warm-reactive autoantibodies, serological investigations are aimed not at an assessment of their pathogenicity, but rather at the critical question of whether obscured clinically significant alloantibodies are present in the panagglutination reactivity typically associated with these antibodies (see Figure 2).

Before addressing this, however, mention must be made of the significance and evaluation of results from direct antiglobulin testing, which is typically performed in tandem or parallel investigations (see Figure 3).

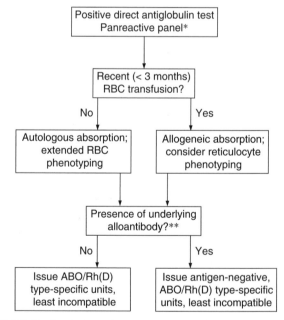

FIGURE 2 Serologic evaluation of patient with suspected warm autoimmune hemolytic anemia. *In some cases, the antibody screen or panel may be negative or will rarely show specificity for a particular RBC antigen (i.e., panel not panreactive), and the direct antiglobulin test results may rarely be negative on routine testing. **In some cases, repeat absorptions or use of enzyme-treated cells may be necessary. In cases of urgent need for transfusion, RBCs may need to be issued prior to completion of these time-consuming procedures.

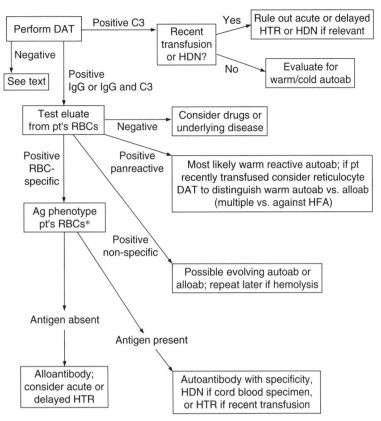

FIGURE 3 Evaluation of positive direct antiglobulin test (DAT). *If pt recently transfused (<3 months), consider antigen phenotyping reticulocyte-enriched RBCs. HTR, hemolytic transfusion reaction; HDN, hemolytic disease of the newborn; HFA, high-frequency antigen; alloab, alloantibody; autoab, autoantibody, and pt, patient.

2. Significance and Workup of a Positive Direct Antiglobulin Test (DAT)

a. General Principles and Implications

The DAT is a serologic technique whereby a 3–5% suspension of a patient's washed RBCs is mixed with a volume of antiglobulin reagent (anti-IgG, anti-C3, or anti-polyspecific), centrifuged immediately, and then suspended by gentle agitation and examined macroscopically for agglutination. Positivity in this test implies *in vivo* red cell sensitization by gamma globulin (antibody) and/or beta globulin (components of complement). A complex relationship exists with respect to the number of IgG molecules on the RBC surface and the degree of positivity. A positive DAT can be associated with multiple causes (Petz and Garrity, 1980) and can be observed in various settings—some in the absence of clinical hemolysis. Thus, it does not necessarily mean shortened RBC survival. The existence of a positive DAT does not automatically imply the pres-

ence of autoantibodies on the RBC either. Indeed, depending on the clinical circumstances, a positive DAT may be associated with clinically significant alloantibodies (e.g., delayed hemolytic transfusion reactions, hemolytic disease of the newborn), or even no RBC-specific antibody at all (i.e., nonspecific adsorption associated with certain drugs, e.g., cephalosporins).

b. The Importance of Clinical History

As in other investigations, pertinent clinical history, especially medications the patient has recently taken, is useful. A screening of other important considerations in the evaluation of a patient with a positive DAT is outlined in Table 3. Clinical history dictates the extent of the workup. If the patient is not exhibiting enhanced RBC destruction, no further studies may be needed. In individuals requiring transfusion and where serum incompatibilities are present, additional studies may be of benefit.

TABLE 3   Considerations in Evaluation of a
Positive DAT

1. Is there evidence of *in vivo* hemolysis?
2. Is there a history of recent transfusions?
3. Does the patient's serum contain unexpected antibodies?
4. What is the patient's current drug therapy?
5. Is there a history of ABO-incompatible plasma transfusion?
6. Is there a history of ALg, ATg, or IVIg administration?

Note. Clinical history dictates extent of workup. ALg, antilymphocyte globulin; ATg, antithymocyte globulin; IVIg, intravenous immunoglobulin.

c. The Role of Elution Techniques

Examination of the specificity of the RBC-associated antibody in a patient with a positive DAT is performed by standard IAT methods, using material removed (eluted) from the RBC surface. This is accomplished via the use of standard methodologies known to disrupt and separate the antibody–antigen complex (Judd, 1994). False negatives and false positives may complicate interpretation. Reactivity in a supernatant fraction obtained from a last wash prior to elution must be assessed to ensure that all unbound antibody has been removed.

d. Special Situations

Additionally, special circumstances surrounding DAT evaluation may exist in cases of suspected immune hemolysis in which the DAT is completely negative or is positive for complement (C3) alone. In the latter situation, after certain clinically significant alloantibodies (e.g., those directed against the Kidd blood group system) have been ruled out, evaluation for certain cold-reactive autoantibodies should be considered.

Techniques directed at evaluating such activity have previously been described; however, in the case of a patient in which paroxysmal cold hemoglobulinuria (PCH) is suspected, additional testing should be done. In this rare DAT-positive form of autoimmune hemolytic anemia (AIHA), IgG cold-reactive autoantibodies bind to RBCs in colder areas of the body, activate, and fix complement to the membrane of RBCs, and then dissociate from RBCs at warmer temperatures, leaving the complement attached but resulting in RBC hemolysis.

This activity has classically been described as a biphasic hemolysin (Donath–Landsteiner antibody). It forms the basis for the serologic test in which a patient's serum only causes hemolysis with RBCs when incubated at 37°C after those cells and serum have had a

prior cold temperature exposure or incubation. Eluates prepared from RBCs from such patients are typically nonreactive. Specificity for the P antigen is most frequently seen.

Another unusual situation involves patients with suspected immune hemolysis who exhibit completely negative findings by conventional DAT and IAT methodologies (Gilliand et al., 1971; Gilliand, 1976). The immune-mediated nature of their disease may become apparent when more sensitive immunoserologic testing is used or when improvement in their clinical status results after institution of immunosuppressive therapy. Antibodies with low-affinity RBC interactions and/or immunoglobulins other than those routinely detected by conventional reagents (e.g., IgM, IgA) may be involved.

3. Workup of Panagglutinin Reactivity in the Plasma

a. Autoadsorptions and the Search for Masked, Clinically Significant Alloantibodies

As discussed above, evaluation of antibody bound to RBCs is a critical component in the workup of suspected immune-mediated hemolysis, especially where autoantibodies are concerned. Of no less equal importance is the examination of antibody in the plasma of such patients. Because of the panreactive nature of warm-reactive autoantibodies, the danger of masked underlying clinically significant alloantibodies cannot be overstated. Depending on whether or not the patient has been recently transfused, various options are available for their discovery. The core of the investigation is centered around the use of the patient's own RBCs acting as affinity adsorption substrates to remove warm reactive autoantibodies.

Clinically significant alloantibodies are left in the serum by virtue of the fact that autologous RBCs lack their corresponding ligand. This clearance of warm-reactive autoantibodies from the serum allows for detection of specific reactivity patterns associated with clinically significant alloantibodies when antibody panels are performed. In practice, multiple adsorptions may be needed to clear the warm-reactive autoantibody from the serum. This may in part be due to coverage of self-antigen sites by warm-reactive autoantibodies, thus decreasing the accessibility of these antigens as targets.

Methods to enhance warm-reactive autoantibody adsorption have been developed. Here use of modified autologous RBCs by various agents, including heat, enzymes, ZZAP, and chloroquine, can increase the efficiency of this process. Treatment of the RBCs by any means raises the possibility of alterations and/or modifications to their antigenic structures. This must be kept in mind when using such approaches, but practically, it appears to be of little concern in this diagnostic setting.

b. Special Concerns in Recently Transfused Patients

In patients recently transfused, autoadsorptions are of limited value and indeed may lead to false assessments regarding the presence or absence of clinically significant alloantibodies. This is because transfused cells possessing the alloantigen may be present, thus removing clinically significant alloantibodies along with warm-reactive autoantibodies. Alternative strategies using allogenic RBCs of known phenotypes, that is, differential adsorption along with titration/dilution analysis, may be of value.

Sometimes the use of less sensitive IAT methods may afford the ability to find crossmatch-compatible blood for patients and/or identify an underlying clinically significant alloantibody. Factors enhancing clinically significant alloantibody activity also increase autoantibody reactivity. Avoidance of such methods where antibody enhancement occurs (e.g., polyethylene glycol [PEG], enzyme-treated RBC substrates, gel test) may allow for clinically significant alloantibody detection and the provision of crossmatch-compatible blood for the patient.

References

Boral, L. I., and Henry, J. B. (1977). The type and screen: A safe alternative and supplement in selected surgical procedures. *Transfusion* **17**, 163–168.

Branch, D. R., and Petz, L. D. (1982). A new reagent (ZZAP) having multiple applications in immunohematology. *Am J Clin Pathol* **78**, 161–167.

Coombs, R. R. A., Mourant, A. E., and Race, R. R. (1945). A new test for the detection of weak and "incomplete" Rh agglutinins. *Br J Exp Pathol* **26**, 255.

Dzik, W., Yang, R., and Blank, J. (1986). Rabbit erythrocyte stroma treatment of serum interferes with recognition of delayed hemolytic transfusion reactions [Letter]. *Transfusion* **26**, 303–304.

Gilliand, B. C. (1976). Coombs-negative immune hemolytic anemia. *Semin Hematol* **13**, 267–275.

Gilliand, B. C., Baxter, E., and Evans, R. S. (1971). Red-cell antibodies in acquired hemolytic anemia with negative antiglobulin serum tests. *N Engl J Med* **285**, 252–256.

Judd, W. J. (1994). "Methods in Immunohematology," 2nd ed. Montgomery Scientific, Durham, NC.

Marks, M. R., Reid, M. E., and Ellisor, S. S. (1980). Adsorption of unwanted cold autoagglutinins by formaldehyde-treated rabbit erythrocytes [Abstract]. *Transfusion* **20**, 629.

Oberman, H. A., Barnes, B. A., and Steiner, E. A. (1982). Role of the crossmatch in testing for serologic incompatibility. *Transfusion* **22**, 12–16.

Petz, L. D., and Garrity, G. (1980). "Acquired Immune Hemolytic Anemias." Churchill Livingstone, New York.

Widmann, F. K. (1997). Early observations about the ABO blood groups. *Transfusion* **37**, 665–667.

SECTION IV

SPECIALIZED COMPONENT PROCESSING OR TESTING

14

Leukoreduced Products

WALTER H. DZIK

Massachusetts General Hospital
Harvard Medical School
Boston, Massachusetts

Leukocyte reduction of blood has become an established technology for the prevention of several complications of transfusion. In some countries of Western Europe and in Canada, all blood components are leukoreduced ("universal leukoreduction"). In the United States, leukoreduction is used for specific indications described below. The practicing physician should understand how blood is leukoreduced, the current standards for leukoreduction, and the indications for leukoreduced blood.

I. DESCRIPTION

Leukocyte reduction technology refers to methods to decrease the number of residual donor leukocytes in cellular blood components. Both packed red blood cells (PRBCs) and platelets may be leukoreduced. The current American Association of Blood Banks (AABB) standards and U.S. FDA guidelines for leukoreduction are as follows: *Leukoreduced whole blood (WB), leukoreduced PRBCs, and leukoreduced apheresis platelets should contain no more than 5×10^6 residual white blood cells (WBCs) per unit.* Leukoreduced platelet concentrates derived from whole blood should contain no more than 0.83×10^6 WBCs per unit. Thus, 6 units of pooled whole-blood-derived platelet concentrates should contain no more than 5×10^6 WBCs per pool.

Leukocyte reduction can be accomplished by several different techniques, and at three distinct time points.

A. Prestorage Leukoreduction

PRBCs leukoreduced shortly after collection, during component preparation, or within the first 5 days of storage are referred to as prestorage leukoreduced PRBCs. In practice, most products are leukoreduced within 24 hours of collection. By virtue of the design of certain apheresis machines that allows for the collection of platelet products with low numbers of donor leukocytes, apheresis-derived platelets are leukoreduced at the time of collection ("process leukoreduction"). Platelets derived from whole blood ("platelet concentrates") may also be leukoreduced prior to storage by passing the platelet concentrates through a leukoreduction filter. There are several advantages of prestorage leukoreduction which are described in Section II, below.

B. In-Laboratory Leukoreduction

Both PRBCs and platelets (either apheresis-derived or pooled concentrates) may be leukoreduced by filtration in the laboratory (blood bank) prior to release for transfusion.

C. Bedside Leukoreduction

PRBCs and platelets (either apheresis-derived or pooled concentrates) may be leukoreduced by filtration at the time of transfusion to the patient. This method is considered the least beneficial, as it is: (1) accomplished later in the product storage period, allowing accumulation of cytokines, (2) has no associated quality control mechanisms, and (3) requires specialized training of a large number of health professionals who will be administering the products.

II. COLLECTION, STORAGE, AND HANDLING—ADVANTAGES OF PRESTORAGE LEUKOREDUCTION

There is an increasing trend toward the use of prestorage leukoreduced blood components. Potential advantages of prestorage leukoreduction over bedside leukoreduction include prevention of cytokine accumulation in platelets during storage and greater consistency and quality control of leukoreduction processing.

III. SPECIAL PREPARATION AND PROCESSING

Most leukoreduction is accomplished by filtration. Leukoreduction blood filters consist of a nonwoven web of surface-treated synthetic microfibers that have been compressed to a controlled functional pore size. These filters remove leukocytes from PRBCs and platelets by several mechanisms, including barrier size exclusion, cell adhesion to the filter media, and biologic interactions between leukocytes, plasma proteins, and the filter media. Because of fundamental differences in design, *leukoreduction filters intended for use with PRBCs cannot be interchanged with filters designed for platelets.* In addition, bedside leukocyte reduction may not result in a successfully leukoreduced component if the filter is not used as intended by the manufacturer, if excessive pressure is applied during the filtration, or if the filter is rinsed after use.

IV. ABO AND Rh COMPATIBILITY

The usual rules governing ABO and Rh compatibility that apply to non-leukoreduced blood components apply to leukoreduced blood.

V. ORDERING AND ADMINISTRATION

When a patient is assigned to receive leukoreduced blood products, both PRBCs and platelets should be leukoreduced (Dzik, 1996). Fresh frozen plasma (FFP) does not require leukoreduction, as the number of residual donor WBCs in unmodified FFP is less than

the number of WBCs in a leukoreduced PRBC unit; this may change, however, as filter technology advances. In general, the restriction to receive leukoreduced blood components is assigned to a patient and not for a particular single transfusion. Thus, when a patient has a clinical indication to receive leukoreduced blood (see "Indications" below), the requesting physician should arrange for instructions to be added to the blood bank record of the patient so that subsequent PRBCs and platelets will automatically be leukoreduced. For obvious reasons, leukoreduction filters should not be used with therapeutic granulocyte transfusions (see Chapter 8) or during infusion of bone marrow or peripheral blood stem cells (see Chapters 18 and 19).

VI. INDICATIONS AND CONTRAINDICATIONS

A. Indications

Leukocyte reduction is used to reduce the risks of complications of allogeneic blood transfusion (see Table 1).

1. Febrile Nonhemolytic Transfusion Reactions (FNHTRs; See Chapter 30)

a. PRBCs

Patients who have received multiple transfusions, have undergone organ transplantation, or have had multiple pregnancies may develop antibodies to foreign human leukocyte antigens (HLAs) or leukocyte surface antigens. Many such patients will experience recurrent FNHTRs at the time of PRBC transfusion. The use of leukoreduced PRBCs is extremely effective for the prevention of such reactions and abolishes the need to premedicate patients with antipyretics. Because leukocyte reduction is so effective for the prevention of

TABLE 1   Advantages to Leukoreduction of Blood Components

Reduces the incidence of febrile nonhemolytic transfusion reactions

Reduces the incidence of transfusion-transmitted leukocyte-associated infectious diseases (e.g., CMV) in at-risk recipients

Reduces the frequency of primary HLA alloimmunization

Possibly reduces transfusion-related immunomodulation (TRIM)

FNHTRs to PRBCs, clinicians should give special attention to patients who experience fever or chills to leukoreduced PRBCs, as these may represent signs of a more serious transfusion reaction such as hemolysis or bacterial contamination.

b. Platelets

As in the case of PRBCs, patients who have developed antibodies to foreign HLAs or leukocyte surface antigens may experience FNHTRs to either apheresis-derived or whole-blood-derived platelets. Leukocyte reduction is very effective for reducing the overall frequency of such reactions. However, some platelet units that have not been leukoreduced prior to storage may continue to result in FNHTRs because pyrogenic cytokines (released from donor leukocytes during storage) accumulate in some, but not all, platelet units. These cytokines—principally interleukin 1 (IL-1) and TNF-alpha—are able to pass through the leukoreduction filter, and if present in high enough concentration, may produce fever in the recipient. Prestorage leukoreduction of platelets prevents cytokine accumulation. In addition, patients with HLA sensitization may continue to experience FNHTRs to prestorage leukoreduced platelets when the transfused incompatible platelets are cleared by the recipient's alloantibodies. The failure to gain a satisfactory post-transfusion platelet increment is an important clinical clue to the development of such allosensitization.

2. Primary HLA Alloimmunization (See Chapter 24)

HLA molecules on cotransfused leukocytes may cause HLA alloimmunization, which can lead to platelet refractory states and potential difficulty with solid organ or bone marrow transplant acceptance and maintenance. *Thus, for patients awaiting transplantation and for patients requiring long-term blood component support (including oncology and bone marrow transplant patients), leukoreduced blood components are indicated to reduce the risk of primary HLA alloimmunization.* Leukocyte reduction is less effective for the prevention of secondary HLA alloimmunization among individuals who have been previously sensitized to HLA antigens by virtue of prior transfusions, transplantations, or pregnancies. The national Trial to Reduce Alloimmunization to Platelets (TRAP) compared the development of HLA alloimmunization among patients randomized to receive leukoreduced pooled platelets, leukoreduced apheresis platelets, ultraviolet B light-treated pooled platelets, and unmodified pooled platelets (control; TRAP, 1997). Compared with the control group, patients assigned to any of the other groups had a signif-

TABLE 2 Patient Subgroups for Whom Leukoreduced Blood Products are Recommended

Patients with serious congenital immune deficiency syndromes
HI V-positive, CMV-negative individuals
Very low-birth-weight (< 1200 g) infants
Bone marrow transplant patients
CMV-negative recipients of CMV-negative solid organ allografts
Patients requiring long-term blood product support
Patients awaiting renal, heart, or lung transplantation

icantly lower incidence of HLA alloantibody sensitization. There was no difference between leukoreduced pooled platelet concentrates and leukoreduced apheresis-derived platelets, demonstrating that the number of donor leukocytes—not the number of donor platelets—is the main determinant of primary HLA sensitization.

3. Transfusion-Transmitted Cytomegalovirus (CMV) among At-Risk Recipients (See Chapters 16 and 35)

Certain patients are at risk for significant clinical morbidity from transfusion-transmitted CMV infections. These patients are listed in Table 2. For these patient groups, the incidence of transfusion-transmitted CMV may be reduced either by using blood components collected from CMV-seronegative donors or by using leukoreduced blood collected from CMV-positive or -negative blood donors (Dzik, 1996; Hillyer *et al.*, 1994). The residual risk of transfusion-transmitted CMV is ~ 3% with either CMV-seronegative or leukoreduced blood products.

Patients who are CMV-seropositive generally develop CMV disease from reactivation of their endogenous CMV infection during the course of their medical treatment. These patients are currently considered unlikely to derive medical benefit from the use of leukoreduced components.

4. Potential Effect on Transfusion-Related Immunomodulation (TRIM)

A large body of indirect evidence suggests that non-leukoreduced blood transfusions may result in some degree of immunosuppression. Clinically, this effect manifests as an increase in postoperative wound infections. Additionally, some evidence suggests that there is an effect on natural tumor suppressor mechanisms.

The mechanism by which blood transfusions induce mild global immunosuppression is not understood. A modification in the TH1/TH2 ratio toward a state of "tolerance" has been postulated.

Despite clinical studies suggesting an effect of transfusions on the incidence of postoperative bacterial infections, little experimental evidence exists regarding the effect of transfusion on recipient neutrophil function. Experimental studies in rodents have induced transplantation tolerance by localization of donor-type HLA proteins in the recipient's thymus. Other studies have demonstrated T-cell anergy resulting from exposure to donor cells expressing HLA antigens in the absence of necessary costimulatory molecules. As noted above, polarization (by unknown mechanisms) of the recipient immune response towards a T_H2 phenotype has also been suggested to result from transfusion. Some studies in animals have suggested that transfusion induces suppressor T-cells in the transfusion recipient.

There is also controversy about whether or not the incidence of TRIM may be reduced by transfusion of leukoreduced blood components (Blajchman and Vamvakas, 1999). Some studies of patients undergoing colorectal cancer surgery, gastrointestinal surgery, and cardiac surgery have shown a reduction in TRIM with the use of leukoreduced blood components, while other studies have not (Jensen *et al.*, 1996; Tartter *et al.*, 1998; van der Watering *et al.*, 1998; Blajchman and Vamvakas, 1999). The expected benefits include potential decreases in the incidence of postoperative wound infections and in the rate of recurrence of tumors after primary surgical resections.

While the mechanism and magnitude of TRIM remain to be fully elucidated, these concepts have contributed to the movement toward universal leukoreduction, as TRIM represents a potential adverse effect of transfusion in all recipients.

B. Contraindications

1. Transfusion-Associated Graft-versus-Host Disease (TA-GVHD)

Leukocyte reduction is not an adequate or appropriate method to prevent TA-GVHD, as there are enough residual lymphocytes in leukoreduced products to induce this complication. Gamma irradiation of cellular blood components is the only approved method for prevention of TA-GVHD.

2. Bedside Leukoreduction for Patients on Angiotensin Converting Enzyme (ACE) Inhibitors Who Have Experienced Hypotensive Reactions to Prior Transfusions

Some patients receiving ACE inhibitor therapy will experience hypotensive reactions at the time of transfusion of blood through a bedside leukoreduction filter (Mair and Leparc, 1998). These reactions are currently considered to result from the elaboration of bradykinin during filtration. ACE inhibitors interfere with the breakdown of the infused bradykinin, which can sometimes reach concentrations that induce hypotension. Prestorage or in-laboratory leukoreduction does not result in hypotensive reactions, because the delay in time between filtration and infusion allows for the breakdown of bradykinin. Thus, as most leukoreduced products undergo prestorage leukoreduction, this complication is no longer a significant risk in most patients.

References

Blajchman, M. A., and Vamvakas, S., Eds. (1999). "Immunomodulation and Blood Transfusion." AABB Press, Bethesda, MD.

Dzik, W. H. (1996). Leukoreduced blood components: Laboratory and clinical aspects. In "Principles of Transfusion Medicine" (E. C. Rossi, T. L. Simon, G. S. Moss, and S. A. Gould, Eds.), 2nd ed., pp. 353–374. Williams and Wilkins, Baltimore.

Dzik, S., Aubuchon, J., Jeffries, L., *et al.* (2000). Leukocyte reduction of blood components: Public policy and new technology. *Transfusion Med Rev* 14, 34–52.

Hillyer, C. D., Emmens, R. K., Zago-Navaretti, M., *et al.* (1994). Methods for the reduction of transfusion-transmitted cytomegalovirus infection: Filtration versus the use of seronegative donor units. *Transfusion* 34, 929–934.

Jensen, L. S., Kissmeyer-Nielsen, P., Wolff, B., *et al.* (1996). Randomised comparison of leucocyte-depleted versus buffy coat poor blood transfusion and complications after colorectal surgery. *Lancet* 348, 841–847.

Mair, B., and Leparc, G. F. (1998). Hypotensive reactions associated with platelet transfusions and angiotensin converting enzyme inhibitors. *Vox Sang* 74, 27–30.

Tartter, P. I., Mohandas, K., Aza, P., *et al.* (1998). Randomized trial comparing packed red cell blood transfusion with and without leukocyte depletion for gastrointestinal surgery. *Am J Surg* 176, 462–466.

Trial to Reduce Alloimmunization to Platelets (TRAP) Study Group (1997). Leukocyte reduction and ultraviolet B irradiation of platelets to prevent alloimmunization and refractoriness to platelet transfusion. *N Engl J Med* 337, 1861–1869.

Van der Watering, L. M. S., Hermans, J., Houbiers, J. G. A., *et al.* (1998). Beneficial effects of leukocyte depletion of transfused blood on postoperative complications in patients undergoing cardiac surgery: A randomized clinical trial. *Circulation* 97, 562–568.

15

CMV and Other Virus-Safe Products

Leukoreduction and Virus Inactivation

JOHN D. ROBACK

Department of Pathology and Laboratory Medicine
Emory University Hospital Blood Bank
Atlanta, Georgia

Technical advances during the last two decades have markedly decreased the risk of viral transmission by blood transfusion. In fact, the probability of contracting human immunodeficiency virus (HIV) from a blood transfusion has recently been calculated as 1 in 493,000 transfusions, but the rate could be as low as 1 in 2,778,000 (Schreiber *et al.*, 1996). For perspective, this level of risk is considerably lower than the estimated risk of being killed in a motor vehicle accident, and more comparable to that of extremely low-probability events such as drowning in a bathtub (Lee *et al.*, 1998).

The safety of today's blood supply is due primarily to current serological screening methodologies, which after years of refinements are exceptionally sensitive and specific for detection of infectious, pathogenic viruses in donated blood. With the introduction of nucleic acid amplification technology (NAT)-based testing in 1999, the ability to identify blood donations from individuals with early ("window phase") viral infections, which would otherwise be missed by current serological techniques, has been significantly enhanced. Screening to prevent viral transmission is discussed in detail in Chapter 2.

Because infectious disease testing is imperfect, the prevention of infectious disease transmission requires additional, complementary approaches to remove or inactivate viral contaminants in blood components prior to transfusion. For example, filtration of blood components prior to transfusion is currently used clinically to decrease the incidence of cytomegalovirus (CMV)

transmission. The roles of leukofiltration and other methodologies in preventing infectious disease transmission and maintaining the safety of the blood supply are the subject of this chapter.

I. PRODUCT DESCRIPTION

Detailed descriptions of standard blood components and derivatives are discussed in more detail in Chapters 4–10. The blood components described below have undergone additional screening or processing steps to further reduce the risk of viral transmission.

A. Cytomegalovirus-Seronegative Units

Blood donations are not routinely tested for anti-CMV antibodies to identify donors previously infected with CMV. However, many transfusion services employ CMV serology in order to maintain an inventory of CMV-seronegative components, which historically have been regarded as the gold standard for reducing the incidence of transfusion-transmitted CMV (TT-CMV) infections in selected patient populations (see Table 1; Hillyer *et al.*, 1990, 1999). Transfusion of at-risk patients with CMV-seronegative blood components has an associated TT-CMV incidence of 0–3% (Bowden *et al.*, 1986; Miller *et al.*, 1991), as compared to an incidence of 13–37% with unscreened blood compo-

TABLE 1 CMV-Seronegative Patients for Whom CMV-Safe
Blood Products[a] are Recommended

Immunocompromised	Immature immune systems
Bone marrow transplant recipients	Premature and newborn infants
Solid organ transplant recipients	
HIV infected and AIDS patients	
Patients with congential immunodeficiency states	

[a] Also recommended for those patients in whom CMV serologic status is unknown.

nents (Miller *et al.*, 1991; Yeager *et al.*, 1981). As described below, the incidence of TT-CMV with leukofiltered, non-CMV-screened blood components is similar to the low TT-CMV incidence seen with CMV-seronegative units, leading some authorities to recommend that seronegative and leukofiltered ("CMV-safe") units can be used interchangeably for at-risk patient populations (Bowden *et al.*, 1995).

B. Leukoreduced Units

Leukoreduction of packed red blood cell (PRBC) and platelet units, usually accomplished by filtration, decreases the incidence of adverse transfusion outcomes, including alloimmunization to human leukocyte antigens (HLAs), febrile nonhemolytic transfusion reactions, transfusion-mediated immunosuppression, and transmission of leukocyte-associated infectious agents such as CMV (Hillyer *et al.*, 1994, 1999). Leukoreduction can be performed either in the blood center prior to transfusion (prestorage leukoreduction) or at the time of transfusion (bedside leukoreduction). Multiple clinical trials have demonstrated similar low incidences of TT-CMV with either seronegative components or leukoreduced, unscreened components (see Chapter 14; Bowden *et al.*, 1995; Luban *et al.*, 1987).

Although not yet confirmed by experimentation, there is reason to believe that leukoreduction may also be effective in mitigating the transfusion transmission of other WBC-associated infectious agents, including Epstein-Barr virus (EBV), human herpes virus-8 (HHV-8), and *Ehrlichial* species that cause human granulocytic ehrlichiosis. In addition, B-cells are believed to be important for some aspects of the pathogenesis of prion-mediated spongiform encephalopathies, such as Creutzfeldt-Jakob disease (CJD). While there is not yet convincing evidence that CJD can be transfusion transmitted, it has been suggested that B-cell removal by filtration could reduce the possibility of such an event (Brown, 1997). Leukoreduction methodologies and indications are discussed in more detail in Chapter 14.

C. Solvent/Detergent-Treated Frozen Plasma

Solvent/detergent (SD) plasma is prepared by pooling 2500 ABO-group specific plasma fractions from volunteer blood donations, and treating the pool with the solvent tri(n-butyl)phosphate (TNBP) and detergent Triton X-100 to inactivate any contaminating lipid-enveloped viruses, including HIV and hepatitis C virus (HCV). After removal of the solvent and detergent, the plasma pool is filtered, aliquoted into 200-ml units, and frozen for storage. SD plasma contains a minimum of 1.8 mg/ml of fibrinogen and 0.7 U/ml of coagulation factors V, VII, VIII, IX, X, XI, and XIII, representing concentrations similar to those seen in fresh frozen plasma (FFP). However, the largest von Willebrand multimers are removed from SD plasma by the filtration process; this may prove efficacious in the treatment of FFP. The SD process is discussed in more detail below.

II. COLLECTION, STORAGE, HANDLING, AND SPECIAL PROCESSING

The standard techniques for collection, storage, and handling of blood components are described in more detail in Chapters 3–10. Likewise, standard serological and NAT-screening methodologies used to prevent transfusion transmission of HIV, hepatitis B virus (HBV), HCV, and human T-cell lymphotrophic virus (HTLV) are discussed in Chapter 2. The following section describes components and plasma derivatives rendered "viral-safe" through special collection, screening, and/or treatment methodologies.

A. Leukoreduction Filtration to Decrease Viral Transmission

Passenger leukocytes in RBC and platelet components can transmit infectious agents to the transfusion recipient (Hillyer *et al.*, 1994, 1999). A number of techniques are available to remove leukocytes from blood components. These methodologies, which vary in efficacy, include simple centrifugation and buffy coat removal, washing, freezing with subsequent deglycerolization, and filtration. Of these techniques, filtration is used most frequently because it is simple, rapid, and cost effective, and reproducibly reduces leukocyte counts by at least 3-$\log_{10}$. Prestorage filtration is becoming the international standard due to clinical advantages of leukocyte removal within 24 hours of blood collection and to improved standardization and quality control as compared to bedside leukoreduction (Anonymous, 1998).

1. Leukoreduction for CMV

While a policy of transfusing CMV-seronegative blood components to at-risk patients can markedly reduce the incidence of TT-CMV, CMV-seronegative units are often in short supply. Peripheral blood leukocytes of the monocyte/macrophage lineage are believed to be important to the pathogenesis of TT-CMV since they carry latent CMV DNA, and can support CMV replication, including the production of infectious virions (Soderberg-Naucler et al., 1997a, b; Stoddart et al., 1994). The logistical complications of providing seronegative units, along with the observations that CMV is highly leukocyte-associated in peripheral blood, led to interest in determining whether leukoreduction could render blood units "CMV-safe." Several clinical trials have demonstrated clinically significant reduction in the incidence of TT-CMV with leukoreduction of CMV-unscreened components (Bowden et al., 1995; Luban et al., 1987). These findings have led to the conclusion that leukoreduced components are essentially equivalent to CMV-seronegative components with respect to the transfusion transmission rate of CMV disease (Bowden et al., 1995; Hillyer et al., 1994).

2. Leukoreduction for Other Herpes Viruses

In addition to CMV, other human herpes viruses can also cause significant morbidity and mortality. EBV infection, for example, can produce B-cell lymphoproliferative diseases, including Burkitt's lymphoma and post-transplantation lymphoproliferative disorder (PTLD). HHV-8, the etiologic agent of Kaposi's sarcoma in AIDS patients, also participates in the pathogenesis of body-cavity-based lymphomas (BCBLs; Hillyer et al., 1999). While the clinical impact of these herpes viruses is clear, transmission of these viruses by blood transfusion has proven more difficult to document and study. Although EBV transfusion transmission has been demonstrated (Alfieri et al., 1996), approximately 90% of the population is EBV-seropositive, which leaves only small numbers of at-risk

seronegative patients to study. In contrast, less than 20% of the population is seropositive for HHV-8. However, the evidence regarding HHV-8 transfusion transmission is currently inconclusive (Blackbourn et al., 1997; Operskalski et al., 1997). Nonetheless, both of these viruses primarily reside within leukocytes, and leukofiltration would be expected to reduce the titer of these viruses in transfused blood components.

3. Leukoreduction for Prion-Mediated, Transmissible Spongiform Encephalopathies (TSEs)

CJD and new-variant CJD (nvCJD) are TSEs mediated by proteinacious infectious particles known as prions. Both CJD and nvCJD are incurable, rapidly progressive, fatal diseases. Despite the fact that up to 1 in every 1.2 million blood donors may have CJD, current evidence suggests that CJD is not transfusion transmitted (Brown, 1998). There have been no documented cases of prion disease transmission by blood transfusion, nor have epidemiologic studies demonstrated an increased incidence of CJD in heavily transfused patient populations, including hemophiliacs (Dodd and Sullivan, 1998). Nonetheless, the possibility of TSE transmission by transfusion cannot be completely excluded, although if it occurs the efficiency is likely to be extremely low, as has been seen in transfusion transmission of experimental prion infections in animals (Brown, 1998). In human whole blood units experimentally infected with prions and in blood from prion-infected mice, prions appear to associate with leukocytes as well as other blood constituents, including platelets and RBCs (Brown et al., 1998). In addition, B-lymphocytes appear to be important for the neuroinvasiveness of scrapie, a prion disease of animals. These findings have led to the suggestion that leukoreduction of blood components may decrease the potential risk of prion transmission by blood components (Brown, 1997). However, because the degree of association between prions and leukocytes is not known, and transfusion transmission of prion diseases has not been documented, filtration for the sole purpose of preventing transfusion transmission of TSEs cannot be recommended at this time.

4. Leukoreduction for Other Infectious Agents

Among leukocyte subsets, monocytes harbor *Ehrlichia chaffeensis*, the first identified human ehrlichial agent (Dawson et al., 1991). Neutrophils are reservoirs for the human granulocytic ehrlichiosis agent, which is, or is closely related to, E. equi and E. phagocytophilia (Hodzic et al., 1998; Klein et al., 1997). CD19$^+$ B-cells, as well as T-cells and NK-cells, can be infected by EBV (Jaffe, 1995; Kanegane *et al.*, 1998), while HHV-8 replication has been observed in CD19$^+$ B-cells *in vitro* (Blackbourn *et al.*, 1997). The human T-cell lymphotropic viruses HTLV-I and HTLV-II preferentially infect CD4$^+$ and CD8$^+$ T-cells *in vivo*, respectively (Cereseto *et al.*, 1996). In addition, other infectious agents, such as *Toxoplasma gondii* are also likely to display distinct WBC tropism. Thus, based on these considerations, leukoreduction filtration could be effective in preventing transfusion transmission of these infectious agents. However, efficacy has not yet been documented in the clinical setting.

5. Adverse Effects of Leukoreduction

Leukofiltration can infrequently produce untoward effects. For example, case reports have indicated that bedside leukoreduction has produced marked hypotensive reactions in a very small subset of transfusion recipients. Hypotensive reactions are discussed in more detail in Chapters 7, 14, and 30.

B. Viral Inactivation

The capacity of serologic and NAT screening methodologies to identify potentially infectious blood components is limited by factors that include the time interval (window) required to mount a host immune response and lower limits of detection sensitivity. In addition, only recognized viral pathogens can be identified, while emerging infectious agents cannot be detected. Leukoreduction, while effective for infectious agents that reside primarily in leukocytes, like CMV, has little impact on viruses that exist in a stable, infectious form in plasma, like HIV and HCV. For these reasons, further reductions in the risk of transfusion-transmitted infections will require strategies that complement screening and leukoreduction, including approaches to inactivate viruses in blood components prior to transfusion. The attractiveness of broad-spectrum sterilization methodologies derives from the fact that they should be equally effective against common, rarely occurring, and even currently unidentified pathogenic microorganisms that contaminate blood components.

1. Solvent/Detergent (SD) Plasma and Coagulation Factor Concentrates

Many of the pathogenic viruses transmissible by blood transfusion, such as HIV and HCV, are encapsu-

lated within a lipid envelope. Treatment of plasma and plasma derivatives with a combination of a solvent (such as ethyl ether or TNBP) and a detergent (Tween 80 or Triton X-100) disrupts lipid bilayers, markedly inhibiting infectivity of contaminating lipid-enveloped viruses. Plasma proteins, in contrast, are extremely stable to SD treatment. After chromatography to remove the solvent and detergent reagents, the plasma is filtered, which removes residual leukocytes, bacteria, and the largest von Willebrand factor multimers. Solvent/detergent plasma is subsequently dispensed into 200-ml units and frozen to preserve coagulation factor activity during storage.

Solvent/detergent treatment efficiently inactivates all enveloped viruses tested, including HIV, HBV, and HCV. It has been estimated that with SD treatment of coagulation factor concentrates, less than 1 in 10^{16}, 10^{13}, and 10^6 will contain infectious HIV, HBV, and HCV, respectively. In contrast, viruses without lipid envelopes, such as hepatitis A virus (HAV) and parvovirus B19, are resistant to SD treatment. HAV and B19 have been transmitted by SD-treated factor concentrates (Klein *et al.*, 1998; Pehta, 1996). However, HAV and B19 transmission by SD plasma is less likely than for factor concentrates.

During the SD plasma manufacturing process, ABO group-specific plasma is first pooled from 2500 individual volunteer blood donations. The plasma in the pool from HAV- and B19-immune donors contributes high levels of neutralizing antibodies to the pool, leading to inactivation of contaminating HAV and B19 viruses that may have been introduced by a minority of the plasma donations. In addition, SD plasma pools are now tested by the manufacturer for the presence of HAV and B19 viral nucleic acids; contaminated pools are discarded prior to release. The Food and Drug Administration (FDA) granted approval for the use SD plasma in 1998. Prior to its introduction in the United States, SD plasma was used routinely in a number of European countries. More than 3 million units of SD plasma and 11 million units of SD-treated plasma derivatives, such as antihemophilia factor concentrates, have been transfused through 1997 without reported transmission of lipid-enveloped viruses (Klein *et al.*, 1998).

2. Heat-Treated Plasma Derivatives

A different approach to viral inactivation involves heat treatment of derivatives. Various heat-treatment protocols have been implemented, including heating in solution (pasteurization, usually at 60°C for 10 hours), heating previously freeze-dried plasma derivatives (80°C), and heating in hot vapor at 60°C (Mannucci et al., 1992). Pasteurization of IVIg preparations leads to marked reductions in viral load (Chandra et al., 1999). Total levels of viral inactivation have been calculated to be a minimum of $\sim 15\text{-log}_{10}$ for HIV, and $\sim 6\text{-log}_{10}$ for a model nonenveloped virus. HBV and HCV are more resistant to these inactivation procedures than is HIV. Parvovirus B19, which causes erythema infectiosum (fifth disease) and transient aplastic anemia, is even more resistant to heat treatment. For example, in one study patients receiving either factor VIII or IX concentrates treated with dry heating in combination with solvent/detergent did not seroconvert for HIV, HBV, or HCV. In addition, there was no seroconversion for HAV, which is likely inactivated by the heating step, since it is unaffected by SD treatment. However, over half of the patients (8 of 15) developed anti-B19 antibodies and were positive for B19 DNA by PCR analysis (Santagostino et al., 1997), demonstrating the resistance of this virus to currently employed virucidal techniques.

3. Experimental Approaches

While techniques such as SD treatment and pasteurization are effective and approved for sterilizing plasma and plasma-derived products, the FDA has not yet approved methods for inactivating viral contaminants in PRBC or platelets components. Thus, the technologies for cellular component sterilization described below remain experimental (reviewed by Ben-Hur et al., 1996). Each of these approaches relies on the use of a chemical compound that, when activated by light of the appropriate wavelength or by a pH shift, can irreversibly inactivate contaminating pathogens.

Psoralens, heterocyclic planar compounds, intercalate into the DNA of viruses, bacteria, and eukaryotic cells. After exposure to UV-A light (320–400 nm), activated psoralens can cross-link DNA, blocking subsequent replication (Lin et al., 1997; Margolis-Nunno et al., 1997). In experimentally contaminated platelet units, these agents are highly effective in blocking replication of cell-free HIV virions, as well as cell-associated HIV that is either actively replicating or latent. However, psoralen reagents are not yet effective for sterilization of RBC units, because UV-A light is efficiently absorbed by hemoglobin.

The porphyrin-like phthalocyanines, in contrast, are activated by longer wavelength red light (650–700 nm), obviating interference from hemoglobin. These agents have been shown to inactivate viruses as well as parasites, including *Plasmodium falciparurn* and *Trypanosoma cruzi* (Lustigman and Ben-Hur, 1996). However, activated phthalocyanines appear to target the lipid envelope, and nonenveloped viruses may be resistant to sterilization with this agent. Phenothiazine derivatives have also been effectively used with RBC units (Wagner et al., 1998).

Under the proper experimental conditions, platelet and erythrocyte function are not significantly affected by these treatments. Blood component sterilization is an area of active investigation, and it is expected that viral inactivation methodologies of this type will eventually be approved as a complement to the currently used screening and leukoreduction approaches to decrease viral transmission.

C. Recombinant Coagulation Factors

Advances in molecular biological techniques over the last decades now allow coagulation factors, such as factor VIII, to be expressed and purified *in vitro*, virtually eliminating concerns of infectious disease transmission that still plague factor VIII concentrates prepared from human plasma. For example, parvovirus B19, which is nonenveloped, appears to be transmitted by plasma-derived factor VIII concentrates, even after solvent/detergent treatment (VanAken, 1997). Transgenic animal "bioreactors" may also prove useful in producing large amounts of recombinant coagulation factors (Lubon et al., 1996). In fact, it has been estimated that the entire yearly requirement for factor VIII in the United States could be met by recombinant human factor VIII (rhfVIII) produced by a single transgenic pig (Lubon et al., 1996). In multiple clinical trials for patients with hemophilia A, rhfVIII has been shown to be effective, with hemostatic activity profiles similar to those of plasma-derived factor VIII (VanAken, 1997). Development and testing of other recombinant coagulation factors for clinical use is currently underway. A complete consideration of coagulation factor concentrates available for transfusion is beyond the scope of this chapter, but is discussed in more detail in Chapter 9.

III. ABO AND Rh COMPATIBILITY

Issues of ABO and Rh compatibility as they relate to transfusion of PRBCs, platelets, and plasma products are discussed in more detail in Chapters 4–8.

IV. ORDERING AND ADMINISTRATION

Ordering and administration of blood components is discussed in more detail in Chapters 4–8 and 22.

V. INDICATIONS AND CONTRAINDICATIONS

Indications and contraindications for the transfusion of blood components are discussed in more detail in Chapters 4–8.

References

Alfieri, C., Tanner, J., Carpentier, L., *et al.* (1996). Epstein-Barr virus transmission from a blood donor to an organ transplant recipient with recovery of the same virus strain from the recipient's blood and oropharynx. *Blood* **87**, 812–817.

Anonymous. (1998). The use and quality control of leukocyte-depleted cell concentrates. *Vox Sang* **75**, 82–92.

Ben-Hur, E., Moor, A. C., Margolis-Nunno, H., *et al.* (1996). The photodecontamination of cellular blood components: Mechanisms and use of photosensitization in transfusion medicine. *Transfusion Med Rev* **10**, 15–22.

Blackbourn, D. J., Ambroziak, J., Lennette, E., *et al.* (1997). Infectious human herpesvirus 8 in a healthy North American blood donor. *Lancet* **349**, 609–611.

Bowden, R. A., Sayers, M., Flournoy, N., *et al.* (1986). Cytomegalovirus immune globulin and seronegative blood products to prevent primary cytomegalovirus infection after marrow transplantation. *N Engl J Med* **314**, 1006–1010.

Bowden, R. A., Slichter, S. J., Sayers, M., *et al.* (1995). A comparison of filtered leukocyte-reduced and cytomegalovirus (CMV) seronegative blood products for the prevention of transfusion-associated CMV infection after marrow transplant. *Blood* **86**, 3598–3603.

Brown, P. (1997). B lymphocytes and neuroinvasion. *Nature* **390**, 662–663.

Brown, P. (1998). Donor pool size and the risk of blood-borne Creutzfeldt-Jakob disease. *Transfusion* **38**, 312–315.

Brown, P., Rohwer, R. G., Dunstan, B. C., *et al.* (1998). The distribution of infectivity in blood components and plasma derivatives in experimental models of transmissible spongiform encephalopathy. *Transfusion* **38**, 810–816.

Cereseto, A., Mulloy, J. C., and Franchini, G. (1996). Insights on the pathogenicity of human T-lymphotropic/leukemia virus types I and II. *J Acquir Immune Defic Syndr Hum Retrovirol* **13**(Suppl 1), S69–75.

Chandra, S., Cavanaugh, J. E., Lin, C. M., *et al.* (1999). Virus reduction in the preparation of intravenous immune globulin: *in vitro* experiments. *Transfusion* **39**, 249–257.

Dawson, J. E., Anderson, B. E., Fishbein, D. B., *et al.* (1991). Isolation and characterization of an Ehrlichia sp. from a patient diagnosed with human ehrlichiosis. *J Clin Microbiol* **29**, 2741–2745.

Dodd, R. Y., and Sullivan, M. T. (1998). Creutzfeldt-Jakob disease and transfusion safety: Tilting at icebergs? *Transfusion* **38**, 221–223.

Hillyer, C. D., Snydman, D. R., and Berkman, E. M. (1990). The risk of cytomegalovirus infection in solid organ and bone marrow transplant recipients: Transfusion of blood products. *Transfusion* **30**, 659–666.

Hillyer, C. D., Emmens, R. K., Zago-Novaretti, M., and Berkman, E. M. (1994). Methods for the reduction of transfusion-transmitted cytomegalovirus infection: Filtration versus the use of seronegative donor units. *Transfusion* **34**, 929–934.

Hillyer, C. D., Lankford, K. V., Roback, J. D., *et al.* (1999). Transfusion of the HIV seropositive patient: Immunomodulation, viral reactivation, and limiting exposure to EBV (HHV-4), CMV (HHV-5) and HHV-6, 7, and 8. *Transfusion Med Rev* **13**, 1–17.

Hodzic, E., Ijdo, J. W., Feng, S., *et al.* (1998). Granulocytic ehrlichiosis in the laboratory mouse. *J Infect Dis* **177**, 737–745.

Jaffe, E. S. (1995). Nasal and nasal-type T/NK cell lymphoma: A unique form of lymphoma associated with the Epstein-Barr virus. *Histopathology* **27**, 581–583.

Kanegane, H., Bhatia, K., Gutierrez, M., *et al.* (1998). A syndrome of peripheral blood T-cell infection with Epstein-Barr virus (EBV) followed by EBV-positive T cell lymphoma. *Blood* **91**, 2085–2091.

Klein, M. B., Miller, J. S., Nelson, C. M., and Goodman, J. L. (1997). Primary bone marrow progenitors of both granulocytic and monocytic lineages are susceptible to infection with the agent of human granulocytic ehrlichiosis. *J Infect Dis* **176**, 1405–1409.

Klein, H. G., Dodd, R. Y., Dzik, W. H., *et al.* (1998). Current status of solvent/detergent-treated frozen plasma. *Transfusion* **38**, 102–107.

Lee, D. H., Paling, J. E., and Blajchman, M. A. (1998). A new tool for communicating transfusion risk information. *Transfusion* **38**, 184–188.

Lin, L., Cook, D. N., Wiesehahn, G. P., *et al.* (1997). Photochemical inactivation of viruses and bacteria in platelet concentrates by use of a novel psoralen and long-wavelength ultraviolet light. *Transfusion* **37**, 423–435.

Luban, N. L., Williams, A. E., MacDonald, M. G., *et al.* (1987). Low incidence of acquired cytomegalovirus infection in neonates transfused with washed red blood cells. *Am J Dis Child* **141**, 416–419.

Lubon, H., Paleyanda, R. K., Velander, W. H., and Drohan, W. N. (1996). Blood proteins from transgenic animal bioreactors. *Transfusion Med Rev* **10**, 131–143.

Lustigman, S., and Ben-Hur, E. (1996). Photosensitized inactivation of plasmodium falciparum in human red cells by phthalocyanines. *Transfusion* **36**, 543–546.

Mannucci, P. M., Schimpf, K., Abe, T., *et al.* (1992). Low risk of viral infection after administration of vapor-heated factor VIII concentrate. International Investigator Group. *Transfusion* **32**, 134–138.

Margolis-Nunno, H., Bardossy, L., Robinson, R., *et al.* (1997). Psoralen-mediated photodecontamination of platelet concentrates: Inactivation of cell-free and cell associated forms of human immunodeficiency virus and assessment of platelet function *in vivo*. *Transfusion* **37**, 889–895.

Miller, W. J., McCullough, J., Balfour, H. H., Jr., *et al.* (1991). Prevention of cytomegalovirus infection following bone marrow transplantation: A randomized trial of blood product screening. *Bone Marrow Transplant* **7**, 227–234.

Operskalski, E. A., Busch, M. P., Mosley, J. W., and Kedes, D. H. (1997). Blood donations and viruses. *Lancet* **349**, 1327.

Pehta, J. C. (1996). Clinical studies with solvent detergent-treated products. *Transfusion Med Rev* **10**, 303–311.

Santagostino, E., Mannucci, P. M., Gringeri, A., *et al.* (1997). Transmission of parvovirus B19 by coagulation factor concentrates exposed to 100 degrees C heat after lyophilization. *Transfusion* **37**, 517–522.

Schreiber, G. B., Busch, M. P., Kleinman, S. H., and Korelitz, J. J. (1996). The risk of transfusion-transmitted viral infections. The Retrovirus Epidemiology Donor Study. *N Engl J Med* **334**, 1685–1690.

Soderberg-Naucler, C., Fish, K. N., and Nelson, J. A. (1997a). Interferon-gamma and tumor necrosis factor-alpha specifically induce formation of cytomegalovirus permissive monocyte-derived macrophages that are refractory to the antiviral activity of these cytokines. *J Clin Invest* **100**, 3154–3163.

Soderberg-Naucler, C., Fish, K. N., and Nelson, J. A. (1997b). Reactivation of latent human cytomegalovirus by allogeneic stimulation of blood cells from healthy donors. *Cell* **91**, 119–126.

Stoddart, C. A., Cardin, R. D., Boname, J. M., *et al.* (1994). Peripheral blood mononuclear phagocytes mediate dissemination of murine cytomegalovirus. *J Virol* **68**, 6243–6253.

VanAken, W. G. (1997). The potential impact of recombinant factor VIII on hemophilia care and the demand for blood and blood products. *Transfusion Med Rev* **11**, 6–14.

Wagner, S. J., Skripchenko, A., Robinette, D., *et al.* (1998). Preservation of red cell properties after virucidal phototreatment with dimethylmethylene blue. *Transfusion* **38**, 729–737.

Yeager, A. S., Grumet, F. C., Hafleigh, E. B., *et al.* (1981). Prevention of transfusion-acquired cytomegalovirus infections in newborn infants. *J Pediatr* **98**, 281–287.

C H A P T E R

16

Irradiated Components

RAYMOND L. COMENZO
DAVID L. WUEST
Memorial Sloan-Kettering Cancer Center
New York, New York

Scientific interest in graft-versus-host (GVH) reactions began 50 years ago when the recovery of animals after splenic cell transplants was found to be complicated by a wasting or "runting" syndrome soon to be observed in humans after allogeneic bone marrow transplantation (BMT; Billingham, 1967–1968; Billingham *et al.*, 1954, 1962). In the mid-1960s a similar syndrome was observed after allogeneic blood transfusion in neonates and children with congenital immunodeficiency syndromes (Hathaway *et al.*, 1965, 1967). Subsequently, transfusion medicine specialists have developed an appreciation for the etiology of transfusion-associated graft-versus-host disease (TA-GVHD) and for the prophylactic measures required to prevent it.

I. TRANSFUSION-ASSOCIATED GRAFT-VERSUS-HOST DISEASE

TA-GVHD is caused by transfusion of allogeneic blood components containing immunocompetent lymphocytes that are not rejected by the recipient's immune system. These allogeneic lymphocytes engraft, clonally expand, and mount an immune attack on the recipient's bone marrow, skin, liver, and gastrointestinal tract, causing TA-GVHD. TA-GVHD is fatal in the majority of cases.

A. Genetics

Tissue compatibility between human recipients and donor grafts is controlled by the genes of the major histocompatibility complex on chromosome 6. The proteins encoded by these genes are called human leukocyte antigens (HLAs), and the HLA system is vastly complex and polymorphic, meaning that related individuals who are HLA-identical share haplotypes that have the same genes in them, while unrelated individuals who are HLA-identical likely have different versions of the same genes (polymorphisms).

B. Etiology

The development of TA-GVHD is influenced by:

1. The immunocompetence of the recipient
2. The degree of HLA similarity between donor and recipient
3. The number of functional lymphocytes infused

C. Patients at Risk for TA-GVHD

1. Those who have defective cell-mediated immunity and cannot, therefore, reject histoincompatible lymphocytes
2. Those receiving "HLA-matched" transfusions of platelets from unrelated donors
3. Those receiving cellular blood products from first- or second-degree relatives, where the donor may be an HLA match and also homozygous for an HLA haplotype for which the recipient is heterozygous; in this situation, there is no foreign HLA antigen on donor lymphocytes, and the recipient's immune system cannot recognize the shared tissue type of the donor as foreign (cannot reject donor cells), but donor lymphocytes can react to recipient tissues and cause TA-GVHD

D. Epidemiologic Considerations

The risk of such an occurrence in the heterogeneous Caucasian population of the United States has been calculated to be low (1 in 20,000 to 40,000), and reported cases of TA-GVHD in the USA remain even rarer than those odds would suggest (Wagner and Flegel, 1995). In contrast, however, in more homogeneous populations such as the Japanese, where the risk of TA-GVHD to immunocompetent individuals was first described, the likelihood that donors will be homozygous for an extended HLA haplotype is much greater.

E. Lymphocyte Dose

The number of immunocompetent lymphocytes required to cause TA-GVHD is not known. Following allogeneic marrow transplantation, evidence for GVHD is not found if the dose of T-cells is less than 10^5 per kilogram recipient weight (Kernan et al., 1986). In transfusing packed red blood cells (PRBCs) or platelets, leukoreduction filters routinely achieve such low levels of T-cells. Interestingly, the intensity of the mixed lymphocyte reaction (a classical cellular assay for alloreactivity) can be decreased by either reducing the number of unirradiated responding cells or increasing the dose of gamma irradiation to responding cells (Dzik et al., 1993). At the levels currently in clinical use, however, leukoreduction per se does not prevent TA-GVHD, as one case report has shown (Akahoshi et al., 1992).

II. CLINICAL COURSE OF TA-GVHD

TA-GVHD begins 1 to 2 weeks after transfusion of cellular components containing immunocompetent lymphocytes into individuals at risk (Parkman et al., 1974; Brubaker, 1983; Juji et al., 1989; Kruskall et al., 1990; Thaler et al., 1989; Anderson and Weinstein, 1990). Fever usually heralds its onset, followed in 2 to 3 days by an erythematous maculopapular rash on the face and trunk (von Fliedner et al., 1982; Wagner et al., 1989; Moncharmont et al., 1988).

A. Skin

A skin biopsy may reveal epidermal cells at the junction of the dermis that display "balloon degeneration," often surrounded by mature lymphocytes of donor origin (if HLA-typed) in a pattern of "satellite necrosis" (Kaye et al., 1984). These skin changes often worsen with disease progression.

B. Gastrointestinal Tract and Liver

As TA-GVHD affects the gastrointestinal tract and liver, it causes massive diarrhea and acute hepatocellular injury with dramatic elevations in hepatic enzymes. Because TA-GVHD often is unrecognized early in its course, biopsies of the gastrointestinal tract, particularly rectal biopsies, may be helpful in making the diagnosis although the pathologic findings of "exploding" crypt cells (due to necrosis) and lymphocytic infiltration of the lamina may also be seen in AIDS enteropathy, CMV colitis, and inflammatory bowel disease (Sale et al., 1979). Liver damage caused by TA-GVHD differs from that caused by GVHD due to allogeneic BMT, since TA-GVHD patients do not usually receive dose-intensive therapy and rarely have the obstructive cholestasis of veno-occlusive disease as part of their picture. Liver damage in TA-GVHD patients is associated with marked elevations in hepatic enzymes due to severe hepatocellular injury.

C. Bone Marrow

TA-GVHD affects the marrow, causing hypoplasia and associated cytopenias. Unlike GVHD associated with allogeneic marrow transplantation, a situation in which the engrafted marrow is of donor origin, the pancytopenia in TA-GVHD occurs as the result of histoincompatibility between donor lymphocytes and host marrow, and it is the marrow destruction of TA-GVHD that causes its fulminance.

D. Diagnosis and Treatment

When fever, rash, diarrhea, and liver injury appear in susceptible patients following transfusion of cellular blood components that have not been irradiated, TA-GVHD should be considered a diagnostic possibility. The mortality of TA-GVHD exceeds 90% and, as one may infer, there is no effective therapy, even if the diagnosis is made early. Steroids, antithymocyte globulin, and even stem cell transplant have been tried, all without success (Brubaker, 1994). The few patients who have survived TA-GVHD have eventually succumbed to chronic GVHD. *The best and only approach to TA-GVHD is prevention.*

III. A POLICY OF PREVENTION

In order for an institutional policy of blood component irradiation to fulfill its preventive function, patients at risk for TA-GVHD must be reliably identified prospectively.

TABLE I   Conditions for Which Blood Component
Irradiation is Recommended

Allogeneic and autologous hematopoietic progenitor
  cell transplantation

Congenital immunodeficiency syndromes

Directed-donor components

HLA-matched apheresis platelets

Intrauterine transfusions

Very low-birth-weight neonates

Solid organ transplantation

Immunosuppressive cancer therapy

Hodgkin's disease

Lymphoma

Leukemia

Aplastic anemia

A. Patients at Risk

Certain categories of patients clearly and uniformly require irradiated cellular components (see Table 1), such as:

1. Those receiving HLA-matched apheresis platelets
2. Allogeneic and autologous bone marrow, blood stem cell, or cord blood transplant patients
3. Those with congenital immunodeficiencies
4. Those with cancer, Hodgkin's disease, chronic lymphocytic leukemia receiving fludarabine, and other hematologic malignancies actively undergoing intensive chemotherapy or radiation
5. Those receiving intrauterine or neonatal transfusions due to hemolytic disease of the newborn, alloimmune thrombocytopenia, or anemia of prematurity
6. Those receiving directed donations from first- or second-degree relatives—since irradiation policies require clarity and fail-safe logic, it is rational to require that all directed-donor blood components, regardless of degree of relationship known, be irradiated, a policy in common practice

B. Patients Currently Not Considered at Risk

It is worth noting that patients with HIV disease, multiple trauma, viral infections, and advanced age are not considered appropriate candidates for irradiated components at this time, although 30% of transfusion services irradiate cellular components for patients with AIDS. However, the evolution of the literature on TA-GVHD has been one in which case reports predominate, and, therefore, the indications for irradiation are matters of vigilance, discussion, and debate, and

tend to increase with time. It is the responsibility of the transfusion service to maintain and review the indications, components, validation procedures, and, where appropriate, the dosimetry involved in component irradiation.

C. Components Requiring Irradiation

Currently the standard prophylactic measure for preventing TA-GVHD is to gamma irradiate cellular blood components prior to their use (Anderson and Weinstein, 1990).

Blood components that must be irradiated for categories of patients at risk include the following:

1. PRBCs, granulocytes, platelets, platelet-rich plasma, freshly separated plasma, and buffy coats used to treat neonatal sepsis (Betzhold and Hong, 1979).
2. Frozen deglycerolized PRBCs can also be a source of immunocompetent lymphocytes capable of causing TA-GVHD, and therefore must be irradiated for indicated categories of patients (Brubaker, 1984; Crowley et al., 1974; Kessinger et al., 1987).
3. HLA-matched apheresis platelets must always be irradiated.

D. Components Currently Not Requiring Irradiation

Coagulation factor concentrates and albumin do not require irradiation, nor has TA-GVHD been described with infusion of fresh frozen plasma (FFP) or cryoprecipitate, although there have been reports of leukocyte and hematopoietic progenitor cell viability in thawed FFP (Bernvil et al., 1994; Wieding et al., 1994). Interestingly, solid organ grafts can contain passenger lymphocytes that may cause a spectrum of syndromes including GVHD; however, solid organ grafts are not routinely irradiated (Burdick et al., 1988).

E. Components That May Be Irradiated

Donor lymphocyte infusions (DLIs) may or may not be irradiated, depending on the indication for their use; those used to treat viral infections should be irradiated, while those used as therapy for disease recurrence post-allograft should not be irradiated (Camitta et al., 1994; Verdonck et al., 1998).

F. Components That Should Not Be Irradiated

Cells transfused with the expectation of engraftment to rescue marrow function should not be irradiated. Allogeneic and autologous blood or marrow hematopoietic stem cells and cord blood stem cells should *never* be irradiated, and the procedures for processing these components should specifically state as much.

IV. TECHNICAL ASPECTS OF BLOOD COMPONENT IRRADIATION

Three sources exist for gamma irradiation: linear accelerators, cobalt-60, and cesium-137. Although some centers continue to use linear accelerators in radiation oncology departments, this method is suboptimal because it is difficult to be confident about the adequacy of radiation dosing for each component. Specially constructed blood irradiators are expensive, but do provide uniformity of source (cesium-137), and certification of dose measurement is provided for containers designed to accommodate a specific volume of blood components.

A. Dose

Although some variation in dosing still exists, with centers using anywhere from 15 Gy (1500 rads) to 50 Gy, the data with respect to the ability of lymphocytes to proliferate in mixed lymphocyte cultures or in response to mitogens indicate that a small fraction of T-lymphocytes retain some proliferative capacity even at 30 Gy. Current standards require a dose of no less than 25 Gy to the central midplane of the irradiation field, and this dose appears to be appropriate for clinical indications. Approaches to quality control and assurance continue to be developed for component irradiation, and unit tags or indicators of appropriate irradiation have become commercially available (Moroff *et al.*, 1997; Hillyer *et al.*, 1993).

B. Dosimetry

The dosimetry of the instrument itself is critical also, and calibration is required to ensure that the length of exposure of the components is adjusted to the age of the source, a proviso the manufacturers may build into the software that accompanies the instrument. Moreover, the technical specialists involved in monitoring blood component irradiators are required to perform regular interval checks ("leak tests") to ensure that the instrument is not emitting radiation into the surrounding area, an event that has yet to be reported from a blood center or transfusion service (Brubaker, 1994).

C. Effects on Nonlymphoid Blood Cells

The effects of gamma irradiation on nonlymphoid blood cells remain a matter of research interest.

1. Platelets have been shown to be quite radioresistant, with few if any changes in morphology, volume, soluble mediator release (a sign of activation), function, or aggregation (Moroff *et al.*, 1986). Important enzymatic pathways and responses to key aggregants were unchanged even after platelets were irradiated to several hundred Gy and stored beyond the normal 5-day period.

2. Granulocytes and their bactericidal capacity, oxidative functions, and chemotactic responsiveness also appear to be unaffected by doses of gamma irradiation far in excess of those required for the prevention of TA-GVHD.

3. PRBCs have been thought to be minimally affected by irradiation, perhaps because they are the least metabolically active formed elements of the blood. However, elevations in potassium due to membrane leakage after irradiation have been identified in PRBC units (Hillyer *et al.*, 1991). Still, initial concerns regarding pediatric transfusion of small aliquots of blood have been proven to be unfounded, but PRBCs used in exchange transfusions and in neonatal large-volume transfusions (> 50 ml) should be washed to eliminate the excess potassium that may cause cardiovascular instability in the neonate. Postirradiation storage of PRBCs is limited to 28 days, leading most centers to irradiate as close to the time of issue as possible (FDA, 1993).

D. Other Considerations

Gamma irradiation of blood components does not affect passenger viruses, such as cytomegalovirus (CMV), and has no role in preventing transfusion-transmitted disease or other transfusion-related side effects such as febrile nonhemolytic transfusion reactions or allosensitization.

References

AABB (1991). Blood Products Advisory Committee addresses full agenda during January meeting. *Blood Bank Week* **8**, 1–2.

Akahoshi, M., Takanashi, M., Masuda, M., *et al.* (1992). A case of transfusion-associated graft-versus-host disease not prevented by white cell-reduction filters. *Transfusion* **32**, 169–174.

Anderson, K. C., and Weinstein, H. J. (1990). Transfusion-associated graft-versus-host disease. *N Engl J Med* **323**, 315–321.

Anderson, K. C., Goodnough, L. T., Sayers, M., *et al.* (1991). Variation in blood component irradiation practice: Implications for prevention of transfusion-associated graft-versus-host disease. *Blood* **77**, 2096–2102.

Bernvil, S. S., Abdulatiff, M., Al-Sedairy, S., *et al.* (1994). Fresh frozen plasma contains viable progenitor cells—should we irradiate? *Vox Sang* **67**, 405.

Betzhold, J., and Hong, R. (1978). Fatal graft-versus-host disease after a small leukocyte transfusion in a patient with lymphoma and varicella. *Pediatrics* **62**, 63–66.

Billingham, R. E. (1967–1968). The biology of graft-versus-host reactions. *Harvey Lect* **62**, 21–78.

Billingham, R. E., Brent, L., and Medawar, P. B. (1954). Quantitative studies on transplantation. Immunity. I and II. *Proc R Soc* **143**, 43–80.

Billingham, R. E., Defendi, V., Silvers, W. K., and Steinmuller, D. (1962). Quantitative studies on the induction of tolerance of skin homografts and on runt disease in neonatal rats. *J Natl Cancer Inst* **28**, 365–435.

Brubaker, D. B. (1983). Human post-transfusion graft-versus-host disease. *Vox Sang* **45**, 401–420.

Brubaker, D. B. (1984). Fatal graft-vs-host disease occurring after transfusion with unirradiated normal donor red cells in an immunodeficient neonate. *Plasma Ther Transfusion Technol* **5**, 117–125.

Brubaker, D. B. (1994). Transfusion-associated graft-versus-host disease. In "The Scientific Basis of Transfusion Medicine" (K. C. Anderson and P. Ness, Eds.), pp. 544–579. Saunders, Philadelphia, PA.

Burdick, J. F., Vogelsang. G. B., Smith, W. J., *et al.* (1988). Severe graft-versus-host disease in a liver-transplant recipient. *N Engl J Med* **318**, 689–691.

Camitta, B., Chusid, M. J., Starshak, R. J., and Gottschall, J. L. (1994). Use of irradiated lymphocytes from immune donors for treatment of disseminated varicella. *J Pediatr* **124**, 593–596.

Crowley, J. P., Skrabut, E. M., and Valeri, C. R. (1974). Immunocompetent lymphocytes in previously frozen washed red cells. *Vox Sang* **26**, 513–517.

Dzik, W. H., and Jones, K. S. (1993). The effects of gamma irradiation versus white cell reduction on the mixed lymphocyte reaction. *Transfusion* **33**, 93–100.

FDA (1993). Recommendations regarding license amendments and procedures for gamma irradiation of blood products. Memorandum, July 22.

Hathaway, W. E., Githens, J. A., Blackburn, J. R., Fulginiti, V., and Kempe, C. H. (1965). Aplastic anemia, histiocytosis, and erythroderma in immunologically deficient children. *N Engl J Med* **273**, 953–955.

Hathaway, W. E., Fulginiti, V. A., Pierce, C. W., *et al.* (1967). Graft-vs-host reaction following a single blood transfusion. *JAMA* **201**, 1015–1020.

Hillyer, C. D., Tiegerman, K. O., and Berkman, E. M. (1991). Evaluation of the red cell storage lesion after irradiation in filtered packed red cell units. *Transfusion* **31**, 497–499.

Hillyer, C. D., Hall, J. M., Lackey, D. A., and Wazer, D. E. (1993). Development of a colorimetric dosimeter for quality control of blood units and irradiators. *Transfusion* **33**, 898–901.

Juji, T., Takahashi, K., Shibata, K., *et al.* (1989). Post-transfusion graft-versus-host disease in immunocompetent patients after cardiac surgery in Japan [letter]. *N Engl J Med* **321**, 56.

Kaye, V. N., Neumann, P. M., and Kersey, J., *et al.* (1984). Identity of immune cells in graft-vs-host disease of the skin: Analysis using monoclonal antibodies by indirect immunofluorescence. *Am J Pathol* **116**, 436–443.

Kernan, N. A., Collins, N. H., Juliano, L., *et al.* (1986). Clonable T lymphocytes in T-cell-depleted bone marrow transplants correlate with development of graft-v-host disease. *Blood* **68**, 770–773.

Kessinger, A., Armitage, J. O., Klassen, L. W., *et al.* (1987). Graft-versus-host disease following transfusion of normal blood products to patients with malignancies. *J Surg Oncol* **36**, 206–209.

Kruskall, M. S., Alper, C. A., and Yunis, E. J. (1990). HLA-homozygous donors and transfusion-associated graft-versus-host disease [letter]. *N Engl J Med* **322**, 1005–1006.

Leitman, S. F., and Holland, P. V. (1985). Irradiation of blood products: Indications and guidelines. *Transfusion* **25**, 293–300.

Moncharmont, P., Souillet, G., Rigal, D., *et al.* (1988). Post-transfusion graft versus host disease: Report of three cases. *Ann Pediatr* **35**, 247–251.

Moroff, G., George, V. M., Siegl, A. M., and Luban, N. L. C. (1986). The influence of irradiation on stored platelets. *Transfusion* **26**, 453–456.

Moroff, G., Leitman, S. F., and Luban, N. L. C. (1997). Principles of blood irradiation, dose validation, and quality control. *Transfusion* **37**, 1084–1092.

Ohto, H., and Anderson, K. C. (1996). Survey of transfusion-associated graft-versus-host disease in immunocompetent recipients. *Transfusion Med Rev* **10**, 31–43.

Parkman, R., Mosier, D., Umansky, I., *et al.* (1974). Graft versus host disease after intrauterine and exchange transfusion for hemolytic disease of the newborn. *N Engl J Med* **290**, 359–363.

Sale, G. E., Shulman, H. M., McDonald, G. B., and Thomas, E. D. (1979). Gastrointestinal graft-versus-host disease in man: A clinicopathological study of the rectal biopsy. *Am J Surg Pathol* **3**, 291–299.

Thaler, M., Shamiss, A., Orgad, S., *et al.* (1989). The role of blood from HLA-homozygous donors in fatal transfusion-associated graft-versus-host disease after surgery. *N Engl J Med* **321**, 25–28.

Verdonck, L. F., Petersen, E. J., Lokhorst, H. M., *et al.* (1998). Donor leukocyte infusions for recurrent hematologic malignancies after allogeneic bone marrow transplantation: Impact of infused and residual donor T-cells. *Bone Marrow Transplant* **22**, 1057–1063.

Von Fliedner, V., Higby, D. J., and Kim, U. (1982). Graft-versus-host reaction following blood product transfusion. *Am J Med* **72**, 951–956.

Wagner, J. E., Vogelsang, G. B., and Beschorner, W. E. (1989). Pathogenesis and pathology of graft-vs-host disease. *Am J Pediatr Hematol Oncol* **11**, 196–212.

Wagner, F. F., and Flegel, W. A. (1995). Transfusion-associated graft-versus-host disease: Risk due to homozygous HLA haplotypes. *Transfusion* **35**, 284–289.

Wiedling, J. U., Vehmeyer, K., Dittman, J., *et al.* (1994). Contamination of fresh-frozen plasma with viable white cells and proliferable stem cells. *Transfusion* **34**, 185–186.

Wuest, D. L. (1996). Transfusion and stem cell support in cancer treatment. *Hemetol Oncol Clin N Am* **10**, 397–429.

17

Washed and Volume-Reduced Components

RAYMOND L. COMENZO
DAVID L. WUEST

Memorial Sloan-Kettering Cancer Center
New York, New York

Reactions of patients to the transfusion of packed red blood cells (PRBCs) and other blood components are commonly categorized as febrile, allergic, volume-related, or not related to transfusion. Washing and/or volume-reducing cellular blood components are manipulations designed to remove presumed or potential offending plasma proteins or to make the infused volume compatible with the patient's size or cardiovascular status.

I. WASHING FOR TRANSFUSION REACTIONS

Cellular components are often washed after severe, recurrent allergic reactions have been observed (see Table 1).

A. Washed Cellular Blood Components Cause Fewer Reactions in General

In one large study performed in the early 1980s, transfusion reactions were carefully charted in over 10,000 patients randomized to receive saline-washed or packed red blood cells. Those receiving washed red blood cells had significantly fewer overall transfusion reactions, both febrile nonhemolytic transfusion reactions (FNHTRs) (reduced from 19 to 4) and allergic ones (reduced from 10 to 4), $p < 0.03$ (Goldfinger and Lowe, 1981). Washing, however, decreases the PRBC or platelet content of the product, and therefore is only used as indicated in Table 1.

B. Saline Washing

The saline washing of blood components is labor-intensive and expensive, and incurs cellular losses of 10 to 20%, a factor particularly relevant to dosing platelets (Comenzo *et al.*, 1992). At this time the routine prospective use of washed components is neither reasonable nor practical.

C. Allergic Reactions

Allergic reactions may range from limited hives and itching to whole body rash, wheezing, and near-anaphylaxis, and often represent reactions to plasma proteins in the blood components infused with antibodies in the recipient. By removing plasma with washing, the incidence of such reactions can be markedly reduced. Subsequent to severe allergic reactions patients should receive saline-washed red blood cells or ACD–saline-washed platelets. Washing is an effective method for removing plasma proteins, microaggregates, and cytokines that may accumulate during storage (Heddle *et al.*, 1994).

D. Febrile Nonhemolytic Transfusion Reactions

FNHTRs are often secondary to infusion of either incompatible leukocytes or pyrogenic cytokines (Heddle *et al.*, 1994). Leukoreduction filters remove white blood cells (WBCs) that often cause FNHTRs. The removal via washing of cytokines that accumulate during storage has also been been demonstrated to reduce

TABLE 1   Indications for Washed Cellular
Blood Components

Severe, recurrent allergic reactions
Recurrent FNHTRs despite leukoreduction
IgA deficiency when IgA-deficient products are unavailable

the frequency of FNHTRs, but due to the widespread use of leukoreduction, washing is not utilized for this indication.

II. PROSPECTIVE USE OF WASHED COMPONENTS

Washing may be used prospectively in the case of IgA-deficient patients, if IgA-deficient products are not available (see Table 1).

A. IgA-Deficient Patients with Anti-IgA

Patients who are IgA-deficient and have anti-IgA (due to prior exposure to IgA as a result of transfusion or pregnancy) can experience life-threatening anaphylactic transfusion reactions. It is recommended that IgA-deficient patients receive IgA-deficient blood components, such as IgA-deficient plasma. However, IgA-deficient donors are rare. Thus, frozen, deglycerolized PRBCs are likely to be safe for IgA-deficient patients, since they contain so little native plasma protein. When neither IgA-deficient donors nor frozen PRBCs are available, liquid PRBCs may be used if washed twice with a liter of saline per PRBC unit (T'oth et al., 1998).

If platelets are needed emergently for IgA-deficient patients and IgA-deficient components or donors are unavailable, platelets may also be transfused if washed several times prior to transfusion. However, it is important to note that significant platelet losses result from platelet washing and patients may still react to washed platelets (because platelets sequester immunoglobulins).

B. Complement-Sensitive Disease Processes

Patients with complement-sensitive disease processes such as paroxysmal cold hemoglobinuria, paroxysmal nocturnal hemoglobinuria, or cold agglutinin disease have been historically transfused with washed cellular components in order to avoid infusing complement. The scientific validity of this practice remains controversial.

III. PLATELET WASHING

Platelets as well as PRBCs may be washed on automated cell washers without functional compromise, a fact that has clear-cut implications with respect to the management of allergic transfusion reactions. However, platelet viability and function appear to be affected by the composition of the wash solution, and the number of platelets in the product is substantially reduced following washing.

IV. OTHER APPLICATIONS OF WASHED COMPONENTS

A. Cryopreserved Hematopoietic Stem Cells

The cryopreservation of hematopoietic stem cells requires the use of the intracellular cryoprotectant dimethylsulfoxide (DMSO) at concentrations of 5 to 10%. In addition, when thawed, stem cell components often contain red blood cell hemolysate and other microaggregate debris. In pediatric patients, patients with pulmonary compromise, and patients who have reacted to stem cell infusion, the use of a method for washing stem cell components is reasonable, although the standard of care remains immediate post-thaw infusion of unmanipulated cells.

B. Shed Blood from Surgical Wounds

At some centers, postoperatively shed blood from surgical wounds may be employed for transfusion in an attempt to minimize the use of allogeneic blood. See Chapter 4 for more information regarding the use of cell saver instruments.

C. Bypass Surgery in Infants

Different substances may be removed by washing. For example, the use of washed red blood cells as a cardiopulmonary bypass priming solution in infants significantly attenuates the increase in blood glucose concentration observed during cardiopulmonary bypass, a difference of potential clinical relevance (Hosking et al., 1990).

D. Renal Failure

Washing can also reduce the amount of potassium infused to patients with renal failure who have small blood volumes and/or require massive transfusion.

V. VOLUME REDUCTION

The modification of component volume is common practice in pediatric transfusion because components are dosed not in units but in milliliters per kilogram and are often administered by syringe or "pedi unit." In adults, units of PRBCs and other components can be split or centrifuged to allow excess plasma to be extracted manually in order to reduce the volume infused to patients at risk of volume overload due to severe congestive heart failure or cardiomyopathy. Volume reduction to deplete plasma may also be employed in the transfusion of single-donor platelets that are ABO incompatible and contain high-titer isoagglutinins. Such manipulations are not routine and therefore may require the approval of the transfusion specialist. In addition, these manipulations take technical time and are usually not appropriate in emergencies.

References

Blaylock, R. C., Carlson, K. S., Morgan, J. M., *et al.* (1994). In vitro analysis of shed blood from patients undergoing total knee replacement. *Am J Clin Pathol* **101**, 365–369.

Comenzo, R. L., Malachowski, M. E., and Berkman, E. M. (1992). Determining the dose of platelets for transfusion. In "Clinical Decisions in Platelet Therapy" (S. R. Kurtz and D. B. Brubaker, Eds.), pp. 1–17. AABB, Bethesda, MD.

Goldfinger, D., and Lowe, C. (1981). Prevention of adverse reactions to blood transfusion by the administration of saline-washed red blood cells. *Transfusion* **21**, 277–280.

Heddle, N. M., Klama, L., Singer, J., *et al.* (1994). The role of the plasma from platelet concentrates in transfusion reactions. *N Engl J Med* **331**, 625–628.

Hosking, M. P., Beynen, F. M., Raimundo, H. S., *et al.* (1990). A comparison of washed red blood cells versus packed red blood cells (AS-1) for cardiopulmonary bypass prime and their effects on blood glucose concentration in children. *Anesthesiology* **72**, 987–990.

Kalmin, N. D., and Brown, D. J. (1982). Platelet washing with a blood cell processor. *Transfusion* **22**, 125–127.

Pineda, A. A., Zylstra, V. W., Clare, D. E., *et al.* (1989). Viability and function of washed platelets. *Transfusion* **29**, 524–527.

T'oth, C. B., Kramer, J., Pinter, J., Th'ek, M., and Szab'o, J. E. (1998). IgA content of washed red blood cell concentrates. *Vox Sang* **74**, 13–14.

SECTION V

HEMATOPOIETIC STEM CELLS AND RELATED CELLULAR PRODUCTS

18

Bone-Marrow-Derived Hematopoietic Progenitor Cells

STEVEN R. SLOAN

Children's Hospital Boston
Harvard Medical School
Boston, Massachusetts

Bone marrow (BM) is one source of hematopoietic stem cells. Bone-marrow-derived stem cell products contain stem cells and hematopoietic progenitor cells that are committed to particular hematopoietic lineages. Hence, the term "hematopoietic progenitor cells" (HPCs) is more accurate than "stem cells" and will be used in this chapter.

There are two categories of BM transplants (BMTs). Autologous BMTs are donated by the person who will receive the transplant (the steps in an autologous BM-HPC transplant are depicted in Figure 1). Allogeneic BMTs are donated by a person other than the recipient. Currently, almost all autologous HPC transplants are collected from peripheral blood (see Chapter 11), while allogeneic HPC transplants may be collected from peripheral blood (see Chapter 11), umbilical cord blood (see Chapter 12), or bone marrow. Except where stated, information in this chapter refers to allogeneic BMT.

BMTs are still relatively new when compared with transfusions of other commonly used blood products and derivatives. Procedures and practices for collection and preparation vary among institutions. Various organizations, including the Food and Drug Administration (FDA), Foundation for the Accreditation of Hematopoietic Cell Therapy (FAHCT), and the American Association of Blood Banks (AABB), have either recently adopted standards or are planning to do so in the near future. FAHCT was started with input from several different organizations having interest and expertise in HPC transplants. Most of these standards focus on the quality of the environment, manpower, and procedures for HPC collection and processing, and do not specify product content or usage.

I. PRODUCT DESCRIPTION

A. Cellular Constituents

In addition to HPC, BM-derived cell collections contain all of the cellular components that are normally present in BM. Mature and immature RBCs, lymphocytes, myeloid cells, and platelets are present in BM-derived cell collections. Some of these cells, especially T lymphocytes, can contribute to graft-versus-host disease (GVHD) and/or a graft-versus-tumor effect. In addition, other cells may contaminate the BM collection. Of particular concern is the fact that *malignant cells may contaminate the autologous BM harvest of an oncology patient*. Because some of these other types of cells are not therapeutic and may even harm the patient, BM products may undergo additional processing to remove unwanted cells or to preferentially select for HPCs (see Chapter 11).

B. Characterization of HPC Cell Content

Traditionally, the quantity of useful cells in a BM-HPC product is determined by measuring the number of mononuclear cells. The number of mononuclear cells may roughly correlate with the number of HPCs and stem cells. Stem cells compose about 1/100,000 BM mononuclear cells, but this proportion can vary

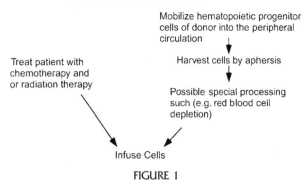

Allogeneic PB-HPC Transplant

FIGURE 1

significantly among donors. In general, BM-HPCs collected for autologous transplants should contain approximately $1-3 \times 10^8$ mononuclear cells/kg recipient body weight, and BM-HPCs collected for allogeneic transplants should contain approximately $2-4 \times 10^8$ mononuclear cells/kg recipient body weight, with some variation allowed for different clinical situations. For example, some institution-specific guidelines might indicate that BM-HPCs for HLA-identical sibling allogeneic transplants for aplastic anemia should contain at least 3×10^8 cells/kg recipient body weight; BM-HPCs for unrelated donors should also contain 3×10^8 cells/kg recipient body weight; and BM-HPCs for HLA-identical allotransplants for leukemia, hemoglobinophathies, or inborn errors of metabolism should contain at least 2×10^8 cells/kg recipient body weight (Atkinson, 1998). The National Marrow Donor Program (NMDP) in the United States requires 2.0×10^8 mononuclear cells/kg recipient body weight for allogeneic BM-HPC transplants.

Most HPCs express the CD34 protein on their surfaces. CD34$^+$ cells compose $1-5\%$ of peripheral blood mononuclear cells following mobilization, but many cells in addition to HPCs express the CD34$^+$ antigen (Siena $et\ al.$, 1993).

Most institutions use additional means to enumerate the HPCs in BM products, such as granulocyte-macrophage colony-forming unit (CFU-GM) assays or CD34$^+$ cell counts by flow cytometry. Approximately $0.1-1 \times 10^4$ CFU-GM/kg patient body weight are needed to ensure timely engraftment. However, CFU-GM assays usually provide retrospective data only, since they take 2 weeks to complete. Hence, most institutions rely primarily on CD34$^+$ cell counts to quantify HPCs, with $2-5 \times 10^6$ CD34$^+$ cells/kg patient body weight usually being required for timely engraftment. These values must be determined by each institution and may depend on the type of disease being treated and whether

the cells are intended for an autologous transplant, a matched, related allogeneic transplant, or an unrelated allogeneic transplant. These assays are also performed for peripheral-blood-derived HPCs (PB-HPCs), and the target number of CD34$^+$ cells is similar. These assays are described in Chapter 11.

Unprocessed BM-HPCs contain everything that was collected from the donors marrow. This product will be in a volume that ranges from 500 to 2000 ml.

C. Anticoagulants and Additives

BM-HPC collections contain plasma, anticoagulant, and additional buffered solution. The anticoagulant may be heparin, acid-citrate-dextrose (ACD), citrate-phosphate-dextrose (CPD), or CPD-adenine (CPDA-1). ACD should be used if the marrow will be kept in liquid form for an extended period of time. BM-HPCs collected at some institutions contain tissue culture media that contain electrolytes and buffers and may contain vitamins and/or minerals. Another practice is the addition of a buffered electrolyte solution instead of tissue culture media. Tissue culture medium provides conditions for long-term cell growth but is not needed for short-term storage of BM-HPCs. Because tissue culture medium is not currently approved for human use in the United States, FDA-approved solutions such as Normosol or other infusion-grade solutions are more commonly used.

Human plasma contaminates collections of BM-HPCs. While this plasma is not usually specifically removed from the final BM-HPC product, the plasma may be removed during other processing, such as CD34$^+$ selection. Additional processing may result in the addition of other substances. For example, some products depleted of RBCs may contain hetastarch or albumin.

D. Labeling

Labeling requirements are determined by accrediting organizations such as the AABB and the FAHCT and regulatory agencies such as the FDA. The label should contain the name and address of the processing facility unless it is an unrelated allogeneic transplant, the unit's unique numeric or alphanumeric identifier, the date and time (and time zone if applicable) of collection, the volume of the product, the ABO group and Rh type of the donor, the type and volume of additives such as anticoagulants and cryprotectants, a biohazard label, the expiration date, manipulation method(s), and the recommended storage temperature range. In addition, the label must contain the phrases "Human Bone Marrow," "Properly Identify Intended

Recipient and Component or Unit," and "Warning. This Product May Transmit Infectious Agents." An HPC product intended for an autologous transplant must contain the phrase, "For Autologous Use Only," on the label. An HPC product intended for an allogeneic transplant must have the phrase, "For Use By Intended Recipient Only," printed on the label. Units from unrelated donors must be labeled with the donor registry name and unique donor registry number. The most recent results of infectious disease tests must be on the label or must be communicated to the transplant facility by some other documented means. A biohazard label must be attached to the product if infectious disease tests indicate that the donor is likely to have been infected with HIV, hepatitis B virus (HBV), or hepatitis C virus (HCV).

II. COLLECTION AND STORAGE

A. Donor Evaluation

1. HLA Compatibility

The human leukocyte antigen (HLA) types of potential allogeneic bone marrow donors are determined so that the patient will receive an optimal match (for review see Little *et al.*, 1998). The major HLA system encodes for class I (A, B, and C) and class II (DR, DP, and DQ) cell surface molecules that are critical to the immune response. There are multiple alleles for each HLA locus, resulting in significant diversity in the population. Patients transplanted with donor HPCs that are HLA incompatible with the patient are more likely to develop GVHD and more likely to have delayed immune reconstitution. However, HLA-incompatible transplants are more likely to produce beneficial graft-versus-leukemia (GVL) effects than are HLA-compatible transplants.

Because of the risks associated with GVHD and delayed immunologic recovery, *transplant donors are usually selected to maximize HLA compatibility*. This is accomplished by attempting to identify people who share identical HLA-A, HLA-B, and HLA-DR genes with the patient. Some institutions may also attempt to match HLA-C genes. Matching HLA-DP and HLA-DQ genes is not important. Up to one mismatched allele may be acceptable. The specific mismatch is important. Two alleles may present only minor mismatches if they are in the same cross-reactive group (CREG). Other mismatched alleles may cause severe GVHD. Minor histocompatibility antigens (mHag) can also contribute to GVHD. These antigens are not normally measured or used for identifying donors. Transplants from rela-

tives pose a lower risk from mHag incompatibility when compared to transplants from unrelated donors, but mHag may cause GVHD in transplant recipients, including transplants from HLA-identical siblings. Traditionally, sera that react against specific HLA antigens have been used to type patients and potential donors. For a variety of reasons, serologic techniques are error prone and fail to discriminate between some related alleles. DNA-based techniques that are based on the polymerase chain reaction (PCR) have been advancing in use in recent years. These techniques provide greater precision and greater discrimination between related alleles.

Family members are often the most likely histocompatible donors. There is a 97–99% chance that an identical twin sibling is HLA identical (this percentage is not 100% because of the relatively high frequency of chromosomal recombination near the HLA genes). Theoretically, there is a 25% chance that any one nonidentical sibling of a patient is HLA identical with the patient. Certain HLA genes are inherited preferentially, however, and thus there is actually a 35% chance that a child with leukemia, for instance, and any one sibling are HLA identical (Chan *et al.*, 1982). Other family members may also be histocompatible.

If no compatible family members are identified, then national donor registries can be accessed. The National Marrow Donor Program (NMDP) in the United States is the largest registry. They can request registries in other nations to search for donors if none are identified in the NMDP. The registries contain HLA typing information that is usually based on serologic typing and often only includes HLA-A, HLA-B, and sometimes HLA-DR typing information. Potential donors then need to be typed using PCR-based methods for all alleles that the transplant center normally considers important for matching.

2. Medical History

Once a potential donor is identified, his or her health is assessed by history and physical and laboratory tests. This evaluation has two purposes. One purpose is to determine whether a donation will be safe for the donor. The second purpose is to determine whether the cells can be safely administered to the patient recipient. Up to 0.4% of bone marrow donors develop life-threatening complications; this may be reduced to 0.1–0.2% if older donors, donors with cardiovascular disease, and obese donors are excluded (Buckner *et al.*, 1994). FAHCT also requires a pregnancy test for all female donors of childbearing potential. Donor risk behavior and health are also evaluated to determine the likelihood of infectious disease transmission to the patient. Standards do not specify definitive exclusion criteria. However, infectious disease tests routinely performed for whole blood donation should be performed on BM donors no more than 30 days prior to donation (Chapter 2). In addition to these tests, FAHCT requires cytomegalovirus (CMV) testing of donors. Individuals excluded from whole blood donation may be acceptable BM donors. BMT programs determine the acceptability of donors on a case-by-case basis if there is risk of infectious disease transmission.

Potential autologous BMT candidates are screened using criteria similar to those described in Chapter 19 for peripheral-blood-derived HPCs. BM fibrosis can make BM donation difficult or impossible. In addition, some protocols attempt to evaluate the BM for presence of residual disease, though no technique is sensitive enough to assure that the marrow is totally devoid of malignant cells.

B. Collection Procedure

Bone marrow is collected in the operating room from anesthetized donors. The donor usually receives general anesthesia but may receive spinal or epidural anesthesia. Syringes are used to aspirate bone marrow from the posterior iliac crests and pelvic rim. In some patients, BM is also aspirated from the anterior iliac crests and sternum (Thomas and Storb, 1970). The harvested BM is placed into a sterile container with an anticoagulant (usually heparin) and tissue culture media or buffered normal saline solution. After collection, the BM is filtered through stainless steel mesh screens to remove clots, bone fragments, fat, and fibrin, and is transferred to a blood bag or other sterile container. The most serious risks to the BM donor are those related to anesthesia. Most donors experience fatigue, pain at the collection site, and lower back pain that lasts for a few weeks (Stroncek et al., 1993). Acetaminophen with codeine, or a similar narcotic medication, for a few days following the harvest usually alleviates the pain.

C. Laboratory Evaluation of Bone Marrow Harvest

Various quality control measurements may be made from aliquots of the BM product. Many institutions measure mononuclear cell concentrations and CD34$^+$ cell concentrations. Culture of BM product for bacterial and fungal contamination is recommended. In addition, some institutions perform colony-forming unit (CFU) assays. In most cases, the results of cultures and CFU assays will not be available until several days after the transplant, but these assays can provide data for the quality of the laboratory's processing and may help in the treatment and diagnosis of a patient whose engraftment is delayed or who is septic. Though none of these assays is specifically required by any organization standards, accrediting organizations do require that there be an ongoing monitoring of the quality of the work performed in the laboratory.

D. Storage and Transport

Allogeneic bone marrow is usually stored for 2 to 36 hours at room temperature (not to exceed 37°C or drop below 2°C). While storage time should be minimized, HPCs that have been stored at 4–25°C for 24–36 hours have been successfully transplanted (Lasky et al., 1986). If the donor and patient are at the same institution, then the BM is usually harvested and transplanted on the same day. If the donor and patient are in different locations, then the BM is usually shipped from the donor's location to the recipient's location within hours after the harvest. A courier from the patient's institution will often personally transport the cells to ensure that the cells are kept at an acceptable temperature.

E. Cryopreservation

Allogeneic BM-HPCs are usually not cryopreserved for fear of damaging the cells and reducing the chances of engraftment. However, allogeneic BM-HPCs have been successfully transplanted following cryopreservation. If a BM-HPC collection is cryopreserved, it is stored using the same techniques used to store cryopreserved PB-HPCs (see Chapter 19). Autologous BM-HPC collections are usually cryopreserved using these same techniques.

III. SPECIAL PREPARATION AND PROCESSING

A. T-Cell Depletion

Allogeneic BM-HPCs may be depleted of T-cells. This is done to reduce the chances of severe GVHD. However, some studies suggest that beneficial graft-versus-cancer effects may also be reduced, and some T-cell depletion techniques may decrease the chance that the BM will engraft.

There are several techniques that may be used to deplete allogeneic BM-HPC harvests of T-cells. These techniques are identical to those used to deplete T-cells from PB-HPCs (see Chapter 19). Certain methods of T-cell depletion that utilize antibodies against T and NK cells are not usually performed on BM-HPCs, however, because of increased risk of engraftment failure and increased risk of cancer recurrence (Marmont *et al.*, 1991). In contrast, physical techniques such as counterflow elutriation, lectin separation, and E-rosette depletion are sometimes used to remove T-cells from bone marrow. Of these methods, counterflow elutriation is most commonly used.

Counterflow centrifugation elutriation separates cells based on cell size and density. Elutriation is used to separate lymphocytes from most of the other cells. $CD34^+$ hematopoietic progenitor cells vary in size and density. While the smaller, more primitive $CD34^+$ cells segregate with the lymphocytes, at least half of the $CD34^+$ cells can be separated from the lymphocytes (Gee and Lee, 1998). The lymphocyte-depleted cell population is then transplanted. Because the transplanted product may lack the most immature small $CD34^+$ cells, long-term engraftment may be at risk. For this reason, some institutions may additionally select $CD34^+$ cells from the lymphocyte fraction and add these back to the lymphocyte-depleted cells (Gee and Lee, 1998).

T-cell-depleted products may contain additional solutions, antibodies, or other chemicals depending on the T-cell depletion method. BM-HPCs that have been depleted of T-cells by counterflow elutriation contain fewer lymphocytes of all types. BM-HPCs that have been depleted of T-cells by $CD34^+$ selection contain fewer RBCs, platelets, and WBCs of all types, although a large proportion of the $CD34^+$ cells are retained. $CD34^+$-selected products are enriched for HPCs and are usually contained in a few milliliters.

B. RBC and Plasma Depletion

Allogeneic BM-HPCs may contain RBCs or plasma that is incompatible with the patient. This is an indication for depletion of RBCs or plasma as described in Section IV, below.

Plasma and RBCs may be removed from BM-HPCs. BM-HPC product that has been depleted of RBCs will contain fewer RBCs than unmanipulated BM-HPCs. The laboratory performing the manipulation determines a satisfactory level of depletion. Usually, this is about 90% depletion of the RBCs. In addition, RBC-depleted products may contain hetastarch or albumin, depending on the procedure that was used. Plasma may also be removed from BM-HPC products.

C. Purging and Positive Selection

Autologous BM-HPC transplants can be processed, purged, and/or selected using essentially the same techniques as described in Chapter 11.

IV. ABO AND Rh COMPATIBILITY

Selection of a BM-HPC donor and product for transplantation is made primarily based on an HLA match. RBC incompatibility between donor and recipient is thus considered secondarily. ABO and Rh types of the donor and recipient should be determined in advance of allogeneic transplants to anticipate management of incompatibilities.

A. Approach to Major RBC Incompatibility

A major blood group incompatibility occurs when naturally occurring antibodies (anti-A, anti-B, anti-A,B) in the patient's plasma react with donor RBCs. In addition, tests of the patient's serum with donor RBCs may reveal the presence of other unexpected clinically significant antibodies against antigens on donor RBCs (Kell, Duffy, Kidd, etc.). If clinically significant antibodies react against RBCs by routine pretransfusion testing (see Chapters 19 and 20), then the RBCs should be removed from the BM. Acceptable methods for RBC depletion include hydroxyethyl starch (HES) sedimentation or automated cell separators. In addition, some cell selection methods that are designed to purify the stem cell population, such as methods that select for $CD34^+$ cells, also remove RBCs. HES sedimentation, automated cell separators, and repeated sedimentation and dilution with compatible RBCs remove most, but not all, of the incompatible RBCs. Hence, these procedures mitigate but do not eliminate the risk of hemolytic transfusion reactions associated with BM-HPC transplants. In addition to removing RBCs from BM, some transplant programs remove isoagglutinins from the patient by plasmapharesis (Atkinson, 1998). This should rarely, if ever, be necessary, however, because RBC depletion techniques can remove enough RBCs to eliminate the risk of severe hemolytic transfusion reactions in this patient population.

1. Hydroxyethyl Starch Sedimentation

HES sedimentation has been commonly used to remove incompatible RBCs from BM harvests (Dinsmore *et al.*, 1983). HES added to BM causes the RBCs to form rouleaux and accelerates their sedimentation. The sedimented RBCs are drained from the bottom of the bag, leaving the rest of the BM harvest, including the HPCs, in the original bag. Laboratories often find it difficult to remove enough RBCs while retaining most of the WBCs cells using HES sedimentation (Adkins *et al.*, 1998). This method results in infusion of HES to the patient, which does not typically have significant side effects.

2. Cell Separator

Many transplant laboratories now use a cell separator to remove RBCs from BM harvests. Cell separators use the same principles as apheresis machines to separate the blood's components and selectively remove specified components. During this processing, albumin, saline, and ACD are often added to the product.

3. Sedimentation and Dilution

Repeated sedimentation and dilution with compatible RBCs is a lengthy process because of the time required for RBCs to sediment in the absence of HES. This technique is not used by many processing laboratories but some laboratories have diluted products with compatible RBCs in conjunction with HES sedimentation or automated cell separators.

B. Approach to Minor RBC Incompatibility (ABO and Other)

The ABO group of the BMT donor and recipient may have a minor ABO incompatibility, in which plasma from the donor contains antibodies against ABO antigens on the recipient's RBCs. Additionally, the donor's plasma may contain unexpected antibodies against other RBC antigens that are present on donor RBCs (e.g., Kell, Duffy, or Kid). These antibodies would be discovered when performing antibody screens on the donor's blood.

Although a minor ABO incompatibility does not usually cause immediate, significant hemolysis, *substantial delayed hemolysis can occur in group A or B patients receiving ABO-incompatible BM-HPCs* (Petz, 1998). This hemolysis, which is due to stimulation and proliferation of donor-derived B lymphocytes, usually develops 5 to 16 days following the transplant. Some drugs, such as methotrexate, that are used as a prophylaxis for GVHD inhibit B-lymphocytes and reduce the chances of severe hemolysis. However, other anti-GVHD drugs such as cyclosporin and FK506 do not inhibit B-lymphocytes and do not prevent severe hemolysis. Although massive hemolysis occurs in no more than 10–20% of susceptible patients, it can be abrupt and severe, resulting in substantial declines in hemoglobin and hematocrit and subsequent renal failure. The direct antiglobulin test (DAT) will usually be positive prior to the onset of significant hemolysis and can be used to identify patients at risk for massive hemolysis. However, not all patients with a positive DAT will develop massive hemolysis. Treatment consists of empiric use of corticosteroids, hydration, and transfusion with blood products that are compatible with the group of the donor and the original blood group of the recipient.

If donor plasma contains clinically significant antibodies that react with the patient's RBCs, then one should consider removing plasma from the BM harvest to reduce the chance of clinically significant hemolysis of the patient's RBCs (Lasky *et al.*, 1983). Plasma can be removed by centrifuging the BM collection and removing the plasma layer with a plasma extractor. Alternatively, a cell separator can be used to remove plasma.

C. RBC Crossmatches

Many institutions crossmatch donor RBCs with patient plasma and donor plasma with patient RBCs, even though current accreditation standards do not explicitly require this testing in all cases. This testing can serve as a check to verify whether there is a need for RBC or plasma depletion of the BM product. If the cells are incompatible, then the product should be labeled as being incompatible. If a minor incompatibility exists, this product may be safely administered. If a major incompatibility exists, this product may be safely administered following RBC depletion. The person performing the transplant should verify that the product has been depleted of RBCs.

D. Rh Compatibility

Rh blood group mismatches do not impair engraftment, reduce patient survival, or increase the risk of GVHD. Some Rh(D)-positive donors who have received Rh(D)-negative BM have developed anti-D antibodies. While these patients' RBCs may undergo increased hemolysis, Rh(D)-negative RBCs should eventually repopulate their circulatory systems as the donor BM engrafts.

V. ORDERING AND ADMINISTRATION

When considering an allogeneic BM-HPC transplant, a donor with a compatible HLA type must be identified. Family members, especially siblings, are usually the best potential donors, and their HLA types are the first to be determined. If no donor is identified in the patient's family, searches for potential donors can be made through national and international BMT registries. High-resolution HLA types of potential donors are determined to identify histocompatible donors. If a compatible donor is identified, his or her health status is assessed. Autologous BM-HPCs and autologous PB-HPCs are ordered and administered in the same way (see Chapter 11).

If an allogeneic BMT is planned, efforts must be coordinated among the processing laboratory, the physicians harvesting the bone marrow, the patient, and the donor. This is made even more complex if the donation is made at a location that is different from the patient's location. In those cases, the bone marrow is harvested at a hospital that serves as the collection facility, and a courier, often a member of the transplant team from the patient's hospital, transports the BM product back to the patient's hospital.

Prior to administration of the BM-HPCs, proper identification of the patient and product is critical to ensure that the patient is receiving the correct BM-HPCs. BM-HPCs are usually infused at least 24 hours after completion of chemotherapy to prevent the cyto-toxic effects of chemotherapy from damaging the in-fused cells. BM-HPCs can be immediately infused fol-lowing radiotherapy, however. The patient should be well hydrated prior to infusion and is often premedi-cated with acetaminophen, an antihistamine, and a corticosteroid. Oxygen and antianaphylaxis treatment such as epinephrine should be available. Allogeneic BM-HPCs that have been cryopreserved is adminis-tered using procedures that are similar to those used for cryopreserved PB-HPCs. BM-HPCs are adminis-tered intravenously, usually through a central venous catheter, without any filters. HPC preparations must not be irradiated. Cells may be administered rapidly by intravenous push or, as opposed to BM-HPCs that have been cryopreserved with DMSO, cells that have not been cryopreserved may be infused over several hours. It may be advisable to initiate the infusion slowly to observe for any adverse reactions and then subse-quently accelerate the infusion rate. The patient should be closely monitored and vital signs should be taken periodically during the infusion (as for any blood prod-uct infusion) due to the risks of allergic, anaphy-lactic, hemolytic, or febrile nonhemolytic transfusion reactions.

VI. INDICATIONS

HPCs are usually administered to patients whose own hematopoietic (BM) system is defective. Although a specific disease may directly cause the patient's BM defect, toxic cancer treatment is a more frequent cause of the BM damage. HPC transplants have also been used for the treatment of immunodeficiencies, autoim-mune diseases, and genetic disorders.

A. Autologous

Autologous BM-HPC transplants are indicated for all clinical cases in which autologous stem cell trans-plants are indicated (see Chapter 11). Autologous BM-HPC transplants are usually performed to reconstitute the hematopoietic system of a patient who has received myeloablative chemotherapy and/or radiation therapy to treat a malignancy. These malignancies are most often solid tumors that are responsive to chemotherapy and/or radiation therapy. Autologous transplants are performed for breast cancer, acute lymphocytic leuke-mia (ALL), non-Hodgkin's lymphoma, and Hodgkin's disease. Autologous BM-HPC transplants have also been used for acute myelocytic leukemia (AML), chronic myelocytic leukemia (CML), chronic lympho-cytic leukemia (CLL), multiple myeloma, myelodysplas-tic syndrome, ovarian cancer, small cell lung tumors, malignant melanoma, colon cancer, gastric cancer, germ-cell tumors, and childhood solid tumors such as neuroblastoma, soft tissue sarcoma, Ewing's sarcoma, pediatric glioma, Wilms' tumor, and pediatric germ-cell tumors.

B. Allogeneic

Allogeneic BM-HPC transplants are indicated for reconstitution of the hematopoietic system of patients whose BM has been destroyed by cancer therapy. In cancer cases, allogeneic BM-HPC transplantation is only considered for patients whose disease has re-sponded to chemotherapy and/or radiation therapy. Common indications for allogeneic BM-HPC trans-plants include leukemia, non-Hodgkin's lymphoma, AML, CML, ALL, multiple myeloma, and myelodys-plastic syndrome. Allogeneic BM-HPC transplants have also been used for a variety of other malignancies such as Hodgkin's disease, CLL, juvenille CML, non-Hodg-kin's lymphoma, neuroblastoma, rhabdomyosarcoma, and (rarely) breast cancer. In almost all of these dis-eases, patients can receive either autologous or allo-geneic stem cell transplants, though outcomes differ. However, almost all patients with breast cancer or ovarian cancer who receive a HPC transplant receive an autologous HPC transplant rather than an allo-geneic one.

Allogeneic BM-HPC transplants can also be per-formed for other disorders involving BM cells, such as aplastic anemia, Fanconi's anemia, thalassemias, sickle cell disease, and other metabolic diseases (O'Marcaigh and Cowan, 1997). Currently, transplants are only warranted for severe cases of these disorders with well-defined restrictions for BMT.

Allogeneic HPC transplants may also prove useful in the treatment of severe cases of autoimmune diseases such as rheumatoid arthritis, systemic lupus erythematosus, systemic sclerosis, and antiphospholipid antibody syndrome (Marmont, 1998). Animal models suggest that autoimmune diseases are due at least in part to a disorder intrinsic to the hematopoietic system (van Bekkum, 1993). By replacing the hematopoietic system, allogeneic transplants have cured autoimmune diseases coexisting in patients transplanted for hematologic or oncologic diseases (Killick, 1998). Autologous transplants may also prove useful in treating some of the same autoimmune diseases that can potentially be treated by allogeneic HPC transplants. Though autoimmune diseases usually recur in patients who receive autologous transplants, this may not be the case if T-cells are depleted from the HPCs since autoimmune disorders are thought to be mediated by T-cells (Killick, 1998).

VII. CONTRAINDICATIONS

A. Donor Issues

The physical condition of the donor may make BM collection and associated anesthesia especially dangerous. The importance of these risks when deciding whether to harvest BM from a prospective donor depends on whether the BM-HPCs are intended to be used for an autologous or an allogeneic transplant. Only minimal risk is acceptable when harvesting BM from a healthy allogeneic donor, while some additional risk may be acceptable when harvesting BM for an autologous transplant.

In addition to the general health of the donor, a prospective autologous donor's BM needs to be evaluated. Marrow fibrosis in the autologous donor is a contraindication because BM harvests are impossible in some patients with bone marrow fibrosis. Also, autologous transplants for malignancy are usually not considered in patients who have biopsy-proven evidence of cancer present in their BM. This evaluation, however, can overlook minimal disease, which may be important. For example, immunohistochemistry studies have shown micrometastases of breast cancer cells in 17–60% of marrow harvested from patients who were not thought to have BM metastasis by conventional testing techniques. Unfortunately PB-HPCs may not offer an advantage, as studies have shown 10–78% of PB-HPCs can be contaminated by malignant cells. Furthermore, some studies have shown that micrometas-

tases correlate with poor outcome, though it is unknown if the poor outcomes are due to malignant cells contaminating the transplants or to the more advanced stage of disease in these patients. Regardless, most transplant programs currently collect autologous BM only from those patients who have healthy bone marrow as determined by conventional techniques.

Allogeneic BM harvests from healthy donors are contraindicated when there is significant risk to the donor associated with the BM collection and anesthesia. Donors who are obese, older, or have cardiovascular or pulmonary disease are the ones at increased risk (Buckner et al., 1994). In addition, allogeneic donors who test positive for infectious diseases such as hepatitis B, hepatitis C, or CMV pose an increased risk to recipients, and the transplant physicians must decide on a case-by-case basis whether to transplant BM from these donors. Most transplant physicians will not transplant BM from donors who test positive for HIV.

B. Recipient Issues

BMT is a potentially dangerous treatment that is contraindicated in some patients who are at especially high risk for adverse outcomes. The risk–benefit analysis must consider the fact that the risk associated with transplant depends on the relationship between the donor and the patient. Allogeneic BM-HPC transplants from unrelated donors are the riskiest BMTs and autologous BM-HPC transplants are the safest. Allogeneic BM-HPC transplants from HLA-identical siblings are of intermediate risk.

Each transplant program must establish its own guidelines. For example, some institutions set age limits for transplant, but age may not actually represent an independent risk factor for adverse outcomes (Siegel, 1999; Atkinson, 1998). A sample guideline might include an age limit of 65 years for autologous transplants, 55 years for HLA-identical sibling transplants, and 50 years for unrelated allogeneic transplants (Atkinson, 1998). Other contraindications may include organ dysfunction as indicated by serum creatinine < 2.8 mg/dl, serum bilirubin > 2.4 mg/100 ml, PaO_2 < 70 mm Hg, a left ventricular ejection fraction < 50%, active infection, or a Karnofsky performance score < 70%. These may not be absolute contraindications however. For example, patients with renal failure have been successfully transplanted.

VIII. EXPECTED RESPONSE

BM-HPCs should reconstitute the hematopoietic system. Engraftment, measured as a neutrophil count $\geq 0.5 \times 10^9$/liter, usually occurs between 8 and

30 days post-transplant. Platelet ($\geq 20 \times 10^9$/liter) and RBC (reticulocyte $> 1.55\%$) engraftment usually follows neutrophil recovery. Mean time for neutrophil and platelet engraftment following autologous BM-HPC transplant ranges from 11 to 14 days and 17 to 23 days, respectively (Schmitz *et al.*, 1996). Engraftment kinetics depend on the condition of the supporting BM stroma, the dose of HPC infused, the underlying disease, and post-transplant GVHD prophylaxis. Autologous HPC transplants engraft more rapidly than allogeneic stem cell transplants. Growth factors administered after the transplant can speed engraftment of neutrophils (Gisselbrecht *et al.*, 1994; Stahel *et al.*, 1994). Up to 35% of BM-HPC transplants fail to engraft.

Some experimental protocols involve the use of non-myeloablative treatments followed by allogeneic HPC treatments. With these therapies, no significant period of severe neutropenia or severe thrombocytopenia normally occurs.

In the long term, the patient's hematopoietic system should be completely replaced by the donor's hematopoietic system. For patients with a history of leukemia who are transplanted, failure of donor HPCs to completely and permanently replace the patient's hematopoietic system foreshadows a higher chance of leukemia relapse. The extent of BM replacement is measured by "chimerism analysis." Chimerism analysis determines the phenotype and/or genotype of the hematopoietic cells in the transplant recipient. The blood type and the HLA type of the patient should change to the donors blood and HLA types. If there are no HLA differences, microsatellite DNA markers can be used for chimerism analysis.

Graft-versus-leukemia effect, a potential beneficial effect of allogeneic BM-HPC transplants, is described in Chapter 13.

IX. POTENTIAL ADVERSE EFFECTS

A. Reaction to Cryopreservative, Nonimmune Hemolysis

Acute adverse reactions to transplants of cryopreserved BM-HPCs include reactions to cryopreservatives and are described in Chapter 11.

B. Acute Reactions to Blood Product Transfusion

Acute adverse reactions to allogeneic BMT infusions include hemolytic reactions, allergic reactions (mild or anaphylactic), reactions to rapid volume changes, febrile nonhemolytic reactions, fluid overload reactions, sepsis, and endotoxic shock. These reactions are associated with the same signs and symptoms described in Chapters 28–31. Management of these reactions differs from management of identical reactions that can occur with transfusions of more traditional blood components, due to the fact that BM-HPCs are usually irreplaceable. For this reason, the patient is normally treated for signs and symptoms of the reaction as described in Chapters 28–30 and 40, but the infusion is continued if at all possible. In some cases the infusion may be temporarily halted, but the infusion should be restarted as soon as the patient can tolerate it. If the patient is not being prophylactically treated with antimicrobial therapy, then a febrile reaction could be an indication to commence such therapy.

Signs and symptoms such as flank pain, hypotension, hematuria, and dyspnea should be investigated. Some of these signs and symptoms could be due to a hemolytic transfusion reaction, or due to reactions to DMSO and lysed RBCs contained in cryopreserved products. Whenever a suspected hemolytic transfusion reaction occurs, the label on the BM-HPCs should be rechecked immediately to confirm that the correct product is being infused. A direct antiglobulin test can be performed on the patient's RBCs. Additional tests may include retyping RBCs from the patient and the BM-HPCs, performing antibody screens on serum or plasma from the patient and the BM-HPCs (or BM donor), and performing crossmatches between donor RBCs and patient's serum as well as between patient's RBCs and donor's plasma or serum. These tests will provide additional evidence as to the correct identity of the BM-HPCs, and may suggest that the patient's or donor's RBCs are undergoing hemolysis. In these cases, further depletion of RBCs or plasma in the BM-HPCs or plasmapheresis of the patient may be warranted.

Substantial delayed hemolysis can develop 5–16 days after a BM-HPC transplant complicated by a minor ABO incompatibility (Petz, 1998). Massive hemolysis can be abrupt and severe, resulting in substantial declines in hemoglobin and hematocrit, as well as renal failure. Treatment consists of corticosteroids, hydration to maintain adequate renal blood flow, and transfusion with blood products that are compatible with both the blood group of the donor and the original blood group of the recipient.

C. Chronic Adverse Effects

1. Graft-versus-Host Disease

GVHD is a potentially serious adverse reaction of allogeneic BM-HPC transplants. T lymphocytes derived from allogeneic donor BM can recognize the patient's cells as foreign and attack those cells. The skin, gas-

TABLE 1   Graft Content of HPC Sources

	Bone marrow	Allogeneic peripheral blood	Umbilical cord blood
Nucleated cells ($\times 10^8$)	241 ± 131^a	351 ± 156^a	10.9 ± 4.9^a
CD34$^+$ cell frequency	$1.20\% \pm 0.6\%^b$	$0.74\% \pm 0.34\%^a$	$0.29\% \pm 0.17\%^b$
CD34$^+$ cells ($\times 10^8$)	250 ± 344^a	259 ± 188^a	5.0 ± 4.0^c
CFU-GM ($\times 10^5$)	193 ± 84.4^a	65.8 ± 34.6^a	$5.7 \pm 6/3^c$
CD3$^+$ cell frequency	$33.4\% \pm 12.9\%^a$	$35.7\% \pm 30.4\%$	$12.1\% \pm 5.2\%^a$

[a] University of Minnesota/Fairview-University Medical Center Cell Therapy Clinical Laboratory quality control database.
[b] Knapp *et al.* (1995).
[c] Kogler *et al.* (1996).

trointestinal tract, and liver are the principal targets of this attack. By definition, acute GVHD occurs within 100 days of the transplant, but most frequently occurs around the time of BM engraftment. Risk factors for development of GVHD include transplantation from unrelated donors, HLA-mismatched donors, multiparous female donors, and older recipients. Cutaneous symptoms may include erythema, a maculopapular rash, bullous lesions, and epidermal necrosis. Liver manifestations may include increased conjugated bilirubin and/or serum transaminases, hepatomegaly, and right upper quadrant tenderness. Gastrointestinal manifestations may include diarrhea, nausea, vomiting, and cramping. Chronic GVHD, which by definition arises more than 100 days post-transplant, resembles in its presentation collagen vascular diseases: multiple organ systems are affected. These include the skin, mouth, eyes, sinuses, gastrointestinal tract, liver, lungs, vagina, and the muscular, nervous, urologic, hematopoietic, and lymphoid systems.

2. Infectious Diseases

Allogeneic transplants may transmit the same infectious diseases that can potentially be transmitted by blood transfusions (Chapters 31–35).

X. ALTERNATIVE HPC SOURCES

The main alternatives to transplants of BM-HPCs are PB-HPCs and umbilical-cord-blood-derived HPCs. HPC contents of these alternatives are shown in Table 1. These are described in detail in Chapters 11 and 12. PB-HPCs engraft more quickly than BM-HPCs, resulting in decreased lengths of neutropenia and thrombocytopenia (Schmitz *et al.*, 1996). For this reason, almost all autologous HPC transplants are collected from peripheral blood.

Although PB-HPCs offer several short-term advantages over BM-HPCs, allogeneic BM-HPC transplants are still performed because of possible increased risk of GVHD associated with PB-HPC transplants. PB-HPCs contain nearly 10 times more T lymphocytes than BM-HPCs (Dreger *et al.*, 1994), and T-cells are the principal mediators of GVHD. Several studies with limited numbers of patients suggest that there is no increased incidence of severe acute GVHD in patients who receive allogeneic PB-HPCs (Hagglund *et al.*, 1998) compared to allogeneic BM-HPCs. In contrast, some studies have shown an increased incidence and severity of chronic GVHD following PB-HPC transplants (Scott *et al.*, 1998). These studies have followed a limited number of patients for a limited time. Furthermore, drug therapy or T-cell depletion of PB-HPC may overcome this problem (Chapter 11). Currently, the possible increased risk of chronic GVHD associated with PB-HPC transplants must be taken into account when considering this alternative to BM-HPCs.

Umbilical cord blood-derived HPCs have primarily been used for allogeneic (unrelated and sibling) transplants of pediatric patients. In the future, many more patients may be candidates for transplants of umbilical-cord-blood-derived HPCs. Chapter 20 contains more information regarding the potential advantages and disadvantages of umbilical cord blood for transplantation.

References

Adkins, D., Johnston, M., Walsh, J., *et al.* (1998). Hydroxyethylstarch sedimentation by gravity ex vivo for red cell reduction of granulocyte apheresis components. *J Clin Apher* **13**, 56–61.

Atkinson, K. (1998). "The BMT Data Book: A Manual for Bone Marrow and Blood Stem Cell Transplantation." Cambridge Univ. Press, Cambridge/New York.

Buckner, C. D., Petersen, F. B., and Bolonese, B. A. (1994). Bone marrow donors. In "Bone Marrow Transplantation" (E. D.

Thomas, S. J. Forman, and K. G. Blume, Eds.), p. 259. Blackwell Scientific, Boston.

Chan, K. W., Pollack, M. S., and Braun, D., *et al.* (1982). Distribution of HLA genotypes in families of patients with acute leukemia. Implications for transplantation. *Transplantation* **33**, 613–615.

Dinsmore, R. E., Reich, L. M., Kapoor, N., *et al.* (1983). ABH incompatible bone marrow transplantation: Removal of erythrocytes by starch sedimentation. *Br J Haematol* **54**, 441–449.

Dreger, P., Haferlach, T., Eckstein, V., *et al.* (1994). G-CSF-mobilized peripheral blood progenitor cells for allogeneic transplantation: Safety, kinetics of mobilization, and composition of the graft. *Br J Haematol* **87**, 609–613.

Gee, A. P., and Lee, C. (1998). T-cell deplation of allogeneic stem-cell grafts. In "The Clinical Practice of Stem-Cell Transplantation" (J. Barrett and J. G. Treleaven, Eds.), Vol. 2, pp. 474–505. Isis Medical Media, Oxford.

Gisselbrecht, C., Prentice, H. G., Bacigalupo, A., *et al.* (1994). Placebo-controlled phase III trial of lenograstim in bone-marrow transplantation. *Lancet* **343**, 696–700. [Published erratum appears in *Lancet* **343**, 804]

Hagglund, H., Ringden, O., Remberger, M., *et al.* (1998). Faster neutrophil and platelet engraftment, but no differences in acute GVHD or survival, using peripheral blood stem cells from related and unrelated donors, compared to bone marrow. *Bone Marrow Transplant* **22**, 131–136.

Killick, S. (1998). Autoimmune disorders. In "The Clinical Practice of Stem-Cell Transplantation" (J. Barrett and J. G. Treleaven, Eds.), Vol. 1, pp. 305–312. Isis Medical Media, Oxford.

Knapp, W., Strobl, H., *et al.* (1995). Molecular characterization of CD34* human hematopoietic progenitor cells. *Ann Hematol* **70**, 281–296.

Kogler, G., Callejas, J., *et al.* (1996). Hematopoietic transplant potential of unrelated cord blood: Critical issues. *J Hematother* **5**, 105–116.

Lasky, L. C., Warkentin, P. I., Kersey, J. H., *et al.* (1983). Hemotherapy in patients undergoing blood group incompatible bone marrow transplantation. *Transfusion* **23**, 277–285.

Lasky, L. C., McCullough, J., and Zanjani, E. D. (1986). Liquid storage of unseparated human bone marrow. Evaluation of hematopoietic progenitors by clonal assay. *Transfusion* **26**, 331–334.

Little, A. M., Marsh, S. G., and Madrigal, J. A. (1998). Current methodologies of human leukocyte antigen typing utilized for bone marrow donor selection. *Curr Opin Hematol* **5**, 419–428.

Marmont, A. M. (1998). Stem cell transplantation for severe autoimmune diseases: Progress and problems. *Haematologica* **83**, 733–743.

Marmont, A. M., Horowitz, M. M., Gale, R. P., *et al.* (1991). T-cell depletion of HLA-identical transplants in leukemia. *Blood* **78**, 2120–2130.

O'Marcaigh, A. S., and Cowan, M. J. (1997). Bone marrow transplantation for inherited diseases [see comments]. *Curr Opin Oncol* **9**, 126–130.

Petz, L. D. (1998). Hemolysis associated with transplantation. *Transfusion* **38**, 224–228.

Schmitz, N., Linch, D. C., Dreger, P., *et al.* (1996). Randomised trial of fllgrastim-mobilised peripheral blood progenitor cell transplantation versus autologous bone-marrow transplantation in lymphoma patients. *Lancet* **347**, 353–357.

Scott, M. A., Gandhi, M. K., Jestice, H. K., *et al.* (1998). A trend towards an increased incidence of chronic graft-versus-host disease following allogeneic peripheral blood progenitor cell transplantation: A case controlled study. *Bone Marrow Transplant* **22**, 273–276.

Siegel, D. S., Desikan, K. R., Mehta, J., *et al.* (1999). Age is not a prognostic variable with autotransplants for multiple myeloma. *Blood* **93**, 51–54.

Siena, S., Bregni, M., and Gianni, A. M. (1993). Estimation of peripheral blood CD34 + cells for autologous transplantation in cancer patients. *Exp Hematol* **21**, 203–205.

Stahel, R. A., Jost, L. M., Cerny, T., *et al.* (1994). Randomized study of recombinant human granulocyte colony-stimulating factor after high-dose chemotherapy and autologous bone marrow transplantation for high-risk lymphoid malignancies. *J Clin Oncol* **12**, 1931–1938.

Stroncek, D. F., Holland, P. V., Bartch, G., *et al.* (1993). Experiences of the first 493 unrelated marrow donors in the National Marrow Donor Program. *Blood* **81**, 1940–1946.

Thomas, E. D., and Storb, R. (1970). Technique for human marrow grafting. *Blood* **36**, 507–515.

Van Bekkum, D. W. (1993). BMT in experimental autoimmune diseases. *Bone Marrow Transplant* **11**, 183–187.

19

Peripheral-Blood-Derived Hematopoietic Progenitor Cells

STEVEN R. SLOAN

Children's Hospital Boston
Harvard Medical School
Boston, Massachusetts

Peripheral blood (PB) is one source of hematopoietic stem cells (HPCs). Peripheral-blood-derived stem cell products contain stem cells and hematopoietic progenitor cells that are committed to particular hematopoietic lineages. Hence, the term "hematopoietic progenitor cells" (HPCs) is more accurate than "stem cells" and will be used in this chapter. HPC transplants are used to treat bone marrow failure secondary to chemotherapy, immunodeficiency, autoimmunity, and certain genetic diseases. Progenitor cells, including stem cells, are the critical cells in peripheral-blood-derived HPC (PB-HPC) transplants. Committed, partially differentiated HPCs are probably responsible for the initial circulating leukocytes and platelets following a transplant of PB-HPCs. However, the cells that contribute to long-term multilineage reconstitution of the hematopoietic system are the pluripotent stem cells that have the capacity for self-renewal.

There are two categories of PB-HPC transplants. Autologous PB-HPC transplants (see Figure 1) are donated by the person who will receive the transplant and are usually cryopreserved. Allogeneic PB-HPC transplants (see Figure 2) are donated by a person other than the patient and are usually not cryopreserved. Currently, most autologous and many allogeneic HPC transplants are collected from peripheral blood. Some allogeneic HPC transplants use HPC collected from bone marrow (BM) or umbilical cord blood (Chapters 12 and 17).

PB-HPC transplants are relatively new, and processing procedures vary between institutions as do clinical protocols. Regulatory agencies and accrediting organizations have recently adopted, or are still in the process of developing, standards and regulations for PB-HPC transplant programs. The same agencies that regulate and accredit bone marrow transplants (BMTs) also regulate and accredit PB-HPC transplants (see Chapter 17).

I. PRODUCT DESCRIPTION

A. Cellular Constituents

In addition to progenitor cells, PB-HPC collections contain other hematopoietic cells. Although the PB-HPC collection procedure is designed to enrich for mononuclear cells, all PB-HPC collections contain granulocytes, erythrocytes, and platelets. Certain methods of collection remove most of the platelets in the product. Lymphocytes and immature myeloid cells are also present in PB-HPC harvests. Furthermore, malignant cells may contaminate autologous PB-HPC collections from oncology patients. Some of the non-HPC cells are not therapeutic and may even harm the patient. Hence, PB-HPCs may be further processed to remove these unwanted cells.

The final volume infused to the patient depends on the HPC concentration, the target number of mononuclear and/or HPCs per kilogram desired for transplant, and the size of the patient. The final volume of PB-HPCs infused is usually a few hundred milliliters.

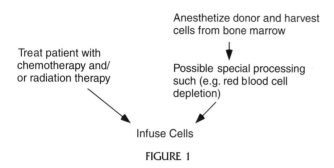

Allogeneic BM-HPC Transplant

Treat patient with chemotherapy and/ or radiation therapy

Anesthetize donor and harvest cells from bone marrow

Possible special processing such (e.g. red blood cell depletion)

Infuse Cells

FIGURE 1

B. Characterization of HPC Content

Different assays can be used to assess the number of HPCs in a PB-HPC product. To determine whether a HPC donor has donated a sufficient number of cells, total mononuclear cell and CD34$^+$ cell concentrations are measured within hours or days of the collection. Total mononuclear and CD34$^+$ cell measurements do not measure HPCs but rather measure the concentrations of cell populations, including many different cells, such as HPCs and true stem cells. Biological growth assays such as the colony-forming unit (CFU) assay and the long-term culture-initiating cell (LTCIC) assay may also be performed to measure the quantity of HPCs in a PB-HPC product.

1. Characterization of Mononuclear Cell Content

The mononuclear cell count is one measure that may help determine whether sufficient cells are present in the PB-HPC collection to result in a timely engraftment of the patient's hematopoietic system. Though some studies have suggested that the mononuclear cell count does correlate with engraftment, this correlation

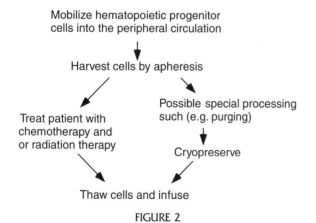

Autologous PB-HPC Transplant

Mobilize hematopoietic progenitor cells into the peripheral circulation

Harvest cells by apheresis

Possible special processing such (e.g. purging)

Treat patient with chemotherapy and or radiation therapy

Cryopreserve

Thaw cells and infuse

FIGURE 2

has not been observed in other studies (Roberts *et al.*, 1993). Mononuclear cell counts are easily performed on automated hematology analyzers or hemocytometers, and several centers use this testing method. The target dose of mononuclear cells usually ranges from 2 to 6×10^8 cells/kg, but each transplant program must establish its own guidelines. Target doses may depend upon the source of the cells (e.g., unrelated allogeneic transplants usually require higher doses than autologous transplants), or on the patient's diagnosis. Most laboratories cryopreserve cells at a concentration of 2×10^7 to 8×10^8 mononuclear cells/ml.

2. Characterization of CD34$^+$ Cell Content

The number of CD34$^+$ cells in the PB-HPC product is the most widely used measurement to predict whether sufficient T-cells are present for timely engraftment of the transplanted cells. CD34 is a cell surface protein that is expressed on most stem cells and many other immature hematopoietic cells. CD34$^+$ cells compose 1–5% of peripheral blood mononuclear cells following mobilization. CD34$^+$ cell counts are determined by flow cytometry. This procedure requires technical expertise and judgment, and, as for the CFU assays, results among laboratories may not correlate well. Standardized approaches for these measurements have been recently developed and should help improve interlaboratory reproducibility (Keeney *et al.*, 1998). Several, but not all, studies suggest that a minimum number of CD34$^+$ cells must be transplanted to ensure rapid engraftment. The target number of CD34$^+$ cells depends on whether autologous cells, allogeneic cells from a related donor, or allogeneic cells from an unrelated donor are transplanted. The target number of cells to transplant must be determined by each transplant program but studies suggest that that a dose of $2–5 \times 10^6$ CD34$^+$ cells/kg is adequate to ensure multilineage engraftment in a timely fashion (Weaver *et al.*, 1995).

3. Characterization of CFU Content

Colony-forming unit assays can be used as an indirect measure of HPCs. This method identifies and counts HPCs based on their ability to proliferate and give rise to more mature hematopoietic cells. Cells from the PB-HPC product are cultured in semisolid media, and the types of cell colonies that grow from individual immature HPCs are identified and counted. CFU culture conditions are not completely standardized, and interpretation of results is somewhat subjective. Not surprisingly, results from CFU assays can vary significantly among laboratories. While some studies have shown a correlation between CFU assay results and engraftment speed, other studies have revealed no such correlation (To *et al.*, 1992). Each transplant program determines whether it will strive for a specific minimum or target CFU dose. Typically, a minimum dose of 0.5×10^5 to 5×10^5 CFU-GM (CFU-granulocyte macrophage)/kg is chosen. In most cases, CFU assays cannot be used to determine whether additional PB-HPC collections are necessary for any individual patient, because colonies cannot be scored until approximately 2 weeks after plating the cells. However, CFU assays may be used for data analysis and for monitoring quality of various aspects of the transplant program. Other methods have been developed to measure the most immature progenitor cells, but these assays are complex and are not performed in most clinical transplant laboratories.

C. Anticoagulants and Additives

PB-HPCs contain plasma, anticoagulants, and additional buffered solutions. In most cases, acid-citrate-dextrose formula A (ACD-A) or a similar anticoagulant is added. Human plasma is present in all PB-HPC products whether or not the product is cryopreserved. Cryopreserved PB-HPC products usually contain 10–20% of a protein solution such as plasma or human albumin and a cryoprotectant solution consisting of 10% dimethylsulfoxide (DMSO) or 5% DMSO, 6% hydroxyethyl starch (HES), and albumin. Cryopreserved PB-HPCs usually contain a buffered electrolyte solution such as Normosol or another infusion-grade solution. Some institutions use tissue culture media that also contain vitamins and/or minerals, but such use is discouraged because tissue culture media are not currently approved for infusion to humans in the United States. RBC-depleted products may contain additional albumin.

D. Labeling

PB-HPCs are labeled in the same manner as BM-HPCs (see Chapter 18), except that they contain the phrase "Peripheral Blood Progenitor Cells" instead of "Human Bone Marrow."

II. COLLECTION

A. Donor Evaluation and Preparation

1. Physical Requirements

a. Autologous Transplants

Autologous donors are subjected to the same risks of anticoagulants and volume shifts that are normally associated with apheresis procedures (see Chapter 37).

Autologous donors will often need venous access for an extended period of time for PB-HPC collections, chemotherapy, PB-HPC transplants, intravenous hydration, intravenous drugs, and transfusions. Hence, an autologous donor will often have a central venous catheter through which PB-HPCs can be collected.

b. Allogeneic Transplants

Allogeneic donors undergo the same apheresis risks as autologous donors. Allogeneic donors undergo the same screening (history and infectious disease testing) as for BM donation (see Chapter 17). Relative contraindications to the collection procedure include hemodynamic instability, symptomatic anemia, evidence of active infections, and recent ingestion of Angiotensin Converting Enzyme (ACE) inhibitors in the donor. Unlike autologous donors, allogeneic donors do not need venous access for an extended period of time for multiple purposes, and the HPC collection can be performed by peripheral venous access.

2. Mobilization

The donor is usually treated with cytokines and/or chemotherapy prior to collection in order to augment the number of HPCs in the peripheral circulation. This process is called "mobilization" of the HPCs. Mobilization treatment causes an increased peripheral white blood cell (WBC) count that predominantly consists of myeloid cells at all stages of development. HPCs are the most critical cells to mobilize, and the number of HPCs correlates with the number of CD34$^+$ cells. While CD34$^+$ cells normally represent approximately 0.1% of PB mononuclear cells, mobilization can increase this proportion to more than 1% (Stadtmauer *et al.*, 1995).

Donors are treated with mobilization medications daily for approximately 4–7 days prior to collection by apheresis. Mobilization kinetics vary significantly between donors and can depend on the mobilization regimen and previous chemotherapy that the donor has received. Though some institutions use the WBC count to determine when to begin apheresis, the peripheral blood CD34$^+$ cell count may be a better indicator to determine the optimal day to begin peripheral cell collection (Stadtmauer *et al.*, 1995). Target cell counts vary. Examples include protocols that commence leukapheresis when WBCs $\geq 8.0 \times 10^9$ cells/liter, when WBCs $\geq 1.0 \times 10^9$ cells/liter, or when the CD34$^+$ cell count reaches 10×10^6/liter (Haas *et al.*, 1994). If cell counts reveal that the patient's HPCs are mobilizing poorly, different mobilization drugs may be considered. If a patient's cells mobilize poorly with granulocyte colony-stimulating factor (G-CSF), mobilization may improve with G/GM-CSF. Growth factors, such as G-CSF, can cause unpleasant side effects such as bone pain, fever, and malaise. The pain and fever can usually be treated successfully with acetaminophen and/or mild narcotics.

a. Autologous Transplants

HPCs from autologous donors can be mobilized with cytotoxic chemotherapeutic agents and/or growth factors. The choice of drugs will depend on a variety of factors, including the type of tumor and prior exposure to chemotherapeutic drugs.

b. Allogeneic Transplants

While HPCs can be mobilized with cytotoxic chemotherapeutic agents and/or growth factors, chemotherapy is not usually used to mobilize cells from an allogeneic donor because of risks associated with administering chemotherapy to individuals. Though G-CSF is the most widely used growth factor used for mobilization, other growth factors, including GM-CSF, interleukin-1 (IL-1), IL-3, IL-8, IL-11, and stem cell factor (SCF), also mobilize HPC (Mauch *et al.*, 1995). Cytokine combinations like G-CSF + SCF, G-CSF + GM-CSF, or IL-11 + SCF may improve mobilization, and some institutions use growth factor combinations or a growth factor–chemotherapy combination to mobilize HPCs (Mauch *et al.*, 1995).

3. HLA Compatibility—Allogeneic Donors

HLA matching is the principal means of choosing an allogeneic HPC donor. Issues concerning HLA matching for allogeneic PB-HPC transplants and allogeneic BM-HPC transplants are identical and are discussed in Chapter 17.

B. Collection Procedure

PB-HPCs are collected by leukapheresis. The mononuclear cells are collected using some of the same equipment that can be used for therapeutic apheresis (see Chapters 36–38). The leukapheresis procedure is conducted to specifically collect mononuclear cells. In many cases, one apheresis is sufficient, though 2–3 collections or more are sometimes needed. The yield of HPCs will depend on the type of apheresis equipment that is used, the efficiency of HPC mobilization in the patient, the length of the apheresis procedure, and the total volume of blood processed.

C. Laboratory Evaluation

Laboratory evaluation includes quantitation of HPCs in the product and assessment of the sterility of the product. Cellular contents are quantified indirectly by measuring mononuclear cells and CD34$^+$ cells, as described in Section I. In addition, a microscopic examination of trypan-blue-stained cells is often performed to determine cellular viability. Sterility is usually determined by culturing products for bacteria and fungi. Proliferation assays such as CFU measurements may also be performed to retrospectively measure HPC content (see Section I). Cell counts and sterility measurements are often made before and after cryopreservation (see Section III.D), and before and after special manipulations (see Section III).

D. Storage and Transport

Autologous products are usually cryopreserved prior to storage (see Section III.C). At the end of cryopreservation, cells are maintained in liquid nitrogen or in −80°C freezers. Many centers store cells in liquid nitrogen tanks because cells may be stored longer than in −80°C freezers and because freezers are more likely to fail. Liquid nitrogen tanks can fail as well, and all freezers and liquid nitrogen tanks that contain HPCs should be regularly monitored and have alarm systems. Frozen cells can be stored for years. Cells stored for 5 years in liquid nitrogen and cells stored at −80°C for over 2 years have been successfully transplanted (Rowley, 1992).

Hepatitis viruses and fungi can contaminate liquid nitrogen storage tanks, contaminate the HPCs stored in the tanks, and infect transplant recipients (Hernandez-Navarro *et al.*, 1995). Double bags and/or storage in the vapor phase of the liquid nitrogen tank may help prevent these contamination problems, but this has not been proven.

PB-HPCs should be administered soon after being thawed. Cryopreserved autologous PB-HPCs are usually transplanted to a patient who is being treated in the institution that is storing the cells. In these cases the PB-HPCs need only be transported from the storage area to the patient treatment area. Cells may be transported in liquid nitrogen and thawed at the bedside, or they may be thawed immediately prior to transport to the patient's room. If cryopreserved cells need to be transported to another institution, then they are transported frozen in liquid nitrogen or packed in dry ice.

Allogeneic PB-HPCs are not usually cryopreserved and are stored and transported the same as BM-HPCs (described in Chapter 17).

III. SPECIAL PREPARATION

Most of the special preparation and processing that is performed on PB-HPCs is designed to purify the

HPCs. While PB-HPCs collected by apheresis contain many different cells, the cells responsible for reconstituting the hematopoietic system are the HPCs. Additional cells and components collected during leukapheresis can be dangerous to the patient. Autologous peripheral blood cells collected from a cancer patient may contain tumor cells that could contribute to relapse of disease. Allogeneic peripheral blood cells contain T-cells that may cause graft-versus-host disease (GVHD). Additionally, allogeneic PB-HPCs may contain RBCs that could be hemolyzed by antibodies in the patient's plasma and/or may contain plasma containing antibodies that will hemolyze the patient's RBCs. Thus, transplant laboratories perform a variety of procedures to purify the PB-HPCs. The processing may involve depletion of RBCs, lymphocytes, or malignant cells, or positive selection for CD34-expressing HPCs. These processed products are in reduced volumes and usually have different cell populations when compared to minimally processed HPCs.

A. RBC and Plasma Depletion

Autologous PB-HPCs do not usually need RBC or plasma depletion. Major or minor RBC incompatibilities or unexpected alloantibodies may necessitate depletion of RBCs or plasma from PB-HPCs intended for allogeneic transplant. The indications for depleting PB-HPCs of RBCs or plasma and the techniques used to deplete them are identical to the indications and techniques that apply to BM-HPCs (see Chapter 17).

B. T-Cell Depletion

Allogeneic PB-HPC transplants can cause GVHD. The increased risk of GVHD associated with allogeneic PB-HPC transplants compared to BM-HPC transplants is related to higher concentrations of T-cells in PB-HPC collections. Though the data are still limited, several studies have found an increased risk of chronic (rather than acute) GVHD associated with allogeneic PB-HPC transplants as compared to BM-HPC transplants (Storek et al., 1997). Attempts to reduce this risk have focused on techniques to reduce the concentration of T-cells in PB-HPC collections by positive selection for HPCs or negative selection and removal of T-cells. Clinical trials are needed to substantiate whether reduction in GVHD results from this manipulation.

Several methods can be used to specifically deplete lymphocytes or T lymphocytes from HPC products. Although CD34$^+$ cell selection (discussed below) is a common way to remove T-cells, several other methods have been used. Counterflow centrifugal elutriation, a method that removes most lymphocytes, is more frequently used to deplete T-cells from bone marrow and

is described in Chapter 17. Soybean agglutinin (SBA) added to the HPC collection will agglutinate all but the progenitor cells. The agglutinated cells are then removed on a 5% bovine serum albumin gradient. Alternatively, SBA-coated plastic can be used to selectively remove all but the HPCs by panning. T lymphocytes rosette around sheep RBCs, and this method can be used to selectively remove T-cells. The rosettes either sediment or can be removed by density-gradient centrifugation.

Most other methods to remove T-cells utilize monoclonal antibodies that recognize antigens that are specifically expressed on lymphocytes, T-cells, or a subset of T-cells. A variety of antibodies have been used, including the "CAMPATH" series of rat monoclonal antibodies that recognize CDw52, an antibody that recognizes the T-cell receptor heterodimer. Other antibodies that specifically recognize CD2, 3, 4, 5, 6, or 8 have been used as well (Gee and Lee, 1998). These antibodies can be used to separate cell populations or to lyse the cells targeted by the antibodies. To separate cells targeted by the antibody, the antibody must be physically bound to a solid matrix such as paramagnetic microspheres, paramagnetic nanoparticles, or relatively large plastic surfaces used for panning. Antibodies can lyse the target cells either by activating complement that is added to the cells or by utilizing complement present in the plasma that is collected with the HPC. Alternatively, some investigators have conjugated a toxin, such as ricin, directly to the monoclonal antibody.

C. Selection

Tumor cells can contaminate autologous collections of bone-marrow-derived HPCs (BM-HPCs) or PB-HPCs. BM-HPCs often have more cancer cells than PB-HPCs. For most malignancies, the danger posed by these contaminating tumor cells has not been fully established, and the risk may depend on the type and stage of the cancer. However, malignant cells contaminating cell collections from patients with acute myelogenous leukemia (AML) are known to increase the risk of disease recurrence, and there is a theoretical risk posed by contaminating malignant cells from other types of cancer (Gorin et al., 1991). Some strategies reduce the concentration of malignant cells by positively selecting HPCs, while other strategies attempt to directly remove malignant cells.

1. Positive Selection for CD34$^+$ Cells

One of the most common manipulation techniques performed on HPC products is selection of HPCs expressing CD34. Most stem cells and many committed

HPCs express the CD34 antigen, which can be used as a marker for selection. This process can be used to reduce tumor cells in autologous transplants or to reduce T-cells in allogeneic transplants. While methods vary, the first step often involves separation of the mononuclear cells from the RBCs and neutrophils by density-gradient centrifugation. Monoclonal anti-CD34 antibodies bound to a solid-phase matrix are then used to selectively adsorb the CD34$^+$ cells. The bound cells are then eluted from the solid matrix.

Several studies have shown that CD34$^+$-selected cells can reconstitute hematopoiesis in the BM. A dose of at least 1.2×10^6 CD34$^+$-selected cells per kilogram appears necessary for rapid platelet recovery, and some institutions transplant double or triple that amount (Shpall *et al.*, 1997). With a sufficient dose, the bone marrow will be reconstituted with CD34$^+$-selected cells as rapidly as with total PB-HPC transplants. This is an area of intensive investigation, and minimal cell dosages have not been determined.

Selection of CD34$^+$ cells would not benefit autologous transplants for patients with tumors that express CD34. CD34 is normally expressed on capillary endothelium and stromal cell precursors in the BM, and tumors derived from these cells frequently express CD34 (Fina *et al.*, 1990). CD34 is also expressed on vascular tumor cells, including those of angiosarcoma, hepatic hemangioendothelioma, and Kaposi's sarcoma (Fina *et al.*, 1990). Additionally, about 40% of AMLs, 65% of pre-B acute lymphoblastic leukemias (ALLs), and 1–5% of acute T-cell lymphoid leukemias express CD34 (Borowitz *et al.*, 1990). CD34 has also been reported to be expressed in a few cases in squamous cell lung carcinoma, neuroblastomas, and Ewing's sarcoma cells.

CD34$^+$-selected cells are usually suspended in a small volume of buffered normal saline solution and contain few if any RBCs, platelets, or mature leukocytes, but may contain additional proteins that were added during processing. If the cells are frozen, 10% DMSO and at least 10–20% protein solution (plasma or albumin) will usually be included in the final infused product.

2. Negative Selection (Purging)

Though CD34$^+$ cell selection is the most common way to remove tumor cells, many other techniques can be used to purge tumor cells from autologous HPC collections. Most of these techniques are investigational and exact approaches vary, depending on the particular research protocol and the type of malignancy. Many of the techniques begin with a physical purification technique such as density-gradient cen-

trifugation or counterflow centrifugal elutriation. Additional techniques to purge tumor cells include methods that use heat, cell cultures, cytotoxic effector cells, cytotoxic drugs, molecular-biology-based molecules, and antibodies.

A variety of cytotoxic drugs have been used to purge tumor cells. The most widely used drugs are 4HC and mafosfamide. Purging with these drugs reduces the risk of disease relapse in patients with AML. While these purging protocols can destroy committed HPCs and prolong engraftment times, pluripotent progenitors survive the purging protocols (Rowley and Davis, 1991). A variety of other drugs have been suggested. Some of these are photoactive drugs that are added to the HPCs, which are then exposed to fluorescent light.

Molecular-biology-based agents include antisense DNA oligonucleotides and ribozymes that are designed to inhibit expression of specific genes that promote cancer cell growth. Only a few preliminary trials using these approaches have been reported (Bishop *et al.*, 1997).

Antibodies that recognize tumor cells can be used to kill those cells in HPC harvests or to physically remove those cells from the HPCs (Gee, 1995). Antibodies and complement can be used to lyse cells, or antibodies conjugated to a toxin can be used to kill T-cells. Antibodies fixed to solid plastic surfaces, colloids, or magnetic particles can be used to selectively remove malignant cells.

D. Cryopreservation

Unlike most allogeneic PB-HPC collections, most autologous PB-HPC collections are cryopreserved. Cryopreservation is designed to preserve the mononuclear cells for extended periods of time. Cryopreservation is not designed to alter the cell populations in the PB-HPC collection and therefore is not considered a manipulation of the product by accrediting agencies such as the Foundation for the Accreditation of Hematopoietic Cell Therapy (FAHCT). The cell concentrate collected by apheresis is centrifuged, the plasma is removed, and the WBCs are resuspended in a cryopreservation solution at a concentration usually ranging between 2×10^7 and 8×10^8 cells/ml. Two techniques are commonly used, though some variations have been described.

Commonly, the cryopreservation solution is designed so that the final product contains 10% DMSO and at least 10–20% human protein solution, such as plasma or albumin. The remainder of acellular volume consists of a buffered saline solution such as tissue culture media, Normosol, or another infusion-grade solution. Tissue culture medium is not approved for human infusion in the United States and its use is discouraged for this purpose. After resuspension, the cell solution is dispensed into freezing bags. The bags are chilled in a controlled-rate freezer at -1 to $-2°C/min$ until they reach approximately $-50°C$, and then at $-5°C/min$ until they reach about $-90°C$. The bags are then transferred to a liquid nitrogen storage tank. Cells stored for 5 years in liquid nitrogen have been successfully transplanted (Rowley, 1992).

In another common technique, the cryopreservation solution is designed so that the final product contains albumin and 5% DMSO + 6% hydroxyethyl starch (Stiff, 1991). With this technique, the freezing bags are placed horizontally in a $-80°C$ freezer for freezing and storage.

E. Thawing

Cryopreserved cells must be thawed prior to transplantation. Cells are thawed rapidly by immersing the product in a 37°C water bath. Rapid thawing can lead to bag breakage, and various strategies have been used to minimize this problem. Bags that have a low breakage rate should be chosen, and some centers double-bag the HPCs. Products exposed to water baths due to bag breakage have been safely infused.

Cell death due to thawing can be minimized by infusing cells as soon as possible. Thus cells are usually thawed immediately prior to infusion in the same room with the patient and are often infused rapidly. Each bag of cells is sequentially thawed and infused, minimizing waste in the event of a patient reaction. Another approach is to thaw the cells in the laboratory and then remove the DMSO by centrifugation. While this method avoids the unpleasant side effects of DMSO, the cells are exposed to DMSO for a longer period of time. In this case, all of the cells to be infused are thawed simultaneously, making it difficult to temporarily discontinue the transplant if the patient experiences an adverse reaction.

IV. ABO AND Rh COMPATIBILITY

ABO and Rh typing should be performed on the donor and patient. In addition, FAHCT requires that the donor be typed at the time of collection, as described in Chapter 17. The PB-HPC label should contain the donor's ABO and Rh type. This helps ensure that the HPC will be administered to the intended recipient. Crossmatches should be performed on allogeneic transplants, and incompatible products should be depleted of RBCs or plasma using the same criteria

and essentially the same techniques described in Chapter 17 and above. If RBCs are depleted, they should be depleted prior to cryopreserving the HPCs. If plasma in the PB-HPCs is incompatible with the patient's RBCs, the incompatible plasma can be removed and compatible plasma can be added during the cryopreservation processing procedures.

In some allogeneic transplants, the donor's plasma contains isohemagglutinins against the recipient's RBCs. This occurs when the donor is group O and the recipient is group A, B, or AB, when the donor is group A and the recipient is group B, or when the donor is group B and the recipient is group A. Although these are known as "minor" ABO incompatibilities, their effects can be serious. *While a minor ABO incompatibility does not usually cause immediate significant hemolysis, massive delayed hemolysis can occur* (Salmon et al., 1999). This hemolysis, which is caused by stimulation and proliferation of donor-derived B lymphocytes, usually develops 5–16 days following the transplant and can be more severe than analogous reactions seen with BM-HPC transplants. Some drugs such as methotrexate that are used as prophylaxis for GVHD inhibit B lymphocytes and reduce the chances of severe hemolysis. However, other anti-GVHD drugs such as cyclosporin and FK506 do not inhibit B lymphocytes and do not prevent severe hemolysis. Although massive hemolysis occurs in no more than 10–20% of susceptible patients, it can be abrupt, severe, and fatal. The direct antiglobulin test (DAT) will usually be positive prior to the onset of significant hemolysis and can be used to identify patients at risk for massive hemolysis, but not all patients with a positive DAT will develop massive hemolysis. Treatment consists of empiric use of corticosteroids, hydration, and transfusion with blood products that are compatible with both the blood group of the donor and the original blood group of the recipient. Severe cases may be additionally treated with methotrexate and RBC exchange transfusions.

V. ORDERING AND ADMINISTRATION

The process for selection of the donor (e.g., HLA matching) is the same as that for BM-HPC discussed in Chapter 17. PB-HPCs are infused intravenously using essentially the same methods used for transplantation of BM-HPCs. The patient should be well hydrated prior to infusion and is often premedicated with acetaminophen, an antihistamine, and a corticosteroid. Oxygen and an antianaphylaxis treatment such as epinephrine should be available. HPCs are administered intravenously, usually through a central venous catheter, without any filters; gamma irradiation of HPCs is contraindicated. The patient should be closely monitored and vital signs should be taken periodically during the infusion (as for any blood product infusion) due to the risks of allergic, anaphylactic, hemolytic, or febrile nonhemolytic transfusion reactions.

VI. INDICATIONS

A. Autologous Transplants

An autologous PB-HPC transplant is usually performed to reconstitute the hematopoietic system of a patient who has received myeloablative chemotherapy and/or radiation therapy to treat a malignancy. Patients with solid tumors who are transplanted with HPCs usually receive autologous PB-HPC transplants (Horowitz and Rowlings, 1997). The tumors must be responsive to chemotherapy and/or radiation therapy. In decreasing order, the most common diseases treated with autologous transplants in North America in 1995 were breast cancer, non-Hodgkin's lymphoma, Hodgkin's disease, multiple myeloma, and ovarian cancer.

B. Allogeneic Transplants

Like autologous transplants, allogeneic PB-HPC transplants are usually performed to reconstitute the hematopoietic system of a patient who received myeloablative chemotherapy and/or radiation therapy to treat a malignancy. Allogeneic transplants are usually preferred for patients with malignancies of the hematopoietic system such as AML, chronic myeloblastic leukemia (CML), and acute lymphoblastic leukemia (ALL). Autologous HPC collections from patients with these diseases are less likely to contain sufficient HPCs and will likely contain malignant cells.

In addition, allogeneic transplants have been used for a variety of other diseases, including aplastic anemia, immune deficiencies, inherited disorders of metabolism, Fanconi's anemia, sickle cell disease, and thalassemia. Severe autoimmune diseases may also potentially be treated with allogeneic transplants of HPCs (see Chapter 17).

VII. CONTRAINIDICATIONS

The contraindications for PB-HPC transplants are similar to the contraindications for BM-HPC transplants. However, PB-HPC donors do not undergo general anesthesia during collection, and thus this risk is eliminated. As with BM-HPC transplants, the risk-benefit analysis depends on the relationship between the donor and the patient. Allogeneic PB-HPC transplants from unrelated donors have the highest risks, while autologous PB-HPC transplants are the safest transplants. Allogeneic PB-HPC transplants from HLA-identical siblings are of intermediate risk. Transplant protocols are designed to consider the age, organ function, and Karnofsy performance score of the patient.

VIII. EXPECTED RESPONSE

PB-HPCs usually reconstitute the recipient's hematopoietic system within a few weeks to a month. Peripheral blood neutrophil counts should rise to $\geq 0.5 \times 10^9$/liter between 8 and 30 days post-transplant, with a mean time of 11–17 days. Peripheral blood platelet counts should rise to $\geq 20 \times 10^9$/liter between 6 and 140 days post-transplant, with a mean time of 9–31 days (Russell and Miflin, 1998). The time to engraftment depends on the number of CD34$^+$ cells infused, the condition of the supporting bone marrow stroma, and the underlying disease. Autologous transplants engraft more rapidly than allogeneic transplants. Administration of a myeloid growth factor (G-CSF or GM-CSF) shortens the time of neutropenia but not of thrombocytopenia (Russell and Miflin, 1998). With allogeneic transplants, the hematopoietic system should be eventually completely replaced by the donor's hematopoietic system. This can be measured by chimerism analysis (see Chapter 17).

IX. POTENTIAL ADVERSE EFFECTS

A. Acute Adverse Reactions

Some acute adverse reactions are specifically associated with cryopreserved products. DMSO in cryopreserved products usually causes an unpleasant taste and odor. In addition, headache, flushing, nausea, vomiting, cramping, diarrhea, bradycardia, hypertension, hypotension, and dyspnea can occur (Zambelli *et al.*, 1998). Other severe reactions that have been reported include anaphylactic-type reactions, neurologic complications, and signs and symptoms of leukostasis. The severity of these adverse reactions varies significantly between patients, but higher DMSO doses are more likely to result in more severe reactions. The LD$_{50}$ for dogs receiving DMSO p.o. is > 10 g/kg, and some institutions infuse no more than 1 g/kg/day into human patients. This corresponds to about 500 ml of cells per day in a 70-kg patient. Other institutions continue the transplant as long as the patient tolerates it. Furthermore, lysis of RBCs during cryopreservation and thawing can lead to hemoglobinuria and hemoglobinemia. Because the HPCs are usually irreplaceable, the infusion is typically continued. Severe reactions should be investigated as described in Chapter 17.

Acute reactions associated with the infusion of a large volume of HPCs can occur. The signs and symptoms are identical to volume overload associated with transfusions (see Chapter 30). CD34$^+$-selected cells have a small total volume, and acute reactions of any kind are uncommon. The approach to acute reactions to these blood products is discussed in Chapter 17.

Substantial delayed hemolysis can develop 5–16 days after an allogeneic PB-HPC transplant complicated by a minor ABO incompatibility (Salmon *et al.*, 1999). Massive hemolysis can be abrupt, severe, and fatal. Treatment consists of corticosteroids, hydration to maintain adequate renal blood flow, and transfusion with blood products that are compatible with both the blood group of the donor and the original blood group of the recipient. Severe cases may also be treated with methotrexate and RBC exchange transfusions.

B. Chronic Adverse Reactions

Chronic adverse reactions that occur with allogeneic BM-HPC transplants (see Chapter 17) can also occur with allogeneic PB-HPC transplants. Compared with BM-HPC transplants, allogeneic PB-HPC transplants have a higher risk of severe chronic GVHD, though this may not be true for transplants of PB-HPCs that have been depleted of T-cells (Storek *et al.*, 1997). Allogeneic PB-HPC transplants can also potentially transmit the same infectious diseases transmissible by other types of blood transfusions (Chapters 31–35).

X. ALTERNATIVE HPC SOURCES

The main alternatives to PB-HPC transplants are BM-HPC transplants and umbilical cord blood transplants. These are described in Chapters 12 and 17, respectively. PB-HPCs engraft more quickly than BM-HPCs, resulting in decreased intervals of neutropenia and thrombocytopenia following PB-HPC transplants (Schmitz *et al.*, 1996). For this reason, almost all autologous HPC transplants are collected from peripheral blood.

BM-HPCs or PB-HPCs can be used for allogeneic transplants of HPCs. PB-HPCs provide faster engraftment but have a higher risk of chronic GVHD, as discussed in Chapter 17 and this chapter. To avoid the potential increased risk of chronic GVHD, BM-HPCs may be chosen. Because allogeneic PB-HPC transplants are relatively new, the data concerning the long-term outcomes of these transplants are limited compared to BM-HPC transplants. Umbilical-cord-blood-derived stem cells have primarily been used for allogeneic transplants of pediatric patients. In the future, many more patients may be candidates for transplantation of umbilical-cord-blood-derived HPCs, as discussed in Chapter 12.

References

Beaujean, F., Bernaudin, F., Kuentz, M., *et al.* (1995). Successful engraftment after autologous transplantation of 10-day cultured bone marrow activated by interleukin 2 in patients with acute lymphoblastic leukemia. *Bone Marrow Transplant* **15**, 691–696.

Bishop, M. R., Jackson, J. D., Tarantolo, S. R., *et al.* (1997). Ex vivo treatment of bone marrow with phosphorothioate oligonucleotide OL(1)p53 for autologous transplantation in acute myelogenous leukemia and myelodysplastic syndrome. *J Hematother* **6**, 441–446.

Borowitz, M. J., Shuster, J. J., Civin, C. I., *et al.* (1990). Prognostic significance of CD34 expression in childhood B-precursor acute lymphocytic leukemia: A Pediatric Oncology Group study. *J Clin Oncol* **8**, 1389–1398.

Fina, L., Molgaard, H. V., Robertson, O., *et al.* (1990). Expression of the CD34 gene in vascular endothelial cells. *Blood* **75**, 2417–2426.

Gee, A. (1995). Purging of peripheral blood stem cell grafts. *Stem Cells* **13** (Suppl. 3), 52–62.

Gee, A. P., and Lee, C. (1998). T-cell depletion of allogeneic stem-cell grafts. In "The Clinical Practice of Stem-Cell Transplantation" (J. Barrett and J. G. Treleaven, Eds.), Vol. 2, pp. 474–505. Isis Medical Media, Oxford.

Gorin, N. C., Labopin, M., Meloni, G., *et al.* (1991). Autologous bone marrow transplantation for acute myeloblastic leukemia in Europe. Further evidence of the role of marrow purging by mafosfamide. European co-operative group for bone marrow transplantation (EBMT). *Leukemia* **5**, 896–904.

Haas, R., Mohle, R., Fruhauf, S., *et al.* (1994). Patient characteristics associated with successful mobilizing and autografting of peripheral blood progenitor cells in malignant lymphoma. *Blood* **83**, 3787–3794.

Hernandez-Navarro, F., Ojeda, E., Arrieta, R., *et al.* (1995). Single-centre experience of peripheral blood stem cell transplantation using cryopreservation by immersion in a methanol bath. *Bone Marrow Transplant* **16**, 71–77.

Herrmann, R. P., O'Reilly, J., Meyer, B. F., *et al.* (1992). Prompt haemopoietic reconstitution following hyperthermia purged autologous marrow and peripheral blood stem cell transplantation in acute myeloid leukaemia. *Bone Marrow Transplant* **10**, 293–295.

Horowitz, M. M., and Rowlings, P. A. (1997). An update from the International Bone Marrow Transplant Registry and the Autologous Blood and Marrow Transplant Registry on current activity in hematopoietic stem cell transplantation. *Curr Opin Hematol* **4**, 395–400.

Keeney, M., Chin-Yee, I., Weir, K., et al. (1998). Single platform flow cytometric absolute CD34 + cell counts based on the ISHAGE guidelines. International Society of Hematotherapy and Graft Engineering. *Cytometry* **34**, 61–70.

Mauch, P., Lamont, C., Neben, T. Y., et al. (1995). Hematopoietic stem cells in the blood after stem cell factor and interleukin-11 administration: Evidence for different mechanisms of mobilization. *Blood* **86**, 4674–4680.

Roberts, M. M., To, L. B., Gillis, D., et al., (1993). Immune reconstitution following peripheral blood stem cell transplantation, autologous bone marrow transplantation and allogeneic bone marrow transplantation. *Bone Marrow Transplant* **12**, 469–475.

Rowley, S. D. (1992). Hematopoietic stem cell cryopreservation: A review of current techniques. *J Hematother* **1**, 233–250.

Rowley, S. D., and Davis, J. M. (1991). The use of 4-HC in autologous purging. In "Bone Marrow Processing and Purging: A Practical Guide" (A. P. Gee, Ed.), pp. 247–262. CRC Press, Boca Raton.

Russell, N. H., and Miflin, G. (1998). Peripheral blood stem cells for autologous and allogeneic transplantation. In "The Clinical Practice of Stem-Cell Transplantation" (J. Barrett and J. Treleaven, Eds.), Vol. 1, pp. 453–472. Isis Medical Media, Oxford.

Salmon, J. P., Michaus, S., Hermanne, J. P., et al. (1999). Delayed massive immune hemolysis mediated by minor ABO incompatibility after allogeneic peripheral blood progenitor cell transplantation. *Transfusion* **39**, 824–827.

Schmitz, N., Linch, D. C., Dreger, P., et al. (1996). Randomised trial of fllgrastim-mobilised peripheral blood progenitor cell transplantation versus autologous bone-marrow transplantation in lymphoma patients. *Lancet* **347**, 353–357. [Published erratum appears in *Lancet* **347**, 914]

Shpall, E. J., LeMaistre, C. F., Holland, K., et al. (1997). A prospective randomized trial of buffy coat versus CD34-selected autologous bone marrow support in high-risk breast cancer patients receiving high-dose chemotherapy. *Blood* **90**, 4313–4320.

Stadtmauer, E. A., Schneider, C. J., Silberstein, L. E. (1995). Peripheral blood progenitor cell generation and harvesting. *Semin Oncol* **22**, 291–300.

Stiff, P. J. (1991). Simplified bone marrow cryopreservation using dimethysulfoxide and hydroxyethylstarch as cryoprotectants. In "Bone Marrow Processing and Purging: A Practical Guide" (A. P. Gee, Ed.), pp. 341–350. CRC Press, Boca Raton.

Storek, J., Gooley, T., Siadak, M., et al. (1997). Allogeneic peripheral blood stem cell transplantation may be associated with a high risk of chronic graft-versus-host disease. *Blood* **90**, 4705–4709.

To, L. B., Roberts, M. M., Haylock, D. N., et al. (1992). Comparison of haematological recovery times and supportive care requirements of autologous recovery phase peripheral blood stem cell transplants, autologous bone marrow transplants and allogeneic bone marrow transplants. *Bone Marrow Transplant* **9**, 277–284.

Weaver, C. H., Hazelton, B., Birch, R., et al. (1995). An analysis of engraftment kinetics as a function of the CD34 content of peripheral blood progenitor cell collections in 692 patients after the administration of myeloablative chemotherapy. *Blood* **86**, 3961–3969.

Zambelli, A., Poggi, G., Da Prada, G., et al. (1998). Clinical toxicity of cryopreserved circulating progenitor cells infusion. *Anticancer Res* **18**, 4705–4708.

20

Umbilical Cord Blood Stem Cells

SCOTT R. BURGER

Department of Laboratory Medicine and Pathology*
University of Minnesota
Minneapolis, Minnesota

As described in Chapters 18 and 19, bone marrow (BM) and, more recently, cytokine-stimulated peripheral blood (PB) have been the principal sources of hematopoietic stem and progenitor cells (HPCs) for transplantation. Umbilical cord blood (UCB), also termed placental or fetal blood, contains transplantable stem and progenitor cells and may possess significant clinical advantages in transplantation (see Table 1). Unlike BM and PB HPCs, UCB cells are collected as a by-product of uncomplicated pregnancy. Because special mobilization and extensive collection processes are not required, UCB cells can be collected for storage in a bank of transplantable cells (Harris, 1996; Rubinstein *et al.*, 1993). There is also evidence suggesting that allogeneic UCB transplants are associated with less frequent and less severe graft-versus-host disease (GVHD) than equivalent BM- or PB-HPC transplants (Rubinstein *et al.*, 1998; Kurtzberg *et al.*, 1996). There are, however, several potential and actual problems associated with UCB transplants. These include: (1) the limited number of cells obtained in UCB collections, (2) the inability to perform follow-up collections, (3) the frequency of microbially contaminated collections, (4) the unknown incidence of GVHD and graft-versus-leukemia (GVL) effect, and (5) questions about the long-term immune competence of marrow reconstituted from UCB cells (Kögler *et al.*, 1996).

First proposed as a potential source of clinically useful HPCs in the 1980s, UCB transplantation has rapidly gained acceptance in transplantation, with over 1200 UCB cell transplants having been performed. In

allogeneic UCB transplantation, UCB cells collected at delivery of a presumably healthy infant are used as the source of HPCs. These may be sibling-directed allogeneic UCB transplants, in which a patient is transplanted using UCB cells collected at birth of an HLA-compatible sibling, or allogeneic-unrelated, in which banked unrelated UCB cells are selected for transplant based on HLA compatibility. Autologous UCB transplants are a theoretical possibility, in which UCB cells, collected at delivery and cryopreserved, could be used in the event the donor requires a transplant later in life. To date, only allogeneic UCB cell transplants have been reported, and these will be the focus of this chapter.

I. PRODUCT DESCRIPTION

A. Cellular Constituents

UCB grafts contain several cell populations necessary for a favorable clinical outcome. The most primitive HPCs are defined by their ability to self-renew as well as proliferate, while lineage-committed and uncommitted progenitors are capable only of proliferation and differentiation. Self-renewing stem cells are extremely long-lived, while short-lived progenitor cells can rapidly give rise to large numbers of functional immune and blood cells. In HPC transplantation, progenitor cells provide short-term hematopoietic and immune function after myeloablation, while stem cells more gradually give rise to a new, long-lived and self-perpetuating hematopoietic system. Ancillary cell pop-

---

* Present address: Merix Bioscience, Durham, North Carolina.

TABLE 1 Established and Potential Advantages of UCB Banking and Transplantation Compared to BM Transplant

Established advantages

- Graft available for transplant immediately after match is found
- Long-term availability of cryopreserved grafts
- UCB collected from ethnically diverse populations increases likelihood of matches
- Decreased incidence of infectious agents

Potential advantages

- Lower incidence of GVHD
- Less severe GVHD, if it occurs
- May permit more extensively HLA-mismatched transplants than BMT

ulations, particularly T-lymphocytes, also function in the graft, mediating some engraftment, GVHD, and antitumor effects (Truitt and Johnson, 1995).

B. HPC Content

Assays for these different cell populations are not completely satisfactory. For practical purposes, broadly defined HPCs are more commonly enumerated. Because HPCs are nucleated and, more specifically, are mononuclear cells, these types of cells are enumerated using automated cell counters. Expression of the CD34 cell surface molecule currently is the most widely accepted marker for stem and progenitor cells. CD34$^+$ cells commonly are measured by flow cytometry, or more recently by laser-scanning cytometry. The progenitor assay, which identifies clonogenic HPCs by their ability to produce myeloid and erythroid colonies (CFU-GM and BFU-E, respectively), also may be used. Various long-term cell-culture-based assays can be used to distinguish stem and progenitor cells, but time constraints make many of these methods impractical for clinical cell engineering. It is therefore difficult to precisely determine the dose of stem and progenitor cells in a typical UCB transplant. UCB grafts are typically described in terms of nucleated cell dose, CD34$^+$ cell dose, and sometimes CFU-GM dose. T-lymphocytes also are commonly measured in UCB grafts as CD3$^+$ cells. The cellular composition of a typical UCB graft is described in Table 2.

C. Anticoagulants and Additives

UCB products contain autologous plasma, anticoagulant, and electrolyte solution. Autologous plasma present in the UCB is collected with the cells, and may

or may not be removed from the final product. Anticoagulants currently used include heparin, acid-citrate-dextrose (ACD), citrate-phosphate-dextrose (CPD), or CPD-adenine (CPDA-1). Because the final volume of the UCB specimen being collected is not precisely known in advance, citrate-based anticoagulants can be problematic, due to the blood:citrate volume ratio needed for effective anticoagulation. Many UCB banks prefer citrate-based anticoagulants, in volumes over 20 ml, capable of anticoagulating all but the largest-volume UCB products. Heparin provides effective anticoagulation across a broader range of anticoagulant concentrations, but is not universally used.

The electrolyte solution used is most often a tissue culture medium, typically added after collection is completed. Because tissue culture media are not approved for clinical use and employ buffer systems that do not function in room air, clinical-grade balanced electrolyte solutions are increasingly being developed and used (Burger et al., 1999).

D. Labeling

Product labels for UCB specify the product name, "Human Cord Blood," the collection date and time, cryopreservation date and time, the unit number, identifying information for the collection and processing facilities, additives such as anticoagulant, antibiotics, electrolyte solution, and cryoprotectant, the product volume and total nucleated cell content, and ABO and Rh type of the donor. Labels may vary depending on whether the UCB product is being stored in an unrelated allogeneic bank or as a directed-donor product. In the latter case, labels may also include identifying information for the donor and recipient and recipient ABO and Rh type.

II. COLLECTION, STORAGE, AND HANDLING

A. Collection for Establishment of a Cell Bank for Transplantation

UCB may be collected for transplant into a specific patient, typically a sibling, or for storage in a bank. Ability to bank cells is a unique advantage of UCB cells. Banked cells ensure that the graft is available and is screened and tested for infectious disease. A bank of UCB cells also can be established such that ethnic diversity of its contents represents prospective patient populations, potentially increasing the likelihood of obtaining matches. Several UCB banks have been established, some in cooperation with the NIH, to provide cells for unrelated allogeneic transplant (Wagner and Kurtzberg, 1998; Harris, 1996; Rubinstein *et al.*, 1993). Private, for-profit cord blood banks also have been established to provide autologous or sibling-directed cells in case transplant is needed later in life. Concerns regarding feasibility and appropriateness of autologous UCB transplants are discussed below.

B. Collection Process

1. Donor Criteria and Evaluation

During pregnancy, the mother must meet specific criteria for the UCB to be collected and banked. Her medical, family, and obstetric history must be suitable, particularly with regard to risk of infectious disease exposure, likelihood of uncomplicated delivery, and family history of genetic disease that might be transmitted by UCB. The mother must give informed consent for her newborn child to donate blood remaining in the umbilical cord and placenta after delivery, for this blood to be cryopreserved, for her and her child's medical history to be reviewed, for her blood to undergo infectious disease testing, and for possible genetic testing of the UCB in certain circumstances. Ideally, the expectant mother should be approached early in pregnancy and provided with information about cord blood collection and, if relevant, donation to a cord blood bank.

2. Collection Procedure

UCB is collected as soon as possible after delivery, with the informed consent of the mother and the cooperation of the obstetric team. Properly performed, the UCB collection process does not harm the infant or interfere with childbirth. Proper collection methods are critical, in that additional cord blood cells cannot be obtained by prolonging the collection or reharvesting at a later time.

UCB may be collected by closed or open methods. Closed collection is performed using anticoagulant-containing syringes to collect blood from the umbilical vein. This method generally results in excellent cell yield, with infrequent microbial contamination. Open collection is performed by draining the transected umbilical cord into a sterile container to which anticoagulant has been added in advance. This typically results in acceptable cell yield but a higher incidence of microbial contamination. In one study, open collection was associated with a bacterial contamination rate of 12.5%, compared to only 3.3% for closed collection (Bertolini *et al.*, 1995).

The time of umbilical cord clamping has a significant influence on the volume of cord blood collected. Umbilical cords clamped within 30 seconds of vaginal delivery produced, in one study of 378 deliveries, UCB collections of 77 ± 23 ml, while delayed cord clamping reduced UCB volume by half. Umbilical cord clamping within 30 seconds of delivery was not associated with adverse effects in the newborn (Bertolini *et al.*, 1995).

UCB may be collected before or after delivery of the placenta. If collected before placental delivery, the obstetric staff collects blood from the undelivered placenta immediately after clamping the cord. This does, however, require the attention of obstetric staff at a busy moment. Alternatively, the delivered clamped or tied placenta may be brought to a nearby collection area where UCB is collected by either closed or open methods. The quality of the UCB product may be compromised, however, as cell yield also decreases with time after umbilical cord clamping and transection, likely due to cord blood clotting. Collection of UCB postplacental delivery produced UCB products with volumes 10-20% lower than those collected with the placenta *in utero* (Bertolini *et al.*, 1995).

3. Donor Infectious Disease Testing

Infectious disease testing for UCB is performed using maternal blood. Testing the UCB product itself is more likely to yield false-negatives, in that the antiviral antibodies for which testing is performed do not always cross the placenta. Testing maternal blood also minimizes the UCB product volume used in testing. Screening includes anti-HIV-1/2, HIV-1 p24 antigen, anti-HTLV-I/II, anti-hepatitis B core antigen, hepatitis B surface antigen, anti-hepatitis C, ALT, and syphilis testing.

4. Laboratory Evaluation of UCB

Tests performed on the UCB product itself typically include product volume, nucleated cell count and differential, percent $CD34^+$ cells, ABO and Rh type, HLA class I and II type, and microbial culture. Other tests may include cell viability, percent $CD3^+$ cells, and a progenitor assay to determine CFU-GM content. Cellular characteristics of typical UCB specimens are shown in Table 2.

C. Storage and Transport

UCB products must be transported to the processing laboratory for further processing and cryopreservation. This entails a period of nonfrozen storage. UCB may be transported and stored with only the anticoagulant and medium or electrolyte solution used in collection, or may be supplemented with additional anticoagulant, plasma, albumin, electrolyte solutions, or tissue culture medium. Optimal additives and storage conditions for UCB have not been established. Successful HPC storage has been reported at 4°C and at ambient temperatures for varying durations, some over 48 hours (Burger et al., 1999), but these studies have not specifically addressed UCB-derived HPCs. One study of UCB cells collected into CPD anticoagulant demonstrated that cells could be maintained at 4°C for 24 hours with essentially unchanged numbers of mononuclear cells and CFU-GM (Tron de Bouchony et al., 1993). CFU-GM were less well maintained after 48 hours at 4°C, however, and were poorly preserved after only 24 hours at 22°C. In practice, UCB is rarely stored for longer than 36 hours without cryopreservation.

III. SPECIAL PREPARATION AND PROCESSING

A. Mononuclear Cell Enrichment

UCB cells may be cryopreserved without additional processing. Although simple, this retains a comparatively large volume of cells not known to function in engraftment, such as granulocytes and RBCs. Cryopreserved and thawed granulocytes are predominantly lysed, which may cause adverse effects in the recipient. Numerous cryopreserved and thawed RBCs also lyse. These cells also may not be ABO-compatible with the recipient. If cells are to be banked, then precryopreservation processing also may be performed to minimize the volume of product stored. For these reasons, UCB is usually processed prior to cryopreservation. A buffy coat may be prepared by centrifugation with a density agent, such as Ficoll or Percoll (Denning-Kendall et al., 1996). This is a comparatively simple process that effectively removes plasma, granulocytes, and RBCs, while retaining mononuclear cells, including hematopoietic stem and progenitor cells. Alternatively, RBCs may be removed, commonly by hydroxyethyl starch (HES) or gelatin sedimentation (Denning-Kendall et al., 1996; Rubinstein et al., 1995). This method also is relatively simple, although it does not remove granulocytes. Cell separation by density centrifugation may be most effective when performed on fresh specimens (Bertolini et al., 1995; Broxmeyer et al., 1989).

B. Cryopreservation

UCB is commonly cryopreserved. Cells are stored in flattened plastic bags designed to maximize heat trans-

fer and to withstand the stresses associated with cryopreservation and thawing. Small-volume (1–2 ml) aliquots of the product, as well as samples of maternal and possibly neonatal blood, also are cryopreserved in cryovials or in sample bags in the event that these specimens require testing not performed prior to freezing. The maximum storage duration of cryopreserved UCB cells is not known, but may be longer than 10 years (Broxmeyer and Cooper, 1997).

Programmed (rate-controlled) freezing is the historical standard for clinical cell cryopreservation, although evidence increasingly indicates that nonprogrammed (mechanical) freezing may be acceptable (Hernandez-Navarro et al., 1998). A cryopreservative agent intended to minimize cryogenic toxicity to the cells is added immediately prior to cryopreservation. The most commonly used cryoprotectant is dimethyl sulfoxide (DMSO), at a final concentration of 10% v/v. Because DMSO can cause adverse reactions when infused, some laboratories cryopreserve using a lower DMSO concentration—most commonly a mixture of 5% DMSO with 6% hydroxyethyl starch (Donaldson et al., 1996). Cryopreserved UCB cells are stored in either liquid nitrogen (liquid or vapor phase) or, rarely, in a mechanical freezer.

C. Thawing

Cryopreserved UCB cells are thawed prior to infusion, most commonly by immersion in a 37°C sterile saline bath. The thawed product, containing intact cells, lysed fragments, and cryoprotectant, may be infused without further manipulation. Because UCB transplants often are performed for smaller patients, the relative dose of DMSO cryoprotectant may be larger than desired, and the adverse effects of DMSO may be more severe. Some laboratories choose to wash UCB cells immediately after thawing to reduce the dose of DMSO administered with the product. The thawed product is centrifuged, the DMSO-rich supernatant removed, and the washed cells resuspended in an infusion solution, for example, dextran-albumin (Rubinstein et al., 1995).

D. HPC Selection and Expansion

Like other HPCs, UCB cells may be further purified by CD34$^+$ cell selection. UCB cell CD34$^+$ selection is often used as a processing step preparatory to gene manipulation and cell expansion (Briddell et al., 1997). Because UCB collection cannot be repeated to increase the number of cells obtained, the potential for ex vivo expansion of UCB-derived stem and progenitor cells will be of obvious importance. UCB cells also have potential applications in gene therapy. Populations of UCB committed progenitor cells (CFU-GM) and more primitive near-stem cells have been expanded in culture, in some cases generating $> 10^7$ CFU-GM from as little as 15 ml of UCB (Koller *et al.*, 1998). In one study, UCB CD34$^+$ cells were transduced with an ADA-containing retrovirus and transplanted into patients with ADA deficiency, with sustained expression of the ADA gene (Kohn *et al.*, 1995). Trials of selected, expanded, and transduced UCB cells can be expected to proliferate.

IV. ABO AND Rh COMPATIBILITY

Transplantation with a less-than-perfectly matched HLA-compatible product can induce life-threatening GVHD, in which functional T-lymphocytes in the UCB graft recognize the immunocompromised recipient as nonself (Ferrara and Krenger, 1998). Also, an HLA-matched graft may not be ABO- and Rh-compatible if the product is cryopreserved without further processing. Given the potentially deadly consequences of GVHD, ABO- or Rh-incompatible, but HLA-matched, products sometimes are transplanted. The UCB transplant may involve major ABO incompatibility, in which RBCs in the product are incompatible with the recipient's plasma antibodies, or minor ABO incompatibility, in which donor plasma antibodies can react with the recipient's RBCs. Clinically significant RBC incompatibility can be prevented by cell processing procedures that remove RBCs and plasma, as described above.

V. ORDERING AND ADMINISTRATION

A. Identification of Transplantable UCB Product

The principal parameters considered in selecting a specific UCB product for transplant include the cell content and HLA compatibility of the products available. The minimum transplantable dose of UCB cells has not yet been established. In the largest study to date, of 562 UCB transplants, 86% of patients receiving a cell dose greater than 5.0×10^7 nucleated cells/kg engrafted by day 42 (Rubinstein *et al.*, 1998). Similarly, the degree of HLA incompatibility tolerable in allogeneic UCB transplant is not yet known. The majority of UCB transplants reported have been matched for 6/6 HLA antigens, although two-antigen mismatched UCB transplants have been described.

B. Infusion Procedure

Patients about to receive a transplant of UCB cells should receive preinfusion medications approximately 30 minutes prior to infusion. Because adverse events associated with infusion of cryopreserved hematopoietic cells include fever, chills, nausea, vomiting, and hypertension (Kessinger *et al.*, 1990), premedications should include an antipyretic, an antihistamine, and an antiemetic.

Cells should be infused as soon as possible after thawing, or after washing, if performed. Post-thaw or postwash durations of greater than 60 minutes should be avoided. Cells may be infused through the patient's central catheter, infusing by gravity feed as rapidly as tolerated. The practice of filtering UCB products is controversial. Filtration can remove large cell aggregates and cellular debris produced after thawing. The possibility of trapping HPCs in the filter is of concern, however. If products are filtered, care should be taken to use a filter size of 80 or 170–210 μm. *Leukodepletion filters must* not *be used*, as these will effectively remove the cells to be transplanted, with potentially fatal consequences.

VI. INDICATIONS AND CONTRAINDICATIONS

A. Indications

Indications for UCB transplant include most malignant and nonmalignant disorders currently treated with HPC transplantation. Diseases that have been treated with UCB-derived cells include acute and chronic lymphocytic leukemia, acute and chronic myelogenous leukemia, non-Hodgkin's lymphoma, neuroblastoma, myelodysplastic syndrome, aplastic anemia, Fanconi's anemia, severe combined immunodeficiency, osteopetrosis, Hurler's syndrome, Wiskott–Aldrich syndrome, adrenoleukodystrophy, and Blackfan–Diamond syndrome.

Unlike BM- or PB-HPCs, if the dose obtained is inadequate, additional UCB cells cannot be obtained by prolonging the collection or harvesting at a later time. For this reason, the majority of UCB transplants have been performed in patients of comparatively small body weight (e.g., children or adults of small stature).

Despite increasing popularity of commercial autologous UCB banking, a theoretical resource if transplant is needed later in life, autologous UCB transplants have not yet been reported. Evidence that leukemic cells are present in fetal and neonatal blood of patients later diagnosed with leukemia at ages 9 and 10 years (Ford *et al.*, 1997) suggests that autologous UCB transplantation may be an extremely inappropriate therapy in certain situations. Also, autologous UCB banking may be excessively costly, given its lack of proven efficacy.

B. Contraindications

Established contraindications to UCB transplant include inadequate cell dose, incompatible HLA-match, and disease stage or state inappropriate for transplantation.

VII. EXPECTED RESPONSE

Like any other HPC transplant, stem and progenitor cells from UCB are expected to engraft and reconstitute the patient's hematopoietic system. The patient is at risk of life-threatening infectious complications during the preengraftment neutropenic interval. The rapidity with which engraftment occurs is therefore of great importance. In the largest study to date, median time to neutrophil engraftment was 28 days, with 81% of patients demonstrating neutrophil engraftment by day 42. Median time to platelet engraftment was 90 days, with 85% of patients showing platelet engraftment by day 180 (Rubinstein et al., 1998). Time to engraftment appears primarily associated with the nucleated cell content of the graft.

Graft-versus-malignant cell effect, also known as the graft-versus-leukemia (GVL) effect or graft-versus-tumor effect, is an additional beneficial effect of allogeneic transplantation (Truitt and Johnson, 1995). This phenomenon appears to be mediated by donor T-lymphocytes. In that UCB lymphocytes appear immunologically immature, it is not clear whether UCB transplants will provide significant graft-versus-malignant cell effect. No conclusions can yet be reached on this point, based on the UCB transplant trials performed to date.

VIII. ADVERSE EVENTS

Adverse events associated with transplantation may be related to conditioning chemo- and radiotherapy, or with the cellular graft. The former will not be discussed here.

A. Acute Adverse Events

Cell-related adverse events may occur at the time of infusion. Patients may experience reactions at-tributable to the DMSO cryoprotectant or to lysed cells in the product, or may experience transfusion reactions, as with any other blood product. It is not always possible to distinguish DMSO-related symptoms from those of a transfusion reaction. Symptoms experienced by some patients during infusions include fever, chills, nausea, vomiting, hypertension, tachycardia, dyspnea, and headache. The incidence of these symptoms is greatly diminished if premedications are administered. For example, one study, in which acetaminophen premedication was not used at infusion of cryopreserved thawed marrow, reported an incidence of infusion-associated fever and chills of 65 and 67%, respectively (Kessinger et al., 1990), while a similar trial with acetaminophen described only 8% of patients developing fever and 13% experiencing chills. Patients may also experience adverse events related to fluid overload, depending on volume infused and the patient's volume tolerance.

GVHD is one of the deadliest complications of allogenoic hematopoietic transplantation. Severe (grades II–IV) acute GVHD occurs in as many as 64% of patients transplanted with BM matched for 6/6 HLA antigens (Kernan et al., 1993). It is noteworthy that *GVHD occurs less frequently, and in milder form, in allogeneic UCB cell transplantation than in equivalently matched allogeneic BM- or PB-HPC transplantation.* In one large study, incidence of grade II-IV acute GVHD was only 27% for 6/6 matched UCB transplants. Not surprisingly, mismatched UCB transplants were associated with a higher incidence of GVHD, with 49% of patients receiving ≥ 2 mismatched antigen grafts developing severe acute GVHD (Rubinstein et al., 1998). That two-antigen mismatched UCB transplants are successful at all is striking, however, particularly considering that half of patients transplanted with these grafts developed only mild GVHD.

B. Delayed or Chronic Adverse Events

Failure to engraft is a potential adverse event in any transplant of HPCs. Engraftment is influenced by multiple factors, and failure to engraft may be attributable to numerous etiologic influences. In one multivariate analysis, the single most important factor influencing neutrophil engraftment in pediatric patients with acute leukemia was a nucleated cell dose of $> 3.7 \times 10^7$ cells/kg (Gluckman et al., 1997).

Disease relapse is also a major source of morbidity and mortality in hematopoietic transplantation for malignant disease. In the largest study of UCB transplants, 9% of patients transplanted for leukemia relapsed within 100 days of transplant, and 26% relapsed by 1 year post-transplant (Rubinstein et al., 1998).

There are some questions regarding the quality of immune function of BM derived from UCB cells. UCB appears enriched for primitive hematopoietic stem and near-stem cells, but UCB-derived cells also appear immunologically more primitive (Cohen and Madrigal, 1998). This may explain observations that transplanted UCB cells give rise to GVHD less frequently, and with less severity, than other HPC sources. Whether the immune systems reconstituted by UCB cells eventually develop and maintain normal adult function is currently being studied.

IX. ALTERNATIVES

There are two alternatives to transplantation with UCB-derived stem and progenitor cells—transplantation with BM-HPCs or mobilized PB-HPCs.

Allogeneic unrelated UCB banks are growing in number and size, in keeping with increasing use of UCB in hematopoietic transplantation. UCB likely will continue to be banked for directed-donor (sibling-related) UCB transplants, particularly when a family member has already been diagnosed with a disorder potentially treatable by UCB transplant. UCB banking for potential autologous transplant appears increasingly unjustifiable, as discussed above. Although numerous questions regarding use of UCB remain to be answered, particularly about the long-term outcomes of UCB transplants, UCB may become the established source of HPCs for transplantation.

References

Bertolini, F., Lazzari, L., Lauri, E., et al. (1995). Comparative study of different procedures for the collection and banking of umbilical cord blood. J Hematother 4(1), 29–36.

Briddell, R. A., Kern, B. P., Zilm, K. L., et al. (1997). Purification of CD34+ cells is essential for optimal ex vivo expansion of umbilical cord blood cells. J Hematother 6(2), 145–150.

Broxmeyer, H. E., and Cooper, S. (1997). High-efficiency recovery of immature haematopoietic progenitor cells with extensive proliferative capacity from human cord blood cryopreserved for 10 years. Clin Exp Immunol 107(1), 45–53.

Broxmeyer, H. E., Douglas, G. W., Hangoc, G., et al. (1989). Human umbilical cord blood as a potential source of transplantable hematopoietic stem/progenitor cells. Proc Nat Acad Sci USA 86(10), 3828–3832.

Burger, S. R., Hubel, A. H., and McCullough, J. (1999). Development of an infusible-grade solution for non-cryopreserved hematopoietic cell storage. Cytotherapy 1(2), 123–133.

Cohen, S. B., and Madrigal, J. A. (1998). Immunological and functional differences between cord and peripheral blood. Bone Marrow Transplant 21(3), S9–12.

Denning-Kendall, P., Donaldson, C., Nicol, A., et al. (1996). Optimal processing of human umbilical cord blood for clinical banking. Exp Hematol 24(12), 1394–1401.

Donaldson, C., Armitage, W. J., Denning-Kendall, P. A., et al. (1996). Optimal cryopreservation of human umbilical cord blood. Bone Marrow Transplant 18(4), 725–731.

Ferrara, J. L., and Krenger, W. (1998). Graft-versus-host disease: The influence of type 1 and type 2 T cell cytokines. Transfusion Med Rev 12(1), 1–17.

Ford, A. M., Pombo-de-Oliveira, M. S., McCarthy, K. P., et al. (1997). Monoclonal origin of concordant T-cell malignancy in identical twins. Blood 89(1), 281–285.

Gluckman, E., Rocha, V., Boyer-Chammard, A., et al. (1997). Outcome of cord-blood transplantation from related and unrelated donors. Eurocord Transplant Group and the European Blood and Marrow Transplantation Group. N Engl J Med 337(6), 373–381.

Harris, D. T. (1996). Experience in autologous and allogeneic cord blood banking. J Hematother 5(2), 123–128.

Hernandez-Navarro, F., Ojeda, E., Arrieta, R., et al. (1998). Hematopoietic cell transplantation using plasma and DMSO without HES, with non-programmed freezing by immersion in a methanol bath: Results in 213 cases. Bone Marrow Transplant 21(5), 511–517.

Kernan, N. A., Bartsch, G., Ash, R. C., et al. (1993). Analysis of 462 transplantations from unrelated donors facilitated by the National Marrow Donor Program. N Engl J Med 328(9), 593–602.

Kessinger, A., Schmit-Pokorny, K., Smith, D., and Armitage, J. (1990). Cryopreservation and infusion of autologous peripheral blood stem cells. Bone Marrow Transplant 5(1), 25–27.

Knapp, W., Strobl, H., Scheinecker, C., et al. (1995). Molecular characterization of CD34+ human hematopoietic progenitor cells. Ann Hematol 70(6), 281–296.

Kögler, G., Callejas, J., Hakenberg, P., et al. (1996). Hematopoietic transplant potential of unrelated cord blood: Critical issues. J Hematother 5(2), 105–116.

Kohn, D. B., Weinberg, K. I., Nolta, J. A., et al. (1995). Engraftment of gene-modified umbilical cord blood cells in neonates with adenosine deaminase deficiency. Nature Med 1(10), 1017–1023.

Koller, M. R., Manchel, I., Maher, R. J., et al. (1998). Clinical-scale human umbilical cord blood cell expansion in a novel automated perfusion culture system. Bone Marrow Transplant 21(7), 653–663.

Kurtzberg, J., Laughlin, M., Graham, M. L., et al. (1996). Placental blood as a source of hematopoietic stem cells for transplantation into unrelated recipients. N Engl J Med 335(3), 157–166.

Rowley, S. D., Zuehlsdorf, M., Braine, H. G., et al. (1987). CFU-GM content of bone marrow graft correlates with time to hematologic reconstitution following autologous bone marrow transplantation with 4-hydroperoxycyclophosphamide-purged bone marrow. Blood 70(1), 271–275.

Rubinstein, P., Rosenfield, R. E., Adamson, J. W., and Stevens, C. E. (1993). Stored placental blood for unrelated bone marrow reconstitution. Blood 81(7), 1679–1690.

Rubinstein, P., Dobrila, L., Rosenfield, R. E., et al. (1995). Processing and cryopreservation of placental/umbilical cord blood for unrelated bone marrow reconstitution. Proc Nat Acad Sci USA 92(22), 10119–10122.

Rubinstein, P., Carrier, C., Scaradavou, A., et al. (1998). Outcomes among 562 recipients of placental-blood transplants from unrelated donors. N Engl J Med 339(22), 1565–1577.

Stroncek, D. F., Fautsch, S. K., Lasky, L. C., et al. (1991). Adverse reactions in patients transfused with cryopreserved marrow. Transfusion 31(6), 521–526.

Sutherland, D. R., Keating, A., Nayar, R., *et al.* (1994). Sensitive detection and enumeration of CD34$^+$ cells in peripheral and cord blood by flow cytometry. *Exp Hematol* **22**(10), 1003–1010.

Tron de Bouchony, E., Pelletier, D., Alcalay, D., *et al.* (1993). Hematopoietic progenitor content of fetal cord blood collected using citrate-phosphate-dextrose:influence of holding temperature and delays. *J Hematother* **2**(2), 271–273.

Truitt, R. L., and Johnson, B. D. (1995). Principles of graft-*vs.*-leukemia reactivity. *Biol Blood Marrow Transplant* **1**(2), 61–68.

Wagner, J. E., and Kurtzberg, J. (1998). Banking and transplantation of unrelated donor umbilical cord blood: Status of the National Heart, Lung, and Blood Institute-sponsored trial. *Transfusion* **38**(9), 807–809.

Webb, I. J. (1997). Umbilical cord blood as a source of progenitor cells to reconstitute hematopoiesis. *Transfusion Med Rev* **11**(4), 265–273.

21

Mononuclear Cell Preparations

MARK A. VALLECA
JOAN DEBELAK
DIANE S. KRAUSE
Department of Laboratory Medicine
Yale University School of Medicine
New Haven, Connecticut

This chapter focuses on the use of mononuclear cell (MNC) preparations for immunotherapy, including unmanipulated cells as well as cells processed *in vitro* to serve specific immunologic functions (e.g., cytotoxic T-lymphocytes and dendritic cells). Transfusion of MNCs is a relatively new practice; therefore, many of the indications for usage as well as the protocols for cell preparation are still under development. The infusion of donor cells is referred to in several ways, including donor leukocyte infusion (DLI), adoptive immunotherapy, and allogeneic cell therapy. MNCs can be used for several different purposes in the postallogeneic transplant setting. Donor MNCs have been effective in inducing remission when there is relapse of malignancy, and are also useful in fighting infections in the post-transplant setting.

I. PRODUCT DESCRIPTION

A. Cellular Constituents

1. Cells for Donor Leukocyte Infusion

Mononuclear cell preparations for DLI are products in which the majority of the cells are mononuclear, including monocytes, lymphocytes, natural killer cells, hematopoietic precursors of all lineages, and hematopoietic stem cells. The percentage of each cell type is dependent on the source of the cells and their intended use. Peripheral blood MNC apheresis from mobilized donors for collection of $CD34^+$ hematopoietic stem and progenitor cells (HSCs and HPCs) is covered in Chapter 19. The cells that are critical for DLI are the cells of the immune system, including predominantly T-lymphocytes ($CD2^+$, $CD3^+$, $CD4^+$, and $CD8^+$ lymphocytes). These cells are present in normal HPC collections, but their use as a DLI–MNC preparation is due to their role in immune function.

2. Cytotoxic T-Lymphocytes (CTLs)

$CD8^+$ cytotoxic T-lymphocytes (CTLs) are activated in response to antigen presentation by cells that coexpress major histocompatibility (MHC) antigens. Donor-derived cytotoxic T-lymphocytes (CTLs) that are targeted against specific antigens are referred to as antigen-specific CTLs. CTLs effectively destroy target cells, including virus-infected cells and tumor cells. CTLs are present in peripheral blood and bone marrow, and can be expanded *ex vivo* to obtain sufficient therapeutic doses. In this chapter, antigen-specific CTLs will be covered separately from MNCs for DLI.

3. Dendritic Cells (DCs)

DCs are potent antigen-presenting cells that are capable of inducing strong T-cell-mediated immunity (Banchereau and Steinman, 1998). Since DCs are exceedingly rare in peripheral blood ($< 0.5\%$ of MNCs), they are often cultured *ex vivo* from monocyte precursors ($CD14^+$) or from hematopoietic stem cells ($CD34^+$). DCs currently are used in the autologous setting as immunotherapy for cancer (or, less frequently, for infectious diseases). DCs are pulsed *ex vivo* with tumor (or pathogen) antigens, and then reintroduced into the patient to induce a T-cell-mediated antigen-specific response, thereby acting as a cellular vaccine to induce remission and/or prevent recurrence.

B. Anticoagulants

If the MNCs are collected by apheresis or by collection of peripheral blood, the most common anticoagulant used is citrate. Sodium citrate chelates calcium to prevent activation of platelets and the clotting cascade. After the cells are infused into the patient, the citrate is rapidly diluted and metabolized by the liver. There is currently no indication for removing citrate from a cellular product. Because CTLs and DCs are grown *in vitro* and do not contain either platelets or activated coagulation factors, no anticoagulant is used.

C. Labeling

As for all other cellular products to be administered to patients, proper labeling must include the patient's name and medical record number, the recipient's name and additional identifying information, the name of the product, the volume of the product, the volume and name of anticoagulant, the total dose and specific cells present in the product, the expiration date, the proper storage conditions, and any relevant infectious warnings if the donor tests positive for infectious markers.

II. COLLECTION PROCESS

A. Donor Evaluation

Donors must be in good physical condition, afebrile, and without malignancy. The donor will be questioned regarding any risk behaviors for infection with transfusion-transmitted diseases (TTDs). As for any other donor evaluation, the donor must be adequately informed of the risks of donating, and the options available if they do not donate. After being informed, and prior to collection of the cells, the donor must sign informed consent for the donation. Donors must be screened for TTDs including HIV (anti-HIV antibodies, p24 antigen), HTLV (antibodies), hepatitis B (antibodies against surface protein and core, hepatitis B surface antigen), and hepatitis C (antibodies). Although positive hepatitis serology may not be an absolute contraindication to donation, if the attending physician chooses to use a donor with positive hepatitis results, a

written statement of this decision must be signed by the attending physician and must be included in the patient's permanent record. This signed statement must also be supplied to the cell collection and processing facility prior to collection of the cells. Positive serology for HIV or HTLV is an absolute contraindication to the use of that donor's cells.

B. Collection Procedures

Due to their relatively low cell density, MNCs can be isolated from other nucleated cells by leukapheresis, Ficoll separation, or buffy coat preparation. MNCs are most often collected by leukapheresis (3–12 liters processed) in the same manner as for PB-HPC collections; however, the donor does not need to receive mobilization drugs (except in the case of CD34$^+$-derived DCs) as sufficient numbers of MNCs can be obtained at steady state. The yield is approximately 1×10^9 MNCs for each liter of blood processed, and approximately 50% (30–70%) of the collected cells will be CD3$^+$. Generally, doses up to 10^8/kg can be collected in one apheresis procedure; however, this is dependent in part on the relative weights of the donor and recipient. If the donor is small (e.g., a child) and the recipient is large, then more than 1 apheresis is more likely to be necessary to obtain an adequate dose. In donors with poor venous access, a central venous catheter is used for apheresis. In the more unusual case, in which lower doses of MNC are requested (e.g., 1×10^6/kg), an adequate number of MNCs can be obtained from the simple collection of whole blood.

C. Laboratory Evaluation

The cells collected will be tested for cell count and bacterial or fungal contamination, and depending on the cells being requested, flow cytometric analysis will be performed. For CTLs, the T-cells will be counted; and based on the T-cell concentration, the absolute volume of cells to infuse will be determined in the cell processing facility. For DCs (prior to *ex vivo* expansion), the number of precursors will be determined by CD14 or CD34 staining. Since DCs lack a unique marker, the cell processing facility will use at least two markers to identify and enumerate the *ex vivo* expanded DCs.

III. SPECIAL PREPARATION

A. Manipulation

1. DLI

For DLI, the MNCs are usually collected, processed, and administered immediately. Short-term storage (fewer than 8 hours) at 4–22°C is not detrimental to the cells. If the cells are to be used at a later time, they are cryopreserved. DLIs are usually administered without extensive manipulation in the laboratory. However, as the major deleterious effect of DLI is the induction of graft-versus-host disease (GVHD) in the recipient, efforts are ongoing to develop methods that effectively separate effector cells that induce GVL (graft versus leukemia) from those that cause GVHD, or to select for antigen-specific cells (see Section III.D below). For example, studies have been performed in which CD8$^+$ T-cells have been removed so that only CD4$^+$ T-cells are infused. In one such study, the GVL effect was still induced by the remaining CD4$^+$ cell population (Giralt et al., 1995). It is not yet clear whether this approach will be adequate, as some other studies suggest that the same T-cell subpopulations cause both GVL and GVHD. Some investigators have modified DLI preparations by gamma irradiation, suicide gene insertion, or using serial, escalating doses.

2. CTLs and DCs

The generation and expansion of these cells ex vivo is described in Section III.D below.

B. Cryopreservation

1. DLI

Though rarely necessary for DLI, cells can be cryopreserved for future use. This is particularly important if the appropriate donor cannot easily travel to the collection facility. In this case, the initial cell collection (e.g., for allogeneic PB-HSCs) is also a good time to collect peripheral T-cells for potential future use. The cells are cryopreserved as described for PB-HPCs in Chapter 19; that is, they are frozen at approximately 1°C per minute in a cryopreservation solution containing 10% DMSO.

2. CTLs and DCs

DCs are not usually cryopreserved. However, CTLs and DCs can be grown from precursors that have been previously frozen.

C. Thawing for Infusion

Thawing of MNCs is performed similarly to cryopreserved PB-HPCs. The cells are thawed quickly in a 37°C water bath either near the patient's bedside or in the cell processing facility. If the cells are thawed at the bedside, they are infused immediately upon thawing in order to dilute the DMSO and prevent cell clumping. If

cells are thawed in the laboratory, then the DMSO is removed by washing the cells in a saline solution.

D. Selection and Expansion

1. CTLs

There are no standard methods for the expansion of CTLs. Currently under development are techniques to produce antigen-specific cells that react only against the target cells of interest, which would be cancer cells (for relapse) or Epstein–Barr virus (EBV)- or cytomegalovirus (CMV)-infected cells (when the desired effect is to treat a viral infection). This approach could be used to induce GVL without triggering GVHD, and is currently being tested for a variety of tumor-specific antigens. This is referred to as adoptive immunotherapy via CTLs generated and "educated" ex vivo (Cardoso et al., 1997; Linsley et al., 1991).

Few centers in the United States are currently set up to perform the complex manipulations required to generate educated CTLs. Therefore, when required, these cells are usually sent to those laboratories for processing on a clinical trial basis.

2. DCs

As for CTLs, there is no standard method for preparing DCs. Since there is much heterogeneity in the DC lineage, it is very likely that different types of DCs will be best suited for particular immunotherapeutic purposes. Generally, there are three potential sources: peripheral blood DCs (used without expansion), monocyte (CD14$^+$ cell)-derived DCs, and CD34$^+$-derived DCs. Monocytes are typically cultured in media supplemented with granulocyte-macrophage colony-stimulating factor (GM-CSF) and IL-4; CD34$^+$ cells are grown in media with GM-CSF and TNF—with or without other growth factors such as Flt3 ligand and stem cell factor (SCF). DCs take approximately 7 days to differentiate in vitro from precursors.

After direct isolation from blood, or after growth from precursors, DCs are pulsed with relevant tumor antigens. The processed antigen is then presented to the recipient's T-cells by molecules of the major histocompatibility complex. Optimal methods for antigen delivery to DCs are still under development. Antigen can be "fed" to DCs in many forms: whole cells, recombinant viruses, peptides, or nucleic acids. DCs exist in distinct maturational stages and only immature DCs will actively take up antigen for processing and presentation. Likewise, only mature DCs are capable of potently stimulating T-cells, so cells must be matured, either in vitro or after reintroduction into the patient, for maximal immunostimulatory effect.

IV. ABO AND Rh COMPATIBILITY

A. DLI

ABO and Rh typing should be performed on the donor and patient, but *HLA compatibility is more critical than ABO and Rh compatibility*. Whether to deplete RBCs or plasma from MNCs used for DLI is determined based on the ABO and Rh compatibility of the donor and recipient and the volume of RBCs or plasma in the product. If there is a major ABO incompatibility and the donor RBCs will be lysed by the recipient, the RBC content of the infused product usually is minimized to 10–20 ml RBCs. This may not require cell processing because MNCs collected by apheresis generally contain less than 15 ml RBCs. If the donor plasma has antibodies against receptor RBCs (minor incompatibility), the plasma can be removed and the cells resuspended in saline.

In cases of Rh(D) incompatibility in which the recipient is Rh(D) negative, and the donor is Rh(D)-positive, the RBC content of the product may be minimized, and the administration of Rh immune globulin should be considered to prevent sensitization to the Rh(D) antigen.

B. CTLs and DCs

In CTL and DC preparations, there are no RBCs present. Therefore, ABO and Rh compatibilities are not an issue.

V. ORDERING AND ADMINISTRATION

A. DLI

1. Written Order

Orders for DLI must be written by the patient's physician with the following items specified: donor identification, recipient information, dose of desired cell population, required processing (e.g., CD4 selection, expansion of antigen-specific cells), and whether to infuse the cells fresh or after cryopreservation. Both the patient and the donor must be informed of the risks and benefits of the procedure, and each should sign a consent form documenting the discussion.

2. Dosage

The dosage of DLI is dependent on the indications for infusion and degree of HLA compatibility between donor and recipient. The dose is generally given as the number of T-lymphocytes (which all express the marker CD3) per recipient body weight in kilograms. For HLA-identical transplants, the range reported in the literature is 1×10^6 to 1×10^8 CD3$^+$ cells/kg recipient body weight, with the latter, higher doses used for GVL effect. Lower doses are used either when the donor and recipient are not HLA matched (to decrease the risk of GVHD) or when the DLI is intended for the delivery of virus-specific cells, as they are likely to be present at a high frequency in the donor T-cell population.

3. Timing of Administration

The ideal timing and frequency of administration of donor leukocytes are not yet well established, and vary from institution to institution. In general, however, DLIs are not administered for at least 6 weeks after allogeneic transplantation in order to decrease the risk of GVHD.

4. Volume

The volume of cells infused varies based on the total cell dose required. In general, an apheresis product has a WBC concentration of about 3×10^8 cells per milliliter, of which about 70% are T-lymphocytes, giving a lymphocyte concentration of approximately 2×10^8 per milliliter. If the average 70-kg patient requires a lymphocyte dose of 10^8 T-cells/kg, then this would require about 30 ml of an apheresis product. If a lower dose of 10^6 T-cells/kg were requested, this could be collected from fewer than 70 ml of total whole blood from the donor. There is rarely a need to either concentrate or dilute these cells prior to administration.

5. Patient Preparation and Infusion

The cell product is generally administered immediately after processing. The patient is premedicated with an antipyretic and an antihistamine, and the cells are then administered through a standard 170-μm RBC filter. As with infusion of any blood product, the cells should initially be administered slowly while the patient is closely monitored for adverse reactions. Once it is clear that the patient is tolerating the infusion without difficulty, the rate can be increased. No optimal rate is defined, but the infusion should be completed within 4 hours. If the product is cryopreserved, the cells are thawed prior to infusion as described for PB-HPCs in Chapter 19.

B. CTLs or DCs

Because use of antigen-specific CTLs and DCs developed *in vitro* is still in the developmental stage, the cells are usually administered at the institution where the cells are produced. Patient consent is obtained and infusion is performed as for DLI.

VI. INDICATIONS AND CONTRAINDICATIONS

There are no absolute indications for DLI or the preparation of CTLs and DCs, as their use is still experimental. Because the use of these cell populations is not standard, all clinical protocols that include infusion of these populations should first be submitted to the local institutional review board (IRB) for approval.

A. DLI for Relapse of Malignancy

Absolute indications for DLI are not yet defined. The immunologic antitumor effect, referred to as graft-versus-leukemia, is mediated by donor T-cells, and can contribute to increased long-term survival after allogeneic bone marrow transplant (BMT) or PB-HPC transplant for some malignancies. For example, DLI may be administered for recurrence of malignancy (chronic myeloblastic leukemia [CML] and acute myelogenous leukemia [AML]) in patients who received T-cell-depleted allogeneic grafts in order to prevent GVHD (Hale and Waldmann, 1994).

1. CML

DLI has been used successfully in the treatment of patients with relapse of primary malignancy after allogeneic transplantation (Kolb *et al.*, 1990; Mehta *et al.*, 1995; Porter *et al.*, 1994; Slavin *et al.*, 1996). The best, most consistent, results are observed in CML, suggesting that these tumor cells are more immunogenic than other types of leukemia cells (Kolb, *et al.*, 1990). In a retrospective study of 140 patients from multiple institutions, complete remission was induced in 60% of CML patients by DLI. Improved outcome (75–80% complete remission) was observed more often in patients with cytogenetic relapse than in patients who had developed blastic or accelerated phase CML (15–30% remission).

2. Other Diseases

In patients with lymphoma and acute leukemia, DLI has been less effective, with only about 15% of patients achieving complete remission. DLI-induced remission has also been reported for relapsed non-Hodgkin's lymphoma and multiple myeloma. It is still controversial whether DLI should be used for relapsed AML or acute lymphoblastic leukemia (ALL), as the chance of remission is low and the risk of causing GVHD is quite high.

It has been suggested that since DLI-induced remission can take from 1 to 5 months to occur, more rapidly growing malignancies may not be amenable at relapse to current DLI protocols. It is unclear why the time to remission is so long and why there is so much interpatient variation. It is likely that during this time interval, the cells necessary for the GVL response are homing to sites within the immune system (e.g. spleen and lymph nodes), responding to the relevant alloantigens, and expanding *in vivo* until they reach a threshold level at which a clinical effect is detectable.

B. DLI for Infectious Complications

DLI has also proven to be highly successful for the treatment of post-transplant infectious complications, particularly those caused by Epstein–Barr virus (EBV) and cytomegalovirus (CMV).

1. EBV

Over 90% of people in the United States have been infected with Epstein–Barr virus and have circulating B-cells harboring latent EBV. EBV infection is controlled in the normal host by virus-specific T-lymphocytes that lyse infected B-cells. Normally, at least 1 in 50,000 circulating peripheral blood T-cells is specific for EBV-associated antigens such as EBNA 1–6 and the latency membrane proteins LMP1 and LMP2; these circulating T-cells act to maintain EBV latency. Immunosuppression, either by T-cell depletion of the allogeneic graft prior to transplantation or by administration of immunosuppressive drugs to prevent or treat GVHD, can have the deleterious effect of permitting reactivation of latent donor B-cell-derived EBV. This is why EBV lymphoproliferative disease (EBV-LPD) is one of the most common complications after allogeneic transplantation.

The transfer of peripheral blood T-cells from the allogeneic donor (i.e., DLI) can be a safe and effective way to reconstitute cellular immunity against EBV after allogeneic marrow transplantation (O'Reilly *et al.*, 1997). In one early study, 5 of 5 patients with EBV-associated immunoblastic lymphoma after allogeneic bone marrow transplantation had complete remissions after infusion of donor-derived leukocytes (Papadopoulos *et al.*, 1994). The dose of T-cells required for this effect is less than that required for treatment of tumor recurrence because of the high frequency of EBV-specific T-cells in the donor's peripheral blood. In one small study, 3 patients who received less than 10^6/kg donor leukocytes for treatment of EBV-related lymphoproliferative disease experienced complete resolution of EBV-related disease (Mackinnon *et al.*, 1995).

2. CMV

Cytomegalovirus infection, another life-threatening complication of allogeneic transplantation, occurs due to insufficient CMV-specific CD8$^+$ cells. CMV-related complications occur in approximately 20–50% of recipients of T-cell-depleted allogeneic transplants, and usually arise within the first 1–2 months post-transplantation. The risk of CMV-related complications has decreased with advances in antiviral medications, especially ganciclovir and foscarnet. The risk of CMV infection is also decreased by prophylactic administration of DLI to restore cell-mediated immunity in the transplant recipient (see below).

C. Prophylactic DLI

Several clinical trials suggest that prophylactic infusion of donor MNCs after transplant can decrease the risk of tumor recurrence. Optimal therapy may be to give allogeneic HPCs depleted of T-cells immediately following high-dose chemotherapy, and then, after a time delay of 6–8 weeks, administer donor leukocytes (DLI) to prevent tumor relapse and infectious complications. However, even though delayed infusion of lymphocytes after establishment of the chimeric state is less likely to cause GVHD than infusion of lymphocytes at the time of stem cell infusion, there is still significant risk of inducing GVHD (van Rhee *et al.*, 1994).

D. Contraindications of DLI

Patients who are already experiencing GVHD should not receive DLI. In addition, patients who are receiving drugs to suppress the immune system will not respond to DLI, as the immunological effects of the MNCs will be inhibited *in vivo* by the drugs.

E. Indications for CTLs

1. T- and B-Cell Malignancies and Melanoma

Tumor-specific antigens have been identified for many malignancies. For example, in T- and B-cell malignancies, there is a clone-specific rearrangement in the T-cell receptor and immunoglobulin genes, respec-

tively. Using DCs developed *ex vivo* or cell lines that express these antigens, one can generate tumor-specific CTLs. Current research is focused on generating host-derived tumor-specific CTLs for many malignancies, including lymphoma, melanoma, and leukemia, in order to prevent or treat recurrence with the patients' own cells.

2. Viral Infections

For viral infections, a similar approach can be taken. Infusion of CMV-specific clones of CD8$^+$ T-cells has been used successfully to reconstitute cellular immunity against CMV after allogeneic marrow transplantation, and EBV-specific CTLs generated against multiple different EBV-associated antigens have also been effective in the treatment and prevention of EBV-related post-transplant complications (Walter *et al.*, 1995).

F. Indications for DCs

While initial results are promising, the use of DCs for immunotherapy is still highly experimental and limited to patients for whom no other standard regimens are likely to succeed (reviewed in Timmerman and Levy, 1999). Since the first use of DCs to treat lymphoma (Hsu *et al.*, 1996, 1997), there have been (or are currently) clinical trials for DC-based immunotherapy of multiple myeloma, malignant melanoma, sarcoma, neuroblastoma, and renal cell carcinoma, as well as lung, colorectal, pancreatic, and prostate cancer. DCs have also been used for the treatment of HIV (Kundu *et al.*, 1998).

VII. EXPECTED RESPONSE

A. Induction of Relapse

DLI-induced remission of relapsed disease usually takes between 1 and 5 months to occur, but has been reported to occur as late as 8 months postinfusion. The effect of donor-specific CTLs and DCs similarly takes up to 6 months to occur, although theoretically the effect of CTLs is likely to be a little faster than that induced by DCs, as the initial lymphocyte priming step has already occurred.

B. Treatment of Viral Infection

DLI-induced resolution of EBV or CMV infection may occur within 10 days to 4 weeks.

C. Evaluation of Response

Patients are generally evaluated at least weekly for signs of remission of malignancy or resolution of EBV- or CMV-related symptoms. Effective treatment for relapsed malignancy also correlates with the elimination of mixed chimerism (i.e., loss of recipient-derived cells in the peripheral blood and bone marrow). In addition, patients should be carefully assessed for GVHD weekly by complete physical exam, including evaluation of skin for rash, specific questions regarding onset of upper or lower gastrointestinal problems, and laboratory evaluations including but not limited to liver function tests.

VIII. ADVERSE REACTIONS

A. Acute Transfusion Reaction

There is rarely an immediate transfusion reaction to MNCs, CTLs, or DCs. Febrile and allergic reactions may occur.

B. Graft-versus-Host Disease

The major causes of morbidity and mortality associated with DLI include GVHD (acute and chronic) and pancytopenia. Of patients who obtain complete remission of CML following DLI, approximately 80% develop grade II–IV GVHD. Dose alone cannot be used to predict the induction of GVL versus GVHD.

C. Pancytopenia

Pancytopenia post-DLI occurs from 1 to 5 months after administration of the donor cells, and often occurs just prior to the beneficial clinical GVL response (Giralt *et al.*, 1995). Pancytopenia may be due to depletion of host-derived tumor cells and normal HPCs from the marrow space. In one study that examined the degree of chimerism in patients relapsing post-transplant, only those patients who had greater than 10% host-derived CD34$^+$ cells in their BM prior to DLI experienced significant aplasia after DLI (Keil *et al.*, 1997). This suggests that chimerism studies should be performed prior to DLI, and that those patients with a high degree of host hematopoiesis should receive both DLI to induce a GVL effect and infusion of donor HPCs for BM reconstitution, in order to shorten or prevent the period of aplasia.

IX. ALTERNATIVES AND NEW DEVELOPMENTS

For the treatment of relapsed malignancy, alternative treatments include administration of additional chemotherapy or treatment with cytokines such as IL-2 that stimulate the immune system. In order to obtain the GVL benefit of DLI and decrease the risk of GVHD, another approach has been to insert inducible "suicide" genes into the donor leukocytes that, when activated, kill the infused leukocytes and thereby abort GVHD. In early clinical trials, this approach has been effective (Bonini *et al.*, 1997). For treatment of viral infections post-transplant, antiviral agents and cytokines to boost the function of the immune system are also used.

Additional cell populations that have been isolated for adoptive immunotherapy for relapsed malignancy include IL-2 activated tumor infiltrating lymphocytes (TILs) or activated lymphocytes derived from tumor-draining lymph node cells (Rosenberg, 1992; Yoshisawa *et al.*, 1991). TILs that have been further modified *ex vivo* by the introduction of cytokine genes that enhance the immune reaction against tumor cells *in vivo* have also been under development (for review, see Hwu and Rosenberg, 1994).

References

Banchereau, J., and Steinman, R. M. (1998). Dendritic cells and the control of immunity. *Nature* **392**, 245–252.

Bonini, C., Ferrari, G., Verzeletti, S., *et al.* (1997). HSV-TK gene transfer into donor lymphocytes for control of allogeneic graft-versus-leukemia. *Science* **276**, 1719–1724.

Cardoso, B. A., Seamon, M. J., Afonso, H. M., *et al.* (1997). Ex vivo generation of human anti-pre-B leukemia-specific autologous cytolytic T-cells. *Blood* **90**, 549–561.

Giralt, S., Hester, J., Huh, Y., *et al.* (1995). CD8-depleted donor lymphocyte infusion as treatment for relapsed chronic myelogenous leukemia after allogeneic bone marrow transplantation. *Blood* **86**, 4337–4343.

Hale, G., and Waldmann, H. (1994). Control of graft-versus-host disease and graft rejection by T-cell depletion of donor and recipient with Campath-1 antibodies. Results of matched sibling transplants for malignant diseases. *Bone Marrow Transplant* **13**, 597–611.

Horowitz, M. M., Gale, R. P., Sandel, P. M., *et al.* (1990). Graft-versus-leukemia reactions after bone marrow transplantation. *Blood* **75**, 555–562.

Hsu, F. J., Benike, C., Fagnoni, F., *et al.* (1996). Vaccination of patients with B-cell lymphoma using autologous antigen-pulsed dendritic cells. *Nat Med* **2**, 52–58.

Hsu, F., Caspar, C., Czerwinski, D., *et al.* (1997). Tumor-specific idiotype vaccines in the treatment of patients with B-cell lymphoma—long-term results of a clinical trial. *Blood* **89**, 3129–3135.

Hwu, P., and Rosenberg, S. (1994). The genetic modification of T-cells for gene therapy: An overview of laboratory and clinical trials. *Cancer Detect Prevent* **18**, 43–50.

Keil, F., Haas, O., Fritsch, G., *et al.* (1997). Donor leukocyte infusion for leukemic relapse after allogeneic marrow transplantation: Lack of residual donor hematopoiesis predicts aplasia. *Blood* **89**, 3113–3117.

Kolb, H., Mittermüller, J., Clemm, C., *et al.* (1990). Donor leukocyte transfusions for treatment of recurrent chronic myelogenous leukemia in marrow transplant patients. *Blood* **76**, 2462–2465.

Kundu, S. K., Engleman, E., Benike, C., *et al.* (1998). A pilot clinical trial of HIV antigen-pulsed allogeneic and autologous dendritic cell therapy in HIV-infected patients. *AIDS Res Hum Retro-viruses* **14**, 551–560.

Linsley, P., Brady, W., Grosmaire, L., *et al.* (1991). Binding of the B-cell activation antigen B7 to CD28 costimulates T-cell proliferation and interleukin 2 mRNA accumulation. *J Exp Med* **173**, 721.

Mackinnon, S., Papadopoulos, E. B., Carabasi, M. H., *et al.* (1995). Adoptive immunotherapy using donor leukocytes following bone marrow transplantation for chronic myeloid leukemia: Is T-cell dose important in determining biological response? *Bone Marrow Transplant* **15**, 591–594.

Mehta, J., Powles, R., Singhal, S., *et al.* (1995). Cytokine-mediated immunotherapy with or without donor leukocytes for poor-risk acute myeloid leukemia relapsing after allogeneic bone marrow transplantation. *Bone Marrow Transplant* **16**, 133–137.

Mellman, I., Turley, S. J., and Steinman, R. M. (1998). Antigen processing for amateurs and professionals. *Trends Cell Biol* **8**, 231–237.

O'Reilly, R., Small, T., Papadopoulos, E., *et al.* (1997). Biology and adoptive cell therapy of Epstein-Barr virus-associated lymphoproliferative disorders in recipients of marrow allografts. *Immunol Rev* **157**, 195–216.

Papadopoulos, E., Ladanyi, M., Emanuel, D., *et al.* (1994). Infusions of donor leukocytes to treat Epstein-Barr virus-associated lymphoproliferative disorders after allogeneic bone marrow transplantation. *N Engl J Med* **330**, 1185–1191

Porter, D., Mark, S., McGarigle, C., Ferra, J., and Antin, J. (1994). Induction of graft-versus host disease as immunotherapy for relapsed chronic myeloid leukemia. *N Engl J Med* **330**, 100–106.

Rosenberg, S. (1992). The immunotherapy and gene therapy of cancer. *J Clin Oncol* **10**, 180–199.

Slavin, S., Naparstek, E., Nagler, A., *et al.* (1996). Allogeneic cell therapy with donor peripheral blood cells and recombinant human interleukin-2 to treat leukemia relapse after allogeneic bone marrow transplantation. *Blood* **87**, 2195–2204.

Timmerman, J. M., and Levy, R. (1999). Dendritic cell vaccines for cancer immunotherapy. *Annu Rev Med* **50**, 507–529.

Van Rhee, F., Lin, F., Cullis, J., *et al.* (1994). Relapse of chronic myeloid leukemia after allogeneic bone marrow transplant: The case for giving donor leukocyte transfusions before the onset of hematologic relapse. *Blood* **83**, 3377–3383.

Walter, E., Greenberg, P., Gilbert, M., *et al.* (1995). Reconstitution of cellular immunity against cytomegalovirus in recipients of allogeneic bone marrow by transfer of T-cell clones from the donor. *N Engl J Med* **333**, 1038–1044.

Yoshisawa, K., Chang, A., and Shu, S. (1991). Specific adoptive immunotherapy mediated by tumor-draining lymph node cells sequentially activated with anti-CD3 and IL-2. *J Immunol* **147**, 729–737.

SPECIALIZED TRANSFUSION SITUATIONS

Approach to Acute Bleeding and Massive Transfusion

DOUGLAS P. BLACKALL

Department of Pathology and Laboratory Medicine
UCLA Medical Center
Los Angeles, California

The acutely bleeding patient represents a special challenge for clinicians, who must stem the flow of blood and replace fluid losses. If blood loss exceeds the patient's ability to compensate for the acute loss of oxygen-carrying capacity, then packed red blood cell (PRBC) transfusions will be initiated. The blood bank will be required to provide these units in a safe but timely manner to support patient needs. With continued blood loss, platelets, fresh frozen plasma (FFP), and cryoprecipitate may also be required.

The term "massive transfusion" is used to designate patients who have been transfused with at least 1 blood volume in a 24-hour period. The patient who is massively transfused is at heightened risk for the pathologic consequences of hemorrhagic shock, for example, acidosis and coagulopathy. The rapid transfusion of blood products may contribute to or complicate this state. In this chapter, an approach to the support of the acutely bleeding patient will be discussed, with special emphasis on the massively transfused patient.

I. THE ACUTELY BLEEDING PATIENT

A. Maintain Intravascular Volume

There is a wide variety of medical, surgical, and obstetric conditions that can result in acute bleeding and have the potential to progress to the point of "massive transfusion." Regardless of cause, however, in the patient experiencing acute hemorrhage, *the maintenance of intravascular volume is initially much more*

important than the maintenance of oxygen-carrying capacity (Crosson, 1996). Therefore, volume replacement therapy in acutely bleeding patients is initially accomplished by the infusion of asanguinous fluids (Reiner, 1998). Two types of these solutions are routinely used in this setting:

1. Crystalloid solutions such as normal saline (0.9% NaCl) and lactated Ringers solution (containing physiologic concentrations of sodium, potassium, and calcium buffered with lactate)
2. Colloid solutions such as albumin (5% and 25% preparations) and plasma protein fraction (83% albumin and 17% globulins)

There has been a great deal of controversy regarding the best solution to use in the initial phase of resuscitation, but the preponderance of evidence *favors the use of crystalloid solutions over colloid solutions* (Stehling, 1994). However, as roughly 80% of infused salt solutions equilibrate into the extravascular space (with only 20% of the infused volume remaining in the intravascular space), a volume of crystalloid 3–4 times that of the patient's estimated blood loss must be infused to maintain intravascular volume.

B. Treat the Source of Bleeding

After aggressive fluid resuscitation has begun, evaluation and treatment of the source of bleeding should ensue. As long as intravascular volume is maintained, experimental and clinical investigations have demonstrated that large-volume PRBC loss is generally well tolerated (Crosson, 1996).

C. Maintain Oxygen-Carrying Capacity

Once blood loss exceeds ~ 30% of blood volume, RBC mass must be replenished in order to maintain adequate tissue oxygenation. The clinical decision to transfuse PRBCs in the setting of acute bleeding must be tailored to the needs of the individual patient. General considerations in this decision-making process include the rate of RBC loss, the oxygen requirements of the patient, and the adequacy of tissue perfusion.

When an increase in oxygen-carrying capacity is required, PRBCs are the component of choice. It is worthy to note that PRBCs stored in additive solutions (AS-1, AS-3, AS-5) have a volume of 300–350 ml and a reduced hematocrit of 50–60%. As such, there is decreased viscosity, which allows for more rapid flow.

D. Close Laboratory Monitoring

In addition to a thorough clinical assessment, all patients who are experiencing ongoing acute bleeding will require close laboratory monitoring. As blood is lost and replaced with crystalloid and PRBCs, serial hemoglobin and hematocrit values should be correlated with the patient's clinical response. These values should then be used to guide further transfusion therapy. Platelet count, fibrinogen level, prothrombin time (PT), and partial thromboplastin time (PTT) should also be assessed to determine the need for additional hemostatic blood components (platelet products, cryoprecipitate, and FFP, respectively).

Algorithms for choosing hemostatic blood products for transfusion to the patient with ongoing bleeding have been published elsewhere (Fakhry *et al.*, 1996; Reiner, 1998). In general, if microvascular bleeding is evident and there is no obvious source of a surgically correctable lesion, then the transfusion of a hemostatic blood component should be considered. Quantitative laboratory values that may serve as general "triggers" for the institution of blood component therapy in the face of microvascular bleeding include a platelet count below $75,000/\mu l$, a fibrinogen level below 100 mg/dl, and PT or PTT values exceeding 1.5 times control.

E. Replace Hemostatic Components

The two most important reasons for deficiencies of platelets, fibrinogen, and coagulation factors are dilution (i.e., depletion, when only PRBCs and asanguinous fluids are given as replacement) and consumption (as with disseminated intravascular coagulation [DIC]).

There is a significant redundancy in total body platelet mass and coagulation factor content that exceeds the levels normally required to achieve hemostasis. In most cases, more than 1 blood volume needs to have been replaced with PRBCs before dilutional coagulopathy becomes an overriding concern (Leslie and Toy, 1991; Murray *et al.*, 1988).

Consumptive processes are more difficult to predict and are more closely correlated to the patient's underlying clinical condition than to the volume of blood lost or transfused (Cosgriff *et al.*, 1997; Ferrara *et al.*, 1990; Harvey *et al.*, 1995; Phillips *et al.*, 1987). Nonetheless, DIC needs to be properly diagnosed and treated. In this dynamic process, the trend in laboratory values is more valuable than a single value.

II. THE MASSIVELY TRANSFUSED PATIENT

Massive transfusion has been defined in the clinical literature in a number of different ways, including: (1) replacement of a patient's entire blood volume within 24 hours, (2) replacement of 50% of the circulating blood volume within 3 hours, and (3) the transfusion of more than 10 units of whole blood or 20 units of PRBCs in a 24-hour period (Crosson, 1996). Alternatively, a working definition that has been adopted to prospectively identify patients at risk for massive transfusion early in the course of bleeding is the transfusion of 4 units of blood within 1 hour, with anticipation of ongoing usage (Crosson, 1996). Regardless of the definition used, the acutely bleeding patient who loses blood rapidly and requires large amounts of blood and blood components over a short period of time represents a particular challenge to the clinical staff and the blood bank.

As with any patient who is experiencing acute bleeding, the end goal is to provide adequate intravascular fluid support to prevent hemorrhagic shock as well as the cellular and end-organ damage that can result from tissue hypoperfusion (Phillips *et al.*, 1994). Thus, treatment of the patient with massive hemorrhage begins with vigorous fluid resuscitation to maintain intravascular volume. This is followed by the restoration of tissue oxygenation through the transfusion of PRBCs and the replacement of lost and/or consumed hemostatic factors using plasma-derived blood components and platelets. As mentioned previously, crystalloid solutions are the replacement fluid of choice in the initial phase of resuscitation. Close monitoring of the patient's volume, hemodynamic, and clinical statuses are essential components of the management of the massively bleeding patient.

A. Administration of Fluids and Components

Initial resuscitation fluids are given through 2 large-bore (> 16-gauge) intravenous access lines. As volume replacement therapy is initiated, baseline laboratory values should be obtained, including electrolyte levels, a complete blood count, and coagulation screening tests. A separate sample should always be sent to the blood bank. As a general rule, hemodynamic normalization and clinical stabilization following a 1- to 2-liter infusion of crystalloid solution usually indicates that the patient has not sustained a significant blood loss and that future blood loss will be minimal (Phillips et al., 1994). In contrast, the patient who only transiently responds, or fails to respond, to an aggressive fluid infusion has likely either suffered a massive hemorrhage, or blood loss is rapid and ongoing (Phillips et al., 1994). This situation requires immediate PRBC transfusion and attention to the site of continued bleeding.

B. Choice of Resuscitative Fluid or Component

The choice of PRBC component used for urgent transfusion will depend on local availability and the nature of the clinical situation. Under routine and nonemergent conditions, pretransfusion compatibility testing is undertaken in the blood bank before PRBC components are issued for transfusion. This testing entails ABO grouping, Rh (D antigen) typing, and an antibody screen for the detection of unexpected blood group antibodies (e.g., Rh, Kell, Kidd, and Duffy blood group systems). Once this testing is complete, PRBC units can be crossmatched with a patient's serum or plasma in preparation for anticipated transfusion. The time necessary to perform pretransfusion testing (including crossmatching) in order to provide specific components is included in Table 1.

1. Use of Group O PRBCs in the Emergent Setting

On occasion, the time required to provide crossmatch-compatible blood for transfusion may be longer than the hemodynamically unstable, massively hemorrhaging patient can safely wait. In this setting, abbreviated or no compatibility testing may be warranted. In this clinical situation, uncrossmatched group O PRBCs may be transfused. Initially, Rh-negative units are commonly provided, in order to prevent immunological sensitization to the D antigen for those recipients who are Rh-negative. Rh(D) sensitization is of particular concern in Rh-negative women of childbearing age, for whom the production of an anti-D antibody could have significant ramifications in future pregnancies (i.e., hemolytic disease of the newborn).

As only 6% of the population is of the group O, Rh-negative blood type, PRBCs of this phenotype are frequently in short supply. As such, it has become common practice in some centers to emergently transfuse uncrossmatched group O, Rh-positive PRBCs to

TABLE 1 Blood Products Used during Massive Transfusion

Blood product	Availability[a]	Testing required	Comments
O, uncrossmatched PRBC[b]	5 minutes	None	Emergency use
ABO type-specific, uncrossmatched PRBC[b]	10–15 minutes	ABO grouping	Emergency use
Crossmatched PRBC (negative screen)	30–60 minutes	ABO, RH, antibody screen	Blood compatible by crossmatch
Crossmatched PRBC (positive screen)	90 minutes to several hours or longer	ABO, RH, antibody screen; +/− special serological studies	Good communication essential to prevent excessive delays
Platelets	20 minutes	No special testing[c]	Pooled concentrates[d] or apheresis product
Fresh frozen plasma	45 minutes	No special testing[e]	Thawing time required
Cryoprecipitate	15–20 minutes	No special testing	Thawing time required

[a] Does not include transportation time.
[b] Rh-negative units generally reserved for females of childbearing potential; Rh-positive units commonly used for males and older females.
[c] ABO-compatible platelets generally provided, when available; Rh-negative platelets provided to Rh-negative females of childbearing potential.
[d] Refers to a pool of platelet concentrates derived from multiple whole blood donations.
[e] ABO-compatible plasma generally provided and routinely available.

male patients and postmenopausal females. This is an extremely safe and clinically efficacious practice (Schwab *et al.*, 1986). It is also highly relevant in the trauma setting, as 70–80% of all victims are young males (Reiner, 1998).

2. Use of ABO Type-Specific PRBCs

Another way to minimize the overuse of valuable group O PRBC units in the emergency transfusion setting is through the provision of ABO type-specific blood. Providing type-specific PRBCs has the benefit of quickly providing the patient with PRBCs prior to the completion of crossmatching. The only increased risk above that in which group O units are used is the erroneous provision of an ABO-incompatible unit to the recipient. The risk of this occurring (with the subsequent probability of an acute hemolytic transfusion reaction) is exceedingly small, but is more likely to occur in high-energy, acute care settings such as operating rooms, emergency departments, and intensive care units.

Aside from this risk, the only other inherent risk of providing type-specific PRBCs or, for that matter, any uncrossmatched PRBC unit is the development of a delayed hemolytic transfusion reaction. This occurs when recipients have been previously sensitized to clinically significant blood group antigens (e.g., Rh, Kell, Kidd, Duffy). Unlike acute hemolytic transfusion reactions, delayed reactions typically result in extravascular hemolysis, occur days after the acute period, and are generally non-life threatening. In other words, the small risk of a delayed hemolytic transfusion reaction is far outweighed by the immediate benefits of PRBC replacement in the massively bleeding, unstable patient.

C. Hemostatic Component Therapy

In addition to the need for PRBCs and crystalloid solutions, the massively transfused patient is at high risk for hemostatic defects requiring the transfusion of plasma-derived components and platelets. The time required to prepare hemostatic components for transfusion is included in Table 1. As previously mentioned, the two most common causes for hemostatic component deficiencies are the dilutional effect of massive transfusion and consumption secondary to the patient's underlying clinical condition (e.g., severe traumatic injury complicated by DIC). However, a wide variety of additional factors may contribute to the development of coagulopathy during transfusion. For example, hypothermia and acidosis effect platelet function and the activity of coagulation cascade enzymes. In addition, certain conditions—including heparin/coumadin us-

age, liver disease, and uremia—may contribute to bleeding (Johnston *et al.*, 1994; Rohrer and Natale, 1992; Valeri *et al.*, 1987). Regardless of the cause, the massively transfused patient is at high risk for the development of coagulopathy. In those cases where there is a clear-cut deficiency of a hemostatic factor, as assessed by appropriate laboratory testing coupled with ongoing evidence of bleeding, replacement therapy is warranted.

1. Thrombocytopenia

Thrombocytopenia is the most commonly encountered hemostatic abnormality in massively transfused patients (Reed *et al.*, 1986). Significant thrombocytopenia frequently occurs in the setting of massive transfusion (particularly after 15–20 units of PRBC replacement), and there is an inverse relationship between the number of units transfused and a patient's platelet count (Counts *et al.*, 1979; Leslie and Toy, 1991). Nonetheless, there is so much patient-to-patient variability in platelet count, irrespective of transfusion volume, that *there is no clear justification for the prophylactic transfusion of platelets to the massively transfused patient in order to prevent microvascular bleeding* (Reed *et al.*, 1986).

Frequent platelet counts correlated with clinical evidence of bleeding should serve as the primary rationale for platelet transfusion. The "trigger" at which platelet transfusion should be considered cannot be precisely defined, primarily because of the complex and variable pathophysiology that accompanies massive transfusion. However, it is widely accepted that a circulating platelet count of $50,000/\mu$l will provide adequate hemostasis for even major surgical procedures (Stehling *et al.*, 1996). However, some authorities believe that this may be an inappropriately low value for the institution of platelet transfusion in the setting of massive hemorrhage, as this patient group is continually under the threat of dilutional thrombocytopenia and rapid platelet consumption. In addition, the patient's clinical state may predispose him or her to platelet dysfunction. Given these factors, some authors feel that it is judicious to maintain a platelet count above $100,000/\mu$l in the setting of massive transfusion (Stehling *et al.*, 1996).

2. Coagulopathy

If the massively transfused patient continues to bleed despite an adequate platelet count, then a coagulation factor deficit (including fibrinogen deficiency) may be the cause. The PT and PTT should be used as a guide for FFP administration. Interestingly though, the risk of microvascular bleeding does not correlate with a prolonged PT or PTT until these values exceed 1.5 times control (Leslie and Toy, 1991). Laboratory-determined deficiencies of fibrinogen also do not correlate with bleeding or risk of bleeding until values fall below 75–100 mg/dl (Murray *et al.*, 1988; Ciavarella *et al.*, 1987). As with platelet transfusions, there is no clear justification for the prophylactic use of FFP or cryoprecipitate in the setting of massive transfusion (Martin *et al.*, 1985).

In principle, it is entirely appropriate to base the transfusion of hemostatic blood components on clinical criteria and frequently determined laboratory values. However, there are at least three factors that militate against an overreliance on transfusion algorithms.

First, in the setting of massive transfusion, there is often no good correlation between specific hemostatic laboratory values and continued bleeding. There is an increasing awareness that prolonged hypotension and accompanying tissue injury and acidosis may profoundly effect the overall ability of the patient to appropriately form a blood clot (Cosgriff *et al.*, 1997; Ferrara *et al.*, 1990; Velmahos *et al.*, 1998). Coupled with this is the deleterious effect that hypothermia (a frequent problem in the massively transfused patient) has on platelet function and, to a lesser extent, coagulation protein function (Johnston *et al.*, 1994; Rohrer and Natale, 1992; Valeri *et al.*, 1987). Thus, in some instances, the ability to achieve adequate hemostasis will be more dependent on the correction of underlying abnormalities related to the clinical condition of the recipient than to the normalization of specific quantitative abnormalities of individual hemostatic factors.

A second factor that militates against an overreliance on transfusion algorithms is the potential for rapid change in the clinical condition of the critically ill patient. Although it is important to obtain frequent laboratory assessments of coagulation and use them in making transfusion decisions, there will always be a delay in the ability to use a given laboratory test to transfuse a patient. This results from a number of factors, including the time required to initiate a laboratory order, the time required for testing and the transmission of data, the time required to initiate a transfusion order, and the time required by the blood bank to process the order into a transfusable blood product. Given the potential for delays that may adversely effect a critically ill patient, clinicians may only be able to use laboratory values as a rough guide to transfusion, relying to a greater extent on the clinical status of the patient.

Finally, the extreme rate of exsanguination of some patients may make it almost infeasible to rely on laboratory-based algorithms for transfusion. In these select patients, a formula-based transfusion protocol may have some merit. Such a protocol might include the transfusion of 2–4 units of FFP and 1 pool of random-donor platelets or an apheresis-derived platelet product for every 10 units of PRBCs transfused. In dire clinical situations, transfusion by formula or protocol may help the blood bank to efficiently and effectively meet the needs of an exsanguinating patient, with less reliance on the rapid interpretation and filling of individual orders from the operating room. This can sometimes lead to serious breakdowns in communication at a time when optimal communication is essential.

III. COMPLICATIONS OF MASSIVE TRANSFUSION

Complications of massive transfusion are primarily due to: (1) the anticoagulant–preservative solutions used for blood storage, (2) the biochemical changes that occur during blood storage, and (3) the temperature at which blood is stored and transfused. These parameters can lead to definable metabolic consequences that can adversely impact the patient receiving high-volume, rapid blood transfusion. Still, patient complications due to massive transfusion are not easily or frequently demonstrated. In fact, there is now significant data to demonstrate that the patient's underlying clinical condition (including extent of injury, length of hypotension, acid–base balance, and temperature) is much better correlated with poor outcome than are the metabolic effects of massive transfusion (Cosgriff *et al.*, 1997; Ferrara *et al.*, 1990; Harvey *et al.*, 1995; Phillips *et al.*, 1987; Velmahos *et al.*, 1998).

The metabolic effects of massive transfusion, including electrolyte disturbances, a decline in 2,3-diphosphoglycerate levels, and hypothermia, are discussed below.

A. The Biochemical Storage Lesion

The potential metabolic consequences of massive transfusion are most closely related to the changes that PRBCs undergo during refrigerated storage (i.e., the "storage lesion"). Specific parameters include:

1. Hyperkalemia

Stored blood commonly has an elevated potassium concentration due to the slow release of potassium from RBCs. By 3–4 weeks, potassium concentrations of approximately 80 mEq/liter are measured in PRBCs (Reiner, 1998). Fortunately, there is only a small volume of plasma in PRBC units, so this represents a small potassium dose. However, during massive transfusion, this may be a significant potassium load. Thus, hyperkalemia is a potential complication of massive transfusion. Although hyperkalemic morbidity and mortality have been reported following rapid, high-volume transfusion, these appear to be uncommon complicating events (Jameson *et al.*, 1990).

Paradoxically, hypokalemia may be the more common accompaniment to massive transfusion (Carmichael *et al.*, 1984; Wilson *et al.*, 1992). This is related to acidosis and the general shock state, as well as the initiation of normal physiologic compensatory mechanisms (e.g., increases in aldosterone, antidiuretic hormone, and steroid hormones).

Given the variability and unpredictability of circulating potassium concentrations, close monitoring of potassium levels is important in the setting of massive transfusion.

2. Citrate Toxicity

Another metabolic consequence of blood storage that may negatively impact the massively transfused patient is citrate toxicity. Citrate is universally used as an anticoagulant for blood storage. It prevents clotting through the chelation of ionized calcium. The potential therefore exists that a patient could be rendered, at least temporarily, hypocalcemic. In reality, this is only likely to occur when infusion rates exceed 100 ml/min, as citrate is rapidly metabolized into sodium bicarbonate by the liver. This may affect acid–base balance, as discussed below (Rudolph and Boyd, 1990).

Patients with hepatic dysfunction may be at greatest risk for the clinically significant effects of hypocalcemia. Given the variability in response to citrate infusion, calcium levels should be monitored closely, with supplementation provided only as necessary. The prophylactic use of calcium solutions is generally not required and can lead to hypercalcemia, with the possibility of adverse clinical events, including bradycardia, hypotension, and death.

3. Acid–Base Abnormalities

These are also possible complications of massive transfusion. The pH of PRBCs decreases over time and may reach levels of approximately 6.5 by the expiration date of the unit (Vengelen-Tyler, 1996). This acid load may contribute to an initial metabolic acidosis during resuscitation. This is quickly countered, however, by the effect of citrate infusion, as citrate is normally metabolized to sodium bicarbonate. In fact, it is more common for adequately resuscitated, massively transfused patients to exhibit a slight metabolic alkalosis during the recovery phase following massive transfu-

sion (Driscoll *et al.*, 1987). Persistent acidosis is much more likely the result of uncontrolled hemorrhage (i.e., severe hypovolemia and poor tissue perfusion) rather than the result of PRBC transfusion. As with potassium and calcium levels, pH values should be monitored closely. Sodium bicarbonate supplementation should not be given empirically during massive transfusion.

4. A Decline in PRBC 2,3-Diphosphoglycerate Levels

At the expiration period of a PRBC unit, less than 10% of normal 2,3-DPG remains (Vengelen-Tyler, 1996). Lowered 2,3-DPG levels are associated with increased hemoglobin oxygen affinity and the theoretical potential for impaired oxygen delivery to peripheral tissues. In practice, however, 2,3-DPG levels are rapidly restored once blood is transfused (Beutler and Wood, 1969). There is no clear evidence that the changes in 2,3-DPG levels that accompany blood storage are clinically significant factors to patients undergoing massive transfusion.

B. Hypothermia

Hypothermia is an important complication that may accompany the rapid infusion of refrigerated blood and may have significant clinical ramifications for the massively transfused patient. Above all, hypothermia has profound effects on myocardial contractility and general cardiac function. It has been shown to adversely effect the function of platelets and coagulation factor proteins, and it also increases the affinity of hemoglobin for oxygen (Johnston *et al.*, 1994; Reiner, 1998; Rohrer and Natale, 1992; Valeri *et al.*, 1987). Patients at particular risk for hypothermia include severe trauma victims, the aged, those with significant and prolonged hypotension, and those exposed to open abdominal or thoracic operative procedures. To counter the adverse consequences of hypothermia, high-flow blood-warming devices are now available which can quickly provide blood warmed from 4 to 37°C. In addition, the ambient temperature in operating rooms in which patients are massively transfused can be elevated, and intravenous fluids and inspired gases can be warmed. As a final note, it should be noted that all blood warmers require careful maintenance and monitoring to ensure that blood is not excessively heated (i.e., above 42°C), as this may result in hemolysis.

IV. ALTERNATIVES TO ALLOGENEIC BLOOD TRANSFUSION

As previously mentioned, the transfusion of any blood product carries certain definable risks to which the massively transfused patient is not immune and, in fact, may be at particular risk. As such, methods to minimize a patient's exposure to allogeneic blood products may be particularly beneficial in the patient undergoing massive transfusion. Currently available alternatives to allogeneic blood transfusion include the donation of autologous blood prior to elective surgery and the use of cell salvage or cell saver devices. An effective RBC substitute is not currently available for clinical use but may have the greatest impact on clinical practice in the setting of massive transfusion.

References

Beutler, E., and Wood, L. (1969). The in vivo regeneration of red cell 2,3-diphosphoglyceric acid (DPG) after transfusion of stored blood. *J Lab Clin Med* **74**, 300–304.

Carmichael, D., Hosty, T., Kastl, D., *et al.* (1984). Hypokalemia and massive transfusion. *South Med J* **77**, 315–317.

Ciavarella, D., Reed, R. L., Counts, R. B., *et al.* (1987). Clotting factor levels and the risk of diffuse microvascular bleeding in the massively transfused patient. *Br J Haematol* **67**, 365–368.

Cosgriff, N., Moore, E. E., Sauaia, A., *et al.* (1997). Predicting life-threatening coagulopathy in the massively transfused trauma patient: Hypothermia and acidosis revisited. *J Trauma* **42**, 857–862.

Crosson, J. T. (1996). Massive transfusion. *Clin Lab Med* **16**, 873–882.

Driscoll, D. F., Bistrian, B. R., Jenkins, R. L., *et al.* (1987). Development of alkalosis after massive transfusion during orthotopic liver transplantation. *Crit Care Med* **15**, 905–908.

Fakhry, S. M., Messick, W. J., and Sheldon, G. F. (1996). Metabolic effects of massive transfusion. In "Principles of Transfusion Medicine" (E. C. Rossi *et al.*, Eds.), pp. 615–625. Williams and Wilkins, Baltimore.

Faringer, P. D., Mullins, R. J., Johnson, R. L., and Trunkey, D. D. (1993). Blood component supplementation during massive transfusion of AS-1 red cells in trauma patients. *J Trauma* **34**, 481–487.

Ferrara, A., MacArthur, J. D., Wright, H. K., *et al.* (1990). Hypothermia and acidosis worsen coagutopathy in the patient requiring massive transfusion. *Am J Surg* **160**, 515–518.

Goodnough, L. T., Brecher, M. E., Kanter, M. H., *et al.* (1999). Medical progress: Transfusion medicine. I. Blood transfusion. *N Engl J Med* **340**, 438–447.

Hakala, P., Lindahl, J., Alberty, A., *et al.* (1998). Massive transfusion exceeding 150 units of packed red cells during the first 15 hours after injury. *J Trauma* **44**, 410–412.

Harvey, M. P., Greenfield, T. P., Sugrue, M. E., *et al.* (1995). Massive blood transfusion in a tertiary referral hospital: Clinical outcomes and haemostatic complications. *Med J Aust* **163**, 356–359.

Jameson, L. C., Popic, P. M., and Harms, B. A. (1990). Hyperkalemic death during use of a high-capacity fluid warmer for massive transfusion. *Anesthesiology* **73**, 1050–1052.

Johnston, T. D., Chen, Y., and Reed, R. L. (1994). Functional equivalence of hypothermia to specific clotting factor deficiencies. *J. Trauma* **37**, 413–417.

Ketcham, E. M., and Cairns, C. B. (1999). Hemoglobin-based oxygen carriers: Development and clinical potential. *Ann Emerg Med* **33**, 326–337.

Kivioja, A., Myllynen, P., and Rokkanen, P. (1991). Survival after massive transfusions exceeding four blood volumes in patients with blunt injuries. *Am Surg* **57**, 398–401.

Leslie, S. D., and Toy, P. T. C. Y. (1991). Laboratory hemostatic abnormalities in massively transfused patients given PRBCs and crystalloid. *Am J Clin Pathol* **96**, 770–773.

Martin, D. J., Lucas, C. E., Ledgerwood, A. M., *et al.* (1985). Fresh frozen plasma supplement to massive PRBC transfusion. *Ann Surg* **202**, 505–511.

Michelsen, T., Salmela, L., Tigerstedt, I., *et al.* (1989). Massive blood transfusion: Is there a limit? *Crit Care Med* **17**, 699–700.

Murray, D. J., Olson, J., Strauss, R., *et al.* (1988). Coagulation changes during packed red cell replacement of major blood loss. *Anesthesiology* **69**, 839–845.

Phillips, G. R., Kauder, D. R., and Schwab, C. W. (1994). Massive blood loss in trauma patients: The benefits and danger of transfusion therapy. *Postgrad Med* **95**, 61–62, 67–72.

Phillips, T. F., Soulier, G., and Wilson, R. F. (1987). Outcome of massive transfusion exceeding two blood volumes in trauma and emergency surgery. *J Trauma* **27**, 903–910.

Reed, R. L., Ciavarella, D., Heimbach, D. M., *et al.* (1986). Prophylactic platelet administration during massive transfusion: A prospective, randomized, double-blind clinical study. *Ann Surg* **203**, 40–48.

Reiner, A. P. (1998). Massive transfusion. In "Perioperative Transfusion Medicine" (B. D. Spiess, R. B. Counts and S. A. Gould, Eds.), pp. 351–364. Williams and Wilkins, Baltimore.

Rohrer, M. J., and Natale, A. M. (1992). Effect of hypothermia on the coagulation cascade. *Crit Care Med* **20**, 1402–1405.

Rudolph, R., and Boyd, C. R. (1990). Massive transfusion: Complications and their management. *South Med J* **83**, 1065–1070.

Sawyer, P. R., and Harrison, C. R. (1990). Massive transfusion in adults: Diagnosis, survival and blood bank support. *Vox Sang* **58**, 199–203.

Schwab, C. W., Shayne, J. P., and Turner, J. (1986). Immediate trauma resuscitation with type O uncrossmatched blood: A two-year prospective experience. *J Trauma* **26**, 897–902.

Stehling, L. (1994). Fluid replacement in massive transfusion. In "Massive Transfusion" (L. C. Jefferies and M. E. Brecher, Eds.), pp. 1–15. AABB, Bethesda.

Stehling, L. C., *et al.* (1996). Practice guidelines for blood component therapy: A report by the American Society of Anesthesiologists Task Force on Blood Component Therapy. *Anesthesiology* **84**, 732–747.

Valeri, C. R., Cassidy, G., Khuri, S., *et al.* (1987). Hypothermia-induced reversible platelet dysfunction. *Ann Surg* **205**, 175–181.

Velmahos, G. C., Chan, L., Chan, M., *et al.* (1998). Is there a limit to massive blood transfusion after severe trauma? *Arch Surg* **133**, 947–952.

Vengelen-Tyler, V. (1996). "Technical Manual" 12th Ed. American Association of Blood Banks, Bethesda.

Wilson, R. F., Binkley, L. E., Sabo, F. M., *et al.* (1992). Electrolyte and acid-base changes with massive blood transfusions. *Am Surg* **58**, 535–544.

23

Evaluation of the Bleeding Patient

KAREN S. ROUSH

Department of Pathology and Laboratory Medicine
Emory University School of Medicine
Atlanta, Georgia

ELEANOR S. POLLAK

University of Pennsylvania Medical Center
Philadelphia, Pennsylvania

This chapter addresses the diagnostic workup of patients who present with or develop bleeding diatheses in the hospital or clinic settings. Organized evaluation followed by timely, efficacious therapy is essential for proper patient management. Acute massive bleeding due to surgical or traumatic sources of injury are discussed in Chapter 22 and will not be reviewed here.

Bleeding may be caused by abnormalities in any part of the normal hemostatic system, which includes blood vessels, platelets, and coagulation pathways. Dysfunction may occur in one or more parts of the hemostatic system and may include blood vessel fragility, quantitative and/or qualitative platelet dysfunction, and coagulation pathway abnormalities. Coagulation pathway abnormalities are complex and are usually in the form of functional factor deficiencies due to congenital absence, antibody inhibitors, or decreased synthesis.

A thorough medical and family history is imperative and can facilitate diagnosis and intervention through increasing or decreasing suspicion of relevant clinical diagnoses (see Table 1). Pertinent questions help differentiate bleeding due to platelet versus coagulation factor abnormalities. Bleeding due to thrombocytopenia or dysfunctional platelets manifests primarily as mucosal bleeds, which include easy bruising, prolonged or excessive bleeding following tooth extractions and other minor injury, frequent nosebleeds, or excessive menstrual bleeding. In contrast, clotting factor deficiencies often present as joint or deep tissue bleeding. History of transfusion may also be relevant, and previous laboratory evaluation may provide additional information.

I. NORMAL HEMOSTASIS

A brief overview of normal hemostasis is included below as a preface for the laboratory diagnosis of the bleeding patient. First, in order to prevent uncontrolled bleeding, platelet plug formation, also known as primary hemostasis, must occur (see Figure 1). Platelet plugs require qualitatively normal platelets in adequate quantities. While the literature does not provide an absolute count below which platelets are insufficient to allow platelet plug formation, some patients with counts as low as $10,000/\mu l$ may not manifest obvious hemorrhage (Gmur *et al.*, 1991).

Primary hemostasis is followed by secondary hemostasis, which, through an enzymatic cascade of coagulation protein activation, ultimately leads to a stable fibrin clot (see Figure 2). Secondary hemostasis is a complex interaction among numerous factors that are separated somewhat arbitrarily into intrinsic and extrinsic pathways. These pathways do not operate independently *in vivo*, but are useful in the laboratory for specific factor dysfunction, deficiency, or inhibitor diagnosis.

II. LABORATORY EVALUATION OF THE BLEEDING PATIENT

By far the most commonly performed tests in the evaluation of bleeding patients are the platelet count, prothrombin time (PT), activated partial thromboplastin time (aPTT), and bleeding time (BT).

TABLE 1 Patient Workup and Screening: Bleeding History and Physical Exam

Physical exam

Bruising. Presence or absence. If present, where? Legs and arms or any areas not likely to have sustained minor trauma? How large? Bruises with appearance of occurrence at same or different times?

Signs of active bleeding problem. Mucosal surface bleeding (epistaxis, oral, anal bleeding) generally associated with platelet defects or von Willebrand's disease. Retroperitoneal, joint bleeding generally associated with coagulopathies.

General signs associated with potential bleeding problem. Splenomegaly? Hepatomegaly? Adenopathy? Delayed wound healing?

History

Any prior abnormal tests associated with a bleeding problem? If so, identify what tests were abnormal. When, where, and under what conditions were the tests performed so that prior records may be evaluated whenever possible.

Bruising. When does bruising occur (spontaneous bruising or only after injury)? Historical data about onset of bruising. Where and how large were the bruises?

Bleeding manifestations.

What is the site where bleeding occurs? Important to differentiate mucosal bleeding such as epistaxis and gum bleeding (suggestive of platelet problem) from joint bleeding (suggestive of a coagulopathy such as deficiencies of factor VIII or IX).

What is bleeding associated with? Spontaneous? What kind of injury? Dental extractions or surgery? Trauma?

When does the bleeding occur in relation to provoking event? Immediately (suggestive of a platelet problem or vWD), after several hours (suggestive of a coagulopathy), or after initial healing (suggestive of a wound repair problem such as FXIII deficiency)?

Frequency of bleeding problem?

Change in nature, duration, and occurrence of bleeding problem?

How much blood is lost with bleeding? Be as quantitative as possible by specific questioning. If nose bleeds are a problem, ask how many tissues are needed to stop the bleeding. Was packing necessary?

Prior surgeries including dental extractions? Was any excess bleeding noted at these times?

Menstrual history? For how long has excessive bleeding been occurring? How much blood loss? Quantitate number of feminine napkins or tampons needed/day for what length of time.

Blood transfusions? Was a blood transfusion ever necessary? If so, under what conditions (surgery, trauma)?

Family history. Any members of the biological family have bleeding problems? If so, what relatives (include distant biological relatives and infant deaths)? In males and females or predominantly one gender?

Other medical problems. Acute or chronic illnesses? Nutritional status? Vitamin K deficiency is associated with low activity of clotting proteins. Folate, vitamin B 12, and severe iron deficiency may be associated with platelet problems. Liver problems (associated with a decreased production of most coagulation and anticoagulation proteins)? Malignancy? Hematological malignancies may be associated with thrombocytopenia. Solid tumors can be associated with DIC. Paraproteins may be associated with interference of clotting-based laboratory tests.

Medications. Clotting factor synthesis is affected by coumadin and some broad-spectrum antibiotics (interfere with vitamin K availability). Platelet function is affected by aspirin, NSAIDs, numerous aspirin-containing over-the-counter medications, and chemotherapeutic drugs. Take a thorough history for ANY drug consumed by the patient for at least 2 weeks prior to testing.

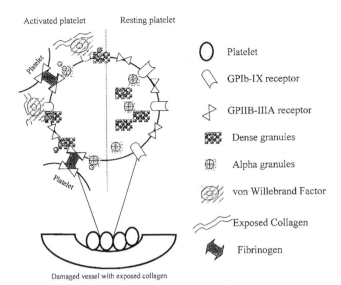

FIGURE 1 Primary hemostasis. When there is disruption of the endothelium, displaying subendothelial substances such as collagen, platelets become attached through von Willebrand factor and release the contents of their alpha and dense granules. Alpha granules contain platelet fibrinogen, platelet-derived growth factor, vWF, FV, P-selectin, β-thromboglobulin, and heparin-neutralizing platelet factor 4 (PF4). Dense granules contain ADP, ATP, 5-hydroxytryptamine, and calcium. The goal of primary hemostasis is formation of a platelet plug.

A. Platelet Count

Normal primary hemostasis requires an adequate number of platelets in addition to normal qualitative platelet function. The platelet count itself does not provide any information about qualitative function. A normal reference range for the platelet count is between 150,000 and 400,000/μl. At platelet counts below 100,000/μl, the BT (see below), an *in vivo* test of hemostasis, usually becomes prolonged beyond the normal range. Relatively normal hemostasis without spontaneous bleeding may occur with platelet counts as low as 50,000/μl, while prolonged bleeding due to minor trauma may occur at counts between 30,000 and 50,000/μl. Finally, spontaneous bleeding may occur with platelet counts less than 30,000/μl, though serious bleeding complications are uncommon with physiologically normal platelets at counts greater than 10,000/μl.

B. Prothrombin Time and the Activated Partial Thromboplastin Time

The PT and the aPTT are laboratory tests used to assess a patient's ability to form a fibrin clot by screening for defects in the coagulation cascade. The PT assesses the extrinsic pathway of coagulation and the aPTT assesses the intrinsic pathway.

1. PT

The PT assesses the tissue factor pathway, also known as the extrinsic pathway. Prolongation of the clotting time in this assay is due to decreased activity of FVII or one of the common pathway factors, including FII, FV, FX, and fibrinogen, all of which participate in fibrin formation in both the intrinsic and the extrinsic systems.

A lupus anticoagulant is an antiphospholipid antibody which is commonly seen in the presence of systemic lupus erythematosis, but may also be seen in other disorders. This is an *in vitro* effect. In fact, the presence of a lupus anticoagulant may be seen as a risk factor for thrombotic and not hemorrhagic disorders or may be an incidental nonpathological finding. Specialized testing is required to rule out true coagulopathies in the presence of a lupus anticoagulant.

Liver disease of several different forms may be associated with a prolonged PT because most clotting factors are synthesized either exclusively or predominantly in the liver. The PT is usually prolonged prior to a prolongation of the aPTT due to the greater sensitivity of the PT to mild and combined decreases in proteins produced in hepatocytes.

2. aPTT

The aPTT tests abnormalities of contact activation factors as well as intrinsic pathway clotting factors. Contact activation factors are FXII, prekallikrein, and high-molecular-weight kininogen. While deficiencies of these factors produce a grossly prolonged aPTT (often > 150 seconds), this is not associated with bleeding disorders. Deficiencies of the intrinsic factors, which include FXI, FIX, and FVIII, are associated with bleeding tendencies. As mentioned earlier, the aPTT may also be prolonged due to a common pathway factor deficiency; however the PT is more sensitive to deficiencies of proteins in the common pathway.

Clinical disorders associated with a prolonged aPTT include hemophilia A (FVIII deficiency), hemophilia B (FIX deficiency), some cases of von Willebrand disease (vWD), and FXI deficiency. Isolated clotting factor deficiencies do not usually affect the PT or aPTT until the activity level of the factor is below 50%. As the factor level decreases, the abnormal coagulation test becomes progressively more prolonged. Combined deficiencies of factors have a multiplicative effect on the PT and aPTT.

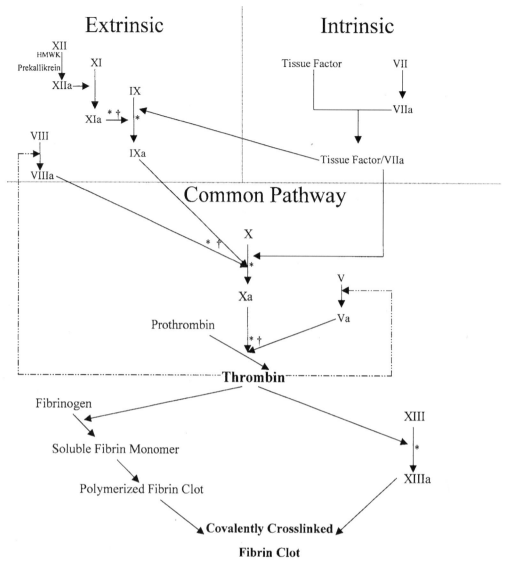

FIGURE 2 Secondary hemostasis—the coagulation cascade. Coagulation is somewhat arbitrarily divided into an extrinsic and intrinsic pathway, which converge into a final common pathway. *In vivo* this separation does not occur, but the division is relevant in *in vitro* analysis and facilitates diagnosis. The ultimate goal of the cascade is to provide a covalently cross-linked fibrin clot. Note that thrombin positively feeds back to activate factors V and VIII. *, requires calcium; †, requires phospholipid.

The most common cause of prolonged aPTT is heparin, which is present either by intentional administration to treat underlying clotting predisposition or by sample contamination from an indwelling catheter flushed with heparin to maintain patency. While the laboratory is capable of identifying heparin's effect on the aPTT, a specimen free of heparin is needed for a meaningful aPTT result.

3. Prolonged PT and/or aPTT

The most important follow-up test for a prolonged PT or aPTT is repetition of the test on a new sample.

Preanalytical variables, including improper blood collection, clotted specimen, underfilled tube, delayed transport of the tube to the laboratory, or improper or untimely plasma separation, may produce falsely prolonged PT and aPTT results. Confirmed prolongation of the PT or aPTT warrants prompt evaluation with additional laboratory testing.

C. Bleeding Time

The BT is a controlled, *in vivo* test of primary hemostasis and reflects platelet function and vascular and connective tissue integrity. Settings in which a BT may provide useful information include evaluation of suspected platelet function disorders or other abnormalities relating to primary hemostasis, or assessing platelet functionality in the setting of uremia. Patients who have a platelet count less than $100,000/\mu l$ will likely have an abnormal BT due to the low platelet count, obviating its usefulness. Patients who have taken aspirin should not be evaluated with a BT since aspirin causes unpredictable variability in the test. Furthermore, the literature suggests that a prolonged BT due to aspirin does not correlate well with bleeding after surgical challenge. vWD is often associated with a prolonged BT, but more specific tests make a BT unnecessary for this diagnosis.

To give meaningful results, it is critical that experienced personnel perform the BT. It should be administered under a constant blood pressure of 40 mm Hg with a standard incision on the volar surface of the forearm and then blotted at 30-second intervals.

III. BLEEDING DUE TO THROMBOCYTOPENIA

Thrombocytopenia is usually defined as a platelet count of less than $150,000/\mu l$. Thrombocytopenia occurs as a result of any of four primary causes: (1) decreased platelet production, (2) decreased platelet survival, (3) platelet sequestration, or (4) platelet dilution. One of the most common causes of suspected thrombocytopenia is pseudothrombocytopenia, in which the subnormal laboratory value is due to *in vitro* EDTA-induced platelet clumping (Silvestri *et al.*, 1995). Thus, examination of the peripheral blood smear is essential prior to making the diagnosis of thrombocytopenia.

A. Disorders Associated with Decreased Platelet Production

1. Congenital Disorders

A decreased platelet count can be associated with congenital abnormalities of platelets, some of which have microscopically visible changes in platelet morphology. In the "gray platelet syndrome," platelets lack alpha granule proteins and appear gray due to hypogranularity. Large or giant platelets may be seen in a number of congenital anomalies, including the May–Hegglin anomaly, Bernard–Soulier syndrome, and Alport syndrome. An additional congenital disorder is the thrombocytopenia-absent radius (TAR) syndrome.

2. Drugs

Any patient who presents with thrombocytopenia should be questioned about therapeutic, as well as recreational, drug use since several commonly used agents have been associated with poor platelet production. These include, but are not limited to, cytoreductive chemotherapeutic agents, interferon, thiazide diuretics, ethanol, and cocaine.

3. Bone Marrow Infiltration

Thrombocytopenia may occur as a result of marrow replacement by tumor, fibrosis, granulomatous disease, and lipid histiocytes. Bone marrow aspirates and/or biopsies are necessary for evaluation.

4. Ineffective Thrombopoiesis

Several states occur that affect platelet production or impair megakaryocyte and platelet survival before platelet release from the marrow. These are collectively referred to as ineffective thrombopoietic disorders and include:

a. Constitutional Disorders

Congenital disorders of the bone marrow include Fanconi's anemia and amegakaryocytic thrombocytopenia, usually present early in childhood, and are associated with thrombocytopenia.

b. Nutritional Deficiency

Notably, vitamin B12, folic acid, and severe iron deficiencies may lead to defective platelet production.

c. Paroxysmal Nocturnal Hemoglobinuria (PNH)

PNH is an acquired stem cell disorder that affects the erythropoietic, granulocytic, and thrombopoietic cell lines and results in thrombocytopenia in approximately two-thirds of patients. PNH is more commonly associated with anemia, which is caused by a reduction in complement defense proteins ultimately leading to complement-mediated intravascular lysis.

d. Infection

Viral infections including rubella, CMV, EBV, and HIV may be associated with inadequate thrombopoiesis leading to thrombocytopenia.

B. Decreased Platelet Survival

Autoimmune, isoimmune, drug-mediated, or infection-related antibodies may lead to destruction of platelets. Syndromes associated with immune-mediated destruction of platelets include but are not limited to autoimmune thrombocytopenic purpura (ITP), heparin-associated thrombocytopenia (HIT), acquired thrombotic thrombocytopenic purpura (TTP), HIV-associated thrombocytopenia, post-transfusion purpura

(PTP), and neonatal alloimmune thrombocytopenia (NAIT). Additionally, immune-mediated thrombocytopenia may be seen along with other symptomatology in systemic disease. Nonimmunologic mechanisms including microangiopathic hemolytic anemias, giant hemangiomas, or disseminated intravascular coagulation (DIC) are also responsible for shortened platelet survival. The primary nonimmune cause of decreased platelet survival is DIC, which is discussed in detail later in this chapter.

1. Autoimmune Thrombocytopenic Purpura (ITP)

Previously referred to as idiopathic thrombocytopenia purpura, ITP is a clinical disorder resulting in immune-mediated platelet destruction and splenic sequestration. Acute ITP is most commonly seen in young children following a viral illness. In 80% of childhood cases, ITP is a mild self-limited disease. Chronic ITP most commonly occurs in women of childbearing age and has an insidious onset. Unfortunately, ITP is largely a diagnosis of exclusion since there are no definitive diagnostic features or laboratory findings. The diagnosis of ITP is made when unexplained thrombocytopenia occurs with normal or increased megakaryocytes in the bone marrow when other causes of thrombocytopenia have been excluded.

a. Clinical and Laboratory Manifestations

Clinically, symptoms of ITP include petechiae, ecchymoses, and mucosal bleeding. Platelet antibody screening tests may be positive for antibodies to GPIIb-IIIa or GPIb-IX, which are the platelet fibrinogen receptor and the platelet vWF receptor, respectively.

b. Management

The management of ITP consists of immunosuppressive therapy which may include various combinations of intravenous immune globulin (IVIg) or intravenous anti-Rh(D), steroids, splenectomy, and, in specific situations, platelet transfusion. Steroids and splenectomy are primary treatments in chronic ITP, while IVIg is first-line therapy for acute ITP. Therapeutic guidelines were developed and published by the American Society of Hematology in 1996 (George *et al.*, 1996).

Unless an ITP patient has severe thrombocytopenia and life-threatening bleeding, platelet transfusions are generally withheld. However, platelet transfusions are NOT contraindicated per se in ITP, and in cases of life-threatening bleeding may be the only available therapeutic option (Carr *et al.*, 1986). Some authors have advocated timing administration of platelets to follow IVIg (Baumann *et al.*, 1986). Prophylactic platelet transfusions may be appropriate in the following situations: (1) prior to splenectomy, especially if the patient is actively bleeding, (2) when the platelet count is less than $10,000/\mu l$, or (3) prior to Cesarean section when the platelet count is less than $50,000/\mu l$.

2. Heparin-Associated Thrombocytopenia (HIT)

While many drugs are associated with immune-mediated platelet destruction, the most notable example is heparin. HIT occurs in approximately 5% of patients receiving unfractionated heparin and <1% patients receiving low-molecular-weight heparin.

a. Clinical and Laboratory Manifestations

Historically, there are two major classifications of HIT, type I and type II. Type I is the most common, occurs rapidly after heparin administration and leads to mildly decreased platelet counts. Type II typically occurs 5–14 days following heparin administration, may result in severe venous and arterial thromboses, and is caused by autoantibodies formed against the heparin–PF4 complex.

b. Management

Heparin, including heparin flushes, should not be administered to any patient suspected of having HIT. The heparin serotonin release assay and an ELISA assay are available to identify antiplatelet antibodies if present. It is important to know the limitations of these assays since not all patients have antiplatelet antibodies (type I) and specificities of the antibodies may not be unique. Therefore, heparin administration should be withheld from a patient with a negative antibody screen when there is a strong clinical suspicion of HIT. Other methods of anticoagulation should be utilized in patients with HIT.

3. Thrombotic Thrombocytopenic Purpura (TTP)

The pathogenesis and treatment of TTP are reviewed in detail in Chapter 39.

a. Clinical and Laboratory Manifestations

Thrombocytopenia, fever, microangiopathic hemolytic anemia, altered mental status, and renal dysfunction characterize TTP. Patients with TTP are at high risk for bleeding due to extremely low platelet counts. TTP has long been considered a disease of exclusion; however, there is now evidence that TTP may be caused by decreased activity of vWF-cleaving metalloprotease. In most cases this is due to the formation of an antibody to the protease. Ultralarge vWF multimers cannot be cleaved without normal protease activity, and this leads to the formation of platelet thrombi in the patient's microvasculature, ultimately causing end-organ ischemia.

b. Management

Although the TTP patient is often markedly thrombocytopenic, *platelet transfusions are considered contraindicated in TTP* because platelet administration provides additional substrate for the continuing formation of platelet thrombi. The treatment of choice for acquired TTP is plasma exchange using fresh frozen plasma (FFP) or cryosupernatant as the replacement fluid. This procedure removes the anti-vWF-cleaving metalloprotease antibody while providing normal protease. Plasma exchange for TTP is often used in conjunction with a variety of immunosuppressive therapies. (See Chapter 39.)

4. HIV-Associated Thrombocytopenia

Thrombocytopenia occurring in the setting of HIV is a multifactorial disorder, caused by impaired platelet production and increased platelet destruction. Impaired production is secondary to HIV-infected megakaryocytes and increased destruction is caused by antibodies directed against the GpIIb–IIIa complex. Platelet transfusions are occasionally necessary to maintain hemostasis in this patient population.

5. Systemic Diseases

Some collagen vascular disorders, particularly systemic lupus erythematosis (SLE), may be associated with platelet destruction. This is most often due to antibody-mediated destruction.

6. Disseminated Intravascular Coagulation (DIC)

DIC, an acquired life-threatening consumptive coagulopathy, is the most common emergent cause of increased platelet destruction requiring immediate attention. A separate section devoted to laboratory diagnosis and management of DIC is included later in this chapter.

C. Platelet Dysfunction

1. Disorders of Platelet Function

a. Congenital Disorders

These include Bernard–Soulier syndrome, Glanzmann's thrombasthenia, storage pool disease, and platelet activation defects, and may necessitate transfusion of platelets. In Bernard–Soulier syndrome, GPIb-IX, a platelet receptor for vWF, is either absent or dysfunctional. In Glanzmann's thrombasthenia, GPIIb-IIIa, a platelet receptor for fibrinogen, is absent or dysfunctional. In storage pool disease, platelet granules are either absent or dysfunctional and improperly store or secrete platelet granule contents. Treatment for bleeding episodes in all of these disorders may include platelet transfusion regardless of platelet count. Treatment with recombinant factor VIIa has also been effective in some cases.

b. Acquired Platelet Dysfunction

Certain drugs may interfere with platelet function by interaction with the platelet membrane or its receptors. These drugs include some antidepressants, penicillins, antihistamines, alcohol, dextran, hydroxyethyl starch (HES), agents that affect the prostaglandin pathway, and drugs that alter phosphodiesterase activity. Commonly used drugs that affect the prostaglandin pathway are aspirin, NSAIDs, indomethacin, furosemide, and verapamil, while examples of those that alter phosphodiesterase activity are caffeine, dipyridamole, and theophyllline. These drugs, such as aspirin and NSAIDSs, are associated with either irreversible or reversible platelet dysfunction, respectively. As has been previously emphasized, a detailed medication history that includes prescription drugs, over-the-counter drugs, and recreational drugs is vital to accurate diagnosis. Treatment of drug-related platelet dysfunction consists primarily of drug discontinuation with subsequent *in vivo* platelet function tests for documentation of restored platelet function.

c. Uremia

Patients who have uremia frequently experience bleeding due to platelet dysfunction. While many mechanisms for this acquired defect have been proposed, no current consensus exists. Primary therapy for uremic bleeding consists of dialysis and red blood cell transfusion to raise the hematocrit above 30%. If dialysis is not a treatment option or is unsuccessful, desmopressin vasopressin (DDAVP) may be administered. While platelet transfusions are only temporarily effective in uremic patients, they may be necessary to effectively treat life-threatening bleeding.

d. Acute and Chronic Liver Disease

These disorders may be associated with dysfunctional platelets, either due to the impact of fibrin degradation products (FDPs) on platelet response or due to altered platelet metabolism. Treatment with DDAVP and platelet concentrates should be considered in a patient with recalcitrant bleeding despite near-normal platelet counts.

e. Mechanical Destruction

Cardiopulmonary bypass (CPB) surgery is associated with hemodilutional deficiencies of platelets and coagulation factors that usually resolve within 12–72 hours postsurgery. Platelet function is also adversely affected, which significantly impairs platelet aggregation and prolongs the bleeding time. The vast majority of bleeding post-CPB surgery that is not due to incomplete surgical hemostasis is due to platelet dysfunction (Woodman and Harker, 1990). Platelet transfusions in CPB surgery may be necessary if there is uncontrolled bleeding, but prophylactic platelet transfusions for CPB surgery are not recommended.

f. Myeloproliferative Disorders (MPDs)

The MPDs include chronic myelogenous leukemia (CML), essential thrombocythemia (ET), and polycythemia vera (PV). Although the platelet count is often elevated in these disorders, platelet transfusion may be indicated for bleeding since the platelets are usually functionally abnormal.

g. Dysproteinemias

Multiple myeloma and Waldenstrom's macroglobulinemia may lead to bleeding due to either qualitative platelet dysfunction or interference with coagulation factors via their adsorption to circulating cancer cells or inhibition by abnormal immunoglobulins. Platelet transfusions may be helpful but are generally ineffective in this setting and treatment consists of treating the underlying abnormality.

2. Clinical and Laboratory Manifestations of Platelet Dysfunction

Bleeding is the most common manifestation of congenital or acquired platelet dysfunction. Bleeding disorders due solely to platelet dysfunction generally are associated with a prolonged BT while the PT and aPTT are normal. Specialized tests of platelet function are described below.

3. Diagnosis

Since platelet-type bleeding may occur in the setting of a normal or elevated platelet count, specialized platelet function tests are generally indicated. These detect the ability of patient platelets to aggregate in response to platelet agonists including adenosine diphosphate (ADP), epinephrine, and collagen, and can help confirm the existence of an increased risk for bleeding due to insufficient platelet function (Bick, 1994b). Platelet function tests must be performed and interpreted cautiously by an experienced reference laboratory since numerous preanalytical variables may affect the sensitivity and specificity of testing. Many drugs, most notably aspirin and NSAIDs, affect platelet function, so relaying medication history to the laboratory is essential prior to performance of specialized testing. Abnormal platelet function due to aspirin may be detected by platelet response to arachidonic acid, which will be negative in this setting.

D. Platelet Sequestration

Platelet sequestration occurs in the setting of splenomegaly or hypersplenism, and patients with unexplained thrombocytopenia should be evaluated accordingly. Platelet transfusions given to splenomegalic patients are typically ineffective and have shortened platelet circulation times and inadequate post-transfusion count increments. Transfused platelets, along with the patient's platelets, are rapidly trapped in the enlarged spleen. Splenectomy is often required in these patients in addition to treatment of the underlying disorder.

E. Dilutional Thrombocytopenia

This is a well-documented phenomenon in massively transfused patients who receive greater than 10 units of packed red blood cells (PRBCs) during a 24-hour period, and is also seen following large-volume fluid resuscitation. Treatment for bleeding is platelet transfusion.

IV. ABNORMALITIES OF THE COAGULATION PATHWAY

Detailed descriptions of the bleeding disorders caused by specific deficiencies in clotting factors are found in Chapter 9 and will not be covered in detail here. Information regarding dosing of factor concentrates for the treatment of clotting factor deficiencies may also be found in Chapter 9.

A. Factor VIII Deficiency (Hemophilia A)

The aPTT is prolonged in patients with FVIII deficiency. The severity of bleeding is highly correlated with the in vitro FVIII activity level. Patients with < 1% FVIII activity levels experience severe and frequent bleeds, primarily in the joint spaces. Patients with activity levels of 1–5% are considered to have moderate disease, and those with activity levels of 6–30% are considered to have mild disease. Factor replacement with either monoclonal antibody-purified FVIII concentrates or with recombinant FVIII is standard therapy in most institutions.

B. Factor IX Deficiency (Hemophilia B)

FIX deficiency is clinically indistinguishable from FVIII deficiency; the aPTT is prolonged in FIX deficiency. Recombinant FIX and purified FIX concentrates are available for replacement therapy. FFP administration may be indicated in emergency settings in which FIX concentrate is not immediately available.

C. Von Willebrand Disease (vWD)

The BT is often prolonged in patients with vWD, a disorder caused by a congenital or acquired deficiency of vWF and characterized by mucosal bleeding. A mildly prolonged aPTT occurs in approximately half of all patients with vWD. Ristocetin cofactor activity, which measures plasma levels of vWF, is reduced. Type I vWD patients, which compose 70% of all vWD patients, are treated with DDAVP and/or a FVIII concentrate that contains vWF. Patients with type II vWD have qualitative defects in the vWF such that the response to DDAVP may be variable and contraindicated in some forms of vWD (type IIb) where it causes an increased aggregation of vWF and platelets. In type III vWD associated with a severe deficiency of vWF, severe bleeding episodes most often require treatment with purified FVIII and/or vWF concentrates.

D. Factor VIII Deficiency with Inhibitor

A prolonged aPTT is seen in FVIII activity deficiency due to specific antibody inhibitors. These inhibitors can be identified by laboratory evaluation, and are formed in 7–52% of patients receiving FVIII therapy. Management of bleeding in these patients depends on the titer of the antibody but includes administration of one or more of the following: (1) recombinant FVIIa, (2) porcine FVIII, (3) prothrombin complex concentrates. Human or recombinant FVIII concentrates and DDAVP may be used in patients with low-titer inhibitors. Treatment with porcine FVIII should be reserved for severe bleeding episodes in order to prevent increased inhibitor formation and development of antiporcine FVIII antibodies. Approximately 20% of patients who receive porcine FVIII will form antibodies against it (Hoyer, 1995).

TABLE 2   Causes of DIC

Obstetric
  1. Amniotic fluid embolus
  2. Retained dead fetus
  3. Placental abruption
  4. Eclampsia
Infection
  1. Sepsis
  2. Viremia
Malignancy
  1. Metastases
  2. Leukemia
Tissue damage
  1. Trauma
  2. Snake bite
  3. Thermal or chemical burn
Acute liver disease
Transfusion
  1. Hemolytic transfusion reaction
  2. Massive transfusion

E. Disseminated Intravascular Coagulation (DIC)

DIC is a secondary thrombohemorrhagic disorder resulting in thrombosis, consumption of platelets and coagulation factors, fibrin deposition in the microvasculature, and subsequent hemorrhage (Bick, 1994a; Williams, 1998). DIC can be life threatening, but may resolve quickly and completely with prompt management of the inciting underlying disease. Primary disorders leading to DIC are listed in Table 2.

1. Pathophysiology

The principal impetus leading to DIC is dysregulation of generation and degradation of thrombin, the enzyme responsible for amplifying clot formation. Thrombin has many procoagulant activities, including cleavage of fibrinogen to directly enable fibrin clot formation, and additionally the activation of factors XI and XIII, and cofactors V and VIII. Thrombin is the most potent platelet activator and also enhances cell migration. However, when bound to the endothelial cell receptor thrombomodulin, thrombin is converted to an antithrombotic protein through the activation of anticoagulant protein C, which catalyzes the degradation of the active coagulation cofactors Va and VIIIa. Thrombin also causes the release of tissue plasminogen activator (TPA) from endothelial cells and activates the fibrinolysis inhibitor, thrombin-activatable fibrinolysis inhibitor (TAFI). The principal mechanism opposing thrombins coagulant activities is the serpin, antithrombin (AT), which forms an irreversible complex with thrombin and inactivates it. This catalytic inactivation is dramatically accelerated by heparin. Strategies for treating DIC, such as administration of heparin and/or AT concentrates, are based on the need to dampen the prothrombotic properties of thrombin.

DIC is an end result of various pathological processes that cause widespread fibrin deposition in the microvasculature. The etiologies responsible for DIC can be grouped into two major categories: (1) disorders that lead to generalized damage to the endothelial cell lining, and (2) disorders of tissue damage that release thromboplastin into the circulation. Despite heterogeneity of endothelial cells, the fibrin deposition of DIC affects the microvasculature, both arterial and venous, of nearly all organ systems.

The fibrinolytic response results in the generation of fibrin degradation products (FDPs) and D-dimers, and also in the consumption of AT and plasminogen. Depletion of AT and plasminogen further fuels the ongoing prothrombotic process. The widespread generation of thrombin, in addition to endothelial damage, leads to platelet recruitment, activation, and consumption. In fact, thrombocytopenia is often the earliest and most sensitive, but least specific, laboratory value indicating the onset of DIC. Importantly, FDPs and D-dimers affect *in vitro* measurements of clotting and lead to increased thrombin time, PT, and aPTT.

2. Clinical Presentation

a. Acute DIC

Acute DIC occurs rapidly in response to overwhelming systemic impetus, and most often manifests as a hemorrhagic disorder. Bleeding that begins as petechiae or purpura, bleeding at venipuncture sites, and mucous membrane bleeding may quickly progress to hemoptysis, hematuria, gastrointestinal bleeding, and bleeding from surgical incisions. As the thrombocytopenia becomes increasingly severe, intracranial and deep tissue bleeding may also occur. DIC may also present with thrombotic characteristics, which may include microvascular thrombosis, acral cyanosis, gangrene of the extremities, and cerebrovascular accidents.

b. Chronic DIC

Low-grade, or compensated, chronic DIC occurs in a variety of settings, including cardiovascular disease, renal disease, immunologic disorders, malignancies, hematologic disorders, and inflammatory diseases. Chronic DIC usually occurs when synthesis of clotting proteins and platelets occurs at a faster rate than their degradation, so that chronic DIC is most often thrombotic in nature. Bleeding manifestations in chronic DIC are more subacute than in acute DIC, but may become emergent as DIC progresses.

3. Laboratory Diagnosis

Widespread vascular fibrin deposition characterizes DIC, but there are no definitive laboratory tests to diagnose DIC. Serial laboratory testing, including platelet count, PT, aPTT, fibrinogen, and FDPs, is essential to follow the progression of DIC. Classic acute DIC exhibits decreased fibrinogen, significantly elevated FDPs or D-dimers, decreased platelet count, and elevated PTs and aPTTs. Unfortunately, not all of these values may be abnormal when DIC is present. The frequency of test abnormalities from a combination of several studies prior to 1988 (Bick, 1994a) (410 patients) showed decreased platelet count in 95%, increased FDPs in 75%, fragmented red cells in 70%, an elevated PT and aPTT of 3 and 6 seconds, respectively, in 56%, and decreased fibrinogen in 57%.

The most characteristic laboratory signature of DIC is elevated D-dimers and/or FDPs in the context of serially decreasing fibrinogen levels. Since fibrinogen is an acute phase reactant, the fibrinogen level may acutely rise before decreasing into the subnormal range. Other laboratory abnormalities may include decreases in FV, FVIII, AT, and plasminogen, all due to consumption of these factors.

4. Differential Diagnosis

a. Chronic Liver Disease

Chronic liver disease can lead to a combination of elevated PT, elevated FDPs, decreased platelet count, and decreased levels of fibrinogen. Decreased plasma protein levels, including fibrinogen, are due to deficient hepatic synthesis. Elevated FDPs are due to inefficient clearance of FDPs and activated clotting proteins. Thrombocytopenia is secondary to destruction and sequestration associated with splenomegaly. FDP elevation in liver disease tends to be milder than that in DIC, and serially increasing FDP is not present in liver disease. Chronic DIC, however, is more difficult to differentiate from liver disease.

b. Microangiopathic Hemolytic Anemias

TTP and hemolytic-uremic syndrome (HUS) have overlapping laboratory features with DIC. However, the fibrinogen and PT are usually within normal limits in TTP and HUS, and FDPs are normally only mildly elevated.

c. Primary Fibrinolysis

This disorder is associated with elevated FDPs, but the platelet count is normal.

d. Other Causes

Increased D-dimers and FDPs may be seen in renal disease and sickle cell disease. Rheumatoid arthritis and cryoglobulinemia may create false positive results in latex and agglutination-based testing for fibrinogen-related antigens.

5. Management

The most critical strategy in the treatment of DIC is the correction of the underlying cause. Symptomatic management, while necessary, is merely temporizing.

a. Check Vital Signs

Obtain pertinent clinical values, including blood pressure, heart and respiratory rate, and temperature, to assess supportive care measures needed immediately.

b. Obtain Laboratory Data

Obtain laboratory values to confirm the diagnosis and to guide appropriate usage of blood products. Hemoglobin, platelet count, PT, aPTT, fibrinogen, FDPs and D-dimers, and, in some cases, AT levels may be helpful for initial, as well as serial, assessment. Frequent laboratory testing is necessary due to precipitous changes that may occur.

c. Treat the Underlying Illness

Removing the inciting stimulus is critical in the effective management of DIC.

d. Blood Product Therapy

Clinical symptoms and laboratory values should guide blood product therapy and will depend on whether the patient has a primarily hemorrhagic or thrombotic disorder. Appropriate blood products may be ordered using the following as a general guide:

i. Thrombocytopenia It is recommended that platelet transfusions be administered when the platelet count is less than $20,000/\mu l$ when the patient is not bleeding. The same dose is indicated in a bleeding patient when the platelet count is less than $50,000/\mu l$.

ii. Hypofibrinogenemia Cryoprecipitate may be indicated for patients with fibrinogen concentrations of less than 100 mg/dl.

iii. Elevated PT/aPTT FFP administration should be considered based on the extent of prolongation of the PT and aPTT. The PT is a better indicator of clotting protein levels than is the aPTT, since FVIII, as an acute phase reactant, may mask abnormalities in the aPTT.

iv. Low AT Levels AT concentrates have been used with success in fulminant DIC. Administration of AT may be particularly useful when DIC arises secondary to sepsis. The amount of AT to administer can be calculated based on the patient's initial AT level: AT dose (IU) = (125% − patient's current % AT level) × 0.6 × weight (kg). The patient's AT level should subsequently be monitored at 12-hour intervals to assess for additional AT therapy.

e. Other Therapies

Heparin therapy may be helpful in cases of primarily thrombotic DIC, particularly those associated with purpura fulminans or cancer. Hirudin and low-molecular-weight heparin have also been proposed for use in DIC. Recombinant human activated protein C may be effective in patients with sepsis-related problems.

References

Baumann, M. A., Menitove, J. E., Aster, R. H., *et al.* (1986). Urgent treatment of idiopathic thrombocytopenic purpura with single-dose gammaglobulin infusion followed by platelet transfusion. *Ann Intern Med* **104**, 808–809.

Bick, R. L. (1994a). Disseminated intravascular coagulation. *Med Clin N Am* **78**, 511–541.

Bick, R. L. (1994b). Platelet function defects associated with hemorrhage or thrombosis. *Med Clin N Am* **78**, 577–607.

Carr, J. M., Kruskall, M. S., Kaye, J. A., *et al.* (1986). Efficacy of platelet transfusions in immune thrombocytopenia. *Am J Med* **80**, 1051–1054.

George, J. N., Woolf, S. H., Raskob, G. E., *et al.* (1996). Idiopathic thrombocytopenic purpura: A practice guideline developed by explicit methods for the American Society of Hematology. *Blood* **88**, 3–40.

Gmur, J., Burger, J., Schanz, U., *et al.* (1991). Safety of stringent prophylactic platelet transfusion policy for patients with acute leukemia. *Lancet* **338**, 1223.

Hoyer, L. W. (1995). Factor VIII inhibitors. *Curr Opin Hematol* **2**(5), 365–371.

McMillan, R., and Imbach, P. (1998). Immune thrombocytopenic purpura. In "Thrombosis and Hemorrhage" (J. Loscalzo and A. I. Schafer, Eds.), Chap. 30, pp. 643–664. Williams and Wilkins, Baltimore.

Rutherford, C. J., and Frenkel, E. P. (1994). Thrombocytopenia. *Med Clin NAm* **78**, 555–575.

Silvestri, F., Virgolini, L., Savignano, C., *et al.* (1995). Incidence and diagnosis of EDTA-dependent pseudothrombocytopenia in a consecutive outpatient population referred for isolated thrombocytopenia. *Vox Sang* **68**, 35–39.

Williams, E. (1998). Disseminated intravascular coagulation. In "Thrombosis and Hemorrhage" (J. Loscalzo and A. I. Schafer, Eds.), Chap. 44, pp. 963–987. Williams and Wilkins, Baltimore.

Woodman, R. C., and Harker, L. A. (1990). Bleeding complications associated with cardiopulmonary bypass. *Blood* **76**, 1680–1697.

24

Approach to the Platelet Refractory Patient

ELINE T. LUNING PRAK

Department of Pathology and Laboratory Medicine
University of Pennsylvania School of Medicine
Philadelphia, Pennsylvania

Rapid loss of transfused platelets observed in two or more consecutive platelet transfusions is termed "platelet refractoriness." Hemorrhagic complications in thrombocytopenic patients with platelet refractoriness can be fatal. Furthermore, because these patients do not respond to platelet products, they typically receive a greater number of platelet transfusions and therefore are usually exposed to a larger number of blood donors. The specialized types of blood products and testing that these patients may require are expensive and labor-intensive. If profound hypoproliferative thrombocytopenia persists, patients may need to remain hospitalized. Longer hospital stays for platelet refractory patients translate into both higher healthcare costs and additional complications, such as nosocomial infections. This chapter describes the laboratory evaluation, monitoring, and transfusion management of patients who become refractory to platelet transfusions.

I. DESCRIPTION

Refractoriness to platelet transfusions is manifested by the absence of a significant and sustained rise in the platelet count following platelet transfusion (the "platelet increment"). Loss of transfused platelets can occur by a variety of mechanisms (summarized in Table 1), including sequestration (splenomegaly), consumption (sepsis, disseminated intravascular coagulation [DIC], massive bleeding, fever, drugs, endothelial cell injury or activation), or immune-mediated destruction (Novotny, 1999). Immune-mediated destruction of platelets occurs in idiopathic thrombocytopenic purpura (ITP; see George *et al.*, 1996) and in response to drugs such as heparin (heparin-induced thrombocytopenia; see Arepally and Cines, 1998), or can arise following human leukocyte antigen (HLA) or platelet alloimmunization. Alloantibodies can be passively acquired (as occurs in neonatal alloimmune thrombocytopenia) or can result from active immunization.

The focus of this chapter is on platelet refractoriness due to HLA alloimmunization. Patients who develop antibodies that are directed against foreign HLA molecules are at increased risk of developing platelet refractoriness.

II. INCIDENCE

The reported incidence of platelet refractoriness due to alloimmunization ranges from 2 to nearly 30% in multiply transfused patients (Novotny *et al.*, 1995; Doughty *et al.*, 1994; Abou-Ella *et al.*, 1995). Part of the explanation for the wide range in incidence is the distinction between patients with *de novo* alloimmunization and patients with prior sensitization (anamnestic response). Exposure to white blood cells (WBCs) via pregnancy, transfusion, or transplantation increases the risk of alloimmunization. Typically anti-HLA antibodies are detected within 1–6 weeks of exposure, if at all. In prospective surveys of bone marrow transplant (BMT) patients, *de novo* alloimmunization to HLA is infrequent, in part because many patients have already formed antibodies prior to entry into the study (Abou-Ella *et al.*, 1995; Novotny *et al.*, 1995; TRAP, 1997). Incidence is influenced by transfusion practices as well

TABLE 1 Possible Reasons for a Poor Response to Platelet Transfusion

Nonimmune factors	Immune factors
Splenomegaly (sequestration)	HLA antibodies
Bleeding	Platelet glycoprotein antibodies
Fever or infection	ABO antibodies
Disseminated intravascular coagulation	Autoantibodies
Large patient size	Antibacterial, antifungal agents
Low platelet dose	Heparin
Poor quality of platelets	Circulating immune complexes

(whether or not leukodepletion filters and/or UV-B irradiation are used). The frequency of alloimmunization also may vary with the underlying disease state. A higher frequency of HLA alloimmunization in acute myelogenous leukemia (AML) patients than in acute lymphoblastic leukemia (ALL) patients has been observed, despite both groups of patients receiving similar numbers of platelet and PRBC transfusions (Lee and Schiffer, 1987). Another area of uncertainty is that not all patients with anti-HLA antibodies are necessarily refractory to platelet transfusions. Thus HLA sensitization cannot simply be equated with platelet refractoriness.

III. PATHOPHYSIOLOGY

A. Alloimmunization

How HLA antibody production is stimulated is not entirely known. There are no published examples of naturally occurring anti-HLA antibodies. Primary HLA alloimmunization likely involves the processing and presentation of foreign HLA-A and HLA-B molecules to the recipient's T cells (McFarland, 1996). Platelets cannot carry out this antigen presenting cell (APC) function because they lack class II MHC (major histocompatibility) molecules. Rather, class II MHC molecules are found on leukocytes present in PRBC and platelet transfusions. The observation that primary alloimmunization can be prevented in many patients by the use of leukodepletion filters or UV-B-irradiated platelet and PRBC products is consistent with this model of immune stimulation (Novotny et al., 1995; TRAP, 1997; Slichter, 1998).

In the appropriate costimulatory context, donor APCs activate the recipient's T cells, causing them to elaborate cytokines, proliferate, and assist in B cell responses. Within days to weeks following primary immunization, anti-HLA antibodies can be detected in the serum. When the alloimmunized patient is rechallenged (transfused) with the same or related HLAs, an anamnestic response ensues. Anti-HLA antibodies coat the transfused platelets, facilitating their clearance (opsonization) by phagocytic cells. Antibody titers and reactivity patterns can shift in individuals over time, depending on the nature and timing of their transfusions. Antibodies may be cleared from the circulation as they coat the transfused platelets, for example. The patient's immune status, disease state, and other therapeutic maneuvers (immunosuppression, radiation, plasmapheresis, etc.) can also influence the concentration and binding characteristics of HLA antibodies.

The specificity of HLA antibodies can be used to predict the HLA types of platelets that are most likely to fail in refractory patients. Alternatively, or in addition, the refractory patient's antibodies can be used in platelet crossmatching. However, not every antibody that binds to platelets or even HLA antigens on platelets necessarily results in refractoriness. Indeed, several studies document the presence of anti-HLA antibodies in patients who are not refractory to platelets (Novotny et al., 1995). The titer, heavy chain subclass (the ability to fix complement or carry out other effector functions), and the specificity and avidity of the antibody for antigen likely play a role in determining pathogenicity.

B. Antigenic Targets of Antiplatelet Antibodies

1. Histocompatibility Antigens (HLAs)

The most frequently encountered antigenic targets of antibodies in patients with alloimmune platelet refractoriness are the class I major histocompatibility antigens (human leukocyte antigens), HLA-A and HLA-B. Class I MHC molecules are found on nearly all cells of the body and play a vital role in immune surveillance for intracellular bacteria and parasites as well as for viruses and some tumors. Class I HLA molecules are highly polymorphic. Of the three HLA class I alleles (HLA-A, -B, and -C), A and B are the most relevant to platelet refractoriness. HLA-C is less polymorphic and is expressed at lower levels on platelets than HLA-A and -B (Santoso et al., 1993). Platelets, in addition to expressing their own class I HLA antigens, can adsorb class I HLA molecules from plasma.

2. ABO Antigens

Less commonly, patients will have antibodies that are directed against ABH substances. The blood group

A and B antigens are structurally related oligosaccharides which reside on glycosphingolipids adsorbed from plasma or on platelet surface molecules, including glycolipids and membrane glycoproteins (including Ib, IIa, IIb, IIIa, and PECAM), as well as on the GPI-anchored platelet protein CD109 (Kelton *et al.*, 1998). Platelet ABH antigens occasionally have been implicated in platelet refractoriness, though this is not a universal finding (Sintnicolaas and Lowenberg, 1996). ABO matching may be relevant in products derived from donors who are high ABH expressors or who express A1. Furthermore, the use of ABO-matched platelets may be of benefit in recipients with high titers of anti-A and/or anti-B isohemagglutinins. The use of ABO-matched platelet products is addressed in Chapter 7.

3. Human Platelet Antigens

At least 15 human platelet alloantigen systems have been described and given a bewildering variety of names. For example, the platelet antigen $P1^{A1}$ is also known as Zw^a or HPA-1a (the latter stands for human platelet antigen 1, allele a). Fetal HPA-1a is recognized by maternal antibodies in the majority of cases of neonatal alloimmune thrombocytopenia (NAITP; see Chapter 25). The epitope for this antigen has been mapped to platelet glycoprotein IIIa (also known as CD61, the beta chain of the vitronectin receptor). Additional platelet glycoprotein antigens include HPA-2 to HPA-8, $Gov^{a/b}$, Va^a, Gro^a, Max^a, Oe^a, La^a, and ly^a (Kelton *et al.*, 1998). There are allelic forms of each of these antigens. Alloantibodies are sometimes formed by transfused or otherwise exposed individuals who lack the corresponding allele(s). For example, patients with Glanzmann's thrombasthenia, a hereditary hemorrhagic disorder in which platelet aggregation is defective, lack platelet gpIIb (HPA-3) and gpIIIa (HPA-1). These patients sometimes form antibodies to HPA-1 or HPA-3 when they receive platelet transfusions. This can lead to platelet refractoriness upon subsequent transfusions. In addition to NAITP and platelet refractoriness, anti-HPA antibodies (especially those directed against HPA-1a) have been implicated in post-transfusion purpura. In platelet refractory patients, the frequency of forming alloantibodies to platelet glycoprotein antigens appears to be 10–25% of that of forming HLA antibodies (Doughty *et al.*, 1994).

IV. DIAGNOSIS

Algorithms for the diagnosis and management of platelet refractoriness due to HLA alloimmunization are presented in Figures 1 and 2. The diagnosis of platelet refractoriness due to HLA alloimmunization is suggested by rapid loss of transfused platelets. The diagnosis is confirmed by identifying antibodies which react with HLA or platelet-specific antigens in the patient's serum. A history of pregnancy, PRBC or platelet transfusions, prior refractoriness to platelet transfusion, or transplantation should increase the suspicion of alloimmune platelet refractoriness.

A. Kinetics of Platelet Loss

Rapid loss can be defined as a rise in the platelet count of less than $10,000/\mu l$ within 2 hours of completing a transfusion with a pooled platelet concentrate (PC; $\sim 4.2 \times 10^{11}$ platelets) or a unit of single-donor apheresis-derived platelets (AP; $\sim 5 \times 10^{11}$ platelets; Kruskall, 1997). A more rigorous approach is to calculate the corrected count increment (CCI) (see Box, p. 215). The CCI value that triggers a serologic workup or constitutes refractoriness has not been standardized (Bishop *et al.*, 1992). One-hour CCIs of $< 3000/\mu l$ (equivalent to 3×10^9/liter; Murphy, 1995), $< 5000/\mu l$ (TRAP, 1997), or $< 7500/\mu l$ (Novotny *et al.*, 1995; Kekomaki, 1998) have been cited as indicative of refractoriness. In any case, a low CCI on 2 or more consecutive platelet transfusions should prompt a workup for antibody-mediated platelet destruction (Murphy, 1995).

B. Laboratory Evaluation

Laboratory tests used in the diagnosis and management of platelet refractory patients are summarized in Table 2.

1. Panel Reactive Antibody (PRA): Anti-HLA Antibodies

A panel reactive antibody (PRA) and autocrossmatch can be performed to evaluate for anti-HLA and autoantibodies, respectively. The PRA is a lymphocytotoxicity assay, in which a panel of lymphocytes expressing different HLAs are incubated with the patient's serum. If present in the appropriate concentration, anti-HLA antibodies in the patient's serum bind to some of the lymphocytes in the panel and fix complement, leading to cell lysis, which is scored visually. In the autocrossmatch, the patient's own lymphocytes are used as targets. The autocrossmatch can be difficult to perform in patients with low WBC counts. A positive PRA not only reveals the presence of anti-HLA antibodies, but also gives qualitative insight into the antibody level (amount of lysis) and extent of sensitization (reactivity pattern). If low titers of clinically significant antibodies are suspected, flow cytometry may be used.

2. HPA Antibody Typing Assays

The most widely used assay to detect HPA antibodies is the monoclonal antibody specific immobilization of platelet antigen (MAIPA) assay (Kroll *et al.*, 1998). In this assay platelets are incubated in the presence of the recipient's serum and a murine monoclonal antibody (MoAb) to a defined glycoprotein antigen. Following binding, the platelets are lysed and the immune complexes are attached to microtiter wells coated with

antimouse antibodies. The MAIPA can be falsely negative if the MoAb and the human antibodies compete for binding to related epitopes. Other difficulties include alteration of the epitope during the platelet lysis step and cross-reactivity of the antimouse antibodies for human antibodies or vice versa.

The MAIPA, as well as platelet agglutination assays, can be used to test for platelet alloantibodies and autoantibodies. One simple method to distinguish antibodies binding platelet glycoproteins from those directed against HLA is to pretreat the target platelets with chloroquine or a weak acid to elute HLA from the surface. With the exception of some commercially available kits for commonly encountered alloantibodies

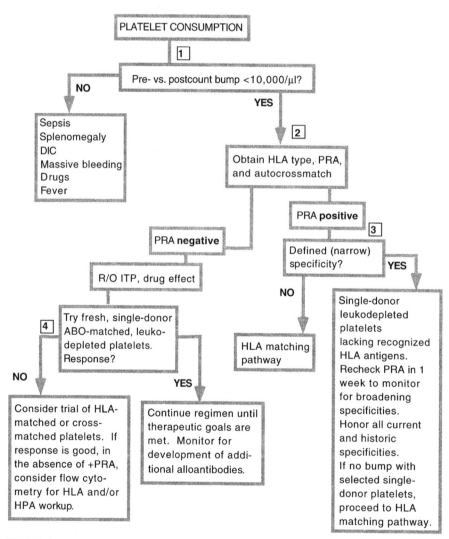

FIGURE 1   Flow chart for the diagnosis and transfusion therapy of platelet refractoriness due to HLA alloimmunization. (See also, Box p. 214.)

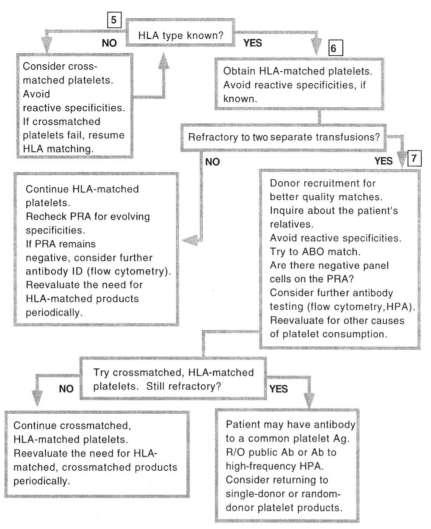

FIGURE 2 HLA platelet pathway for HLA alloimmunized, platelet refractory patients. (See also, Box p. 214.)

(e.g., anti-HPA-1a in NAT), most HPA or HPA antibody assays are performed in specialized reference labs.

3. HPA Typing Assays

There are many methods for typing platelet glycoprotein antigens (HPAs). The concept of typing HPAs in either platelet donors or recipients is analogous to RBC antigen testing. Serologic methods include the enzyme-linked immunosorbent assay (ELISA) and modified antigen capture enzyme-linked immunosorbent assay (MACE). More labor-intensive methods such as immunoprecipitation and immunoblotting may be appropriate for selected antigen systems (Gov) or in characterizing newly identified antigens. In addition, whole platelets from the recipient or donor can be labeled with fluorescently tagged antibodies and analyzed microscopically or by flow cytometry (Sintnicolaas *et al.*, 1996).

4. Genotyping Methods

Genotyping methods are now also available for many of the human platelet antigen alleles. Methods include restriction fragment polymorphism (RFLP), allele-specific oligonucleotide hybridization (ASO), polymerase chain reaction (PCR) with allele-specific primers, and single-strand conformational polymorphism analysis (SSCP; reviewed in Kroll *et al.*, 1998). With the

ADDITIONAL NOTES ON FIGURES 1 AND 2

The algorithms presented in Figures 1 and 2 represent examples of how platelet refractory patients may be assessed and treated. The branching points and products used in this or any other platelet algorithm are obviously dependent on the available platelet inventory and on access to specialized testing. For example, in some institutions platelet crossmatching may be relied upon more early and more often than HLA matching, while in other institutions the opposite pattern (HLA matched before crossmatched) of product use may be observed.

Nevertheless, all algorithms for platelet refractoriness share the following aims: (1) Distinguish immune from nonimmune refractoriness. (2) If alloimmune, determine the antibody specificity. (3) If possible, find antigen-appropriate platelets. Initially standard platelet products (such as random-donor or single-donor platelets) are tried. If transfusions of these products fail, more specialized products (such as HLA-matched, selectively mismatched, or crossmatched platelets) are employed. (4) Monitor the adequacy of transfusion therapy and periodically reevaluate the need for specialized platelet products.

Specific comments on decision points correspond to the numbered boxes in the figures.

1. Rapid consumption of transfused platelets (corresponding to a 1-hour CCI of $< 7500/\mu l$ or a rise of less than approximately $10,000/\mu l$ in the platelet count in patients receiving 1 single-donor or 6 random-donor units of platelets) is consistent with immune-mediated destruction. Management for "nonimmune" causes of shortened platelet survival is generally supportive (continue to transfuse with random-donor or single-donor platelets, if appropriate) and directed toward correcting the underlying abnormality. The presence of nonimmune causes of platelet destruction does not rule out immune-mediated destruction. In fact, most patients with immune-mediated destruction will concurrently exhibit one or more nonimmune causes (see text).

2. To evaluate for HLA alloimmunization, a panel reactive antibody (PRA) and autocrossmatch are performed. In patients with autoimmune thrombocytopenia, the autocrossmatch is often positive whereas the PRA is negative. The opposite pattern is typically observed in HLA alloimmunized patients. The HLA type of the recipient is often obtained before myeloablative therapy in anticipation of possible platelet refractoriness (and before the white blood cell count becomes too low to perform the typing).

3. If HLA antibodies are demonstrating, it may be possible to approximate their specificity by analyzing the reactivity pattern and corresponding HLA genotypes of the panel cells. Recipients with a narrowly defined anti-HLA specificity may respond well to single-donor platelets which lack the corresponding HLA antigen and any strongly cross-reactive antigens. This process is analogous to providing an antigen-negative red cell unit to a patient with clinically significant red cell antibodies. On the other hand, broadly sensitized patients will likely require a more stringent HLA matching regimen (see Figure 2).

4. If the PRA is negative, the patient may have autoimmune platelet destruction or may be developing alloantibodies which are not yet circulating at measurable levels. Some patients will respond well to fresh (less than 24 hours old), ABO-matched single-donor platelets. Time (during which antibody titers will likely rise) and response to therapy will usually reveal HLA alloimmunization. A good response to HLA-matched or crossmatched platelets is suggestive of HLA or HPA antibodies, although a trial with less specialized products may be warranted in the continued absence of demonstrable alloantibodies. If available, platelet crossmatching may be especially helpful in the absence of knowledge about the recipient's HLA type or the specificities of the offending antibodies.

5. If crossmatched platelets and the patient's HLA type are unavailable, it may be possible to use the PRA as a guide for selecting antigen-appropriate platelets (see Figure 1 and text). Alternatively or in addition, it may be possible to transfuse platelets with a rare HLA type (i.e., one to which the patient is unlikely to have been sensitized). The selection of such a platelet product is often based on demographic considerations.

6. If the HLA types of the recipient and the donor platelet inventory are known, HLA matching may be attempted. The quality of the HLA match will depend on how frequently the patient's HLA alleles are represented in the donor population.

7. Failure of HLA-matched transfusions can be due to several factors, including (1) a poor HLA match. If the platelet inventory is limited to C- or D-type matches, consider selective antigen mismatching at low expressor alleles (such as HLA-B12) and/or limiting mismatches to antigens against which the recipient has no demonstrable antibody reactivity. Donor recruitment may improve the quality of HLA matches. The recipient may be broadly alloimmunized to HLA and have intra-CRG antibodies or antibodies directed against public HLA determinants (see text). Other factors include (2) HPA antibodies, (3) high-titer anti-A or anti-B isohemagglutinins, (4) antibodies against antigens which appear on platelets as they age during storage, and, perhaps most important, (5) superimposed nonimmune causes of platelet destruction (see Figure 1 and text).

CALCULATION OF THE CORRECTED COUNT INCREMENT (CCI)

$$CCI = \frac{(P_{post} - P_{pre} \text{ per } \mu l) \times BSA \ (m^2)}{N_{transf}(\times 10^{-11})}.$$

P_{pre} is the pretransfusion platelet count (platelets per microliter of anticoagulated blood). P_{post} is the post-transfusion platelet count. BSA is the body surface area in square meters, and can be approximated by

$$BSA \approx [(\text{height (cm)} \times \text{weight (kg)})/3600]^{1/2},$$

or 1.7 m² can be used as the BSA of an average adult. N_{transf} is the number of transfused platelets.

Example

An average-sized adult patient undergoing bone marrow transplantation has a platelet count of 5000 (per μl). One hour after receiving an apheresis-derived unit of platelets ($\sim 5 \times 10^{11}$ platelets), her platelet count has risen to 11,000. What is her CCI?

$$CCI = \frac{(11,000 - 5000) \times 1.7}{5} = 2040.$$

Another measure of adequacy of platelet transfusion is the platelet percent recovery (PPR). The PPR is defined as

$$PPR = \frac{(P_{post} - P_{pre \ per \ \mu l}) \times 10^6 \ (\mu l/\text{liter}) \times BV \ (\text{liter}) \times 100\%}{N_{transf}},$$

where N_{transf} is the number of platelets in the transfused platelet product, and BV is the blood volume in liters (which can be estimated by $0.07 \times$ body weight [kg]).

possible exception of assays for HPA-1 in the setting of NAT, most of the molecular assays are performed in specialized reference labs.

5. Platelet Crossmatching

In selected cases, crossmatching may be used to evaluate the compatibility of a particular platelet product in a platelet refractory patient. In a platelet crossmatching assay, the recipient's serum is tested for its ability to bind to various antigens on platelets from different donors. Antibody binding can be visualized by a solid-phase red cell adherence assay or by flow cytometry (Sintnicolaas and Lowenberg, 1996). Platelet crossmatching may show incompatibility of recipient serum and donor platelets due to antibodies against HLA, ABO, or HPA antigens. Platelet crossmatching can be done in the absence of knowledge about which

antigens the patient's antibodies are recognizing. Thus, a knowledge of the HLA type of the patient or donors is not required, which is suggested to be an advantage of this technique by some. Others use crossmatching to identify suitable platelet donors for platelet refractory patients with high PRAs (see Figures 1 and 2). In such broadly sensitized individuals, platelet crossmatching is often performed on HLA-compatible platelet products to increase the likelihood of finding a crossmatch-compatible unit.

V. CLINICAL COURSE AND SEQUELAE

Platelet refractoriness typically arises in the multi-transfused patient with hypoproliferative thrombocytopenia secondary to a hematologic malignancy or sys-

TABLE 2   Laboratory Assays in the Diagnosis and
Management of Platelet Refractoriness

Antiplatelet antibodies

  PRA (panel reactive antibody)

  Autocrossmatch

  Platelet agglutination assay

  Immunofluorescence assays of platelets

Platelet antigens

  HLA typing of donor units and recipient

  ELISA (enzyme-linked immunosorbent assay)

  MAIPA (monoclonal antibody specific immobilization of
    platelet antigen)

  Genotyping methods: RFLP, ASO, SSCP, etc.

Compatibility testing

  CCI (response *in vivo*)—platelet counts

  Crossmatching—platelet agglutination, flow cytometry

temic chemotherapy or both. Studies in the 1970s reported that approximately 60% of patients receiving platelet transfusions eventually exhibited HLA antibodies. The rate of appearance of cytotoxic HLA antibodies correlated with prior sensitization (those who had been sensitized by pregnancy, previous transfusions, etc., developed antibodies more quickly than those who did not have a history of sensitization). Interestingly, the likelihood of developing HLA antibodies did not correlate with the number of platelet transfusions given. However, there may be a dose response relationship in terms of donor exposures, particularly if this is studied in patients receiving single-donor, rather than multidonor, platelet transfusions (McFarland, 1996).

The failure to respond to platelet transfusions can place the patient at increased risk for uncontrollable, life-threatening hemorrhage, with few if any treatment options (Schiffer, 1991). The risk of hemorrhage gradually increases as the platelet count decreases (see discussion in Chapter 7 of this volume). Only a fraction of patients develop alloimmune refractoriness, and at the moment there are no good predictive tests (apart from a history of sensitization) to distinguish responders from nonresponders. The current practice is to attempt to limit every patient's risk of becoming sensitized through the use of special blood products (leukodepleted single-donor platelets, typically), which is expensive. The cost and use of limited resources escalates in patients who actually become alloimmunized and refractory. Patients who are refractory to platelets receive transfusions that are more numerous and more expensive (Lill *et al.*, 1997).

VI. THERAPY

Therapy for platelet refractoriness depends upon the underlying etiology. Platelet refractoriness due to alloimmunization is approached by avoiding antigens to which the patient has been sensitized.

A. Prevention of Alloimmunization: Minimizing Leukocyte Exposure

The first approach is to try to prevent primary alloimmunization. In a large, multicenter prospective clinical trial (TRAP, 1997), the use of leukodepletion filters or UV-B irradiation reduced the frequency of *de novo* HLA alloimmunization from 13% in controls who received standard PRBC and platelet transfusions to 3–5% in patients who received leukocyte-depleted or UV-B-irradiated products ($< 5 \times 10^6$ WBCs per transfusion). Others have observed an HLA alloimmunization rate of 5–10% in BMT patients receiving filtered and gamma-irradiated PRBC and platelet products (Abou-Elella *et al.*, 1995). In addition to their effects on HLA sensitization, leukodepleted products offer the advantage of fewer febrile transfusion reactions and lower rates of cytomegalovirus (CMV) transmission (see Chapters 15 and 16).

The risk of HLA sensitization also may be lowered by using a lower platelet threshold for transfusion support in patients with hypoproliferative thrombocytopenia who are not bleeding and do not require an invasive procedure ($10,000/\mu l$ instead of $20,000/\mu l$; Kruskall, 1997). The use of single-donor apheresis platelet (AP) units over platelet concentrates (PC) has been advocated as a means to delay the onset of sensitization, although this remains open to debate. Other sources of HLA sensitization, such as pregnancy or transplanted organs, may not be avoidable, however.

B. Use of Antigen-Appropriate Platelet Products

Once primary alloimmunization has occurred, the treatment approach shifts to an effort to find platelet products to which a patient responds (see Figures 1 and 2). This is achieved, when possible, by the use of HLA-compatible and/or crossmatch-compatible platelet products (for product descriptions, see Chapter 7).

1. Single-Donor, Apheresis-Derived Platelets Lacking Specific Antigens

When the specificities of a patient's HLA antibodies are defined, it may be possible to give the patient apheresis-derived platelet units that lack the appropriate antigens (and any closely related antigens), rather than resorting to HLA-matched products. Serologically related HLA antigens are known as splits and cross-reactive groups (CRGs). When selecting a HLA-compatible apheresis-derived platelet product for a patient on the basis of HLA status, it is best to avoid the implicated HLA antigen(s), their splits, and, if possible, also the cross-reactive antigens. If the patient's HLA type is known, one can further attempt to optimize the product by choosing similar antigens to the patient's own antigens. Some have advocated the use of leukodepletion filters and/or UV-B-irradiated platelet products to minimize determinant spreading of HLA reactivity. The utility of these specialized products in patients who have previously been sensitized to foreign HLAs remains controversial (Sintnicolaas *et al.*, 1995; TRAP, 1997).

2. HLA Matching in Patients with High PRAs

Sometimes patients will have antibodies directed against public HLA antigens (these antibodies recognize several different HLAs, for example, HLA Bw4–Bw6 incompatibility), or will have antibodies to HLAs within their own cross-reactive group (intra-CRG antibodies). Public antibodies to high-frequency antigens and intra-CRG antibodies make it difficult to provide antigen-appropriate products. In cases where the PRA is markedly elevated (>80%), the patient will typically require HLA-matched or crossmatched apheresis-derived platelets (see Figure 2). When it is not possible to ascertain an anti-HLA specificity (for example, in a patient with a very high PRA) it can sometimes be helpful to know the HLA types of the negative (unlysed) panel cells. It may be possible to select apheresis-derived platelets with HLA types that are similar to those of the negative panel cells.

It is important to realize that the quality of HLA-matched apheresis-derived platelets (how similar they are to the patient's own HLA type) varies. The quality of the HLA match between the platelet donor and recipient is graded on the basis of serologic similarity of the HLA A and B alleles. As each individual has two HLA-A alleles and two HLA-B alleles, four antigens are considered for the match. Serologic similarity between the platelet donor and recipient is classified based on the number of identical, cross-reactive, and non-cross-reactive HLA antigens (summarized in Table 3).

As shown in Table 3, in an A-type match, all four donor HLA antigens (two HLA-A antigens and two HLA-B antigens) are identical to those of the recipient. In a B1X match, three donor antigens are identical and one is cross-reactive with those of the recipient. In a

TABLE 3 Grading System for HLA Compatibility in Platelet Transfusion

Type	Antigen status			
	Number identical	X	Non-X	Total IDs in donor
A	4	NA[a]	0	4
B1U	3	NA	0	3
B1X	3	1	0	4
B2U	2	NA	0	2
B2UX	2	1	0	3
C	0–3	0–3	1	2–4
D	0–2	0–2	2	2–4

Note. "Number identical" refers to the number of HLA A and B antigens that are identical in the donor and recipient. X, cross-reactive—the recipient and the donor's HLA antigens share serologic similarities. Non-X, non-cross-reactive. "Total IDs in donor" refers to the number of HLA A and B alleles identified in the donor. C and D type matches are distinguished by the presence of one (C) or two (D) non-cross-reactive HLA differences between the donor and the recipient. NA, not applicable.

B1U match, the recipient and the donor have three identical HLA A and B alleles. In B1U or B2U matches, the absence of the fourth (or third and fourth) antigen(s) may be due to homozygosity for one or more of the alleles, or indicate failure to identify the fourth antigen. Because of the possibility of the latter, a B1U match is ranked below an A-type match and a B2U match is ranked below a B1U match. In general, the higher the quality of the HLA match, the better the chances of successful transfusion in a patient with HLA alloimmunization. In the author's experience, C and D matches generally fair poorly in highly sensitized individuals. However, others have reported surprisingly good results (>50% of platelet transfusions with CCI > 7500/μl) with HLA-mismatched (C type) platelet transfusions in alloimmunized (PRA > 60%), platelet refractory patients (Murphy, 1997).

3. HLA Matching and Crossmatching

The availability of high-quality (A, B1U, or B1X) HLA-matched products is dictated by the inventory of the platelet provider and the rarity of the patient's own HLA type in the provider's donor pool. Sometimes relatives or special donors can be recruited to improve the quality of the match. Some have advocated the use of HLA- and HPA-matched products in refractory patients who express antibodies to both classes of antigen (HLA and HPA; Kekomaki, 1998). Simultaneously HLA-matched and crossmatched apheresis-derived platelets can also be used (reviewed in Sintnicolaas and Lowenberg, 1996). If a high-quality HLA match or

TABLE 4 Additional Therapeutic Options for Patients Unresponsive to Standard Therapy

- Intravenous immunoglobulin (IVIg)
- Epsilon aminocaproic acid (EACA)
- High-dose corticosteroids
- Plasma exchange
- Splenectomy
- Cyclosporin A
- Vincristine
- Acid-treated platelets
- Staphylococcal protein A column therapy

crossmatch is unavailable, or the patient's HLA type is not known, the use of apheresis-derived platelets with rare HLA types may be considered. As the expression of HLA class I antigens on platelets is highly variable, it may be possible to limit HLA mismatches to antigens which are expressed at low levels, such as HLA B12, B44, and B45 (McFarland, 1996). Finally, attempts to strip HLA antigens off of the transfused platelets, circumventing the need for HLA-matched apheresis-derived platelets, have met with limited success (Sirchia, 1996).

C. Additional Treatment Options

Even when HLA-"compatible" products are used, platelet transfusions often fail (up to 40–60% of the time) in refractory patients (Schiffer, 1991). This is due, in part, to the simultaneous presence of immunologic and nonimmunologic causes of platelet consumption in many platelet refractory patients (Doughty et al., 1994). In patients who remain refractory to all types of apheresis-derived platelets, the use of standard platelet concentrates (such as pooled platelets) should be considered.

Additional therapeutic options for immunologically mediated platelet refractoriness are shown in Table 4. Most of these additional therapies are believed to work based on their immunomodulatory effects. However, by and large, the evidence supporting their efficacy is limited, beneficial effects are often transient, and therapy is expensive, precluding their routine use (see reviews by Novotny, 1999; Silberman, 1999). Nevertheless, some of these treatments have been used in refractory individuals with life-threatening hemorrhage. Immunosuppressive regimens have been remarkably ineffective in blocking alloimmunization. It is sobering to realize that patients can be alloimmunized during BMT, a time in which they are receiving formidable doses of immunosuppressive and ablative agents. Fu-

ture therapies of platelet refractoriness may include cryopreservation of autologous platelets, the use of platelet substitutes, and the development of more effective platelet growth factors (Novotny, 1999).

References

Abou-Ella, A. A., Camarillo, T. A., Allen, M. B., et al. (1995). Low incidence of red cell and HLA antibody formation by bone marrow transplant patients. Transfusion 35, 931–935.

Arepally, G., and Cines, D. (1998). Heparin-induced thrombocytopenia and thrombosis. Clin Rev Allergy Immunol 16, 237–247.

Bishop, J. F., Matthews, J. P., Yuen, K., et al. (1992). The definition of refractoriness to platelet transfusions. Transfusion Med 2, 35–41.

Doughty, H. A., Murphy, M. F., Metcalfe, P., et al. (1994). Relative importance of immune and non-immune causes of platelet refractoriness. Vox Sang 66, 200–205.

George, J. N., Woolf, S. H., Raskob, G. E., et al. (1996). Idiopathic thrombocytopenic purpura: A practice guideline developed by explicit methods for the American Society of Hematology. Blood 88, 3–40.

Kekomaki, R. (1998). Use of HLA- and HPA-matched platelets in alloimmunized patients. Vox Sang 74, 359–363.

Kelton, J. G., Smith, J. W., Horsewood, P., et al. (1998). ABH antigens on human platelets: Expression on the glycosyl phosphatidylinositol-anchored protein CD109. J Clin Lab Med 132, 142–148.

Kickler, T., Kennedy, S., and Braine, H. (1990). Alloimmunization to platelet specific antigens on glycoproteins IIbIIIa and IbIX in multiply transfused thrombocytopenic patients. Transfusion 30, 622–625.

Kroll, H., Kiefel, V., and Santoso, S. (1998). Clinical aspects and typing of platelet alloantigens. Vox Sang 74, 345–354.

Kruskall, M. S. (1997). The perils of platelet transfusions. N Engl J Med 337, 1914–1915.

Lee, E. J., and Schiffer, C. A. (1987). Serial measurement of lymphocytotoxic antibody and response to nonmatched platelet transfusions in alloimmunized patients. Blood 70, 1727–1729.

Lill, M., Snider, C., Calhoun, L., et al. (1997). Analysis of utilization and cost of platelet transfusions in refractory hematology/oncology patients [Abstract]. Transfusion 37, 26S.

McFarland, J. G. (1996). Platelet immunology and alloimmunization. In "Principles and Practice of Transfusion Medicine" (E. C. Rossi, T. L. Simon, G. S. Moss, and S. A. Gould, Eds.), 2nd ed., Chap. 22. Williams & Wilkins, Baltimore.

Murphy, S. (1995). Preservation and clinical use of platelets. In "Williams Hematology" (E. Beutler, M. A. Lichtman, B. S. Coller, and T. J. Kipps, Eds.), pp. 1643–1649. McGraw Hill, New York.

Murphy, S. (1997). Mismatched platelet transfusions to alloimmunized patients. Blood 15, 1715–1716.

Novotny, V. M. J. (1999). Prevention and management of platelet transfusion refractoriness. Vox Sang 76, 1–13.

Novotny, V. M. J., Doorn R. V., Witvliet, M. D., et al. (1995). Occurrence of allogeneic HLA and non-HLA antibodies after transfusion of prestorage filtered platelets and red blood cells: A prospective study. Blood 85, 1736–1741.

Santoso, S., et al. (1993). The presence of messenger RNA for HLA class I in human platelets for protein biosynthesis. Br J Haemotol 84, 451–456.

Schiffer, C. A. (1991). Prevention of alloimmunization against platelets. Blood 77, 1–4.

Schnaidt, M., Northoff, H., and Wernet, D. (1996). Frequency and specificity of platelet-specific alloantibodies in HLA-immunized haematologic-oncologic patients. *Transfusion Med* **6**, 111–114.

Silberman, S. (1999). Platelets: Preparations, transfusions, modifications, and substitutes. *Arch Pathol Lab Med* **123**, 889–894.

Sintnicolaas, K., and Lowenberg, B. (1996). A flow cytometric platelet immunofluorescence crossmatch for predicting successful HLA matched platelet transfusions. *Br J Haematol* **92**, 1005–1010.

Sintnicolaas, K., van Marwijk Kooij, M., van Prooijen, H. C., *et al.* (1995). Leukocyte depletion of random single donor platelet transfusions does not prevent secondary human leukocyte anti-gen-alloimmunization and refractoriness: A randomized prospective study. *Blood* **85**, 824–828.

Sirchia, G. (1996). HLA-reduced platelets to overcome platelet refractoriness: Can lemons help? *Transfusion* **36**, 388–391.

Slichter, S. J. (1998). Platelet refractoriness and alloimmunization. *Leukemia* **12**, s51–s53.

Trial to Reduce Alloimmunization to Platelets (TRAP) Study Group (1997). Leukocyte reduction and ultraviolet B radiation of platelets to prevent alloimmunization and refractoriness to platelet transfusions. *N Engl J Med* **337**, 1861–1869.

25

Approach to Transfusion in Obstetrics
Maternal and Fetal Considerations

NANCY C. ROSE

Department of Obstetrics and Gynecology
Divisions of Reproductive Genetics and Maternal-Fetal Medicine
University of Pennsylvania Health System
Philadelphia, Pennsylvania

The increasing demands of the fetus, as well as biologic differences between the fetus and the mother, significantly affect the management of a patient during pregnancy. Obstetric considerations, including physiologic adaptations to pregnancy, issues related to the assessment of obstetric hemorrhage, and the need for maternal transfusion, will be addressed here. This chapter will also discuss the diagnosis and treatment of fetomaternal hemorrhage (FMH) and its contribution to the hemolytic disease of the fetus and the newborn, with emphasis on rhesus (Rh) isoimmunization. Finally, the diagnosis and management of both maternal and fetal thrombocytopenias will be reviewed.

I. APPROACH TO MATERNAL TRANSFUSION IN OBSTETRICS

A. Physiologic Adaptations to Pregnancy

1. Plasma

The physiologic changes that occur during pregnancy are designed to protect the patient from the significant amounts of blood loss caused by delivery. At about 10 weeks of gestation, the maternal plasma volume begins to increase until it plateaus at about 30–34 weeks gestation. The mean increase in plasma volume is about 50%, with a greater increase occurring in multiple gestations. It is important to note that patients who develop preeclampsia have a smaller amount of plasma volume expansion during gestation, but also have a greater degree of blood loss at delivery. This is due to the laboratory changes associated with preeclampsia such as thrombocytopenia, a higher risk for placental abruption secondary to hypertension, and the increased risk for uterine atony secondary to the therapeutic use of magnesium sulfate.

2. Red Blood Cells (RBCs)

The erythrocyte mass also begins to increase at about 10 weeks gestation and continues until term. The increase in RBCs helps provide for the increase in oxygen transport to the fetus. The total iron requirement is 700–1400 mg per gestation, the majority of which is used for RBC mass expansion. The fetus derives its iron solely by active transport across the placenta. It is recommended that women be supplemented with 30 mg of elemental iron daily, which is a 100% increase over daily recommendations for nonpregnant women (ACOG, 1993). Since the maternal increase in plasma volume is proportionately more than the increase in RBC mass, the hematocrit decreases during gestation, causing the "physiologic anemia" of pregnancy. This decrease in hematocrit is lowest at about 34 weeks gestation.

3. Platelets

In the majority (> 90%) of women, platelet volume and function are not signficantly altered during pregnancy.

B. Obstetric Hemorrhage

Hemorrhage is the third leading cause of maternal mortality in the United States and is responsible for 30% of all maternal mortality worldwide. Since blood loss is often clinically underestimated by 30–50%, this may delay therapeutic interventions. Estimation of blood volume deficit is sometimes extremely difficult, especially because blood loss may not be obvious (for example, from a high cervical laceration after vaginal delivery, or retroperitoneal bleeding).

1. Antepartum Hemorrhage

Antepartum causes of obstetrical bleeding include placenta previa and placental abruption, although these complications can also present at term. At term, 600 cc/mm of blood circulate through the placental bed. The expected mean blood loss incurred during spontaneous vaginal delivery is about 500 cc; during a cesarean section it is 1000 cc; and during a cesarean hysterectomy it is 1500 cc.

2. Postpartum Hemorrhage

a. Early

Postpartum hemorrhage is defined as blood loss resulting in either a 10% decrease in hematocrit or the need for transfusion. Early postpartum hemorrhage occurs within the first 24 hours after delivery. Early postpartum hemorrhage is far more common than late postpartum hemorrhage, and is associated with a greater degree of blood loss, morbidity, and mortality. The most common cause of early postpartum hemorrhage is uterine atony that occurs either prior to or after delivery of the placenta. Atony (inadequate uterine contraction to decrease blood loss) can be caused by overdistension of the uterus (polyhydramnios, multiple gestation, macrosomia), oxytocin administration, or uterine relaxants (magnesium sulfate, terbutaline), or can be due to infectious causes such as chorioamnionitis. Other causes of uterine atony include retained products of conception, genital tract lacerations, uterine rupture or inversion, placenta accreta, and hereditary coagulopathies.

b. Late

Late postpartum hemorrhage occurs between 24 hours and 6 weeks after delivery. Causes of late postpartum hemorrhage include endometritis, retained products of conception, and hereditary coagulopathies.

C. Management of Obstetric Hemorrhage

1. Antepartum Evaluation

All patients admitted for labor and delivery should have had a type and screen sent to the blood bank and a complete blood count performed. If the suspicion is great enough that the patient may need a transfusion, blood is typed and crossed upon admission. If transfusion is necessary, all pregnant women who test seronegative for cytomegalovirus (CMV) should receive CMV-safe cellular blood components. Patients with antepartum bleeding should also have a Kleihauer–Betke test to assess for the presence and quantity of fetomaternal hemorrhage (see Section II.B.5). An ultrasound is performed to evaluate for placenta previa. Placental abruptions are not generally diagnosed by ultrasound but by the clinical signs of contractions, pain, and vaginal bleeding. Fetal heart rate monitoring is performed concurrently to evaluate the fetus for tachycardia secondary to blood loss, as well as tocodynamometry to evaluate for uterine contractions. A speculum examination is performed to determine the presence of vaginal infections or labor causing rapid cervical dilation and resultant bleeding.

2. Postpartum Evaluation

Perhaps the most important issue regarding postpartum bleeding is its recognition. After noting that the postpartum bleeding is more than commonly expected, concurrent steps are performed to evaluate and treat the patient. Intravenous access should be placed or maintained. Etiology of the blood loss needs to be determined. Evaluation for obstetric trauma is performed by looking for uterine rupture and lower genital tract lacerations, with repairs as needed. The patient is evaluated for retained products of conception with manual extraction or curettage. Finally, the patient is examined for uterine atony, which is characterized by a flaccid uterus unresponsive to standard postpartum oxytocin infusion. If uterine atony is present, the patient is reexamined for retained products, and bimanual uterine massage is performed. Initial medical therapy for atony is continuous intravenous oxytocin —20 mU/liter in 1000 cc normal saline or lactated ringer's solution. If unresolved, 15-methyl PG F2-alpha, 0.25 mg IM into uterus or muscle, can be given every 15–90 minutes, not to exceed 8 doses. Contraindications to 15-methyl PG include asthma, as well as other active pulmonary, renal, or hepatic disease. Methylergonovine, 0.2 mg IM, can be given every 2–4 hours, although it is contraindicated in cases of preeclampsia or chronic hypertension.

More aggressive surgical therapies for postpartum hemorrhage include uterine arterial ligation or hypogastric arterial ligation; both are designed to decrease blood flow to the uterine vessels. Uterine packing and selected arterial embolization have been performed in specific cases. Hysterectomy is employed when all other modalities fail to arrest bleeding.

D. Autologous Blood Donation During Pregnancy

Patients who are higher risk for postpartum hemorrhage may donate autologous blood prior to delivery. These include patients with placenta previa, patients with an anterior placenta that is also covering a prior uterine scar (increasing the risk for placenta accreta and the need for possible hysterectomy), and patients with a history of a prior placenta accreta or a prior postpartum hemorrhage. These patients should be treated predelivery with aggressive iron supplementation consisting of ferrous sulfate BID to TID throughout gestation. Whole blood is donated at least 2 weeks prior to delivery, and a patient hematocrit of at least 34% is required (ACOG, 1994). However, it is often difficult for pregnant women to achieve this hematocrit level due to the physiologic dilution of RBCs secondary to increased plasma volumes.

II. APPROACH TO FETAL HEMOLYTIC DISEASE

A. Fetomaternal Hemorrhage (FMH)

Approximately 75% of women demonstrate evidence of FMH during gestation or at parturition (Bowman *et al.*, 1986). In 60% of cases, the amount of fetal blood in the maternal circulation is < 0.1 ml. Less than 1% of women have more than 5 ml transferred, and only 0.25% of women have greater than 30 cc of fetal blood in the maternal circulation. The incidence and volume of FMH increases as a pregnancy advances. Although FMH can occur spontaneously, complications and procedures that increase a patient's risk for FMH include vaginal bleeding (placenta previa, placental abruption, etc.), prenatal diagnostic procedures (chorionic villous sampling, amniocentesis, umbilical cord blood sampling), external podalic version, multiple gestation, manual extraction of the placenta, and cesarean section. However, the majority of cases of FMH occur in patients without risk factors during an uncomplicated spontaneous vaginal delivery.

B. Hemolytic Disease of the Newborn

Hemolytic disease of the newborn (HDN) is defined as neonatal anemia and hyperbilirubinemia caused by an incompatibility between maternal and fetal erythrocytes. The clinical manifestations occur most often in the first 24–48 hours of life. Very high levels of bilirubin can cause neonatal kernicterus; preterm infants are the most susceptible. Phototherapy is the first line of treatment. Exchange transfusions are sometimes necessary.

1. ABO Incompatibility

Over 60% of HDN is caused by ABO incompatibility; however, many other RBC antibodies can cause HDN. Clinically apparent hemolytic disease of the newborn is mostly confined to the situation in which the mother is group O, generating both anti-A and anti-B antibodies, and the fetus is either group A or B. Approximately 20–25% of all pregnancies are ABO incompatible. Most ABO-incompatible pregnancies have no adverse outcomes, because ABO antibodies are generally of the IgM class and therefore do not cross the placenta. Since ABO incompatibility does not cause fetal anemia, it is regarded as a neonatal rather than obstetric disease. Group A and B sensitizations do not require prior exposure to RBCs through prior gestation or transfusion. Therefore, in contrast to Rh(D) sensitization, ABO disease can affect a firstborn child. Because ABO incompatibility can also occur in subsequent gestations, it should be documented in the patient's medical record. However, since it is usually a mild form of hemolytic disease, checking maternal antibody titers and analyzing amniotic fluid for hemolysis prior to delivery is not routinely indicated.

2. Erythroblastosis Fetalis

Erythroblastosis fetalis defined by fetal edema, neonatal hyperbilirubinemia, and neonatal anemia is caused by maternal RBC antibodies that cross the placenta and hemolyze fetal RBCs. Although many erythrocyte antigens have been described (Weinstein, 1982), only a few are clinically important causes of maternal isoimmunization (see Table 1). The most common is the Rh(D) antigen, followed by the Kell, Kidd, and Duffy antigen systems. The large majority of minor antigen sensitivity is due to prior blood transfusion. The management plan for Rh(D) sensitization (serial titers and amniocenteses) is used for the less common RBC antibodies as well. Furthermore, both the Kell (Lee *et al.*, 1995) and Rh(D) (Le Van Kim

TABLE 1   Hemolytic Disease of the Newborn

Blood group system	Antigens related to hemolytic disease	Severity of hemolytic disease	Proposed management
Lewis*			
Kell	K	Mild to severe with hydrops fetalis	Amniotic fluid bilirubin studies
	k	Mild only	Expectant
	K_0	Mild only	Expectant
	Kp^a	Mild only	Expectant
	Kp^b	Mild only	Expectant
	Js^a	Mild only	Expectant
	Js^b	Mild only	Expectant
	Fy^a	Mild to severe with hydrops fetalis	Amniotic fluid bilirubin studies
	$Fy^{b†}$		
Kidd	Jk^a	Mild to severe	Amniotic fluid bilirubin studies
	Jk^b	Mild to severe	Amniotic fluid bilirubin studies
MNSs	M	Mild to severe	Amniotic fluid bilirubin studies
	N†		
	S	Mild to severe	Amniotic fluid bilirubin studies
	s	Mild to severe	Amniotic fluid bilirubin studies
	U	Mild to severe	Amniotic fluid bilirubin studies
	Mi^a	Moderate	Amniotic fluid bilirubin studies
	Mt^a	Moderate	Amniotic fluid bilirubin studies
	Vw	Mild only	Expectant
Lutheran	Lu^a	Mild only	Expectant
	Lu^b	Mild only	Expectant
Diego	Di^a	Mild to severe	Amniotic fluid bilirubin studies
	Di^b	Mild only	Expectant
Xg	Xg^a	Mild only	Expectant
Public Antigens	Yt^a	Moderate to severe	Amniotic fluid bilirubin studies
	Lan	Mild only	Expectant
	Ge	Mild only	Expectant
	Co^a	Severe	Amniotic fluid bilirubin studies
Private antigens	Batty	Mild only	Expectant
	Becker	Mild only	Expectant
	Berrens	Mild only	Expectant
	Biles	Moderate	Amniotic fluid bilirubin studies
	Evans	Mild only	Expectant
	Gonzales	Mild only	Expectant
	Good	Severe	Amniotic fluid bilirubin studies
	Heibel	Moderate	Amniotic fluid bilirubin studies
	Hunt	Mild only	Expectant
	Jobbins	Mild only	Expectant
	Radin	Moderate	Amniotic fluid bilirubin studies
	Rm	Mild only	Expectant
	Ven	Mild only	Expectant
	Wright	Severe	Amniotic fluid bilirubin studies
	Zd	Moderate	Amniotic fluid bilirubin studies

* Not a proven cause of hemolytic disease; † not a cause of hemolytic disease. Reproduced with permission.

et al., 1992) antigens have been genetically character-ized in order to make available molecular antenatal diagnosis. The remaining RBC antigens have not been genotyped, and therefore prenatal diagnosis to deter-mine direct fetal antigen type cannot be performed during gestation for RBC antigens other than Rh(D) and Kell.

3. Rh(D) Antigen

The Rh(D) antigen, the most common cause of erythroblastosis fetalis, has recently been genetically characterized as residing on the short arm of chromo-some 1. The genes coding for the Rh locus lie in tandem, with the first coding for Rh(CcEe) followed by Rh(D). Patients that are Rh(D)-negative lack the D locus on both alleles. The Rh gene complex is de-scribed by various combinations of these genes. There are multiple allelic variations of Rh(D) that exist. About 15% of Caucasians, 8% of African-Americans, and 1–2% of Asians are Rh(D)-negative. About 60% of those that are Rh(D)-positive are heterozygous for the Rh(D) gene; 40% are homozygous.

Another Rh(D) allelic variant is the Du variant, also known as "weak D." The erythrocytes of most Du-positive individuals have a quantitative decrease in the amount of D antigen sites on their erythrocytes. There-fore, some of the D antigen is expressed, and these patients are treated as if they are Rh(D)-positive (they are not given Rh immune globulin prophylaxis).

Theoretically, a Rh(D)-negative Caucasian female has about a 60% chance of having a Rh(D)-positive fetus. As many as 30% of Rh(D)-negative women will be physiologic nonresponders; these are women who do not become sensitized when exposed to Rh(D)-positive blood. Also, ABO incompatibility is protective against Rh sensitization. It has been suggested that ABO-incompatible cells are more rapidly cleared from the circulation and therefore are not available to incite an antigen-mediated antibody response.

4. Rh Immune Globulin (RhIg)

RhIg is passively administered antibody that pre-vents active Rh(D) immunization, also known as anti-body-mediated immune suppression (AMIS). The pre-cise mechanism of AMIS is not understood but is thought to be due to central inhibition. Theoretically, fetal erythrocytes are coated with the RhIg and are thus filtered out of the circulation, suppressing the primary immune response to Rh(D). To be effective, RhIg must be given prior to the development of Rh(D) antibody. The standard dose of RhIg is 300 μg, which should protect against a 30 cc fetomaternal (whole

blood) hemorrhage. However, for first trimester preg-nancy losses or terminations or chorionic villous sam-pling in which only a very small volume of FMH is suspected, a 50-μg dose of RhIg is sufficient. Once the second trimester is reached, only the 300-μg dose should be used.

Routine Rh prophylaxis should be given within 72 hours of suspected FMH. If for some reason RhIg is not given to the patient within 72 hours of suspected FMH or delivery, it may be given up to 28 days after the event in an attempt to avoid Rh sensitization (Jackson and Branch, 1996). See Chapter 10 (Table 5) for RhIg indications, dosing schedule, and relative con-traindications.

5. Management of the Rh(D)-Negative, Unsensitized Pregnant Woman

At the first prenatal visit, every pregnant patient should have a blood type and antibody screen per-formed. These tests should be repeated with each new pregnancy, since new antibodies may form. A Rh(D)-negative, Du-negative, Rh(D) antibody-negative patient receives her first RhIg prophylaxis at 28 weeks gesta-tion. When the patient is admitted for delivery, a repeat antibody screen is performed. When the baby is delivered, its blood type is determined. If the patient has no Rh(D) antibody and the neonate is Rh(D)-posi-tive, a second dose of 300 μg of RhIg is given, usually immediately following, but always within 72 hours of delivery.

After delivery, mothers are screened for FMH by the erythrocyte rosette test. The rosette test identifies Rh(D)-positive (fetal) cells in the maternal circulation. The maternal RBC sample is first incubated with human anti-D antibody, and then indicator Rh(D)-positive cells are added. The indicator cells will form agglutinates (rosettes) around the antibody-bound Rh(D)-positive fetal cells if present. If the rosette test is positive, a Kleihauer–Betke test is used to quantitate the number of fetal cells in the maternal circulation. This technique relies on differences between fetal and adult hemoglobin resistance to acid elution. Alternative methods to quantify FMH include the enzyme-linked antiglobulin test (ELAT) and flow cytometry. Only 1% of deliveries have a greater than 30 cc (whole blood) FMH, and in these cases, RhIg doses are increased accordingly to protect against Rh(D) sensitization.

6. Management of the Rh(D)-Immunized Gestation
a. Antibody Titers

If a patient's antibody screen is positive for anti-Rh(D), an antibody titer is performed. A patient with a baseline titer of greater than 1:8 is considered at risk for fetal hemolysis from Rh(D) isoimmunization. If initially less than 1:8, anti-Rh(D) titers are repeated monthly.

b. Paternal Zygosity

Paternal zygosity testing is performed to determine if the father is Rh(D)-positive or -negative, and, if he is Rh(D)-positive, whether he is homozygous or heterozygous for the D allele. If the father is Rh(D)-negative, no further testing or preventive treatment of the mother or fetus is indicated, since the fetus must be Rh(D)-negative. If the father is heterozygous for the Rh(D) allele, the mother has a 50% chance of carrying an Rh(D)-negative fetus. If he is a homozygote for the D allele, then all offspring of these parents will be Rh(D)-positive.

c. Amniocentesis

Once a Rh(D) titer of 1:8 is reached, the patient is offered amniocentesis. If the patient is undergoing prenatal diagnosis for another indication, Rh(D) typing by DNA analysis is performed at the same time. If the patient has no other indication for early prenatal diagnosis, then DNA analysis is performed at the time of the first amniocentesis, which is performed to evaluate hemolysis. If the fetus is Rh(D)-negative by DNA analysis, then no further invasive testing is required. In this situation, serial ultrasound scans are still performed to evaluate for hydrops, in case of a laboratory error. If the fetus is Rh(D)-positive, serial amniocenteses are performed to evaluate the degree of fetal RBC hemolysis. The obstetric history is essential in the management of these patients, in that patients who have had an affected child in the past generally exhibit earlier hemolysis during subsequent gestations. If a patient has a history of a prior hydropic fetus, then the chance that the next fetus will be hydropic is greater than 80% (Jackson and Branch, 1996).

d. Amniotic Fluid Analysis

Bevis (1956) originally noted that spectrophotometric determinants of amniotic fluid bilirubin correlated with the degree of fetal hemolysis. Using a semilogarithmic plot, the optical density curve of normal amniotic fluid is linear between 525 and 375 nm. Bilirubin causes a shift in the spectrophotometric density with a wavelength peak at 450 nm (called the δOD450). Liley (1961) retrospectively correlated δOD450 measurements with neonatal outcome by dividing the graph into three zones (Figure 1). Unaffected and mildly affected fetuses plot in zone 1; moderately affected fetuses plot in zone 2; and severely affected fetuses plot in zone 3. These graphs have been developed for Rh(D)-sensitized patients and not for sensitization to other antigens, and therefore when used to evaluate patients with Kell, Kidd, or Duffy alloantibodies, their data should always be interpreted with caution. Although the Liley data can be interpreted starting at about 28 weeks gestation, multiple groups have extrapolated the curve to evaluate earlier gestations with varied success. Therefore, the majority of amniocenteses to evaluate for fetal hemolysis begin in the third trimester. The timing of repeat amniocenteses depends upon the severity of the hemolysis.

e. Fetal Transfusion

Fetal transfusion corrects anemia, increases fetal oxygenation, and decreases the hematopoietic system, thereby decreasing extramedullary demand, decreasing portal vein pressure, and increasing the volume of the circulation. Fetal transfusion is indicated when either (1) the amniotic fluid reveals a Liley value in zone II or (2) the fetus exhibits (by ultrasound examination) hydrops, ensuring a fetal hemoglobin < 5 g/dl. However, half of fetuses with hemoglobin < 5 g/dl are not hydropic, so ultrasound is not necessarily a good indicator of fetal well-being in this case.

Both intraperitoneal and intravascular umbilical vein transfusion techniques have been described. In the absence of hydrops, 80–100% survival rates have been quoted with each method (Harman et al., 1990). Hydropic fetuses show an increased survival with intravascular transfusions (Grannum et al., 1986). Furthermore, the intravascular approach is now also used for other procedures, and thus clinicians have more experience and better proficiency with the intravascular technique. Further advantages of the intravascular technique include the ability to (1) measure a direct fetal hematocrit, (2) determine the fetal blood type if not already performed, and (3) accurately calculate the amount of blood to transfuse. The blood volume to be transfused has been determined by Nicolaides et al. (1988). The goal is typically to reach a fetal post-transfusion hematocrit of 40–45%. Most centers performing intrauterine and/or neonatal transfusions use less-than-7-day-old group O, Rh(D)-negative PRBCs that are CMV-safe, hemoglobin S-negative, and gamma irradiated.

After transfusion, the decline in fetal hematocrit is dependent upon the rate of hemolysis of the fetal Rh(D) RBCs and the ratio of donor to fetal RBCs. MacGregor et al. (1989) use the following guidelines to predict the fetal hematocrit. On average, the drop in fetal hematocrit is about 1.5%/day. Following multiple transfusions, this decrease is 1.0–1.2%/day. Using these calculations, the interval between transfusion is the maximum time required to maintain the fetal

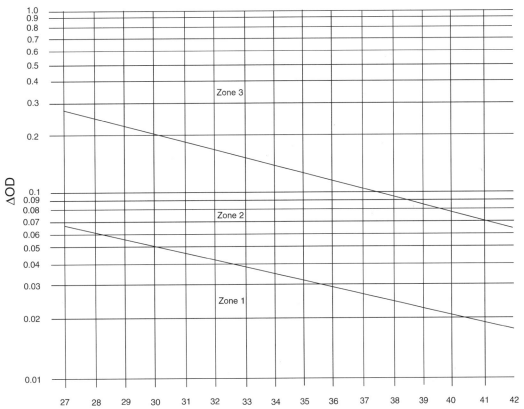

FIGURE 1   Liley graph used to plot degree of sensitization. Reprinted (permission pending) American College of Obsterics and Gynecology. Management of Isoimmunization in Pregnancy. ACOG Technical Bulletin #227, pg 4, 1996[13].

hematocrit above 25–30%. The fetal loss (mortality) rate is about 1–5% for each intrauterine transfusion, with a fetal morbidity of approximately 10–15% (fetal bradycardia, premature labor, and premature rupture of membranes.)

III. THROMBOCYTOPENIA

A. Fetal and Neonatal

The diagnosis of fetal thrombocytopenia is usually made in the neonate, not the fetus. Although there are multiple etiologies for neonatal thrombocytopenia, including sepsis, maternal drug exposure, and maternal disease, neonatal thrombocytopenia without obvious cause warrants an evaluation for neonatal alloimmune thrombocytopenia (NAITP).

1. Neonatal Alloimmune Thrombocytopenia

NAITP is analogous to HDN, but platelets are destroyed instead of RBCs. Unlike HDN secondary to Rh(D) isoimmunization, NAITP can occur in a first

pregnancy. The incidence is 1:2000 deliveries, with the most serious morbidity due to the antenatal intracranial hemorrhage that occurs in 10–15% of cases. Although many different antigens can cause NAITP, the most common is the HPA-1a platelet antigen. Approximately 1–3% of individuals are negative for the HPA-1a antigen. Therefore, if a fetus has an unexplained thrombocytopenia, molecular diagnosis is performed to test maternal HPA-1a antibody status. Once the diagnosis of NAITP due to anti-HPA-1a is made, the parents are tested for HPA-1a antigen. If it is possible that the fetus could be HPA-1a-negative (the father is a heterozygote), then amniocentesis can be offered to test the fetal platelet antigen type.

Once a fetus is determined to be HPA-1a-positive and the mother HPA-1a negative, serial ultrasounds every 2 weeks are performed to evaluate the fetus for intracranial hemorrhage. Late in the second trimester or early in the third, a fetal complete blood count and Kleihauer–Betke test are performed by periumbilical intravascular blood sampling (PUBS) to determine the fetal platelet count. At the end of the procedure, either maternally derived apheresis platelets or randomly donated HPA-1a negative platelets are transfused through the umbilical cord to the fetus.

Various algorithms exist for the management of fetal thrombocytopenia. However, the most common management strategy includes weekly intravenous gamma globulin (IVIg) given to the mother, with serial monthly fetal platelet samplings to be certain that the treatment is effective. If IVIg is ineffective, steroids are sometimes administered to the mother to help increase the fetal platelet counts. A final fetal platelet transfusion is given just prior to delivery; labor is induced by vaginal induction or cesarean section within 1 week of this transfusion. Fetal lung maturity studies are performed at this final umbilical cord sampling to determine the appropriate time for labor induction.

Not infrequently, fetal thrombocytopenia is undetected until after delivery, when the infant demonstrates stigmata of thrombocytopenia such as purpura, or a low platelet count is discovered when a CBC is performed. Neonatal thrombocytopenia is self-limiting, resolving in most cases within 3 weeks. Severely thrombocytopenic or bleeding infants with NAITP should receive maternally derived apheresis platelets or antigen-negative donor platelets. If compatible platelets are not available, random donor platelets, with or without concomitant IVIg administration, should be transfused. Successful treatment of NAITP with IVIg alone has been reported; however, platelet counts often take 48 hours or more to increase.

2. Neonatal Autoimmune Thrombocytopenia

See Section III.B.2, below, for a discussion of maternal ITP and fetal/neonatal autoimmune thrombocytopenia.

B. Maternal

Maternal thrombocytopenia due to autoantibodies directed against platelets is an important diagnosis for maternal management, but it is not certain that this disorder causes antenatal fetal intracranial hemorrhage. Therefore, if this diagnosis is made, then the fetus generally does not undergo PUBS or antenatal evaluations; the mode of delivery is determined by obstetric indications only. If a pregnant woman presents with an incidentally low platelet count (less than $80,000/\mu l$), a thorough search for an etiology should be conducted. The maternal platelet count should be repeated and the peripheral blood smear examined. Evaluation includes a detailed drug history, a detailed family history for inherited disorders (May–Hegglin anomaly), and evaluation for unrecognized medical conditions such as systemic lupus erythematosus (SLE),

antiphospholipid antibody syndrome, HIV, leukemia, preeclampsia, hemolytic uremic syndrome (HUS), and viral and protozoal infections.

1. Benign thrombocytopenia of Pregnancy (BTP)

Approximately 7% of women have unexpectedly low platelets counts during gestation, although the counts are usually greater than $80,000–100,000/\mu l$. Generally, a normal platelet count prior to gestation is documented, and if the patient has no other suggested etiologies for the low platelet count, then no other evaluation is needed (Burrows and Kelton, 1990).

2. Autoimmune Thrombocytopenia (ITP)

ITP is characterized by platelet destruction due to IgG platelet autoantibodies. The main risk during gestation for mothers with ITP is maternal hemorrhage at delivery. Steroids, the mainstay of treatment, are begun if the platelet count declines to less than $50,000/\mu l$. Refractory ITP patients and those intolerant of steroids usually can be managed with IVIg. Cytotoxic drugs with teratogenic potential are contraindicated during the first trimester. Alternative medications used to treat nonpregnant ITP patients should be used in pregnant ITP patients with caution (azathioprine) or should be avoided altogether (danazol and dapsone).

Infants born to mothers with ITP are usually not markedly thrombocytopenic. The frequency of severe fetal thrombocytopenia ($< 50,000$) in cases of maternal ITP is approximately 10%. These fetuses and infants have a much smaller risk ($< 1\%$) of intracranial hemorrhage (antenatally or at delivery) than infants with NAITP. Thus, routine fetal platelet monitoring is not recommended. Furthermore, cesarean section is performed in ITP patients for obstetric indications only. In the postpartum period, neonatal exchange transfusion to remove circulating antibodies or IVIg administration may be effective treatments for severe fetal thrombocytopenia. Neonatal platelet transfusions are reserved for episodes of neonatal hemorrhage.

References

American College of Obstetrics and Gynecology (1993). "Nutrition During Pregnancy." ACOG Technical Bulletin 179. ACOG, Washington, DC.

American College of Obstetricians and Gynecologists (1994). "Blood Component Therapy." ACOG Technical Bulletin 199. ACOG, Washington, DC.

American College of Obstetricians and Gynecologists (1996). "Management of Isoimmunization in Pregnancy." ACOG Technical Bulletin 227. ACOG, Washington, DC.

Bevis, D. C. A. (1956). Blood pigments in haemolytic disease of the newborn. *J Obstet Gynecol Br Emp* **63**, 65.

Bowman, J. M., Pollack, J. M., and Penston, L. E. (1986). Fetomaternal transplacental hemorrhage during pregnancy and after delivery. *Vox Sang* **51**, 117–121.

Burrows, R. F., and Kelton, J. G. (1990). Thrombocytopenia at delivery: A prospective survey of 7615 deliveries. *Am J Obstet Gynecol* **162**, 731–734.

Freda, V. J., Gorman, J. G., Pollack, W., *et al.* (1975). Prevention of rh hemolytic disease—Ten years clinical experience with Rh immune globulin. *N Engl J Med* **292**, 1014–1016.

Grannum, P. A., Copel, J. A., Plaxe, S. C., *et al.* (1986). *In utero* exchange transfusion by direct intravascular injection in severe erythroblastosis fetalis. *N Engl J Med* **314**, 1431–1434.

Harman, C. R., Bowman, J. M., Manning, F. A., *et al.* (1990). Intrauterine transfusion intraperitoneal versus intravascular approach: A case-control comparison. *Am J Obstet Gynecol* **162**, 1053–1059.

Jackson, M., and Branch, D. W. (1996). Isoimmunization in pregnancy. In "Obstetrics, Normal and Problem Pregnancies" (S. Gabbe, Ed.), p. 904. Churchill and Livingstone, London.

Le Van Kim, C., Mouro, I., Cherif-Lazar, B., *et al.* (1992). Molecular cloning and the primary structure of the human blood group RhD polypeptide. *Proc Natl Acad Sd USA* **89**, 10925–10929.

Lee, S., Wu, X., Reid, M., *et al.* (1995). Molecular basis of the Kell (K1) phenotype. *Blood* **85**, 912–916.

Liley, A. W. (1961). Liquor amnii analysis in the management of pregnancy complicated by rhesus sensitization. *Am J Obstet Gynecol* **82**, 1359.

MacGregor, S. N., Socol, M. L., Pielet, B. W., *et al.* (1989). Prediction of hematocrit decline after intravascular fetal transfusion. *Am J Obstet Gynecol* **161**, 1491–1493.

Nicolaides, K. H., Soothill, P. W., Clewell, W. H., *et al.* (1988). Fetal haemoglobin measurement in the assessment of red cell isoimmunization. *Lancet* **1**, 1073–1075.

Weinstein, L. (1976). Irregular antibodies causing hemolytic disease of the newborn. *Ob/Gyn Survey* **31**(8), 587.

26

Transfusion Management of Infants and Children

CATHERINE S. MANNO

The Children's Hospital of Philadelphia
Philadelphia, Pennsylvania 19104

Transfusion of blood products to infants requires an appreciation of infants' smaller intravascular volume as compared to adults and an understanding of their unique immune characteristics. Transfusion principles for the management of children beyond the newborn period are similar to those applied to adult patients, although the smaller intravascular volume of children must be taken into consideration. The blood component most commonly transfused to infants and children is packed red blood cells (PRBCs). For very low-birth-weight (VLBW) infants, PRBCs are most commonly used in treatment of the anemia of prematurity, which is often exacerbated by frequent phlebotomy. A less common indication for newborn transfusion, because effective prophylaxis for Rh hemolytic disease with Rh immune globulin (RhIg) has become widespread throughout the United States, is in the management of hemolytic disease of the newborn (HDN). The indications for PRBC transfusion in older children are diverse, and include support for patients with congenital hemolytic anemias and congenital hypoplastic anemias as well as those treated for cancer or who experience surgical blood loss.

fants must be closely controlled in order to avoid intolerable increases in intravascular volume. In contrast to adults and older children, whose transfusions are ordered in units (or fractions thereof), transfusions for VLBW infants are usually ordered in milliliter quantities. VLBW infants are among the most heavily transfused patient groups; approximately 80% of premature newborns receive at least 1 transfusion, while the average number of transfusions during a hospital stay for prematurity is 8–10 (Strauss, 1991). Several surveys have documented a lack of uniformity in neonatal transfusion practices throughout the United States (Blanchette *et al.*, 1995; Bednarek *et al.*, 1998). A recent study in 6 neonatal intensive care units (NICUs) showed that patient outcomes were the same in NICUs that had conservative transfusion protocols and transfused less blood as in those that had aggressive protocols and transfused more blood (Bednarek *et al.*, 1998). Such findings suggest that adherence to standardized transfusion and phlebotomy guidelines in NICUs might result in fewer transfusions and fewer transfusion-related complications while decreasing associated costs.

I. NEONATAL TRANSFUSION

A. Very Low-Birth-Weight Infants

The blood volume of a newborn is approximately 80–100 ml/kg. The VLBW infant weighs < 1200 g and thus has a blood volume approximating 100 ml. The volume of blood transfusions prescribed for such in-

B. Immune Characteristics

Serum IgG in the neonate is derived from the maternal circulation via placental crossover. Newborn serum may contain IgG antibodies directed against antigens to which the mother has reacted, including those RBC antigens she lacks. In contrast, the large pentameric structure of IgM precludes efficient transfer across the placenta. Low neonatal IgM levels are a reflection of relatively poor neonatal antibody production.

During the first several months of life, infants are usually incapable of making alloantibodies after RBC transfusion (Floss *et al.*, 1986). As a result, once an infant's ABO and Rh type is established and an absence of maternally derived antibodies is demonstrated, repeat pretransfusion testing is not necessary until after 4 months of age, when more efficient antibody formation begins.

C. Anemia of Prematurity

1. Description

Anemia of prematurity is the most common indication for transfusion of a neonate and is a normochromic, normocytic anemia that develops between 2 and 6 weeks of age in infants born before 35 weeks gestation. The mechanism for this anemia lies in the premature infant's poor erythropoietin response to anemia and available oxygen. Poor growth, decreased activity, tachypnea, and tachycardia are sometimes observed when the neonates hemoglobin level falls below 10 g/dl.

2. Treatment

Booster transfusions are given to the neonate with anemia of prematurity to avoid these signs and symptoms and to replace blood losses due to phlebotomy. In addition, exogenous erythropoietin (200 U/kg subcutaneously q.o.d.) given in combination with parenteral iron stimulates erythropoiesis in even the smallest (< 750 g birth weight) infants. Although this regimen does not eliminate the transfusion requirements of VLBW infants, it does result in a significant reduction in the number of transfusion events and in the total volume of blood transfused (Ohls *et al.*, 1997).

D. Isoimmune Hemolytic Anemia

1. Description

Hemolytic disease of the newborn occurs when maternal IgG against a fetal RBC antigen crosses the placenta, causing destruction of the fetal RBCs. The most severe form of HDN may occur in the Rh(D)-positive infant of an Rh(D)-negative mother who has previously formed anti-Rh(D) IgG serum antibodies. The antibodies cross the placenta and causes hemolysis in the fetus (erythroblastosis fetalis). This variety of HDN requires that a mother has been previously sensitized against the Rh(D) antigen, either through previous pregnancy or by a Rh(D)-positive transfusion. The

most severely affected infants in this setting have profound anemia with hyperbilirubinemia, accompanied by organomegaly, heart failure, and anasarca (hydrops fetalis).

2. Treatment

Treatment depends on the severity of hemolysis before and after birth, as well as the degree of postnatal hyperbilirubinemia. In the severely anemic fetus, intrauterine PRBC transfusion may be required. After birth, whole blood exchange transfusion is used to add intact RBCs and remove excess bilirubin. The placenta effectively removes excess bilirubin prior to birth, but high levels of bilirubin are toxic in the newborn period. PRBCs intended for intrauterine transfusion (IUT) or for neonatal exchange transfusion should either be washed or < 5 days old, type O, Rh(D)-negative, and cytomegalovirus (CMV)-safe; for IUTs or transfusion to infants of low birth weight (< 1200 g), PRBCs should also be gamma irradiated to prevent transfusion-associated graft-versus-host disease (TA-GVHD; see Section III and Table 1).

3. Morbidity and Mortality

The estimated risk for Rh-associated HDN is about 1 in 1000 live births. Ten percent of these infants require IUT to correct anemia. Perinatal mortality in infants who receive IUT is approximately 16%. Male fetuses are more severely affected by maternal anti-

TABLE 1   Special Requirements for Intrauterine and Neonatal Transfusions

PRBCs
CMV-safe
Gamma-irradiated
Hemoglobin S negative
Washed or < 5 days old (necessary for large-volume [> 20 ml/kg] or intrauterine transfusions only)
Platelets
Group-specific, volume-reduced/resuspended, or group AB
CMV-safe
Gamma-irradiated
FFP
ABO-compatible plasma or group AB
Granulocytes
ABO/Rh(D)-compatible
CMV-seronegative
Gamma-irradiated

Rh(D) than female fetuses. Males are more likely to have significantly lower hemoglobin levels, to require more IUTs, and to develop hydrops than are female fetuses (Ulm *et al.*, 1999). Male fetuses require more intense antepartum surveillance with regard to maternal alloimmunization to Rh(D). The neurodevelopmental outcome of fetuses who receive intravascular IUT for severe intrauterine hemolysis is similar to that observed in high-risk, VLBW infants.

Approximately 10% of those infants who survive after requiring IUT have neurodevelopmental disabilities (Janssens *et al.*, 1997).

E. ABO Hemolytic Disease

Hemolysis can also occur when the infant's ABO type differs from the mother's, most commonly when a group O mother carries a group A fetus. ABO hemolytic disease can occur with the first pregnancy because anti-A and anti-B are naturally occurring antibodies. Hemolysis due to ABO incompatibility is more common but less severe than Rh hemolysis. Exchange transfusion is only rarely required to treat neonatal hemolysis secondary to ABO incompatibility.

II. BLOOD COMPONENTS

A. Packed Red Blood Cells (RBCs)

1. Additive Solutions

The preservation of PRBCs is improved and the shelf life of the unit prolonged with the addition of extended storage media (also called additive solution, or AS). These solutions contain glucose, adenine, and saline, and may include mannitol and phosphate buffer. Whereas PRBCs preserved in CPDA-1 have a final hematocrit of 75–85%, the addition of AS reduces the final hematocrit to 50–60%. The reduced hematocrit requires that a 10–12% higher volume of AS RBCs be transfused to deliver the same amount of PRBCs present in CPDA-1-preserved PRBCs, a potential problem when transfusing large amounts of PRBCs to very small infants. Centrifugation of a portion of an AS unit to remove extracellular fluid and thus reduce the volume transfused is useful if the patient's intravascular volume must be tightly controlled. It is important to note that transfusion of PRBCs preserved in additive solutions (AS-1, AS-3) has been demonstrated to be safe and effective in VLBW infants in the context of small-volume, booster (5–20 ml/kg) transfusions (Goodstein *et al.*, 1993). The safety of AS PRBCs for large-volume transfusions (exchange transfusion, extra-corporeal membrane oxygenation) has not been definitively established.

2. PRBC Transfusion for Congenital Anemias

Children with certain congenital anemias may require PRBC transfusion during infancy. Inherited anemias due to poor RBC production, such as congenital red cell aplasia (Diamond–Blackfan anemia), are associated with macrocytosis and reticulocytopenia. Patients present with pallor, lethargy, and failure to thrive. Congenital anemias due to increased RBC destruction, which include the red cell membrane defects (e.g., hereditary spherocytosis), enzyme deficiencies (e.g. G-6-PD deficiency), and hemoglobinopathies, are associated with reticulocytosis and specific, individual red cell morphological abnormalities. Infants with increased RBC destruction are often jaundiced and may have splenomegaly. Infants with inherited abnormalities of the beta globin chain (e.g., beta thalassemia major or sickle cell disease) do not usually become symptomatic from anemia until after the first 6 months of life, when decreased or abnormal production of the beta globin chain impedes production of normal adult hemoglobin, Hb A.

Whether the origin of the anemia is hypoplastic or hemolytic, there is no preset laboratory value that triggers transfusion. The goal of transfusion therapy is to allow for optimal growth and development while eliminating the physical manifestations of chronic anemia (respiratory distress and cardiac failure). Long-term, routine PRBC transfusions are required to support some children with chronic anemias. The decision for initiating chronic transfusion therapy must be made on an individual basis, taking into consideration the child's growth activity and individual level of hemolysis.

3. PRBC Transfusions for Sickle Cell Disease (SCD)

There are several acute and chronic indications for transfusion in children with SCD (see Table 2). Acute PRBC transfusions are required to support patients who develop sudden complications of their preexisting anemia, as in episodes of stroke, acute chest syndrome, or aplastic crisis following infection with parvovirus B19 (Cohen *et al.*, 1996). PRBC transfusion intended to raise the preoperative hemoglobin concentration to 10 g/dl is recommended prior to surgery requiring general anesthesia (Vichinsky *et al.*, 1995). Routine long-term PRBC transfusion has been shown to reduce the likelihood of recurrence of stroke in children with SCD (Russell *et al.*, 1984), but chronic PRBC transfusion leads to the major complications of iron overload and RBC alloimmunization. Extended RBC antigen

TABLE 2 Acute and Chronic Indications for Transfusion in Sickle Cell Disease

Acute
  Aplastic crisis
  Acute chest syndrome
  Acute splenic sequestration
  Prior to general anesthesia or eye surgery
  Stroke
  Multiorgan system failure
Chronic
  Recurrent strokes in children
  Complicated pregnancy
  Prevention of first stroke in children
Controversial issues
  Severe pulmonary or cardiac disease
  Priapism
  Intractable pain crises

typing should be completed prior to transfusion in children with SCD, since the provision of antigen-matched PRBCs has been shown to reduce the risk of RBC alloimmunization in patients with SCD (Vichinsky, 1996). The provision of leukocyte-reduced PRBCs to SCD patients requiring chronic transfusion prevents febrile transfusion reactions and mitigates the development of HLA and platelet alloimmunization.

B. Platelets

1. Platelet Transfusions in Infants

Neonatal thrombocytopenia, observed in one-fifth of infants admitted to the intensive care nursery, is more often due to increased platelet destruction or to destruction combined with poor production than to poor platelet production alone. Birth asphyxia is commonly associated with thrombocytopenia that resolves spontaneously (Castle *et al.*, 1985). In one study the risk of neonatal thrombocytopenia was assessed in 80 infants with birth weights < 1500 g (Andrew *et al.*, 1987). Intraventricular hemorrhage (IVH) was observed in 78% of thrombocytopenic ($< 100,000$ μl) infants compared with 48% of non-thrombocytopenic infants. The risk of serious neurological sequelae was significantly higher in the thrombocytopenic survivors than in those who had not had low platelet counts. Based on these and other studies, most clinicians that treat VLBW infants transfuse platelets at higher platelet counts than are recommended for older children and adults. Although early aggressive platelet transfusion does not

eliminate serious bleeding in VLBW infants (Andrew *et al.*, 1993), platelet transfusions are commonly given to maintain the platelet count $> 100,000/\mu$l in the ill VLBW infant and above $50,000/\mu$l the stable VLBW infant. In the patient with little or no platelet destruction, a dose of 10 ml/kg random-donor platelet concentrate should raise the platelet count $50,000/\mu$l when measured directly after transfusion.

Alloimmunization to transfused platelets is rarely observed during infancy; neonatal platelet refractoriness is most often associated with disseminated intravascular coagulation (DIC) or infection. Infants with neonatal alloimmune thrombocytopenia (maternally derived IgG against a platelet antigen on the infant's platelet) will consume random platelet concentrates but retain platelets from a donor who lacks the inciting platelet antigen (the mother or another antigen-negative donor). The mother, who is always antigen-negative may undergo platelet apheresis in order to donate her own platelets for transfusion to the neonate. In most cases, the mother's platelets are both the easiest to obtain and the best product for the affected child.

2. Platelet Transfusions in Children

Platelet transfusions are critical in the management of older children who have thrombocytopenia secondary to poor platelet production or abnormal platelet function. For the thrombocytopenic older child, platelet transfusion is required for bleeding or to prevent bleeding when the platelet count is profoundly decreased ($10,000-20,000/\mu$l). The dose of random-donor platelet concentrates for an older child with poor platelet production is $0.1-0.2$ U of random-donor platelets per kilogram of body weight (Cahill and Lilleyman, 1998). Measuring a platelet increment by drawing a platelet count 15–45 minutes after transfusion is useful in assessing if the dose was appropriate.

C. Plasma

The indication for transfusion of fresh frozen plasma (FFP) in infants is similar to the indication for adults, that is, for the replacement of clotting factors. Bleeding in infancy may be due to a congenital deficiency of a particular factor or to consumption of several or all of them. Plasma contains 1 U/ml of the factors II, V, VII, VIII, IX, X, and XI, as well as the naturally occurring anticoagulants protein C, protein S, and antithrombin III (ATIII). The bleeding infant with a prolonged PT and PTT and bleeding may benefit from 10 ml/kg of FFP administered every 4–6 hours.

D. Factor Concentrates

Concentrated, plasma-derived forms of ATIII, FVIII, and FIX, as well as recombinant FVIII, FIX, and activated FVII, are commercially available. When a bleeding infant is suspected of having hemophilia A or B because of a positive family history or excessive bleeding following circumcision, rapid confirmation of the diagnosis is important, since infusion of recombinant factor concentrates is preferred over transfusion of FFP for replacement of FVIII or FIX.

E. Granulocyte Transfusions

There has been a recent resurgence of interest in granulocyte transfusion, and in particular its role in the management of septic neutropenic patients (Sweetman and Cairo, 1995; see Chapter 8). Premature neonates are at increased risk for bacterial sepsis since they have a limited capacity to increase granulocyte production in response to infection. Cairo et al. (1992) have shown an improvement in the survival of septic neonates following granulocyte transfusion as compared to treatment with IVIg. Granulocytes prepared by apheresis techniques are preferred over buffy coat preparations and should be transfused as quickly as possible following collection (≤ 24 hours).

III. RISKS SPECIFIC TO NEONATES

A. Cytomegalovirus (CMV)

Premature infants born to mothers who are seronegative for CMV antibody are at high risk for CMV-related morbidity or mortality if transfused with cellular blood products from CMV-seropositive donors (Yeager et al., 1981). Infants at highest risk for transfusion-transmitted CMV include those who weigh < 1200 g at birth. The risk of transfusion-transmitted CMV can be substantially reduced through the use of CMV-safe cellular blood products. Leukoreduction via filtration is an effective means of preventing CMV infection in transfused infants (Gilbert et al., 1989; Eisenfeld et al., 1992; Strauss, 1999).

B. Hyperkalemia

PRBC storage in the liquid state is associated with several biochemical alterations that worsen with time. For neonatal transfusion, the most important of these is the leak of potassium from the RBC into the extracellular space, a potential problem with large-volume transfusions of small infants. Case reports of neonatal deaths due to hyperkalemia following exchange transfusion or rapid transfusion following open heart surgery with stored blood (e.g., large-volume transfusions) have prompted many clinicians to request < 5-day-old RBCs for exchange transfusion or after open heart surgery (Blanchette et al., 1984; Hall et al., 1993). Washing PRBCs removes excess extracellular potassium, making washed PRBCs an appropriate substitute if fresh PRBCs are not available. For small-volume transfusions (< 20 ml/kg), however, the actual dose of bioavailable potassium transfused is quite small, and thus "fresh" PRBCs are not necessary for small-volume transfusions (Strauss et al., 1996).

C. Transfusion-Associated Graft-Versus-Host Disease

Neonates at risk for TA-GVHD include VLBW infants, those with congenital cellular immunodeficiencies (for example, severe combined immunodeficiency), and infants who receive blood from a blood relative. Fetuses who require IUT are also at risk. Older children at risk for TA-GVHD include those with cellular immunodeficiencies, patients being treated for malignancy with intensive radiotherapy or chemotherapy, and recipients of bone marrow or solid organ transplants (Manno, 1996). TA-GVHD can be prevented by gamma irradiation of blood products with a dose of 2500 cGy.

IV. REDUCING DONOR EXPOSURES

VLBW infants usually require lengthy hospital stays and repeated small-volume transfusions. Efforts to decrease the exposure of these infants to multiple donors are enhanced by using 1 U of PRBCs for 1 infant's exclusive use. A sterile connecting device allows for preparation of small-volume aliquots from a single PRBC unit (Strauss et al., 1996). In addition, filter–syringe sets have been developed for the purpose of removing PRBCs from the original unit and efficiently delivering them to the infant recipient (Chambers, 1995).

V. AUTOLOGOUS DONATION IN CHILDREN

Autologous transfusion is not routinely performed prior to surgery in children since venous access, adequate volume of donation, and absence of physical or psychological distress to the nonconsenting child cannot always be guaranteed. Successful pediatric autologous transfusion programs have been reported in the context of abdominal, orthopedic, and cardiac surgery. Pediatric autologous transfusion programs work better (less discarded blood, appropriate blood transfusions) if the likelihood of postoperative transfusion is high and when the donor has received preoperative iron supplementation to reduce perioperative anemia. Contraindications to autologous transfusion in infants and children are the same as those applied to adult patients. Patients with limited cardiac reserve and those with decreased oxygen saturation due to cardiomyopathy or pulmonary disease should be carefully scrutinized before donation. Patients with hemoglobinopathies, bleeding disorders, or sepsis are not suitable candidates.

References

Andrew, A., Vegh, C. C., Kirpalani, H., et al. (1993). A randomized, controlled trial of platelet transfusions in thrombocytopenic premature infants. *J Pediatr* **123**(2), 285–291.

Andrew, M., Castle, V., Saigal, S., et al. (1987). Clinical impact of neonatal thrombocytopenia. *J Pediatr* **110**, 457–464.

Bednarek, F. J., Weisberger, S., Richardson, D. K., et al. (1998). Variations in blood transfusions among newborn intensive care units. SNAP II Study Group. *J Pediatr* **133**, 589–590.

Blanchette, V. S., Gray, E., Hardie, M. J., et al. (1984). Hyperkalemia following exchange transfusion: Risk eliminated by washing red cell concentrates. *J Pediatr* **105**, 321–324.

Blanchette, V. S., Kuhne, T., Hume, H., et al. (1995). Platelet transfusion therapy in newborn infants. *Transfusion Med Rev* **9**, 215–230.

Cahill, M. R., and Lilleyman, J. S. (1998). The rational use of platelet transfusions in children. *Semin Thromb Hemost* **24**, 567–575.

Cairo, M. S., Worcester, C. C., Reicker, R. W., et al. (1992). Randomized trial of granulocyte transfusions versus intravenous immune globulin for neonatal neutropenia and sepsis. *J Pediatr* **120**, 281–285.

Castle, V., Andrew, M., Kelton, J., et al. (1985). Frequency and mechanism of neonatal thrombocytopenia. *J Pediatr* **108**, 749–755.

Chambers, L. (1995). Evaluation of a filter-syringe set for preparation of packed cell aliquots for neonatal transfusion. *Am J Clin Pathol* **104**, 253–257.

Cohen, A. R., Norris, C. F., and Smith-Whitley, K. (1996). Transfusion therapy for sickle cell disease. In "New Directions in Pediatric Hematology" (S. M. Capon and L. A. Chambers, Eds.). American Association of Blood Banks, Bethesda, MD.

Eisenfeld, L., Silver, H., McLaughlin, J., et al. (1992). Prevention of transfusion-associated cytomegalovims infection in neonatal patients by removal of white cells from blood. *Transfusion* **32**, 205–209.

Floss, A. M., Strauss, R. G., Goeken, N., et al. (1986). Multiple transfusions fail to provoke antibodies against blood cell antigens in human infants. *Transfusion* **26**, 419–422.

Gilbert, G. L., Hayes, K., Hudson, I. L., et al. (1989). Prevention of transfusion-acquired cytomegalovirus infection in infants by blood filtered to remove leukocytes. *Lancet* **1**, 1228–1231.

Goodstein, M. H., Locke, R. G., Wlodarczyk, D., et al. (1993). Comparison of two preservative solutions for erythrocyte transfusions in newborn infants. *J Pediatr* **123**, 783–788.

Hall, T. L., Barnes, A., Miller, J. R. et al. (1993). Neonatal mortality following transfusion of red cells with high plasma potassium levels. *Transfusion* **33**, 606–609.

Janssens, H. M., de Haan, M. J., van Kamp, I. L., et al. (1997). Outcome for children treated with fetal intravascular transfusions because of severe blood group antagonisms. *J Pediatr* **131**, 340–342.

Kim, H. C., Dugan, N. P., Silber, J. H., et al. (1994). Erythrocytapheresis therapy to reduce iron overload in chronically transfused patients with sickle cell disease. *Blood* **83**, 1136–1142.

Manno, C. S. (1996). What's new in transfusion medicine? *Pediatr Clin N Am* **43**, 793–808.

Ohls, R. K., Harcum, J., Schibler, K. R., et al. (1997). The effect of erythropoietin on the transfusion requirements of preterm infants weighing 750 gm or less: A randomized, double blind, placebo controlled study. *J Pediatr* **131**, 653–655.

Russell, M. O., Goldberg, H. I., Hodson, A., et al. (1984). Effect of transfusion therapy on arteriographic abnormalities and recurrence of stroke in sickle cell disease. *Blood* **63**, 162–169.

Strauss, R. G. (1991). Transfusion therapy for neonates. *Am J Dis Child* **148**, 904–911.

Strauss, R. G. (1999). Leukocyte-reduction to prevent transfusion-transmitted cytomegalovirus infections. *Pediatr Transplant* **3**, 19–22.

Strauss, R. G., Burmeister, L. F., Johnson, K., et al. (1996). AS-1 red cells for neonatal transfusion: A randomized trial assessing donor exposure and safety. *Transfusion* **36**, 873–878.

Sweetman, R. W., and Cairo, M. S. (1995). Blood component therapy and immunotherapy in neonatal sepsis. *Transfusion Med Rev* **3**, 251–259.

Ulm, B., Svolba, G., Ulm, M. R., et al. (1999). Male fetuses are particularly affected by maternal alloimmunization to D antigen. *Transfusion* **39**, 169–173.

Vichinsky, E. (1996). Transfusion therapy. In "Sickle Cell Disease: Basic Principles and Clinical Practice" (S. H. Emburg, R. P. Hebbel, N. Mohandas, M. H. Steinberg, Eds.). Lipincott-Raven, Philadelphia, PA.

Vichinsky, E. P., Haberkern, C. M., Neumayr, L., and the Preoperative Transfusion in Sickle Cell Disease Study Group (1995). A comparison of conservative and aggressive transfusion regimens in the perioperative management of sickle cell disease. *N Engl J Med* **333**, 206–213.

Yeager, A. S., Grumet, F. C., Hafleigh, E. B., et al. (1981). Prevention of transfusion-acquired cytomegalovirus infections in newborn infants. *J Pediatr* **98**, 281–286.

27

Approach to the Immunocompromised Patient

HUA SHAN

Division of Transfusion Medicine Services
Department of Pathology
Johns Hopkins Medical Institutions
Baltimore, Maryland

There are several important issues to consider in transfusion therapy of the immune-compromised patient. In this chapter, patients regarded as immune-compromised will include those with severe congenital immunodeficiencies, severe acquired immunodeficiencies (including patients with leukemia, lymphoma, and certain solid organ neoplasms), recipients of solid organ and bone marrow transplants (BMTs) and patients with human immunodeficiency virus type 1 (HIV-1) infection.

Immunocompromised allogeneic blood transfusion recipients differ from immunucompetent transfusion recipients in their susceptibility to developing transfusion-associated graft-versus-host disease (TA-GVHD), transfusion-transmitted infectious diseases, including cytomegalovirus (CMV) infection (TT-CMV), and transfusion-related immunomodulation (TRIM).

Practice guidelines (Klein, 1996; Przepiorka *et al.*, 1996a, b) for the transfusion management of each of these immunocompromised patients are reviewed below, and specific conditions associated with increased risk for TA-GVHD are listed in Table 1.

I. POTENTIAL COMPLICATIONS OF TRANSFUSION IN THE IMMUNOCOMPROMISED ALLOGENEIC TRANSFUSION RECIPIENT

A. Transfusion-Associated Graft-Versus-Host Disease

1. Description and Incidence

TA-GVHD is a rare complication of allogeneic transfusion (Przepiorka *et al.*, 1996a; Williamson, 1998a,

b). Only 200 cases of TA-GVHD have been published; however, it is likely that more cases have gone unreported due to lack of recognition and absence of definitive diagnostic criteria. TA-GVHD is almost universally fatal ($> 90\%$ mortality rate).

TA-GVHD is caused by replication-competent donor lymphocytes in cellular blood products that engraft and lead to immune-mediated destruction of host tissues. The development of TA-GVHD reflects the inability of the recipient's immune system to reject the transfused, and thus foreign, lymphocytes. Although extremely rare, cases of TA-GVHD have been reported in apparently immunocompetent individuals (Juji *et al.*, 1989). The vast majority of cases occur in patients with severe immunodeficiency, either congenital or acquired (Sanders and Graeber, 1990; Greenbaum, 1991; Przepiorka *et al.*, 1996a).

2. Patients at Risk

Patients at risk of developing TA-GVHD are listed in Table 1. Many of these conditions will be discussed in following sections in this chapter.

3. Diagnosis

a. Clinical

The symptoms of TA-GVHD are similar to those of GVHD that develops following hematopoietic progenitor cell (HPC) transplantation, and include an erythematous maculopapular skin rash, cutaneous erythema and bullae, diarrhea, hepatomegaly, severe pancytopenia, and fever. In contrast to GVHD seen with HPC

TABLE 1   Patients at Increased Risk for TA-GVHD

Congenital immmunodeficiency syndromes
  Severe combined immunodeficiency (SCID)
  Adenosine deaminase (ADA) deficiency
  Wiskott–Aldrich syndrome
  DiGeorge syndrome
  MHC class I and/or II deficiencies
  Omenn syndrome
  Ataxia telangiectasia
  T-cell activation defects
  Bruton's agammaglobulinemia
Acquired immunodeficiency syndromes
  Leukemia
  Hodgkin's disease
  Non-Hodgkin's disease
  Bone marrow or stem cell transplant (allogeneic and
    autologous)
  Solid organ transplant patients on intensive
    immunosuppressive therapy
  Other patients on potent immunosuppressive therapy
  Certain solid tumors
    Immunoblastic sarcoma
    Rhabdomyosarcoma
    Neuroblastoma
    Glioblastoma
    Aplastic anemia and other bone marrow failure states
  AIDS (possible indication)
Fetuses/neonates
  Intrauterine transfusions
  Neonatal transfusions of patients who underwent intrauterine
    transfusions
  Neonatal exchange transfusions
  Maternal platelet transfusions for neonatal alloimmune
    thrombocytopenia
  Premature infants weighing < 1500 g at birth
Pregnant women
Recipients of specific blood components
  Granulocyte infusions
  HLA-matched or crossmatch-compatible platelets
  Blood products donated by a blood relative of the recipient

transplantation, TA-GVHD also results in severe marrow aplasia, with bleeding due to thrombocytopeinia, and infection due to neutropenia. Symptoms of TA-GVHD can begin at any time from 2 days to up to a month post-transfusion of nonirradiated cellular blood products (Brubaker, 1983; Anderson and Weistein, 1990). TA-GVHD may be confused with severe viral infection or drug reaction. Indeed, these conditions may coexist.

b. Laboratory

Early in the course of TA-GVHD, the patient may develop marked elevation in liver enzymes. Skin and liver biopsies show cellular degeneration in the presence of a mononuclear cell infiltrate. Pancytopenia develops later in the course of TA-GVHD, with hypocellularity and fibrosis of the bone marrow in the presence of a mononuclear cell infiltrate. Definitive diagnosis rests with the determination of differences in HLA types between the donor and the recipient, and (if possible) in the identification of donor lymphocytes within the affected tissues of the recipient.

c. Clinical Course

TA-GVHD usually occurs in the setting of an immunocompromised host.

Symptoms of TA-GVHD develop within 8–10 days following transfusion of the implicated product. Death is nearly universal (> 90%) and occurs in 3 to 4 weeks post-transfusion, usually as a result of complications of bone marrow failure.

4. Prevention

There is no effective treatment for TA-GVHD, and as stated above, its mortality exceeds 90% (Greenbaum, 1991). Thus, prevention of TA-GVHD is of paramount importance. The only currently available method of preventing TA-GVHD is by gamma irradiation of cellular blood products (Klein, 1996; Przepiorka *et al*., 1996a)—2500 cGy is the recommended dose, and quality control and validation measures are recommended. The lowest number of replication-competent lymphocytes in a blood component capable of causing TA-GVHD is unknown; of note, transfusion of leukoreduced cellular blood products has been associated with TA-GVHD (Akahoshi *et al*., 1992). Leukoreduction of cellular blood components is *not* adequate to prevent TA-GVHD in susceptible patients.

B. Transfusion-Transmitted Cytomegalovirus Infection

1. Description

CMV is a member of the human herpes virus family (also known as HHV-5). A high percentage of the general population, as well as the blood donor population, has been infected by CMV and has developed CMV-specific antibodies (CMV seropositive; Ho, 1990).

Numerous early studies established blood transfusion as a cause of CMV infection (Kreel *et al.*, 1960; Prince *et al.*, 1971; Lamberson, 1985). Post-transfusion CMV infection can result from primary infection in previously seronegative recipients or from reactivation of a previously latent infection, or secondary infection, in previously seropositive recipients. In secondary infection, a previously seropositive patient may become infected with a different CMV strain (also known as second-strain infection). The incidence of transfusion-associated primary CMV infection has been estimated to be 1 to 3% per unit of blood products (Ho, 1990). There is no reliable method to predict a recipient's risk for transfusion-mediated CMV reactivation or secondary infection.

Although TT-CMV has been the most widely studied, recent data suggest that other transfusion-transmitted viral infections, especially transfusion-transmitted human herpes virus (other than CMV) infections (TT-HHV), can also be of clinical significance. Human herpes viruses that have been implicated in human diseases include Epstein–Barr virus (HHV-4) and potentially HHV-8. TT-HHV can cause significant clinical problems in immunocompromised patients (Hillyer *et al.*, 1999).

2. Patients at Risk

The morbidity and mortality of CMV infection depend on the immune status of the patient, underlying disease, route of infection, and size of viral inoculum. TT-CMV is associated with significant morbidity and mortality in immunocompromised patients, especially those who are CMV-seronegative prior to transfusion (Meyers *et al.*, 1986; Sayers, 1996).

Other populations at risk for TT-CMV include fetuses and some neonates (Yeager, 1974; Przepiorka *et al.*, 1996b), and therefore these should receive CMV-"safe" cellular blood products. CMV-seronegative pregnant women should also receive CMV-safe cellular blood products (Przepiorka *et al.*, 1996b).

3. Prevention

Efforts aimed at preventing TT-CMV infection are of critical importance because all available treatments for this condition are limited in their efficacy. To reduce the risk of TT-CMV infection in high-risk populations, CMV-safe cellular blood products, which are either CMV-seronegative or leukoreduced unit from blood donors with undetermined CMV antibody status, are indicated (Bowden *et al.*, 1986, 1995; Miller *et al.*, 1991; Przepiorka *et al.*, 1996b; Klein, 1996; Narvios

et al., 1998). A more detailed discussion of TT-CMV is provided in Chapters 16 and 36.

Since most clinically significant HHVs in the blood reside in leukocytes, leukoreduced cellular blood products may be effective in preventing TT-HHV as well as other leukocyte-associated virus transmission by transfusion. More studies in this area are needed before definitive recommendations can be made.

C. Transfusion-Related Immunomodulation (TRIM)

The immunomodulatory effects of allogeneic cellular blood transfusion may lead to reactivation of viral infection in immunocompromised patients (Adler *et al.*, 1985; Busch *et al.*, 1992). Although the mechanism and severity of TRIM are a subject of debate, there is increasing evidence that transfused allogeneic leukocytes alter the host immune system, resulting in what appears to be immunosuppression (Blumberg and Heal, 1998; Hillyer *et al.*, 1999). This effect may have more clinical significance in immunocompromised patient populations.

Since TRIM is thought to be mediated through allogeneic donor lymphocytes, removal of these donor leukocytes therefore may be effective in decreasing TRIM (Jensen *et al.*, 1992, 1995, 1996).

II. CONGENITAL IMMUNODEFICIENCY SYNDROMES

Patients with congenital immunodeficiencies are at risk for TA-GVHD and include those conditions listed in Table 1.

A. Prevention of TA-GVHD

Patients with known or suspected severe congenital immunodeficiencies should receive irradiated cellular blood products. Most TA-GVHD cases reported in congenital immunodeficient patients are associated with severe T-cell deficiencies or severe T-cell and B-cell combined deficiencies. TA-GVHD has not been reported in patients with isolated agammaglobulinemia or neutrophil disorders such as chronic granulomatous disease (Greenbaum, 1991; Przepiorka *et al.*, 1996a). Since it is difficult to evaluate the exact risk for TA-GVHD for each different immunodeficient condition, it is prudent to provide irradiated cellular blood products to prevent potential TA-GVHD in any patient for whom a diagnosis of a severe immunedeficient condition is either established or suspected.

B. Prevention of TT-CMV

CMV-seronegative patients with severe congenital immunodeficiencies are susceptible to developing devastating primary TT-CMV. CMV-safe cellular blood products are indicated for all CMV-seronegative patients diagnosed or suspected of severe immunodeficiencies, including T- or B-cell deficiencies, combined deficiencies, and deficiencies of other branches of the immune system. Since CMV-seropositive, severely immunocompromised patients are also at increased risk for high mortality and morbidity associated with reactivation of previously latent infection or second strain infection, clinicians may consider supporting these patient with CMV-safe cellular blood products.

III. ACQUIRED IMMUNODEFICIENCY SYNDROMES

Patients may develop acquired immunodeficiency secondary to malignant diseases, cancer therapies (drugs and irradiation), immunosuppressive treatment required for organ transplant or other medical conditions, HIV infection, and BMT. Specific syndromes are discussed below.

A. Patients with Leukemia and Lymphoma

The overall risk for TA-GVHD in cancer patients has been estimated to be 0.1 to 1.0% (Von Fliedner et al., 1982). This risk may increase in patients with hematological malignancies, especially those receiving immunosuppressive and cytotoxic treatment (Zulian et al., 1995; Przepiorka et al., 1996a). Thus, patients with leukemia and lymphoma should receive irradiated cellular blood products to prevent TA-GVHD.

CMV-seronegative patients in this group should receive CMV-safe cellular blood products, especially those who are likely candidates for allogeneic marrow transplant. Although the risk of significant CMV reactivation and/or second strain infection exists, the indication for CMV-safe cellular blood products in CMV-seropositive patients in this group remains a subject of debate.

B. Patients with Bone Marrow Failure States

TA-GVHD has been reported in aplastic anemia patients (Brubaker, 1983; Przepiorka et al., 1996a).

Patients with aplastic anemia and other marrow failure states should receive irradiated cellular blood products. CMV-seronegative patients with marrow failure states who are likely candidates for allogeneic marrow transplantation should receive CMV-safe products (Meyers et al., 1986; Przepiorka et al., 1996b).

C. Patients with Certain Solid Organ Tumors

Patients with some types of solid organ tumors are considered to be at increased risk for TA-GVHD, including immunoblastic sarcoma, rhabdomyosarcoma, neuroblastoma, and glioblastoma (Von Fliedner et al., 1982; Anderson et al., 1991; Przepiorka et al., 1996a). These patients should receive irradiated cellular blood products. Irradiated cellular blood products should be considered for all cancer patients who are immunosuppressed because of chemotherapy or irradiation. CMV-safe cellular blood products may need to be considered for all CMV-seronegative patients receiving immunosuppressive therapy, especially when shown to be susceptible to opportunistic infections (Przepiorka et al., 1996b).

D. Recipients of Solid Organ Transplant

Immunosuppressed organ transplant recipients should receive irradiated cellular blood products to prevent TA-GVHD (Jamieson et al., 1991; Przepiorka et al., 1996a). CMV-seronegative patients who are likely candidates for organ transplantation should receive CMV-safe cellular blood products (Przepiorka et al., 1996b; Sherman and Ramsey, 1996).

For patients undergoing kidney transplantation, there has been much debate regarding the role of pretransplant blood transfusion (Sherman and Ramsey, 1996). Pretransplant transfusion in kidney transplant patients has been associated with induction of a graft protection effect. Sensitization to human leukocyte antigens (HLAs) and increased graft failure have also been linked to pretransplant transfusion in this patient population. These are complex issues which require further study.

E. Patients Receiving Immunosuppressive Therapy

Irradiated cellular blood products are indicated for all patients receiving potent immunosuppressive therapy. Primary TT-CMV as well as reactivation of latent CMV infection has been documented in patients on potent immunosuppressive therapy for autoimmune diseases or other indications (e.g., prevention of alloimmunization; Dowling *et al.*, 1976; Suassuna *et al.*, 1993). Thus, it has been recommended that CMV-safe cellular blood products are indicated for those patients on immunosuppressive therapy who are CMV-seronegative and already at risk for opportunistic infections (Przepiorka *et al.*, 1996b).

IV. PATIENTS WITH HIV INFECTION

Significant hematologic and immunologic abnormalities are prominent features of HIV-1 infection, with pancytopenia present in the majority of patients with advanced disease (Sloand and Groopman, 1997; Moses *et al.*, 1998; Hillyer *et al.*, 1999). HIV-1 viruses compromise an individual's immune defense system, thereby increasing one's susceptibility to opportunistic infections and malignant diseases. By similar mechanisms, a compromised immune status also renders a patient more vulnerable to various adverse consequences secondary to blood transfusion.

A. TA-GVHD in HIV-1 Infected Patients

For unclear reasons TA-GVHD has not been associated with transfusion in HIV-1-injected patients (Greenbaum, 1991; Przepiorka *et al.*, 1996a; Hillyer *et al.*, 1999). It is possible that donor lymphocytes (CD4-positive T-cells) become infected by HIV and are subsequently rendered immunologically incompetent. Another explanation is that TA-GVHD may be underreported in this patient population, because the symptoms of TA-GVHD are similar to those in HIV-1-infected patients without TA-GVHD. It has been proposed that HIV-1-infected patients, especially those with a history of opportunistic infections, should receive irradiated cellular blood products (Przepiorka *et al.*, 1996a; Hillyer *et al.*, 1999).

B. CMV-Safe Cellular Blood Products

Primary TT-CMV in CMV-seronegative HIV-1-infected patients can be devastating. All CMV-seronegative HIV-infected patients should receive CMV-safe cellular blood products. The majority of HIV-1-infected patients (about 90%) are CMV-seropositive. For these patients, post-transfusion CMV reactivation and/or second strain infection can also lead to significant morbidity and mortality (Spector *et al.*, 1984; Britt and Alford, 1996; Mocarski, 1996). Therefore, CMV-safe cellular blood products may also be warranted for these HIV-infected patients, especially those with a history of opportunistic infections

(Sloand *et al.*, 1994; Przepiorka *et al.*, 1996b; Hillyer *et al.*, 1999).

C. Leukoreduced Cellular Blood Products

HIV-1-infected patients are also at higher risk for transfusion-transmitted infections from other HHVs (e.g., EBV and possibly HHV-8) as well as other infectious agents such as *Toxoplasma gondii* (Siegel *et al.*, 1971; Hillyer *et al.*, 1999). Since these agents are highly leukocyte associated, leukoreduction may minimize transmission of these infectious agents by transfusion. This has not, as yet, been studied clinically.

Since immunocompromised HIV-1-infected patients may be more susceptible to transfusion-associated immunomodulation mediated by leukocytes, leukoreduced cellular blood products in this patient population may have an added benefit of decreasing the immunosuppressive effect of transfusion (Hillyer *et al.*, 1999).

V. BONE MARROW TRANSPLANT (BMT) PATIENTS

The diseases for which BMT has been applied include leukemias, lymphomas, congenital severe immunodeficiencies, aplastic anemia, certain solid tumors, congenital metabolic disorders, some autoimmune disorders, and AIDS. Appropriate transfusion support for BMT patients is an essential element of a successful BMT program and is critical to the care of BMT recipients. Many patients require blood transfusion support before and after BMT. Most of them are immunocompromised due to a combination of underlying diseases, preparative chemotherapy prior to transplantation, and the post-transplant marrow aplasia and hypoplasia period. The most frequently used blood products in BMT patients are platelet products and packed red blood cells. On rare occasions, granulocyte products are also indicated.

A. Prevention of TA-GVHD

Potent immunosuppressive therapy is often used in allogeneic marrow transplantation patients to permit engraftment of donor marrow cells. High-dose chemotherapy and/or radiation therapy in autologous BMT candidates also significantly compromise a patient's immune defense system. Therefore both allogeneic and autologous BMT patients are at higher risk of developing TA-GVHD (Slichter, 1988; Greenbaum, 1991). Irradiated cellular blood products are indicated for all BMT recipients. Irradiated blood products should be used beginning at the initiation of the pretransplant conditioning therapy (Petz, 1996; Przepiorka *et al.*, 1996a).

B. Prevention of TT-CMV and TT-HHV

It has been estimated that 28% of CMV-seronegative marrow transplant recipients with CMV-seronegative marrow donors develop CMV infection during the marrow transplant period, and that these cases were attributable to CMV transmission by blood products (Meyers *et al.*, 1986). CMV infection in marrow transplant patients carries a significant mortality rate, especially when patients develop CMV pneumonia (Meyers *et al.*, 1986). The efficacy of CMV-safe cellular blood products in preventing TT-CMV in the CMV-seronegative marrow transplant patient population has been reported in several studies, with a residual TT-CMV risk of < 3% (Bowden *et al.*, 1986, 1995; Miller *et al.*, 1991; Narvios *et al.*, 1998).

CMV-safe cellular blood products are indicated for CMV-seronegative allogeneic BMT recipients with CMV-seronegative donors, as well as for all CMV-seronegative patients who are likely candidates for allogeneic marrow transplantation (Petz, 1996; Przepiorka *et al.*, 1996b). CMV-safe cellular blood products should also be used for CMV-seronegative candidates and recipients of autologous or syngeneic BMT (Przepiorka *et al.*, 1996b). The duration for which BMT recipients should receive CMV-safe cellular products is still controversial, but certainly includes a period of up to 1 year following BMT (Przepiorka *et al.*, 1996b). The need for CMV-safe cellular blood products in CMV-seropositive BMT patients is still under debate.

Similar to HIV-1-infected patients, BMT patients are also at increased risk for significant clinical consequences from other transfusion-transmitted viral infections (e.g., EBV). Use of leukoreduced cellular blood products may decrease transmission of these viruses by transfusion.

References

Adler, S. P., Bagget, J., *et al.* (1985). Transfusion-associated cytomegalovirus infections in seropositive cardiac surgery patients. *Lancet* 2, 743–746.

Akahoshi, M., Takanashi, M., *et al.* (1992). A case of transfusion-associated graft-versus-host disease not prevented by white cell-reduction filters. *Transfusion* 32, 169–172.

Anderson, K. C., and Weistein, H. J. (1990). Transfusion-associated graft-versus-host disease. *N Engl J Med* 323, 315–321.

Anderson, K., Goodnough, L. T., *et al.* (1991). Variation in blood component irradiation practice: Implications for prevention of transfusion-associated graft-versus-host disease. *Blood* 77, 2096–2102.

Bishop, J. F., McGrath, K., *et al.* (1988). Clinical factors influencing the efficacy of pooled platelet transfusions. *Blood* 71, 383–387.

Blumberg, N., and Heal, J. M. (1998). Blood transfusion immunomodulation. The silent epidemic. *Arch Pathol Lab Med* 122, 117–119.

Bowden, R. A., Sayers, M., *et al.* (1986). Cytomegalovirus immune globulin and seronegative blood products to prevent primary cytomegalovirus infection after marrow transplantation. *N Engl J Med* 314, 1006–1010.

Bowden, R. A., Slichter, S. J., *et al.* (1995). A comparison of filtered leukocyte-reduced and cytomegalovirus (CMV) seronegative blood products for the prevention of transfusion-associated CMV infection after marrow transplant. *Blood* 86, 3598–3603.

Britt, W. J., and Alford, C. A. (1996). Cytomegalovirus. In "Fields Virology" (B. N. Fields, D. M. Knipe, P. M. Howley, *et al.*, Eds.), 3rd ed., pp. 2493–2523. Lippincott-Raven, Philadelphia, PA.

Brubaker, D. B. (1983). Human posttransfusion graft-versus-host disease. *Vox Sang* 45, 410–420.

Buckner, C. D., Clift, R. A., *et al.* (1978). ABO-incompatible marrow transplants. *Transplantation* 26, 233–238.

Busch, M. P., Lee, T. H., *et al.* (1992). Allogeneic leukocytes but not therapeutic blood elements induce reactivation and dissemination of latent human immunodeficiency virus type 1 infection: Implications for transfusion support of infected patients. *Blood* 80, 2128–2135.

Champlin, R. B., Horowitz, M. M., *et al.* (1989). Graft failure following bone marrow transplantation for severe aplastic anemia: Risk factors and treatment results. *Blood* 73, 606–613.

Decost, S. D., Boudreaux, C., *et al.* (1990). Transfusion-associated graft-vs-host disease in patients with malignancies: Report of two cases and review of the literature. *Arch Dermatol* 126, 1324–1329.

Dowling, J. N., Saslow, A. R., *et al.* (1976). Cytomegalovirus infection in patients receiving immunosuppressive therapy for rheumatic disorders. *J Infect Dis* 133, 399–408.

Greenbaum, B. H. (1991). Transfusion-associated graft-versus-host disease: Historical perspective, incidence and current use of irradiated blood products. *J Clin Oncol* 9, 1889–1902.

Hatley, R. M., Reynolds, M., *et al.* (1991). Graft-versus-host disease following ECMO. *J Pediatr Surg* 26, 317–319.

Hillyer, C. D., Lankford, K. V., *et al.* (1999). Transfusion of the HIV-seropositive patient: Immunomodulation, viral reactivation, and limiting exposure to EBV (HHV-4), CMV (HHV-5), and HHV-6, 7, and 8. *Transfusion Med Rev* 13, 1–17.

Ho, M. (1990). Epidemiology of cytomegalovirus infections. *Rev Infect Dis* 12, S701–S710.

Jamieson, N. V., Joysey, V., *et al.* (1991). Graft-versus-host disease in solid organ transplantation. *Transplant Int* 4, 67–71.

Jensen, L. S., Andersen, A. J., *et al.* (1992). Postoperative infection and natural killer cell function following blood transfusion in patients undergoing elective colorectal surgery. *Br J Surg* 79, 513–516.

Jensen, L. S., Grunnet, N., *et al.* (1995). Cost-effectiveness of blood transfusion and white cell reduction in elective colorectal surgery. *Transfusion* 35, 719–722.

Jensen, L. S., Kissmeyer-Nielsen, P., *et al.* (1996). Randomized comparison of leukocyte-depleted versus buffy-coat-poor blood transfusion and complications after colorectal surgery. *Lancet* 348, 841–845.

Juji, T., Takahashi, Y., *et al.* (1989). Post-transfusion graft-versus-host disease in immunocompetent patients after cardiac surgery in Japan. *N Engl J Med* 321, 56.

Klein, H. G. (1996). "Standards for Blood Banks and Transfusion Services." American Association of Blood Banks, Bethesda, MD.

Klumpp, T. R. (1991). Immunohematologic complications of bone marrow transplantation. *Bone Marrow Transplant* 8, 159–170.

Kreel, I., Zarroff, L. I., *et al.* (1960). A syndrome following total body perfusion. *Surg Gynecol Obstet* **111**, 317–321.

Lamberson, H. V. (1985). Cytomegalovirus (CMV). The agent, its pathogenesis and its epidemiology. In "Infection, Immunity and Blood Transfusion" (R. Y. Dodd, and L. F. Barber, Eds.), pp. 149-173. A. R. Liss, New York.

Linden, J. V., and Pisciotto, P. T. (1992). Transfusion-associated graft-versus-host disease and blood transfusion. *Transfusion Med Rev* **6**(2), 116–123.

Meyers, J. D., Floumoy, N., *et al.* (1986). Risk factors for cytomegalovirus infection after human marrow transplantation. *J Infect Dis* **153**, 478–488.

Miller, W. J., McCullough, J., *et al.* (1991). Prevention of cytomegalovirus infection following bone marrow transplantation: A randomized trial of blood product screening. *Bone Marrow Transplant* **7**, 227–234.

Mocarski, E. S. J. (1996). Cytolomegalovirus and their replication. In "Fields Virology" (B. N. Fields, *et al.*, Eds.), 3rd ed., pp. 2447-2492. Lippincott-Raven, Philadelphia, PA.

Moses, A., Nelson, J., *et al.* (1998). The influence of human immunodeficiency virus-1 on hematopoieses. *Blood* **91**, 1479–1495.

Narvios, A. B., Przepiorka, D., *et al.* (1998). Transfusion support using filtered unscreened blood products for cytomegalovirus-negative allogeneic marrow transplant recipients. *Bone Marrow Transplant* **22**, 575–577.

Petz, L. D. (1996). Bone marrow transplantation. In "Clinical Practice of Transfusion Medicine" (L. D. Petz, S. N. Swisher, S. Kleinman, R. K. Spence, and R. G. Strauss, Eds.), 3rd ed., pp. 757–782. Churchill Livingstone, New York.

Prince, A. M., Szmuness, W., *et al.* (1971). A serological study of cytomegalovirus infection associated with blood transfusion. *N Engl J Med* **284**, 1125–1131.

Przepiorka, D., LeParc, G. F., *et al.* (1996a). Use of irradiated blood components, practical parameter. *Am J Clin Pathol* **106**, 6–11.

Przepiorka, D., LePark, G. F., *et al.* (1996b). Prevention of transfusion-associated cytomegalovirus infection. Practice parameter. *Am J Clin Pathol* **106**, 163–169.

Sanders, M. R., and Graeber, J. E. (1990). Posttransfusion graft-versus-host disease in infancy. *J Pediatr* **117**, 159–163.

Sayers, M. H. (1996). Cytomegalovirus and other herpesviruses. In "Clinical Practice of Transfusion Medicine" (L. D. Petz, *et al.*, Eds.), 3rd ed., pp. 875-889. Churchill Livingstone, New York.

Sherman, A. A., and Ramsey, G. (1996). Solid-organ transplantation. In "Principles of Transfusion Medicine" (E. C. Rossi, T. L. Simon, G. S. Moss, and S. A. Gould, Eds.), 2nd ed., pp. 635-640. Williams & Wilkins, Baltimore.

Siegel, S. E., Lunde, M. N., *et al.* (1971). Transmission of toxoplasmosis by leukocyte transfusion. *Blood* **37**, 388–394.

Slichter, S. J. (1986). Prevention of platelet alloimmunization. In "Transfusion Medicine: Recent Technological Advances" (K. Murawski and F. Peetoom, Eds.), pp. 83–116. A. R. Liss, New York.

Slichter, S. J. (1988). Transfusion and bone marrow transplantation. *Transfusion Med Rev* **2**, 1.

Slichter, S. J. (1996). Transfusion support in bone marrow transplantation. In "Principles of Transfusion Medicine" (E. C. Rossi, *et al.*, Eds.), 2nd ed., pp. 521–535. Williams & Wilkins, Baltimore.

Sloand, E. M., and Groopman, J. E. (1997). Transfusion in HIV disease. *Curr Opin Hematol* **4**, 449–454.

Sloand, E., Kumar, P., *et al.* (1994). Transfusion of blood components to persons infected with human immunodeficiency virus type 1: Relationship to opportunistic infections. *Transfusion* **34**, 48–53.

Spector, S. A., Hirata, K. K., *et al.* (1984). Identification of multiple cytomegalovirus strains in homosexual men with acquired immunodeficiency syndrome. *J Infect Dis* **6**, 953–956.

Storb, R., Thomas, E. D., *et al.* (1980). Marrow transplantation in thirty "untransfused" patients with severe aplastic anemia. *Ann Intern Med* **92**, 30–36.

Suassuna, J. H., Machado, R. D., *et al.* (1993). Active cytomegalovirus infection in hemodialysis patients receiving donor-specific blood transfusions under azathioprine coverage. *Transplantation* **56**, 1552–1554.

Von Fliedner, V., Higby, D. J., *et al.* (1982). Graft-versus-host reaction following blood transfusion. *Am J Med* **72**, 951–959.

Williamson, L. M. (1998a). Transfusion associated graft versus host disease and its prevention. *Heart* **80**, 211–212.

Williamson, L. M. (1998b). Transfusion-associated graft-versus-host disease: New insights and a route towards therapy? *Transfusion Med* **8**, 169–172.

Yeager, A. S. (1974). Transfusion-acquired cytomegalovirus infection in newborn infants. *Am J Dis Child* **128**, 478–483.

Zulian, G. B., Roux, E., *et al.* (1995). Transfusion-associated graft-versus-host disease in a patient treated with Cladribine (2-chlorodeoxyadenosine): Demonstration of exogenous DNA in various tissue extracts by PCR analysis. *Br J Haematol* **89**, 83–89.

TRANSFUSION REACTIONS

Acute and Delayed Hemolytic Transfusion Reactions

DEBORAH A. SESOK-PIZZINI

University of Pennsylvania Medical Center
Philadelphia, Pennsylvania

Acute and delayed hemolytic transfusion reactions result from immune-mediated destruction of transfused incompatible packed red blood cells (PRBCs). Acute hemolytic transfusion reactions, while relatively rare, can be devastating. The initial symptoms of fever and chills are seen with both acute and delayed hemolytic (as well as in febrile nonhemolytic) transfusion reactions, transfusion-related acute lung injury (TRALI), and transfusion of bacterially contaminated blood. Early evaluation of any suspected reaction is necessary to determine the cause and ensure appropriate management (Sazama, 1990). Delayed hemolytic transfusion reactions are generally less severe and associated primarily with extravascular hemolysis. The pathophysiology, signs, symptoms, and steps for evaluation and management of these reactions are presented here. Nonimmune hemolysis of transfused cells is discussed in Chapter 31.

I. DESCRIPTION

A. Acute Hemolytic Transfusion Reactions (AHTRs)

Acute hemolytic transfusion reactions, by definition, occur within 24 hours of transfusion. Administration of ABO-incompatible blood accounts for most AHTRs. Incompatibility is usually due to clerical errors, which are most often due to misidentification of patients or specimens, or to system errors in the release or administration of products (Sazama, 1990). The acute hemolysis following the transfusion of ABO-incompatible blood results in a rapid intravascular hemolysis. Complement activation and intravascular hemolysis may

also occur with antibodies against non-ABO antigens. Alloantibodies causing hemolysis directed against these other blood group antigens often result in a less severe extravascular hemolysis.

B. Delayed Hemolytic Transfusion Reactions (DHTRs)

Delayed hemolytic transfusion reactions are caused by accelerated destruction of transfused PRBCs by a recipient's immune response directed against an antigen on the donor RBCs. There are two types of DHTRs (Vamvakas, 1995). The first type of reaction occurs several weeks following transfusion and is caused by a new antibody generated in the recipient (a primary response). The second type of DHTR occurs within hours or days after transfusion of a previously immunized patient and results in extravascular hemolysis due to a secondary antibody (anamnestic) response to an antigen on donor RBCs. DHTRs are usually mild reactions, although serious complications may occasionally be observed.

II. INCIDENCE

A. Acute HTR

ABO-incompatible transfusion occurs in approximately 1:6000 to 1:33,000 transfused PRBC units. The estimated fatalities due to acute hemolytic reactions range from 1:500,000 to 1:800,000 per year (DeChristopher and Anderson, 1997).

B. Delayed HTR

DHTRs occur in approximately 1:4000 transfused PRBC units (DeChristopher and Anderson, 1997). The distinction between DHTRs and delayed serologic transfusion reactions (DSTRs) is made on the basis of detectable hemolysis (i.e., DSTRs show only serologic signs of the development of an RBC antigen alloantibody, without the presence of clinically evident symptoms of hemolysis). One study reported the incidence of DHTR or DSTR as 1:1899 transfused PRBC units, with a DHTR:DSTR ratio of 36:64 (Vamvakas et al., 1995).

III. PATHOPHYSIOLOGY

A. Acute HTRs

The volume of incompatible donor RBCs in circulation in the recipient will often determine the severity of the reaction. Other factors such the potency of the anti-A or anti-B in the recipient's plasma will also influence the outcome following an incompatible PRBC transfusion (Mollison, 1993). In general, immediate HTRs from transfusion of incompatible PRBCs are due to the binding of naturally occurring IgM antibody (anti-A or anti-B) to donor RBCs. Alternatively, other blood group antibodies, either IgM or complement-fixing IgG, may also be involved in acute HTRs. Hemolytic antibodies, in addition to anti-A and anti-B, include anti-PP_1, P^k, anti-Vel, and occasional Lewis antibodies. Kidd antibodies can also lyse RBCs and are often of the complement-fixing IgG class (e.g., IgG3, IgG1; Mollison, 1999).

In addition to RBC lysis, the antibody–RBC complex activates a series of systemic pathways that may lead to serious sequalae. These pathways include neuroendocrine response, complement activation, coagulation activation, and cytokine effects (Anstall, 1996).

1. Neuroendocrine Response

Hageman factor is activated in response to the antigen–antibody and complement complex present on the donor RBCs. The kinin system generates bradykinin in response to activated Hageman factor (factor XIIa). The release of bradykinin increases capillary permeability and arteriolar dilation, resulting in a fall in systemic blood pressure. A sympathetic nervous system response from hypotension causes a rise in norepinephrine and other catecholamines. These factors in turn produce vasoconstriction in renal, splanchnic, pulmonary, and cutaneous capillaries.

2. Complement Activation

The interaction of antibody and antigen on the RBC surface activates the complement cascade. Activation depends on factors such as the type of immunoglobulin class or IgG subclass, the thermal range of the antibody, antigenic density and distribution, and the antibody titer. With most non-ABO hemolytic reactions, complement activation is incomplete and hemolysis is primarily extravascular.

3. Coagulation Activation

Activated Hageman factor and circulating RBC stroma initiate both the intrinsic and the extrinsic coagulation pathways. Activation of the coagulation cascade leads to generation of thrombin and often to some degree of disseminated intravascular coagulation (DIC). The degree of DIC produced is dependent on the volume of immune complexes bound to the RBCs following complement activation. This in turn is dependent on the volume of incompatible RBCs infused.

4. Cytokine Effects

Cytokines are released from white blood cells (WBCs) in response to RBC antibody–antigen complex formation. Accumulating evidence supports a central role for cytokines in the pathophysiology of acute HTRs. Under the influence of IL-1 and TNF cytokines, the endothelial cells alter the expression of surface adhesion molecules to favor thrombosis (Davenport, 1995).

B. Delayed HTRs

DHTRs are generally caused by a secondary (anamnestic) immune response to a repeat exposure to an RBC antigen. The first exposure may have been due to a prior transfusion or pregnancy. Antibodies in the Kidd system (anti-Jk^a and anti-Jk^b) and Rh system (anti-E and anti-C) often fall to undetectable serum antibody levels weeks to months after the primary sensitization (Ramsey and Larson, 1988). These antibodies will not be detectable by subsequent antibody screen or crossmatch. Following reexposure to the antigen through additional transfusions, the immune system may quickly regenerate large amounts of IgG antibody. The IgG antibody titer increases over 5–14 days and coats the donor RBCs. This results in a gradual decrease in hematocrit, as the donor RBCs are removed by macrophages in the extravascular space. More rarely, an overt intravascular hemolysis occurs when the antibody titer (Kell, Kidd, and Duffy systems) is sufficiently high and complement is fully activated. The difference between intravascular and extravascular phagocytosis of antibody and/or complement-labeled RBCs depends on the degree of activation of the complement system. If complement is not fully activated, then C3 convertase is formed, and the RBCs coated with antibody and complement are removed by the macrophage/reticuloendothelial (RE) system (Mollison, 1993).

The immune response to an antigen may also be primary, either producing a DHTR or a DSTR several weeks following the transfusion. The probability of forming an alloantibody after the transfusion of 1 unit of crossmatch-compatible PRBCs is less than 1% (Ness et al., 1990). However, should the antibody form, there is a possibility that hemolysis may ensue.

IV. DIAGNOSIS

A. Acute HTRs

1. Clinical Evaluation at the Bedside

The most common initial clinical sign of an acute HTR is fever (temperature rise $> 1°C$; Pineda et al., 1978). The fever may be accompanied by chills. Other potential signs and symptoms of an acute HTR are listed in Table 1, all resulting from acute intravascular hemolysis. A patient may report burning or itching along the course of the vein as well as lower back pain. In the anesthetized patient, generalized oozing from surgical or intravenous lines, hemoglobinuria and unexplained hypotension may indicate an acute HTR.

TABLE 1   Signs and Symptoms of an Acute Hemolytic Transfusion Reaction

Cardiac	Renal	Hematological	Generalized
Chest pain	Hemoglobinuria	Anemia	Fever, chills
Hypotension	Oliguria	Generalized oozing	Flushing
Hypertension	Anuria	Uncontrolled bleeding	Nausea or vomiting
Tachycardia		DIC	Dyspnea
			Pain at infusion site
			Uriticaria
			Back, flank, or abdominal pain

In a patient who shows clinical signs and symptoms suggestive of an acute HTR, the transfusion should immediately be stopped (see Table 2). The initial evaluation should focus on recording vital signs (blood pressure, temperature, pulse), patient complaints, and urine output. The intravenous line should remain open with a slow infusion of normal saline. The unit label and the patient identification data should be confirmed and the following specimens should be collected and immediately sent to the laboratory for evaluation (Table 3): (1) an EDTA (purple-topped tube) for a direct antiglobulin test (DAT), (2) a red-topped tube for ABO and Rh confirmation, repeat antibody screening, and crossmatch if needed, and visual inspection for hemolysis and icterus, and (3) first voided urine after the event for detection of hemoglobinuria. The blood unit bag should also be returned to the laboratory for inspection and possible bacterial culture (AABB, 1999b). If an ABO-incompatible transfusion is suspected, basic coagulation parameters (PT, PTT, FDPs) should also be drawn (a blue-topped tube).

2. Laboratory Findings in Acute HTR

The immediate laboratory evaluation steps for a suspected acute HTR are as follows: (1) repeat clerical check, (2) visual inspection for hemolysis or icterus, and (3) direct antiglobulin test on postreaction sample

TABLE 2   Treatment of Acute Hemolytic Transfusion Reactions

First, STOP THE TRANSFUSION

Hypotension

1. Normal saline (simple hypotension)
2. Dopamine, 1 g/kg body weight per hour (contraindicated in patients with significant volume deficits)

Renal failure

1. Monitor urine output, BUN, creatinine, electrolytes
2. Treat hypotension to help prevent renal ischemia
3. Maintain urine output of 100 ml/hr for 24 hours
4. Furosemide (20–100 mg over 12 hours) if needed
5. Dialysis to manage fluid and electrolyte imbalances (hyperkalemia)—consultation with nephrology

DIC

1. Monitor coagulation parameters (PT, PTT, fribinogen, FDPs, platelet count)
2. Early use of heparin if not contraindicated at a loading dose of 40 IU/kg body weight, followed by 5000–6000 IU over 10–12 hours for an average 70-kg adult
3. Blood component transfusion for patients with active bleeding (fresh frozen plasma, platelets as indicated)

TABLE 3   Approach to Suspected Acute Transfusion Reactions

1. **Stop the transfusion** and keep the IV line open

2. Monitor vital signs, urine output, patient complaints

3. Evaluate for evidence of a severe reaction (e.g., acute hemolytic transfusion reaction, septic shock, anaphylaxis)

4. Check patient's identification and medical record number against component labels to ensure correct unit was transfused

5. Laboratory testing

    a. Clerical check

    b. Direct antiglobulin test (EDTA—purple-topped tube)

    c. Visual inspection for hemolysis or icterus, and repeat ABO/Rh, antibody screen, crossmatch (clotted—red-topped tube)

    d. Coagulation studies as needed (citrate—blue-topped tube)

    e. Urinalysis for hemoglobinuria

6. Review results of blood bank workup and evaluation prior to subsequent transfusions

7. Continue to monitor patient and provide appropriate medical support

---

TABLE 4   Tests Performed in Blood Bank to Evaluate Suspected DHTRs

1. Antibody screen (indirect antiglobulin test [IAT] or indirect Coombs' test) to detect any new RBC alloantibodies

2. Direct antiglobulin test (DAT or direct Coombs' test) to detect IgG or complement on patient's RBCs

3. IAT on eluate containing any alloantibodies coating patient's RBCs to determine specificity

4. Antigen phenotyping of patient's RBCs and recently transfused PRBCs to compare to alloantibody specificity in serum and eluate

---

(AABB, 1999b). If any of these tests are abnormal, and a patient has clinical signs and/or symptoms suspicious of acute HTR, additional steps include a repeat ABO and Rh typing of patient and unit and a repeat antibody screen and crossmatch (AABB, 1999a). Laboratory findings in acute HTRs include hemoglobinemia, hemoglobinuria, hyperbilirubinemia, low haptoglobin, and elevated lactate dehydrogenase (LDH). The DAT may be positive for IgG, C3, or both. The DAT may rarely be negative if all the transfused RBCs are lysed following transfusion of a few milliliters of incompatible blood. The antibody screen (IAT) may or may not be positive, depending on the amount and specificity of antibody in the serum. The ABO and Rh testing may show a mixed field, suggesting two RBC populations are present.

B. Delayed HTRs

1. Evaluation of a Suspected DHTR

The most common characteristics of DHTRs are fever, anemia, and mild jaundice resulting from extravascular hemolysis. These symptoms may be observed anywhere from 3 to 10 days following transfusion. Sometimes a DHTR will present as a failure to adequately respond to PRBC transfusions. In rare cases, features of overt intravascular hemolysis, such as transient hemoglobinemia and hemoglobinuria, may be observed (Vamvakas et al., 1995). Unlike acute HTRs, acute renal failure is uncommon. Signs or symptoms of DHTR should be followed by serologic testing in the blood bank for further evaluation (Table 4).

2. Laboratory Findngs in DHTR

Since DHTRs are frequently subclinical, the diagnosis may first be suspected because of a newly positive direct antiglobulin test. The antibody may be present on donor RBCs even when serum antibody is not detectable (i.e., negative antibody screen). A newly positive DAT following recent transfusion warrants an investigative workup to determine the specificity of the antibody coating the RBCs (AABB, 1999b). It may require hours or days for the antibody titer to be sufficiently high to be detectable in the patient's serum. The antibody may be eluted from the transfused RBCs to provide evidence for the presence of a DHTR or a DSTR. If a new alloantibody is identified coating the transfused RBCs (e.g., DSTR), the patient should be monitored for signs and symptoms of a DHTR. Once the antibody is detected in the recipient's serum, crossmatches will be incompatible if the donor PRBCs have the corresponding antigen. Thus, PRBC units negative for the corresponding antigen will be selected for future transfusions.

V. CLINICAL COURSE AND SEQUELAE

A. Acute HTRs

The most severe complications of acute HTRs are hypotension, renal failure, and DIC. Hypertension may also be observed, but much less frequently than hypotension. Hypotension may be severe and even progress to shock. Renal failure is due to hypotension, reactive vasoconstriction, and thrombin occlusions in renal microvasculature, leading to patchy ischemia of segments of renal tubules. The majority of patients will recover normal renal function over several days to 3 weeks. Some patients may require hemodialysis. Activation of the fibrinolytic system and coagulation cascade may result in varying degrees of DIC with microthombi and ischemic damage to organs and tissues.

There may be consumption of fibrinogen, platelets, factors V and VIII, and generation of fibrin degradation products (FDPs). In these cases, hemorrhage, hemoglobinuria, and extensive blood loss may be observed.

In general, the mortality associated with acute HTRs depends on the volume of incompatible PRBCs transfused. However, a fatal outcome may result from the transfusion of even small amounts of incompatible blood. Aggressive treatment of hypotension, DIC, and renal failure will help prevent organ failure and death in patients experiencing acute HTRs.

B. Delayed HTRs

Treatment of DHTRs is rarely necessary because of the mild nature of the reactions. However, the patient's hemoglobin levels should be monitored, and if hemolysis is significant, PRBC transfusions may be required. The PRBC units chosen for transfusion should lack the antigen corresponding to the new alloantibody. Additionally, monitoring the patient's urine output, BUN, and creatinine and examining coagulation parameters will help detect signs of rare complications, including DIC and renal failure.

Although DHTRs are considered mild and DSTRs are asymptomatic, there have been fatalities reported following delayed hemolysis. In 50 cases of DHTRs reported from the Mayo Clinic, deaths occurred from alloantibodies to anti-Jk[a], anti-D, and anti-K (Pineda et al., 1978). Fatalities caused by DHTRs due to anti-Fy[a], anti-Jk[a], anti-E, and anti-C have also been reported (Brecher, 1996). The FDA cited 26 fatalities from DHTRs during the years 1976–1985 (Sazama, 1990).

VI. MANAGEMENT

A. Acute HTRs

Therapy should focus on management of hypotension or shock, prevention or therapy of DIC, and maintenance of renal perfusion. Infusion of normal saline may be sufficient to manage simple hypotension, but more severe hypotension will require vasopressor therapy. Dopamine, at an infusion rate of 1 g/kg of body weight per hour, is the agent of choice because it produces selective vasodilatation to the kidneys. This dosage should maintain a systolic pressure of 100 mm Hg or higher (Anstall, 1996). Dopamine is contraindicated in patients with a significant volume deficit.

Early use of heparin is a consideration after an ABO-incompatible transfusion if no other contraindi-

cation exists and the baseline coagulation parameters are only marginally elevated. In adults, an intravenous heparin loading dose of 40 IU/kg body weight, followed by 5000–6000 IU over 10–12 hours, is an appropriate initial therapy (Anstall, 1996). Heparin therapy may be problematic for intraoperative and postoperative patients due to risk of hemorrhage.

Treatment of hypotension will help prevent renal ischemia that, if untreated, might lead to renal failure. A urine output of 100 ml/hr for 24 hours should be maintained in an adult. Many favor the use of furosemide in low-to-moderate doses (20–100 mg over 12 hours) over mannitol (20% solution up to a total dose of 100 g over 24 hours) for diuresis (Goldfinger, 1977). Mannitol is not recommended as a primary therapy because it increases urine output without correcting the underlying ischemia. Furosemide will both promote diuresis and improve renal cortical perfusion. After massive hemolysis, the kidney's ability to excrete high concentrations of potassium may also be impaired. In this situation, acute dialysis is necessary to prevent life-threatening cardiac arrhythmias. If acute tubular necrosis (ATN) occurs, proper fluid and electrolyte balance must be maintained. Dialysis may also help correct these fluid and electrolyte imbalances. Consultation with a nephrologist is advised (see Table 3).

B. Delayed HTRs

If a DHTR is suspected due to decline in hemoglobin following PRBC transfusion, then serologic testing should be pursued (see Table 4). The patient should also be monitored for symptoms of anemia and signs of hemolysis (i.e., an increase in LDH and indirect bilirubin, and a decrease in haptoglobin). If laboratory tests confirm a DSTR, the primary care team is alerted to monitor and evaluate for DHTR. Confirmation of DHTR may help avoid unnecessary diagnostic testing for other causes of declining hemoglobin.

VII. PREVENTION

A. Acute HTRs

The primary prevention of acute HTRs requires elimination of clerical errors in the process of blood administration (e.g., mislabeled blood samples, transfusion of blood to wrong patient). Therefore, identifying the specific step in which an error occurred may help prevent future errors and accidents (Linden, 1999). Self-assessment practices help to identify failures with procedure compliance. For example, the AABB Quality Program provides detailed guidelines for establishing a total quality management program designed for continued quality improvement. Process control of operations through validated standard operating procedures, standardized training, and assurance of reliable functioning equipment and reagents are all technical quality control points built into the blood bank (Galel and Richards, 1997). Extension of process control to include pretransfusion collection and blood administration will help reduce the incidence of errors and accidents.

B. Delayed HTRs

The prevention of DHTRs involves the initial selection of donor PRBC units that lack the corresponding alloantibody in the patient's serum. Routine laboratory testing of the patient's serum will detect these alloantibodies if the titer is sufficiently high. Some blood banks issue written notification of alloantibodies to be used by clinicians and patients to guide future transfusions. Also, AABB Standards mandate the review of permanent records of a patient's clinically significant alloantibodies prior to issuance of PRBCs for transfusion (AABB, 1999a). The antibodies that are historically present in the patient's serum are noted, and antigen-negative PRBCs are selected for transfusion. The rationale for obtaining a new patient sample every 72 hours for hospitalized patients is for detection of potentially newly formed alloantibodies. Each of these practices helps to prevent DHTRs. In addition, a patient's RBC phenotype may be determined and a donor with an antigen profile similar to the patient may be selected for transfusion. This strategy, which limits exposure to foreign antigens and prevents formation of alloantibodies, is used for patients receiving multiple transfusions.

References

AABB (1999a). "Standards for Blood Banks and Transfusion Services," 19th ed. AABB, Bethesda, MD.

AABB (1999b). "Technical Manual," 13th ed. AABB, Bethesda, MD.

Anstall, H. B., and Blaylock, R. C. (1996). Adverse reactions to transfusion—Part I. In "Practical Aspects of the Transfusion Service," pp. 201–228. ASCP Press, Chicago.

Brecher, M. E. (1996). Hemolytic transfusion reactions. In "Principles of Transfusion Medicine" (E. C. Rossi, T. L. Simon, G. S. Moss, et al., Eds.). Williams and Wilkins, Baltimore, MD.

Davenport, R. D. (1995). The role of cytokines in hemolytic transfusion reactions. Immunol Invest 24, 319–331.

DeChristopher, P. J., and Anderson, R. R. (1997). Risks of transfusion and organ and transplantation: Practical concerns that drive practical policies. AJCP 107, S2–S11.

Galel, S. A., and Richards, C. A. (1997). Practical approaches to improve laboratory performance and transfusion safety. AJCP 107, S43–S49.

Goldfinger, D. (1977). Acute hemolytic reactions—A fresh look at pathogenesis and considerations regarding therapy. Transfusion 17, 85–98.

Linden, J. V. (1999). Errors in transfusion medicine. Arch Pathol Lab Med 123, 563–565.

Mullison, P. L., Engelfreit, C. P., and Contreras, M. (1993). Blood Transfusion in Clinical Medicine. Blackwell Scientific Publication, London.

Ness, P. M., Shirey, R. S., and Buck, SA. (1990). The differentiation of delayed serologic and delayed hemolytic transfusion reactions: Incidence, long-term serologic findings, and clinical significance. Transfusion 30, 688–693.

Pineda, A. A, Brzica, S. M., Jr., and Taswell, H. F. (1978). Transfusion reaction. An immunologic hazard of blood transfusion. Transfusion 18, 1–7.

Ramsey, G., and Larson, P. (1988). Loss of red cell alloantibodies over time. Transfusion 28, 162–165.

Sazama, K. (1990). Report of 355 transfusion-associated deaths 1976–1985. Transfusion 30, 583–590.

Vamvakas, E. C., Pineda, A. A., Reisner, R., et al. (1995). The differentiation of delayed hemolytic and delayed serologic transfusion reactions: Incidence and predictors of hemolysis. Transfusion 35, 26–32.

Febrile Nonhemolytic Transfusion Reactions

ANNE F. EDER

Department of Pathology and Clinical Laboratories
The Children's Hospital of Philadelphia
Philadelphia, Pennsylvania

Febrile nonhemolytic transfusion reactions (FNH-TRs) are suspected in recipients of blood components who experience an increase in temperature of 1°C or more in the absence of other identifiable causes for the fever. The reaction may occur either during or 1–2 hours following the transfusion. Associated symptoms may include chills or rigors and rarely nausea and vomiting. The reaction is self-limited, but may be extremely uncomfortable for the patient. The clinical importance of FNHTRs lies in distinguishing them from more serious, even life-threatening, transfusion reactions. Because fever is a nonspecific sign of several different types of transfusion reactions, a diagnosis of FNHTR must not be made until other causes, including acute hemolytic transfusion reactions (HTRs) and transfusion of bacterially contaminated units, are excluded (Table 1). In a hospital setting, differentiation of a FNHTR from other possible causes of transient temperature increases such as concurrent infection or drug reaction is often difficult.

I. INCIDENCE

FNHTRs are the most commonly encountered transfusion reactions. The incidence of FNHTRs varies depending on the blood component transfused and its preparation method, as well as on the clinical history of the recipient. Overall, FNHTRs occur in approximately 0.5–2.0% of units transfused (DeChristopher and Anderson, 1997; Miller and AuBuchon, 1999). FNHTRs are more likely to occur following transfusion of platelets than packed red blood cells (PRBCs), and they are more common following transfusion of pools of random donor platelet units than single-donor apheresis platelet units (Miller and AuBuchon, 1999). FNHTRs are also more frequently observed in individuals with a history of repeated transfusions and in women with a history of multiple pregnancies.

II. PATHOPHYSIOLOGY

The sine qua non of FNHTRs is fever, which is ultimately caused by the action of pyrogenic cytokines (e.g., IL-1, IL-6, TNF) on the thermoregulatory center of the anterior hypothalamus. The following mechanisms have been proposed to account for the generation of cytokines in association with blood component transfusion (Miller and AuBuchon, 1999; Stack *et al.*, 1996):

1. Leukocytes from the donor unit may be stimulated by anti-HLA or other antibodies in the recipient to produce cytokines following transfusion into the recipient.
2. Leukocytes (or endothelial cells) in the recipient of the transfusion may be stimulated directly or indirectly by transfused foreign cells or plasma constituents to produce cytokines.
3. Leukocytes in the donor unit may produce and release cytokines into the blood bag during blood component storage that are subsequently transfused into the recipient.

These models are not mutually exclusive, rather all have been supported by various observations and experimental results.

TABLE 1   Transfusion Reactions: Differential Diagnosis of Fever and Chills

- Febrile nonhemolytic transfusion reaction (FNHTR)
- Acute hemolytic transfusion reaction
- Delayed hemolytic transfusion reaction
- Transfusion-related acute lung injury (TRALI)
- Bacterial contamination
- Unrelated to transfusion (i.e., drug reaction, concurrent infection)

A. Activation of Donor Leukocytes by Preformed Antibodies in the Recipient

Buttressing the first model is the observation that previous exposure to blood components, either through transfusion or pregnancy, increases the risk of FN-HTRs. This observation implicates sensitization to foreign antigens and development of antibodies that may react against transfused leukocytes, resulting in the release of pyrogenic cytokines. Approximately 10% of FNHTRs are associated with the presence of anti-HLA antibodies in the recipient (Miller and AuBuchon, 1999). The risk of FNHTRs is correlated with the number of leukocytes in transfused units. A minimum threshold of approximately 5×10^8 leukocytes per unit of PRBCs must be present for FNHTRs to occur in most recipients (Miller and AuBuchon, *et al.* 1999).

This model fails to account for several observations in sensitized individuals. Antibodies implicated in FN-HTRs may have specificity for antigens other than HLA and may be directed against granulocyte or platelet antigens. In these cases, the recipient's antibody is directed against a cell that cannot produce pyrogenic cytokines. In addition, leukocyte reduction does not prevent all FNHTRs, suggesting that transfusion of a critical number of leukocytes is not necessary for some febrile reactions to occur.

B. Activation of Recipient Leukocytes by Transfused Substances

The second model posits the activation of recipient's leukocytes to release cytokines directly or indirectly by cells or plasma constituents in the donor unit. Antigen–antibody complexes formed by reaction of the recipient's antibodies with their target blood component antigens, whether platelet, granulocyte, or plasma proteins, may cause complement activation. Production of C5a may then stimulate the release of pyrogenic cytokines from recipient monocytes.

This mechanism fails to account for FNHTRs occurring in individuals who have never been transfused or pregnant, because without prior sensitization, antibodies directed against HLA, granulocyte, or platelet antigens are not produced by the recipient. An alternative mechanism described by this model could account for this observation, in that antibodies in the plasma of the transfused unit that are directed against HLA or granulocyte antigens (also referred to as leukoagglutinins) may directly or indirectly stimulate cytokine production by recipient monocytes. The majority of FNHTRs, however, are not associated with the presence of antibodies in either the recipient or the donor unit.

C. Cytokines in the Donor Unit

The third pathogenic model for FNHTRs, production and accumulation of pyrogenic cytokines during component storage, may account for a significant proportion of FNHTRs. Increased concentrations of cytokines in blood components are correlated with the initial leukocyte count, the length of storage, and the storage temperature. FNHTRs are more common after the transfusion of platelets than after the transfusion of RBCs, even though the latter may contain more leukocytes (Miller and AuBuchon, 1999). Cytokine production by leukocytes is favored in platelet units because they are stored at room temperature (20–24°C), whereas RBCs are stored at 4°C, and cellular metabolism, necessary for active synthesis of cytokines, is inhibited at the lower storage temperature of RBCs. Moreover, the risk of FNHTRs increases as the length of storage increases and corresponds to the exponential increase in cytokine concentration observed with storage time.

II. DIAGNOSIS

FNHTR, a diagnosis of exclusion, cannot be made until other potentially life-threatening transfusion reactions, such as acute HTRs, bacterial contamination of the unit, or other medical causes of fever (such as a preexisting infection or a drug reaction), are eliminated as possibilities (Table 1).

A. Clinical Evaluation

Clinical evaluation to determine the cause of a transfusion reaction should ensue after the initial steps have been taken to manage a suspected acute transfusion reaction (see Section V, Management). An account of the events surrounding the transfusion should be obtained from the patient or transfusionist. The onset, duration, and severity of symptoms surrounding the transfusion should be documented. The clinical events preceding the transfusion should be investigated, such as the administration of medications or other medical interventions. The patient's clinical history, including the history of the present illness, temperatures recorded during the current hospitalization, concurrent medical conditions, and indication for transfusion, should be reviewed. The patient's transfusion history, including any previous adverse reactions to blood components, should be obtained. Symptoms suggestive of other types of transfusion reactions (i.e., back pain and hemoglobinuria for acute HTRs and hives and urticaria for allergic reactions) should be documented. The patient's vital signs before the transfusion started should be compared to those obtained at the time the transfusion was discontinued, and should be monitored until the fever and other symptoms resolve.

B. Laboratory Evaluation

Laboratory evaluation of suspected transfusion reactions to blood components should include the following steps (Vengelen-Tyler, 1999): (1) check the accuracy of the clerical transfusion record; (2) check for hemolysis by visually inspecting the pre- and post-transfusion serum samples for discoloration and examining a post-transfusion urine specimen for free hemoglobin; and (3) compare the pretransfusion and the post-transfusion direct antiglobulin (Coombs') test, also known as the DAT. If results from any one of these tests is positive or suspicious, or if the clinical presentation is strongly suggestive of a serious transfusion reaction, additional laboratory evaluation is required to rule out the possibility of an acute HTR or bacterial contamination of the transfused unit (see Chapters 28 and 35). If the initial laboratory evaluation (see 1–3 above) is negative and the patient's presentation and clinical course are consistent with a FNHTR, additional laboratory tests are not required.

IV. CLINICAL COURSE AND SEQUELAE

FNHTRs are typically benign, and usually resolve completely without sequelae. The fever and other symptoms abate within 1–2 hours after the transfusion in most cases. Persistent fever lasting more than 18–24 hours is likely to be unrelated to the transfusion (Stack *et al.*, 1996).

TABLE 2   Approach to Suspected Acute
Transfusion Reactions

1. Stop the transfusion and keep the IV line open
2. Monitor symptoms, vital signs, urine output
3. Evaluate for evidence of a severe reaction (e.g., acute hemolytic transfusion reaction, septic shock, anaphylaxis)
4. Check patient's identification and medical record number against component labels to ensure correct unit was transfused
5. Return product to blood bank along with blood samples (6.b and 6.c, below) and completed transfusion record indicating vital signs, symptoms
6. Laboratory testing
   a. Clerical check
   b. Direct Antiglobulin Test (EDTA—purple-topped tube)
   c. Visual inspection for hemolysis or icterus. Repeat ABO/Rh, antibody screen, crossmatch, if indicated. (Clotted—red-topped tube)
   d. Coagulation studies as needed (Citrate—blue-topped tube)
   e. Urinalysis for hemoglobinuria
7. Review results of blood bank workup and evaluation prior to subsequent transfusions
8. Continue to monitor patient and provide appropriate medical support

V. MANAGEMENT

A. General Management of Suspected Transfusion Reactions

Because the clinical presentation of fever and chills may represent either a life-threatening HTR or a benign FNHTR, the first step in the management of adverse transfusion reactions is to discontinue the transfusion (see Table 2; Vengelen-Tyler, 1999). The intravenous line should be kept open with normal saline. The patient's identification and component labels should be checked to ensure the transfused component was intended for the recipient. The bag of blood and post-transfusion blood and urine samples from the patient should be sent to the blood bank (or other appropriate laboratory) for the tests and investigation previously described. The remaining component should not be transfused, even after the transfusion investigation is complete. Although immune-mediated hemolysis may be ruled out within an hour, investigation of possible bacterial contamination of the unit requires a Gram stain and culture, and cannot be ruled out quickly enough to allow for resuming transfusion of the same unit. Moreover, no routine laboratory tests are currently available to identify plasma constituents such as cytokines, leukoagglutinins, or other biological mediators that may be responsible for the reaction. If transfusion is still required following resolution of the reaction, another unit should be issued.

B. Management of Symptoms

The signs and symptoms of FNHTRs are generally self-limited; however, the patient may experience significant anxiety and discomfort. Antipyretic medication may be used to treat fever. For most adult patients, acetaminophen (325–650 mg orally) is preferred over aspirin because it does not affect platelet function. Shaking chills or rigors may be treated with meperidine (25–50 mg intravenously; Davenport, 1999). Antihistamines are not indicated unless the patient also shows signs or symptoms of an allergic reaction.

VI. PREVENTION

A. Premedication

FNHTRs may be prevented or ameliorated by premedicating the patient with antipyretics (e.g., acetaminophen). The major disadvantage of premedicating patients receiving transfusions is that antipyretics may mask immune-mediated hemolysis or bacterial contamination, as fever may be the first recognizable sign associated with these other transfusion reactions as well. Consequently, routine use of antipyretics prior to transfusion is not appropriate. Premedication is recommended for individuals who have: (1) experienced repeated FNHTRs or (2) experienced FNHTRs after receiving leukocyte-reduced blood components.

B. Leukocyte-Reduced Blood Components

In many cases, FNHTRs may be prevented by transfusing leukocyte-reduced blood components. The American Association of Blood Banks and the Food and Drug Administration define a "leukocyte reduced" component as one prepared by a method that reduces the residual leukocyte number in the final product to less than 5×10^6 (Standards Committee, 2000). In addition to preventing FNHTRs, this extent of leukocyte reduction may also reduce the incidence of HLA alloimmunization, and transmission of cytomegalovirus (CMV) and other cell-associated pathogens. Most methods in current practice surpass this degree of leukocyte reduction. A comprehensive discussion of leukocyte reduced products is found in Chapter 14.

1. Apheresis Collection Methods

Filters may be used to reduce the leukocyte content of pools of random-donor platelet units; alternatively, the leukocyte content of single-donor platelets may be minimized by special apheresis collection methods (Miller and AuBuchon, 1999). Certain collection instruments have the ability to yield apheresis units of platelets with fewer than 1×10^6 leukocytes.

2. Filtration: Prestorage versus Poststorage Leukoreduction

High-efficiency leukoreduction filters are capable of reducing the leukocyte count of blood components from 2×10^9 to less than 1×10^6. Filtration may be performed at the bedside during transfuson of the unit (poststorage leukoreduction) or during or shortly after component collection (prestorage leukoreduction). Poststorage leukocyte reduction prevents most FN-HTRs associated with PRBC transfusion (Miller and AuBuchon, 1999; Stack et al., 1996). The effectiveness of poststorage leukocyte reduction in preventing FN-HTRs associated with platelet transfusion is more controversial (Miller and AuBuchon, 1999; Stack et al., 1996).

Poststorage leukocyte reduction is only effective for reducing FNHTRs due to leukocyte alloimmunization and is ineffective in preventing FNHTRs associated with storage-generated cytokines. Prestorage leukoreduction offers the dual advantage of preventing both FNHTRs caused by leukocyte alloimmunization and FNHTRs caused by the accumulation of soluble mediators during storage. Federowicz et al. (1996) demonstrated FNHTRs were less common after transfusion of PRBC units that were leukocyte reduced before storage than after the transfusion of those units leukocyte reduced after storage at the bedside by filtration. The same advantage may be observed for prestorage leukocyte-reduced platelet units (Muir et al., 1995).

Unfortunately, recurrence of FNHTRs with subsequent transfusions cannot be predicted, and only about 10–15% of patients who experience a FNHTR will have a similar reaction to the next transfusion (Stack et al., 1996). An individual who experiences a FNHTR may never experience one again. Prestorage leukoreduction or poststorage leukoreduction adds significantly to the cost of the transfusion. Leukocyte-reduced components generally are not recommended until after the patient experiences at least two FN-HTRs. Prestorage leukocyte reduction, however, is becoming more common in blood banking practice. In the near future, all institutions may exclusively maintain prestorage leukocyte-reduced inventory (universal leukoreduction; see Chapter 14).

References

Davenport, R. D. (1999). Management of transfusion reactions. In "Transfusion Therapy: Clinical Principles and Practice" (P. D. Minz, Ed.), pp. 313–339. AABB Press, Bethesda, MD.

DeChristopher, P. H., and Anderson, R. R. (1997). Risks of transfusion and organ and tissue transplantation—Practical concerns that drive practical policies. *AJCP* **107**, S2–S11.

Federowicz, I., Barrett, B. B., Anderson, J. W., Urashima, M., Popovsky, M. A., and Anderson, K. C. (1996). Characterization of reactions after transfusion of cellular blood components that are white cell reduced before storage. *Transfusion* **36**, 21–28.

Miller, J. P., and AuBuchon, J. P. (1999). Leukocyte-reduced and cytomegalovirus-reduced-risk blood components. In "Transfusion Therapy: Clinical Principles and Practice" (P. D. Minz, Ed.), pp. 313–339. AABB Press, Bethesda, MD.

Muir, J. C., Herschel, L., Pickard, C., and AuBuchon, J. P. (1995). Prestorage leukocyte reduction decreases the risk of febrile reactions in sensitized platelet recipients [Abstract]. *Transfusion* **35**, 45S.

Stack, G., Judge, J. V., and Snyder, E. L. (1996). Febrile and non-immune transfusion reactions. In "Principles of Transfusion Medicine" (E. C. Rossi, T. L. Simon, G. S. Moss, and S. A. Gould, Eds.), 2nd ed., pp. 773–784. Williams & Wilkins, Baltimore.

Standards Committee of the American Association of Blood Banks (2000). "Standards for Blood Banks and Transfusion Services," 20th ed., pp. 32–35. AABB, Bethesda, MD.

Vengelen-Tyler, V., Ed. (1999). "Technical Manual," 13th ed., pp. 577–600. AABB, Bethesda, MD.

CHAPTER

30

Allergic Transfusion Reactions

DEBORAH A. SESOK-PIZZINI
University of Pennsylvania Medical Center
Philadelphia, Pennsylvania

Allergic reactions to blood product transfusion encompasses a broad range of clinical symptoms and range of severity. Allergic reactions include mild urticarial reactions, more serious anaphylactoid reactions, and rare anaphylactic reactions.

I. DESCRIPTION

Mild allergic reactions mediated by IgE are among the most common adverse reactions to transfusion therapy. These mild transfusion reactions, thought to be caused by soluble proteins in donor plasma, may be associated with transfusion of any type of blood component.

Anaphylactic or anaphylactoid reactions are characterized by cardiovascular instability, hypotension, tachycardia, cardiac arrhythmia, cardiac arrest, loss of consciousness, and shock. Anaphylactoid reactions resemble anaphylaxis, but lack demonstrable IgE antibodies. Documented cases of acute anaphylaxis have occurred upon transfusion of blood products to IgA-deficient patients who have high titers of anti-IgA in their sera.

II. INCIDENCE

Mild allergic transfusion reactions are estimated to occur in 1–3% of patient's requiring large plasma transfusions and 0.3% for other blood components (DeChristopher and Anderson, 1997). Acute anaphylaxis has been estimated to occur at a frequency of 1:20,000 to 1:50,000 blood products, while IgA deficiency has an estimated incidence of 1:900 persons.

Anti-IgA antibodies are found in only ~20% of completely IgA-deficient patients. It is these patients who are susceptible to anaphylactic reactions following transfusion.

III. PATHOPHYSIOLOGY

A. Mild Allergic Reactions

These are due to soluble substances, usually proteins, present in the donor's plasma. The interaction of these antigens with preformed IgE can activate mast cells that release bioactive mediators. Histamine, leukotrienes, prostaglandins, and platelet-activating factor (PAF) all contribute to the signs and symptoms of an allergic transfusion reaction. Passive transfer of drugs, IgE antibodies, or vasoactive substances may also incite allergic transfusion reactions.

B. Acute Anaphylactic Reactions

In an acute anaphylactic reaction, C3a and C5a are generated during complement activation and lead to the activation of mast cells and the release of bioactive mediators. These bioactive mediators include histamine, leukotrienes, PAF, prostaglandins, and kinins, all of which contribute to the signs and symptoms of an anaphylactic reaction (Anstall *et al.*, 1993).

C. Other Conditions

Other conditions that can mimic generalized IgE-mediated reactions include reactions associated with angiotensin converting enzyme (ACE) inhibition and transfusion-related acute lung injury (TRALI). Also, histamine, serotonin, and PAF in stored platelets may cause reactions that mimic immediate hypersensitivity reactions.

IV. DIAGNOSIS

A. Clinical Evaluation

Mild allergic reactions may present clinically as hives, generalized urticaria, pruritis, erythema, and flushing. Hoarseness, stridor, wheezing, chest tightness, substernal pain, dyspnea, and cyanosis characterize the more severe allergic reactions. Allergic reactions may also be associated with nausea, vomiting, abdominal pain, and diarrhea. Since other substances, including certain drugs or latex, may also incite allergic reactions, these factors should also be considered in the initial clinical evaluation of a suspected allergic transfusion reaction. Underlying medical conditions such as asthma may mimic an allergic transfusion reaction. Dyspnea may represent a sign of TRALI or of circulatory overload (see Chapter 31).

Symptoms of acute anaphylaxis include bronchospasm, respiratory distress, vascular instability, nausea, abdominal cramps, vomiting, diarrhea, shock, and loss of consciousness. These symptoms typically occur immediately following the infusion of only a few milliliters of the offending blood product.

B. Laboratory evaluation

For simple allergic reactions, there are no readily available tests that will specifically identify the inciting antigen. Only a few allergic reactions will be traced to a food allergy or to the passive transfer of an IgE antibody that interacts with an antigen in the recipient. Reactions characterized by signs and symptoms other than simple, focal urticaria should prompt a thorough transfusion reaction workup (Vengelen-Tyler, 1999). Initial evaluation should include routine steps to rule out an acute hemolytic transfusion reaction (see Chapter 40). Patients with anaphylactic reactions should be: (1) tested for quantitative IgA levels, and (2) evaluated for the presence of anti-IgA antibodies. Recent blood transfusion may falsely elevate serum IgA levels.

V. CLINICAL COURSE AND SEQUELAE

Most allergic reactions are mild and self-limited and cease upon discontinuation of the transfusion and administration of an antihistamine (50 mg of diphen-

hydramine hydrochloride, Benadryl). If untreated, severe anaphylactic reactions may progress to shock, loss of consciousness, and death.

VI. MANAGEMENT

See Table 2 in Chapter 28 for an algorithm describing the approach to suspected transfusion reactions. Mild allergic reactions are best managed by discontinuation of the transfusion and administration of antihistamines. For mild reactions characterized by focal urticaria only, the transfusion usually can be restarted after 15 minutes, following treatment with antihistamines and alleviation of symptoms. Restarting the transfusion is not recommended for more severe allergic reactions. For dyspnea with desaturation, oxygen should be administered. More severe reactions with symptoms of respiratory distress may require immediate intubation and additional urgent treatment. When early symptoms consistent with an acute anaphylactic reaction occur, the transfusion should be immediately stopped. Urgent therapy may include epinephrine (0.3 ml of 1:1000 dilution SQ or IM), or if shock is present, intravenous epinephrine (3–5 ml of a 1:10,000 dilution) should be given (Davenport, 1999). Antihistamines may be beneficial during acute anaphylaxis. Aerosolized or intravenous beta-2 agonists, histamine antagonists, theophylline, and glucagon may be required for those patients with bronchospasm unresponsive to epinephrine. Cases of severe laryngeal edema may require endotracheal intubation. Cardiac instability may require emergent cardiopulmonary resuscitation.

VII. PREVENTION

Patients with a history of mild allergic reactions usually benefit from premedication with an oral or intravenous antihistamine given 30–60 minutes prior to transfusion. Patients with repeated allergic reactions may benefit from premedication with methylprednisolone given several hours prior to transfusion. Patients with repeated or significant allergic reactions may require plasma-reduced cellular products. Plasma proteins may also be removed by washing PRBCs (Davenport, 1999). For patients documented to be IgA deficient and have anti-IgA antibodies, products from IgA-deficient donors identified through rare donor registries should be provided. If these products are not available, frozen/deglycerolized saline-suspended PRBCs or thoroughly washed PRBCs may be transfused. Washing of blood products has been effective at removing donor plasma IgA immunoglobulin.

Premedication with antihistamines, steroids, and epinephrine may help prevent or diminish severe allergic reactions due to causes other than IgA deficiency. There is, however, no combination of premedications that will entirely prevent a subsequent anaphylactic reaction. Emergency medical treatment must be anticipated for any patient with a history of anaphylactic reactions.

Special consideration must be given patients with antibodies to IgA, as well as those who have selective IgA deficiency and require intravenous immunoglobulin (IVIG) infusion. The management approach to these patients (e.g., selection of appropriate blood products) is discussed in Chapter 17.

References

Anstall, H. B., Blaylock, R. C., and Craven, C. M. (1993). Immunologic hazards of transfusion of blood components and derivatives. In "Managing Hazards in the Transfusion Service" (K. D. Johnson, Ed.). ASCP Press, Chicago, IL.

Davenport, R. D. (1999). Management of transfusion reactions. In "Transfusion Therapy: Clinical Principles and Practice" (P. D. Mintz, Ed.). AABB Press, Bethesda, MD.

DeChristopher, P. J., and Anderson, R. R. (1997). Risks of transfusion and organ transplantation—Practical concerns that drive practical policies. *AJCP* **107**, S2–S11.

Vengelen-Tyler, V., Ed. (1999). Noninfectious complications of blood transfusion. In "AABB Technical Manual," 13th ed. AABB, Bethesda, MD.

31

Other Noninfectious Complications of Transfusion

PETER J. LARSON

University of Pennsylvania School of Medicine
The Children's Hospital of Philadelphia
Philadelphia, Pennsylvania

In addition to immune-mediated hemolytic reactions, allergic and febrile reactions, and the risks of transfusion-transmitted viral infections, noninfectious complications of transfusion also occur. The transfusionist must be familiar with many potential noninfectious complications of the infusion of blood or blood products, including circulatory overload, transfusion-related lung injury (TRALI), non-immune-mediated hemolysis, citrate toxicity, electrolyte imbalances, hypothermia, cytokine reactions, immune reactions to leukocyte and platelet antigens, iron overload, and transfusion-associated graft-versus-host disease (TA-GVHD; discussed in Chapter 27), each of which can be life-threatening (see Table 1). Some of these may be due to human errors and are therefore preventable. Specific therapies are indicated for the treatment of many of these events and are discussed in this chapter.

The FDA requires that fatalities due to the infusion of blood or blood products be promptly reported (within 7 days) to the Office of Compliance of the Center for Biologics Evaluation and Research (CBER). A compilation of reported fatalities has shown that ~60% of these were preventable (Sazama, 1990, 1992). The FDA also mandates that each institution maintain records documenting adverse reactions to the transfusion of blood products, including a written report of the investigation of each adverse event. Such requirements have patient safety as a primary goal. In an attempt to better define the extent of reactions to blood and components, the FDA has recently encouraged transfusion services to report nonfatal reactions to CBER. Despite these requirements and recommendations, adverse reactions to transfusion are underreported, and many of

these reactions go unrecognized. Incidences of the adverse sequelae of transfusion are thus difficult to determine and vary according to the complexity of care and practice rendered at a particular institution. Estimates of the incidence of these reactions can be speculated from published series of fatalities due to transfusion (see Table 2; Sazama, 1990, 1992).

I. CIRCULATORY OVERLOAD

A. Description

Circulatory overload is defined as the expansion of a patient's intravascular volume, due to the infusion of blood components and/or other fluids, that results in cardiac decompensation.

B. Incidence

This complication can occur at any age but is more likely in elderly patients, infants, or other groups of patients with compromised myocardial function. Estimates are that between 0.1 and 1% of transfusions result in circulatory overload, with the highest incidence occurring in elderly patients undergoing transfusion following orthopedic surgery (Popovsky and Taswell, 1985; Audet *et al.*, 1995). It is likely that the majority of fatalities reported to the FDA as acute lung injury are actually related to pulmonary edema secondary to volume overload (see Table 2; Sazama, 1990; Popovsky, 1996b).

TABLE 1   Noninfectious Complications of Transfusion

- Acute/delayed hemolytic transfusion reactions (HTRs)
- Febrile nonhemolytic transfusion reactions (FNHTRs)
- Allergic/anaphylactic reactions
- Circulatory overload
- Transfusion-related acute lung injury (TRALI)
- Non-immune-mediated RBC hemolysis
- Citrate toxicity
- Electrolyte disturbances (see Chapter 22)
- Hypothermia (see Chapter 22)
- Cytokine reactions
- Post-transfusion purpura (PTP)
- Iron overload
- Transfusion-associated graft-versus-host disease (TA-GVHD; see Chapter 27)

C. Pathophysiology

Circulatory overload can occur with the transfusion of any allogeneic or autologous blood product, and occurs most often with higher-volume products such as plasma or PRBCs. Infusion of oncotically active blood products increases the intravascular blood volume, resulting in increased central venous pressure and pulmonary edema (right-sided cardiac failure) and symptoms of diminished peripheral perfusion, such as agitation or disorientation (left-sided cardiac failure). Anemic patients are less tolerant of increases in blood volume because their hearts are already hyperkinetic and unable to compensate for the increased blood volume (Stack *et al.*, 1996).

D. Diagnosis

1. Clinical

Symptoms and signs include headache, chest tightness, agitation, dyspnea, orthopnea, cyanosis, tachycardia, increased blood pressure, pulmonary or dependent edema, jugular venous distention, cough, rales, and signs of poor cerebral perfusion.

2. Laboratory

Studies may reveal a decrease in O_2 saturation, decreased pAO_2, and characteristic bilateral infiltrates on CXR. Differential diagnosis includes acute hemolytic transfusion reaction (HTR), TRALI, anaphylaxis, sepsis, myocardial infarction, and pneumonia.

E. Clinical Course and Sequelae

The course is similar to that of acute congestive heart failure (CHF). With appropriate management many patients will recover; however, in elderly or severely compromised patients, this adverse event can be fatal.

F. Management and Prevention

When signs of circulatory overload develop, if at all possible, the transfusion should be terminated. Therapy for CHF should be instituted, including fluid restriction, supplemental O_2, administration of diuretics, and morphine. For severe symptoms, mechanical ventilation, intravenous preload and afterload reducers, and beta agonists may be necessary. Phlebotomy with removal of plasma and retransfusion of RBCs, and rotating tourniquets may be useful acutely. Circulatory overload may be prevented by delaying transfusion in patients who are in positive fluid balance until volume reduction can be accomplished (an important discriminator for the development of circulatory overload is pretransfusion volume status) (Audet *et al.*, 1995). Unless necessary to replace loss due to hemorrhage, blood should be infused slowly (2–4 ml/kg/h, or as low as 1 ml/kg/h in patients at risk) (Marriott and Keckwick, 1940). If such rates make infusion of a unit of PRBCs impossible over 4 hours, the unit may be split into smaller volumes in the blood bank prior to issue using a sterile docking device. This also allows for institution of diuretic therapy between the infusion of aliquots.

II. TRANSFUSION-RELATED ACUTE LUNG INJURY (TRALI)

A. Description

TRALI is a syndrome of acute hypoxia due to noncardiogenic pulmonary edema that develops within 1–6 hours following transfusion. The syndrome is clinically indistinguishable from the adult respiratory distress

TABLE 2   Nonviral Transfusion Fatalities Reported to the FDA, 1976–1990

	N
Acute hemolysis	227
TRALI	50
Bacterial contamination	47
Delayed hemolysis	35
Anaphylaxis	8
External hemolysis	5
TA-GVHD	1

syndrome (ARDS), but almost always resolves with supportive therapy within 48–96 hours.

B. Incidence

TRALI has been reported to occur in less than 1 of 5000 transfusions of products containing plasma (Popovsky and Moore, 1985). Approximately 150 cases have been reported since 1985 (Popovsky, 1996b), although this syndrome is likely to be misdiagnosed and perhaps underreported. There is no age or sex prevalence, and most patients have no prior history of reactions to transfusion.

C. Pathophysiology

Evidence from clinical cases and animal models suggests that TRALI is due to the presence of leukoagglutinating antibodies in donor plasma (most often found in donations from multiparous women) that react with recipient neutrophils (Popovsky and Moore, 1985). Less commonly these antibodies are present in the plasma of recipients and react with donor leukocytes. In approximately 50% of cases, the implicated antibodies are directed against HLA-A or HLA-B epitopes. These antibodies fix complement and result in neutrophil aggregation in the microvasculature of the lung. Neutrophils then release oxygen radicals, proteases, and acidic lipids, damaging the respiratory endothelium. Injury to the endothelium results in extravasation of proteins and fluids from the vascular space, causing the syndrome of noncardiogenic pulmonary edema. In 5–15% of cases, no antibody has been detected. Lipid metabolites which accumulate in stored blood may also play a role in this syndrome (Silliman, 1997, 1998).

D. Diagnosis

This syndrome must be distinguished from circulatory overload, bacterial contamination of blood products, and acute anaphylactic reactions.

1. Clinical

Patients with TRALI develop acute respiratory distress, indistinguishable from ARDS, within 1–6 hours following transfusion of a blood product that contains plasma. Symptoms and signs include dyspnea, tachycardia, cyanosis, and severe hypoxemia (paO_2 as low as 30–50 torr). On physical exam, patients will have no signs of cardiac decompensation. The reaction may be associated with moderate fever and hypotension.

2. Laboratory

Severe hypoxemia in the presence of bilateral alveolar and interstitial pulmonary infiltrates is a hallmark of the TRALI. Central venous pressures are normal and pulmonary artery or pulmonary capillary wedge pressures are normal or low, distinguishing this syndrome from circulatory overload.

E. Clinical Course and Sequelae

TRALI resolves in 48–96 hours with supportive care including supplemental oxygen and mechanical ventilation (Popovsky and Moore, 1985). Despite the relatively rapid and complete resolution of TRALI, between 5 and 8% of cases are fatal due to complications of the pulmonary injury (Eastlund et al., 1989).

F. Management and Prevention

If symptoms develop during infusion of a blood product, the transfusion should be terminated. Supportive care includes supplemental O_2 and mechanical ventilation, if indicated. Vasopressors may be necessary for prolonged hypotension. Corticosteroids are felt to have little benefit, and diuresis is not indicated since TRALI alone does not result in elevated cardiac filling pressures.

There are no practical methods to screen donations for antileukocyte antibodies or lipid inflammatory mediators. In addition, up to 2% of blood donors have HLA-reactive antibodies, but TRALI is exceedingly rare. Exclusion of these donors would have a significant impact on the blood supply without preventing many TRALI reactions. The patient–donor combination may be important; therefore, if the implicated blood product is from a directed donor, that donor should not be used again with the same recipient. Some investigators advocate the withdrawal of plasma-containing products collected from donors implicated in cases of TRALI from inventories destined for transfusion (Popovsky et al., 1992). Prestorage leukoreduction may have a role in limiting the incidence of TRALI, as leukoreduction filtration will remove significant numbers of neutrophils having a theoretical impact on the less common reactions due to antineutrophil antibodies in the recipient. In addition, prestorage leukoreduction may reduce the accumulation of lipid mediators during storage.

III. NONIMMUNE HEMOLYSIS OF RBCS

A. Description

This involves the destruction of RBCs within the circulation due to physical injury to the RBC membrane. This may be due to osmotic, thermal, chemical, or mechanical damage to the RBCs. As opposed to immune-mediated intravascular hemolysis, these reactions are rarely life-threatening. All of these are preventable.

B. Incidence

There is no good information regarding incidence, and because of minimal symptoms, nonimmune hemolysis is likely to be underrecognized and underreported.

C. Pathophysiology

The common final pathway to hemolysis is damage to the RBC membrane structures. This can occur due to thermal injury when RBCs are exposed to temperatures greater than 42°C (due to improper blood warming techniques) or lower than 1°C (due to improper storage conditions). Non-isotonic solutions, especially hypo-osmotic solutions (e.g., 5% dextrose), result in intracellular fluid shifts that can cause the membrane to fail. Mechanical injury occurs when RBCs are infused at high flow rates through small-bore catheters, by certain roller apparatus infusion pumps, and through kinked infusion lines or catheters, and when clotting occurs in the blood bag or infusion lines. Clotting can result from bacterial contamination of the unit or due to the concomitant infusion of calcium-containing IV fluids, such as lactated Ringer's solution, which reverse the citrate anticoagulant. In cases of nonimmune hemolysis, there is no evidence of antibody binding or complement fixation to the RBC as determined by the direct antiglobulin test (DAT). Systemic release of inflammatory mediators, such as C5a and C3a, that are responsible for the constellation of signs and symptoms of immune-mediated hemolytic reactions, does not occur.

D. Diagnosis

Nonimmune hemolysis is usually a diagnosis of exclusion. Therefore, other causes of hemolysis, including immune-mediated-hemolytic reactions, bacterial contamination, cold agglutinin disease, drug-induced hemolysis, oxidative stress hemolysis (G6PD), and paroxysmal nocturnal hemoglobinuria, should be considered and ruled out.

1. Clinical

The patient may complain of low back pain or chest discomfort, but more often will be asymptomatic. These reactions can occur in the operating room where they are detected first by free hemoglobin in the plasma and/or the urine. These signs are hallmarks of this reaction.

2. Laboratory

Free hemoglobin may be noted in the plasma of collected blood specimens and may be quantified. Hemoglobin is detected in the urine in the absence of RBCs or RBC casts by urinalysis. The DAT is negative. Examination of the blood bag and the intravenous setup will help determine the etiology. Potassium released from the lysed RBCs can result in hyperkalemia. Severe hemolysis may result in a fall in the hemoglobin level or lack of an expected rise in hemoglobin, and haptoglobin levels will decrease. BUN and creatinine may be elevated if free hemoglobin results in renal injury. LDH and AST will be released from RBCs, raising levels of these enzymes in patient plasma.

E. Clinical Course and Sequelae

On the whole, these hemolytic reactions are relatively benign unless large amounts of free hemoglobin result in damage to the nephron. It is critical that they be rapidly distinguished from a potentially life-threatening acute immune-mediated hemolytic reaction. The cause of hemolysis should be determined prior to initiating further transfusion to prevent more widespread donor RBC destruction. With significant hemolysis, depending on the indication for transfusion, the added requirement for PRBCs may result in additional donor exposures and a resultant small increase in the risk of transmissible infectious diseases and alloimmunization.

F. Management

Management involves terminating the transfusion, ruling out an immune-mediated mechanism, and determining the cause of RBC injury prior to continuing RBC transfusion. Intravenous access is maintained in order to provide adequate hydration for good urine output. Serum potassium levels should be monitored, and intervention for hyperkalemia is instituted if necessary.

G. Prevention

Prevention requires strict adherence to proper storage conditions, the use of approved and controlled blood warming devices, infusion through large-bore infusion sets and catheters at slow rates of flow, the use of pumps approved for PRBC infusions, and the use of approved intravenous fluids. Medications must not be added to blood or components of whole blood.

1. Osmotic or Chemical Lysis

To prevent osmotic or chemical lysis, no drug infusions should be mixed with any blood products at any point outside the intravascular space. For practical purposes, 0.9% (normal) saline should be considered the only intravenous fluid that is compatible with blood products. However, the FDA does allow the use of other tested and approved solutions such as ABO-compatible plasma, 5% albumin or plasma protein fraction, and isotonic solutions that have been demonstrated to be safe and efficacious when mixed with blood. Hypotonic solutions such as 0.45% saline, 0.22% saline, or 5% dextrose, and isotonic solutions containing calcium (e.g., lactated Ringer's solution), should *never* be coinfused with blood products. Lactated Ringer's contains sufficient calcium (3 mEq/liter) to overcome citrate chelation and cause clots to develop.

2. Thermal Lysis

To prevent thermal lysis, blood products should be warmed using only approved devices such as thermostatically controlled water baths, dry heat devices with electric warming plates, and high-volume countercurrent heat exchangers with water jackets (Uhl *et al.*, 1992). Microwave warmers should not be used as they cause uneven heating, resulting in hemolysis. Warmers should not raise temperature above 42°C and should have a visible thermometer and preferably an audible alarm.

3. Mechanical Lysis

To prevent lysis due to shearing, infusion devices must be manufacturer-approved for the infusion of blood products. Many of these are designed for crystalloid alone and can cause hemolysis of cellular products, especially high-viscosity RBC products. Pressure devices for rapid transfusion are available. These should not be used at pressures greater than 300 mm Hg, which may cause the blood bag seams to fail. Large-bore needles should be used for rapid infusion of blood to prevent hemolysis due to high-flow shearing, especially when administering blood under pressure.

IV. CITRATE TOXICITY

A. Description

The most common anticoagulant used in blood and blood components is trisodium citrate. Excessive circulating citrate lowers the level of ionized calcium and results in typical signs and symptoms of hypocalcemia, from mild paresthesias to frank tetany.

B. Incidence

This reaction is common with rapid large-volume infusions of blood products, such as those given during resuscitation maneuvers and during apheresis procedures.

C. Pathophysiology

Citrate functions as an anticoagulant by chelating ionized calcium in whole blood and components. Calcium is required for the function of many proteases in coagulation reactions. Citrate is a normal intermediary in the Krebs (tricarboxylic acid) cycle, and the liver metabolizes excessive amounts of citrate in the circulation. Large infusions of citrate may overwhelm the capacity of the liver to metabolize citrate, with the result that ionized calcium levels fall below normal (< 1.16 mM for adults and < 1.20 mM for children).

D. Diagnosis

1. Clinical

A conscious patient initially may have symptoms of peripheral neural dysfunction and complain of perioral tingling or distal extremity paresthesias. As the plasma level of ionized calcium falls, the autonomic system becomes involved and the patient may develop symptoms of nausea, emesis, or other signs of abnormal gastrointestinal motility; hypotension; and, if severe, cardiac arrhythmias.

2. Laboratory

Ionized calcium levels are well below normal, typically < 0.9 mM in patients with citrate toxicity.

E. Clinical Course and Sequelae

Ionized calcium level improves or returns to normal upon withdrawal of citrate or by slowing the rate of infusion of blood products.

F. Management and Prevention

Slowing the rate of infusion of blood products should reverse or prevent this reaction. In some situations, infusion of calcium in the form of calcium gluconate (10% solution) may be utilized to prevent or reverse the reaction (for example, during apheresis procedures where slowing the rates of blood draw and infusion may greatly prolong the procedure). *Calcium must never be added to blood or blood products, or to intravenous fluids that are infused through the same line used for the infusion of blood products* (see Nonimmune Hemolysis, Section III). Saline flushes should be performed prior to and following calcium infusion, especially if the same catheter is also being used for blood infusion.

V. OTHER ELECTROLYTE DISTURBANCES

A. Description

Perturbations in whole blood pH and serum potassium can occur due to the infusion of blood.

B. Incidence

Electrolyte disturbances are rare, especially with routine elective transfusion.

C. Pathophysiology

Storage of PRBC products results in a slow leakage of potassium. The concentration of potassium in the supernatant plasma of PRBCs is much higher than normal plasma levels at the end of the storage period (>50 mEq/liter); however, the total extracellular potassium is usually no greater than 7 mEq/U (Uhl and Kruskall, 1996). Although stored blood is acidic due to the presence of lactic acid and citrate anticoagulant, patients often develop metabolic alkalosis following massive transfusion due to the metabolism of citrate to bicarbonate. The minimal quantity of plasma in the product, dilution into the extravascular volume, and a slow rate of infusion obviate the need for washing of PRBCs to reduce or remove excess potassium, citrate, or lactate in the supernatant plasma.

D. Diagnosis

1. Clinical

Cardiac arrhythmias may develop with significant hyperkalemia. Metabolic alkalosis can result in neurological symptoms similar to those seen with hypocalcemia.

2. Laboratory

Serum potassium and bicarbonate levels and arterial pH will be abnormal. Widened QRS complex, arrhythmias, and other manifestations of hyperkalemia may be observed by electrocardiogram.

E. Clinical Course and Sequelae

Cardiac arrhythmias and death due to hyperkalemia have been reported from the rapid infusion of older red cell products directly into the central circulation (Hall *et al.*, 1993). These occur most often in infants or small children.

F. Management and Prevention

Reversal of severe hyperkalemia is accomplished by termination of the transfusion and standard interventions such as the administration of bicarbonate, glucose, insulin, or oral potassium-binding resins, or institution of hemodialysis. Impaired renal function may delay the reversal of both metabolic alkalosis and hyperkalemia. Slow infusion of blood products will prevent these complications. Some services recommend the use of PRBC products that are less than 5 days old for rapid large-volume infusion into the central circulation, especially in children.

VI. HYPOTHERMIA

A. Description

A fall in the core body temperature results in metabolic, coagulation, platelet, oxygen transport, and cardiac conduction dysfunction. Severe hypothermia occurs at body temperatures below 34°C.

B. Incidence

Hypothermia occurs primarily in the setting of massive transfusion. Transfusion may exacerbate the hypothermia of shock, trauma, and surgery.

C. Pathophysiology

Blood products such as PRBCs and thawed FFP are stored at 4°C. Rapid infusion of large amounts of these components results in a fall in core body temperature. Enzymatic dysfunction at low body temperatures accounts for some or all of the abnormalities seen with hypothermia (Rohrer and Natale, 1992; Rudolf and Boyd, 1990). Severe hypothermia with temperatures below 34°C can result in platelet dysfunction and prolonged coagulation times, resulting in bleeding, hyperkalemia, and cardiac conduction defects that may result in dysrhythmias and death (Uhl and Kruskall, 1996).

D. Diagnosis

1. Clinical

Patients will exhibit signs of decreased core body temperature, bleeding, and circulatory compromise.

2. Laboratory

Evaluation may disclose metabolic acidosis, elevated serum potassium, prolonged bleeding time, prothrombin time, and activated partial thromboplastin time, dysrhythmias, and evidence of poor tissue oxygenation with elevated serum lactate levels.

E. Clinical Course and Sequelae

Raising the core temperature of the patient will reverse most of the defects. Substantial bleeding has been observed during rewarming.

F. Management and Prevention

Maneuvers to raise the core temperature include warming of blankets and beds, use of heat lamps, and others. Plasma and platelet transfusions may be useful in slowing hemorrhage during rewarming. Preventive measures include the warming of fluids and blood components for intravenous infusion using approved devices. However, except for countercurrent methods of warming, most blood warmers are of limited use in resuscitation and massive transfusion because of flow and heating limitations (Uhl *et al.*, 1992). It should be emphasized that the use of unapproved warming devices such as culinary microwaves will result in uneven warming and red cell lysis (see Section III, Nonimmune Hemolysis). For the same reason, coinfusion of pre-warmed saline should not be attempted.

VII. CYTOKINE REACTIONS

A. Description

These are systemic effects due to the infusion of a variety of vasoactive and pyrogenic mediators of inflammation that contaminate blood products.

B. Incidence

These reactions have only recently been discriminated from nonhemolytic febrile transfusion reactions, and their true incidence is not known.

C. Pathophysiology

White blood cells (WBCs) release cytokines during storage. Specific WBC-derived mediators implicated in these reactions include IL-1, IL-6, and tumor necrosis factor-α (TNF-α), which can cause fever, hypotension, chills, and rigors (Muylle *et al.*, 1993; Heddle *et al.*, 1994). These reactions are more common with platelet concentrates that contain a larger number of contaminating leukocytes (Heddle *et al.*, 1993; Hume *et al.*, 1996). Activation of the complement, kinin, and kallikrein systems may result in the production of C3a, C5a, and bradykinin, which cause hypotension, flushing, bradycardia, and dyspnea (Owen and Brecher, 1994). These reactive substances of anaphylaxis may be generated when the corresponding proteases are activated following exposure to negatively charged surfaces present in the extracorporeal circuits used in hemodialysis, apheresis, and other procedures. Patients who are receiving angiotensin converting enzyme (ACE) inhibitors are at increased risk for reactions, since ACE is important for the degradation of bradykinin.

D. Diagnosis

1. Clinical

Hypotension, flushing, non-bronchospastic pulmonary symptoms, chills, rigors, fever, and other non-specific symptoms of malaise may be present during or following transfusion or procedures involving extracorporeal circulation.

2. Laboratory

More serious reactions such as acute hemolytic reactions, allergic reactions, and TRALI are associated with these same symptoms and should be ruled out by appropriate testing and clinical history. The DAT is negative, and chest x-ray does not show bilateral infiltrates in cytokine reactions.

E. Clinical Course and Sequelae

These reactions resolve with supportive therapy and there are no long-term sequelae.

F. Management and Prevention

Prevention of these reactions is a focus of ongoing studies. Prestorage leukoreduction of blood products reduces the accumulation of inflammatory mediators released during storage. If possible, patients who are to undergo procedures that involve extracorporeal circulation should stop ACE inhibitor therapy at least 24 hours prior to treatment.

VIII. POST-TRANSFUSION PURPURA (PTP)

A. Description

PTP is a self-limiting severe thrombocytopenia that develops within 5–10 days following transfusion of blood components containing platelet antigens (usually PRBCs, whole blood, or platelets, but may rarely occur with plasma).

B. Incidence

The incidence of PTP is likely to be underreported, but may be as high as 9000 cases in the United States per year (MacFarland, 1996).

C. Pathophysiology

PTP results from an antibody directed against a platelet-specific antigen that develops after exposure to the antigen by transfusion in a previously sensitized, antigen-negative individual. Destruction of both antigen-positive (donor) platelets and antigen-negative (autologous) platelets occurs. In 70–90% of cases, the implicated antigen is HPA-1a (PI^{A1}). The mechanism for destruction of the patient's own (autologous) antigen-negative platelets is unclear, but may involve adsorption of soluble allogeneic platelet antigen or antigen–antibody complexes to autologous platelets. Alternatively the syndrome may be the result of an autoreactive antibody that is stimulated following the transfusion of alloantigen (Bussel and Cines, 2000). Previous sensitization to the antigen occurs through transfusion or pregnancy. The majority of cases of PTP occur in women, and the syndrome has not been described in children under the age of 16.

D. Diagnosis

1. Clinical

Purpura is the hallmark of the disorder and bleeding can occur from mucosal surfaces, including the GI and urinary tracts. Mortality from the disorder is ~10% and is usually the result of intracranial hemorrhage that occurs during periods of extreme thrombocytopenia (MacFarland, 1996).

2. Laboratory

Severe thrombocytopenia with platelet counts of $< 10,000/\mu L$ is present in over 80% of cases. Thrombotic thrombocytopenia purpura (TTP) and disseminated intravascular coagulopathy (DIC) are ruled out by normal RBC morphology and normal coagulation assays. Bone marrow is normo- or hypercellular with normal or increased megakaryocytes. Patient serum reveals the presence of a platelet-specific antibody (usually anti-HPA-1a). The patient's platelets, when isolated following recovery, are negative for the implicated antigen.

E. Clinical Course and Sequelae

Most patients with PTP recover spontaneously in 7–48 days; however, due to the high mortality rate from intracranial hemorrhage, treatment is indicated. The syndrome does not usually recur with subsequent transfusion.

F. Management and Prevention

Patients are unresponsive to either antigen-negative or antigen-positive platelets. IVIg (400–500 mg/kg for 1–10 days) may be effective in shortening the duration of thrombocytopenia. If IVIg is not effective, plasma exchange using FFP replacement is indicated. High-dose corticosteroid therapy is often used, although there is no strong evidence for its benefit.

IX. IRON OVERLOAD

A. Description

Chronic transfusion for congenital and acquired anemias can result in systemic iron overload, causing deposition of iron in parenchymal tissues that leads to hepatic, endocrine, and cardiac organ system failure (Cohen, 1987; Hollan, 1987).

B. Incidence

Patients at risk in the United States include patients with thalassemia and patients with sickle cell disease who are managed by chronic transfusion therapy. Other less common chronic diseases such as myelodysplastic syndromes, aplastic anemia, sideroblastic anemia, and Diamond–Blackfan anemia are also managed with PRBC transfusion and these patients may also develop iron overload. After 15–20 years of transfusion therapy, nearly 100% of these patients will develop symptomatic iron overload.

C. Pathophysiology

Each unit of PRBCs contains ~ 250 mg of elemental iron. In the absence of blood loss, the average rate of iron excretion is very small (< 1.0 mg per day) and is accounted for by cellular losses from the gastrointestinal and urinary tracts and skin. Menstruating women lose ~ 1.5 mg of iron per day, and iron loss from pregnancy is about 500 mg. Iron absorption is relatively unaffected by tissue iron load and in some chronic anemias such as thalassemia may actually be increased. The presence of the gene defect that results in hereditary hemochromatosis may also result in a further increase in iron absorption (Feder *et al.*, 1996).

D. Diagnosis

The symptoms and signs of iron overload are usually anticipated because of the awareness of transfusion-related iron overload on the part of physicians who treat patients with chronic anemias.

1. Clinical

Bronze discoloration of the skin is the result of iron deposition. Endocrine manifestations include diabetes mellitus resulting from iron overload in the islet cells of the pancreas. Failure to undergo a pubertal growth spurt and normal sexual maturation are due to pituitary and sexual organ iron deposition. Deposition of iron in the thyroid and parathyroid glands can result in hypothyroidism and hypocalcemia. Iron buildup in the parenchyma of the liver leads to cirrhosis, and iron overload in the heart eventually results in supraventricular and ventricular arrhythmias and CHF.

2. Laboratory

In iron overload, the serum ferritin rises, followed by increased saturation of transferrin. Ferritin > 1000 ng/ml is observed in patients with thalassemia major after transfusion of 20 units of PRBCs. After the transfusion of 100 units of PRBCs, stainable iron is present in heart and liver biopsies. End organ dysfunction is evidenced by glucose intolerance, hypocalcemia, hypothyroidism, cardiac dysrhythmias, and hepatic synthetic dysfunction.

E. Clinical Course and Sequelae

Chronic iron deposition over a period of 15-20 years inevitably leads to cardiac, hepatic, and endocrine failure. Death is usually the result of cardiac or hepatic failure.

F. Management and Prevention

Since iron overload is a chronic complication, a hematologist experienced with the following therapeutic modalities should manage iron removal therapy (Cohen, 1987; Hollan, 1997). Intermittent phlebotomy, when feasible, is the treatment of choice; however, most of these patients are unable to tolerate phlebotomy due to their underlying anemias. Therapy with a parenteral iron chelator (desferrioxamine) can limit or reverse iron overload. Desferrioxamine is typically administered by subcutaneous infusion over 10-12 hours per day. For severe cardiac symptoms, more rapid removal can be effected by intravenous chelation therapy. There is no approved oral iron chelator available. Recent experience in sickle cell disease suggests that chronic partial exchange transfusion therapy (usually performed by apheresis techniques) can limit or reverse iron overload in sickle cell patients (see also Chapter 39; Kim *et al.*, 1992; Singer *et al.*, 1999; Hilliard *et al.*, 1998).

X. TRANSFUSION-ASSOCIATED GRAFT-VERSUS-HOST DISEASE (TA-GVHD)

TA-GVHD is another noninfectious complication of transfusion and is discussed in detail in Chapter 27.

References

Audet, A., *et al.* (1995). Current transfusion practices in orthopedic surgery patients: A multi-institutional study. *Blood* **86** (Suppl.), 853a.

Bussel, J., and Cines, D. (2000). Immune thrombocytopenic purpura, neonatal alloimmune thrombocytopenia, and posttransfusion purpura. In "Hematology: Basic Principles and Practice" (R. Hoffman, *et al.*, Eds.), 3rd ed., pp. 2096-2114. Churchill Livingstone, Philadelphia.

Cohen, A. (1987). Management of iron overload in the pediatric patient. *Hematol Oncol Clin N Am* **1**, 521-544.

Eastlund, T., *et al.* (1989). Fatal pulmonary transfusion reaction to plasma containing donor HLA antibody. *Vox Sang* **57**, 63-66.

Feder, N., *et al.* (1996). A novel MHC class I-like gene is mutated in patients with hereditary haemochromatosis. *Nature Genet* **13**, 399.

Hall, T. L., *et al.* (1993). Neonatal mortality following transfusion of red cells with high plasma potassium levels. *Transfusion* **33**, 606-609.

Heddle, N. M., *et al.* (1993). A prospective study to identify the risk factors associated with acute reactions to platelet and red cell transfusions. *Transfusion* **33**, 794-797.

Heddle, N. M., *et al.* (1994). The role of the plasma from platelet concentrates in transfusion reactions. *N Engl J Med* **331**, 625-628.

Hilliard, L. M., *et al.* (1998). Erythrocytapheresis limits iron accumulation in chronically transfused sickle cell patients. *Am J Hematol* **59**, 28-35.

Hollan, S. R. (1997). Transfusion-associated iron overload. *Curr Opin Hematol* **4**, 436-441.

Hume, H. A., *et al.* (1996). Hypotensive reactions: A previously uncharacterized complication of platelet transfusion? *Transfusion* **36**, 904–909.

Kim, H. C., *et al.* (1992). A modified transfusion program for prevention of stroke in sickle cell disease. *Blood* **79**, 1657–1661.

MacFarland, J. (1996). Posttransfusion purpura. In "Transfusion Reactions" (M. Popovsky, Ed.), pp. 205–229. AABB Press, Bethesda, MD.

Marriott, H., and Kekwick, A. (1940). Volume and rate in blood transfusion for the relief of anemia. *Br Med J* **1**, 1043.

Muylle, L., *et al.* (1993). Increased tumor necrosis factor α (TNFα), interleukin 1, and interleukin 6 (IL-6) levels in the plasma of stored platelet concentrates: Relationship between TNF and IL-6 levels and febrile transfusion reactions. *Transfusion* **33**, 195–199.

Owen, H. G., and Brecher, M. E. (1994). Atypical reactions associated with use of antiotensin-converting enzyme inhibitors and apheresis. *Transfusion* **34**, 891–894.

Popovsky, M. (1996a). Circulatory overload. In "Transfusion Reactions" (M. Popovsky, Ed.), pp. 231–236. AABB Press, Bethesda, MID.

Popovsky, M. (1996b). Transfusion-related acute lung injury (TRALI). In "Transfusion Reactions" (M. Popovsky, Ed.), pp. 167–184. AABB Press, Bethesda, MD.

Popovsky, M., and Moore, S. (1985). Diagnostic and pathogenic considerations in transfusion-related acute lung injury. *Transfusion* **25**, 573–577.

Popovsky, M., and Taswell, H. (1985). Circulatory overload: An underdiagnosed consequence of transfusion. *Transfusion* **25**, 469.

Popovsky, M., *et al.* (1992). Transfusion-related acute lung injury: A neglected, serious complication of hemotherapy. *Transfusion* **32**, 589–592.

Rohrer, M., and Natale, A. (1992). Effect of hypothermia on the coagulation cascade. *Crit Care Med* **20**, 1402–1405.

Rudolf, R., and Boyd, C. (1990). Massive transfusion: Complications and their management. *South Med J* **83**, 1065–1070.

Sazama, K. (1990). Reports of 355 transfusion-associated deaths: 1976 through 1985. *Transfusion* **30**, 583–590.

Sazama, K. (1992). Analysis of Causes of Blood Transfusion Fatalities. Fall 1992 Teleconference, American Society of Clinical Pathologists.

Silliman, C. C., *et al.* (1997). The association of biologically active lipids with the development of transfusion-related acute lung injury: A retrospective study. *Transfusion* **37**, 719–726.

Silliman, C. C., *et al.* (1998). Plasma and lipids from stored packed red blood cells cause acute lung injury in an animal model. *J Clin Invest* **101**, 1458–1467.

Singer, S. T., *et al.* (1999). Erythrocytapheresis for chronically transfused children with sickle cell disease: an effective method for maintaining a low hemoglobin S level and reducing iron overload. *J Clin Apheresis* **14**, 122–125.

Stack, G., *et al.* (1996). Febrile and nonimmune transfusion reactions. In "Principles of Transfusion Medicine" (E. Rossi, *et al.*, Eds.), 2nd ed. Williams & Wilkins, Baltimore.

Uhl, L., and Kruskall, M. (1996). Complications of massive transfusion. In "Transfusion Reactions" (M. Popovsky, Ed.), pp. 301–320. AABB Press, Bethesda, MD.

Uhl, L., *et al.* (1992). A comparative study of blood warmer performance. *Anesthesiology* **77**, 1022–1028.

INFECTIOUS COMPLICATIONS OF TRANSFUSION

32

Hepatitis

DAVID F. FRIEDMAN
The Children's Hospital of Philadelphia
Department of Pediatrics
University of Pennsylvania School of Medicine
Philadelphia, Pennsylvania

The potential for transmission of hepatitis through blood and blood products was recognized prior to the 1950s. Most of the fundamental mechanisms that are in place today to reduce the risk of transmission of infectious disease in the blood supply were developed in response to epidemiologic studies of post-transfusion hepatitis, which in the 1960s was estimated at 1–3% of patients transfused (Aach, 1987; Holland, 1998). Measures such as the use of volunteer blood donors and transmissible disease testing of individual blood donations were implemented to reduce the risk of post-transfusion hepatitis. These measures have subsequently been expanded and refined as understanding of the infectious agents that cause post-transfusion hepatitis has improved, and as other transfusion-transmissible infectious diseases have been identified. The safety of the blood supply has improved markedly, and the risk of post-transfusion hepatitis has been reduced by more than two orders of magnitude.

Several well-characterized viral agents are implicated in transfusion-transmitted hepatitis because they are: (1) known to be transmissible by transfusion, and (2) known to cause acute and/or chronic liver disease. These are the hepatitis viruses A, B, C, and D (HAV, HBV, HCV, and HDV, or delta virus, respectively), the Epstein–Barr virus (EBV), and cytomegalovirus (CMV). Other, more recently described viruses are not included in this discussion because they do not fulfill both criteria given above. In the case of hepatitis E (HEV), transmission through blood transfusion has not been documented, and in the cases of Hepatitis G (HGV; H. J. Alter *et al.*, 1997; M. J. Alter *et al.*, 1997), a role in causing hepatitis has not been demonstrated.

This chapter will review hepatitis viruses HAV, HBV, HCV, and HDV, the viruses that infect hepatocytes specifically and that have the greatest impact on the blood supply. CMV and EBV will be discussed elsewhere.

I. HEPATITIS A (HAV)

A. Incidence

The CDC published an estimate of 80,000 clinical cases of HAV and 134,000 total infections in the United States for 1994 after correction for reporting error, and from 10 to 15 cases per 100,000 population (CDC, 1996b). The incidence varies significantly with age, with the lowest incidence in age groups < 4 years and greater than 40 years, and the highest incidence rates in children 5–15 years, accounting for 30% of all cases. Up to 70% of childhood infections are asymptomatic or subclinical. The incidence of HAV also varies with race/ethnicity, with the highest incidence among Hispanics and Native Americans. HAV is rarely acquired by blood transfusion, with a transfusion-associated risk estimated at less than 1 per 1,000,000 units of blood transfused (Dodd, 1994).

B. Pathophysiology

HAV is a small, single-stranded RNA virus of the picornavirus family. It lacks a lipid envelope, is stable at room temperature and in refrigerated storage, and is resistant to heating.

By far the most common route of transmission of HAV is from person to person by the oral–fecal route. There is no insect vector or animal reservoir for HAV. The virus is endemic in the U.S.A., and local outbreaks are often reported. Commonly identified sources for HAV outbreaks are insular communities, child care centers, common-source food- and water-borne outbreaks involving uncooked or undercooked foods (including shellfish), nosocomial infections, and household members and personal contacts of recent cases. Other risk factors for HAV infection include low socioeconomic status, travel to endemic areas, male homosexual contact, and intravenous drug use (Lemon, 1985; Committee on Infectious Diseases, 1997). About 40% of HAV cases have no identified risk factor, but since nonicteric/asymptomatic infection is common, especially among children younger than 6 years, apparently isolated cases may represent household contacts of unidentified HAV cases.

The incubation period for HAV is 15–50 days, with an average of 25–30 days. Viral shedding lasts 1–3 weeks, with the highest viral load of HAV in stool occurring 1–2 weeks prior to the onset of illness. Viral shedding diminishes after clinical illness starts, and there is no HAV carrier state. Immunity is associated with development of serum anti-HAV IgG and is lifelong (Committee, 1997).

Parenteral transmission of HAV is rare because: (1) the duration of viremia in acute infection is short, (2) potential blood donors with hepatitis A would be quite ill when infectious, and therefore would be excluded from donation, and (3) there is no chronic HAV carrier state. However, rare transmission via blood products (Giacoia and Kaprisin, 1989) and clotting factors (Mannucci et al., 1994) has been reported. Since HAV is heat resistant and has no lipid envelope, infectious virions can withstand certain viral inactivation procedures such as solvent/detergent treatment. Rare outbreaks of HAV transmission in blood products derived from plasma pools have been reported (CDC, 1996a).

C. Diagnosis

1. Clinical Evaluation

The clinical symptoms caused by HAV infection typically start abruptly and include fever, malaise, anorexia, weight loss, pruritis, nausea, abdominal discomfort, dark urine, and jaundice. Dark urine may be the first sign of HAV illness, with jaundice and acholic stools following in 1–2 days. Abdominal tenderness and

moderate hepatomegaly may be present. The likelihood of symptomatic HAV infection is a function of age: about 70% of children have subclinical or anicteric infections, and 70% of adults have icteric infections.

2. Laboratory Evaluation

Elevation of the serum transaminase ALT occurs ∼4 weeks after infection with HAV; the ALT peak is coincident with clinical symptoms, if they occur. Elevation of total and direct serum bilirubin occurs about 1 week following ALT elevation. Diagnosis of acute HAV infection is made by detection of IgM anti-HAV in patient serum obtained during or shortly after the onset of jaundice or clinical illness. IgM anti-HAV may be present as early as 5–10 days following exposure. IgG anti-HAV is detectable shortly after IgM anti-HAV and usually persists for life. IgG anti-HAV without IgM anti-HAV is consistent with either past HAV infection or post-HAV vaccination.

D. Clinical Course and Sequelae

The symptoms of HAV infection are self-resolving, usually lasting less than 2 months, although 10–15% of patients with HAV may have prolonged or relapsing illness for up to 6 months. Many children with HAV have mild symptoms or are asymptomatic. When symptoms persist, malaise, anorexia, and weight loss are the most common.

In the case of both symptomatic and asymptomatic HAV infections, virions are shed in the stool. The highest viral titers in the stool and the peak of infectivity occur during the two weeks before the onset of jaundice or elevation of ALT. Children and infants may shed HAV in the stool for longer periods than do adults—up to several months. However, a chronic HAV carrier state does not occur.

The most common outcome of HAV infection is full recovery. The hepatitis associated with HAV infection is rarely fulminant and is not usually associated with hepatic dysfunction sufficient to cause hypoalbuminemia or coagulopathies. HAV infection is not associated with chronic active hepatitis, cirrhosis, or hepatocellular carcinoma. Lifelong protective immunity to HAV follows infection.

E. Management and Prevention

The management of acute HAV infection consists primarily of supportive care: provision of fluids, analgesics, and antipyretics. To control the spread of HAV infection, hospitalized patients should be isolated to prevent spread by the oral–fecal route for 1 week following the onset of symptoms. Outside the hospital setting, caretakers should observe careful hygiene and hand-washing, especially with diaper changes and food handling. Children with HAV should be excluded from day care or school for 1 week following the onset of symptoms.

Postexposure prophylaxis is available in the form of HAV immune globulin, which is 80–90% effective if given within 2 weeks of exposure. Establishing the time of exposure may be difficult in some situations, such as day care centers, where asymptomatic cases may preceed the symptomatic index case. HAV immune globulin may also be used for preexposure prophylaxis for up to 6 months for individuals traveling to areas where HAV is endemic. Two inactivated HAV vaccines are available, and are highly effective in recipients older than 2 years, with 99–100% seroconversion rates and 94–100% protection from infection after 2 to 3 doses. Vaccination to prevent HAV is recommended for persons living in confined or circumscribed communities with high endemic rates of HAV, persons with preexisting liver disease, men who have sex with men, intravenous drug abusers, and individuals with potential occupational exposures. Others who may benefit from the HAV vaccine include food handlers, child care workers, children in day care centers, all hospital personnel, residents and employees of custodial care institutions, and hemophiliacs (due to rare reported outbreaks associated with factor concentrates; Mannucci et al., 1994).

Screening of blood donors for a history of hepatitis, intravenous drug abuse, foreign travel, and receipt of immune globulin will identify some risk factors for HAV. No specific laboratory screening of blood donations for HAV is currently performed, since there is no chronic carrier state. Screening of blood donations for elevated ALT levels, where performed, may identify some donors who have asymptomatic HAV infection.

II. HEPATITIS B (HBV)

A. Incidence

Hepatitis B virus remains a significant public health problem in the U.S.A. The estimated incidence of acute hepatitis B infection is 200,000–300,000 new infections per year (data through 1991). In addition, there are an estimated 1–1.25 million individuals with chronic HBV infection in the U.S.A., and these carriers are potentially infectious to others. Chronic carriers of HBV are also at risk for long-term sequelae, since as many as 5–8% will develop chronic active hepatitis, which may in turn progress to cirrhosis, liver failure,

and primary hepatocellular carcinoma. About 350 individuals die each year from fulminant HBV infection, 4000–5000 die from chronic liver disease due to HBV, and 1500 die from liver cancer associated with HBV (CDC, 1991a, b).

Risk factors for HBV include: (1) parenteral exposure to human blood, (2) intravenous drug abuse, (3) men who have sex with men, (4) sexual partners of HBV-infected individuals, (5) health care workers, (6) dialysis patients, (7) transfusion recipients, and (8) household contacts of HBV-infected individuals. However, as many as 40% of persons with HBV have no identifiable risk factor. HBV is also transmitted vertically: 40–90% of infants of HBV-infected mothers become infected. Even if infants are not infected during the perinatal period, children of HBV-infected mothers are at high risk of HBV infection by horizontal maternal transmission. The majority of people who acquire HBV congenitally or in the first 5 years of life become chronic carriers; these people are important reservoirs of endemic HBV infection in underdeveloped countries.

The risk for HBV transmission by blood transfusion in the U.S.A. was estimated by the Retrovirus Epidemiology Donor Study (REDS) as 1:63,000 units transfused (95% confidence interval = 1:31,000 to 1:147,000; Schreiber et al., 1996). This estimate was based on a database of over 2.3 million donations from repeat blood donors between 1991 and 1993, with an adjustment for the limitation in estimating the incidence of new HBV infections in blood donors from HBsAg seroconversion rates. HBV accounted for more than 50% of the seroconversions to transfusion-transmissible diseases in the repeat blood donors studied, reflecting the fact that the donor questionnaire for risk factors fails to defer a substantial number of donors with HBV.

B. Pathophysiology

The hepatitis B virus is a double-stranded DNA virus of the hepadnavirus family that replicates through an RNA intermediate. The whole virus, also called the "Dane particle," is 42 nm in diameter, and contains an outer envelope that bears the surface antigen (HBsAg) and a nucleocapsid that contains the genome and bears the core antigen (HBc). The virus is tropic for and replicates in human hepatocytes, and circulates in the bloodstream as whole virions. HBsAg-positive subviral particles may also be present in the blood. The incubation period is typically 1–6 months, but HBsAg and HBV DNA appear in the circulation 2–6 weeks after infection, prior to both evidence of hepatocellular injury and onset of symptoms (Lee, 1997).

Acute HBV infection may follow one of three courses. (1) Fewer than 1% of patients progress to fulminant hepatic failure. A fulminant course is more likely if coinfection with HBV and the hepatitis D virus occurs (see below). (2) The majority of acute HBV infections eventually resolve completely, with clearance of HBV-infected hepatocytes and loss of HBsAg and HBV DNA. Individuals who clear HBV infection in this manner have lifelong immunity to HBV. (3) Three to five percent of acute HBV infections are not resolved and result in chronic infection, in which circulating HBsAg persists. This subgroup may develop chronic active hepatitis with intermittent elevations of the transaminases, hepatic fibrosis, and eventual progression to cirrhosis. Individuals with chronic HBV infection are subject to superinfection with the defective hepatitis D virus (see below).

The host's immune response is a major determinant of the course of HBV infection. The hepatocellular injury associated with the icteric and symptomatic phase of the infection, as well as the progressive fibrosis of chronic hepatitis, is caused by host-cell-mediated immunity directed against HBV-derived epitopes on infected hepatocytes. Essentially all infected individuals develop antibody to HBc, but antibody to HBsAg is associated with protective immunity, and is lacking in the chronic carriers who remain HBsAg-positive (Lee, 1997).

C. Diagnosis

1. Clinical Evaluation

About one-third of patients with acute HBV present with the classic symptoms of acute viral hepatitis: jaundice, dark urine, fatigue, loss of appetite, and abdominal pain. If jaundice is not present, symptoms may be less specific, including malaise, fatigue, nausea, and vomiting. About one-third of cases of HBV infection are asymptomatic. Skin rash and arthritis may occur, as well as polyarteritis nodosa, glomerulonephritis, and vasculitis. Hepatomegaly and abdominal tenderness are usually present; splenomegaly is present in 20% of patients.

2. Laboratory Evaluation

Elevation of the transaminases, most specifically ALT, and direct hyperbilirubinemia (may be > 15 mg/dl), occur 2–6 months after HBV exposure and may persist for 4–6 weeks. These values typically return to normal, except in those patients who become chronic carriers.

The time courses of detectable HBV antigens and nucleic acids, clinical signs and symptoms, and anti-HBV antibodies are well described (Hoofnagle and DiBisceglie, 1991). HBsAg and hepatitis B envelope antigen (HBeAg) appear first, along with HBV DNA, at ~2–6 weeks following exposure, followed by elevation of the transaminases, jaundice, and clinical symptoms within 2–6 months of exposure. The anti-HBc response, first IgM and later IgG, is the earliest and most consistent antibody response and is detectable during the symptomatic phase. The antibody response to HBsAg arises after 5–6 months of exposure in patients destined to clear the infection.

Evaluation of an individual patient for HBV usually includes testing for serum HBsAg, antibody to HBsAg, and IgM anti-HBc. The IgM anti-HBc assists in detecting the "window period" prior to seroconversion with anti-HBsAg. In contrast, screening of the blood supply in the U.S.A. is intended primarily to detect chronic HBV carriers who are persistently positive for HBsAg, so only this test is performed.

Blood donations in the U.S.A. are also screened for HBV core antibody (anti-HBc). Anti-HBc testing was initially instituted as a "surrogate test" to reduce the risk of post-transfusion "non-A, non-B" hepatitis, now known to be hepatitis C, and for which specific testing of the blood supply is now performed. Thus, some have advocated the discontinuation of anti-HBc testing of the blood supply. However, testing for anti-HBc also contributes to the detection of recent HBV infections in at least three situations: (1) asymptomatic blood donors who donate in the window period prior to the appearance of HBsAg, (2) chronic carriers with low levels of HBsAg that may not be detectable with current testing methods, and (3) rare HBV mutants with altered HBsAg epitopes. These three special situations provide a convincing argument to continue anti-HBc testing of the blood supply.

D. Clinical Course and Sequelae

The clinical symptoms of HBV infection and the associated hyperbilirubinemia and elevation of the ALT resolve within 4–6 months after infection in 90–95% of acute HBV infections. Symptomatic illness, with anorexia, malaise, and nausea, may last 2 months or more, requiring a prolonged period of convalescence Fulminant HBV infection, with life-threatening coagulopathy, encephalopathy, cerebral edema, and acute liver failure, occurs in 1% of HBV infections. Patients with fulminant HBV infection may require liver transplantation; however, the usual outcome is that the engrafted liver also becomes infected with HBV.

A chronic carrier state develops in 3–5% of HBV infections. Chronically infected patients may have ongoing hepatic injury. They also have a high risk of developing cirrhosis (50% within 5 years) and of dying from cirrhosis (relative risk is 12–79 times that of healthy individuals). The serious complications of cirrhosis include ascites, portal hypertension, variceal bleeding, and hepatic encephalopathy. These patients also have a markedly elevated risk of developing hepatocellular carcinoma, with a relative risk from 30- to over 100-fold that of normal individuals. About 5% of chronic carriers per year will have spontaneous late seroconversions with formation of anti-HBsAg and cessation of chronic hepatic injury (Lee, 1997).

E. Management and Prevention

The care of patients with acute HBV infection is supportive, with provision of fluids, analgesics, antipyretics, and antiemetics as needed. Universal precautions, as for all patients, are indicated when treating inpatients with identified HBV.

Interferon alpha-2b has been shown to induce sustained remissions of chronic hepatitis in 30–40% of selected chronic HBV carriers (Perrillo et al., 1990). Some of the responders to interferon therapy apparently clear HBsAg from their circulations (Korenman et al., 1991). Interferon therapy, which is given 3 times weekly over 16 weeks, is thought to have both antiviral and immunomodulatory effects, stimulating host-cell-mediated immunity to clear HBV-infected hepatocytes. The value of corticosteroids along with interferon alpha-2b is less clear. Antiviral agents such as nucleoside analogues have also been administered for the treatment of chronic HBV. It appears that antiviral therapy alone may suppress, but not eradicate, HBV, and resistant HBV mutants can arise during antiviral therapy.

Safe and highly effective recombinant vaccines for HBV are available that induce an anti-HBsAg response and confer protective immunity to HBV in 90–95% of vaccine recipients. The HBV vaccine is usually given by 3 intramuscular injections at intervals of 1 month and 6 months from the initial dose (CDC, 1990). Initially, vaccination for preexposure prophylaxis against HBV was recommended only for groups at increased risk for HBV, such as health care workers, IV drug abusers, household and sexual contacts of HBV carriers, prison inmates, chronic transfusion recipients, and others. In 1991, the Immunization Practices Advisory Committee of the NIH published a recommendation favoring universal HBV vaccination of newborns and children (CDC, 1991a). The American Academy of Pediatrics (AAP) recommends 3 doses of recombinant HBV vaccine for newborns: (1) at birth, (2) at 1–3 months of age, and (3) at $\geq$ 6 months of age, but at least 2 months after the second dose. The AAP also recommends a standard course of 3 doses for unimmunized older children (CDC, 1995). Vaccination against HBV does not usually interfere with a person's eligibility to donate blood, since donated blood is not screened for anti-HBsAg, and the HBV vaccine does not typically elicit an anti-HBc response.

Perinatally acquired HBV is a significant public health issue, because both the chance of perinatal infection and the chance of a chronic carrier state developing are high among infants born to HBV-infected mothers. More than 90% of these infections can be prevented if the infant born to an HBV-infected mother receives HBV vaccine and HBV immune globulin (HBIg) within 12 hours after birth. Prenatal and perinatal screening of pregnant mothers for HBsAg is an important part of this prevention program (CDC, 1991a).

The recommendations for postexposure HBV prophylaxis for exposures other than perinatal vary with the type of exposure. In general, a person who has sexual contact with or known blood exposure to a patient with acute HBV is treated with HBIg and the HBV vaccine, while sexual and household contacts of chronic carriers receive the HBV vaccine (CDC, 1991a).

III. HEPATITIS D (DELTA; HDV)

A. Incidence

HDV is a "defective" virus that can replicate only in cells that are infected with HBV. For the most part, the prevalence of HDV infection correlates with the prevalence of chronic HBV infection. In the U.S.A. and in other countries that have a low overall prevalence of chronic HBV, HDV is found in fewer than 10% of HBV carriers. In some European countries with a high prevalence of chronic HBV, the prevalence of HDV infection may be $\geq$ 20% among asymptomatic HBV carriers and $\geq$ 60% among patients with HBV-related chronic liver disease. However, in Southeast Asia and China, HDV is uncommon, though chronic HBV is relatively common. Localized epidemics of HDV infection among chronic HBV carriers have also been described worldwide. The risk of HDV in the blood supply in the U.S.A. is presumed to be significantly lower than the transfusion-transmitted risk of HBV.

Risk factors for HDV are similar to those for HBV, with percutaneous blood exposure being the most common means of transmission. HDV most commonly occurs in IV drug users, hemophiliacs, and chronic transfusion recipients. Sexual and perinatal transmission of HDV is less common than for HBV.

B. Pathophysiology

HDV is a single-stranded RNA virus that requires HBV as a helper virus. The HDV viral particle is coated with HBsAg. The RNA genome of HDV has ribozyme activity and does not require protein enzymes to replicate.

HDV infection can be acquired either as a coinfection with HBV (i.e., simultaneously with the initial HBV infection) or as a superinfection in an individual with chronic HBV infection. These two scenarios of HDV infection have different serological characteristics and clinical outcomes, as described below (Anderson and Ness, 1994).

C. Diagnosis

1. Clinical Evaluation

The clinical presentation of HDV–HBV coinfection is similar to that of HBV infection alone, except that the symptoms are likely to be more severe and the chance of fulminant hepatic failure is higher than for acute HBV alone. HDV superinfection may present as a second episode of acute hepatitis in an HBV carrier or as an acceleration of the progression of chronic HBV infection.

2. Laboratory Evaluation

In patients who are coinfected with HBV and HDV, HDV antigen and HDV RNA appear about 1 week after the appearance of serum HBsAg and remain detectable for a few months. Patients develop IgM anti-HDV, followed by IgG anti-HDV, during the course of the infection, although in about 15% of HDV–HBV coinfection cases, either the IgM or the IgG anti-HDV cannot be detected. Most patients with HDV–HBV coinfection do not develop chronic infection with HBV; the IgG anti-HDV usually declines and becomes undetectable within a year following the acute episode. Thus, many patients with resolved coinfection with HDV and HBV may have no persistent serological marker of the HDV infection.

In contrast, superinfection with HDV in a patient with chronic HBV typically results in chronic infection with HDV as well. HDV antigen appears in the serum 2–3 months after exposure and persists indefinitely, as does chronic HBV antigenemia. The concentration of HBsAg usually declines as HDV antigen appears. Finally, high titers of IgG anti-HDV develop in the serum and persist indefinitely.

The distinction between HDV–HBV coinfection and superinfection can be made in patients with acute hepatitis and anti-HDV antibody based on a test for IgM anti-HBc. If IgM anti-HBc is present, the patient is more likely to have an acute coinfection with both HDV and HBV. If IgM anti-HBc is absent, the patient is more likely to have an HDV superinfection (Hoofnagle and DiBisceglie, 1991).

D. Clinical Course and Sequelae

The acute presentation of HDV–HBV coinfection is similar to the presentation of HBV infection alone, with symptoms appearing 3–13 weeks after exposure; however, symptoms may be more severe than with HBV infection alone. Chronic hepatitis occurs less frequently in persons with HBV–HDV coinfection than with HBV alone, but coinfection incurs a higher risk of fulminant hepatitis (2–20%) compared with HBV alone.

Chronic HBV carriers who acquire HDV superinfection most often develop chronic HDV infection, and 70–80% of these patients will develop chronic liver disease and cirrhosis (a much higher risk than that of chronic HBV carriers).

E. Management and Prevention

Since HDV is dependent on HBV for replication, HDV–HBV coinfection may be prevented by measures to prevent HBV infection, for example, either postexposure prophylaxis with HBIg or preexposure prophylaxis with HBIg and/or HBV vaccine. There are no specific interventions to prevent HDV superinfection in chronic HBV carriers, other than counselling to avoid behaviors which risk exposure to HDV. Treatment of chronic HDV–HBV infection with interferon alpha-2b is thought to be less effective than in the case of HBV.

There is no additional screening of blood donations for HDV. Since chronic carriers of HDV must also be chronic carriers of HBV, screening the blood supply for HBsAg and anti-HBc effectively screens for HDV as well.

IV. HEPATITIS C (HCV)

A. Incidence

Of the chronic infectious diseases transmitted by blood transfusion, HCV is the most common in the U.S.A., exceeding both HBV and HIV infection in prevalence. HCV disease was described first as a form of post-transfusion hepatitis ("non-A, non-B hepatitis"), but was also thought to be the cause of 15–20% of community-acquired chronic hepatitis cases. The CDC estimates that 3.9 million individuals are currently infected with HCV (the highest prevalence in individuals 30–49 years of age) and that there are 8000–10,000 deaths per year from HCV-related chronic liver disease (CDC, 1998). The world's prevalence of HCV infection may be as high as 3% (Ebeling, 1998). The U.S. incidence of new HCV infections has fallen dramatically since the 1980s. In the 1980s, the incidence of new "non-A, non-B" hepatitis infections in the U.S.A. was estimated at 230,000 cases per year; in 1996, 36,000 new HCV cases per year were estimated (CDC, 1998). Part, but not all, of the decline in the incidence of new HCV infections can be attributed to the discovery of HCV in 1988 and to the development of specific tests for this pathogen.

Prior to the availability of the first tests for HCV, an indolent form of chronic hepatitis was recognized as distinct from HAV and HBV infection, and was referred to as non-A, non-B hepatitis (Aach et al., 1981). Both intravenous drug use and therapeutic blood transfusions were significant risk factors for non-A, non-B hepatitis, each associated with about 20% of these cases. Controlled studies of post-transfusion hepatitis gave risk estimates for non-A, non-B hepatitis at 3–5% of transfused patients. When the HCV ELISA test became available, studies of chronic hepatitis demonstrated that about 90% of what had been called non-A, non-B hepatitis could be attributed to HCV infection (Alter et al., 1989; Aach et al., 1991).

The risk of HCV transmission in blood products declined substantially because of "surrogate" testing strategies to reduce the transfusion-transmitted risk of non-A, non-B hepatitis (see Section IIC2 regarding anti-HBc testing), and probably as a beneficial side effect of screening of the blood supply for HIV. Since the start of screening of blood donors for HCV by a single-antigen EIA (enzyme immunoassay) test in 1990, and then by multiantigen EIA tests in 1992, the incidence of transfusion-transmitted HCV has continued to decline to its most recent risk estimate of 1:103,000 per donor exposure (Schreiber et al., 1996). Additional HCV risk reduction, perhaps to the level of 1 per 500,000–1,000,000 units transfused, is estimated now that nucleic-acid-based HCV screening (NAT) of the blood supply has been widely instituted. However, HCV remains a significant public health problem because there remains a large reservoir of HCV-infected individuals.

The most common risk factor for HCV infection is exposure to blood or blood products (IV drug abuse, chronic transfusion, hemophilia prior to 1987, and exposure to transfusions from known HCV-infected donors). Exposures such as chronic hemodialysis, his-

tory of a sexually transmitted disease, needlestick injury, and prior blood transfusion are associated with smaller risks for HCV infection, although these risks are still greater than those in the general population.

HCV can be transmitted vertically: ~6% of fetuses/infants of HCV-infected mothers become infected. Maternal coinfection with HIV increases the chance of vertical HCV transmission to the fetus/infant to ~17% (CDC, 1998). There is debate about the relative importance of sexual transmission of HCV, primarily due to the relatively low prevalence of HCV infection among partners of HCV-infected individuals. HCV can probably be transmitted sexually, but this route of transmission is likely relatively inefficient (CDC, 1998).

B. Pathophysiology

HCV is a lipid-enveloped viral particle, similar to the flavivirus family, with a 9400-base genome consisting of single-stranded RNA. Six genotypes and over 90 subtypes of HCV have been defined based on nucleotide sequence; type 1 accounts for about 70% of HCV infections in the U.S.A. and Europe, but different genotypes predominate in other regions of the world. The viral proteins detected by multiantigen screening tests are nonstructural proteins of the nucleocapsid and a viral protein called NS4, whose function is unknown (Anderson and Ness, 1994). As stated above, HCV is transmitted from human to human primarily through blood exposure.

During the initial phase of HCV infection, the virus replicates rapidly, having a doubling time of less than 1 day. Symptoms related to HCV infection, if they occur, may begin ~6 weeks following exposure, and seroconversion with formation of anti-HCV antibody occurs at 8–10 weeks postexposure. However, only 80% of HCV-infected patients have detectable antibody at 15 weeks after exposure, and 1–3% may not seroconvert for ~6–9 months after exposure. Most HCV infections become chronic, and of patients with chronic HCV infections, the majority will have ongoing hepatic damage. Liver damage from chronic HCV infection usually progresses slowly, but may result in cirrhosis in 20–30 years (CDC, 1998).

C. Diagnosis

1. Clinical Evaluation

Acute HCV infection presents with a milder form of the syndrome similar to that seen with the other hepatitis viruses. Fatigue, malaise, weight loss, nausea, vomiting, and abdominal pain occur in about 50% of HCV acute infections, and these symptoms may last up to 10 weeks. Some acute HCV infections are entirely asymptomatic. Fulminant hepatitis with a fatal outcome has been described with HCV infection, but is rare. Jaundice and dark urine occur in about 25% of cases. The liver may be enlarged or tender at the time of acute HCV infection.

2. Laboratory Evaluation

Elevation of serum ALT is common, but levels above 800 IU/ml occur in less than one-third of patients with acute HCV infection. A fluctuating pattern of mild ALT elevation is the most common biochemical abnormality in acute HCV infection; therefore, a single serum ALT result cannot establish or exclude a diagnosis of HCV infection. Direct serum bilirubin usually does not exceed 10 mg/dl. There is no pattern of laboratory findings that is predictive of either resolution or chronic HCV infection.

The licensed, commercially available test for antibody to HCV is an enzyme immunoassay format using 2 or more recombinant HCV antigens. The sensitivity of the test is > 97% (CDC, 1998), but it has a significant false-positive rate, especially in low-risk populations (e.g., blood donors). The EIA also has a relatively long mean window period to seroconversion, so the EIA typically only becomes positive $\geq$ 70 days following exposure. Positive EIA tests should be "confirmed" with a recombinant immunoblot assay (RIBA); the RIBA uses the same recombinant HCV antigens as the EIA, but is more specific because it can distinguish antibody reactivity to specific HCV antigens from false-positive cross-reactions that occur in the EIA. Testing of all blood donations with the HCV EIA was mandated in 1990. This screening has greatly improved the safety of the blood supply, and has greatly reduced the risk of blood transfusion as a mechanism of HCV transmission.

The reverse-transcriptase–polymerase chain reaction (RT-PCR) assay for HCV RNA is a highly specific, nucleic-acid-based test, and it detects HCV much earlier than does the EIA or the RIBA (i.e., by 1–2 weeks postexposure). RT-PCR is most useful in detecting HCV infection in the following circumstances: (1) during the the "window period" prior to seroconversion, (2) in neonates, in the presence of transplacentally acquired maternal anti-HCV antibodies, and (3) in monitoring HCV viral load to assess effectiveness of therapy. Patients with chronic HCV may be intermittently positive or negative in the RT-PCR assay.

In 1999, the American Red Cross and other blood collection centers began phasing in methods to screen the blood supply for HCV by these nucleic-acid-based tests (NATs), a strategy which should even more dra-

matically reduce the risk of HCV transmission by transfusion.

D. Clinical Course and Sequelae

The symptoms and laboratory abnormalities associated with HCV infection usually abate within 10 weeks of infection. Fulminant hepatic failure resulting from acute HCV infection is rare.

Most HCV infections become chronic, only 15–25% of infections resolving completely. Of the chronic HCV infections, the majority (60–70%) of patients will develop chronic active hepatitis. Liver damage from chronic HCV infection progresses slowly but may result in cirrhosis in 10–20% of patients after 20–30 years (Alter et al., 1992; Seeff et al., 1992; CDC, 1998). There is an increased risk of developing hepatocellular carcinoma for patients with cirrhosis secondary to HCV (Tong et al., 1995). Since the rate of progression to liver disease is slow and variable, the impact of HCV infection on any individual's morbidity and mortality is difficult to estimate. One case-control study with a 18-year follow-up of transfusion-transmitted non-A, non-B hepatitis demonstrated no effect of this form of hepatitis on mortality (Seeff et al., 1992).

Extrahepatic manifestations of HCV infection include cryoglobulinemia, glomerulonephritis, and other immunologically mediated complications.

E. Management and Prevention

The care of patients with acute HCV infection is entirely supportive. There is no therapeutic intervention that has been shown to reduce the likelihood of chronic HCV infection. Infection control measures for blood-borne pathogens are the same as those for HBV.

Since many individuals with chronic HCV may not be aware they are infected, the CDC has recommended several strategies to identify infected individuals, including HCV lookbacks by blood centers, testing of persons with risk factors for HCV, and increased public education campaigns regarding HCV (CDC, 1998).

Chronic HCV hepatitis has been treated with interferon alpha-2b and with a combination of interferon alpha-2b and the antiviral drug ribavirin; these regimens last 24–48 weeks or more. This combination drug approach has superior effectiveness to interferon alone (Davis et al., 1998; McHutchison et al., 1998), but the overall rate of sustained response (clearance of virus, as detected by RT-PCR and normalization of the ALT) is only 30–40%. The response rate for the most prevalent HCV genotype is significantly lower than that for other, less common genotypes. Antiviral treatment is generally indicated for patients with persistently elevated ALTs and histologic evidence of fibrosis, but it may not be indicated for patients with less active hepatic disease. Counseling to reduce alcohol intake is also suggested for individuals with HCV, since alcohol may accelerate the progression of fibrosis (CDC, 1998).

An important and effective measure for prevention of HCV transmission in the population is screening of blood, organ, and tissue donors for HCV. No vaccine for HCV is available, and intravenous immunoglobulin has no value in postexposure prophylaxis against HCV.

References

Aach, R. D. (1987). Primary hepatic viruses: Hepatitis A, hepatitis B, delta hepatitis and non-A, non-B hepatitis. In "Transfusion-Transmitted Viruses: Epidemiology and Pathology" (S. J. Insalaco and J. E. Menitove, Eds.), pp. 17–40. AABB, Arlington, VA.

Aach, R. D., Szmuness, W., et al. (1981). Serum alanine aminotransferase of donors in relation to the risk of non-A, non-B hepatitis in recipients: The Transfusion-Transmitted Viruses Study. N Engl J Med 304, 989–994.

Aach, R. D., Stevens, C. E., et al. (1991). Hepatitis C virus infection in post-transfusion hepatitis. N Engl J Med 325, 1325–1329.

Alter, H. J., Purcell, R. H., et al. (1989). Detection of antibody to hepatitis C virus in prospectively followed transfusion recipients with acute and chronic non-A, non-B hepatitis. N Engl J Med 321, 1494–1500.

Alter, H. J., Nakatsuji, Y., et al. (1997). The incidence of transfusion-associated hepatitis G virus infection and its relation to liver disease. N Engl J Med 336, 747–754.

Alter, M. J., Margolis, H. S., et al. (1992). The natural history of community-acquired hepatitis C in the U.S. The Sentinel Counties Chronic Non-A, Non-B Hepatitis Study Team. N Engl J Med 327, 1899–1905.

Alter, M. J., Gallagher, M., et al. (1997). Acute non-A-E hepatitis in the U.S. and the role of hepatitis G virus infection. Sentinel Counties Viral Hepatitis Study Team. N Engl J Med 336, 741–746.

Anderson, K. C., and Ness, P. M. (1994). "Scientific Basis of Transfusion Medicine. Implications for Clinical Practice." Saunders, Philadelphia, PA.

Busch, M. P. (1998). Prevention of transmission of hepatitis B, hepatitis C, and human immunodeficiency virus infections through blood transfusion by anti-HBc testing. Vox Sang 74 (Suppl. 2), 147–154.

CDC (1990). Protection against viral hepatitis recommendations of the Immunization Practices Advisory Committee (ACIP). Morb Mortal Wkly Rep 39(RR-2), 1–26.

CDC (1991a). Hepatitis B virus: A comprehensive strategy for eliminating transmission in the U.S. through universal childhood vaccination: Recommendations of the Immunization Practices Advisory Committee (ACIP). Morb Mortal Wkly Rep 40(RR-13), 1–19.

CDC (1991b). Public Health Service inter-agency guidelines for screening donors of blood, plasma, organs, tissues, and semen for evidence of hepatitis B and hepatitis C. Morb Mortal Wkly Rep 40(RR-4), 1–17.

CDC (1995). Notice to readers update: Recommendations to prevent hepatitis B virus transmission—U.S. Morb Mortal Wkly Rep 44(30), 574–575.

CDC (1996a). Hepatitis A among persons with hemophilia who received clotting factor concentrate. Morb Mortal Wkly Rep 45, 29–32.

CDC (1996b). Prevention of hepatitis A through active or passive immunization: Recommendations of the Advisory Committee on Immunization Practices. Morb Mortal Wkly Rep 45(RR15), 1–30.

CDC (1998). Recommendations for prevention and control of hepatitis C virus (HCV) infection and HCV-related chronic disease. Morb Mortal Wkly Rep 47(RR-19), 1–39.

Committee on Infectious Diseases of the American Academy of Pediatrics (1997). "Hepatitis A. The Red Book" (G. Peter, Ed.), pp. 237–246. Am. Acad. Pediat., Elk Grove Village, IL.

Davis, G. L., Esteban-Mur, R., et al. (1998). Interferon alfa-2b alone or in combination with ribavirin for the treatment of relapse of chronic hepatitis C. International Hepatitis Interventional Therapy Group. N Engl J Med 339, 1493–1499.

Dodd, R. Y. (1994). Adverse consequences of blood transfusion: Quantitative risk estimates. In "Blood Supply: Risks, Perceptions, and Prospects for the Future" (S. T. Nance, Ed.), pp. 1–24. AABB, Bethesda, MD.

Ebeling, F. (1998). Epidemiology of the hepatitis C virus. Vox Sang 74 (Suppl. 2), 143–146.

Giacoia, G. P., and Kaprisin, D. O. (1989). Transfusion-acquired hepatitis A. South Med J 82, 1357–1360.

Holland, P. V. (1998). "Post-transfusion hepatitis: Current risks and causes. Vox Sang 74, 135–141.

Hoofnagle, J. H., and DiBisceglie, A. M. (1991). Serologic diagnosis of acute and chronic viral hepatitis. Semin Liver Dis 11, 73–83.

Korenman, J., Baker, B., et al. (1991). Long term remission of chronic hepatitis B after alpha-interferon therapy. Ann Intern Med 114, 629–634.

Lee, W. (1997). Hepatitis B virus infection. N Engl J Med 337(24), 1733–1745.

Lemon, S. (1985). Type A viral hepatitis: New developments in an old disease. N Engl J Med 313, 1059–1067.

Mannucci, P. M., Gdovin, S., et al. (1994). Transmission of hepatitis A to patients with hemophilia by factor VIII concentrates treated with organic solvent and detergent to inactivate viruses. Ann Intern Med 120, 1–7.

McHutchison, J. G., Gordon, S. C., et al. (1998). Interferon alfa-2b alone or in combination with ribavirin as initial treatment for chronic hepatitis C. Hepatitis Interventional Therapy Group. N Engl J Med 339, 1485–1492.

Perrillo, R. P., Schiff, E. R., et al. (1990). A randomized, controlled trial of interferon alfa-2b alone and after prednisone withdrawal for the treatment of chronic hepatitis B. N Engl J Med 323, 295–301.

Schreiber, G. B., Busch, M. B., et al. (1996). The risk of transfusion-transmitted viral infections. N Engl J Med 334, 1685–1690.

Seeff, L. B., Buskell-Bales, Z., et al. (1992). Long-term mortality after transfusion-associated non-A, non-B hepatitis. The National Heart, Lung, and Blood Institute Study Group. N Engl J Med 327, 1906–1911.

Tong, M. J., el-Farra, N. S., et al. (1995). Clinical outcomes after transfusion-associated hepatitis C. N Engl J Med 332, 1463–1466.

33

CMV and Other Herpesviruses

JOHN D. ROBACK

Department of Pathology and Laboratory Medicine
Emory University Hospital
Atlanta, Georgia

The human herpesvirus (HHV) family is composed of 8 viral species categorized into 3 subfamilies based on similarities in genome structure, nucleic acid sequences of conserved open reading frames (ORFs), and biological properties (see Table 1). Most members have a commonly employed name, such as cytomegalovirus (CMV), as well as an accompanying HHV designation according to the *Guidelines of the International Committee on Taxonomy of Viruses* (CMV is HHV-5). The *Alphaherpesvirinae* subfamily contains HSV-1 and -2 (HHV-1 and -2) as well as VZV (HHV-3). The *Betaherpesvirinae* subfamily contains CMV and HHV-6 and -7. Finally, the *Gammaherpesvirinae* subfamily contains Epstein–Barr virus (EBV; also HHV-4) and HHV-8 (previously known as Kaposi's sarcoma (KS)-associated herpesvirus).

Inclusion into the *Herpesviridae* family is based on viral structure. Common characteristics include a linear double-stranded DNA genome of 120–230 kb, a surrounding icosadeltahedral nucleocapsid, a dense tegument, and a lipid envelope containing viral glycoproteins. Mature virions of approximately 200 nm in diameter (ranging from 120 to 300 nm) are released from infected cells. Herpesviruses also share biological similarities, including the capacity for both lytic growth and latency. During the former, active viral gene expression occurs, and viral progeny are produced with ensuing host cell destruction. In contrast, latent infections are characterized by maintenance of the viral genome in an otherwise healthy host cell in the absence of extensive lytic viral gene expression and without viral replication. Latent virus may subsequently reactivate to lytic growth. Herpesviruses differ from one another, however, in species specificities, cellular tropism, and pathologic effects of infection.

This chapter discusses the human herpesviruses with emphasis on those leukocytotropic viruses (primarily CMV) where transmission by transfusion (or transplantation) is associated with significant clinical sequelae. In general, the significance of each virus to transfusion medicine is proportional to (1) the viral prevalence in leukocytes of otherwise healthy blood donors, and (2) the percentage of transfusion recipients that are seronegative (nonimmune; Table 2).

I. HERPES SIMPLEX VIRUSES 1 AND 2 (HHV-1, -2)

A. Viral Characteristics

HSV-1 and -2 are distinct viral species, each encoded by largely colinear genomes of ~ 150 kb.

B. Clinical Aspects of Community-Acquired HSV Infections

HSV-1, which is spread by contact with virus from oral vesicular fluid, is usually acquired early in life, with seropositivity rates reaching 80% by adolescence. In contrast, HSV-2 is usually acquired during sexual contact from vesicular fluid in the anogenital area. The rate of seropositivity is very low until adolescence. In most U.S. populations, 10–60% of adults are seropositive for HSV-2.

TABLE 1 The Human Herpesvirus Family

Subfamily	Genus	Species	HHV designation
Alphaherpesvirinae	Simplexvirus	HSV-1, -2	HHV-1, -2
	Varicellovirus	VZV	HHV-3
Betaherpesvirinae	Cytomegalovirus	CMV	HHV-5
	Roseolovirus	HHV-6	
		HHV-7	
Gammaherpesvirinae	Lymphocryptovirus	EBV	HHV-4
	Rhadinovirus	HHV-8	

C. Viral Pathobiology

After replicating in epithelial cells at the mucosal surface, HSV is retrogradely transported in peripheral nerves to neuronal cell bodies in the dorsal root ganglia, where the viral genome remains in a latent state.

D. Significance of HSV Transmission by Blood Transfusion or Organ Transplantation

While these two viruses have been well studied, there is no evidence that transmission by blood or organ transplantation is of clinical concern. Neither virus efficiently infects peripheral blood mononuclear cells (PBMCs). In patients with disseminated disease and viremia, platelet-associated HSV may be found (Forghani and Schmidt, 1983), however these blood donors would almost certainly be deferred based on clinical screening criteria. Moreover, high seroprevalence rates indicate that most potential transfusion recipients are immune to HSV-1 and/or -2.

E. Recommendations

Because there is no evidence of transfusion transmission, standard blood components are acceptable for transfusion (Table 2).

II. VARICELLA–ZOSTER VIRUS (HHV-3)

A. Viral Characteristics

The VZV genome is ~ 125 kb in length, and contains about 70 ORFs encoding putative viral polypeptides.

B. Clinical Aspects of Community-Acquired VZV Infections

Primary VZV infection is the causative agent of varicella (chicken pox). Occurrence of VZV infection peaks during early childhood, and 95% of the U.S. population is seropositive by 18 years of age. After infection, the virus usually enters lifelong latency, but in elderly or immunocompromised patients VZV can reactivate to cause herpes zoster (shingles).

C. Viral Pathobiology

VZV is believed to enter the host via respiratory tract mucosal surfaces, and then to draining lymph nodes. Replication leads to primary viremia and, at the end of the incubation period, a secondary viremia during which VZV can be isolated from PBMCs. Epidermal dissemination and the characteristic vesicular rash follows. The virus subsequently becomes latent in cells of the dorsal root ganglia.

TABLE 2 Transfusion Transmission of Herpesvirus Infections

Viral species	WBC-associated	Seronegative recipients	Risk of TTI	Recommendations for use of special components	
				Seronegative units	Filtered units
HSV-1,2	—	20–40%	—	n/a	n/r
VZV	—	5%	—	n/a	n/r
CMV	+	40%	+	Acceptable	Acceptable
HHV-6	+	5%	(+)	n/a	n/r
HHV-7	+	5%	(+)	n/a	n/r
EBV	+	10%	+	n/a	Possibly acceptable
HHV-8	+	75–95%	(+)	n/a	n/r

Note. TTI, transfusion-transmitted infections; +, documented; —, not documented, or not of significant concern; (+), theoretically; +, possible but not documented; n/a, not generally available; and n/r, not recommended.

33. CMV and Other Herpesviruses

287

D. Significance of VZV Transmission by Blood Transfusion or Organ Transplantation

At the end of the incubation period VZV can be identified in 0.01–0.001% of PBMCs, but is subsequently cleared within 24–72 hours. Although transiently PBMC-associated, transfusion transmission of VZV does not represent a significant clinical problem because most individuals with acute varicella are too young to donate blood, and most transfusion recipients are immune.

E. Recommendations

Because there is no evidence of transfusion transmission, standard blood components are acceptable for transfusion.

III. EPSTEIN–BARR VIRUS (HHV-4)

A. Viral Characteristics

The EBV genome is ~172 kb in length. The two primary EBV variants, EBV-1 and -2, display extensive sequence similarities and can recombine with one another to form a biological gradient of viral variants.

B. Clinical Aspects of Community-Acquired EBV Infections

EBV is transmitted by close contact, and the majority of the adult population (>90%) has been infected with EBV based on serologic investigations. While primary EBV infections are largely asymptomatic, EBV can cause acute heterophile antibody-positive infectious mononucleosis (IM). EBV has also been implicated as an etiologic factor in a number of malignancies, including post-transplantation lymphoproliferative disorder (PTLD), Burkitt's lymphoma, nasopharyngeal carcinoma, and Hodgkin's disease.

C. Viral Pathobiology

While EBV can produce a lytic infection in cells of mucosal epithelium, B-lymphocytes predominantly support latent infections. Eleven viral genes are expressed during latent B-cell infection, including nuclear antigens (EBNA 1, 2, 3A, 3B, 3C, LP), membrane proteins (LMP 1, 2A, 2B), and noncoding nuclear RNAs (EBER 1, 2). These latent gene products disrupt B-cell regulatory mechanisms, leading to the characteristic polyclonal B-cell lymphoproliferation seen in IM. Although the host cellular immune response eventually controls

the lymphoproliferation, EBV persists in a latent state in 1 in every 10^5–10^6 B-cells in immunocompetent individuals. Latent EBV is believed to undergo sporadic reactivation within B-cells to release progeny virus, and the lymphoid system probably represents the true reservoir of this virus.

D. Significance of EBV Transmission by Blood Transfusion or Organ Transplantation

Transmission of EBV by transfusion has been recognized for a number of years. First known as postperfusion mononucleosis, it presented in a similar manner to classic IM in seronegative recipients of EBV-seropositive transfusions. In most cases, the blood donor was found to be in the incubation period for IM with high B-cell viral load (Walsh et al., 1970). More recently, viral genotyping was used to document EBV transmission from a blood donor to a recipient who subsequently developed EBV-driven PTLD (Alfieri et al., 1996). EBV can also be transmitted by organ grafts.

In general, immunodeficient EBV-seronegative recipients of seropositive organ grafts or blood transfusions are at greatest risk to develop rapidly progressive EBV-associated B-cell lymphoproliferations, such as PTLD. For example, in pediatric orthotopic liver transplantation (OLT), PTLD has been identified in 10–20% of patients immunosuppressed with tacrolimus. PTLD encompasses a range of disorders from benign polyclonal B-cell lymphoid hyperplasia to malignant high-grade non-Hodgkin's lymphoma. In pediatric OLT, PTLD is a cause of significant morbidity, including graft loss, as well as mortality rates of up to 20%.

E. Recommendations

Although not yet the subject of controlled trials, leukoreduction of cellular blood components is likely to decrease the incidence of transfusion-transmitted EBV infections in at-risk seronegative patients, such as immunosuppressed recipients of seronegative liver transplants or other immunocompromised individuals. EBV-seronegative units may also be effective in this setting, but are not routinely available.

IV. CYTOMEGALOVIRUS (HHV-5)

A. Viral Characteristics

The CMV genome, at ~230 kb in length, is the largest of any herpesvirus, encoding at least 208 ORFs. Numerous variants have been identified, but differences in clinical behavior are still under investigation.

B. Clinical Aspects of Community-Acquired CMV Infections

CMV can be shed into body fluids, leading to transmission upon close person-to-person contact. Healthy seropositive individuals intermittently shed CMV into saliva, and CMV can also be found in breast milk of 30% or more of seropositive women. Sexual transmission of CMV is supported by the detection of CMV in the genital tracts of up to 35% of seropositive women. Urine, semen, tears, and feces may also contain infectious CMV virions. CMV can be transmitted transplacentally to the developing fetus *in utero*. In fact, fetal infection occurs in up to 50% of pregnancies complicated by primary CMV infection in a previously seronegative mother, and is a significant cause of birth defects, including mental retardation. In contrast, vertical transmission is reduced to approximately 2% of pregnancies in seropositive mothers, indicating a role for the maternal immune system in limiting transplacental CMV infections.

In immunocompetent individuals, primary CMV infections are often asymptomatic, although CMV may cause about 8% of IM cases. In contrast, significant morbidity and mortality can result from primary infections in immunocompromised individuals, including neonates (< 1200 g bodyweight), patients with leukemia, immunosuppressed solid organ and marrow transplant recipients, and HIV+ individuals. Overall, about 50–70% of the population is seropositive, although the incidence can be higher in urban areas.

C. Viral Pathobiology

During primary infection, viral replication occurs in a number of tissues and cell types. The host cell receptor(s) for CMV and the other host cell proteins that facilitate or restrict CMV replication have not been well defined. CMV can infect epithelial and endothelial cells, as well as other cell types such as fibroblasts. With respect to blood transfusion, CMV infection of leukocytes and leukocyte progenitors is of most significance. While CMV has been found to infect T- and B-lymphocytes, NK cells, neutrophils, and monocytes, evidence suggests that the monocyte subset is most important to the pathogenesis of transfusion-transmitted CMV infection (TT-CMV).

Latent CMV DNA can be found in monocytes of healthy, seropositive individuals (Bolovan-Fritts *et al.*, 1999). Interestingly, CMV DNA has been found in peripheral blood monocytes from some seronegative donors (Larsson *et al.*, 1998), leading to the suggestion that occasional seronegative individuals may have been previously infected with CMV. These findings may indicate that serology alone will not detect all CMV-infected blood donors. While monocytes harbor primarily latent CMV, evidence suggests that differentiation of monocytes into macrophages, for example, in the transfusion recipient, allows reactivation of CMV with generation of infectious viral progeny.

PCR can also detect CMV in bone marrow CD34$^+$ cells from healthy seropositive, and also some seronegative, donors in both early and more differentiated hematopoietic progenitor cells (HPCs; Sindre *et al.*, 1998). These results suggest that while peripheral blood monocytes contain latent CMV, the true reservoir for long-term CMV latency in seropositive individuals may be bone marrow HPCs.

D. Significance of CMV Transmission by Blood Transfusion or Organ Transplantation

TT-CMV is currently of more concern than is transfusion transmission of other herpesviruses. First described in 1966, TT-CMV has subsequently been documented as a cause of significant morbidity and mortality in immunocompromised as well as immunocompetent recipients. Current understanding of the pathogenesis of TT-CMV is based on the tropism of the virus for peripheral blood monocytes, which as described above are believed to harbor, transmit, and produce infectious CMV virions after infusion into seronegative transfusion recipients.

For the prevention of TT-CMV, donor history cannot be used to identify CMV carriers, but serology can differentiate between CMV-seropositive and -seronegative donations. In fact, as compared to unscreened blood, the use of seronegative units can decrease the incidence of TT-CMV from 13 to 37% down to ~2.5% in at-risk transfusion recipients (Miller *et al.*, 1991). Unfortunately, less than half of the donor population is CMV-seronegative. In addition, as discussed above, some seronegative units may harbor CMV DNA, which may explain the low rate of TT-CMV in recipients of seronegative units.

Filtration of cellular blood components, which results in a 3–4 log$_{10}$ reduction in monocytes and other leukocytes, removes cell-associated CMV and decreases the incidence of TT-CMV (Hillyer *et al.*, 1994; Bowden *et al.*, 1995). In fact, it has been suggested that filtered units are essentially equivalent to seronegative units with respect to the risk of TT-CMV. Still, protection from TT-CMV is not absolute, since ~2.5% of immunocompromised patients who receive seronegative or bedside-filtered blood components can develop significant CMV-associated disease (Bowden *et al.*, 1995). It is also noteworthy that CMV can be transmitted by transplantation of solid organs. For example, a

50–60% incidence of primary CMV infection has been seen in heart and heart-lung transplant recipients.

Factors related to the recipient's immune system are likely important to the pathogenesis of TT-CMV. Significantly more immunosuppressed than immunocompetent patients develop TT-CMV disease, including retinitis, esophagitis, and pneumonitis, even though the immunosuppressed patients are preferentially transfused with seronegative units. In addition, *in vitro* allogeneic interactions lead to macrophage differentiation, which renders the cells permissive for CMV replication, supporting the possibility that interactions between transfused leukocytes and the recipient's immune system enhance TT-CMV. While the molecular signals have not been completely elucidated, stimulation of monocyte cultures with IFN-γ or TNF-α produces macrophage differentiation and CMV replication, suggesting that some components of the normal host antiviral response may actually increase the frequency of TT-CMV.

E. Recommendations

Standard blood components are not appropriate for transfusion of seronegative immunocompromised patients due to the unacceptably high risk of TT-CMV. Instead, these patients should receive either seronegative blood components or filtered components, which are essentially equivalent in terms of the associated risk of TT-CMV.

V. HHV-6A AND -6B

A. Viral Characteristics

There are two HHV-6 variants, 6A and 6B, which can be distinguished by genome structure, monoclonal antibody immunoreactivities, host cell tropism in culture, and differing spectrums of disease after natural infections. The HHV-6 genome consists of a single unique segment of 143 kb flanked by terminal direct repeats of 8–13 kb, and encodes 102 potential ORFs. There is over a 90% sequence similarity between HHV-6A and -6B.

B. Clinical Aspects of Community-Acquired HHV-6 Infections

HHV-6 DNA is present in saliva from over 90% of apparently healthy individuals, and is likely spread by salivary secretions during close personal contact. Intrauterine transmission has also been documented by identification of HHV-6 DNA in 1.6% of postpartum cord blood samples. Based on serology, HHV-6 is endemic, with at least 70% of children infected during the first year of life, and 90% infected during infancy. HHV-6 is the causative agent for the majority of cases of the childhood illness roseola infantum (sixth disease), which presents with the maculopapular skin rash exanthem subitum (ES; Yamanishi *et al.*, 1988). Over 95% of ES cases are caused by the HHV-6B subtype. HHV-6 also shows a predilection to neuroinvasiveness during primary infection. Although HHV-6A is detected more frequently in cerebrospinal fluid (CSF) than in PBMCs or saliva, there is still no clear disease association for HHV-6A.

HHV-6 infection has been weakly associated with other diseases, including heterophile antibody-negative mononucleosis, hepatitis, chronic fatigue syndrome, hemophagocytic syndrome, encephalitis, Rosai–Dorfman disease, Kawasaki disease, Kikuchi disease, sarcoidosis, and a variety of lymphoproliferative disorders (Braun *et al.*, 1997). HHV-6 infection of oligodendrocytes has also been seen in central nervous system (CNS) plaques in patients with multiple sclerosis. HHV-6 DNA has been identified in malignant cells from non-Hodgkin's lymphoma, Hodgkin's disease, cervical carcinoma, acute lymphocytic lymphoma, and carcinomas of the oral cavity (Braun *et al.*, 1997). Nonetheless, HHV-6 has not yet been shown to play an etiologic role in any of these disorders.

C. Viral Pathobiology

The primary tropism of HHV-6 is now recognized to be CD4$^+$ T-lymphocytes, although it can also infect CD8$^+$ T-cells, NK cells, cells of the monocyte/macrophage lineage, megakaryocytes, epithelial cells, neurons, and oligodendrocytes. The cellular receptor for HHV-6 has not been identified. After the resolution of primary infection, HHV-6 can persist in a latent state for the lifetime of the host. Salivary glands, PBMCs, and cells of the CNS have all been proposed as sites of latency.

In immunocompromised patients, such as immunosuppressed marrow and solid organ transplant recipients or AIDS patients, reactivation of latent HHV-6 has been associated with a spectrum of clinical sequelae including fever, rash, leukopenia, graft rejection, interstitial pneumonitis, and encephalitis. HHV-6 recovered from bone marrow transplant (BMT) patients is almost always variant B. In AIDS patients, HHV-6 may be a cofactor for HIV progression since both can infect CD4$^+$ cells, and like CMV, HSV-1, and EBV, HHV-6 can transactivate the HIV Long Terminal Repeat (LTR). However, the end results of *in vivo* interactions are unclear, since both suppressive and enhancing effects of HHV-6 on HIV replication have been seen (Lusso *et al.*, 1989; Braun *et al.*, 1997). HHV-6 infection of T-cells and NK cells can also increase CD4 expression, possibly rendering the cells permissive for HIV infection.

D. Significance of HHV-6 Transmission by Blood Transfusion or Organ Transplantation

HHV-6 DNA, predominantly variant 6B, has been identified by PCR in PBMCs in up to 90% of healthy blood donors (Cuende *et al.*, 1994). However, HHV-6 transmission by blood transfusion has not yet been clearly documented, probably because the vast majority of recipients are immune. While HHV-6 transmission by marrow and solid organ transplantation has been demonstrated (Lau *et al.*, 1998), reactivation of latent HHV-6 remains a greater problem than primary HHV-6 infection in these transplant populations.

E. Recommendations

Although transfusion transmission of HHV-6 can theoretically occur, it does not represent a significant clinical concern. In the event that it becomes a concern, the use of filtered blood components is likely to mitigate the risk of transfusion transmission.

VI. HHV-7

A. Viral Characteristics

The HHV-7 genome is composed of a single unique segment of ~ 133 kb flanked by 6-kb direct repeat sequences. There is 40% sequence similarity with the HHV-6A and -6B genomes.

B. Clinical Aspects of Community-Acquired HHV-7 Infections

HHV-7 is transmitted primarily by salivary secretions. Although primary infections occur slightly later than for HHV-6, most children are seropositive for HHV-7 by 5 years of age. HHV-7 appears to cause a small number of childhood ES cases. Although a causative role for HHV-7 in pityriasis rosea has been suggested (Drago *et al.*, 1997), other disease associations have not been proven.

C. Viral Pathobiology

HHV-7 was first isolated in 1990 from CD4+ T-lymphocytes of a healthy individual (Frenkel *et al.*, 1990). HHV-7, along with HHV-6A and -6B, are the only human herpesviruses that are primarily T-lymphotropic. HHV-7 also infects salivary glands, and is frequently found in saliva of seropositive patients. The CD4 membrane protein has been identified as the HHV-7 receptor, explaining its T-cell tropism. Studies have suggested that HHV-7 and HIV can reciprocally interfere with one another during infection of CD4+ cells (Lusso *et al.*, 1994).

D. Significance of HHV-7 Transmission by Blood Transfusion or Organ Transplantation

The possibility exists that HHV-7 may be transmitted by blood transfusion, since HHV-7 DNA was found in peripheral blood buffy coats of 66.1% of blood donors (Wilborn *et al.*, 1995). However, most transfusion recipients are immune, and the disease spectrum is likely limited, so the clinical significance of transfusion transmission is currently unclear.

E. Recommendations

These are the same as for HHV-6 (above).

VII. HHV-8

A. Viral Characteristics

The HHV-8 genome is ~ 170 kb in size, and contains at least 81 open reading frames. Gene products of potential interest include homologues to mammalian cyclin D, cytokines (vIL-6, vMIP), cytokine-regulated proteins (vIRF), cytokine receptors (v-GCR), and Bcl-2.

B. Clinical Aspects of Community-Acquired HHV-8 Infections

As compared to the other human herpesviruses, the seroprevalence of HHV-8 is low, with only 1–5% of healthy adults testing seropositive with some assays (Kedes *et al.*, 1996). However, if lytic antigens are used as serological targets, up to 25% of the general population, 90% of HIV-infected homosexual men, and almost all KS patients are seropositive (Lennette *et al.*, 1996).

Most evidence suggests that HHV-8 is sexually transmitted, and seropositivity is rare prior to adolescence. However, HHV-8 can also be detected in salivary and nasal secretions from some seropositive patients, suggesting another potential route of transmission. A possible explanation for the low seroprevalence is inefficiency of horizontal transmission. A recent epidemiologic study of ORF-K1 glycoprotein gene sequences showed that although HHV-8 is an ancient human pathogen (as opposed to a relatively recently acquired one), its penetrance in the population remains low (Zong *et al.*, 1999).

C. Viral Pathobiology

HHV-8 DNA sequences were first identified in biopsies of Kaposi's sarcoma using the technique of representational difference analysis (Chang *et al.*, 1994). DNA sequence analysis revealed significant similarities to herpesvirus saimiri, a herpesvirus of subhuman primates, and EBV. Most evidence suggests that HHV-8 plays a causative role in KS lesions from both AIDS and non-AIDS patients, as well as in body-cavity-based lymphoma (BCBL, or primary effusion lymphoma), and multicentric Castleman's disease (MCD). While currently the subject of debate, it remains unclear whether HHV-8 has an etiologic role in multiple myeloma.

The mechanisms by which HHV-8 could immortalize cells of KS lesions or BCBL remain the topic of intensive investigations. While the viral cyclin D and Bcl-2 homologues may play a role, it has been shown that almost all KS spindle tumor cells are latently infected, with only a minority expressing lytic genes. In BCBL, most cells are infected, once again displaying predominantly latent expression. Surprisingly, in MCD only a few of the cells in the lesion are infected and those all show lytic gene expression. The pathobiological significance of these findings awaits further investigation.

D. Significance of HHV-8 Transmission by Blood Transfusion or Organ Transplantation

HHV-8 DNA can be found in PBMCs of 55% of patients with KS, and HHV-8$^+$ spindle cells can also be isolated from their peripheral blood (Whitby *et al.*, 1995). However, the prevalence of HHV-8 DNA in blood samples from healthy donors is currently unclear. In one reported case, infectious HHV-8 was isolated from PBMCs of a healthy donor (Blackbourn *et al.*, 1997). After activation of CD19$^+$ cells with phorbol ester and IL-6, HHV-8 transmission to naive PBMCs was documented, supporting the possibility of HHV-8 transmission by blood transfusion. However, a study of renal allograft recipients which demonstrated HHV-8 transmission by transplantation of solid organs from seropositive donors also showed that seronegative recipients of seronegative transplants remained seronegative, despite numerous blood transfusions. These data indicate that if HHV-8 is transmitted by blood transfu-

sion, the incidence is very low (Regamey *et al.*, 1998). Thus, while HHV-8 can infect PBMCs of blood donors, and most transfusion recipients are seronegative, HHV-8 transmission by transfusion appears to be a rare event. There is not yet an adequate explanation for these findings.

E. Recommendations

As discussed above, transfusion transmission of HHV-8 is not currently considered a significant clinical concern, and the use of filtered components for the sole purpose of preventing TT-HHV-8 cannot be recommended at this time. However, HHV-8 is present in PBMCs and has been implicated as an etiologic agent in malignancies, and a large seronegative transfusion recipient population exists. Thus, the above recommendation may require modification if transfusion transmission is confirmed in future investigations.

References

Alfieri, C., Tanner, J., Carpentier, L., *et al.* (1996). Epstein-Barr virus transmission from a blood donor to an organ transplant recipient with recovery of the same virus strain from the recipient's blood and oropharynx. *Blood* **87**, 812–817.

Blackbourn, D. J., Ambroziak, J., Lennette, E., *et al.* (1997). Infectious human herpesvirus 8 in a healthy North American blood donor. *Lancet* **349**, 609–611.

Bolovan-Fritts, C. A., Mocarski, E. S., and Wiedeman, J. A. (1999). Peripheral blood CD14(+) cells from healthy subjects carry a circular conformation of latent cytomegalovirus genome. *Blood* **93**, 394–398.

Bowden, R. A., Slichter, S. J., Sayers, M., *et al.* (1995). A comparison of filtered leukocyte-reduced and cytomegalovirus (CMV) seronegative blood products for the prevention of transfusion-associated CMV infection after marrow transplant. *Blood* **86**, 3598–3603.

Braun, D. K., Dominguez, G., and Pellett, P. E. (1997). Human herpesvirus 6. *Clin Microbiol Rev* **10**, 521–567.

Chang, Y., Cesarman, E., Pessin, M. S., *et al.* (1994). Identification of herpesvirus-like DNA sequences in AIDS-associated Kaposi's sarcoma. *Science* **266**, 1865–1869.

Cuende, J. I., Ruiz, J., Civeira, M. P., and Prieto, J. (1994). High prevalence of HHV-6 DNA in peripheral blood mononuclear cells of healthy individuals detected by nested-PCR. *J Med Virol* **43**, 115–118.

Drago, F., Ranieri, E., Malaguti, F., *et al.* (1997). Human herpesvirus 7 in patients with pityriasis rosea. Electron microscopy investigations and polymerase chain reaction in mononuclear cells, plasma and skin. *Dermatology* **195**, 374–378.

Forghani, B., and Schmidt, N. J. (1983). Association of herpes simplex virus with platelets of experimentally infected mice. *Arch Virol* **76**, 269–274.

Frenkel, N., Schirmer, E. C., Wyatt, L. S., *et al.* (1990). Isolation of a new herpesvirus from human CD4+ T cells. *Proc Natl Acad Sci USA* **87**, 748–752.

Hillyer, C. D., Emmens, R. K., Zago-Novaretti, M., *et al.* (1994).

Methods for the reduction of transfusion-transmitted cytomegalovirus infection: Filtration versus the use of seronegative donor units. *Transfusion* **34**, 929–934.

Kedes, D. H., Operskalski, E., Busch, M., *et al.* (1996). The seroepidemiology of human herpesvirus 8 (Kaposi's sarcoma-associated herpesvirus): Distribution of infection in KS risk groups and evidence for sexual transmission. *Nat Med* **2**, 918–924.

Larsson, S., Soderberg-Naucler, C., Wang, F. Z., *et al.* (1998). Cytomegalovirus DNA can be detected in peripheral blood mononuclear cells from all seropositive and most seronegative healthy blood donors over time. *Transfusion* **38**, 271–278.

Lau, Y. L., Peiris, M., Chan, G. C., *et al.* (1998). Primary human herpes virus 6 infection transmitted from donor to recipient through bone marrow infusion. *Bone Marrow Transplant* **21**, 1063–1066.

Lennette, E. T., Blackbourn, D. J., and Levy, J. A. (1996). Antibodies to human herpesvirus type 8 in the general population and in Kaposi's sarcoma patients. *Lancet* **348**, 858–861.

Lusso, P., Ensoli, B., Markham, P. D., *et al.* (1989). Productive dual infection of human CD4+ T lymphocytes by HIV-1 and HHV- 6. *Nature* **337**, 370–373.

Lusso, P., Secchiero, P., Crowley, R. W., *et al.* (1994). CD4 is a critical component of the receptor for human herpesvirus 7: Interference with human immunodeficiency virus. *Proc Natl Acad Sci USA* **91**, 3872–3876.

Miller, W. J., McCullough, J., Balfour, H. H., Jr., *et al.* (1991).

Prevention of cytomegalovirus infection following bone marrow transplantation: A randomized trial of blood product screening. *Bone Marrow Transplant* **7**, 227–234.

Regamey, N., Tamm, M., Wernli, M., *et al.* (1998). Transmission of human herpesvirus 8 infection from renal-transplant donors to recipients. *N Engl J Med* **339**, 1358–1363.

Sindre, H., Tjoonnfjord, G. E., Rollag, H., *et al.* (1996). Human cytomegalovirus suppression of and latency in early hematopoietic progenitor cells. *Blood* **88**, 4526–4533.

Walsh, J. H., Gerber, P., and Purcell, R. H. (1970). Viral etiology of the postperfusion syndrome. *Am Heart J* **80**, 146.

Whitby, D., Howard, M. R., Tenant-Flowers, M., *et al.* (1995). Detection of Kaposi sarcoma associated herpesvirus in peripheral blood of HIV-infected individuals and progression to Kaposi's sarcoma. *Lancet* **346**, 799–802.

Wilborn, F., Schmidt, C. A., Lorenz, F., *et al.* (1995). Human herpesvirus type 7 in blood donors: Detection by the polymerase chain reaction. *J Med Virol* **47**, 65–69.

Yamanishi, K., Okuno, T., Shiraki, K., *et al.* (1988). Identification of human herpesvirus-6 as a causal agent for exanthem subitum. *Lancet* **1**, 1065–1067.

Zong, J. C., Ciufo, D. M., Alcendor, D. J., *et al.* (1999). High-level variability in the ORF-K1 membrane protein gene at the left end of the Kaposi's sarcoma associated herpesvirus genome defines four major virus subtypes and multiple variants or clades in different human populations. *J Virol* **73**, 4156–4170.

34

HIV and HTLV

UNA O'DOHERTY

Department of Pathology and Laboratory Medicine
Hospital of the University of Pennsylvania
Philadelphia, Pennsylvania

WILLIAM J. SWIGGARD

Department of Medicine
Hospital of the University of Pennsylvania
Philadelphia, Pennsylvania

Beginning in the early 1980s, the HIV epidemic transformed the practice of transfusion medicine significantly. Still, transfusion practices to reduce the risk of transmission of HIV continue to evolve. The need to reduce the risk of HIV transmission by blood transfusion has resulted in improvements in donor history taking and has accelerated the development of screening tests, not only for HIV, but also for other blood-borne viruses, including human T-lymphotropic virus (HTLV).

HIV and HTLV are retroviruses with similar structures. The risk of disease transmission from transfusion is similar for both viruses and is now extremely low in the United States as the result of continuing efforts to improve the safety of the blood supply. However, as recently as 1983, when AIDS was first emerging as an important new disease, the risk of HIV-1 transmission from blood transfusion was unacceptably high. This chapter reviews the modes of transmission and the pathophysiology of both of these viruses. Also included are recommended transfusion practice guidelines for infected recipients.

I. DESCRIPTION

A. Genomic Organization

Like all retroviruses, HIV (types 1 and 2) and HTLV (types I and II) contain RNA genomes (see Figure 1) that encode structural proteins (gag), enzymes (pol), and receptors (env). The gag genes encode proteins which initiate viral assembly and determine the shape of the virus—the matrix, capsid, and nucleocapsid pro-

teins. The pol enzymes include proteins that convert the RNA genome into DNA (reverse transcriptase), integrate the DNA reverse transcript into the host genome (integrase), and separate the gag and gag–pol gene products, which are initially translated as polyproteins, into functional units (protease). The env gene encodes a receptor required for binding and fusing with target cells.

B. Virion Structure

Figure 2 is a simplified diagram of a retrovirus. Two copies of the RNA genome are encased in the cone-shaped capsid, along with nucleocapsid, reverse transcriptase, and integrase proteins. The pol enzymes within the virion are required for initiation of the viral life cycle and viral assembly. The capsid is enclosed in a shell of matrix proteins that associate with the envelope, a lipid bilayer derived from the plasma membrane of an infected cell. The viral receptor proteins span this envelope.

C. Accessory Genes

HTLV-I and -II (Figure 1) encode 2 unique accessory proteins, tax and rex, which regulate RNA transcription and splicing, respectively. Analogous accessory proteins in HIV-1 and -2, tat and rev, have similar functions. In addition, HIV-1 and -2 express several additional unique accessory proteins, including Vpr, Vif, Nef, and Vpu or Vpx. The functions of these proteins are currently not well understood, but are under intense study.

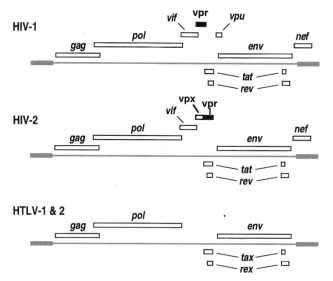

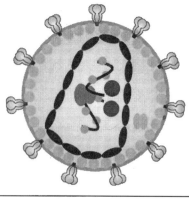

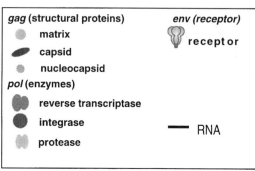

FIGURE 1   Genomic structures of HIV-1 and -2 and HTLV-I and -II. Boxes show coding regions. In addition to the canonical retroviral genes (gag, pol, and env), HTLV-I and -II contain 2 accessory genes, tax and rex. HIV-1 and -2 are unique among retroviruses in possessing 6 accessory genes, vif, vpr, tat, rev, nef, and either vpu (HIV-1) or vpx (HIV-2).

FIGURE 2   Simplified diagram of a retrovirus. Accessory gene products have been omitted for clarity. The current ELISA test (1995) for HIV-1 and -2 tests for antibodies that bind recombinant gag and env proteins immobilized on beads. The ELISA test for HTLV-I and -II tests for antibodies that bind a mixture of a viral lysate and recombinant env proteins. The p24 assay that was implemented in 1996 detects the HIV-1 capsid protein in the serum of individuals.

D. Sequence Homology and Serological Cross-Reactivity

The RNA genomes of HTLV-I and -II are 9 and 8.9 kilobases (kb) in length, respectively, and they share 65% nucleotide homology. An antibody screening test for HTLV-I exposure was implemented in 1988, and cross-reacts with 44–91% of HTLV-II isolates (Anderson et al., 1989).

The genomes of HIV-1 and -2 are both 9.2 kb in length. They share 60% amino acid identity between their gag and pol proteins, and 37% identity between their env receptors (Clavel, 1987). Not all sera from HIV-2-infected individuals cross-react with the recombinant HIV-1 gag and env gene products that are used in the HIV-1 screening ELISA antibody test (Denis et al., 1988). As a result, the screening ELISA test for HIV has been expanded to include HIV-2 antigens, even though this virus is detected in only 3 out of 74 million blood donations in the United States. Based on sequence similarities, HIV-1 isolates are organized into 3 groups, called M (main), N, and O. HIV-1 group M is also divided into 9 subtypes or clades (A through I) based on envelope sequence. The vast majority of cases of HIV infection in the U.S.A. are HIV-1, group M, subtype B. The current ELISA test in the U.S.A. detects antibody responses against all subtypes of group M HIV-1, but not all antibody responses against group

O or N. The first case in the U.S.A. of group O HIV-1 was reported in 1994, and was not detectable by the existing HIV-1 ELISA tests. Most cases of group O HIV-1 infection occur in West Africa and France (Pau et al., 1998). Recently, the European ELISA has been modified to enhance detection of group O HIV-1 antibodies. Continued surveillance is required to determine if the HIV-1 screening tests used in the U.S.A. should be modified to detect group O. Currently, the FDA does not require screening HIV ELISA tests to detect groups O and N.

II. INCIDENCE AND TRANSMISSION RATES FROM BLOOD TRANSFUSIONS

A. HIV Transmission Risk

The exact number of HIV-infected individuals in the U.S.A. is unknown, but the seroprevalence among military recruits is between 1 in 2000 and 1 in 4000 (Renzullo et al., 1995). The prevalence of HIV among blood donors is much lower: only 6 per 100,000 blood donations are positive for anti-HIV antibodies (PHS, 1996). Because donors are screened for antibodies, only individuals with acute HIV infection (i.e., those in the "window period" prior to the appearance of anti-HIV antibodies) should escape detection by screening methods, before implementation of nucleic acid screening for HIV. Among screened repeat donors, the estimated incidence is about 3 per 100,000 donors per year, and is higher among first-time blood donors—7 per 100,000 donors per year (Janssen *et al.*, 1998). Because the window period prior to seroconversion is only an average of 16 days or 0.04 years, the actual risk of HIV transmission has been calculated to be 1 in 2,000,000 to 1 in 200,000 blood transfusions in 1996, before the implementation of nucleic acid testing (NAT; Schreiber *et al.*, 1996). This represents a dramatic improvement over the HIV transfusion-transmission rate that existed in 1983, before HIV screening was introduced.

B. Evolution of Screening Tests for HIV

Screening tests have become increasingly sophisticated, starting with exclusion of individuals with high-risk behavior (implemented in 1983), anonymous self-exclusion after donation, screening for antibodies against HIV using whole virus lysate as the immobilized ELISA antigen (implemented in 1985, with an initial mean window period of 56 days; Petersen *et al.*, 1994), and surrogate testing for hepatitis (implemented in 1986; AuBuchon *et al.*, 1997).

In the mid-1990s, the average window period for HIV antibody tests decreased to 22 days following the introduction of a more sensitive ELISA, in which recombinant gag and env proteins were used as the immobilized antigens (Busch *et al.*, 1995). In 1996, testing for p24, the capsid antigen (see Figure 2), was implemented. The addition of this test reduced the window period to an average of 16 days. There has been considerable debate about the cost effectiveness of this test, since it is predicted to detect only 4 to 6 p24 antigen-positive units per 12 million donated units per year (PHS, 1996). This translates into a cost of $10 million per HIV transmission prevented, or $2.3 million per quality-adjusted life year saved (Busch and Alter, 1996). Nonetheless, this test reduces the number of undetected infectious units by 25%. Screening of donated units for viral RNA reduces the window period further, to an average of only 11 days (Busch *et al.*, 1995). Viral RNA assays should also detect rare patients with antibody-negative HIV infection, one case of which has been documented in the current literature (Schockmel *et al.*, 1996).

TABLE 1 Probability of Transmission of HIV Following Various Types of Exposure

Exposure type	Probability of HIV transmission
Transfusion of HIV-negative[a] blood unit	1/200,000 to 1/2,000,000
Transfusion of HIV-infected blood unit	9/10
Needle stick	1/300
Vertical transmission	1/4
Sexual transmission	
Vaginal intercourse	1/1000 to 1/2000
Anal intercourse	1/31 to 1/1250

[a] Tested negative when screened for HIV infection.

C. HIV Transmission Risks of Other Exposures

Table 1 compares the risk of HIV transmission from a screened blood unit with other exposures, in the absence of antiretroviral therapy (Carpenter *et al.*, 1999; Katz and Gerberding, 1997). This information may be useful for discussion regarding informed consent prior to transfusion of allogeneic blood. The transmission rates of other exposures vary greatly, in part because they depend on the donor's viral load at the time and the volume of infected material to which the recipient is exposed. For example, the risk of HIV transmission from the transfusion of an infected unit of blood is typically much greater than the risk from sexual contact or a needle stick, as the transfusion of an infected unit of blood exposes the individual to a greater volume of infected material. Vertical transmission rates have been dramatically reduced by treating the mother before delivery and/or the newborn after delivery with antiretrovirals. The use of postexposure prophylaxis along with reduced viral loads due to widespread use of highly active antiretroviral therapy should reduce actual HIV transmission rates for all types of exposures.

D. HTLV Transmission Risk

The current incidence of HTLV infection in the United States is 1 in 6250 individuals identified as being HTLV-seropositive per year. Approximately half of these individuals are infected with HTLV-I and half are infected with HTLV-II (Khabbaz *et al.*, 1993). Risk factors for HTLV-I infection include birth in an endemic area (Japan, the Caribbean), having had unprotected sex with someone from an endemic area, injection drug use, and blood transfusion. Injection drug users in the U.S.A. are more likely to have HTLV-II than HTLV-I (Khabbaz *et al.*, 1993).

Transmission from mother to child usually occurs through breast milk (Khabbaz *et al.*, 1993). Sexual transmission is 60-fold more likely from man to woman than woman to man (Khabbaz *et al.*, 1993). The transmission rate of HTLV-I or -II in a recipient of an infected blood unit is between 20 and 60% (Schreiber *et al.*, 1996). Infected leukocytes must be present for transmission to occur, so that the number of leukocytes and the time of storage affect the transmission rate. Noncellular blood products such as fresh frozen plasma (FFP) appear to be incapable of infecting recipients with either HTLV-I or -II (Vrielink *et al.*, 1997). The risk of transmission of HTLV from a screened unit of blood is low (1 in 640,000 units; Schreiber *et al.*, 1996). The average window period of HTLV as measured by the current HTLV-I and -II ELISA test is 51 days (Vrielink *et al.*, 1997).

III. PATHOPHYSIOLOGY

A. Cellular Targets of Infection

1. HIV-1 and -2

HIV-1 and -2 have been shown to infect $CD4^+$ lymphocytes, $CD4^+$ monocytes, macrophages, and dendritic cells. Fusion occurs most efficiently in target cells that express both CD4 and a chemokine receptor, usually utilizing CCR5 for macrophage-tropic strains of HIV, or CXCR4 for T-tropic strains. Virions isolated from individuals who have recently seroconverted utilize predominantly CCR5, suggesting that these virions are transmitted more efficiently. With disease progression, HIV-1 evolves to use an expanded range of receptors, including CXCR4. Within a minority of infected T-cells, HIV DNA integrates into the host genome (Chun *et al.*, 1998). A still smaller subset of infected T-cells actively produces new virions (Natrajan *et al.*, 1999). HIV infection leads to a gradual decline in $CD4^+$ T-cells by unknown mechanisms, leading eventually to AIDS, except in a small number of long-term nonprogressors.

2. HTLV-I and -II

HTLV-I productively infects $CD4^+$ T-cells. *In vivo*, 99% of HTLV-I DNA resides in these cells; however, *in vitro*, productive infection of nonlymphoid cells, including fibroblasts and endothelial cells, has been demonstrated. In addition, HTLV-I has been isolated *in vitro* from EBV-transformed B-cells from patients, suggesting that infection of B-cells may occur. HTLV-II infects predominantly $CD8^+$ T-cells. HTLV transforms

T-cells *in vitro*, and individuals infected with HTLV-I occasionally develop adult T-cell leukemia (ATL; see Section V, Clinical Course and Sequelae). It is not known whether HTLV-I causes this *in vivo* transformation directly; although the leukemic cells contain HTLV-I DNA in their genomes, viral RNA production has never been detected within these cells.

IV. DIAGNOSIS

Tests utilized for screening of blood products for HIV and HTLV are discussed in Chapter 2. The same ELISA tests that are utilized to screen blood products for antibodies against HIV-1 and -2 and HTLV-I and -II are also used to make the primary diagnosis. A positive ELISA test is first repeated in duplicate. If repeatedly positive, the sample is tested for antibodies against HIV or HTLV proteins by a Western blot. The titer that results in a positive ELISA can be utilized to identify patients who have recently seroconverted.

V. CLINICAL COURSE AND SEQUELAE

A. HIV

Primary HIV-1 infection can be asymptomatic or can be associated with a self-limited viral syndrome resembling mononucleosis or influenza. Laboratory abnormalities include transient leukopenia, lymphopenia, and thrombocytopenia. CD4/CD8 ratios may be acutely inverted. During primary infection, there is a high titer of HIV virions and p24 (see Figure 2). After acute infection, HIV-positive individuals can expect to be symptom-free for several years. Their CD4 counts slowly decline until they become immunosuppressed and develop opportunistic infections. The long-term prognosis of HIV-infected individuals has been dramatically improved recently with the introduction of highly active antiretroviral therapy (i.e., "HAART" therapy; see Section VI, Management and Prevention).

B. HTLV-I Associated Diseases

1. Tropical Spastic Paraparesis (TSP)

TSP is a slowly progressive chronic spastic paraparesis, previously referred to as HTLV-I-associated myelopathy (HAM). Symptoms are mostly indicative of upper motor neuron lesions, including lower extremity weakness, hyperreflexia with clonus and Babinski signs, urinary incontinence, and impotence. In addition, patients exhibit sensory disturbances and lower back pain

(Manns *et al.*, 1992). Autopsy studies and MRI scans show lesions in the white matter (especially in the thoracic spinal cord). There has been 1 well-documented case in the U.S.A. of TSP myelopathy after HTLV-I infection acquired by transfusion (Gout *et al.*, 1990). In this case, the myelopathy developed unusually rapidly, just 18 weeks after transfusion. The usual latent period for myelopathy is 6 months to 8 years (Vrielink *et al.*, 1997). Patients with TSP usually have a high viral load and polyclonal HTLV-I integration. The lifetime risk for TSP among HTLV-I-infected patients is low (< 1%; Khabbaz *et al.*, 1993).

2. Adult T-Cell Leukemia/Lymphoma (ATL)

There are several subtypes of ATL, with varying severity and location of symptoms (see Table 2; Khabbaz *et al.*, 1993). All subtypes of ATL are associated with an oligoclonal or clonal population of CD4$^+$ T-cells, each carrying a single integrated copy of HTLV-I. The range of symptoms in ATL is determined by the locations where the malignant CD4$^+$ T-cells accumulate. While the subtypes pre-ATL, smoldering ATL, and chronic ATL can all progress to acute ATL, 50% of the pre-ATL cases spontaneously remit. Prominent lymphadenopathy is seen in lymphoma ATL. These patients have a rapidly progressive course and usually die within 10 months. Acute ATL has the most severe symptoms, including all the signs and symptoms of the other categories, but patients also develop visceral involvement, lytic bone lesions, and sometimes interstitial pneumonitis, and die rapidly. These patients are immunosuppressed and can develop opportunistic infections. Laboratory abnormalities include "flower cells," abnormal lymphocytes with folded nuclei on blood smears, and hypercalcemia.

The lifetime risk for developing ATL with HTLV-I infection is between 2 and 4% (Khabbaz *et al.*, 1993).

The latent period is estimated to be several decades long, since individuals infected with HTLV-I in childhood tend to develop ATL between 40 and 60 years of age (Khabbaz *et al.*, 1993).

In addition, there are several inflammatory diseases associated with HTLV-I, including chronic inflammatory arthropathy, Sjogren's syndrome, polymyositis, uveitis, T-cell alveolitis, and infective dermatitis (Vrielink *et al.*, 1997).

C. HTLV-II-Associated Diseases

HTLV-II has not been clearly associated with any malignancies (Khabbaz *et al.*, 1993). However, its *in vitro* ability to transform T-cells and its sequence homology to HTLV-I suggest it has the potential to cause transformation of T-cells *in vivo*. There is some evidence that HTLV-II may cause a neurologic syndrome similar to TSP (Vrielink *et al.*, 1997).

VI. MANAGEMENT AND PREVENTION

A. HIV Antiretroviral Therapy

In HIV disease, the rate of progression to AIDS (Mellors *et al.*, 1995) and the number of opportunistic infections (Mouton *et al.*, 1997) have been greatly reduced with the introduction of multidrug, highly active antiretroviral therapy (HAART). Furthermore, at least partial recovery of the immune system has been demonstrated in HIV-infected individuals who have achieved persistently undetectable viral burdens while on HAART (Komanduri *et al.*, 1998). Individuals who have had a significant exposure to HIV (including those exposed by transfusion) should receive antiretroviral therapy, since there is evidence that rapid institution of this therapy decreases the probability of infection (Car-

TABLE 2 Stages of ATL

Stage	Signs and symptoms	Laboratory tests	Median survival
Pre ATL	none	slight lymphocytosis	
Smoldering ATL	Skin lesions	marrow involvement	
Chronic ATL		lymphocytosis	
Lymphoma ATL	lymphadenopathy		10 months
Acute ATL	all of the above and hepatosplenomegaly, interstitial pneumonitis, opportunistic infections	lymphocytosis, neutrophilia, eosinophilia, ↑LDH, hypercalcemia, hyperbilirubinemia, lytic bone lesions	6 months

penter *et al.*, 1999; Katz *et al.*, 1997). Theoretically, individuals treated with HAART during primary infection should have a better prognosis, since there is evidence that the initial "set point" of viral load determines prognosis (Mellors *et al.*, 1995).

B. Transfusion Guidelines for HIV+ and HTLV+ Patients

Very few studies address transfusion guidelines for HIV+ individuals (Hillyer *et al.*, 1999). Blood products given to HIV-infected patients should be leukoreduced by filtration, thereby decreasing exposure to transfusion-transmissible herpesviruses. Additional benefits from leukoreduction are likely, since exposure to allogeneic leukocytes has been shown to increase HIV replication and to increase the likelihood of opportunistic infections, and may reactivate several herpesviruses, such as CMV (Hillyer *et al.*, 1999).

In general, it is not necessary for HIV-positive individuals to receive irradiated products. However, since HIV-positive patients are immunosuppressed, they are theoretically at risk for transfusion-associated graft-versus-host disease (TA-GVHD). Two extensive reviews of TA-GVHD failed to report a single case among HIV-infected individuals (Hillyer *et al.*, 1999).

Appropriate transfusion practices for HTLV-infected individuals are even less clear (Vrielink *et al.*, 1997). Certainly, patients with ATL are immunosuppressed, and therefore at risk for TA-GVHD, as well as at risk for viral illnesses from transfusion-transmissible herpesviruses; thus, ATL patients should receive blood products that are irradiated and filtered. If ATL patients or HIV-infected individuals are CMV-seronegative, they should receive CMV-safe blood products. Because leukoreduction filters reduce the risk of CMV transmission dramatically, to levels equivalent or nearly equivalent to that of CMV-seronegative blood products (Hillyer *et al.*, 1999), it may no longer be necessary to give CMV-seronegative blood products, provided that leukoreduction filters are used.

References

Anderson, D., Epstein, J. S., Lee, T. H., *et al.* (1989). Serological confirmation of human T-lymphotropic virus type I infection in healthy blood and plasma donors. *Blood* **74**, 2585–2591.

AuBuchon, J. P., Birkmeyer, J. D., and Busch, M. P. (1997). Safety of the blood supply in the United States: Opportunities and controversies. *Ann Intern Med* **127**(10), 904–909.

Busch, M. P., and Alter, H. J. (1996). Response to: Cost-effectiveness of p24 antigen testing in preventing transfusion transmission of human immunodeficiency virus infection. *Transfusion* **36**, 382–383.

Busch, M. P., Lee, L. L. L., Satten, G. A., *et al.* (1995). Time course of detection of viral and serologic markers preceding human immunodeficiency virus type 1 seroconversion: Implications for screening of blood and tissue donors. *Transfusion* **35**, 91–97.

Carpenter, C. C. J., Fischl, M. A., Hammer, S. M., *et al.* (1999). Antiretroviral therapy for HIV infection in 1996: Recommendations of an international panel. *JAMA* **276**, 146–154.

Chun, T. W., Engel, D., Berrey, M. M., *et al.* (1998). Early establishment of a pool of latently infected, resting Cd4(+) T-cells during primary HIV-1 infection. *Proc Natl Acad Sci USA* **95**, 8869–8873.

Clavel, F. (1987). HIV-2, the West African AIDS virus. *AIDS* **1**, 135–140.

Connor, R. I., Sheridan, K. E., Ceradini, D., *et al.* (1997). Change in coreceptor use correlates with disease progression in HIV-1-infected individuals. *J Exp Med* **185**, 621–628.

Denis, F., Leonard, G., Sangare, A., Gershy-Damet, G., Rey, J., Soro, B., Schmidt, D., Mounier, M., Verdier, M., Baillou, A., and Barin, F. (1988). Comparison of 10 enzyme immunoassays for detection of antibody to human immunodeficiency virus type 2 in West African sera. *J Clin Microbiol* **26**, 1000–1004.

Gout, O., Baulac, M., Gessain, T., *et al.* (1990). Rapid development of myelopathy after HTLV-I infection acquired by transfusion during cardiac transplantation. *N Engl J Med* **322**, 383–388.

Hillyer, C. D., Lankford, K. V., Roback, J. D., *et al.* (1999). Transfusion of the HIV-seropositive patient: Immunomodulation, viral reactivation, and limiting exposure to EBV (HHV-4), CMV (HHV-5), and HHV-6, 7, and 8. *Tranfusion Med Rev* **13**, 1–17.

Janssen, R. S., Satten, G. A., Stramer, S. L., *et al.* (1998). New testing strategy to detect early HIV-1 infection for use in incidence estimates and for clinical and prevention purposes. *JAMA* **280**, 42–48.

Katz, M. H., and Gerberding, J. L. (1997). Postexposure treatment of people exposed to the human immunodeficiency virus through sexual contact or injection-drug use. *N Engl J Med* **336**, 1097–1100.

Khabbaz, R. F., Fukuda, K., Kaplan, J. E., *et al.* (1993). Guidelines for counseling persons infected with human T-lymphotropic virus type I (HTLV-1) and type II (HTLV-II). *Ann Intern Med* **118**, 448–454.

Komanduri, K. V., Viswanathan, M. N., Wieder, E. D., *et al.* (1998). Restoration of cytomegalovirus-specific CD4+ T-lymphocyte responses after ganciclovir and highly active antiretroviral therapy in individuals infected with HIV-1. *Nature Med* **4**, 953–956.

Manns, A., Wilks, R. J., Murphy, E. L., *et al.* (1992). A prospective study of transmission by transfusion of HTLV-I and risk factors associated with seroconversion. *Int J Cancer* **51**, 886–891.

Mellors, J. W., Kinsley, L. A., Rinaldo, C. R. J., *et al.* (1995). Quantitation of HIV-1 RNA in plasma predicts outcome after seroconversion. *Ann Intern Med* **122**, 573–579.

Mouton, Y., Alfandari, S., Valette, M., *et al.* (1997). Impact of protease inhibitors on AIDS-defining events and hospitalizations in 10 French AIDS reference centres. *AIDS* **11**, F101–F105.

Natrajan, V., Bosche, M., Metcalf, J. A., *et al.* (1999). HIV-1 replication in patients with undetectable plasma virus receiving HAART. Highly active antiretroviral therapy. *Lancet* **353**, 119–120.

Pau, C. P., Hu, D. J., Spruill, C., *et al.* (1996). Surveillance for human immunodeficiency virus type 1 group O infections in the United States. *Transfusion* **36**, 398–400.

Perelson, A. S., Neumann, A. U., Markowitz, M., *et al.* (1996). HIV-1 dynamics in vivo: Virion clearance rate, infected cell life-span, and viral genration time. *Science* **271**, 1582–1586.

Petersen, L. R., Satten, G. A., Dodd, R., *et al.* (1994). Duration of time from onset of human immunodeficiency virus type 1 infectiousness to development of detectable antibody. *Transfusion* **34**, 283–289.

PHS (1996). U.S. Public Health Service guidelines for testing and counseling blood and plasma donors for human immunodeficiency virus type 1 antigen. *Morb Mortal Wkly Rep* **45**, 1.

Renzullo, P. O., McNeil, J. G., Wann, Z. F., *et al.* (1995). United States Military Medical Consortium for Applied Retroviral Research. Human immunodeficiency virus type-1 seroconversion trends among young adults serving in the United States Army. *J Acquir Immune Defic Syndr* **10**, 177–185.

Schockmel, G. A., Balavoine, J. F., Yerly, S., *et al.* (1996). Persistent lack of detectable HIV-1 antibody in a person with HIV infection. *Morb Mortal Wkly Rep* **45**, 181–185.

Schreiber, G. B., Busch, M. P., Kleinman, S. H., *et al.* (1996). The risk of transfusion-transmitted viral infections. *N Engl J Med* **334**, 1685–1690.

Vrielink, H., Zaaijer, H. L., and Reesink, H. W. (1997). The clinical relevance of HTLV type I and II in transfusion medicine. *Tranfusion Med Rev* **11**, 173–179.

35

Other Transfusion-Transmitted Infections

JONATHAN D. KURTIS

*Department of Pathology & Laboratory Medicine
and International Health Institute
Brown University
Providence, Rhode Island*

FRANK J. STROBL

*University of Pennsylvania Health System
Philadelphia, Pennsylvania*

In the United States, protozoal infections are infrequent complications associated with transfusion therapy, with a transmission rate of less than 1 in 1,000,000 (Shulman, 1994). Increases in immigration from and travel to parasite-endemic areas will continue to pose a small but significant risk to the domestic blood supply. Donor screening by history and physical exam remain the cornerstone of efforts to prevent such infections. The principal protozoal infections with the potential for transmission by transfusion are malaria, babesiosis, American trypanosomiasis (Chagas' disease), and toxoplasmosis. Other parasitic infections such as microfilariasis and leishmaniasis have not been associated with transfusion-transmitted infections in the U.S.A. Although not parasitic infections, ehrlichiosis, Lyme disease, syphilis, parvovirus B19, and prion disease (Creutzfeldt-Jakob) will also be discussed in this chapter.

I. MALARIA

A. Description

Human malarial infections are caused by four *Plasmodium* species, *P. falciparum*, *P. ovale*, *P. vivax*, and *P. malariae*, and result in 300–500 million infections and over 2 million deaths each year. In endemic areas (Africa, regions of Asia, South America, Middle East), malaria parasites are inoculated into the human bloodstream as motile sporozoites that require an initial developmental stage within hepatocytes prior to a cyclical infection of erythrocytes (merozoites and schizonts).

In *P. vivax* and *P. ovale* infections, liver stage parasites may enter a dormant phase (hypnozoites) that can reactivate and cause blood stage infection years after the initial infection. Malarial organisms survive cyropreservation of PRBCs with glycerol and remain viable in PRBCs for at least 1 week at 4°C (Guerrero *et al.*, 1983).

B. Incidence

On average, 3 cases of transfusion-transmitted malaria occur in the U.S.A. each year.

C. Pathophysiology

As in mosquito-transmitted malaria, the principal physiologic insult in transfusion-transmitted malaria results from the infection and subsequent destruction of erythrocytes. *Plasmodium ovale* and *P. vivax* merozoites preferentially infect reticulocytes and young erythrocytes, *P. malariae* infects older erythrocytes, and *P. falciparum* infects erythrocytes of all ages. Merozoites bind to the surface of erythrocytes (via the Duffy blood group chemokine receptor for *P. vivax*), are internalized, and begin a cycle of asexual reproduction leading to the formation of schizonts. When the schizont is mature, the erythrocyte ruptures, releasing multiple merozoites that infect additional erythrocytes. This erythrocytic cycle produces parasitemia in which, in the case of *P. falciparum*, 50% of all erythrocytes can be infected; this can result in marked anemia via intravascular RBC destruction as well as splenic sequestration. In addition to RBC destruction, impaired erythropoesis may occur (Jootar *et al.*, 1993). *Plasmodium falciparum* also results in occlusion of small-caliber blood vessels. *Plasmodium falciparum*-infected RBCs express parasite-encoded proteins in knoblike structures on their cell surfaces. These parasite proteins bind to vascular endothelial receptors, resulting in the sequestration of infected RBCs in the microcirculation. By direct occlusion and possibly dysregulated local immune responses, the sequestered parasites contribute to the clinical syndromes of malaria.

D. Diagnosis

The diagnosis of malaria is most commonly made by examination of both thick and thin blood films. Both molecular and immunologic detection methods are available, but not widely employed.

E. Clinical Course and Sequelae

Guerrero and colleagues documented 26 cases of transfusion-transmitted malaria in the U.S.A. between 1972 and 1981, with an estimated risk of 0.25 cases per 1,000,000 transfused units (Guerrero *et al.*, 1983). They identified 9 patients with *P. malariae*, 8 with *P. falciparum*, 8 with *P. ovale*, and 1 with *P. vivax*. *Plasmodium falciparum* and *P. ovale* infection became clinically evident 12 days after transfusion, while the incubation period for *P. malariae* was 35 days. Signs and symptoms included intermittent fever, chills, diaphoresis, malaise, headache, and nausea. In two cases, subclinical infection was diagnosed on routine blood smear. Four of the 26 cases resulted in fatalities attributable to malaria infection, a case fatality rate of 15%.

F. Prevention and Management

The potential for reactivation of infection (*P. vivax* and *P. ovale*) and asymptomatic infection (*P. malariae* and occasionally *P. falciparum*) forms the basis for the deferral of prospective blood donors. Travelers to malaria-endemic areas who have not contracted malaria are deferred from donation for 1 year following return to a nonendemic region. Travelers who have contracted malaria and immigrants from endemic areas are deferred for 3 years.

In addition to screening donors by travel history, proposed strategies for the prevention of transfusion-transmitted malaria include serologic screening of donors, examination of blood films from donors, and (in endemic areas) chemoprophylaxis of transfusion recipients. Because of the low incidence of transfusion-trans-

mitted malaria in the U.S.A. these approaches are not cost-effective.

As in mosquito-acquired malaria, treatment of transfusion-transmitted malaria is complicated by the high prevalence of drug-resistant parasites, particularly for *P. falciparum*. Current drug treatment recommendations should be obtained from the Malaria Branch of the CDC. Because transfusion-transmitted malaria does not involve liver stage parasites, drugs active only against blood stage parasites are required for therapy. In addition, RBC exchange transfusions may be employed as therapy.

II. BABESIOSIS

A. Description

Babesiosis is caused by several RBC parasites of the genus *Babesia*. *Babesia* is transmitted by the bite of an *Ixodes* tick. Endemic areas in the U.S.A. include New England, the Midwest, and the Pacific Northwest. Transfusion-associated babesiosis has been linked to platelet, PRBC, and frozen PRBC units (Popovsky *et al.*, 1988), and is primarily caused by *B. microti* (Herwaldt *et al.*, 1997).

B. Incidence

Human babesiosis is the second most commonly reported transfusion-transmitted parasitic infection after malaria (Shulman, 1994). Although the seroprevalence of *B. microti* in Northeastern U.S.A. has been estimated to be as high as 10%, the seroprevalence in all U.S. blood donors approximates 1% (Leiby, 1998). However, this estimate should be viewed as conservative, as the sera were collected in the winter, not in the summer, when tick bites more frequently occur.

C. Pathophysiology

As in malaria, the principal physiologic insult in babesiosis results from the infection and subsequent destruction of RBCs. Transfusion-transmitted parasites bind to the surface of RBCs, are internalized, and begin a cycle of asexual reproduction leading to the formation of numerous merozoites. When the merozoites are mature, the RBC ruptures, releasing multiple merozoites which infect additional RBCs. This RBC cycle produces parasitemia that, in immunocompromised and asplenic individuals, can infect 25% of all RBCs and results in marked anemia from intravascular destruction. Unlike mosquito-transmitted malaria, *Ba*-

besia does not have an obligate intrahepatic stage, nor does it sequester in vascular beds.

D. Diagnosis

Diagnosis is made by identification of parasites on blood film.

E. Clinical Course and Sequelae

The majority of tick-transmitted *Babesia* infections are asymptomatic, with a minority resulting in severe disease, including hemolysis, renal failure, and death. Immunocompromised and asplenic individuals often have more severe infections. Transfusion-transmitted babesiosis typically presents with fatigue, malaise, and weakness (91%); fever (91%); shaking chills (77%); and diaphoresis (69%; White *et al.*, 1998). Due to concurrent illnesses, severe disease may be more common in transfusion-transmitted *babesiosis* than in tick-acquired infection.

F. Prevention and Management

Due to the low incidence of babesiosis, donor screening by history and physical exam remains the most cost-effective approach for decreasing contamination of the U.S. blood supply. However, because of the frequency of asymptomatic infections, this approach often fails to identify infected donors. Surprisingly, potential donors residing in endemic areas have similar seropositivity compared to donors in nonendemic areas (3.7% vs 4.7%); therefore, deferment based only on geographic location is not effective (Popovsky *et al.*, 1988). Individuals with a history of babesiosis are indefinitely deferred due to the possibility of chronic infection. Treatment of symptomatic infection includes quinine sulfate, clindamycin, and occasionally RBC exchange transfusion.

III. AMERICAN TRYPANOSOMIASIS (CHAGAS' DISEASE)

A. Description

American trypanosomiasis, or Chagas' disease, is caused by infection with the protozoal parasite *Trypanosoma cruzi*. In endemic areas of Mexico and Central and South America, *T. cruzi* is transmitted by the reduviid bug via postprandial defecation into the bite site or surrounding mucus membranes. *Trypanosoma cruzi* parasites remain viable in stored PRBCs for several weeks (Bruce-Chwatt, 1972) and were first impli-

cated in transfusion-transmitted Chagas' disease in 1952 (Wendel, 1998). All blood products contaminated with *T. cruzi*, with the exception of lyophilized plasma and plasma derivatives, are potentially infectious.

B. Incidence

Trypanosoma cruzi remains a significant problem in transfusion therapy for endemic countries (Schmunis *et al.*, 1998). A recent seroepidemiologic study of potential blood donors in Los Angeles found that 1311 (39.5%) of 3320 donors were identified by questionnaire as potentially at risk for transmission of *T. cruzi* (Shulman *et al.*, 1997). Seven donors (0.2%) were reactive by an enzyme immunoassay, and 6 of these 7 were positive in a radioimmunoprecipitation assay. This report underscores the potential for transfusion-transmitted Chagas' disease in America as immigration from endemic countries increases.

C. Pathophysiology

In arthropod-transmitted Chagas' disease, acute infections present as inflammation around the site of inoculation (often the eye, causing the so-called Romaña's sign). Transfusion-transmitted Chagas' disease, however, lacks this local inflammation. Both modes of transmission can result in acute constitutional symptoms resembling the common cold. After a latent period of up to 10 years, the parasite and/or the host's immune response to the parasite may damage the neural tissue of the host's heart and GI tract, resulting in dilated cardiomyopathy, megacolon, and megaesophagus.

D. Diagnosis

Only in the first months of acute disease can *T. cruzi* be demonstrated on a blood smear. Xenodiagnosis (feeding of uninfected reduviid bugs on patients followed by microscopic examination of the bug's gut) is positive in up to 40% of chronic cases. Serodiagnosis can identify anti-*T. cruzi* antibodies in both the acute (IgM) and chronic (IgG) stages.

E. Clinical Course and Sequelae

The symtoms of acute Chagas' disease occur after an incubation period of 20 to 40 days. Fever is the most common (and frequently only) symptom but may be accompanied by lymphadenopathy and hepatosplenomegaly. Cardiac arrhythmia, decreased cardiac ejection fraction, and pericardial effusion occur rarely. Central nervous system involvement, including seizures, myoclonus, and fatigue, has been reported, with menin-gitis and meningoencephalitis occuring very rarely in immunocompromised patients. However, 20% of indi-viduals infected with *T. cruzi* via blood transfusion are asymptomatic until progression to chronic disease. The chronic phase of Chagas' disease typically occurs years to decades later, with the development of dilated car-diomyopathy, megacolon, and megaesophagus.

F. Prevention and Management

In endemic areas, serologic screening represents the most practical approach to preventing transfusion-transmitted Chagas' disease. In the U.S.A. donor screening by history is performed, and any potential donor with a history of Chagas' disease is indefinitely deferred.

Trypanosoma cruzi parasites can be inactivated by treating infected blood units with Gentian violet, or WR6026 (Chiari *et al.*, 1996). There is currently no effective therapy for chronic Chagas' disease, but para-sites can be eliminated during acute infection with Nifurtimox.

IV. TOXOPLASMOSIS

A. Description

Toxoplasma gondii is an obligate intracellular para-site that is usually transmitted to humans by consump-tion of undercooked infected meat. Alternatively, in-gestion or inhalation of infected cat feces can also cause human infection. In adult immunocompetent in-dividuals, infection is usually self-limited and asymp-tomatic. In clinically apparent cases, toxoplasmosis mimics infectious mononucleosis, with generalized lym-phadenopathy, prominent cervical lymph nodes, and low-grade fever. In immunosuppressed individuals, *T. gondii* infection can lead to encephalitis and myocardi-tis. Congential infection may result in chorioretinitis, hepatitis, and encephalitis, particularly if infection oc-curs before the third trimester of pregnancy.

B. Incidence

Toxoplasma gondii is a WBC-associated parasite, and can survive for several weeks in stored whole blood (Raisanen, 1978). Toxoplasmosis has been reported as an unusual complication of transfusion therapy in im-munocompromised patients receiving granulocyte WBC transfusions (Kimball *et al.*, 1965; Roth *et al.*, 1971; Siegel *et al.*, 1971). No case of *T. gondii* infection associated with PRBC transfusion has been reported.

The use of filters for WBC reduction in platelets and PRBCs should theoretically reduce the risk of trans-mission of *T. gondii*.

C. Pathophysiology

Illness typically presents 3–4 weeks after transfusion with a mononucleosis-like syndrome, including fever, rash, lymphadenopathy, and, in severe cases, hepati-tis, meningoencephalitis, pneumonitis, and myocarditis. Pathology results from cellular invasion of the parasite and direct cellular toxicity, as well as from the exuber-ant host inflammatory response.

D. Diagnosis

Toxoplasmosis is diagnosed by serologic tests and by demonstration of the parasite in host tissues.

E. Clinical Course and Sequelae

The only well-documented case of granulocyte-transmitted toxoplasmosis (Siegel *et al.*, 1971) resulted in a fatal outcome (Roth *et al.*, 1971). Extrapolation from naturally acquired infection suggests a 50% cure rate for immunosuppressed individuals treated with pyrimethamine.

F. Prevention and Management

Clinically significant transfusion-transmitted toxo-plasmosis thus far has been reported only in recipients of granulocyte transfusions. Irradiation of granulocyte products may reduce the risk of transmission. For RBC and platelet transfusions, the use of leukoreduction filters (as *T. gondii* is WBC-associated) should theoret-ically reduce the risk of transmission. Clinically signifi-cant toxoplasmosis is treated with pyrimethamine plus sulfadiazine.

V. LEISHMANIASIS

Visceral leishmaniasis (kala-azar) is caused by *Leish-mania donovani*, normally transmitted by the bites of sand flies. The parasite is found in circulating WBCs and thus can also be transmitted by blood transfusion. Several cases of transfusion-transmitted leishmaniasis have been reported outside of the U.S.A. Individuals diagnosed with visceral leishmaniasis should be perma-nently deferred from donating blood. Since no case of transfusion-transmitted leishmaniasis has been docu-mented in the U.S.A., routine screening by history or serology of blood donors to detect asymptomatic carriers is not currently performed.

VI. MICROFILARIASIS

Microfilariasis in humans is caused by *Brugia malayia* (endemic to Asia), *Loa loa* (Africa), *Wuchereria bancrofti* (Asia, Africa, Central America), *Mansonella perstans* (Africa), and *Mansonella ozzardi* (Central and South America). Microfilaria survive in blood stored at 4°C for several weeks. Once transfused, microfilaria can persist in the recipient's circulation for days (*L. loa*, *W. bancrofti*) to years (*M. perstans*, *M. ozzardi*). Fortunately, transfused microfilaria do not develop into adult worms, and thus transfusion-transmitted microfilariasis is a self-limited disease characterized by fever, headache, rash, and inflammatory reactions in host tissues. In the U.S.A., no documented cases of transfusion-transmitted microfilariasis have been reported. Thus, as with leishmaniasis, routine screening by history or serology for asymptomatic donors infected with microfilaria is not performed in the U.S.A. However, individuals diagnosed with microfilariasis should be permanently deferred from blood donation.

VII. SYPHILIS

The spirochete *Treponema pallidum* is the infectious agent that causes syphilis. Treponema pallidum survives at most for 5 days in blood stored at 4°C. Only rare cases of transfusion-transmitted syphilis have been documented. Despite a very high false-positive rate in blood donors, performance of serologic testing for syphilis continues to be mandated by the FDA. The rapid-plasma reagin (RPR) test is the most commonly used screening test for syphilis. Donor units positive for syphilis by serologic screening tests should not be used for allogeneic transfusion.

VIII. EHRLICHIOSIS AND LYME DISEASE

Human granulocytic ehrlichiosis (*Ehrlichia* sp.) and Lyme disease (*Borrelia burgdorferi*) are transmitted to humans by ticks (*Ixodes scapularis* and *Ixodes dammini*, respectively). The seroprevalence of *Ehrlichia* sp. in blood donors in endemic areas of the Northeast ranges from 0.5 to 3.5%. *Borrelia burgdorferi* survives in PRBCs stored at 4°C and plasma stored at −18°C for at least 45 days and in platelets stored at room temperature for at least 6 days (Badon *et al.*, 1989). To date, no case of transfusion-transmitted ehrlichiosis or Lyme disease has been documented. In fact, over 20 recipients of blood products obtained from donors infected with *B. burgdorferi* have failed to demonstrate serologic or clinical evidence of Lyme disease. However, potential donors with a history of Lyme disease or ehrlichiosis should only be accepted if they are completely asymptomatic and have received a full course of appropriate antibiotic therapy.

IX. PARVOVIRUS B19

Parvovirus B19, a non-lipid-enveloped DNA virus, is the cause of the childhood illness erythema infectiosum (fifth disease). In immunocompromised adults, parvovirus B19 can cause severe, sometimes fatal, anemia. The virus infects and lyses RBC precursors in the bone marrow. Approximately 30 to 60% of normal blood donors have antibodies to parvovirus B19 (Luban, 1994). Viremia only occurs during the initial phase of infection. The incidence of viremia in blood donors is estimated to range from 1:3000 to 1:40,000 (Prowse *et al.*, 1997). Transfusion transmission of parvovirus B19 has been reported most commmonly with pooled clotting factor concentrates and pooled plasma. Heat inactivation and solvent/detergent treatment both fail to reliably inactivate parvovirus B19. Rare transmissions with single-donor PRBC, platelet, and plasma blood components have also been reported. Currently, routine screening of blood donors in the U.S.A. for parvovirus B19 is not performed.

X. CREUTZFELDT−JAKOB DISEASE (CJD)

Creutzfeldt−Jakob disease is a spongiform encephalopathy caused by prions (proteinaceous infectious particles). No case of transfusion−transmitted prion disease has been reported in humans, despite the experimental transmission of prion diseases via blood transfusions in animal models (Brown, 1995). In studies with up to 20 years of follow-up, not 1 of nearly 2000 transfusion recipients has acquired CJD from a blood donor who later died of CJD. Nevertheless, individuals at increased risk for CJD are excluded from donating in the U.S.A. This includes individuals with a family history of CJD or individuals who have received human tissue products known to transmit CJD, such as dura mater allografts and pituitary-derived human growth hormone.

In 1996, a new variant of CJD (nvCJD) was described in the U.K. This variant affects younger individuals and is thought to be related to the consumption of cattle suffering from bovine spongiform encephalopathy (BSE; Turner and Ironside, 1998). Because nvCJD may be more transmissible than CJD, the possibility exists that this agent will be transmissible by transfusion. Based on this hypothetical concern, the FDA has issued revised donor deferral guidelines that disallow donation (1) by individuals who have spent more than a total of 6 months in the U.K. between the years 1980 and 1996 (the years of the BSE epidemic), and (2) by individuals who have ever received bovine insulin originating in the U.K.

References

Badon, S. J., Fister, R. D., and Cable, R. G. (1989). Survival of *Borrelia burgdorferi* in blood products. *Transfusion* **29**, 581–583.

Brown, P. (1995). Can Creutzfeldt-Jakob disease be transmitted by transfusion? *Curr Opin Hematol* **2**, 472–477.

Bruce-Chwatt, L. J. (1972). Blood transfusion and tropical disease. *Tropic Dis Bul* **69**, 825–862.

Chiari, E., Oliveira, A. B., Prado, M. A., *et al.* (1996). Potential use of WR6026 as prophylaxis against transfusion-transmitted American trypanosomiasis. *Antimicrob Agents Chemother* **40**, 613–615.

Gerber, M. A., Shapiro, E. D., and Krause, P. J. (1994). The risk of acquiring Lyme disease or babesiosis from blood transfusion. *J Infect Dis* **170**, 231–234.

Guerrero, I. C., Weniger, B. G., and Schultz, M. G. (1983). Transfusion malaria in the United States, 1972–1981. *Ann Intern Med* **99**, 221–226.

Herwaldt, B. L., Kjemtrup, A. M., Conrad, P. A., *et al.* (1997). Transfusion-transmitted babesiosis in Washington State: First reported case caused by a WA1-type parasite. *J Infect Dis* **175**, 1259–1262.

Jootar, S., Chaisiripoomkere, W., Pholvicha, P., *et al.* (1993). Suppression of erythroid progenitor cells during malarial infection in Thai adults caused by serum inhibitor. *Clin Lab Haematol* **15**, 87–92.

Kimball, A. C., Kean, B. H., and Kellner, A. (1965). The risk of transmitting toxoplasmosis by blood transfusion. *Transfusion* **5**, 447–451.

Leiby, D. A. (1998). Short topic 115. Transfusion-transmitted diseases (bacteria & parasites): Latest trends in transfusion-transmitted parasitic infections. In "The Compendium," 51st Annual AABB Meeting, pp. 332–336. AABB, Bethesda.

Luban, N. L. (1994). Human parvoviruses: Implications for transfusion medicine. *Transfusion* **34**, 821–827.

Popovsky, M. A., Lindberg, L. E., Syrek, A. L., *et al.* (1988). Prevalence of Babesia antibody in a selected blood donor population. *Transfusion* **28**, 59–61.

Prowse, C., Ludlam, C. A., and Yap, P. L. (1997). Human parvovirus B19 and blood products. *Vox Sang* **72**, 1–10.

Raisanen, S. (1978). Toxoplasmosis transmitted by blood transfusions. *Transfusion* **18**, 329–332.

Roth, J. A., Siegel, S. E., Levine, A. S., *et al.* (1971). Fatal recurrent toxoplasmosis in a patient initially infected via a leukocyte transfusion. *Am J Clin Pathol* **56**, 601–605.

Schmunis, G. A., Zicker, F., Pinheiro, F., *et al.* (1998). Risk for transfusion-transmitted infectious diseases in Central and South America. *Emerging Infect Dis* **4**, 5–11.

Shulman, I. A. (1994). Parasitic infections and their impact on blood donor selection and testing. *Arch Pathol Lab Med* **118**, 366–370.

Shulman, I. A., Appleman, M. D., Saxena, S., *et al.* (1997). Specific antibodies to Trypanosoma cruzi among blood donors in Los Angeles, California. *Transfusion* **37**, 727–731.

Siegel, S. E., Lunde, M. N., Gelderman, A. H., *et al.* (1971). Transmission of toxoplasmosis by leukocyte transfusion. *Blood* **37**, 388–394.

Turner, M. L., and Ironside, J. W. (1998). New-variant Creutzfeldt-Jakob disease: The risk transmission by blood transfusion. *Blood Rev* **12**, 255–268.

Wendel, S. (1998). Transfusion-transmitted Chaga's disease. *Curr Opin Hematol* **5**, 406–411.

White, D. J., Talarico, J., Chang, H. G., *et al.* (1998). Human babesiosis in New York State: Review of 139 hospitalized cases and analysis of prognostic factors. *Arch Intern Med* **158**, 2149–2154.

36

Bacterial Contamination of Blood Products

BRUCE S. SACHAIS

Hospital of the University of Pennsylvania
Philadelphia, Pennsylvania

Since the risk of transfusion-transmitted viral disease has declined in recent years, the risk of bacterial contamination of blood components is now receiving more attention. Septic transfusion reactions due to bacterial contamination of blood products are unusual but serious complications of transfusion. Severe hypotension and even death may result from transfusion of accumulated organisms or endotoxins in contaminated units. Methods to further reduce the incidence of bacterial contamination of blood products and to detect contaminated units prior to transfusion are under investigation. This chapter describes: (1) the incidence of contaminated platelet products, PRBCs, and hematopoietic stem cells (HSCs), (2) the commonly implicated organisms, and (3) the evaluation and management of reactions due to contaminated blood products.

I. DESCRIPTION

Bacterial contamination of blood components occurs despite adherence to careful donor screening and rigorous collection and processing procedures. Sources of bacterial contamination may include the donor venipuncture site or unsuspected donor bacteremia. Alternatively, bacteria may be introduced into the blood component during processing or manipulation. The infusion of a component that contains bacteria may: (1) cause no obvious signs and symptoms, (2) cause fever with evidence of sepsis, or (3) cause septic shock and death. The initial signs and symptoms associated with septic transfusion reactions may also be observed in the more common acute transfusion reactions; thus, laboratory testing is required to confirm suspicion of

bacterial contamination. Septic transfusion reactions are more commonly associated with platelet transfusion than PRBC transfusion, due to the presence of significant volumes of plasma and storage at room temperature of platelet products. Septic transfusion reactions may also be seen following transfusion of autologous blood products.

II. INCIDENCE

Accurate determination of the incidence of bacterial contamination is complicated by the variety of microbial culture methodologies and variability in reporting among institutions. Transfusion-related bacterial sepsis due to product contamination is believed to be underreported (CDC, 1997; Hoppe, 1992; Klein *et al.*, 1997).

A. Incidence of Fatal Transfusion-Related Bacterial Sepsis

It is required that all transfusion-related fatalities be reported to the Food and Drug Administration (FDA). Between 1986 and 1991, 29 of 182 (11%) transfusion-related deaths reported to the FDA were determined to be a result of bacterial sepsis (Wagner, 1997). Of these, 21/29 bacterial sepsis-related deaths occurred as a result of platelet transfusion. It has been estimated that the risk of death from a bacterially contaminated platelet product is 1 per 1,000,000 transfused units, while the risk of death from bacterially contaminated PRBCs is 1 per 10,000,000 transfused units (Wagner, 1997).

B. Incidence of Nonfatal Transfusion-Related Bacterial Sepsis

1. PRBCs

The incidence of nonfatal transfusion-related bacterial sepsis is more difficult to assess than fatal reactions, since milder signs and symptoms of sepsis may go unrecognized. *Yersinia enterocolitica* is the most common bacterial organism causing PRBC contamination and subsequent donor-related sepsis. Using strict criteria, the CDC identified 10 cases of *Y. enterocolitica* sepsis between March 1991 and November 1996 (CDC, 1997). The mortality rate for these cases was 50%, with 5 of the 10 patients dying primarily as a result of *Y. enterocolitica* sepsis. Two of these 10 patients received contaminated autologous blood transfusions, illustrating the fact that autologous units also are at risk for bacterial contamination. The U.S. General Accounting Office (GAO) has estimated the rate of transfusion reactions due to bacterial contamination (e.g., *Yersinia*) to be 1 per 500,000 for PRBC transfusions (CDC, 1997).

2. Platelets

Bacterial contamination of platelets, based on the bacterial culture of units in inventory not ultimately transfused, has been reported to occur at a rate of 0.5 to 12 per 10,000 platelet units. Contamination occurs more often for platelet concentrates (PCs) than single-donor apheresis-derived platelets (APs; Blajchman, 1995; Morrow *et al.*, 1991; Wagner *et al.*, 1994; Yomtovian *et al.*, 1993). The incidence of bacterial contamination is directly correlated with increased storage time of the units (Braine *et al.*, 1986; Simon *et al.*, 1983). In one study, contamination rates were 12 per 10,000 for units stored for 5 days, and 1.8 per 10,000 for platelet units stored for 4 days or less (Yomtovian *et al.*, 1993).

This observation led the FDA, in 1985, to reverse a 1983 policy that allowed platelets to be stored for up to 7 days, thus permitting platelets to be stored up to 5 days only. The U.S. GAO has estimated the rates of bacteria-related transfusion reactions to be 0.6 per 1000 pooled PCs (CDC, 1997).

III. ETIOLOGY

The specific organisms responsible for transfusion-related sepsis, as well as the most common sources of contamination, are different for each blood product. This is largely due to differences in storage and processing of PRBCs, platelets, and HPCs.

A. PRBCs

PRBCs are stored at 4°C. Accordingly, the most common organisms associated with septic transfusion reactions following PRBC transfusion are the "psychrophilic" or "cold-loving" bacteria, *Y. enterocolitica* and *Pseudomonas fluorescens* (Wagner *et al.*, 1994; see Table 1). Donors infected with *Y. enterocoiltica* may have *Y. enterocolitica* bacteremia at the time of blood donation, but are usually asymptomatic, though some may report mild gastrointestinal illness within the month prior to donation (CDC, 1997; Wagner *et al.*, 1994). Thus, these donors would not necessarily be eliminated by predonation screening questions. *Yersinia enterocolitica* may proliferate in the PRBC unit during storage such that the product may contain up to 10^8 colony-forming units (CFUs)/ml after several days (Wagner, 1997). Transfusion of a *Yersinia*-contaminated unit may cause rapid development of life-threatening gram-negative sepsis in the recipient.

TABLE 1 Organisms Most Commonly Associated with Transfusion-Related Bacterial Sepsis

Product	Storage	Most common organisms	Less common organisms
Packed RBCs	4°C	*Yersinia enterocolitica*	*Pseudomonas putida*
		Pseudomonas fluorescens	*Treponema pallidum*
Platelets	24°C	*Staphylococcus epidermidis*	*Serratia marcescens*
			Staphylococcus aureus
			Bacilus cereus
Progenitor cells (HPCs)	−70°C	*Staphylococcus epidermidis*	*Staphylococcus aureus*
		Proprionibacterium sp.	*Pseudomonas* sp.

TABLE 2 Sources of Bacterial Contamination

- Asymptomatic donor bacteremia
- Skin at the phlebotomy site
- Introduction during processing of blood products

TABLE 3 Signs and Symptoms Associated with Transfusion-Related Bacterial Sepsis

- Fever
- Chills and rigors
- Hypotenslon
- Nausea and vomiting
- Abdominal cramps
- Dyspnea

B. Platelets

In contrast to PRBCs, platelets are stored with constant agitation at room temperature and are thus more frequently contaminated with bacteria. As shown in Table 1, the most common organisms contaminating platelets are rapidly growing aerobic organisms. *Staphylococcus epidermidis* is responsible for approximately 25% of septic transfusion reactions related to platelet transfusions (Wagner *et al.*, 1994). Potential sources for these common skin bacteria include the donor's cutaneous phlebotomy site and contamination during processing of the product (Table 2).

C. Hematopoietic Stem Cells (HSCs)

Because of the high number of HSC transplants, the incidence of bacterial contamination has been determined for both bone marrow (BM) harvests and peripheral blood stem cell (PBSC) preparations. The bacterial contamination rate of BM harvests ranges from ~1 to 7%, while the rate for PBSC preparations is lower, ranging from ~0 to 2% (Padley *et al.*, 1996; Wagner, 1997). Over 50% of the reported cases were due to *S. epidermidis*, while *Pseudomonas* species accounted for 5% of all cases (Wagner, 1997; Table 1). *Staphylococcus epidermidis* may be present at the donor's cutaneous venipuncture site (i.e., introduced during phlebotomy). Alternatively, this organism may be introduced during processing of the HSC product. In contrast, *Pseudomonas* contamination likely occurs due to either contaminated liquid nitrogen storage tanks or contaminated water baths used to thaw the HSC product (Fountain *et al.*, 1997; Padley *et al.*, 1996). As these products are frozen or used immediately after processing, there is no time for bacteria to multiply. Thus, few patients who receive culture-positive HSCs have adverse reactions, and no deaths due to sepsis from HSCs have been reported (Wagner, 1997).

D. Plasma

Plasma is stored frozen, and upon thawing may be stored refrigerated for no more than 24 hours. As a consequence, transfusion-related bacterial sepsis from plasma products is exceedingly rare. Cases that have been reported are attributed to contaminated water baths, often by *Pseudomonas* species.

IV. DIAGNOSIS AND CLINICAL PRESENTATION

A. Clinical Presentation

The clinical presentation of a septic transfusion reaction typically consists of fever, chills, rigors, hypotension, nausea, and vomiting (see Table 3). Less common signs and symptoms include headache, dyspnea, pain at the infusion site, and chest pain. The increase in body temperature is often $\geq 2°C$, but may be less, particularly in patients premedicated with acetaminophen. The clinical presentation of a septic transfusion reaction depends upon the immune status of the patient, pretransfusion administration of medications such as antibiotics and antipyretics, the bacterial load, and the amount of endotoxin in the blood product, if present. Fever and chills may begin during or soon after the transfusion, while hypotension may be delayed by 1 or more hours following start of the transfusion. Thus, the initial clinical presentation of septic transfusion reactions, febrile nonhemolytic transfusion reactions (FNHTRs), acute hemolytic transfusion reactions (AHTRs), and transfusion-related acute lung injury (TRALI) may be similar. *Septic transfusion reactions are often misclassified as FNHTRs*; such a misinterpretation may lead to less than optimal patient therapy (Wagner, 1997). Therefore, it is critical to consider the possibility of bacterial contamination in patients with the above-mentioned symptoms.

1. Clinical Evaluation

When a transfusion reaction is suspected, the transfusion should be discontinued and the patient evaluated (see Table 2, Chapter 28). Blood cultures should be collected prior to antibiotic administration if possible. If antibiotic treatment has been initiated, cultures may still be of value, if positive, and therefore should still be performed. Patients with suspected septic reactions should be monitored for shock, respiratory distress, and disseminated intravascular coagulation (DIC) and given supportive therapy.

2. Laboratory Evaluation

Following the initial steps to rule out an AHTR (see Table 2, Chapter 28), laboratory evaluation for suspected septic transfusion reactions should include visual examination and microbial culture of the transfused blood products remaining in the bag. Thus, implicated products should be returned to the blood bank as soon as possible. Care should be taken to avoid contamination of the product after the transfusion is terminated and during transport to the laboratory.

The product should be visually inspected for clots or discoloration, followed by Gram stain and culture (both aerobic and anaerobic). Determination of the source of the organism and its significance may be difficult when common skin organisms such as *S. epidermidis* are identified. If an organism is isolated from a pooled PC, then each of the individual platelet bags should be cultured to determine which product represents the source of contamination. Negative culture of a product that has been properly processed indicates that the component was likely not heavily contaminated, if at all.

V. CLINICAL COURSE AND SEQUELAE

Signs and symptoms of a septic transfusion reaction (see Table 3) often begin during transfusion or shortly thereafter, although presentation as far out as 2 weeks post-transfusion has been reported (Morduchowicz *et al.*, 1991). Severe reactions typically manifest themselves in the recipient as septic shock. Even with appropriate management, the morbidity and mortality from transfusion-related bacterial sepsis is high. The clinical course may range from a limited febrile response (in patients who rapidly clear the bacteria) to severe sepsis complicated by respiratory distress, DIC, and death (Morduchowicz *et al.*, 1991).

VI. MANAGEMENT AND PREVENTION

Management of patients with suspected septic transfusion reactions should include broad-spectrum antibiotic therapy following collection of blood cultures. Identification of specific organism(s) and antimicrobial testing to determine susceptibility or resistance may impact this antibiotic therapy. Treatment of hypoten-

sion, renal failure, respiratory distress, and DIC may be required.

Specific, sensitive detection systems are needed to identify contaminated blood components prior to release from blood bank inventories. Approaches which have been considered include automated cultures, chemiluminescent probes for bacterial nucleic acids, Gram stain, and measurement of glucose consumption (Klein *et al.*, 1997). Alternatively, steps may be taken to inactivate or remove contaminating bacteria during the collection process or during storage. Methods under investigation include diversion of the initial small volume of donor blood from the collection container during phlebotomy, prestorage leukocyte reduction, use of antibiotics in storage bags, and photochemical treatment of collected units (Ben-hur *et al.*, 1996).

A cooperative study by the American Association of Blood Banks, the American Red Cross, the Center for Disease Control and Prevention, and the Department of Defense, known as the BaCon study, has been performed to better define the problem of bacterial contamination of the blood supply and to develop strategies to decrease the incidence of contamination and transfusion-related sepsis.

References

Ben-hur, E., Moor, A. C., Margolis-Nunno, H., *et al.* (1996). The photodecontamination of cellular blood components: Mechanisms and use of photosensitization in transfusion medicine. *Transfusion Med Rev* **10**, 15–22.

Blajchman, M. A. (1995). Bacterial contamination of blood products and the value of pre-transfusion testing. *Immunol Invest* **24**, 163–170.

Braine, H. G., Kickler, T. S., Charache, P., *et al.* (1986). Bacterial sepsis secondary to platelet transfusion: An adverse effect of extended storage at room temperature. *Transfusion* **26**, 391–393.

CDC (1997). Red blood cell transfusions contaminated with *Yersinia enterocolitica*—United States, 1991–1996, and initiation of a national study to detect bacteria-associated transfusion reactions. *Morb Mortal Wkly Rep* **46**, 553–555.

Fountain, D., Ralston, M., Higging, N., *et al.* (1997). Liquid nitrogen freezers: A potential source of microbial contamination of hematopoietic stem cell components. *Transfusion* **37**, 585–591.

Hoppe, P. A. (1992). Interim measures for detection of bacterially contaminated red cell components. *Transfusion* **32**, 199–201.

Klein, H. G., Dodd, R. Y., Ness, P. M., *et al.* (1997). Current status of microbial contamination of blood components: Summary of a conference. *Transfusion* **37**, 95–101.

Krishnan, L. A., and Brecher, M. E. (1995). Transfusion-transmitted bacterial infection. *Hematol Oncol Clin N Am* **9**, 167–185.

Morduchowicz, G., Pitlik, S. D., Huminer, D., *et al.* (1991). Transfusion reactions due to bacterial contamination of blood and blood products. *Rev Infect Dis* **13**, 307–314.

Morrow, J. F., Braine, H. G., Kickler, T. S., *et al.* (1991). Septic reactions to platelet transfusions. A persistent problem. *JAMA* **266**, 555–558.

Padley, D., Koontz, F., Trigg, M. E., *et al.* (1996). Bacterial contamination rates following processing of bone marrow and peripheral blood progenitor cell preparations. *Transfusion* **36**, 53–56.

Simon, T. L., Nelson, E. J., Carmen, R., *et al.* (1983). Extension of platelet concentrate storage. *Transfusion* **23**, 207–212.

Wagner, S. (1997). Transfusion-related bacterial sepsis. *Curr Opin Hematol* **4**, 464–469.

Wagner, S. J., Friedman, L. I., and Dodd, R. Y. (1994). Transfusion-associated bacterial sepsis. *Clin Microbiol Rev* **7**, 290–302.

Yomtovian, R., Lazarus, H. M., Goodnough, L. T., *et al.* (1993). A prospective microbiologic surveillance program to detect and prevent the transfusion of bacterially contaminated platelets. *Transfusion* **33**, 902–909.

THERAPEUTIC APHERESIS

37

Overview and Practical Aspects of Therapeutic Apheresis

FRANK J. STROBL
University of Pennsylvania Health System
Philadelphia, Pennsylvania

Therapeutic apheresis is the separation and selective removal of a component (plasma, platelets, RBCs, or WBCs) of a patient's blood that contains a specific or suspected pathogenic agent (antibodies, immune complexes, abnormal RBCs, malignant WBCs, platelets, protein-bound drug or toxin). When cells are selectively removed, the procedure is termed therapeutic cytapheresis, or more specifically, leukocytapheresis, thrombocytapheresis, or erythrocytapheresis. When plasma is removed and then replaced by either albumin or fresh frozen plasma (FFP), the procedure is called plasmapheresis or therapeutic plasma exchange (TPE). In most cases, therapeutic apheresis is primarily utilized as an adjunctive or secondary therapy. Only for a few conditions (e.g., thrombotic thrombocytopenic purpura (TTP)) is TPE considered first-line treatment.

I. APHERESIS SERVICES

A. Regulations and Guidelines

Apheresis services should adhere to regulations outlined in the AABB *Standards for Blood Banks and Transfusion Services* (Klein, 1999) and the *Code of Federal Regulations* (CFR, 1998). The Food and Drug Administration (FDA) and the American Society for Apheresis also publish guidelines and recommendations.

B. Personnel

All personnel participating in apheresis activities should be familiar with these regulations and guidelines, and it should be documented that they are qualified by training and experience to perform apheresis. These individuals can be medical technologists, nurses, or physicians.

C. Procedures

Apheresis device manufacturers' instructions should be consulted for specific techniques. These instructions typically supply detailed information and protocols. Each facility performing apheresis must have its own readily available manual containing detailed descriptions of each procedure (standard operating procedure, or SOP).

D. Facilities

Apheresis should only be performed in facilities that provide the trained personnel, equipment, and medications necessary to manage serious reactions or problems.

II. PATIENT EVALUATION

A. Physician

Therapeutic apheresis requires experienced medical knowledge and judgment. Both the primary physician and the apheresis physician should evaluate the patient for treatment. The apheresis physician should make the final determination regarding appropriateness of apheresis and suitability of the patient. The apheresis evaluation should include a medical history, physical

TABLE 1   Essential Components of Apheresis Orders

- Patient name and unique identifier
- Date
- Procedure
- Frequency and duration of treatment
- Vascular access
- Premedications
- Amount of whole blood to be processed
- Anticoagulant
- Replacement fluids: Type and amount (liters)
- Laboratory data to be collected
- Monitoring and nursing care

TABLE 2   Essential Components of Apheresis Record

- Patient name and unique identifier
- Date
- Diagnosis
- Procedure
- Vascular access
- Premedications
- Amount of whole blood processed
- Anticoagulant
- Replacement fluids: Type and amount (liters)
- Vital signs
- Instrument identification number
- Reagent/supply lot numbers
- Adverse reactions

examination, and review of medications, allergies, and laboratory data.

B. Orders

Prior to initiating apheresis, a written physician order should be placed in the patient's chart including (1) type of procedure, (2) frequency and duration of treatments, (3) type of vascular access, (4) type and dose of premedications, (5) amount of whole blood processed, (6) type and dose of anticoagulant, (7) type and amount of replacement fluids, (8) laboratory data to be obtained during therapy, and (9) patient monitoring and nursing care (Table 1).

C. Informed Consent

Informed consent must be obtained. At a minimum, this should include a discussion of the nature of the procedure, its anticipated benefits, its potential risks, and existing alternatives.

D. Records

A record of each apheresis procedure should be maintained and kept on file in the apheresis unit. These records should include the patient's name and medical record number, the date, the diagnosis, the type of procedure, the type of vascular access, the blood volume processed, any premedications given, the replacement fluids administered, vital signs before, during, and after the procedure, instrument identification numbers, and reagent and supply lot numbers (Table 2).

III. INSTRUMENTATION

A. Centrifugation

Currently available apheresis instruments require the use of specific disposable software, which includes sterile collection chambers, tubing, and bags. Depending on the procedure and instrument used, the time to complete the procedure ranges from 1 to 3 hours. The most widely used apheresis instruments withdraw blood from the patient, immediately mix it with anticoagulant, pump the anticoagulated blood into bowl or belt chambers, and separate the blood components by centrifugation according to their specific gravities. The desired component is harvested into a collection or waste bag, while the remaining components are returned to the patient. Centrifugation-based apheresis instruments can be separated into two basic types: intermittent flow centrifugation (IFC) or continuous flow centrifugation (CFC).

1. Intermittent Flow Centrifugation

Intermittent flow centrifugation procedures are performed in cycles. Withdrawal, separation, and reinfusion of a specific volume of blood completes 1 cycle. The cycles are repeated until the desired volume of blood is processed or the desired quantity of harvested product is obtained. The advantage of IFC apheresis is that the same needle can be used for blood withdrawal and reinfusion and thus only requires 1 venipuncture. The disadvantage is that the extracorporeal volume is generally greater than that with CFC. This is particularly important in children and elderly patients who may have small total blood volumes. An example of an IFC instrument is the Haemonetics Corporation model V50.

2. Continuous Flow Centrifugation

In CFC apheresis, the blood is withdrawn, processed, and reinfused simultaneously. Unfortunately, this requires 2 venipuncture sites; one for withdrawal and one for reinfusion. However, since the CFC procedures are uninterrupted, the time to complete the procedure is less than that with IFC instruments. Examples of CFC instruments include the COBE Spectra, Fenwal CS3000 Plus, and Fresenius AS 104.

B. Membrane Filtration

Blood components can also be separated by passage through membranes with specific pore sizes. Plasma constituents smaller than the designated pore size pass through the membrane into the filtrate. These instruments have the membranes arranged as flat plates or hollow fibers. Membrane separation of plasma is rarely used for therapeutic apheresis.

C. Adsorption

Affinity chromatographic methods are available to selectively remove certain solutes with a high degree of specificity. Using these methods, the whole blood is usually passed through a centrifuge or a hollow fiber membrane filter before the separated plasma is passed through columns containing inert matrices to which specific sorbents or ligands (e.g., staphylococcal protein A, dextran sulfate, antibodies, charcoal) are attached. The sorbent or ligand selectively absorbs the desired solute (e.g., pathogenic antibody or immune complexes, LDL, bile acids) from the plasma. Specific absorption methods are described below.

1. Staphylococcal Protein A

Staphylococcal protein A has a strong and specific affinity for the Fc fragments of IgG1, IgG2, and IgG4, and various IgG-containing immune complexes. Staphylococcal protein A columns are currently licensed in the United States and have shown some efficacy in the treatment of refractory autoimmune thrombocytopenia and rheumatoid arthritis.

2. Dextran Sulfate

The Liposorber LA-15 instrument marketed by the Kaneka America Corporation is an example of a dextran sulfate absorption system that is FDA approved for use in patients with familial hypercholesterolemia.

D. Manual Apheresis

Apheresis can also be performed manually using refrigerated centrifuges and commercially available kits containing sterile plastic bags and tubing. The procedure is inexpensive compared to the use of sophisticated apheresis instruments. However, manual apheresis is significantly more labor intensive and time consuming, but is occasionally warranted (e.g., in emergent situations if automated devices are not available).

E. Photopheresis and Extracorporeal Photochemotherapy (ECP)

In ECP, leukapheresis is performed and the mononuclear cells which have been exposed to a photosensitizing agent are treated with ultraviolet A (UV-A) light before being returned to the patient. The photosensitizing agent is usually orally administered 8-methoxypsoralen (8-MOP), which is given several hours prior to the procedure. The oral formulation is typically administered at a dose of 0.6 mg/kg, 2 hours before starting the leukapheresis, in order to achieve a plasma level of psoralen between 50 and 200 ng/ml. As 8-MOP absorption in the gastrointestinal tract is extremely variable, it is recommended that plasma 8-MOP levels be measured at each procedure to ensure correct dosing. Alternatively, the 8-MOP can be added to the leukapheresis component directly. This extracorporeal formulation requires substantially smaller amounts of 8-MOP and reduces the side effects of 8-MOP, including nausea and light sensitivity. Upon irradiation with UV-A light, 8-MOP undergoes a conformational change, binding to one or both strands of leukocyte DNA. This covalent linkage is thought to prevent normal DNA function and replication in treated cells (Gasparro *et al.*, 1989). Reinfusion of the treated cells may also induce a clone-specific suppressor cell response against the abnormal lymphocytes (Khavari *et al.*, 1988).

Photopheresis has been found to be efficacious in treating cutaneous T-cell lymphoma/leukemia (CTCL). It has also been used with some success in autoimmune diseases, chronic solid organ transplant rejection, and the treatment of graft-versus-host disease (GVHD).

IV. ANTICOAGULANTS

A. Acid-Citrate-Dextrose

The most common anticoagulant used in apheresis is acid-citrate-dextrose (ACD). The citrate ion of ACD anticoagulates by chelating calcium ions and blocking calcium-dependent clotting factor reactions. The advantage of citrate is that the liver quickly metabolizes citrate to bicarbonate. In patients with renal disease, however, clearance of bicarbonate may be impaired and thus lead to or exacerbate metabolic alkalosis.

B. Heparin

Heparin can also be used either alone or in combination with citrate. The advantage of using heparin alone or with reduced doses of citrate is the avoidance of citrate-related side effects such as hypocalcemia.

V. REPLACEMENT FLUIDS

A. Crystalloids

Normal saline is used to prime the apheresis circuit and to maintain intravascular fluid volume and patency of the intravenous line when needed. In pediatric patients, where the extracorporeal volume is greater than 10–15% of the total blood volume, red blood cells are used to prime the apheresis system. Although crystalloid solutions like normal saline are inexpensive, their use as replacement fluids during daily or every-other-day plasma exchange regimens is severely limited due to rapid decreases in oncotic pressure and fluid shifts into the extravascular space. The volume of replacement crystalloid required is typically 2 to 3 times the volume of plasma removed. Thus, it is typical to use albumin in addition to saline as a replacement fluid.

B. Albumin

Albumin (5% solution in normal saline) is the most commonly used replacement fluid to maintain intravascular volume and normal oncotic pressure during therapeutic plasmapheresis.

C. Fresh Frozen Plasma (FFP)

Due to the risk of transfusion-transmitted diseases, FFP is only indicated in patients with thrombotic microangiopathies (e.g., TTP) or for replacing coagulation factors in patients undergoing repeated daily plasmapheresis procedures. The need for supplemental FFP replacement should be based on laboratory parameters such as prothrombin time, activated partial thromboplastin time, and fibrinogen levels during the course of apheresis therapy.

D. Colloid Starches

Concerns over the increased cost of albumin and disease transmission associated with FFP have led to the sporadic use of colloid starches (pentastarch or hetastarch) for partial or full replacement during plasma exchange. Unfortunately, life-threatening reactions have occurred on rare occasions when colloidal starch was used as replacement fluid during plasma exchange. The use of colloidal starches (3% or 6% solutions) during plasma exchange should only be considered in patients who suffer clinically significant reactions to albumin replacement.

VI. VASCULAR ACCESS

A. Peripheral

For most patients requiring only infrequent or a limited number of apheresis procedures, the antecubital veins are adequate for vascular access. A 16-gauge or larger needle is recommended for both access and return lines.

B. Central

Occasionally the antecubital veins may be very difficult to access. In this situation, or when patients are critically ill, an indwelling central catheter can be placed in the subclavian, femoral, or internal jugular vein. Placement of a central catheter should also be considered when apheresis is to be performed repeatedly over a protracted period of time.

1. Catheter Types

Various commercially available, double-lumen catheters allow the withdrawal and return of blood simultaneously at high flow rates (50–200 ml/min). These catheters typically have large lumens and rigid-wall construction. Central catheters of 10 French (FR) or greater in size, and 15 cm or greater in length, are adequate for most adults. Except for femoral vein placement, these central catheters can be maintained for weeks if necessary.

2. Catheter Care

Strict catheter care is imperative to prevent infection of the catheter and the surrounding site. Flushing the catheter with heparinized saline (100 to 5000 U/ml) immediately after the procedure is critical to prevent clotting. If the catheter does become clotted, urokinase carefully injected into only the catheter lumens often lyses the clot and restores patency.

3. Catheter Complications

When placed in the subclavian or internal jugular veins, the tip of the central catheter should remain in the superior vena cava. Cardiac irritability can occur if the catheter tip is placed in the right atrium secondary to direct physical contact or indirectly from the infusion of cold or citrated replacement fluids. A chest x-ray should be obtained following the placement of an internal jugular or subclavian central venous catheter to ensure proper location and confirm the absence of an inadvertent hemothorax or pneumothorax. Other complications of venous access catheters include venous thrombosis, arterial puncture, arteriovenous fistula formation, hemorrhage, or air embolism. Fistulas and grafts are alternative sites for vascular access.

VII. ADVERSE EFFECTS OF APHERESIS

Apheresis is considered a safe procedure, but complications are not uncommon. Unfortunately, it is often difficult to determine whether the adverse effect was caused by the procedure or coincidentally by the underlying disease. Rare severe complications include cardiac arrhythmias or arrest, respiratory arrest, seizures, and even death.

A. Citrate Toxicity

Citrate, the most commonly used anticoagulant during apheresis, chelates calcium and thus lowers the patient's ionized calcium level. The degree to which ionized calcium levels are reduced reflects the rate at which the citrate is returned to the patient and how quickly the citrate is metabolized and/or excreted. Initial symptoms of hypocalcemia due to citrate toxicity include perioral paresthesias, tingling, chills, and a heightened sense of vibrations. Hyperventilation, hypothermia, hypomagnesemia, and FFP replacement exacerbate citrate toxicity. These mild adverse effects are usually controlled by reducing the proportion of citrate added to whole blood, reducing the blood flow rate, or interrupting the procedure until the symptoms subside.

Oral calcium supplementation before and during apheresis may also help to prevent or minimize the symptoms of citrate toxicity. If inadequately controlled, however, citrate toxicity can progress to muscle twitching, hypotension, tetany, nausea, vomiting, chest pain, cardiac arrhythmias, and seizures. These severe side effects of citrate toxicity can be avoided with careful monitoring of the ionized calcium level and parenteral administration of calcium, if needed. Intravenous infusion of 10% calcium gluconate (10 cc) over 20 to 30 minutes provides up to 4.65 mEq of calcium. Calcium gluconate (8 cc of a 10% solution) added to 1 liter of 5% albumin-saline and infused during the TPE provides up to 3.7 mEq of calcium. Calcium should never be added to blood components for reinfusion.

B. Allergic Reactions

Allergic reactions characterized by urticaria, flushing, oral mucosa swelling, dyspnea, wheezing, and hypotension are primarily observed in patients receiving FFP as replacement during TPE. Less frequently, similar reactions have been reported with albumin, plasma protein fraction, and colloid starch replacement. These reactions usually respond to pre- and postmedication with antihistamines and corticosteroids. If anaphylaxis occurs, epinephrine should be administered.

C. Vasovagal Episodes

Reactions characterized by pallor, nausea, vomiting, diaphoresis, bradycardia, hypotension, and syncope can occur during apheresis. If a vasovagal reaction occurs, it is helpful to interrupt the procedure, infuse intravenous fluids, and place the patient in the Trendelenburg position until the symptoms resolve.

D. Anaphylactoid Reactions

Angiotensin converting enzyme (ACE) inhibitors prevent the metabolism of bradykinin generated during apheresis. Elevated levels of bradykinin have been shown to cause anaphylactoid reactions characterized by severe hypotension, flushing, and respiratory distress. Patients should be closely monitored and, if possible, apheresis should be delayed for 24 to 48 hours after the last dose of ACE inhibitor.

E. Infections

Apheresis patients may be at increased susceptibility to infections due to their underlying condition, prescribed immunosuppressive drugs, or decreased levels of immunoglobulins and complement while undergoing intensive apheresis regimens. Bacterial infection associated with apheresis usually originates from the vascular catheter. Viral transmission (HIV, hepatitis A, B, and C, parvovirus B19, HTLV-I/II) is associated exclusively with the reinfusion of allogenic blood components during apheresis. In patients who will receive large volumes of FFP as replacement fluid, virally inactivated, solvent/detergent plasma should be considered.

F. Drug Removal

Therapeutic plasma exchange may lower the blood levels of medications, especially those that circulate bound to proteins and have long plasma half-lives. It is common practice to withhold the administration of drugs scheduled 1 hour prior to or during apheresis until after the procedure is finished. It may be preferable to withhold medications given once or twice daily until after the apheresis procedure. For drugs given 3 or more times a day, administration can usually proceed as scheduled.

G. Hemolysis and Anemia

Hemolysis is most often caused by a mechanical problem with the equipment, such as a kink in the plastic tubing. Although the currently available apheresis instruments perform a final rinseback, a significant number of red cells remain in the instrument at the completion of the procedure. Thus, repeated apheresis procedures in patients with compromised bone marrow function may result in a progressive anemia.

H. Hypothermia

Because of concerns over patient comfort and the risk of arrhythmias, a blood warmer should be considered especially when reinfusion is performed through a central line.

I. Hypovolemia

Hypovolemia and associated hypotension may occur during apheresis when the extracorporeal blood volume exceeds 15% of the patient's total blood volume. Extreme care to prevent hypovolemia must be taken when performing apheresis in children or elderly patients. Intermittent flow devices with large extracorporeal volumes (> 500 ml) are more apt to cause hypovolemia. Continuous flow devices typically have smaller extracorporeal volumes. Hypovolemia can also occur if the return flow is inadvertently diverted away from the patient due to operator or instrument error. Therefore, it is important to maintain accurate and continuous records of the volumes removed and returned during all procedures.

J. Hypervolemia

Hypervolemia or fluid overload may be problematic in patients with cardiac or renal insufficiency. Plasma exchanges to remove volume are not recommended. Acceptable alternatives for reducing volume include diuretics or hemodialysis.

K. Respiratory Distress

Respiratory distress during or following apheresis can be secondary to volume overload and congestive heart failure, pulmonary embolus, anaphylactic reactions, ACE inhibitors, or transfusion-related acute lung injury (TRALI).

VIII. ONE-COMPARTMENT MODEL OF PLASMA EXCHANGE

The "one-compartment model" (the assumption that the substance of interest is primarily intravscular with little or no reequilibration with the extravascular space) has gained wide acceptance for predicting the removal of most pathogenic substances during therapeutic plasma exchange.

A. Mathematical Model

Assuming a one-compartment model of distribution, a formula can be used to estimate the fraction of a substance remaining in the intravascular space at anytime during a plasma exchange procedure,

$$y_t = y_0 e^{-x},$$

where y_t is the concentration at time t, y_0 represents the starting concentration, e is the natural logarithm, and x is the number of plasma volumes exchanged at time (t) during the procedure. A plot of this formula can be used to predict the disappearance of a substance from the intravascular space according to the number of plasma volumes removed during the plasma exchange (see Figure 1). The curve is almost identical for plasma exchanges performed by either discontinuous or continuous flow instruments. This formula can also be used to predict the reduction or removal of the patient's red blood cells during a RBC exchange. The removal of 1 plasma volume results in an overall net exchange of between 60 and 70%. The efficiency of

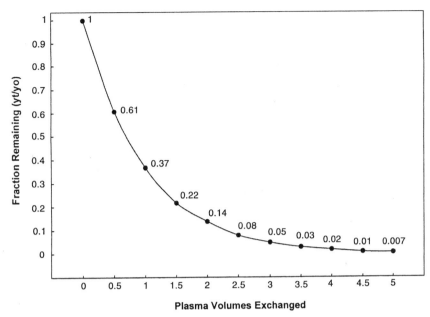

FIGURE 1   Depletion of soluble substances from the intravascular space during plasma exchange according to the one-compartment model of distribution. Note that the highest efficiency during therapeutic plasma exchange occurs in the first 1.0 to 1.5 plasma volumes processed.

exchange decreases dramatically as additional volumes of plasma are removed due to progressive dilution by the replacement fluids during the procedure. Thus, the most effective therapeutic plasma exchange procedures remove only 1.0 to 1.5 plasma volumes.

B. Limitations

In reality, the efficiency of plasma exchange depends on the proportion of the pathogenic substance in the intravascular compartment compared to that in the extravascular compartment, the amount of the patient's total plasma volume removed, and the rate at which the intravascular component is replenished by either synthesis or reequilibration. If the pathogenic agent is an IgG antibody, the amount of IgG in the intravascular compartment is usually only about 50% of the total body IgG. The remainder is in the extravascular compartment. Because IgG reequilibration readily occurs between the extravascular and intravascular compartments, it is unusual to produce more than a transient reduction in the level of circulating IgG following a single plasma exchange. In this case, frequent daily or every-other-day plasma exchanges are usually required. Although IgM antibodies are often synthesized at a rapid rate, they are found primarily in the circulation and are therefore more effectively removed during a single plasma exchange.

IX. EFFECT OF PLASMA EXCHANGE ON NORMAL BLOOD CONSTITUENTS

It should be remembered that during plasma exchange, normal plasma constituents are removed along with the pathogenic agent. Daily procedure schedules, in particular, can be expected to deplete many normal plasma constituents, including the following:

A. Electrolytes

Excluding calcium and magnesium, decreases in the levels of most electrolytes and other small molecules (e.g., potassium, chloride, glucose, bicarbonate) can be expected to be small and transient because of rapid reequilibration with extravascular and intracellular sources.

B. Proteins

The removal of normal plasma proteins, including coagulation factors, complement components, and immunoglobulins, can also be reasonably estimated using the one-compartment model. Recovery of baseline plasma protein levels after apheresis is dependent primarily on net synthesis and secondarily on reequilibration. The levels of most plasma proteins, including coagulation factors and complement components, return to preexchange levels within 24 to 72 hours. On the other hand, the recoveries of some immunoglobulin levels to baseline have been shown to take several weeks (Woods and Jacobs, 1986). The transient hypocoagulability following plasma exchange with albumin is typically not associated with clinically significant bleeding complications.

C. Cellular Constituents

The platelet count can also drop by as much as 33% following a single plasma exchange (Woods and Jacobs, 1986), and greater than 75% following several daily therapeutic plasma exchanges (Keller *et al.*, 1979). As with decreased coagulation factor levels, decreased platelet counts following plasma exchange are not associated with bleeding complications. The platelet count normally returns to baseline within 2 to 3 days. The potential effects of plasma exchange on RBCs are outlined in Section VII.G.

References

"Code of Federal Regulations: 21 CFR 640" (1998). U.S. Government Printing Office, Washington, DC.

Gasparro, F., Dall' Amico, R., Goldminz, D., *et al.* (1989). Molecular aspects of extracorporeal photochemotherapy. *Yale J Biol Med* **62**, 579–594.

Keller, A., Chirnside, A., and Urbaniak, S. J. (1979). Coagulation abnormalities produced by plasma exchange on the cell separator with special reference to fibrinogen and platelet levels. *Br J Haematol* **42**, 593–603.

Khavari, P., Edelson, R., Lider, O., *et al.* (1988). Specific vaccination against photoinactivated cloned T cells. *Clin Res* **36**, 662A.

Klein, H., Ed. (1999). "Standards for Blood Banks and Transfusion Services," 18th ed. AABB, Bethesda.

Woods, L., and Jacobs, P. (1986). The effect of serial therapeutic plasmapheresis on platelet count, coagulation factors, plasma immunoglobulin, and complement levels. *J Clin Apheresis* **3**, 124–128.

38

Therapeutic Plasma Exchange

Rationales and Indications

CHESTER ANDRZEJEWSKI

Department of Pathology
Baystate Health System
Springfield, Massachusetts
and
Tufts University School of Medicine
Boston, Massachusetts

Therapeutic plasma exchange (TPE or plasmapheresis) is the most frequently requested and performed extracorporeal therapeutic hemapheresis intervention. In its simplest sense, this "blood cleansing" procedure results in the removal of pathologic substances from the circulating plasma.

I. RATIONALES FOR INTERVENTION

Rationales for performing TPE vary. They can be broadly categorized into immune-related and non-immune-related reasons.

A. Immune-Related

1. Immunoglobulins

The greatest application of TPE is in the treatment of immune-mediated diseases, whether autoreactive or alloreactive in nature. Here the removal of circulating immunoglobulins presumed to be pathologic contributors to the patient's disease state serves as the major rationale for its use. The class of the immunoglobulins (e.g., IgM, IgG) involved is important not only from a disease pathogenic perspective, but also as it impacts the efficacy of the exchange process itself.

a. IgM

IgM molecules (MW = 900,000 daltons [Da]) have lower volumes of distribution in the body, with their predominant though not exclusive localization in the intravascular space. Because of this, processes involving IgM antibodies are more immediately impacted by a single TPE procedure than those involving IgG antibodies.

b. IgG

IgG antibodies, due to their lower molecular weight (150,000 Da), freely diffuse into the extravascular spaces throughout the body. In IgG-mediated disease processes, a strategy using a series of TPE procedures over an extended time period might offer greater benefit in decreasing the total pathologic antibody burden. In such cases a longer interprocedural interval would allow for diffusion of IgG molecules from the affected target tissue into the intravascular space, thus resulting in the additional capture and elimination of these substances through the exchange process.

2. Immune Complexes

Besides removing free antibody, TPE may also decrease bound antibody circulating as immune complexes. The beneficial effect of this directly impacts selected disease processes by interruption of the sequence of immune complex deposition in sensitive microvascular networks. Further, removal of both free and complexed antibody populations as well as antigens may alter the ratio of these constituents, thus potentially affecting immune complex lattice formation and configuration by impacting the *in vivo* precipitin curve of the immune reactants.

3. Other Humoral Mediators

Other humoral immune system mediators and reactants may also be removed during TPE. Complement proteins are depleted according to expected removal kinetics; however, their rapid rates of synthesis result in little to no change in their plasma concentrations unless intensive daily TPE protocols are used. Immune system regulatory proteins such as lymphokines may also be removed. Theoretically, removal of such agents could alter immunoregulatory balances and result in disease improvement or exacerbations.

B. Non-immune-related

Non-immune rationales for performing TPE include the elimination of a poison or toxic metabolite, or reduction of excessively elevated plasma constituents affecting intravascular viscosity.

II. INDICATIONS FOR TREATMENT

A. Initial Concerns

No apheresis procedure should be undertaken without a thorough review of the patient's medical history and a clear understanding of treatment objectives. The indications or rationale for performance of the intervention should be identified. Although a definitive diagnosis at the time of the initiation of therapy is ideal, occasionally a "therapeutic trial" of TPE may be requested, typically where a final diagnosis is awaiting confirmation. In such settings, re-evaluation of the situation and of the need for TPE is critical as additional data are required. Other considerations in the treatment plan are outlined in Table 1.

B. Derivation of Categorizations and Guidelines

Guidelines for the use of TPE in the treatment strategies of various disorders are predicated on information obtained from a combination of sources including but not limited to case reports, anecdotal experience, case controlled clinical trials, and consensus opinion statements published by various professional and governmental organizations. The latter typically involve the assembly of a group of practitioners with substantial experience in the field who critically review the current literature, analyze available information, and offer recommendations for the use of TPE in various clinical settings. In this respect, both the American Society for Apheresis (ASFA) and the American Association of Blood Banks (AABB) have issued

TABLE 1. Treatment Plan Considerations Involved in TPE Interventions

- Clinical diagnosis/history and therapy/outcome objectives
- TPE indication and/or rationale
- Selection of hemapheresis equipment
- Volume/duration/frequency targets
  —Per procedure
  —For entire series
- Vascular access considerations
- Choice of replacement fluids
- Physiologic/adverse reaction monitoring
- Coordination of pre/intra/inter-procedural lab testing and drug/blood administration
- Concurrent and post medications
- Use of ancillary equipment PRN (As Needed)
  —Blood warmers
  —Fetal monitors

guidance regarding the value of TPE in the treatment of a variety of disorders (Strauss *et al.*, 1993; AABB Hemapheresis Committee, 1992). Periodic reexamination and reassessment of the recommendations by such groups are ongoing initiatives (McLeod, 2000).

From such studies, a guidelines classification system consisting of four categories of potential benefits of TPE has evolved (see Table 2). Although much agreement exists in the placement of many of the disorders in such categorizations, differences do exist. A comparison of the different classification systems as they relate to specific disorders is presented in Table 3. Diseases are arranged according to their presumed immunologic basis.

A discussion of some of the more commonly encountered disorders treated with TPE follows. For a discussion of other less commonly encountered disorders where TPE has been used, or for more detail regarding any of the more commonly treated conditions, the reader is referred to a comprehensive text on the subject (McLeod *et al.*, 1997).

C. Neurologic Diseases

Neurologic diseases represent an important group of disorders for which TPE is requested. Illnesses affecting both the peripheral nervous system and the central nervous system are included. The majority of these have an autoimmune basis with both acute and chronic disorders represented. Historically this group of illnesses was one of the first areas where TPE was demonstrated to be of benefit in their management.

TABLE 2 Comparison of Consensus Group Category Classification Schemes for TPE Indications

Category	AABB	ASFA
I	Standard and acceptable in certain circumstances, including primary therapy	Standard and acceptable; valuable as primary or first-line adjunct therapy
II	Sufficient evidence to suggest efficacy; acceptable therapy on an adjuvant basis	Generally accepted; supportive to other more definitive treatment strategies
III	Inconclusive evidence for efficacy; uncertain risk:benefit ratio	Reported evidence insufficient to establish efficacy
IV	Lack of efficacy in controlled trials	Lack of efficacy in controlled trials

They remain one of the most prevalent and frequently consulted on group of disorders seen by apheresis specialists.

Typically, these disorders have a prominent antibody-mediated component to their pathogenesis, although as in many autoimmune conditions, contributions from cellular immune processes cannot be totally excluded. Various protocols have been advanced to optimize the impact and timing of apheresis intervention in different disorders (Ciavarella *et al.*, 1993). Frequently TPE is used in conjunction with other immune-modulating agents, for example, corticosteroids, immunosuppressive drugs, and/or intravenous immunoglobulin (IVIg). Antibody rebound phenomena (i.e., heightened synthesis and expression of pathologic antibodies post-TPE) can occur, especially in patients receiving TPE as sole therapy.

1. Myasthenia Gravis and Related Disorders

a. Description

A prototypical autoantibody-mediated disorder involving the peripheral nervous system, myasthenia gravis is characterized clinically by sporadic muscular fatigability and weakness. Symptoms are usually more severe later in the day than earlier. Characteristic of the disorder is an immunologic attack targeting the alpha subunit of the acetylcholine receptor (AChR) located at the neuromuscular junction, leading to impairment in muscle contraction. Skeletal muscles throughout the body may be affected; however, involvement of those innervated by cranial nerves is particularly worrisome.

b. Role of TPE

Intervention with TPE can lead to a rapid decrease of pathologic autoantibodies in the circulation, with marked improvement in patient symptomatology even after only 1 or 2 exchanges. Such results may also be observed in patients negative for anti-AChR antibodies, suggesting the possibility of other pathogenic antibody populations. Although the use of TPE in this setting has never been the subject of a controlled trial, it is widely accepted to be of benefit in certain circumstances. These situations include: (1) initial presentation of selected patients with severe respiratory impairment, dysphagia, or compromised locomotion, especially before other therapies have become effective; (2) individuals unresponsive to convential therapies; (3) episodes of acute and precipitous deterioration (myasthenic crisis); (4) pre- and postsurgical interventions, especially thymectomy; and (5) during introduction of corticosteroid therapy, when many patients may experience a clinical decline (Raphael *et al.*, 1996).

c. Technical Considerations

Various TPE treatment protocols may be used as dictated by clinical circumstances, with more aggressive daily schedules followed by a gradual taper for severely compromised patients. Generally, for less severe disease activity a course of 3-6 treatments over a 10- to 14-day period can suffice. Occasional patients may benefit from a "maintenance" course occurring intermittently over an extended period of time. In all cases replacement fluids of choice are 5% albumin and crystalloid solutions.

2. Guillain–Barré Syndrome

a. Description

Acute polyneuropathies, best exemplified by the Guillain-Barré syndrome (GBS), have TPE as a primary therapeutic consideration in patients who do not spontaneously recover from the initial immunologic assault. Primarily involving the peripheral nervous system, patients present with parasthesias and bilateral

TABLE 3   Summary[a] of Most Recent AABB and ASFA Guidelines for Therapeutic Plasma Exchange in Selected Disorders According to their Presumed Immunologic Basis

Clinical condition	Category (AABB)	Classification (ASFA)	Clinical condition	Category (AABB)	Classification (ASFA)
Immune: autoimmune/neurologic			Immune: alloimmune		
Acute inflammatory polyneuropathy/GBS	I	I	ABO-incompatible organ or marrow transplantation	III	II
Chronic inflammatory demyelinating polyneuropathy/CIDP	II	I	Coagulation factor inhibitors	I	II
Myasthenia gravis	I	I	Hemolytic disease of the newborn	III	III
Lambert–Eaton myasthenic syndrome	NC[b]	II	Post-transfusion purpura	I	I
Multiple sclerosis	III	III	Transfusion refractoriness (RBC, PLT, HLA)	III	III
Paraneoplastic CNS syndromes	NC	III	Renal transplant rejection	IV	IV
Amyotrophic lateral sclerosus (ALS)	IV	IV			
Schizophrenia	IV	IV			
Immune: autoimmune/hematologic			Nonimmune: metabolic		
Coagulation factor inhibitors	I	II	Drug overdose/poisoning	II	III
Hemolytic anemia			Hepatic failure: acute/fulminant	IV	III
—Cold related	II	III	Hypercholesterolemia (homozygous/familial)	I	II
—Warm related	III	III			
Pure red cell aplasia	III	III	Refsum's disease	I	I
Immune thrombocytopenicpurpura (ITP)	IV	II[c]	Thyroid storm	III	NC
Immune: autoimmune/miscellaneous[d]			Unknown: miscellaneous		
Goodpasture's syndrome	I	I	Cryglobulinemia	I	II
Rapidly progressive glomerulonephritis	II	II	Dysproteinemia/hyperviscosity	I	II
SLE/nephritis	IV	IV	Hemolytic uremic syndrome (HUS)	II	III[e]
Arthritis, rheumatoid	IV	IV	Thrombotic thrombocytopenic purpura (TTP)	I	I
Polymyositis/dermatomyositis	IV	III			
Progressive systemic sclerosis (PSS)	III	III			
Raynaud's phenomenon	NC	III			
Systemic lupus erythematosus (SLE)	NC	III			
Systemic vasculitis	II	III			
Bullous pemphigoid	NC	NC[e]			
Pemphigus vulgaris	II	NC[e]			
Psoriasis	IV	NC			
Thyroid disease (Graves' disease)	NC	NC[f]			

[a] Adapted and modified with permission from: H. G. Owen and M. E. Brecher (1997), Management of the Therapeutic Apheresis Patient, in *Apheresis: Principles and Practice* (B. C. McLeod, Senior Ed.), pp. 223–249, AABB Press, Bethesda. Category classifications derived from AABB Hemapheresis Committee (1992), Strauss *et al.* (1993), and McLeod (2000).

[b] Not categorized as such in current classifications.

[c] Treatment with protein A column apheresis, not TPE.

[d] Renal, rheumatologic, dermatologic, endocrine conditions.

[e] Previously identified as Category II (Strauss *et al.*, 1993).

[f] Previously identified as Category III (Strauss *et al.*, 1993).

extremity weakness in a predominantly distal to proximal progression. Autonomic nervous system involvement may also occur, presenting challenges during the apheresis procedure in the control of hemodynamic parameters. Severe cases can progress to respiratory distress requiring ventilatory support and ICU admissions. Several variants from the basic clinical presentation can be encountered (e.g., the Fisher variant characterized by reflexia, ataxia, and ophthalmoplegia; Rostami and Sater, 1997).

Histopathologically widespread focal demyelination and lymphocytic infiltration of the nerve tissue is observed. Various pathways to the autoreactive process probably exist, as evidenced by the diversity of associated prodromal conditions. Antecedent infectious processes or illnesses are commonly reported, with both viral and bacterial agents thought to be involved. Immunologic activation via the vaccination process (strikingly observed with particular product lots of swine flu vaccine in the 1970s) may also lead to the syndrome. Based on the various specificities of antibodies found in GBS patients, a number of myelin-derived antigens have been suggested as targets in a potentially cross-reactive immunologic attack triggered by and directed toward a foreign structure presented to and normally handled by the host's immune apparatus.

b. Role of TPE

Several well-controlled clinical trials have demonstrated the efficacy of TPE in GBS in shortening recovery time, reducing disability, and in halting progression to more severe involvement (Osterman et al., 1984; the Guillain–Barré Syndrome Study Group, 1985). Best results are obtained in starting TPE as near to the onset of clinical symptomatology as possible. In contrast to myasthenia gravis, however, the patient's clinical improvement during the apheresis treatment phase is not typically dramatic, sometimes requiring weeks to months to return to premorbid baseline parameters.

c. Technical Considerations

Apheresis protocols derived from the clinical trials experience target exchanges of 200-250 ml plasma removal per kilogram of body weight (5–6 total exchanges) accomplished over a 7–14 day interval. Albumin in combination with saline is the replacement fluid mix of choice. A repeat or truncated series of exchanges may be of value in select cases (e.g., relapses or where improvement or stabilization is delayed). IVIg, either alone or in conjunction with TPE, may benefit certain subsets of patients; however, predicting which patients are candidates for the various interventions prior to their institution remains a challenge (Rostami and Sater, 1997).

3. Chronic Inflammatory Demyelinating Polyneuropathy

a. Description

Chronic polyneuropathic conditions constitute another category of disorders where TPE has been applied. Within the group, immune-mediated and non-

immune-mediated examples are seen. Chronic inflammatory demyelinating polyneuropathy (CIDP), representative of the former, is a syndrome of unknown etiology characterized by the insidious and progressive onset of motor and sensory peripheral nerve deficits. Clinical symptoms (e.g., parasthesias, hyporeflexia, bilateral symmetric limb weakness) develop over weeks, months, or years with a slowly worsening or relapsing course. Nerve conduction abnormalities can be observed showing widespread but patchy focal demyelination. Histopathological tissue examinations from nerve biopsy is significant for demyelination. The immunologic nature of the disorder remains to be fully clarified; however, cellular as well as humoral immune system elements appear to be involved, (Harati, 1997).

b. Role of TPE

No single most effective treatment for all patients with CIDP is known. Corticosteroid therapy, immunosuppressive drugs, and IVIg infusion all have been used. TPE may be useful in patients who have failed to respond to corticosteroids and immunosuppressant agents (Dyck et al., 1986). It may also be of benefit in patients who are nonambulatory and as alternative therapy for disease resistant to tapering of prednisone dosages.

c. Technical Considerations

No particular TPE protocol has been identified as best; however, a series of 6 exchanges over a 2-week period followed with a tapering schedule of 2 procedures per week for another 2–4 weeks (total TPE exchanges, 10–14) may be used. Albumin and saline are the recommended replacement fluids.

4. Paraproteinemic Peripheral Neuropathy

a. Description

Peripheral neuropathy occurring in the context of a paraprotein merits separate consideration. Although binding of a monoclonal protein to structures within nerve tissue may occur, leading to immunologic-induced damage, nonimmunologic neuropathology mediated by such paraproteins may also occur via an ischemic basis secondary to hyperviscosity.

b. Role of TPE

In either situation, reduction of the paraprotein load by medication and/or TPE may be of benefit (Dyck, et al., 1991; Hillyer and Berkman, 1996). Treatment of the underlying B-cell malignancy should occur prior to any TPE since improvement may result.

c. Technical Considerations

Intervention with TPE should target reduction of the pathologic accumulation of the paraprotein and resolution of symptoms. This may be accomplished in a single episode or a short series of exchanges. Albumin and saline are the recommended replacement fluids.

D. Hematologic Disorders

Hematologic disorders constitute another major focus of activity for apheresis specialists (McLeod *et al.*, 1993). Here immune-mediated disorders predominate. Nonimmune problems, or conditions with a presumed immunologic basis awaiting further clarification, may also be seen. Such disorders may be part of a multisystem pathological complex, and their arbitrary assignment to certain organ categories is dependent on the individual's perspectives.

1. Thrombotic Thrombocytopenic Purpura

a. Description

One such challenging multisystem disorder is thrombotic thrombocytopenic purpura (TTP). Referrals of patients with TTP typically occur through hematologists, because of the striking hematologic abnormalities exhibited. Neurologic dysfunction, however, can be a prominent component to the clinical presentation. Acute as well as chronic or relapsing variants of TTP can be encountered.

Though beyond the scope of this chapter, TTP results from lack of activity of a serum protease that normally cleaves ultralarge von Willebrand factor (ULvWF) multimers. This lack of enzyme activity is most often due to the presence of an IgG antibody against the vWF-cleaving protease. With inactivation of the vWF-cleaving protease, ULvWF multimers accumulate within patient plasma, causing platelet aggregation and resultant ischemia (Furlan *et al.*, 1998; Tsai and Lian, 1998).

TPE with plasma as replacement fluid both removes the ULvWF multimers and the IgG antibody to the protease and replaces normal vWF-cleaving protease in the TTP patient.

b. Role of TPE

Until the recognition of the beneficial role of plasma infusion therapy, and subsequently the enhancement of its administration and therapeutic effect via the plasma exchange process, TTP patient mortality rates approached 90%. The application of TPE as a major first line therapeutic component in the treatment of TTP patients, along with advances in clinical management, has resulted in significant improvements in both patient morbidity and mortality (Rock *et al.*, 1991).

Upon diagnosis of TTP, TPE therapy should begin as early as possible and continue on a daily basis, using disease activity marker levels and their value trends to judge therapeutic response. Such markers include clinical status, platelet count, serum lactate dehydrogenase (LDH), hemoglobin/hematocrit, and extent of schistocytosis on peripheral blood smear exam.

Single plasma volume exchange with fresh frozen plasma (FFP) as the replacement fluid should continue on a daily basis until normalization of the platelet count and preferably until the level of LDH has declined to within normal limits. Once these goals are achieved, many practitioners continue daily TPE for another 24-48 hours before stopping TPE or segueing into a TPE taper. Abrupt cessation of TPE in TTP patients should be avoided, since relapses may ensue.

c. Technical Considerations

Options regarding replacement fluids for TTP are limited to plasma products other than albumin. Cryosupernatant and pooled plasma (solvent/detergent-treated) offer alternatives to FFP as replacement fluids for patients with chronic or relapsing forms of TTP. When patients are unresponsive to TPE using FFP, these other plasma product options should be considered. Due to decreased vWF multimer content, a wider role for application of cryosupernatant and solvent/detergent-treated plasma in initial treatment strategy for TTP may be justified. Using solvent/detergent-treated plasma would also reduce the risk of viral transmission.

Regarding other blood product support, patients may require PRBC transfusions, especially early in the disease process. Platelet transfusions are essentially contraindicated in TTP, due to the possibility of exacerbation of the thrombotic process. The use of platelet transfusion in TTP is reserved for episodes of life-threatening hemorrhage.

2. Post-transfusion Purpura

In the management of another thrombocytopenic disorder, post-transfusion purpura (PTP), TPE has a more limited adjunctive role. This is a rare immune-mediated condition resulting in a sudden and severe decline in platelets after exposure to blood products. Although exact immunopathologic mechanisms remain to be fully elucidated, antibodies directed against the platelet specific antigen HPA-1a (Pl^{A1}) on glycoprotein IIIa appear critical to development of PTP. Stimulation of these antibodies can occur either by pregnancy or prior transfusion with Pl^{A1} positive products. Reexposure to such products results in the rapid destruction not only of the transfused antigen-positive blood but, paradoxically, the patient's autologous Pl^{A1} negative platelets as well. The participation of soluble antigen in immune complexes with attachment to autologous platelets and/or the development of a cross-reactive autoantibody may account for such a finding. High-dose IVIg is the primary therapy for the disorder, with TPE reserved for the minority of unresponsive patients (Drew, 1997).

3. Autoimmune (Idiopathic) Thrombocytopenic Purpura (ITP)

In ITP, IgG autoantibodies directed against platelet antigens cause accelerated platelet destruction. In children ITP typically occurs following a viral infection and is usually acute and self-limited. In adults, a chronic course of ITP is often seen. Treatment with corticosteroids is first-line therapy. For steroid-unresponsive cases, splenectomy may be performed. High-dose IVIg therapy and immunosuppressive agents can also be used. In severe refractory patients with active or imminent life-threatening bleeding, platelet transfusions may offer temporary effectiveness. Due to shortened platelet survival in ITP, however, platelet transfusions offer little in the routine management of patients in chronic settings.

TPE has a very limited role in the management of these patients. An exception, however, may be AIDS-related ITP, where TPE in conjunction with IVIg (Stricker, 1991) has occasionally been beneficial. In such cases, as well as in severe refractory ITP patients, the use of extracorporeal affinity absorption column technology (i.e., staphylococcal protein A) may also be an option (McCullough, 1998). For alloimmune platelet refractoriness, no definitive benefit to TPE has been shown.

4. Red Cell Alloimmunization

Another area where TPE has only limited application is in treating both alloimmune and autoimmune hemolytic anemias. In the former, TPE has been used: (1) in the management of patients who receive a bone marrow transplant from an ABO-incompatible donor; (2) as part of the management of alloimmunization to various RBC antigens in the setting of hemolytic disease of the newborn to decrease maternal antibody levels, and (3) as a strategy to enhance transfusion support in patients with antibodies to high-frequency RBC antigens. Although TPE decreases alloantibody levels in such settings, its use is limited by advances in the management of these conditions (e.g., red cell depletion of harvested marrow and intrauterine transfusions). Its continued value in patients sensitized to high-frequency RBC antigens may be an exception. As an extension of this area, TPE has also been re-

ported as a component in the management of ABO-unmatched or ABO-incompatible organ transplantation to treat episodes of post-transplantation hemolysis and/or humoral rejection of allograft organs possessing these antigenic structures (Drew, 1997).

5. Autoimmune Hemolytic Anemia

Case reports of the use of TPE have been described for cold-reactive, warm-reactive, and mixed cold/warm-reactive types of autoimmune RBC destruction. Here, however, because of the usually successful outcomes with conventional therapies applied to these disorders, TPE has been relegated a minor role. Its greatest use may be with cold-reactive autoantibody-mediated disease, where standard therapy is of modest impact and the IgM nature of the antibody allows for its efficient removal via TPE from the intravascular space. Used in concert with steroids and other immunosuppressive agents, however, it is sometimes difficult to evaluate the precise contribution of TPE in these settings. This is especially true for warm-reactive autoantibody-mediated processes.

Nevertheless, it would not be unreasonable to add TPE to the list of therapeutic considerations in patients with severely active hemolysis and marked transfusion dependence who are unresponsive to conventional drug therapy. The use of TPE as a temporizing measure while awaiting the attainment of therapeutic thresholds for immunosuppressive agents or in attempts to modify drug dosing schedules due to a patient's medication intolerance may also be appropriate. In any situation, however, TPE used in combination therapy strategies with steroids and/or other immune-modulating agents increases the likelihood of an effect. The application of pulse dose synchronization therapy, that is, daily sessions of TPE followed by pulses of cyclophosphamide and prednisone (Silva et al., 1994) or IVIg (Hughes and Toogood, 1994), may be useful in some patients.

6. Pure Red Cell Aplasia

In select patients with pure red cell aplasia who have no bone marrow transplantation options or who are unresponsive to conventional immunosuppressive regiments, TPE may have a role if antibodies to early RBC precursors are present (Messner et al., 1981).

7. Clotting Factor Inhibitors

In patients with clotting factor inhibitors, TPE has an adjunctive role. TPE may be used when uncontrollable bleeding occurs or in preparation for surgery in order to lower clotting factor inhibitor titers, thus allowing for normal hemostatic activity following factor infusion. TPE may have value in this regard for both alloimmune and autoimmune factor inhibitors. TPE for the treatment of factor inhibitors is often used in combination with a variety of other immunomodulatory and/or immunosuppressive regimens.

8. Dysproteinemias

Defined as a derangement of the protein content of the blood dysproteinemia typically involves immunoglobulins. The pathway leading to this derangement may occur in either neoplastic or non-neoplastic conditions. These disorders include multiple myeloma, macroglobulinemia. monoclonal gammopathy of unknown significance (MGUS), and cryoglobulinemia. Although such antibodies can act in an immunologically specific manner, resulting in coagulopathies and neuropathies, nonimmunological problems associated with their plasma concentration can be dramatic and acutely life-threatening. Hyperviscosity, hypervolemia, and renal failure are the main presentations in this regard. Depending on the particular syndrome (e.g., hyperviscosity), 1 or 2 treatments may suffice, whereas in renal disease, a series of daily exchanges over 7 days may be warranted. Customization of the therapeutic regimen is key to the successful management of these challenging conditions.

E. Miscellaneous Conditions

Aside from neurologic and hematologic diseases seen by apheresis specialists, other conditions in rheumatology, nephrology, dermatology, and endocrinology may occasionally be encountered. Autoimmune conditions predominate, with overlap involvement of multiple systems frequently seen.

1. Rheumatologic and Renal

Recognized Category I and II indications (see Table 2) for TPE include Goodpasture's syndrome, rapidly progressive glomerulonephritis, and systemic vasculitis (primary and secondary forms; Kiprov *et al.*, 1993). Results from studies using TPE in the treatment of lupus nephritis have been disappointing (Euler *et al.*, 1996). TPE alone is not justified in the treatment of systemic lupus erythematosus (SLE). The value of TPE synchronized with cytotoxic agents in certain SLE-associated crisis situations is controversial (Euler *et al.*, 1994; Wallace *et al.*, 1998). Generally, patients unresponsive to standard drug therapy or intolerant of high-dosage regimens may be candidates for TPE during severe disease exacerbations (Wallace, 1999).

2. Dermatologic

Two autoimmune diseases involving the skin, bullous pemphigoid and pemphigus vulgaris, have been treated successfully with TPE. This has typically been in the context of adjunct therapy to first-line corticosteroids and/or cytotoxic drug administration (Yamada *et al.*, 1997).

3. Metabolic and Endocrine

Major indications for the use of TPE in this group of diseases involve patients with homozygous familial hypercholesterolemia, Refsum's disease, and drug overdoses and poisonings (Kasprisin *et al.*, 1993). Methods of selective removal of various lipoproteins and cholesterol by various extracorporeal ligand-specific extraction methods have also been developed and used in the selling of familial hypocholesterolemia. With respect to poisoning and drug overdose, TPE is of a more acute and limited application. Toxic agents that are highly plasma protein-bound are the best candidates for removal. Other considerations favoring TPE in this setting would include poor lipid solubility, low extravascular tissue affinity, and small volumes of extravascular distribution.

References

AABB Hemapheresis Committee. (1992). "Guidelines for Therapeutic Hemapheresis." AABB, Bethesda.

Ciavarella, D., Wuest, D., Strauss, R. G., *et al.* (1993). Management of neurological disorders. *J Clin Apher* **8**, 242–257.

Drew, M. J. (1997). Therapeutic plasma exchange in hematologic diseases and dysproteinemias. In "Apheresis: Principles and Practice" (B. C. McLeod, Sr. Ed.), pp. 307–333. AABB Press, Bethesda.

Dyck, P. J., Daube, J., O'Brien, P., *et al.* (1986). Plasma exchange in chronic inflammatory demyelinating polyradiculoneuropathy. *N Engl J Med* **314**, 461–665.

Dyck, P. J., Low, P. A., Windebank, A. J., *et al.* (1991). Plasma exchange in polyneuropathy associated with monoclonal gammopathy of undetermined significance. *N Engl J Med* **325**, 1482–1486.

Euler, H. H., Schroeder, J. O., Harten, P., Zeuner, R. A., and Gutschmidt, H. J. (1994). Treatment-free remission in severe systemic lupus erythematosus following synchronization of plasmapheresis with subsequent pulse cyclophosphamide. *Arthritis Rheum* **37**, 1784–1794.

Euler, H. H., Zeuner, R. A., and Schroeder, J. O. (1996). Plasma exchange in systemic lupus erythematosus. *Transfusion Sci* **17**, 245–265.

Furlan, M., Robles, R, Galbusera, M., Remuzzi, G., Kyrle, P. A., Brenner, B., Krause, M., Scharrer, I., Aumann, V., Mittler, V., Solenthaler, M., and Lammle, B. (1998). Von Willebrand factor-cleaving protease in thrombotic thrombocytopenic purpura and the hemolytic-uremic syndrome. *N Engl J Med* **339**, 1578–1585.

Guillain–Barré Syndrome Study Group (1985). Plasmapheresis and acute Guillain-Barré syndrome. *Neurology* **35**, 1096–1104.

Harati, Y. (1997). Chronic immune-related demyelinating polyradiculoneuropathy. In "Neuro-Immunology for the Clinician" (L. A.

Rolak and Y. Harati, Eds.), pp. 229–235. Butterworth-Heinemann, Boston.

Hillyer, C. D., and Berkman, E. M. (1996). Plasma exchange in the dysproteinemias. In "Principles of Transfusion Medicine" (E. C. Rossi, T. L. Simon, G. S. Moss, and S. A. Gould, Eds.), 2nd ed., pp. 569–575. Williams and Wilkins, Baltimore.

Hughes, P., and Toogood, A. (1994). Plasma exchange as a necessary prerequisite for the induction of remission by human immunoglobulin in autoimmune hemolytic anemia. *Acta Haematol* **91**, 166–169.

Kasprisin, D. O., Strauss, R. G., Ciavarella, D., *et al.* (1993). Management of metabolic and miscellaneous disorders. *J Clin Apher* **8**, 231–241.

Kiprov, D. D., Strauss, R. G., Ciavarella, D., *et al.* (1993). Management of autoimmune disorders. *J Clin Apher* **8**, 195–210.

McCullough, J. (1998). "Transfusion Medicine," pp. 453–477. McGraw-Hill, New York.

McLeod, B. C. (2000). Introduction to the third special issue: Clinical applications of therapeutic apheresis. *J Clin Apher* **15**, 1–5.

McLeod, B. C., Strauss, R. G., Ciavarella, D., *et al.* (1993). Management of hematological disorders and cancer. *J Clin Apher* **8**, 211–230.

McLeod, B. C., Price, T. H., and Drew, M. J., Eds. (1997). "Apheresis: Principles and Practice." AABB Press, Bethesda.

Messner, H. A., Fauser, A. A., Curtis, J. E., and Dotten, D. (1981). Control of antibody-mediated pure red cell aplasia by plasmapheresis. *N Engl J Med* **304**, 1334–1338.

Osterman, P. G., Lundemo, G., Pirskanen, R., *et al.* (1984). Beneficial effects of plasma exchange in acute inflammatory polyradiculoneuropathy. *Lancet* **2**, 1296–1299.

Raphael, J. C., Chevretts, S., and Gajdos, P. (1996). Plasma exchange in neurological diseases. *Transfusion Sci* **17**, 267–282.

Rock, G. A., Shumak, K. H., Buskard, N. A., *et al.* (1991). Comparison of plasma exchange with plasma infusion in the treatment of thrombotic thrombocytopenic purpura. *N Engl J Med* **325**, 393–397.

Rostami, A., and Sater, R. A. (1997). Guillain-Barré syndrome. In "Neuroimmunology for the Clinician" (L. A. Rolak and Y. Harati, Eds.), pp. 205–228. Butterworth-Heinemann, Boston.

Silva, V. A., Seder, R. H., and Weintraub, L. R. (1994). Synchronization of plasma exchange and cyclophosphamide in severe and refractory autoimmune hemolytic anemia. *J Clin Apher* **9**, 120–123.

Strauss, R. G., Ciavarella, D., Gilcher, R. O., *et al.* (1993). An overview of current management. *J Clin Apher* **8**, 189–194.

Stricker, R. B. (1991). Hematologic aspects of HIV disease: Diagnostic and therapeutic considerations. *J Clin Apher* **6**, 106–109.

Tsai, H.-M., and Lian, E. C.-Y. (1998). Antibodies to von Willebrand factor-cleaving protease in acute thrombotic thrombocytopenic purpura. *N Engl J Med* **339**, 1585–1594.

Wallace, D. J. (1999). Apheresis for lupus erythematosus. *Lupus* **8**, 174–180.

Wallace, D. J., Goldfinger, D., Pepkowitz, S. H., *et al.* (1998). Randomized controlled trial of pulse/synchronization cyclophosphamide/apheresis for proliferative lupus nephritis. *J Clin Apher* **13**, 163–166.

Yamada, H., Yaguchi, H., Takamori, K., and Ogawa, H. (1997). Plasmapheresis for the treatment of pemphigus vulgaris and bullous pemphigoid. *Ther Apher* **1**, 178–182.

39

Therapeutic Cytapheresis

Red Blood Cell Exchange, Leukapheresis, and Thrombocytopheresis (Plateletpheresis)

JOANNE L. BECKER
Roswell Park Cancer Institute
Buffalo, New York

KIRSTEN ALCORN
Washington Hospital Center
Washington, DC

Cytapheresis removes one or more cellular components of the blood. Common therapeutic cytapheresis procedures include red blood cell (RBC) exchange, leukapheresis, and thrombocytapheresis. These are in contradisctinction to whole blood exchange, which is occasionally used for the treatment of neonatal hyperbilirubinemia and hemolytic disease of the newborn (HDN; see Chapters 25 and 26). The cytapheresis procedure itself may be carried out manually or by automated instruments. Intravascular volume is maintained by either replacement of the removed element with normal packed red blood cells (PRBCs; e.g., RBC exchange), crystalloids, or colloids. As the procedure continues, the efficiency of removal of the cellular component decreases due to the dilutional effect of the replacement fluid.

The indications, adverse events associated with, and technical considerations regarding the three therapeutic cytapheresis procedures are discussed in this chapter.

I. RBC EXCHANGE

In RBC exchange, RBCs are the blood component requiring replacement. RBCs are removed from the patient, and allogeneic PRBCs are transfused. Sickle cell disease is the most commonly cited indication for RBC exchange, but as technology has advanced to its current level of relative ease and safety, the indications

have expanded. RBC exchange transfusion has now been used successfully in the treatment of parasitic infections of RBCs, including malaria and babesiosis, hemolytic transfusion reactions, and carbon monoxide poisoning (see Table 1).

A. Sickle Cell Disease (SCD)

1. Pathophysiology

In SCD, an amino acid mutation in the beta globin gene leads to production of hemoglobin S instead of hemoglobin A. Hemoglobin S is unstable, especially under conditions of relative hypoxia. The hemoglobin S polymerizes, leading to RBC sickling, a decrease in RBC deformability, and therefore a reduced ability of RBCs to pass easily through small vessels. Sickled RBCs become immobilized in the microvasculature, and ischemia of the distal tissue ensues. Patients with SCD suffer numerous vaso-occlusive complications, most of which are directly attributable to this basic defect in the RBCs' ability to circulate effectively.

2. Adverse Events

Some of the most serious adverse sequelae of vasoocclusion in SCD patients include stroke, acute chest syndrome, splenic/hepatic sequestration, and priapism. RBC exchange may be required for the treatment of some of these complications.

TABLE 1   Indications for Red Cell Exchange

- Sickle cell disease (SCD)[a]
- Hemolytic disease of the newborn (HDN)
- Protozoal infections[a]
- Polycythemia vera
- Acute/delayed hemolytic transfusion reactions[a]
- Carbon monoxide poisoning[a]
- Iron depletion

[a] Discussed in this chapter.

a. Stroke

Stroke is one of the most serious adverse events in patients with SCD. Strokes occur in about 10% of sickle cell patients and represent about two-thirds of first adverse events in children. The neurological events include reversible changes, transient ischemic attacks, and both ischemic and hemorrhagic strokes. Children more commonly present with ischemic stroke, and adults are more prone to hemorrhagic stroke. Adults with hemorrhagic strokes have very high mortality (approximately 50%). In both children and adults, the occurrence of a first stroke predicts those patients at high risk for recurrent neurologic events.

Treatment for stroke in SCD should begin as early as possible following onset of symptoms, preferably within 24–48 hours. RBC exchange transfusion with the goal of limiting hemoglobin S to 30% or less is recommended. The most common use of the procedure is a single RBC volume exchange, but other methods, including double RBC volume exchange, are used as well (Pepkowitz, 1997; Ohene-Frempong, 1991). Following the acute event, regular RBC simple or exchange transfusion to maintain the hemoglobin S level under 30% for several years is the most common prophylactic regimen.

b. Acute Chest Syndrome

Acute chest syndrome in SCD patients is another serious, rapidly developing, acute complication of the disease that has high morbidity and mortality. Acute chest syndrome presents with hypoxemia, shortness of breath, chest pain, and opacification of the chest x-ray. These findings often develop quickly, and may be seen in association with other events, including pneumonia and pain crisis. As with other major acute events in SCD, patients with one episode of acute chest syndrome appear to be at increased risk for recurrences. RBC exchange transfusion is indicated as an early intervention in acute chest syndrome, with the goal of lowering hemoglobin S to 30% or less. Patients gener-

ally demonstrate rapid improvement in respiratory symptoms following the RBC exchange procedure.

c. Splenic Sequestration

Sequestration crises may occur in both the liver and the spleen. In both organs, the etiology is simply vaso-occlusion related to the presence of sickled RBCs in the microvasculature. Most patients with SCD undergo gradual infarction of the spleen, resulting in auto-splenectomy by adulthood. However, in young children, the efferent sinusoids may become acutely occluded with sickled RBCs. This obstruction to outflow from the spleen results in rapid pooling of large volumes of blood within the organ. Patients often present with splenomegaly, hypovolemic shock, and anemia. For the initial treatment of splenic sequestration, simple RBC transfusion and crystalloid solutions are recommended (Powell et al., 1992). Transfusion aimed to return the hemoglobin to its baseline value has been recommended traditionally, but as the crisis resolves and previously sequestered blood reenters the circulation, volume overload may occur. RBC exchange can be used for management of the acute volume crisis (Pepkowitz, 1997). Children who have experienced one episode of splenic sequestration are at high risk of recurrent episodes (Powell et al., 1992). To avoid the risks associated with splenectomy in young patients, simple RBC transfusion with or without delayed splenectomy has been used.

d. Hepatic Sequestration

Hepatic sequestration may present similarly to splenic sequestration, or may present as intrahepatic cholestasis. The former may occur concurrently with a splenic sequestration episode, and is treated as such. In intrahepatic cholestasis, RBC sickling and stasis in the hepatic sinusoids results in severe parenchymal damage. In combination with hemolysis, this leads to acute elevation of hepatic enzymes and bilirubin, both direct and indirect; as hepatic synthetic ability is impaired, coagulation factor production may be decreased with subsequent clinical coagulopathy. Intrahepatic cholestasis is often lethal. Patients who survive may suffer chronic hepatic failure. RBC exchange transfusion may be helpful for both acute and chronic hepatic sequestration.

e. Priapism

Male SCD patients are at risk of developing priapism, painful penile erections that may result in significant tissue damage and future erectile dysfunction. Hydration and analgesia are first-line therapy for this disorder. If the priapism has not resolved after approximately 24–48 hours with standard therapy, simple or exchange RBC transfusion should be considered to reduce the risk of further RBC sickling and facilitate resolution. "ASPEN" syndrome is the acronym given to the syndrome of acute neurologic events such as seizure or stroke following RBC exchange transfusion, which has been described in several SCD patients with priapism (Siegel *et al.*, 1993). As a result of the emergence of this syndrome, the indications for RBC exchange transfusion in SCD patients with priapism are unclear at this time.

3. Surgical Prophylaxis

Simple RBC transfusion is recommended prior to surgery for SCD patients undergoing general anesthesia or eye surgery in order to maintain a hematocrit of 28–30% (Koshy *et al.*, 1995). Exchange RBC transfusion is rarely necessary.

4. Complications of Pregnancy

During the last 30 years, SCD has evolved from a disease of childhood to one with a life expectancy of over 40 years. This extended life expectancy has made pregnancy possible for women with SCD. In the not-too-distant past, women with SCD were warned of the dangers of pregnancy and were advised against child-bearing altogether. As management of the disease has improved, pregnancy in the SCD patient has become less dire. Studies have shown that routine simple transfusion throughout pregnancy (i.e., prophylactic transfusion) does not offer significant benefit over transfusions given only when indicated by specific clinical events such as pain crises (Koshy *et al.*, 1988). Therefore, exchange and simple RBC transfusions in pregnant SCD patients are indicated for treatment of pregnancy complications only. Prophylactic transfusion in otherwise normal SCD pregnancies is not recommended.

5. Chronic RBC Exchange Therapy

Many SCD patients who have had serious adverse events such as stroke receive chronic prophylactic simple RBC transfusions in order to reduce the incidence of future similar adverse events. Some centers have advocated the use of chronic RBC exchange instead of chronic simple RBC transfusion (Singer *et al.*, 1999; Hilliard *et al.*, 1998). This change has been recommended for the purpose of reducing iron overload (see Chapter 32), a frequently occurring complication in chronically transfused patients. Studies show that iron overload is in fact reduced (or at least its progression is halted) in most SCD patients on chronic RBC exchange protocols. However, patients on chronic RBC

exchange therapy receive 50–100% more blood products than do patients receiving simple transfusions, increasing their risks of infectious disease exposure and RBC and HLA alloimmunization.

B. Parasitic Infection

1. Malaria

Malaria is an endemic infection in many parts of the world, with high rates of morbidity and mortality. In fact, malaria is reported as the single largest cause of death around the world. The disease is caused by *Plasmodium* protozoa that infect human RBCs. Infection also induces microvascular endothelial changes and a systemic inflammatory response, including cytokine production. In this setting, infected RBCs occlude small blood vessels throughout the body (similar to SCD). In the most severe form of malaria, associated with *P. falciparum* infection, the cerebral blood vessels are affected. Acute organ damage may proceed rapidly with diffuse microvascular occlusion. The role of RBC exchange is discussed in Section I.B.3, below.

2. Babesia

Babesia species are another group of RBC-infecting protozoa. Infected RBCs have decreased membrane deformability and are sequestered in the spleen. In most immunocompetent individuals, the infection is eventually resolved, with mild flulike symptoms. However, in asplenic patients, or in patients who are otherwise immunosuppressed or are coinfected with Lyme disease, babesiosis may result in severe or relapsing illness.

3. Role of Hemapheresis

In severely ill patients with either malaria or babesiosis, RBC exchange is becoming an accepted treatment modality. It removes infected RBCs, reducing the parasitemia. Although studies with adequate data to definitively support the efficacy of the procedure are not available, it appears that high-risk patient groups with either infection may benefit from RBC exchange. For malaria, numerous factors have been cited as poor prognostic indicators (White, 1996), including parasitemia greater than 500,000/ml, seizures, altered consciousness, and end-organ damage, including pulmonary injury. For babesiosis, asplenic and immunosuppressed individuals have a more severe disease course (Matchinger *et al.*, 1993). Because the infested RBCs are often sequestered in the spleen and other microvascular beds, it is difficult to calculate the expected "yield," that is, reduction of parasitemia, based on apparent RBC volume of the patient. For malaria, there is a therapeutic goal of parasitemia of less than 5%. A single RBC exchange may be performed, with close clinical observation following the procedure, with repeat RBC exchanges, if necessary. Using higher volumes may avoid the need for further procedures. Again, observation for clinical improvement and parasitemia is necessary to guide appropriate care.

C. Carbon Monoxide Poisoning

In carbon monoxide poisoning, the CO molecule binds the oxygen-carrying sites of hemoglobin with high affinity. This functionally excludes oxygen from these binding sites, resulting in severe hypoxemia. To prevent death in cases of acute carbon monoxide poisoning, rapid reoxygenation is necessary. This is achieved by having the patient breathe 100% oxygen and by the use of hyperbaric chambers. RBC exchange transfusion is an alternative approach. It is more widely available than hyperbaric chambers, and is relatively quick and easy to initiate. Rather than attempting to alter the hemoglobin–carboxyhemoglobin equilibrium, it removes the patient's RBCs and replaces them with oxygenated donor blood.

D. Hemolytic Transfusion Reactions

1. Acute Hemolytic Reactions

Hemolytic transfusion reactions occur when patients with anti-RBC antibodies are exposed to RBCs that are antigenically different from their own. The classic example of an acute hemolytic transfusion reaction is ABO-incompatible transfusion. ABO antibodies are naturally occurring, and may cause rapid, serious hemolysis, even with minimal exposure to blood from an incompatible ABO group. Although many safeguards are in place in the practice of transfusion medicine, these reactions still occur occasionally. Estimates for the United States suggest that 1 in approximately 33,000 RBC transfusions results in acute hemolytic transfusion reactions due to ABO incompatibility and that 6% of these events are fatal.

2. Delayed Hemolytic Reactions

Delayed hemolytic transfusion reactions occur when antibodies to transfused RBCs cause hemolysis at some time, usually several days, following the transfusion. Although acute hemolytic transfusion reactions are more likely to require immediate intervention, delayed reactions may be clinically severe and require treatment interventions as well.

3. Role of Hemapheresis

The rationale of RBC exchange transfusion for the treatment of hemolytic transfusion reactions is to remove the mismatched donor RBCs. These mismatched cells are replaced with compatible donor RBCs, thereby reducing the risk of further hemolysis. RBC exchange for hemolytic transfusion reactions is not often used, but is an available form of therapy.

E. Technical Considerations

1. General

Once one has decided to proceed with RBC exchange transfusion, one is faced with a new set of considerations: volume to be exchanged, choice of products to be used for replacement, and evaluation of response. Mathematically, the plateau of efficiency for RBC exchange transfusion is approximately 1 RBC volume.

To calculate 1 RBC volume, there are several formulas available. The simplest is to multiply the hematocrit by an estimated blood volume of 70 ml/kg. For example, 1 red cell volume for a 70-kg patient with a hematocrit of 25% would be calculated as

$$70 \text{ kg} \times 70 \text{ ml/kg} \times 0.25 = 1225 \text{ ml.}$$

Other variables that may be considered in the calculation of a RBC volume for purposes of exchange transfusion may include patient gender and weight, preservative (i.e., concentration or hematocrit) of the replacement RBCs, anticoagulant volume ratio, desired postexchange hematocrit, and "fraction of cells remaining." Fortunately, most exchange transfusions are now done by automated apheresis equipment with computer programs capable of making the required calculations based on several such variables. Experience and familiarity with the equipment, as well as attention to clinically useful evaluation of each procedure, are extremely valuable.

2. Adverse Effects

Replacement RBCs used in exchange transfusion carry with them all the risks associated with usual transfusions, notably including transfusion reactions, alloimmunization, and disease transmission.

Other risks of exchange transfusion are those that are generic to apheresis procedures and the use of large-bore indwelling central venous catheters. These risks include citrate toxicity, infection, thrombosis, and bleeding, and are discussed at length in Chapter 36.

3. Replacement RBCs

a. Leukoreduction

Leukoreduced RBCs are recommended for use in RBC exchange procedures. Leukoreduction for RBC exchange transfusion should be performed in the laboratory, as bedside filtration is not rapid enough to accommodate the high flow rate of the automated procedure. Leukoreduction minimizes the risks of febrile nonhemolytic transfusion reactions and CMV transmission, and decreases HLA sensitization and the immunomodulatory effects of transfusion (see Chapter 14).

b. Compatibility

In patients who are undergoing RBC exchange but are not otherwise expected to need chronic transfusion support, for example, as treatment for malaria, no special attention beyond ABO and Rh compatibility is generally paid to RBC phenotype matching of the patient with units used for the exchange procedure.

However, for SCD patients, the importance of careful PRBC unit selection for all transfusions, including exchange transfusions, has been amply demonstrated. Alloimmunization is a primary concern in these patients who are chronically transfused and depend on availability of compatible blood. Alloimmunization also increases with increasing RBC exposure. Exposure to Rh, Kell, and Kidd antigens is especially immunogenic. Most SCD patients are of African descent, and possess RBC antigenic phenotypes that differ significantly from Caucasian populations. Major sickle cell treatment centers recommend performing an extended RBC antigenic phenotype for all SCD patients, and transfusing RBCs matched for Rh and Kell antigens (at the least). Some have shown that the incidence of alloimmunization is greatly reduced by transfusing PRBCs as completely phenotypically matched as possible, including Rh, Kell, Kidd, Duffy, and S blood group antigens.

c. Hemoglobin S-Negative

In addition to considering extended antigenic matching of RBC units for SCD patients, hemoglobin S-negative units should also be used. Several tests to detect the ability and propensity of RBCs to sickle under certain conditions are easily available and relatively rapid to perform. For practical purposes, this excludes blood donated by those with sickle trait (heterozygotes for hemoglobin S). The rationale for this exclusion is straightforward: to avoid providing replacement RBCs that may sickle under stressful physiologic conditions.

II. LEUKAPHERESIS

A. Leukocytosis

Patients with hematologic malignancies can present with symptoms caused by increased numbers of circulating WBCs, platelets, or RBCs.

1. Definition

Hyperleukocytosis, defined as a white cell count greater than $100,000/\mu l$, is often associated with increased morbidity and mortality in patients with leukemic processes. When symptomatic, hyperleukocytosis is considered a medical emergency. A WBC count of greater than $100,000/\mu l$ is a generic value. In patients with AML or CML (described below) in blast crisis, a white count of less than $50,000/\mu l$ can be enough to cause symptoms, whereas in CLL (also described below), patients can be asymptomatic despite a WBC count greater than $100,000/\mu l$.

2. Pathophysiology

The predominant manifestations of hyperleukocytosis are a result of hyperviscous blood leading to vascular obstruction, which induces tissue hypoxia and/or rapid cellular destruction. The central nervous system (CNS) and lungs are the most common sites for vascular obstruction to cause symptoms, but effects on other organ systems are seen. The CNS symptoms of vascular obstruction include stupor, dizziness, blurry vision, delirium, and ataxia. Focal deficits can be elicited and retinal hemorrhages may be present. Respiratory symptoms of obstruction include dyspnea, tachypnea, and hypoxemia, with the presence of ausculatory rales. A chest x-ray will often show bilateral interstitial infiltrates. Rapid cellular destruction, also called tumor lysis syndrome, can induce fever; increased uric acid, phosphate, and potassium; and decreased calcium. These metabolic disturbances can lead to renal failure, cardiac arrythmias, and death.

Patients also typically present with anemia and thrombocytopenia due to marrow suppression. These patients are at risk of increased symptoms of vascular obstruction if transfusions occur in the presence of elevated WBC counts.

3. Role of Leukapheresis

In patients with hyperleukocytosis, there are three methods of reducing the WBC count: hydroxyurea, induction chemotherapy, and leukapheresis. Leukapheresis is the most rapid method of acutely lowering a patient's WBC count. The effectiveness and duration of peripheral leukocyte depletion by leukapheresis are predominantly dependent upon two factors: the efficiency of the WBC removal and the rate at which WBCs are added to the peripheral circulation from the marrow. In patients with electrolyte imbalances or renal insufficiency, leukapheresis can be a temporizing measure to reduce the WBC count while the patient is being prepared to receive chemotherapy, or it can be used as an adjuvant to chemotherapy to reduce the effects of cytolysis. The use of leukapheresis to reduce the symptoms of tumor lysis is gaining favor among hematologists. Leukapheresis does not alter the underlying disease but only modifies the course of the disease.

B. Acute Myelogenous Leukemia (AML)

Patients with AML have large blasts that are less deformable than mature granulocytes. Symptoms of hyperleukocytosis are most common in this disease, even with WBC counts less than $50,000/\mu l$. With this increase in WBCs, these patients experience anemia, thrombocytopenia, and coagulopathies including disseminated intravascular coagulation (DIC).

1. Leukapheresis and Transfusion

Because the large blasts increase the blood viscosity, PRBC and platelet transfusions, which further increase the blood viscosity and risk of vascular stasis, can be dangerous in these patients. Leukapheresis can decrease WBCs so that RBCs and platelets can be safely transfused (Murray et al., 1996).

2. Leukapheresis and Chemotherapy

Leukapheresis has been suggested as a precursor to induction chemotherapy in cases where the WBC count is greater than $100,000/\mu l$. By reducing the amount of cytolysis, the systemic side effects of induction chemotherapy are moderated, and it is hoped that the patients will have a better prognosis. Unfortunately, studies thus far have not been able to support this hypothesis (Porcu et al., 1997; Bunin and Pui, 1985).

3. Leukapheresis and Pregnancy

There is one situation in which leukapheresis is the preferred therapy. In patients with AML diagnosed during the first trimester of pregnancy, leukapheresis has been used to control the effects of the disease until the second trimester, when chemotherapy will no longer cause developmental defects in the fetus.

C. Acute Lymphocytic Leukemia (ALL)

The treatment of hyperleukocytosis in ALL is different than that in AML. The blasts in ALL tend to be smaller and more malleable than those in AML. This makes the WBCs less likely to cause symptomatic leukostasis. Studies in the pediatric population have not shown any significant differences in the outcomes of patients who received leukapheresis prior to induction as compared with those who did not. Despite this, leukapheresis is still considered a valid therapy for patients experiencing symptomatic leukostasis and as a "debulking" procedure in those patients with hyperuricemia and renal insufficiency prior to initiation of chemotherapy (Bunin and Pui, 1985; Maurer et al., 1988).

D. Chronic Myelogenous Leukemia (CML)

1. Blast Crisis

Patients with CML in blast crisis have similar risks of leukostasis as patients with AML, and should be treated similarly. Leukostasis is less common in CML because the WBCs are more mature and have more deformability.

2. Chronic Phase

For chronic phase CML, regularly scheduled leukapheresis procedures have been used and can effectively decrease the peripheral WBC count and associated organomegaly (lymphadenopathy, splenomegaly, and hepatomegaly). The effects are transitory; thus, regularly scheduled leukapheresis is required (Morse et al., 1966). When leukapheresis has been compared to chemotherapy for this disease, chemotherapy has been found to be more effective in creating a stable remission (Hadlock et al., 1975). CML has been successfully treated during pregnancy with leukapheresis (Strobl et al., 1999; Caplan et al., 1978).

E. Chronic Lymphocytic Leukemia (CLL)

Because mature lymphocytes are small and maintain their deformability, symptomatic hyperleukocytosis is uncommon in CLL. Leukapheresis has been shown to be effective in removing these cells, but in most patients chemotherapy has been found to be a better method of controlling the lymphocyte count. In those patients who cannot tolerate chemotherapy, leukapheresis has been used to reduce the WBC count, controlling constitutional symptoms; but similarly to CML, it does not change the course of the disease (Han and Rai, 1990; Goldfinger *et al.*, 1980).

F. Technical Considerations

1. Selection of Patients

Since the ability to remove the WBCs is related to the ability to separate the cells, one way to determine whether leukapheresis will be effective is to measure the leukocrit of the peripheral blood. In normal patients, the cytocrit (hematocrit + leukocrit) is equal to the hematocrit. In patients with elevated WBC counts, the leukocrit can be a significant percentage of the cytocrit. If the leukocrit is not elevated, the ability of apheresis to remove WBCs may be limited.

2. Instrumentation

Centrifugation is the most efficient method of differentiating the WBC layer from the RBCs and plasma, and most instruments for performing leukapheresis are based upon this principle. The procedure can be performed in either a continuous or discontinuous flow format. Because of the smaller extracorporeal volume and faster procedure time, a majority of institutions now perform the procedure in the continuous format.

3. Sedimenting Agents

Hydroxyethyl starch (HES) is a RBC sedimenting agent that is used to improve the collection of granulocytes, bands, and metamyelocytes from normal donors. It causes the RBCs to "rouleaux" and pack more tightly. The cellular layers created with centrifugation using HES are better defined, allowing the WBCs to be more easily identified and removed. HES can be useful in patients with symptoms who do not have extremely high WBC counts, and in those patients with a predominance of moderately mature, malignant WBCs. HES is available in two formulations: 6% hespan and 10% pentastarch. The use of sedimenting agents carries with it additional risks, such as volume expansion in the recipient, and unknown risks associated with metabolism of the product that have not yet been completely defined.

4. Volume Processed

Two blood volumes or 10 liters of whole blood, whichever is greater, should be the starting point for leukapheresis in adults. In patients who tolerate the procedure well, additional volumes can be processed. In general, the collection rate is from 3 to 8 ml/min, dependent upon the inlet rate and the WBC count Manufacturers' documents state that up to 15 ml/min can be collected, but most procedures have no need to progress more than 8 ml/min. The product removed in the procedure ranges from 100 cc of cells in plasma to much greater amounts. A majority of patients will not require volume repletion at the end of the procedure. In those patients for whom the collection is greater than 500 ml, some suggest repletion with a small volume of colloid.

5. Goal

The goal of leukapheresis is to reduce the peripheral WBC count. This action can decrease the acute symptoms of hyperleukocytosis and/or prevent symptoms of cytolysis. Some investigators are using leukapheresis in patients without symptoms as a modality to reduce cytolysis during induction chemotherapy. In these cases, they are typically requesting that the patient's leukocyte count be decreased to less than $100,000/\mu l$ before chemotherapy is begun.

6. Monitoring

The best monitors for adequacy of leukapheresis are improvement of the patient's symptoms. WBC pre- and postcounts and leukocrits can also be helpful. An additional measure that can be used is the percent reduction in WBC count. This is determined by calculating the WBC count in the collection, dividing by the patient's circulating WBC mass, and then multiplying by 100. When this percentage is known, and the patient's pre- and postcounts are compared, the effectiveness of the procedure in reducing the patient's WBC count can be seen.

7. Number of Procedures to Be Performed

It is hoped that a single procedure will achieve the desired goal of decreasing the WBC count, but in some patients, 1 procedure will not be adequate. The decision to perform additional procedures should be based upon the desired goals, the improvement or lack of improvement in symptoms, and the change in peripheral WBC count. When symptoms dictate, the procedure can be performed after chemotherapy has started. In this situation, the WBCs will become more fragile and likely to break, so the patient should be watched carefully during the procedure to make sure that the mechanical cell lysls does not cause serious sequelae.

8. Complications

a. Citrate Toxicity

Patients will be receiving sodium citrate as an anti-coagulant throughout the procedure. The use of citrate will decrease the patient's circulating ionized calcium level and may lead to citrate toxicity. The treatments for this include slowing the procedure to reduce the rate of citrate infusion, decreasing the amount of anti-coagulant used, and giving the patient calcium supplements orally or intravenously.

b. Cell Loss

Depending upon the ability of the procedure to separate the WBCs from the RBCs, the patient may have varying degrees of RBC or platelet loss during the procedure. This is most obvious when multiple leuka-phereses are performed, and is in addition to the anemia and thrombocytopenia that are generally present in these patients. If cell loss during collection of peripheral blood progenitor cells is used as a guide, patients we lose approximately 20% of circulating platelets and 0.3 g/dl Hb with each leukapheresis procedure.

III. THROMBOCYTOPHERESIS

Primary or essential thrombocytosis is a chronic myeloproliferative disease with megakaryocyte hyperplasia and a platelet count greater than $600 \times 10^9/\mu l$ in the absence of an increased RBC mass. Secondary thrombocytosis occurs in response to a stimulus such as iron deficiency anemia. Thrombocytosis can present as thrombosis, with signs and symptoms of vascular insufficiency or of occlusive events; conversely, it can present with hemorrhage.

The short-term course of thrombocytosis can be improved by using plateletpheresis to reduce the peripheral platelet count in those patients experiencing thrombotic or hemorrhagic symptoms. Similar to leukapheresis, the effectiveness of plateletpheresis is dependent both on the efficiency of platelet removal and on the rate at which platelets are being released into the peripheral circulation. Again, this procedure can alter the symptoms of the disease, but does not alter its course. Thrombocytosis is more effectively treated by chemotherapy (Goldfinger *et al.*, 1979; Pearson, 1991).

The procedure of plateletpheresis is the same as that used for donor platelet collection, with the exception of the number of blood volumes processed. At least 2 blood volumes should be processed to begin reduction of the platelet count in thrombocytosis. Because the volume of plasma removed in this procedure will be approximately 500 cc per blood volume, volume replacement beyond that supplied by the citrate anticoagulant will be required for this procedure. There should be minimal to no RBC loss during platelet-pheresis.

References

Bunin, N. J., Pui, C. H. (1985). Differing complications of hyper-leukocytosis in children with acute lymphoblastic or acute non-lymphoblastic leukemia. *J Clin Oncol* **3**, 1590–1595.

Caplan, S. N., Coco, F. V., and Berkman, E. M. (1978). Management of chronic myelocytic leukemia in pregnancy by cell pheresis. *Transfusion* **18**, 120–124.

Goldfinger, D., Thompson, R., Lowe. C., et al. (1979). Long-term plateletpheresis in the management of primary thombocytosis. *Transfusion* **19**, 336–338.

Goldfinger, D., Capostagno, V., Lowe, C., et al. (1980). Use of long-term leukapheresis in the treatment of chronic lymphocytic leukemia. *Transfusion* **20**, 450–454.

Hadlock, D. C., Fortuny, I. E., McCullough, J., et al. (1975). Continuous flow centrifuge leucapheresis in the management of chronic myelogenous leukemia. *Br J Haematol* **29**, 443–453.

Han, T., and Rai, K. R. (1990). Management of chronic lymphocytic leukemia. *Hematol Oncol Clin N Am* **4**, 431–445.

Horn, A., Freed, N., and Pecora, A. A. (1987). Sickle cell hepatopathy: Diagnosis and treatment with exchange transfusion. *J Am Osteopath Assoc* **87**, 130–133.

Koshy, M., Burd, L., Wallace, D., et al. (1988). Prophylactic red-cell transfusions in pregnant patients with sickle cell disease. A randomized cooperative study. *N Engl J Med* **319**, 1447.

Koshy, M., et al. (1995). Surgery and anesthesia in sickle cell disease. *Blood* **86**, 3676–3684.

Matchinger, I., Telford, S. R., Inducil, C., et al. (1993). Treatment of babesiosis by red blood cell exchange in an HIV-positive, splenectomized patient. *J Clin Apher* **8**, 78–81.

Maurer, H. S., Steinhertz, P. G., Gaynon, P. S., et al. (1988). The effect of initial management of hyperleukocytosis on early complications and outcome of children with acute lymphoblastic leukemia. *J Clin Oncol* **6**, 1425–1432.

Morse, E. E., Carbone, P. P., Freireich, E. J., et al. (1966). Repeated leukapheresis in patients with chronic myelocytic leukemia. *Transfusion* **6**, 175–182.

Murray, J. C., Dorfman, S. R., Brandt, M. J., et al. (1996). Renal venous thrombosis complicating acute myeloid leukemia with hyperleukocytosis. *J Pediatr Hematol Oncol* **1918**, 327–330.

O'Callaghan, A., O'Brien, S. G., Ninkovic, M., et al. (1995). Chronic intrahepatic cholestasis in sickle cell disease requiring exchange transfusion. *Gut* **37**, 144–147.

Ohene-Frempong, K. (1991). Stroke in sickle cell disease: Demographic, clinical, and therapeutic considerations. *Semin Hematol* **28**, 213–219.

Pearson, T. C. (1991). Primary thombocythemia: Diagnosis and management. *Br J Haematol* **78**, 145–148.

Pepkowitz, S. (1997). Red cell exchange and other therapeutic alterations of red cell mass. In "Apheresis: Principles and Practice" (B. C. Mcloed, T. H. Price, M. I. Drew, Eds.). AABB Press, Bethesda, MD.

Porcu, P., Danielson, C. F., Arazi, A., et al. (1997). Therapeutic leukapheresis in hyperleucocytic leukaemias: Lack of correlation between degree of cytoreduction and early mortality rate. *Br J Haematol* **98**, 433–436.

Powell, R. W., Levine, G. L., Yang, Y. M., *et al.* (1992). Acute spleniuc sequestration crisis in sickle cell disease: Early detection and treatment. *J Pediatr Surg* **27**, 215–219.

Sheehy, T. W., Law, D. E., and Wade, B. H. (1980). Exchange transfusion for sickle cell intrahepatic cholestasis. *Arch Int Med* **140**, 1364–1365.

Siegel, J. F., Ricj, M. A., and Brock, W. A. (1993). Association of sickle cell disease, priapism, exchange transfusion, and neurological events: ASPEN syndrome. *J Urol* **150**, 1480–1482.

Strobl, F. J., Voelkerding, K. V., and Smith, E. P. (1999). Management of chronic myeloid leukemia during pregnancy with leukapheresis. *J Clin Apher* **14**, 42–44.

White, N. J. (1996). The treatment of malaria. *N Engl J Med* **335**, 800–806.

SECTION X

CONSENT, QUALITY, AND RELATED ISSUES

40

Informed Consent for Transfusion

CHRISTOPHER P. STOWELL

Massachusetts General Hospital
Boston, Massachusetts

Although obtaining informed consent for medical and surgical treatment is very familiar, the application of this process to transfusion has developed largely in the last 15 years. In this chapter, the basis for informed consent and its necessary elements will be reviewed, particularly as they relate to transfusion. In addition, the components of a well-developed informed consent policy will be described. The topic of informed consent for transfusion has been treated in detail by Stowell (1997).

I. BASIS FOR INFORMED CONSENT

The concept of informed consent has its basis in both law and ethics. From these two sources the concept has developed that physicians may not make decisions for patients, but must provide them with the information necessary to make intelligent decisions about treatment. In the 1980s, dramatic escalation in the perception of infectious risks of transfusion shattered the image of blood as a benign and life-giving therapy. The increased awareness of these significant hazards intensified the legal and ethical imperative to obtain informed consent for transfusion. As a result, obtaining informed consent for transfusion has become common practice in many areas in the United States (Eisenstaedt *et al.*, 1993).

A. Legal Basis

The legal origins of informed consent lie in the corpus of case law which has established the principle that an individual "has a right to determine what shall

be done with his own body" (Schloendorff v. Society of New York Hospital, 1914) and must consent to medical intervention. The legal concept of informed consent has further evolved with the application of negligence theories to allow recovery in situations where a patient has given consent for an intervention but the physician has failed to adequately disclose its risks or possible outcomes.

B. Ethical Basis

The ethical origin of our concept of informed consent is based on our regard for the rights of patients, which include individual autonomy and self-determination. Patients have the right to make decisions about medical intervention that may affect their well-being. Implicit in this right is the need for access to sufficient information to be able to make a reasoned choice, and the freedom to consent to or refuse the intervention without coercion or bias.

II. STATUTORY AND REGULATORY REQUIREMENTS AND RECOMMENDATIONS

In addition to the legal and ethical rationale for obtaining informed consent for transfusion, there are also statutory and regulatory requirements. Although there are no federal regulations pertaining to informed consent for transfusion, statutes have been passed in California, New Jersey, and Pennsylvania which make it mandatory for physicians to obtain informed consent for transfusion and for procedures for which transfusion may be required. The Joint Commission on Accreditation of Healthcare Organizations (JCAHO) also has a requirement of informed consent for transfusion for accreditation (JCAHO, 1995). The American Association of Blood Banks (AABB) supports the application of informed consent to transfusion (AABB, 1994), although it is not a standard.

III. ELEMENTS OF INFORMED CONSENT

Informed consent is a process and not a document. It consists of a dialogue between patient and physician during which the physician discloses to the patient the information needed to make a sound decision about the proposed transfusion. Disclosure does not need to be exhaustive; however, it should provide the patient with the information a reasonable person would want to have before making a decision about undergoing a transfusion. This information must be provided in an appropriate manner, taking into consideration the patient's facility with language, education level, and health care experience. A number of key elements must be present to ensure that patients are given a valid opportunity to make an informed choice about their health care.

A. Explanation

In the case of transfusion, the explanation of the procedure is relatively straightforward and consists of describing the blood component to be transfused and the reason it is being administered.

B. Benefits

The benefits anticipated from the transfusion must be explained to patients so that they understand the therapeutic goals and can assess their importance. Patients may place different values on relieving fatigue and averting cardiac ischemia, for example, which might affect their tolerance for risk.

C. Risks

The hazards of transfusion must be reviewed with the patient. It is not necessary to review every single possible complication associated with blood transfusion. The risks which must be described are those material to a reasonable person attempting to make a decision about transfusion by balancing them against the anticipated benefits. In general, common complications of transfusion (e.g., urticarial and febrile, non-hemolytic transfusion reactions) should be described, as well as those which may eventuate in serious outcomes (e.g., death, organ failure, chronic illness). Risks specific to a particular patient (e.g., volume overload in a patient with congestive heart failure) should also be disclosed. Although the risks of viral infection from transfusion have been reduced to extremely low levels, the history of these transfusion-transmitted diseases and the high level of awareness among the public warrant discussing these complications explicitly. Conveying the magnitude of risks to patients can be difficult, particularly when such risks are very low. Numerical representation of risks may not be helpful to some patients. In some instances, comparing the frequency of a transfusion complication to other kinds of events (e.g., risk of death in a car accident), or explaining risks in terms of numbers of occurrences within a cohort easily imagined by the patient (e.g., number of people employed in a facility, number of people living in a town), may be useful.

D. Alternatives

The physician must also present the patient with any applicable alternatives to allogeneic transfusion. Not all of these alternatives may be suitable or available; hence it is not necessary to present all of them as options. Although blood donation by friends and family may be available, the physician should avoid representing this option as an alternative to allogeneic transfusion or suggesting that it is "safer," since it might be construed as an implied warranty.

E. Patient Decision-Making

The patient must be afforded an opportunity to question the physician in order to clarify points of information. The physician may find it useful to query the patient as key points are discussed in order to assess patient comprehension. The patient may decide to consent to the intervention or refuse it. Patients may refuse transfusion on religious grounds or for other reasons. The wishes of a competent adult to refuse even a life-saving transfusion must be honored. Similarly, if a surrogate or patient advocate refuses permission for transfusion and there is evidence that this decision reflects the wishes the patient expressed when competent, the decision must usually be honored. There are circumstances, however, when refusal of consent for transfusion may be challenged. Particularly in a situation where the well-being of a fetus or child is threatened, a court order may be obtained to transfuse a pregnant woman or a child.

F. Documentation

Once informed consent has been obtained, the physician should document that fact in the patient's record. This documentation may entail the use of a specially designed form but can just as easily consist of a note in the patient's chart. The key elements that should be included are that the procedure was described; the risks, benefits, and alternatives were discussed; an opportunity to ask questions was afforded; and the patient gave consent. The note or form should be signed by both the physician and the patient. In a situation where the patient has refused transfusion, the physician may wish to document in more detail that the patient understands the possible consequences of not being transfused and that the physician advised the patient that the transfusion was in his or her best interest.

IV. WHO MAY GIVE CONSENT?

Local regulations and standard practices that specify who may give informed consent for medical interventions usually also apply to informed consent for transfusion. In general, any competent adult may give consent for transfusion. Local practice and hospital policy may govern who may give consent for minors and incompetent adults, and the role of surrogates, patient advocates, and guardians. In many states, emancipated minors may give consent for medical treatment and transfusion. Minors may be considered emancipated if they are married, have children, or serve in the armed forces.

V. EXCEPTIONS

The law recognizes that, in emergent situations, it may not always be possible to obtain informed consent before initiation of medical procedures, including transfusion. In situations where life or a critical organ is threatened with immediate harm, a physician may elect to transfuse without consent. The rationale for this decision should be documented in detail in the patient's record.

VI. ROLE OF THE TRANSFUSION SERVICE AND HOSPITAL

Although the obligation to obtain informed consent rests with the physician ordering the transfusion, the hospital also has obligations to the patient, which the transfusion service may help to execute. The transfusion service should play a role in developing or reviewing hospital policies for informed consent for transfusion.

A. Duty of the Transfusion Service and Hospital

Hospitals have generally not been held responsible for obtaining informed consent for transfusion from their patients. However, there have been situations in which a hospital was held liable for a physician's failure to obtain informed consent based on the theory of "corporate negligence" (Johnson v. Sears, Roebuck & Co., 1992). Under this theory, the hospital is liable for failing to oversee the care of its patients by not requiring its medical staff to obtain informed consent, despite the fact that the principal obligation to obtain informed consent remains with the physician. A hospital may also be found to be liable under the theory of "respondeat superior" (Howell v. Spokane & Inland Empire Blood Bank, 1990), which holds that an employer is responsible for the acts of its employees. If courts begin to interpret the JCAHO requirement for informed consent for transfusion to represent the standard of care, then hospitals may also be found liable for not meeting that standard if they do not require their medical staff to obtain documented informed consent.

Therefore, hospitals where documented informed consent for transfusion is not routinely obtained should carefully consider their liability and the extent to which they may be seen as not fulfilling their duty to their patients. Hospitals with an informed consent policy in place should review it periodically to verify that it continues to meet legal, ethical, and regulatory standards.

B. Process for Development of Informed Consent Policies and Procedures

In the situation where the hospital does not have a requirement for informed consent for transfusion, the transfusion service may need to take the lead in developing a policy. The following are needed: (1) commitment of the hospital executive body to institute a policy for informed consent for transfusion, (2) pertinent information, such as applicable state and local statutes and the requirements of regulatory and accrediting organizations, and (3) the development of an institutional policy with the input of administrators, transfusion medicine specialists, physicians and nurses involved in transfusion therapy, legal counsel, and/or an ethicist or patient advocate.

Once the parameters for an informed consent policy have been established by the committee, responsibility for writing procedures can be delegated to members of the committee or via usual hospital channels. The procedures should be reviewed by the committee and legal counsel and approved by the responsible body within the hospital.

C. Components of Informed Consent Policies and Procedures

1. *Format*. The format and approval process should be consistent with other hospital policies and procedures. Other relevant policies (e.g., informed consent, transfusion) should be cross-referenced and checked for consistency.

2. *Statement of purpose*. The ethical, legal, and regulatory reasons for developing the informed consent policy should be summarized, and applicable statutes, regulations, and standards should be cited.

3. *Applicability*. The policy should delineate the sites within the hospital where it applies (e.g., outpatient clinics vs. inpatient wards), which blood components and derivatives require consent (e.g., red blood cells, intravenous immunoglobulin), and the frequency with which consent must be obtained (e.g., once per admission, for each transfusion, for a series of transfusions).

4. *Who may give consent*. The policy must indicate who may give informed consent for transfusion and in which circumstances parents, guardians, surrogates, and patient advocates may give consent.

5. *Who may obtain consent*. Although the physician ordering the transfusion should ordinarily obtain consent from the patient, in a complex health care facility different professionals may play a role in the decision to transfuse, the education of the patient, and the discussion of risks, benefits, and alternatives. If some parts of the informed consent process are delegated, the policy must specify who may undertake each task.

6. *Information for the patient*. Although the physician ordinarily determines what information should be conveyed to the patient, the hospital may wish to indicate in the policy what the key elements should be.

7. *Transfusion procedures*. The policy must describe how the transfusionist should verify that pretransfusion consent has been obtained and whether or not that verification should be documented.

8. *Documentation*. The policy and procedures need to delineate the requirements for documentation of informed consent. The method for documenting informed consent (e.g., using a special form or making a note in patient's record) should be stipulated.

9. *Quality assurance and medical staff review*. A system for assessing compliance with the hospital policy for obtaining documented informed consent must be developed. Provision for medical staff review of the results of the assessments must be established. A mechanism for periodic review of the policy should also be set up.

VII. ROLE OF THE PHYSICIAN

A. Duty to Inform

The duty of the physician to the patient includes the full and accurate disclosure of the information the patient needs in order to make an intelligent choice about transfusion. In part, this duty consists of describing the nature of the transfusion, and reviewing the material risks, benefits, and alternatives, including withholding the transfusion. In discharging this duty, the physician must assess the patient's ability to grasp this information and present it in such a way that the patient can assimilate and understand it. The physician is also obligated to hold this discussion in a noncoercive atmosphere in which the patient feels that he or she may freely exercise the right to choose. The physician should certainly convey his or her advice, but in a manner which is not overbearing or leading. Finally, the physician has a duty to respect the decision of the patient.

B. Duty to be Knowledgeable

In order to provide the patient with the most accurate information about the risks, benefits, and alternatives to transfusion, the physician has a duty to be well informed. Physicians who transfuse patients infrequently should consult with local experts in transfusion medicine for current information and resources.

References

AABB (1994). Association Bulletin 94-3: Informed Consent for Transfusion. AABB, Bethesda.

Eisenstaedt, R. S., Glanz, K., Smith, D. G., and Derstine, T. (1993). Informed consent for transfusion: A regional hospital survey. *Transfusion* 33, 558–561.

Howell v. Spokane & Inland Empire Blood Bank, 785 P.2d815, 823 (Wash. 1990).

JCAHO (1995). "Comprehensive Accreditation Manual for Hospitals." JCAHO, Oakbrook Terrace.

Johnson v. Sears, Roebuck & Co., 832 P.2d (N.M. App. 1992).

Schloendorff v. Society of New York Hospital, 211 N.Y. 125, 105 N.E. 92 (1914).

Stowell, C. P., Ed. (1997). "Informed Consent for Blood Transfusion." AABB, Bethesda.

CHAPTER

41

Home Transfusion

CHERIE S. EVANS

Northern California Region
American Red Cross Blood Services
Oakland, California

Provision of medical services in the home, including transfusion, is not a new concept. In 1947, Montifiore Hospital in New York City established its "Hospital without Walls" program in which a mobile, coordinated team of nurses, therapists, and physicians would travel to patients' homes to provide medical care (Koren, 1986). By the 1970s, there were published reports of successful home transfusion programs supporting hemophiliacs (Rabiner *et al.*, 1970). Despite this, and largely due to concerns about liability, transfusion medicine essentially became restricted to hospitalized patients. Over the last decade, there has been renewed interest in providing transfusion services in the home. The rationale for home transfusion and description of such a service are reviewed below.

I. RATIONALE FOR HOME TRANSFUSION

A. Increasing Safety and Sophistication of Outpatient Care

The type of medical care provided in outpatient settings has become increasingly sophisticated over the last 2–3 decades. Dialysis is now routinely performed in outpatient centers and patient homes. Patients also receive infusion therapies, such as parenteral nutrition, and chemotherapy for malignancy and infection without leaving home. Agencies providing such care were among the early proponents of incorporating transfusion into the acceptable array of high-technology home therapies.

B. Aging Population

A growing segment of the population is elderly. This is a population for whom transportation may be expensive, difficult, or even dangerous. If they are in otherwise stable condition, transfusions may appropriately be provided at home.

C. Psychosocial Indications

1. Pediatric Patients

It is not surprising that hemophiliacs were among the first to be transfused at home. This patient population is largely pediatric. Hospitals are particularly disruptive for children.

2. Chronic Illness

Patients with chronic medical conditions often have difficulty with the enforced dependence and loss of control that comes with hospitalization. This is not only psychologically difficult but can lead to setbacks in the maintenance of their routine treatment regimen. Treatment provided in the home, on the other hand, helps reinforce a sense of independence and self-worth in chronically ill individuals of any age.

D. Demand for Cost Containment

1. Diagnosis Related Group (DRG)

The DRG system of reimbursement forced hospitals to reevaluate their practices to lower inpatient costs.

One common mechanism is to limit the length of stay. The result has been an increased need for outpatient services to provide the continuing care that was previously supplied within the hospital.

2. Managed Care

Capitated contracts are now the norm in many parts of the country. This places both physicians and hospitals under tremendous pressure to provide care in the least expensive setting possible. It is assumed that care in the home will be less expensive than that in a hospital, although this has been debated.

II. POLICIES FOR HOME TRANSFUSION SERVICES

While many of the policies and procedures routinely employed for inpatient transfusion are applicable to home transfusion, some are not. In addition, there are unique aspects of a home service that must be addressed before performing these procedures.

A. Patient Selection

Only patients for whom the expected benefits of home transfusion outweigh the perceived increase in risk should be chosen. Most home transfusion programs do not accept a request based on convenience alone. In addition, Medicare requires "homebound" status before allowing reimbursement.

1. Stable Chronic Condition

Recognizing that a transfusionist in the home does not have the same access to the sophisticated medical facilities that would be immediately at hand in a hospital setting, patients with acute and/or unstable conditions should not be transfused at home. The stability of their cardiorespiratory system and fluid balance are particularly important. Typical acceptable conditions are listed in Table 1.

2. Prior Transfusion

An important requirement for home transfusion eligibility is that the patient has previously received similar transfusions in a hospital without a reaction (Snyder, 1994). Others believe that this confers no real added protection (Evans, 1994). The two most likely reactions to result in an immediate transfusion fatality are ABO-related hemolytic reactions and anaphylaxis.

TABLE 1   Typical Diagnoses of Home Transfusion Patients

1. AIDS/ARC
2. Anemia of chronic renal disease
3. Hemoglobinopathies
4. Anemia associated with malignancy
5. Chronic gastrointestinal bleeding
6. Refractory anemia
7. Myelosclerosis, myelodysplasia, and myelophthisic states
8. Bone marrow failure
9. Thrombocytopenia
10. Coagulation factor deficiency

3. Measures to Reduce the Occurrence of ABO-Incompatible Hemolytic Reactions

The ABO-incompatible transfusion is still the most common reason for an immediate transfusion fatality in any setting (Sazama, 1990). To avoid this complication, home transfusion agencies have employed a number of strategies, including the following examples.

a. Group O Packed Red Blood Cells (PRBCs)

Group O PRBCs are used for the first transfusion when a patient is new to the service. Subsequently, the record of the ABO grouping from the first transfusion can be used as a second, independent check on the patient's ABO group and group-specific cells can be used.

b. Check Sample Procedure

A "Check Sample Procedure" is used when transfusion is ordered on a patient new to the service. An additional sample is drawn on a separate occasion from the sample provided for compatibility testing. Typically, the nurse draws the sample on the day of transfusion and brings it to the compatibility-testing laboratory where the ABO group is verified before the units are released. This can usually be accomplished in the time required to complete other documentation. This method conserves valuable group O cells by providing a second, independent verification of blood group prior to the first transfusion. All subsequent transfusions can be verified against the known historical group. This may not be a realistic option when the patient's home is distant from the laboratory.

4. Clinically Significant Alloantibodies

It is not uncommon for chronically transfused patients to become alloimmunized to red cell antigens. If a patient has a well-documented transfusion history and, if antibodies are present, the specificities are known and all other clinically significant common antibodies have been excluded, then selection of units negative for the complementary antigens can be acceptable if the home transfusion service chooses to allow it.

B. No Prior Transfusion Reaction

Some services consider a history of a transfusion reaction a contraindication to home transfusion. Others are willing to assess the nature of the previous reaction and make a case-by-case decision regarding the acceptance of the patient.

C. Site Suitability

Even when a suitable patient is identified, the order should not be acted upon without prior evaluation of the home in which the transfusion will occur. While every effort is taken to prevent a transfusion reaction, one still might occur and the agency needs to know in advance whether the home environment would allow such an emergency to be handled appropriately. The following are required: telephone access, rapid emergency assistance, and the presence of a second adult.

D. Component Selection

Even a minor reaction when giving a transfusion in the home is much more problematic than it would be in a hospital. In addition to consideration of premedicating patients to avoid minor reactions, the component chosen can also reduce the likelihood of a reaction.

PRBCs and platelets are the most frequently given components, although cryoprecipitate and fresh frozen plasma (FFP) may also be appropriate for home transfusion. Cellular components should be prestorage leukoreduced. If platelets are to be given there are two advantages to using plateletpheresis products rather than platelet concentrates. First, a plateletpheresis unit is easier to transfuse than 6–8 individual platelet concentrates. Second, plateletpheresis units are leukoreduced in the process of collection.

E. Premedication

Most physicians premedicate patients with an antihistamine and/or antipyretic. These medications are relatively innocuous and will prevent minor febrile and/or urticarial transfusion reactions that are more troublesome in the home setting than in the hospital.

The order for such premedication should be included on the physician's order for transfusion.

F. Number of Units Transfused

1. Two-Unit Limitation

There are published protocols for home transfusion that limit the number of units to no more than 2 units of packed PRBCs in any 24-hour period due to concerns about volume overload (DePalma et al., 1986). This limitation is somewhat conservative but does allow for standardization of the service's policy. On the other hand, even 2 units may be too much for an individual patient. Such patients are often better served with 1 unit each 24 hours, infused over 4 hours and followed up by an assessment of fluid status the next day.

2. Customized Service

Conversely, a service may choose to involve its Medical Director in making case-by-case decisions whenever an unusual number of components is ordered. For example, a chronically transfused AIDS patient who is young, volume depleted, and in good cardiorespiratory condition might safely receive routine home transfusions of 3 or even 4 units at a time. The reduction in cost and time expended may be significant.

III. LEGAL AND REGULATORY ISSUES

Unlike hospital-based transfusion, there are no established uniform standards or federal regulations for home transfusion. To date, only the State of New York has promulgated regulations that directly address home transfusion (Kasprisin, 1994).

A. State Regulations

Although states are not likely to have explicit regulations for home transfusion, anyone setting up a home transfusion service would be well advised to carefully review their state's inpatient transfusion regulations. Many of these regulations are likely to apply to home transfusions.

B. AABB Standards

Home transfusion "Guidelines" were published by the American Association of Blood Banks' (AABB) Scientific Section Coordinating Committee and cover the basic aspects of such a service (Beck and Grindon, 1986).

C. Qualification and Continuing Education of Staff

In the home, the transfusionist is operating almost independently and is therefore likely to be held to a higher level of professional accountability. It is important that a transfusion service consider this when determining what qualifications should be required when hiring staff and how continuing education needs are to be managed and documented.

1. Qualification of Staff

Many agencies employ only nurses with recent experience in acute, preferably intensive or emergency, care. Some have required more advanced training, such as a nurse practitioner. The requirements should ensure the agency employs only individuals with sufficient experience with transfusion and its potential complications to operate independently.

2. Continuing Education

Regardless of the nurse's original experience, continuing education that specifically relates to transfusion medicine needs to occur and be documented. This is particularly important when a nurse is not going to be performing transfusions frequently.

D. Physician Supervision

1. Order

A physician must order the transfusion, and most services require the order be documented in writing.

2. Special Orders

Every order must specify what components are to be given, when, for what indication, and so on; however, some physicians also provide "standing orders" for use in the case of a transfusion reaction.

3. Supervision

While not a regulatory requirement most home transfusion services require that the ordering physician or other designated physician be immediately available by telephone for consultation in case of complications.

E. Contract

Discussion of the elements of a contract for home transfusion services is beyond the scope of this chapter.

If there are two different organizations involved (typically a hospital or regional blood center and a home nursing agency) then a contract is essential. It needs to spell out the specific responsibilities of each of the parties and the remedies to be employed should those responsibilities not be met.

IV. ORGANIZATIONAL APPROACHES

There are two major approaches to home transfusion. There are advantages and disadvantages to both approaches.

A. Regional Blood-Center-Based Service

Many regional blood centers have established contractual relationships with home nursing agencies to provide compatibility testing and blood components that the agency will transfuse in the patient's home. This is a logical extension of the blood center's core business.

1. Advantages

Many different physicians and/or agencies may wish to provide home transfusion services. By centralizing everything but the administration of the transfusion into a blood center, the service is uniform throughout the community. There is only one set of forms, protocols, and so on, for the staff to deal with regardless of how many different hospitals the ordering physicians represent. Blood centers routinely provide units that are CMV-safe, leukoreduced, irradiated, and otherwise treated; therefore special processing orders are unlikely to create inventory problems for them. Similarly, packaging and shipping blood is part of a blood center's daily operation. Blood centers operate in accordance with current Good Manufacturing Practices of the Code of Federal Regulations, which means their processes are validated and they have a quality control mechanism to ensure units are properly transported.

2. Disadvantages

A blood center is unlikely to have a preexisting transfusion committee and establishing one based on the traditional model may be difficult.

B. Hospital-Based Services

Under the pressure of DRGs and managed care, hospitals are also forming partnerships with home health agencies in order to remain the care provider even after patients are discharged to their homes.

1. Advantages

Hospitals already have an established mechanism for quality assurance of inpatient transfusions that they may directly apply to their home program. It requires little more than establishing appropriate audit criteria for transfusion in the home setting.

2. Disadvantages

In general, a hospital will only provide transfusions ordered by the members of that hospital's medical staff. As a result, nursing agencies working with physicians from several different hospitals are likely to end up using different forms and protocols, increasing the opportunity for errors.

V. PROCEDURE FOR HOME TRANSFUSION

A. Physician Orders, Patient Identification, and Transportation

Transfusions may only be performed on the order of a physician licensed in the state in which the transfusion is to be performed. This order must contain a minimum of information, including patient identification, specific location, diagnosis and indication for transfusion, blood components to be transfused, transfusion history of the patient, and medications allowed or required.

Informed consent should be obtained before any elective transfusion. In some states, such as California, the law requires it (California Health and Safety Code, 1991).

Once the order is received and deemed appropriate, a sample must be drawn for compatibility testing. If the home has not been previously assessed it can be done at the same time with the understanding that, if the home is not appropriate, the sample may not be drawn. In labeling this sample, the critical process of identification begins.

Mistakes in identification of patient, sample, or unit continue to be the most common reasons for immediate transfusion fatalities. This is no less true in the home setting despite the fact that the same person who drew the sample may also be the one performing the transfusion. There are three critical control points in regard to identification: identification of the patient, the sample, and the unit(s) to be transfused.

All blood components must be transported within specific temperature ranges. This is generally achieved by using containers that have been validated to ensure that a specific number of units will remain within the acceptable temperature range while being transported under reasonably predictable conditions for a specified period of time.

B. Administration

The equipment and protocols used to perform the transfusion in the home are not much different from those employed in the hospital. The patient is usually monitored much more intensely than in a hospital as the nurse remains with the patient through the entire transfusion and for as long as an hour afterward. Vital signs are monitored and documented in the patient record every 5 minutes during the first 15 minutes and every 15 minutes thereafter. The use of blood warmers or infusion pumps for home transfusions is not recommended.

Immediate transfusion reactions almost always occur during the transfusion or within the first hour after it. This is the reason for keeping a clinically knowledgeable person and a second individual present in the home for 30 to 60 minutes after the procedure.

VI. QUALITY ASSURANCE

In view of the fact that home transfusion is thought to have a higher degree of risk than inpatient transfusion, at least in terms of liability, an effective quality assurance program for such a service is essential.

A. Review of Transfusion Appropriateness

The Medical Director and/or Quality Assurance Committee is responsible for developing criteria for home transfusion for their service. Subsequent review is also required to document that these criteria are being met, and corrective action taken if they are not.

B. Multiple-Organization Services

In the most common model for home transfusion service delivery, the transfusion is ordered and performed by individuals who do not work directly for the facility performing the compatibility testing and providing the blood components. As a result there is potential for confusion about where the responsibility for transfusion review resides.

C. Monitoring and Enforcing Compliance

This presents a significant challenge for many home transfusion services in which multiple organizations are involved. Since the ordering physicians are often not directly affiliated with either the blood center or the nursing agency, the usual mechanism of linking staff privileges to satisfactory peer review may not be available.

D. Transfusion Reactions

In the event of an adverse transfusion reaction the blood center or hospital must, at a minimum, be informed. In some cases this will also include active consultation about management and/or involvement in the serologic workup of a suspected hemolytic reaction.

References

Beck, M. L., and Grindon, A. J. (1986). Home transfusion therapy. *Transfusion* 26, 296–298.

Blood Products Advisory Committee (1998). 60th Meeting Minutes, September 17–18.

California Health and Safety Code. The Paul Gann Blood Safety Act (1991), § 1645.

DePalma, L., and Snyder, E. L. (1986). Medical aspects of home transfusion. In "Home Transfusion Therapy" (E. L. Snyder and J. E. Menitove, Eds.), Chap. 4, pp. 25–40. AABB, Arlington.

Evans, C. S. (1994). Out-of-hospital transfusions: A regional blood center approach. In "Out-of-hospital Transfusion Therapy" (J. Fridey and L. Issitt, Eds.), Chap. 2, pp. 27–41. AABB, Bethesda.

Kasprisin, D. O. (1994). Are the marshals coming to dodge city? In "Out-of-Hospital Transfusion Therapy" (J. Fridey and L. Issitt, Eds.), Chap. 6, pp. 83–109. AABB, Bethesda.

Koren, M. J. (1986). Home care—Who cares [Ed.]. *N Engl J Med* 26, 296–298.

Rabiner, S. F., *et al.* (1970). Home transfusion for patients with hemophilia A. *N Engl J Med* 283, 1011–1015.

Sazama, K. (1990). Reports of 335 transfusion-associated deaths: 1976–1985. *Transfusion* 30, 583–590.

Sazama, K. (1994). Legal considerations for out-of-hospital transfusions. In "Out-of-hospital Transfusion Therapy" (J. Fridey and L. Issitt, Eds.), Chap. 5, pp. 73–81. AABB, Bethesda.

Snyder, E. L. (1994). Medical aspects of home blood transfusion: A view from a hospital. In "Out-of-Hospital Transfusion Therapy" (J. Fridey and L. Issitt, Eds.), Chap. 1, pp. 1–26. AABB, Bethesda.

42

Lookback and Recipient Notification, and Product Recalls and Withdrawals

K. SAZAMA

Department of Laboratory Medicine
University of Texas M.D. Anderson Cancer Center
Houston, Texas

"Lookback" refers to a process of retrospective investigation of either donors or recipients triggered by: (1) recognition of disease in a recipient, causing a "looking back" for all donors of transfused components, or (2) newly discovered serologic evidence of infection of a donor, resulting in tracing and notification of all recipients of prior donations (usually for a set period of time prior to the documented donor seroconversion).

Federal guidance and regulations require that certain transfusion recipients be notified when the donor of their transfused components is found to have seroconverted for selected viruses. Today, human immunodeficiency virus (HIV) and hepatitis C virus (HCV) seroconversion require such notifications, referred to as part of "lookback" in the existing HIV regulations (Title 21 CFR 620.46-7, 1999). However, the roots of lookback arose much earlier, prior to hepatitis B virus (HBV) testing, when some recipients were noted to acquire hepatitis after transfusion, causing hospitals to notify the donor center to "look back" for the implicated donors of the recipients transfusions (Title 21 CFR 606.170(a), 1999). This donor lookback requirement has never been rescinded.

I. VIRUSES FOR WHICH DONOR LOOKBACK AND RECIPIENT NOTIFICATION IS REQUIRED

There are 6 transfusion-transmitted viruses for which serologic testing is available and is routinely used to prevent disease transmission. These viruses are HIV types 1 and 2 (HIV-1/2), HBV, HCV, and human T-lymphotropic virus, types I and II (HTLV-I/II). In the case of HIV, U.S. government regulations specify exactly how such recipient notification must be done. For HCV notification, the FDA has published a series of guidances, including proposed regulations, describing the process of recipient notification. At present, recipient notification for possible HTLV or HBV transmission is not required (FDA, 1996; Table 1).

II. DONOR LOOKBACK

Before 1985 (the date when specific HIV testing began), lookback was almost always initiated from the hospital transfusion service back to the donor center (donor lookback).

A. Role of the Treating Physician

In the past, it was up to the treating physician to recognize illness (usually hepatitis) and inform the hospital transfusion service medical director. Together they assessed the transfusion recipient for possible transfusion-transmitted illness and decided whether to notify the donor center.

B. Role of the Transfusion Service

Until 1985, the only lookback process that occurred was when a transfusion recipient became symptomatic

TABLE 1 Transfusion-Transmitted Viruses and Lookback

Virus	Lookback and Recipient Notations
HIV	Mandatory: Federal Regulations: 21 CFR 610.46-47 and 42 CFR 482
HCV	
HTLV	Not Required: FDA Guidance: 7/19/96
HBV	Not Required: FDA Guidance: 7/19/96

with a viral illness (usually hepatitis B) that could have been transmitted via a transfused blood component (otherwise known as "adverse reaction" by the FDA). As a result of recognition of recipient illness and its possible association with a prior transfusion, the medical director of the transfusion service was required by federal regulations to initiate communication with the donor center that had supplied the transfused blood.

C. Role of the Donor Center

After the above events, the donor center would then trace the donors of the units transfused to the infected recipient(s) to determine whether any of the donors had developed a similar illness or (after 1972) had developed a positive test for hepatitis B. If a donor tested positive for hepatitis B, the donor center deferred that donor indefinitely. Because no system of notifying other transfusion services was ever established as a routine practice, other transfusion recipients who might have received components donated by the infected donor were never notified of the possibility of transfusion-transmitted infection. Often, however, the donor could not be identified unequivocally, so more than one donor would be implicated. All such donors would be identified as "suspicious" for HBV, and usually were put into some kind of an "alert" category; however, each could continue to donate until a second recipient of blood components from this donor developed hepatitis. If any of the previously suspected donors was also a donor to the second ill recipient, then this donor would be permanently deferred as well, even if the donor continued to be free of overt illness and tested negative for hepatitis B.

III. RECIPIENT "LOOKBACK" (RECIPIENT NOTIFICATION AFTER DONOR LOOKBACK)

Recipient lookback, or, more precisely, recipient notification after donor lookback, occurs when a donor is found to be positive for one or more transfusion-transmissible infections for which recipient notification is required or recommended. (Donor deferral may occur without recipient notification, as is the case when donor seroconversion for HBV occurs.)

A. Role of the Donor Center

The blood collecting facility investigates prior donations made by the seroconverting donor for the past 12 months, and occasionally for as long as donor records exist (as is the case for HCV lookback). It notifies the transfusion services to which *prior* donations were distributed (not the current donation, which is discarded or used for research or manufacturing purposes and never transfused). The blood center must communicate the seroconversion information to the transfusion service within 30 (for HIV) or 45 (for HCV) calendar days.

For HCV recipient lookbacks, the donor center investigates donations as far back as records are available. For example, if records existed as far back as 1/1/88, donor centers had until 3/23/00 to communicate this information to transfusion services. If donation records existed prior to 1/1/88, then donor centers had until 9/30/00 to communicate with transfusion services. The mechanism for such communication from a donor center is either via a biological recall or a market withdrawal letter to the transfusion service (Table 1).

1. Biological Recalls and Market Withdrawals

The means by which a donor center communicates with a transfusion service about possible problems with collected components is by written statements—either biological recalls or market withdrawals. FDA regulations (Title 21 CFR 7.40, 1999) describe the steps a blood collecting organization may voluntarily undertake to inform transfusion services of a potential risk from an already distributed component. Blood collectors may initiate recalls on their own or at the request of the FDA. The FDA encourages use of voluntary recalls rather than official action, stating that recalls are "an alternative to an FDA-initiated court order for removing or correcting violative, distributed products . . . ."

a. Biological Recalls

The recall communication must clearly indicate, preferably in bold red letters, that it is a "Biological Recall," the identity of the product and the unit number, the concise reason for which recall is indicated,

TABLE 2   Health Hazard Evaluation for Biological Recall

The health hazard evaluation includes determining:

(1) whether any disease or injuries have already occurred from use of this product;

(2) whether any existing conditions could contribute to a clinical situation that could expose humans . . . to a health hazard . . . ;

(3) assessment of the hazard to various segments of the population . . . who are expected to be exposed to the product being considered, with particular attention paid to the hazard to those individuals who may be at greatest risk;

(4) assessment of the degree of seriousness of the health hazard to which the populations at risk would be exposed;

(5) assessment of the likelihood of occurrence of the hazard;

(6) assessment of the consequences (immediate or long-range) of occurrence of the hazard.

and what action the transfusion service should take regarding the recalled component(s), with a request to report back to the collecting organization whether any untransfused component still exists. If the component has not been transfused, it is quarantined and either destroyed by the transfusion service or returned to the collecting facility for final disposition. Recalls are indicated when "products . . . present a risk of injury or gross deception or are otherwise defective."

All recalls are subject to a health hazard evaluation and recall classification (Table 2). On basis of a health hazard evaluation, a recall is classified as Class I, II, or III, with Class I indicative of the most serious health hazard.

b. Market Withdrawals

Another communication option for any product thought to be deficient in some manner but which does not meet the requirements for a biological recall is a "market withdrawal." The FDA district office works with the collecting organization to ensure that a market withdrawal is appropriate (i.e., to make certain that a biological recall—a more serious problem—is not indicated).

B. Role of the Transfusion Service

1. Notification Requirements for HIV

When a donor becomes HIV positive, federal regulations require that either the hospital or the treating physician (or both) notify the recipient of any component that was donated within 12 months of the last negative donation by that same donor. The transfusion service must inform the treating physician, who may choose to notify the patient (or decline to do so). If the treating physician declines, the hospital, usually through the transfusion service medical director, must notify the recipient. This notification must occur within 8 weeks of receiving the letter from the blood center, with at least 3 attempts to notify the recipient documented in that time period. If the recipient has died or been adjudged incompetent since the transfusion, a legal representative or relative must be notified.

2. Notification Requirements for HCV

When a donor becomes HCV positive, federal regulations require that the hospital and transfusion service notify either the transfusion recipient or the treating physician (requesting that she or he notify the transfusion recipient).

For HCV lookbacks, the transfusion service must notify either the treating physician or the recipient within 12 months of receiving the letter from the blood center, with at least 3 documented attempts to notify the recipient during that time period. If the recipient has died, no attempts at notification are required. If the recipient is alive but adjudged incompetent, then a relative or legal representative must be informed. The transfusion recipient must be offered testing and counseling as part of the notification process.

3. Notification Requirements for Viruses Other Than HIV or HCV, and Other Recalls or Market Withdrawals

Currently, no recipient notification is required when a donor seroconverts to anti-HTLV or develops HBsAg or anti-HBc. Similarly, when positive preliminary tests for any transfusion-transmitted agent are not confirmed (confirmatory testing either is not done, is negative, or is indeterminate), or when donor history suggests that the components should not have been distributed or there is some variance in performance of testing, donor centers may issue reports to transfusion services. Transfusion service medical directors then must decide how to handle these communications. One way of dealing with these letters is to involve the transfusion committee, risk management specialists, the hospital ethics committee, medical specialists, and public health officials in discussions regarding the implications of notification to recipient health. Because of the public's demand for a zero-risk blood supply, other situations may be subjected to national review and new guidance in the future.

C. The Role of the Treating Physician

Treating physicians are expected to conduct the actual recipient notification. For HIV, the notification must be made within 8 weeks of the information being received by the transfusion service, with 3 documented attempts to reach the patient. If the patient is deceased or adjudged incompetent, the next of kin or guardian for the recipient must be reached. For HCV, using current methodologies, the time period is 12 months, with the same 3-attempt requirement, and with notification of relative or guardian if recipient has been found incompetent. No notification is required for deceased recipients.

The existing regulations for both HIV and HCV lookback acknowledge that there may be reasons why the treating physician is unable to notify, or is uncomfortable with notifying, the recipient. In cases where the treating physician does notify the patient, the conditions in the next section must be met. When the treating physician cannot be located, declines to notify, or is unable to complete the required notifications, the hospital, usually via the transfusion service medical director, must comply as described above.

When notification is suggested for reasons other than confirmed seroconversion, the role of the treating physician becomes less well defined. When, as can occur, the transfusion committee determines that the treating physician should be informed about circumstances other than HIV- or HCV-confirmed seroconversion, the treating physician may find it helpful to discuss the situation with the transfusion service medical director and other medical colleagues, as well as with local public health officials, prior to notification.

D. Contents of Recipient Notification

The federal regulations for both HIV and HCV lookbacks require that recipient notification include at least the following information:

1. Basic explanation to the recipient of the need for HIV or HCV testing and counseling.
2. Sufficient oral or written information so that the transfusion recipient can make an informed decision about whether to obtain HIV or HCV testing and counseling.
3. A list of programs or places where the patient can obtain HIV or HCV testing and counseling, including any requirements or restrictions the program may impose (Title 21 CFR 620.46-7, 1999).

Whoever actually contacts the recipient should establish and use a system for documentation of both the efforts to locate the transfusion recipient and the actual content of such notification and discussion, when it occurs.

References

Food and Drug Administration (1996). "Recommendations for the Quarantine and Disposition of Units from Prior Collections from Donors with Repeatedly Reactive Screening Tests for Hepatitis B Virus (HBV), Hepatitis C Virus (HCV) and Human T-Lymphotropic Virus Type 1 (HTLV-1)." U.S. Food and Drug Administration, Center for Biologics Evaluation and Research, Bethesda.
Title 21, Code of Federal Regulations, Part 606.170(a) (1999). U.S. Government Printing Office, Washington, DC.
Title 21, Code of Federal Regulations, Part 620.46-7 (1999). U.S. Government Printing Office, Washington, DC.
Title 21, Code of Federal Regulations, Part 7.40 (1999). U.S. Government Printing Office, Washington, DC.

43

Process Improvement
One Essential of a Quality System

K. SAZAMA

Department of Laboratory Medicine
University of Texas M.D. Anderson Cancer Center
Houston, Texas

Federal guidance in 1995 provided a new framework for blood establishments (both donor centers and transfusion services) to evaluate and improve on the quality of their practices (FDA, 1995). The American Association of Blood Banks (AABB), the volunteer professional association which provides consensus standards for the practice of blood banking and transfusion medicine, translated the federal requirements into a workable and understandable quality system (AABB, 1994, 1996, 1997; Menitove, 1997) that, together with the integration of ISO 9000 principles (ISO, 1996), forms the basis for AABB accreditation, the premier recognition for adherence to the highest standards of transfusion medicine in the world. Moving from quality control and quality assurance to quality systems is neither easy nor inexpensive, but the benefits have been proven in other industries and services around the world.

I. THE GLOBAL COMMITMENT TO QUALITY

The worldwide quality movement is being accepted in healthcare organizations somewhat later than in other service and manufacturing industries. While clinical laboratories have long been accustomed to performing quality control during testing, as mandated by the U.S. Clinical Laboratories Improvements Act of 1967 and quality assurance as defined in the amendments in 1988, the broader quality concepts have only recently been defined and implemented in blood banks and transfusion services (Table 1).

Frequently, these quality concepts are depicted in a pyramid, with quality control and quality assurance forming the base, quality systems the center, and quality management and total quality management (TQM) the top. This visual depiction assists in understanding the relationship between the concepts and the necessity for having a thorough understanding of what each concept requires.

II. QUALITY SYSTEMS DEFINED

Generally, quality systems include the underlying quality control and quality assurance activities, with the added concept of organizational commitment to provide structure, procedures, processes (including quality or, more accurately, process improvements), and resources. The AABB has organized an approach to achieving quality systems in blood banks and transfusion services by defining the 10 quality system essentials (QSEs), choosing language that is consistent with ISO standards and the FDA's quality assurance guideline (Vengelen-Tyler, 1999; Table 2).

III. APPLYING QUALITY SYSTEMS TO ORGANIZATIONAL GOALS FOR QUALITY

The QSEs form the backbone or framework for a well-run organization. In addition, they can be used to understand the interrelationships between these essen-

TABLE 1   Definitions: Quality

Word/phrase	Definition	Organization
Quality control	Activities and controls used to determine the accuracy and reliability of the establishment's personnel, equipment, reagents, and operations	FDA[a]/AABB[b]
	Operational techniques and activities that are used to fulfill requirements for quality	ISO 8402[c]
Quality assurance	Actions planned and taken to provide confidence that all systems and elements that influence the quality of the product are working as expected individually and collectively	FDA/AABB
	Planned and systematic activities implemented within the quality system	ISO 8402
Quality system	The organizational structure, procedures, processes and resources needed to implement quality management	ISO 8402
Quality management	All activities of the overall management function that determine the quality policy, objectives, and responsibilities and implement them by means such as quality planning, quality control, quality assurance, and quality improvement within the quality system	ISO 8402

[a] Guideline for Quality Assurance in Blood Establishments, U.S. Food and Drug Administration, Bethesda, MD, 7/11/95.

[b] AABB Association Bulletin 97-4: Quality program implementation, Bethesda, MD, American Association of Blood Banks, 1997; Quality program supplement, 1996; Quality program, 1994.

[c] International Standard ISO 8401: Quality management and quality assurance—vocabulary. In: ISO 9000 Quality Managemt Compendium, 6th ed., Geneva, Switzerland: International Organization for Standardization, 1996.

TABLE 2   AABB's Quality System Essentials

1. Organization
2. Personnel
3. Equipment
4. Supplier issues
5. Process control, final inspection, and handling
6. Documents and records
7. Incidents, errors, and accidents
8. Assessments: internal and external
9. Process improvement
10. Facilities and safety

tials and the work being performed in any operation within the organization. This is depicted in recent publications (NCCLS, 1999).

A. Quality or Process Improvement in Relation to the Other Elements of Quality Systems

Process improvement is just one element of a quality system, generally limited to acquiring and analyzing current information to decide when system changes may be needed and then using standard methods to reach new solutions that will prevent mistakes or occurrences. There are various published methods, which can be taught in formal training programs variously called "FADE," "PDCA," "Juran's Journey," the "Joiner method," or some similar name. However, the fundamentals of process improvement can be described, as the Joint Commission on Healthcare Organizations (JCAHO) proposed in the early 1980s, in 10 steps (Nevalainen et al., 1998; see Table 3).

1. Transfusion Services Example: Sample Misidentification

Most transfusion services have documented problems with misidentified patient samples, using audit techniques or by random observations. If only random observations are in use, sometimes the error is discovered only when the historic and current ABO/Rh types are discrepant (sometimes only when a patient dies). If the problem is considered significant enough, a process improvement team, using the 10-step approach outlined above, can address this problem to prevent

TABLE 3   The Ten Steps of Process Improvement

1. What do we want to achieve?	The mission
2. Who cares and what do they care about?	Customers and their requirements
3. What are we doing now and how well are we doing it?	Assess the current state
4. What can we do better?	Define the preferred state, problems
5. What prevents us from doing better?	Barriers/root causes
6. What changes could we make to do better?	Develop solutions
7. Can we do changes now?	Implement solution
8. How did we do? Do we need to try again?	Monitor results
9. If it worked, how can we do it every time?	Standardize
10. What did we learn?	Conclude project

recurrence of these errors. Whatever the proposed system revision is, part of process improvement is to plan for and conduct follow-up audits to determine whether the change actually achieved improvement.

B. Utility of Process Improvements

Implicit in process improvement is the recognition that a process should be improved. Interpreting silence (or lack of negative feedback) as evidence of lack of problems can be misleading. To improve an organization, the elements of occurrence reporting and routine assessments, both internal and external, are equally important as the element of process improvement. These are linked activities in a continuum of quality for an organization. The data obtained from these tools, plus the important customer feedback information (often negative) that you receive regularly, can provide you with important information about where to focus process improvement efforts.

In transfusion services, it is easy to limit audits to those areas within direct control. However, observing actual transfusions can be very enlightening, providing specific information for discussions about opportunities for improvements.

C. Benefits of Process Improvements

For blood banks and transfusion services, the use of process improvements is mandatory. For those facilities that are inspected by the FDA, the quality assurance guideline has been used for citations for the past 4–5 years. In addition, the quality system essentials against which every AABB-accredited organization is measured incorporate explicit requirements for process improvement. Today, AABB accreditation signals adherence to not only operational excellence, but also quality system integration. Because AABB accreditation has received status recognition by the JCAHO, additional benefits include the streamlining and combining of hospital-wide performance improvement activities with the process improvements of a quality system.

References

AABB (1994). "Quality Program," Vols. 1 and 2. AABB, Bethesda.

AABB (1996). "Quality Program Supplement." AABB, Bethesda.

AABB (1997). Association Bulletin 97-4: Quality Program Implementation. AABB, Bethesda, MD.

FDA (1995). "Guideline for Quality Assurance in Blood Establishments." U.S. Food and Drug Administration, Bethesda.

ISO (1966). "ISO 9000 Quality Management Compendium," 6th ed. International Organization for Standardization, Geneva, Switzerland.

Menitove, J., Ed. (1997). "Standards for Blood Banks and Transfusion Services," 18th ed. AABB, Bethesda.

NCCLS (1999). "A Quality System Model for Health Care; Approved Guideline." NCCLS, Wayne, PA.

Nevalainen, D. E., Berte, L. M., and Callery, M. F. (1998). "Quality Systems in the Blood Bank Environment," 2nd ed. AABB,

Index

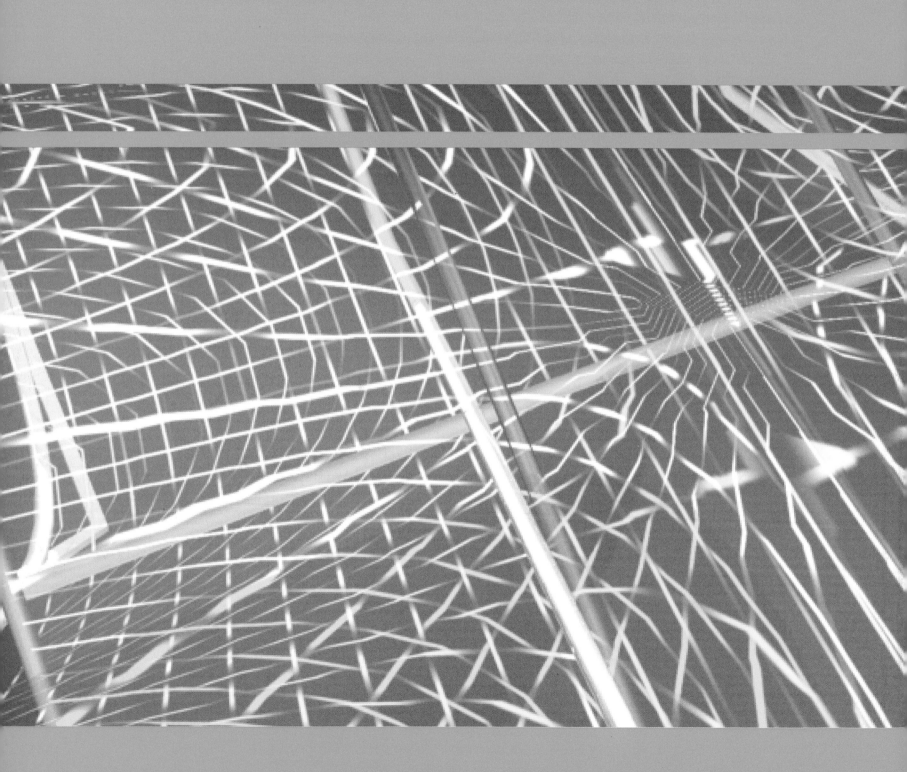

Peter Zellner

HYBRID SPACE

NEW FORMS IN DIGITAL ARCHITECTURE

WITH 644 ILLUSTRATIONS 534 IN COLOUR

Thames & Hudson

Hybrid Space is dedicated to my parents, Yolanda and Peter, Sr.
Their continued advice, support and unquestioning love have carried
me far during the process of putting this book together.

Acknowledgments

Hybrid Space was undertaken at the suggestion of Lucas Dietrich, architecture
and design editor at Thames & Hudson. His combined vision, encouragement,
editing efforts and friendship have inspired me and positively illuminated the
project. The graphic design was admirably executed by Aaron Hayden
and patiently urged along by Catherine Hall. I am also especially grateful to
Neil M. Denari, Robert Elfer, Mark Goulthorpe, Bart Lootsma, Thom Mayne, Marcos
Novak, Kas Oosterhuis, Stephen Perrella, Colin Rofe, Lars Spuybroek, Eul-Ho Suh,
Wade Stevens, Tom Verebes, R. Shane Williamson and Eui-Sung Yi for their continued
support, informed criticism and advice. Finally, *Hybrid Space* could not have been
conceived of without the dedication, bravery, intellect and imagination of all the
architects featured herein – individuals and teams alike. Their efforts continue to
invigorate and encourage me, and I hope that the book introduces their works to a
wide and appreciative audience.

Title page: Stephen Perrella; page 5: dECOi

First published in the United Kingdom in 1999 by
Thames & Hudson Ltd, 181A High Holborn, London WC1V 7QX

British Library Cataloguing-in-Publication Data
A catalogue record for this book is available from the British Library

ISBN 0-500-341737

Printed and bound in Singapore by C.S. Grapics

CONTENTS

FOREWORD
BART LOOTSMA

Within a short space of time the computer has become a widely accepted feature of architecture, both in the design process and in the everyday operation of buildings and the city, and we are constantly aware that the computer's introduction into architecture will eventually have far-reaching consequences. After all, the current revolution is not just about the computer as a tool but about its role in the global telecommunications network.

Since its creation, architects, artists, media designers and theorists have speculated about the ramifications of the computer. It is a theme that invites speculation, experiment and play – but that is not the only reason for the persistent questioning. Today we are aware that we cannot foresee all the implications of the technological revolution. We remember how long it took for people to grasp the social changes of the Industrial Revolution and how another fifty years passed before they started to think about its cultural impact. However sceptical or critical one might be, no one wants to be accused of having "eyes that do not see," as Le Corbusier did in his 1923 *Towards A New Architecture*. On the other hand – at least since Manuel De Landa's recent concept of a nonlinear history – we are aware that innovations and inventions need time to incubate, and that their effects on the organization of society can be completely unexpected. The radicalization of modernity that has been triggered by the computer means that it has become increasingly difficult to fall back on traditions: more than ever, we must reflect on what the future will bring.

Hybrid Space presents a wide spectrum of approaches, from architects who incorporate the computer's techniques into their working methods in a more efficient or exploratory way, to practices that are based on the belief that the computer will dramatically change the nature of architecture, in terms of the design process, as well as on the levels of organization and experience. *Hybrid Space* is not a movement – like many others that have been recently coined – but a concept that helps us better understand and connect the many attempts to establish the computer's role in architecture. The power of this concept and this book is that – analogous to the way Peter Zellner makes an enthusiastic plea for an architecture that might organize and rationalize the gaps, pauses and intervals of pause in an ever-accelerating world – it makes us aware of the many opportunities that exist between and behind design approaches. Instead of trying to validate conventional architectural thinking in a different realm, our strategy today should be to infiltrate architecture with other media and disciplines to produce a new crossbreed. Reducing everything to flows of data and quantities, the computer offers us exactly this possibility.

INTRODUCTION

The cultural and social revolution brought on by telecommunications and information technologies is rapidly transforming the field of architecture. We live in an era of accelerated change, in which data speeds invisibly around the globe and the flow of information has superseded material exchange, and complex digital infrastructures have inscribed themselves within our recognizable mechanical and urban patterns. Metropolitan populations, previously culturally bound and physically localized, have become nomadic and transitory, following the dynamic movements, accumulations and diffusions of international capital investment and diversification.

Against this shifting background, architecture itself is mutating, redefining its boundaries, its essential codings, to adjust to an increasingly supple and volatile world. Architecture is recasting itself, becoming in part an experimental investigation of topological geometries, partly a computational orchestration of robotic material production and partly a generative, kinematic sculpting of space. These new modalities for architecture, the topological-geometric, the industrial-choreographic and the generative-kinematic, are united transversally by the notion of a hybridized spatiality. "Hybrid space" defines an architecture that is produced by breeding ideas or concepts of contrast and heterogeneity – the strong and the weak, the formed and the formless, the real and the virtual – and that evolves through the embodiment of competing identities – unraveling and consuming opposed geometries and spatial postulates. This new architecture organizes the world by arranging the spaces *between* things rather than perpetuating the myth of ideal form. Hybrid space unbinds an architecture of inclusion and absorption, recombination and admixture.

01. Revolutions, Evolutions and Transformations

By the end of the nineteenth century, the impact of the industrial revolution was widely felt in architecture and urbanism. Newly harnessed forms of power and energy distribution were quickly dispersing and dissolving the pre-nineteenth-century city while new manufacturing and assembly processes were transforming the structural logic, appearance and materiality of the most common building forms. The use of steel and reinforced concrete in multistory, free-span structures gave rise to startling new possibilities in construction. In employing the structural steel frame and Otis's mechanical elevator, Chicago architect Louis Sullivan (1856-1924) pioneered the concept particular to urban life in the twentieth century: the vertical city. Nascent communications and media technologies, such as telegraphy, aerial photography and cinema, reshaped humanity's sense of geographic distance, proximity and temporality. More significantly, the advent of the car – a kind of horizontal elevator – brought about an even more radical transformation: an extreme form of urbanity that would spread itself over national networks of highways, secondary roadways and transportation interchanges. In the first

architecture *n*. **1.** The art or science of building; *specif:* the art or practice of designing structures and esp. inhabitable ones. **2.** Formation or construction as or as if the result of conscious act. **3.** Architectural product or work. **4.** A method or style of building. **5.** *computing:* the entire structure of a network, including protocols, operating system, hardware, cabling, interfaces and adapter cards.

hybrid *n. Abbr.* **hyb. 1.** *Genetics.* The offspring of genetically dissimilar parents or stock; especially, the off-spring produced by breeding plants or animals of different varieties, species or races. **2.** Something of mixed origin or composition. **3.** A word whose origins are derived from different languages. (Latin *hybrida*, mongrel).

space *n*. **1.a.** A set of elements or points satisfying specified geometric postulates: non-Euclidean space. **b.** The intuitive three-dimensional field of everyday experience. **c.** The infinite extension of the three-dimensional field. **2.** Broadly, the distance between two points or the area or volume between specified boundaries. **3.** An area provided for a particular purpose: room. **4.** A period or interval of time. **5.** One of the intervals during the telegraphic transmission of a message when the key is open or not in contact. *tr. v.* **1.** To organize or arrange with spaces between. **2.** To separate or keep apart. (Middle English, time interval, from Old French *espace*, from Latin *spatium*, space, distance).

TECHNOLOGIES AND EVOLUTIONARY ARCHITECTURES
SPACE

half of the century, such modern visionaries as Le Corbusier, Frank Lloyd Wright and Mies van der Rohe would seek to devise appropriate architectural forms and urban design strategies to address the technological advances and modes of urban organization.

At the close of our century, it is the information revolution that is metamorphosing architecture and urban design. Digital technologies are transforming the nature and intent of architectural thinking and creativity, blurring the relationships between matter and data, between the real and the virtual and between the organic and the inorganic and leading us into an unstable territory from which rich, innovative forms are emerging. A new time-space vernacular is rescripting the model of the city as cable and satellite connections span massive physical distances along a curved terrestrial geography. Spatial description itself, previously rendered fixed by the freeze-frame of the Albertian perspective is shifting toward a post-representational program defined by the hyperspace of data exchange. Objects, places, people, buildings and cities are no longer framed in the moment but instead approached along multiple and associative routes. Through visual and non-visual means of mobile cognition – satellite-imaging, electron-scanning or heat-sensing – structures and buildings are being set free from a conventional linear viewpoint. Buildings can become less like icons of fixity and immobility and more like inclusive fields of organized materialization.

Paradoxically, while architecture may at last free itself from the shackles of an overdetermined mode of visualization, the building user has become increasingly static. As human cognizance and transience reach around the planetary surface via telecommunications networks, we remain relatively fixed to our points of interface – our workstations, televisions and fax machines. The idea of place has therefore also been recast, as instantaneous data exchange replaces traditional means of mobility. Buildings can now be seen from anywhere at once with the aid of a digital cognition, and, strangely, we are able to perceive everything at once by not moving at all.

It is in this context that today's experimental architects are deploying novel "hard" (manufacturing and material) and "soft" (digital) technologies to engender an architecture of incorporation and conjunction, to test the radical generative and creative potential made possible through computer application. If the seminal avant-garde of the early twentieth century designed an architecture for the Machine Age, then the architects featured in this book are devising transformative, poetic and pragmatic responses to the technologies, urban networks and post-mechanical processes of the Information Age. They are developing spatial routines and urban codings for a world that is at once unfixed and fixed, here-there and there-here, dislocated and located. Theirs is neither a revolutionary nor utopian architecture but an architecture of evolution, contextualization and transmutation. Their researches are triggering a phase-shift in our perception and comprehension of space, materiality and time at the start of a new millennium.

02. Coterminous Territories: the Real and the Virtual

coterminous *adj.* (alter of L. *conterminous, fr. com + terminus* boundary) Having a common boundary.

In the popular cultural construction of the virtual and the real, whether in literature or film, everyday experience is mirrored across an unbridgeable chasm in *another reality*, the threatening, chromed-metallic world of cyberspace. In such films as *The Lawnmower Man* or books like William Gibson's *Neuromancer*, the risk of entering this other world is to fall headlong into an insatiable anti-physical vortex in which we lose the body altogether. According to French theorist Paul Virilio, today's urbanists find themselves in a similarly dangerous position. On the knife's edge between the virtual urbanity of the information machine and the actual urbanity of the city, they are "torn between the permanent requirements of organizing and constructing real space – with its land problems, the geometric and geographic constraints of the center and the periphery – and the new requirements of managing the real time of immediacy and ubiquity."[1] Indeed, in Virilio's "Overexposed City," the flickering expanse beyond the computer interface is an electronic terrain that is eating away the edges of the city.

In such descriptions of our world, the virtual has superseded the real. Yet the virtual – from *virtus*, meaning potential essence or force – has been a philosophical concept for as long as the actual has been manifest in the idea of the city, *polis*. *Virtus* is not an ideal – the Utopian City, for instance – but a bundle of dormant forces awaiting actualization. Its total potential is necessarily never achieved because it offers an infinite number of trajectories instead of singular ideal abstractions or truths.[2] However, the popular insistence of pitching a virtual anti-corporeality and anti-urbanity against an actual corporeality and urbanity has enforced a false split between that which has potential and that which is actualized. This is a false dialectic.

The virtual is real but not actual, ideal but never abstract. Indeed, the two sides of this purported dialectic, the real-actual and the virtual-imaginary are not distinct halves but something akin to oscillating forces in a shifting field, existing not side by side but through and across each other. If we were to assign identities to the real-actual and the virtual-imaginary, we might say that they are at once singular and doubled, like Siamese twins. If they are entities at all, they share functions and space over coterminous territories, or overlapping regions of non-exclusivity. In our cities, there already exist demonstrations of the links between the real and the virtual: the ubiquitous cash machine (ATM), for example, the garish video arcade, even the lowly phone booth all call into play the possibility of a coterminous merging of very real city of bricks and a conceptually experienced "city of bits."[3]

Ironically, the twinning of the virtual and the real in architecture is not a phenomenon specific to our time or technologies. The notion of real spaces enriched by a virtual logic has existed since the seventeenth century, if not earlier. The puzzling forms of the garden maze, for instance, or the infinite reflections of the mirrored gallery, spaces in which vision and reasoning bend and warp according to a virtualized logic of reflection, simulation and distortion, were in many ways precursors of our intermingled electronic-virtual and material-real structures: the actual being recorded in a world network (data-maze) and the virtual as the points of interface (data-mirrors). If in the seventeenth century the real-virtual might have existed only in the mirrored halls and garden mazes of the privileged, today's intertwined real-virtual is more democratically shared across cities and social classes.

The globalized liquid "soft architectures" of digital media flow over, under and through the local, concrete and "hard architectures" of our contemporary cities, creating an indeterminate, "floating" environment, an interface between public and private, collective and subjective, provincial and planetary.

An architecture capable of addressing – or, better yet, choreographing – the dance between the doubled worlds of the real-actual and the virtual-potential is beginning to present itself. Dutch historian and critic Bart Lootsma has written, "Instead of trying to guarantee the eternal life of an existing architecture in a different medium, our strategy today should be the contamination of that architecture with other media and disciplines in order to produce a new and more robust mongrel."[4] Whether "hypersurface," Stephen Perella's investigations into a topology of relational, mediated human agencies [pages 44-53], or "transarchitectures," Marcos Novak's turning-inside-out of cyberspace [pages 126-35], these experimental forms promise to occupy the coterminous territories of the real and the virtual. In them, we may begin to experience a world no longer divided by virtuality but one made rich with spaces of animated potentials and realities.

03. Emergent Dimensions

The new communication system radically transforms space and time . . . Localities become disembodied from their cultural, historical, geographic meaning, and reintegrated into functional networks, or into image collages inducing a space of flows that substitutes for a space of places.[5]

Manuel Castells

Our international telecommunication networks have become characterized by agitated, irreversible super-connections that operate outside conventional human understanding of time and space. We no longer communicate with friends, family or associates exclusively in a particular place; rather, we communicate both in the local context and across time zones and cultures. A seamless virtual geography of informational interchange has replaced locale as an indicator of space and rearranged "natural" temporal sequences along the earth's surface. The globalized liquid "soft architectures" of digital media flow over, under and through the local, concrete and "hard architectures" of our contemporary cities, creating an indeterminate, "floating" environment, an interface between public and private, collective and subjective, provincial and planetary. *Hybrid Space* architects claim this ambient, symbolically rich and multidimensional world-space as an extraordinary context for architectural exploration.

A number of these architects' projects endeavor to instrumentalize the simultaneous identities of the global and the local experience. In the words of Ben van Berkel and Caroline Bos of UN Studio [pages 164-76], "the freedom to assume different identities is an achievement of the condition of endlessness." It is this very sense of endlessness, a free spinning-out of identity and consciousness over the planet's surface, that is becoming a condition in our lives and in our architecture. Confronted not by the possibility of new beginnings but of multiple emergent dimensions of continuous returns and recyclings, architecture can shed the burden of the nostalgia for place and the oppression of a falsely perceived internationalism. As Manuel Castells has pointed out, "networks do more than organizing activities and sharing information. *They are the actual producers, and distributors, of cultural codes.*"[6] In the place of merely representing these projected identities and territories (the global or the local), *Hybrid Space* proposes an approach to architecture that creates new cultural codes and modalities – turbulences or disruptions within the physical and electronic networks that connect our international and local cultures.

Architecture need no longer be generated through the static conventions of plan, section and elevation. Instead, buildings can now be fully formed in three-dimensional modeling, profiling, prototyping and manufacturing softwares, interfaces and hardwares, thus collapsing the stages between conceptualization and fabrication, production and construction.

04. Urban Life

Within our lifetimes we are watching unprecedented deviations from the basic outline of the city. The boundaries between urban conditions, between private and public space, natural and urban space, are blurring. While whole families of urban and architectural types – 1950s skyscrapers and 1960s malls – are becoming marginalized or superannuated, urban forms like featureless information factories, gated exurban estates, anonymous strip malls and hopelessly tangled parking-lot complexes are evolving within the topographies and ecologies of our wired cities. At the never-ending edges of town, urban forms germinate and grow almost instantaneously, appearing in the world as if overnight, fully formed by the forces of global capitalism.

In the post-industrial urbanscape there is no sense of accumulation, history or Colin Rowe's delicious "*bricolage.*" In this environment there are no marvelous accidents, no "collage cities."[7] The logic of these developments is ruthless and relentless; the urban forms it generates are hardly the outcome of municipal mismanagement. Rather, this multicentered, multiperipheral city-suburb represents inevitable and calculated fiscal evolutions in the form of the city – catalyzed by information technologies. Just as the elevator, electricity and the telephone made New York's vertical urbanity possible, contemporary technologies are expanding cities horizontally through new systems of digital and physical infrastructure. Technologies, such as shipping and tracking systems, electronic personal credit and payment records, and data bases, or the global positioning systems (GPS) that are being rapidly integrated into our cars and palmtops, are not mere byproducts of a new urbanity, they are precisely what make living and operating in sprawling agglomerations like Los Angeles-Tijuana or Tokyo-Yokohama possible. Such advances are in fact key neural links in an increasingly sentient urban ambience.

05. Topologies

One definition of topology is the topographic study of a region: the survey and graphical delineation of a place and its configurations, elevations and positions. A topology shows the relationship between things, man-made and natural, on the earth's continuous surface. Topology is also the branch of mathematics that investigates geometric configurations (as a point set) that cannot be altered if subjected to one-to-one transformations by shrinking or enlargement. For instance, if one tied a piece of string around a sphere and shrunk the sphere and the loop of string, eventually the loop would diminish to a point. Shrinking the sphere will not prevent the collapse of the loop; a sphere is therefore not a topological object. On the other hand, if one tied a loop of string around a doughnut (torus) and reduced the doughnut and the loop, the loop would never shrink to a point: the doughnut's hole would prevent the loop from collapsing. A torus is thus a topological object. Topology involves the study of strange surfaces that can be transformed without collapsing or breaking because of the rubbery structure of their surfaces.

A third definition of topology is the anatomy of a particular area of the body, the form and qualities of an organ, for example, or the figure and outline of an organism. As D'Arcy Thompson (1860-1948), classicist, naturalist and biomathematician demonstrated, mathematical functions can be applied to pictures of one living organism to turn it into another organism. Among the most striking examples of these operations were his use of gridded studies of animals or animal parts (shoulder blades or feet, for example) to transform the morphology and appearance from one species to another.[8] This brand of topology divulges the geometry and surface nature of that which is shared in the natural world.

All of these definitions of topology are essential to the computational architectures collected in *Hybrid Space*. These explorations of built form are not based on the pure Euclidean geometries of the sphere, cube or pyramid but instead are often modeled on the torus, the Möbius strip (see Winka Dubbeldam's Yokohama Ocean Liner Terminal on pages 92–93) and the Klein bottle. Unifying disparate spaces in the same way, the earth's surface unifies hill and valley, cliff and plain; these approaches to architecture are relational and may even emulate our own corporeal topologies, the soft geometries of the body's hollows and cavities. In the process, this architecture affirms the paradox of topology: a continuous looping into and out of, back and forth, on a surface without end or beginning, which has neither interior nor exterior, but which is always experienced as a single, strange entity.

06. Firmware: Digital Architectures in Hardware

Computerized design and manufacturing processes have brought about working practices that irrevocably affect the way buildings are assembled, function and behave. Little more than a decade ago, most offices reproduced their architectural drawings and schedules mechanically or by hand, documents were then delivered to consultants for review and updates, before revisions were painstakingly added to working drawings by hand. Today, three-dimensional CAD models can be relayed between workstations or offices, executed in different time zones and endlessly revised without ever leaving the electronic sphere. As computer processing power increases exponentially and advanced manufacturing softwares become more available and less expensive, both large corporate offices and one-person studios will reap the practical benefits of the electronic paradigm shift.

Perhaps the most spectacular (and publicized) example of the extent to which these new technologies are influencing architects' production and aesthetic practices is the captivating use of complex-curve-generation software, digitization devices and numeric command-machining in Frank Gehry's Guggenheim Museum in Bilbao. Using CATIA, an aeronautic and automotive design and manufacturing software, Gehry was able to produce precise three-dimensional models for every facet of the titanium and stone surfaces, as well as the intricate structure of the interior curtain walls and stairways, before directly delivering the design details to Spanish subcontractors in CATIA format.

Architecture need no longer be generated through the static conventions of plan, section and elevation. Instead, buildings can now be fully formed in three-dimensional modeling, profiling, prototyping and manufacturing softwares, interfaces and hardwares, thus collapsing the stages between conceptualization and fabrication, production and construction, numerical data formations and spatial experience. The unique character of handwork and systemic mass production can now commingle in CAD/CAM mode of creation, which can produce series-manufactured, mathematically coherent but differentiated objects, as well as elaborate, precise and relatively cheap one-off components.

Architecture is becoming like "firmware," the digital building of software space inscribed in the hardwares of construction. Soft, complex-curved surfaces modeled in data-space will be transmuted to real space as bent or torqued variable panels, as sheets in steel, copper or plastics, or as kevlar or glass-fiber skins; massive involuted elements designed in data-space will become milled, routed or turned elements in wood or aluminium, or cut as molds for quick-setting resins, rubbers or metals (see Greg Lynn's Embryo House on pages 138-41). Bridging the boundaries between the real-technical and the virtual-technical, firmware will favor a far more malleable relationship between bits, space and matter.

As the French architect, technologist and theorist Bernard Cache has argued, architecture today should be understood as an "electronic technical art," based less in the representation of ideal forms than in the scripting of machining codes and routines for numerically controlled (NC) routers, lasers and water jets. Mark Goulthorpe of dECOi [pages 54-69] suggests that the calculation of space, form and structure will usurp design altogether and eclipse the architect's previously deterministic role. What calculation challenges, he proposes, is "the very distinction of engineer, architect, etc. The separation of entities corresponding to the productive division of elements is precisely what is being called into question. If there are any sacred cows to kill, it is not so much the strict geometry and standardization of components that industrial production has seemed to suggest, but the structures of thought itself, and in particular the linear and rationalizing tendencies that such divisions have championed."[9]

The computer, then, will no longer be merely a production, engineering or facilitation tool under the command of the architect-user but a generating entity with its own virtual intelligence or "knowledge" of the design process; the computer will function as a partner. Architecture is becoming a computational collaborative art based on the choreography of robotic manufacturing, while the architect, freed from the need to continuously invent anew, is becoming more like a choreographer of space and material production.

07. Generative Form

Sed fugit interea, fugit inreparabile tempus
But meanwhile it is flying, irretrievable time is flying
Virgil, *Georgics*, III, 1.284

Time, perhaps once seen as an impediment to building, a source of delay and decay, has assumed a decidedly intimate role in an architecture that engages in a kinematic sculpting of space. Today, time and movement have been instrumentalized in architecture with the aid of powerful animation softwares, which have enabled architects like Greg Lynn [pages 136-49], Marcos Novak and Lars Spuybroek of NOX [pages 110-25] to develop dynamic, mutable and evolving design techniques and new spatial paradigms.

The use of animation software has inscribed duration and motion into static form. Rather than creating an architecture that is essentially the organization of stationary, inert forms, these architects view spatial design as a highly plastic, flexible art in which the building form itself continuously evolves through motion and transformation. With complex time sequences and simulations, forms are no longer defined by the simple parameters of scale, volume and dimension; multivalent and shifting external

> **The investigation and application of technology by architects must consider the ramifications of the potentially reckless and uncritical coercion of technology's powers into architecture. We must remain watchful of the machine's ruinous endgame played out as urban forms, spaces and relations.**

or invisible forces and inclinations can also affect forms. Employing software routines that track time-related factors, such as pedestrian and automotive movement, environmental elements such as wind and sun, urban conditions such as views or site density, these designers are producing buildings in which virtual and real media technologies are inextricably linked. Marcos Novak suggests that mathematical models and generative procedures can be used to build models "derived from the particulars of the real world, from data and processes of the virtual world, or from numerous techniques of capturing the real and casting it into the virtual, motion-capture, for instance. Since time is a feature of the model, if the model is fed time-based data, the form becomes animate, the architecture, liquid."[10]

Some might argue that because architecture is ultimately static, it cannot incorporate or embody kinematics, animation or any other form of movement or transformative energy. But Lars Spuybroek would respond that "Media are a way to inhabit time ... a movement connected with our own movements ... we should keep in mind that architecture was the first machine, *the first medium* to connect behavior and action to time, to place it under the revolving light of the sun, but now, on the other hand, we should not mix up the old history of architecture, its Euclidean mathematics with its new potentials."[11]

08. Crash Culture: Speed Limits, Bombs and End Games

The shift in the twentieth-century image of architecture from the "hard" forms of industrial and military technologies (the biplane, the transoceanic steamliner and the automobile) to a more pliant investigation of broader techno-cultural conditions (the soft technologies of leisure and domesticity or the interface model of the computer) is an ongoing manifestation of the ethics of technology in the aesthetics of building. As it has been put forward by Manuel Castells, technology is ultimately society, and society cannot be understood or represented without its technological tools.

Nevertheless, the investigation and application of technology by architects must consider the ramifications of the potentially reckless and uncritical coercion of technology's powers into architecture. We must remain watchful of the machine's ruinous endgame played out as urban forms, spaces and relations. Recalling early modernism's utopian romance with machine form tells us much about the dangers of an addictive technology fix and the consequences of a technological overdose. Indeed, there is something ominous that lies beneath modernity's play of sleek forms and pure surfaces under light. It was no accident that in the aftermath of the Second World War, Le Corbusier, once the champion of the engineer's aesthetic of cold, naked, polished steel, would reject the accelerated technologies of terrestrial movement and aerial flight. Witness to the spectacular violence brought into the world through the combined efforts of mid-century science and the war machine, Corbusier turned his architecture to the vernacular forms of the Maisons Jaoul (1952-56) and the sacred space of Ronchamp (1950-54).

Today, in a post-industial age, we assume too easily that the more supple technologies of communication and computation are less threatening or less likely to drive us towards a total societal crash. However pliant and mobile the technologies of the Infomation Age might seem, real-time connectivity and interface may be only slightly more subtle in their potential for violence than their the brutal counterparts in the Machine Age. Technological violence, whether manifest in the controlled social manipulation of global capital market movements or the precision of surgical bombing strikes, might be less menacing but the potential for terror is no less powerful, no less awful.

With supercomputing speed estimated to achieve some twelve trillion calculations per second by 2003 and the sum of stored human knowledge to double every seventy-three days in the year 2020,[12] the heart of a highly technologized millennial architecture must lie in the critical relationship between technical speed and the architect's ethical concerns. As Virilio has pointed out, time and space may be at their most useful when not in use. Architecture may serve us best when it helps us to organize the gaps, pauses and intervals of respite in an ever-accelerating world. Sometimes speed limits us, and sometimes limits set us free.

Notes

1. Paul Virilio, *Open Sky,* tr. Julie Rose (New York: Verso, 1997), p. 13.

2. John Rajchman, *Constructions* (Cambridge [MA]: The MIT Press, 1998), p. 116.

3. William Mitchell's term for a virtualized urbanity. See his *City of Bits* (Cambridge [MA]: The MIT Press, 1996).

4. Bart Lootsma, "The Computer as Camera and Projector," *Archis* 11 (1998).

5. Manuel Castells, *The Rise of the Network Society* (London: Blackwell Publishers, 1996), p. 375.

6. Manuel Castells, *The Power of Identity* (London: Blackwell Publishers, 1996), p. 362.

7. Colin Rowe and Fred Koetter, *Collage City* (Cambridge [MA]: MIT Press, 1984).

8. See D'Arcy Thompson, *On Growth and Form: The Complete Revised Edition* (New York: Dover Publications, Inc., 1992 [originally published in 1917]).

9. Mark Goulthorpe, "Man Creates a Tool, Tool Changes Man," lecture at Ove Arup & Partners, London (April 1998).

10. Marcos Novak, "Transarchitectures and Hypersurfaces," *Hypersurface Architecture* (London: *Architectural Design*, 1998), p. 89.

11. From "Where Space gets Lost," interview with Andreas Ruby and Lars Spuybroek.

12. Twelve trillion calculations per second are equivalent to 60,000 times the speed of a standard 350-megahertz PC. Human knowledge is estimated to double every twenty years in 1950 and every five years in 1999.

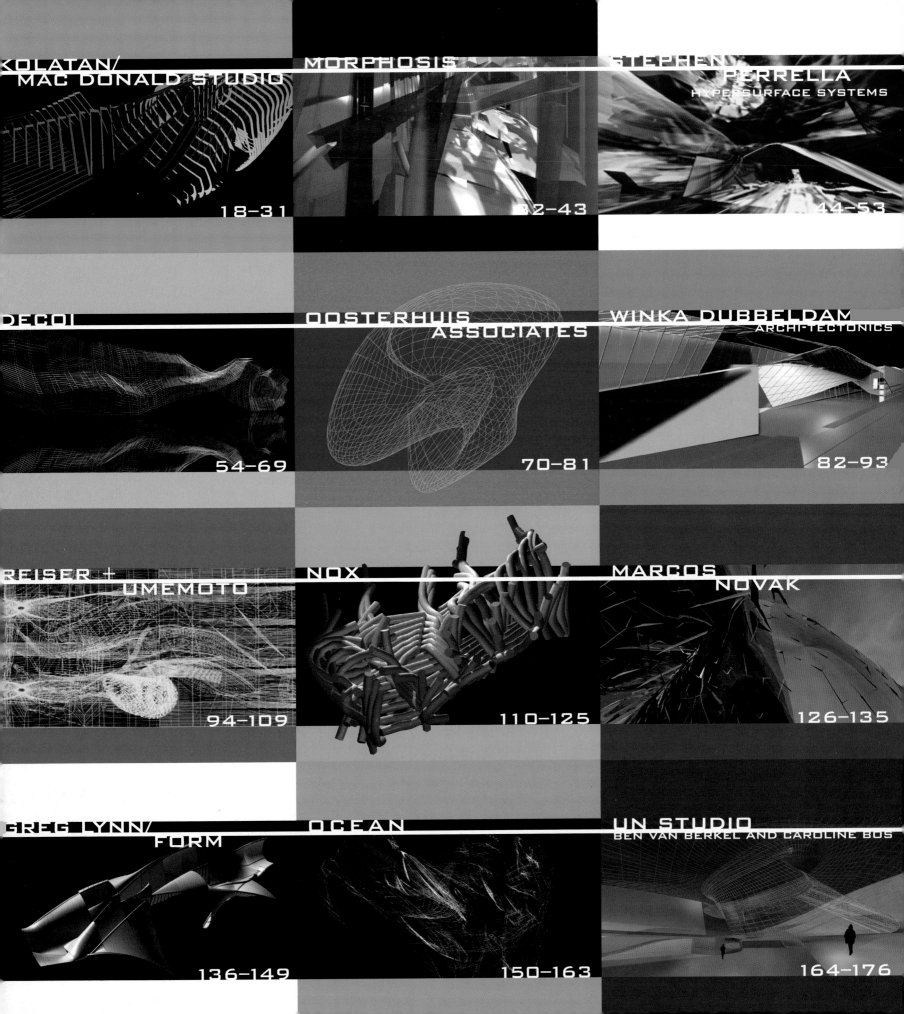

KOLATAN/
MAC DONA

Seamlessly integrating variant forms and conditions,
Kolatan/Mac Donald generate chimerical object-spaces
that challenge traditional architectural idioms and
modes of production.

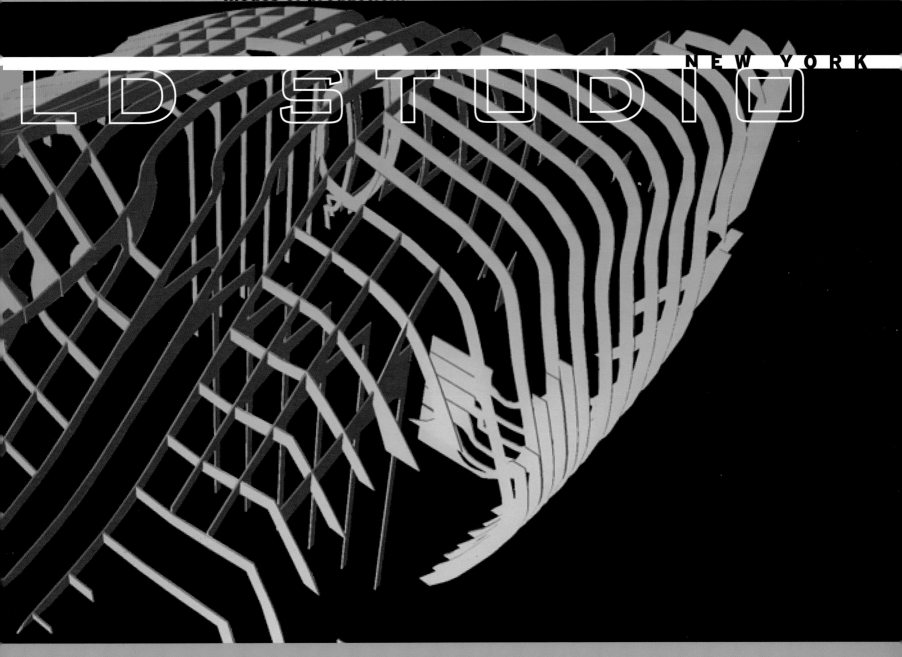

LD STUDIO NEW YORK

YOKOHAMA : SMALE'S HORSESHOE-TRANSFORMATIONS

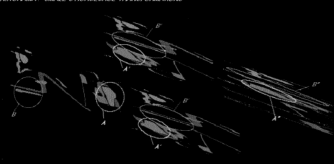

Sulan Kolatan and William J. Mac Donald produce startling composite architectural forms and urban spaces. Their exceptionally flexible and inventive design methodology harnesses the computer's ability to aggregate spatial data, which are tracked and organized in non-hierarchically crosslinked informational indices. These indices are then transformed through a process of data-grafting, in which established spatial types are mapped across and through each other, producing new patterns for non-standard material constructions, moldings and castings. By inducing systematized fields of data into organized material and spatial relations with high-speed computation, Kolatan/Mac Donald has developed the principles behind interlinked and crossreferenced networks into a tool for architectural production.

The architects describe the transposition of data into form as "co-citation mapping," an involved process that charts spatial information – cross-sectional profiles, volumes and dimensions – onto catalogues or databases of dissimilar elements. As an index, co-citation mapping works like any keyword-based library search, registering titles or papers related to the same term. Like such a search model, this mapping technique reveals conceptual connections across categories that might not be immediately apparent. In a house, for instance, the indexing of the sectional profiles of familiar domestic objects might disclose strange alliances between unlike elements – a soap dish and a seat, for example – based on shared morphologies rather than function or scale. The design's next stage is to map such relationships to create a territorial description of associations and disassociations, to demarcate fields of gravitation and repulsion and to coordinate groups of morphologies into clusters that each represent a network of interrelated elements, a matrix of objects. Having no absolute axis, the matrix is subject to frequencies of citation and crossreference, and thus its spatial coordination evolves over time. "These new systems," explain the architects, "are not determined and cannot be understood through a logical extension of the initial parts alone. They are hybrid, but nonetheless seamlessly and inextricably continuous."

Co-citation mapping was employed in the design of the O/K Apartment in New York City, a renovation of two loft apartments that were

Yokohama Ocean Liner Terminal
The terminal's design (1994) was generated through a series of operations in which the local urban fabric was sampled, involuted and stretched into "deep" layers of enfolded skin that radically challenge the conventional relationship of plan to section. A process, called Smale's Horseshoe Transformation, was used to digitally squeeze, fold, stretch, embed and shrink the form, like a mechanical taffy maker's circling arms distend the taffy's surface to become long, thin and intricately self-embedded. The result is a condition, rather than a building per se, defined by the multiple continuities of structure and skin, horizontal and vertical, thickness and thinness. The terminal's functional requirements would be realized within this deep skin in "twisters" (locations of level changes), "movers" (conveyors, escalators, conduits) and "footholds" (places for outlets, brackets, anchors, taps, sockets, locks).

left Smale's Horseshoe Transformation
below left Aerial view of the terminal's roof garden from the sea to the city.
far right Rotated sectional degree cuts of the terminal.
right Aerial views
below Plans
bottom Long section

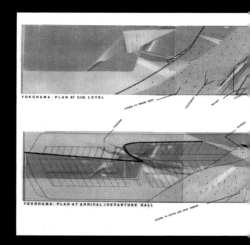

YOKOHAMA: PLAN AT C.I.Q. LEVEL

YOKOHAMA: PLAN AT ARRIVAL/DEPARTURE HALL

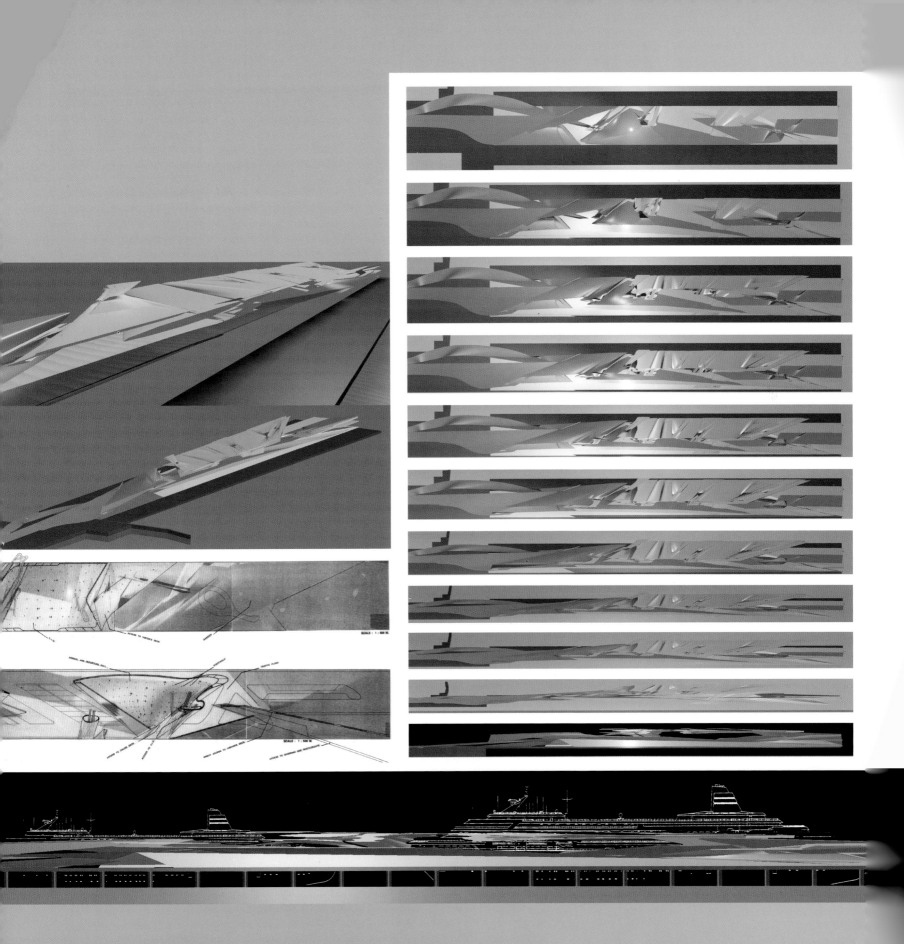

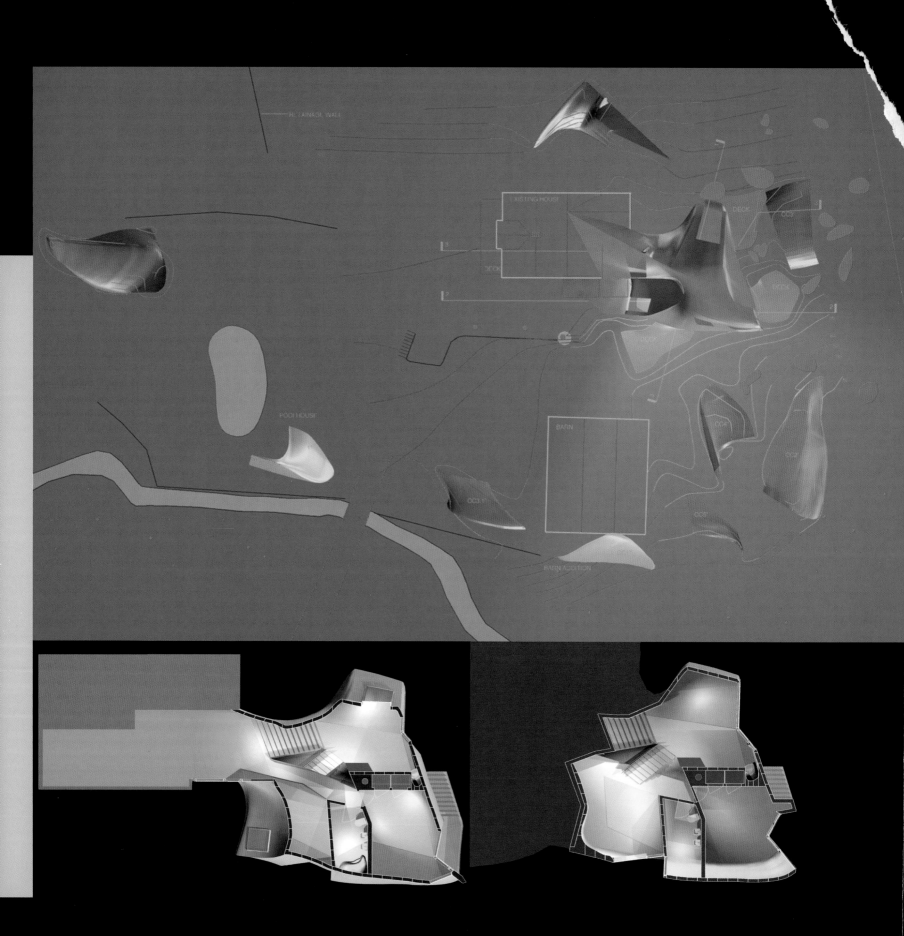

Raybould House
By tracing site/object contour lines at different angles through an existing saltbox house and the surrounding landscape, the architects created a residence in Connecticut (1997-99) that crossreferences and melds into a new hybrid structure.

opposite, above and **below**
Site plan and plans
below Studies of the CNC-manufactured structural rib-frame.
bottom Roof plan showing smooth transition from old roof structure to new house form.

combined by the architects for use as a corporate guesthouse. Kolatan/Mac Donald employed the technique to produce a "domestic scape" – a synthetic residential topography – that was generated by compiling a lexicon of cross-sectional profiles of everyday objects and furnishings, such as a bed, pillow, sink, soap dish, seat, lamp, refrigerator and bath tub, which were then electronically crossreferenced with each other. Disregarding the objects' original scales and categories, the architects combined and recombined the profiles to ascertain their formal and operational similarities. Using this information, the profiles were associated to form entirely new spatial structures, which, interestingly, display aspects of their domestic ancestors in that the initial profiles saved as indexes of specific domestic identities (bed, sink, sofa) are embedded within the new entities. Once linked and surfaced, the hybrid structures were arrayed around the apartment to build up a flowing domestic environment.

Unlike a conventional residential space (the room) or object (a piece of furniture), the domestic scape resists categorical classification because, like a landscape, its identification depends on the presence of inherited idiosyncratic features. It is always situated across the boundaries of pre-existing domestic spaces, sharing qualities of spaces or objects, and thus cannot be understood without making multiple associations to its hybrid taxonomy. One of the overlaps produced in the O/K Apartment, for example, is the seamless transformation between the space shaped by the "bath tub" and the surface of the bedroom floor/wall, forming a single bed-/bathscape across the sleeping and bathing areas. What emerges, finally, is an interior environment that changes incrementally and continuously, a condition that morphs furniture, space and surface.

Kolatan and Mac Donald characterize their spatial productions as chimerical hybrids. The chimera – a she-monster composed of a lion's head, goat's body and dragon's tail – was the ultimate crossbreed in Greek mythology. Chimera can also refer to an individual organ or biological element that consists of genetically diverse tissue, often combined through the process of grafting. "One of the emerging spatial paradigms"

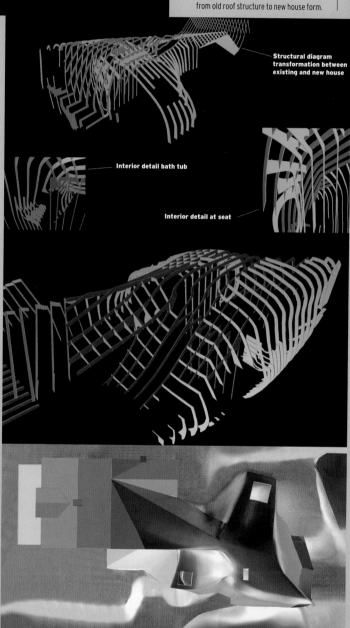

Structural diagram transformation between existing and new house

Interior detail bath tub

Interior detail at seat

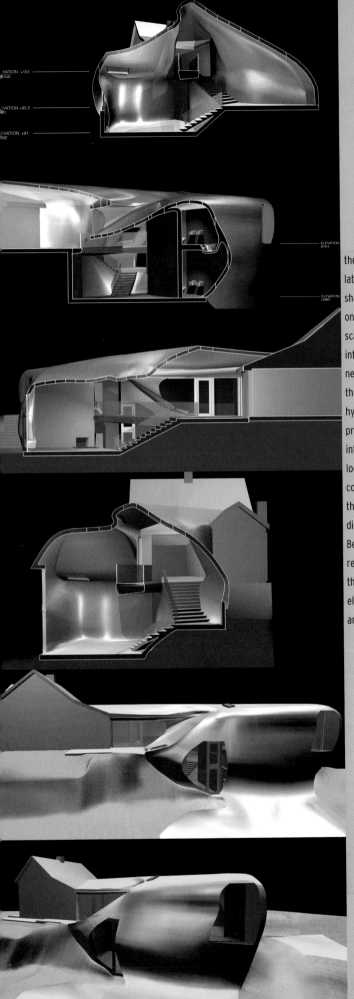

the architects have written, "is that of the network as a system of interrelations between dissipative processes and aggregative structures that shape new spatial patterns and protocols ... Our work focuses in particular on the network model's capacity to facilitate cross-categorical and cross-scalar couplings whereby the initial systems/morphologies are not merely interconnected, but form new hybrid identities. What differentiates this new generation of chimerical hybrids from previous mechanistic ones is the act of transformation." The chimerical diverges from other forms of hybridization, especially mechanical forms like collage, montage or the prosthesis, in crucial ways. While most processes of hybridization integrate diverse elements in a single entity, the constituent parts rarely loose their individual identities; moreover, the singular unity of each component is more pronounced when combined with other elements through superimposition or juxtaposition. In these forms of conjugation differences are enforced rather than incorporated in a complex unit. Because each component in a collage, montage or prosthetic coupling remains discrete, the whole can be broken down and disassembled. Within the chimerical hybrid, on the other hand, a union between distinct elements produced through the logic of the digital machine is seamless and indistinguishable and thus irreversible and irreducible.

In a chimera, the relationship between the parts is not one of interconnection or adjacency. At least, not simply. The limits of the parts, the exact delineations of the thresholds between parts, are not clearly identifiable. Rather, like the result of a successful graft, the border disappears. Locally, the part that was different becomes inextricably bonded with the rest ... We have two primary interests in the chimerical. One has to do with its seeming capability as a concept to help *define* existing phenomena of complex hybridity in which categorically different systems *somehow* operate as a single identity. The second is based on the assumption that the ways in which the chimera are constituted and operate hold clues to a transformatively aggregative model of construction/production. That is to say, an aggregation that becomes more than the sum of its parts, and therefore is not reducible to its constituent elements. Thus, the chimerical has the potential to be both an analytical and methodological tool.

Raybould House

Hybridizing the site/object contour lines along different parameters, Kolatan/Mac Donald generated the new house's profiles, which will become variable templates that can be cut in plywood at 1:1 scale on CNC machines and erected on site. Once assembled, they will form a skeleton for the morphology of the overall building envelope and the hybrid connection between the new residence's integral structure and the frame of the existing house. During construction a spray-on foam will be applied to the rib-frame matrix; once expanded and set in place, the foam insulation will be reduced to conform to the profiles using a machine akin to a lawn mower. Finally, a soft, leathery, aluminiumized polyurethane skin will be applied to the exterior while the interior shell will be finished in a smooth cementious acrylic mix.

left, top to bottom Section and elevation studies.
right Rendered interior and exterior perspectives and sectional studies.

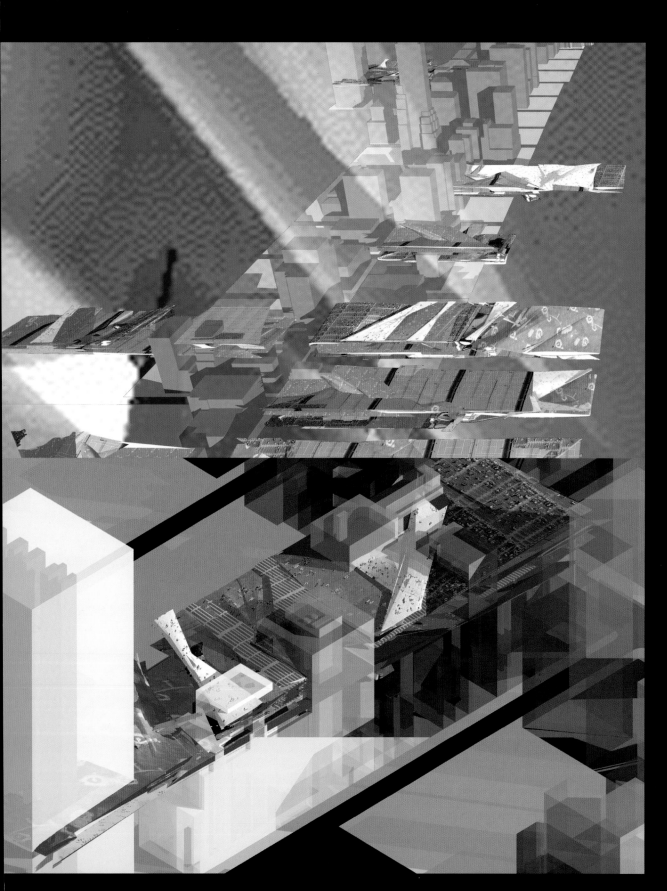

Take 5 on Manhattan Horizontal

Aerial views of Manhattan suggest that the mid-air roofscape constitutes a second urban layer (above street level). Kolatan/Mac Donald's urban design proposal (1997) for 5th Avenue in Manhattan speculates on the potential of this layer to be occupied horizontally. In place of strategies of unilinear continuity along the avenue, the plan is a network based on the notions of hybrid identities, "co-citation" and "soft-sites." In this model, the concept of co-citation permits the construction of relations between non-adjacent sites and the identification of 5th Avenue not only along its own spine but elsewhere in Manhattan. Just like the differentiation between Broadway, Off-Broadway and Off-Off Broadway, the "5th Avenue" identity is exported, while other identities are imported to 5th Avenue proper. Sites on 5th Avenue are termed "soft" if they are underbuilt according to zoning ordinances or the site's 5th Avenue identity is already infused with hybrid elements. The project builds out horizontal skyplane levels, where necessary using transferred air space – reservoirs that establish foothold positions on the ground and connections to existing rooftops and elevators. Rather than assembling small sites and air rights in a package as a single large development site and then extrude its footprint, the proposal builds over existing sites without demolishing them.

above left Overall view of the midtown Manhattan proposal from 23rd to 59th street.
left Sample block isometric at 47th street.
opposite, above View of skyplane-level recreational field in the roofscape of 53rd street.
opposite, below Detail of the Harlem proposal at Marcus Garvey Park.

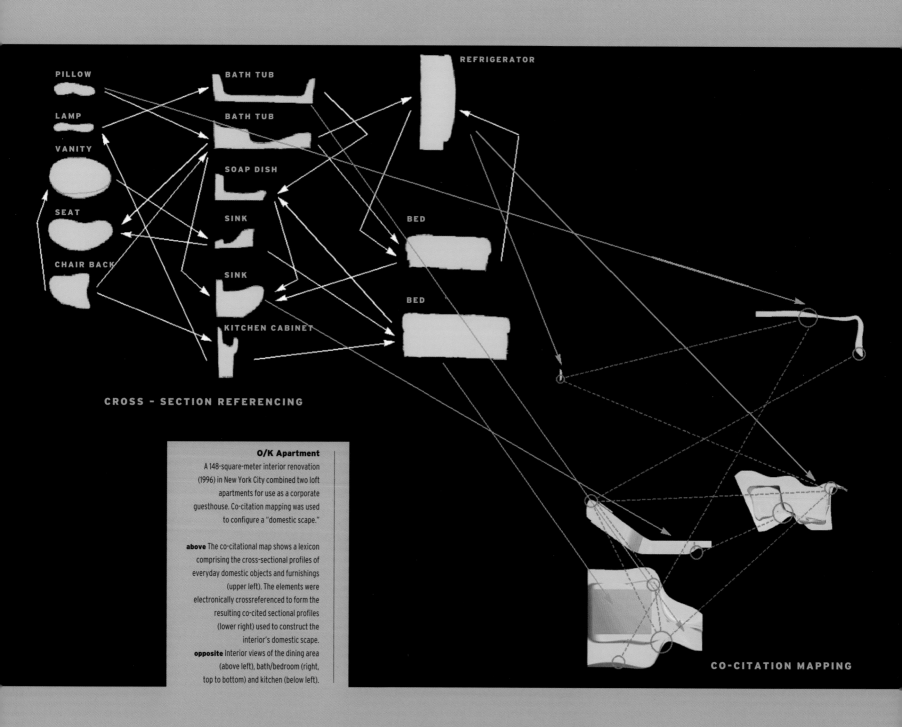

PILLOW

BATH TUB

REFRIGERATOR

LAMP

BATH TUB

VANITY

SOAP DISH

SEAT

SINK

BED

CHAIR BACK

SINK

BED

KITCHEN CABINET

CROSS - SECTION REFERENCING

CO-CITATION MAPPING

O/K Apartment

A 148-square-meter interior renovation (1996) in New York City combined two loft apartments for use as a corporate guesthouse. Co-citation mapping was used to configure a "domestic scape."

above The co-citational map shows a lexicon comprising the cross-sectional profiles of everyday domestic objects and furnishings (upper left). The elements were electronically crossreferenced to form the resulting co-cited sectional profiles (lower right) used to construct the interior's domestic scape.
opposite Interior views of the dining area (above left), bath/bedroom (right, top to bottom) and kitchen (below left).

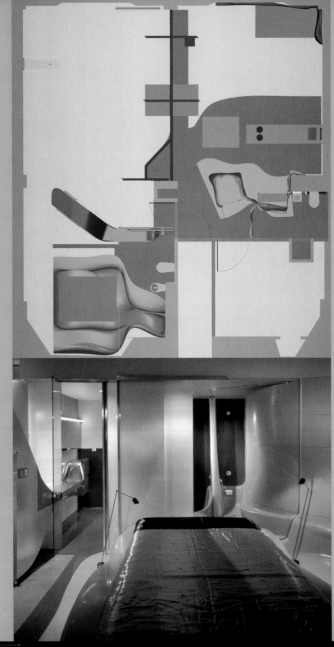

The construction possibilities that arise from the process of seamless coupling is exemplified in the Raybould House, which successfully crossreferences and melds the surrounding landscape and existing saltbox structure into a chimerical hybrid. Tracing contour lines at different angles through the existing house and landscape and hybridizing these lines along defined parameters provided the basis for the new house's form. The profiles generated by the tracing become variable templates that can be cut in plywood on CNC machines and erected on site. Once assembled, they will form a three-dimensional matrix for the building envelope and the skeletal connection between the house's new integral structure and its existing frame, a hybrid of both constructural systems. In later phases of the project, specific sites in the neighboring landscape will also be altered to assume the information extracted from the profiling: The modified sites will correspond to portions of the building and elements of the new house will reappear in the landscape, which – like the domestic objects in the O/K Apartment – assume an embedded history of form. Thus the forming of the new house, the old house and the new landscape are intricately connected as conceptual and constructional couplings and hybridizations.

In their research Kolatan and Mac Donald have discovered unexpected formal and programmatic possibilities through the computer's capacity to be both instrumental and non-deterministic. Seen as a group, the studio's projects suggest a preoccupation with applying the radical potential of digital technologies to architecture, and in the process they overturn traditional or inherited notions of architectural methods and production. Their innovative architectural idiom takes its cue directly from the processing power of the computer and has engendered a range of previously unimaginable fabrications and conditions for which no specific designations have been conceived. Beguiling and fascinating, Kolatan/Mac Donald studio's works are novel object-spaces and amalgams of strangely familiar circumstances that emerge fully formed, as if by the odd logic of the computer's sentience.

O/K Apartment

While being entirely new entities, the forms resulting from co-citation mapping display aspects of the domestic ancestors from which they were cross-referenced. Unlike a standard domestic space (the room) or the domestic object (a piece of furniture), the evolved domestic scape is a fluid, hybrid composition that resists categorical classification, a morphed condition between furniture, space and surface.

left Overall plan of the apartment with bedroom/bathroom 1 (lower left), bathroom 2/kitchen (lower right) and pivoting panels arranged top to bottom in place of the original party wall.
below left View of bedroom area with conjoined bathroom.
right View of entrance area with kitchen and bedroom 1 behind the aluminum wardrobe. Orange pieces are made of fiberglass.

MORPHOSIS
LOS ANGELES

Accelerating from analogue to digital modes of production, Morphosis continues to expand and redefine its organizational principles and formal obsessions.

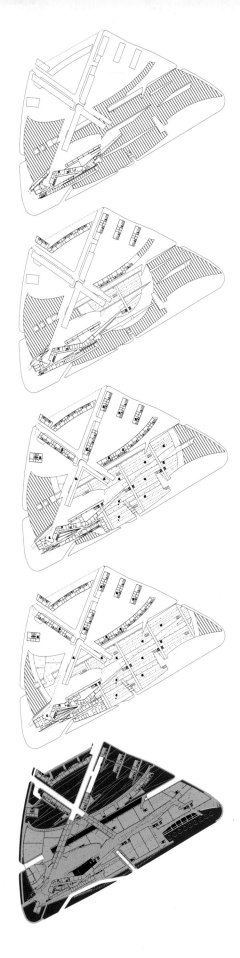

Led by Thom Mayne, Morphosis has long been admired as an architectural practice dedicated to the intensive production of physical drawings and models. For many years drawing was in and of itself the primary architectural project for the firm, particularly in the late 1970s and early 1980s when much of the studio's architecture went unrealized – giving its exquisite drawings and models a presence and importance of their own. In many instances these artifacts did exist independently, in books and in galleries, detached from the architecture they depicted; in other cases, models and drawings were actually completed after a building's construction and served as meditations on architectural systems of organization and processes of fabrication. Less than five years ago the practice still produced buildings in largely the same way that it had done for fifteen years: concepts began as initial sketches, which led to the production of hard-line drawings and/or a model; from there larger and more detailed study models and technical drawings were produced by hand. Recently, however, Morphosis abandoned its well-documented modes of working in favor of a largely digital process of production, and Mayne and his team currently build their drawings exclusively within the space of the computer. Physical models usually follow from virtual models.

For Morphosis today, the act of drawing – outlining a figure – is less an exploratory process of loose inquiry on paper than a means of building directly within the digital medium. Rather than elaborate flat representations, the architects at Morphosis employ the computer to work continuously on a single object – manipulating, rescaling, stretching, amending, subtracting and prying it apart until a design solution is reached. The process, according to Mayne, is like "building, unbuilding, building again ... very direct, very 'physical' ... The other day we asked ourselves if there were any lines left untouched after two years of working on a project. I don't know how many there are, but there are still bones in this work ... from the model produced on that first day ... We just build, construct in one-to-one scale within the virtual space of the computer ... no plan, no section, no elevation ... it's more like shaping clay. I hadn't been interested in this technology until recently. The shift to computers in our office was made inevitable by the nature of practice today."

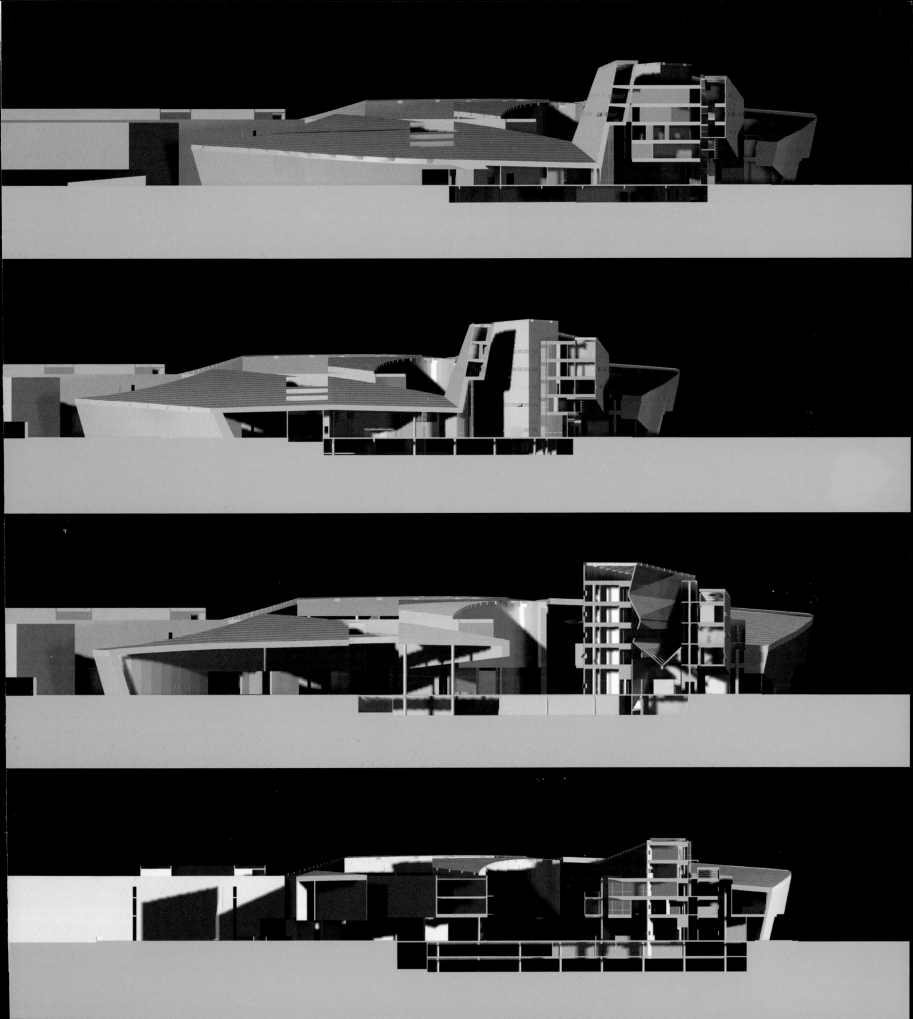

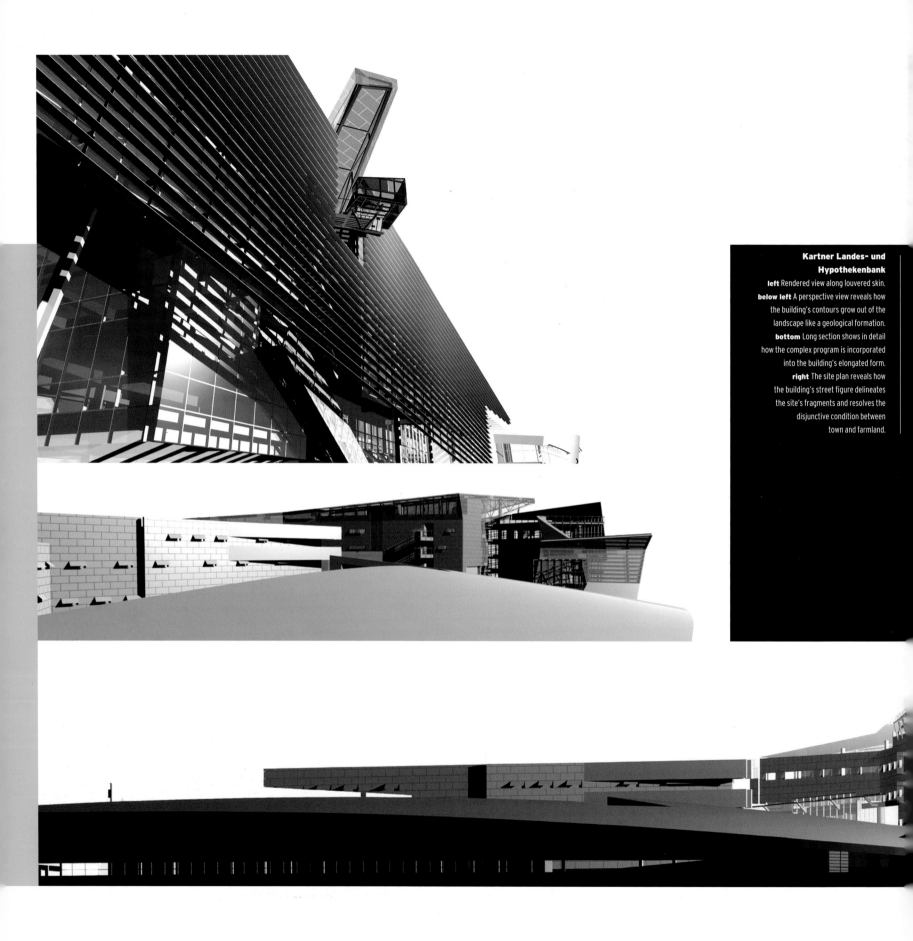

Kartner Landes- und Hypothekenbank
left Rendered view along louvered skin.
below left A perspective view reveals how the building's contours grow out of the landscape like a geological formation.
bottom Long section shows in detail how the complex program is incorporated into the building's elongated form.
right The site plan reveals how the building's street figure delineates the site's fragments and resolves the disjunctive condition between town and farmland.

The shift away from analogue to digital production has been mirrored in the realization of several buildings and significant commissions for large-scale public projects. It could be argued that the challenges, pressures and demands of large-scale construction, not the desire for the development of increasingly sophisticated modes of visualization and representation, have driven the practice's adoption of digital modes of production. But Mayne points out, "When we elected to go with computers we decided to treat them as just another tool ... like an electronic pencil. I had a certain scepticism, anxious to avoid a dependence on technology. Our intention was to continue more or less as we always had. The first task we set on the computer was a simple set of drawings of the Landa House [a recent residential project in Los Angeles]. It was interesting in that the result looked like all of our other line drawings – axonometric sections, elevations – which many people already thought were computer-generated ... except these were outputs of a three-dimensional model. The difference with using computers really isn't in the presentation but rather in the constructional possibilities, however seductive the drawings might be."

These parallel changes – from analogue modes to fully digital production of large-scale constructions – have enabled the studio to leap from directly supervising smaller-scale projects in their local area (the Los Angeles region) to administering much larger and more complex projects in remote locations of Asia and Europe. The shift toward digital production has also helped refine the collaborative rules by which the practice operates. In the Kartner Landes- und Hypothekenbank project in Austria, for instance, Morphosis regularly monitored construction by Internet video links; design decisions based on the data were shared electronically with consultants and the client. Thus, for Morphosis the computer has not only facilitated a new means of production but encouraged an examination of the very processes by which the studio generates its work and distributes project information, both internally and externally.

The ASE Design Center, located in a high-rise housing and retail tower complex in Hsichih, a fast-growing suburb of Taipei, is one of the first generation of Morphosis's projects to be developed exclusively using

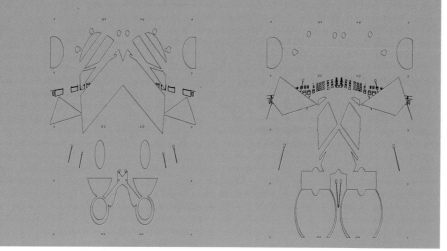

Hippocampus

This conceptual project (1998) translates space into a stream of time in which things and events can be assimilated, linked and manipulated. A microcosm of a city, in which one can live and work, in virtual and real space, the model represents received and frozen elements as a singular memory-event, which fills the brain and gives rise to the spirit of the memory. This is a project about the point at which knowledge ends and darkness begins.

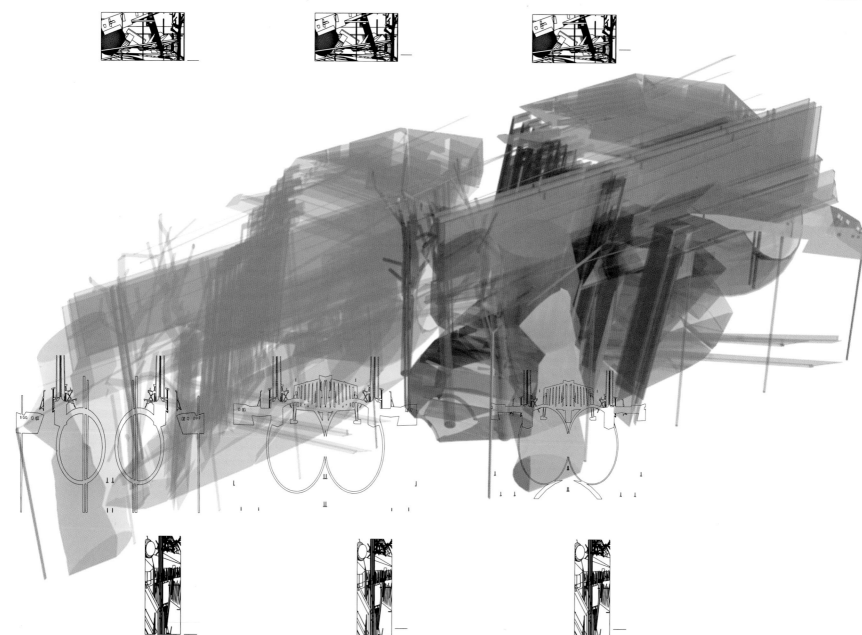

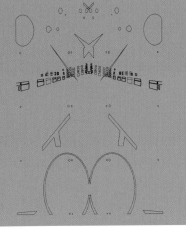
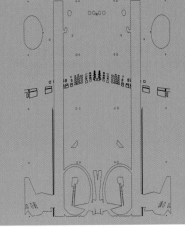
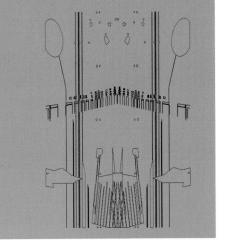

Hippocampus
The campus offers relaxation, socializing and pleasure in all possible combinations and at top speed; it even contains an upside-down pyramid, a "churchy" zone, a 24-hour chat room, an official district, unchartered territory and a bar at the corner.

these pages, top and **bottom** CAT-scan-like sections show the "brain's" layers.
these pages, middle Two transparent renderings of the structure and an interior view of the campus's space, in which virtual and real co-exist.

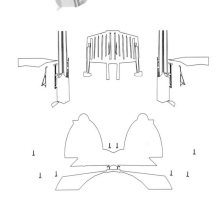

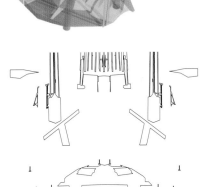

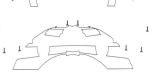

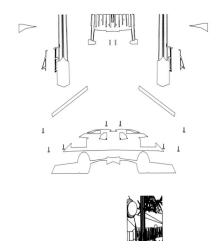

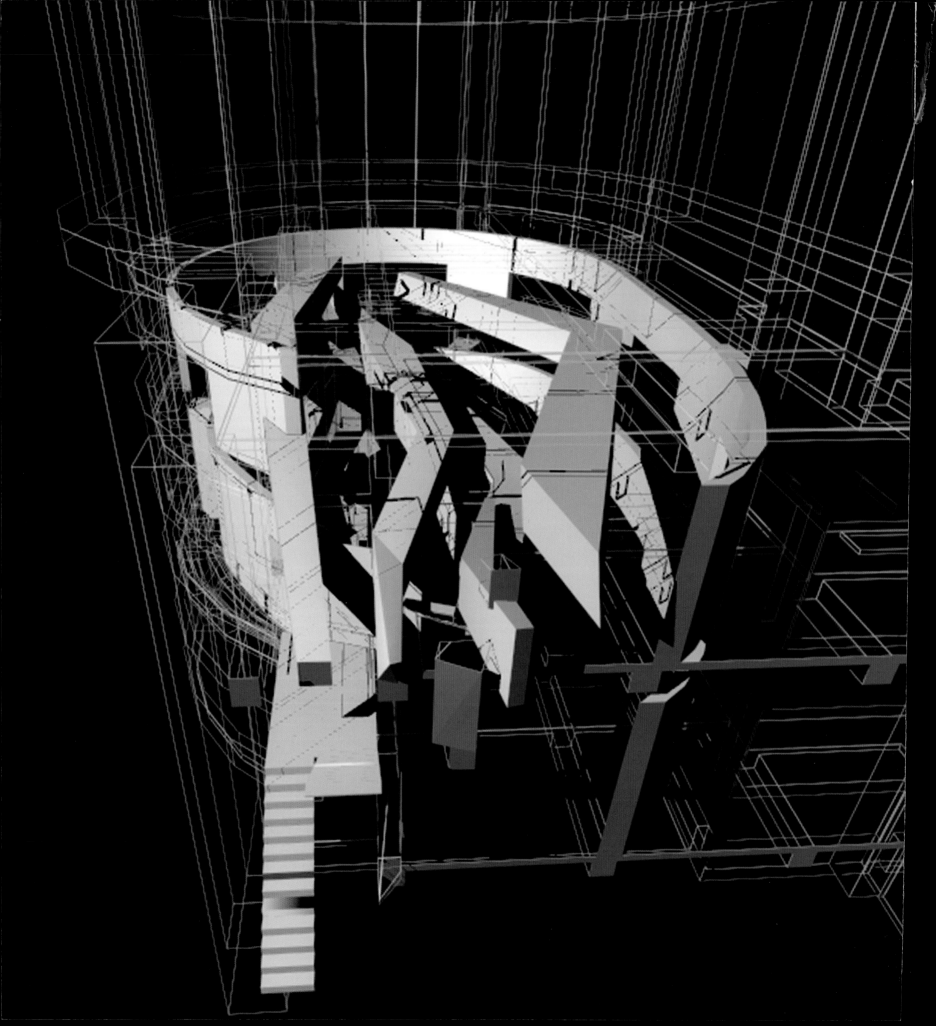

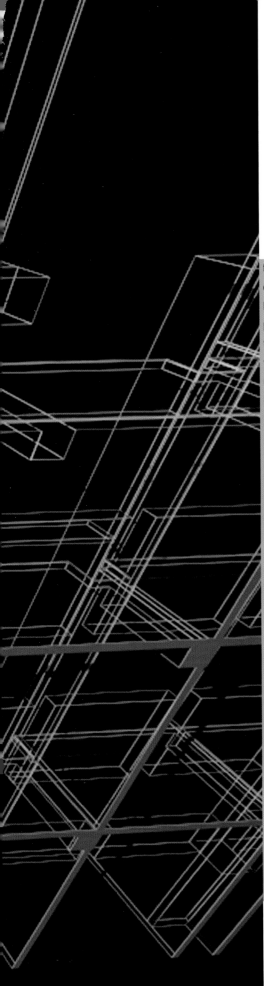

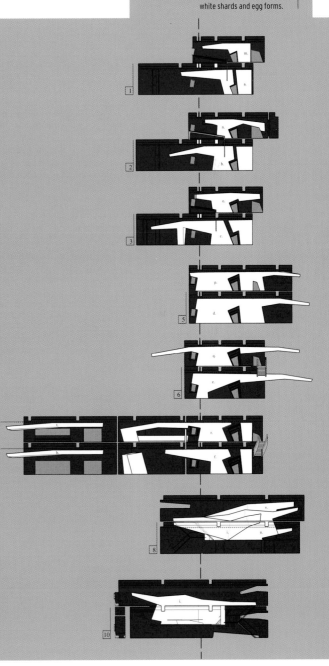
the computer. The studio took on the project with the express intention of overturning its most basic working practice. "Working on a computer," Mayne says, "started to challenge the sequence with which we explore a problem ... we recognized early on that we were building a model and that the model was able to be translated to the convention of drawings that we were accustomed to working with. I am intrigued with this reversal ... the CAT-scan quality of the sections ... still descriptive, analytical, but also expansive and interpretive ... We find that we still have to build physical models because they tell us different things from the virtual models. We developed Taipei solely through computer images, making a physical model only at the very end."

Morphosis's design involved inserting a 2000-square-meter space, interrupted by a random arrangement of columns, into the existing building. Two formal languages, or systems, eggs and shards, were arrayed within the perplexing column grid to rationalize the space and to synthesize its requirements into six distinct egg-shaped zones on two levels that envelop three separate exhibition spaces, two restaurants and a lecture hall. The eggs, fabricated in rusted steel-plate using CNC-machining processes, extend and mimic a rounded swelling at the complex's western end. The design repeats the curved form three times throughout the complex, stacking it to contain the eating, exhibition and lecture areas. The shards, deep walls that rise up to seven meters off the ground floor, are a series of kinked white plaster blade-walls that follow and encase the variable lines of the structural column grid. The shards fly across the complex on east-west "flight patterns" and are punctured by openings that allow for north-south perpendicular movement. A floating school of fish-like fiberglass pods swim parallel to the blade-walls – completing the composition and adding an amusing touch to the pregnant, ovoid spaces.

As with most of Morphosis's work, pre- and post-digital, spatial systems are intermeshed in the Taipei project to accommodate the varied needs of the program. However, unlike earlier projects, in which space was organized through linking regular, repetitive and Platonic formal languages, here a coherent sequence of spaces was generated by randomly interweaving uneven lines of force (the shards) and variant geometries

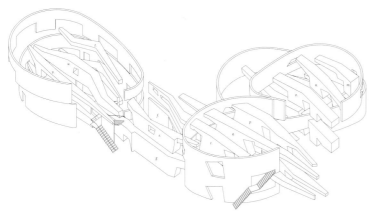

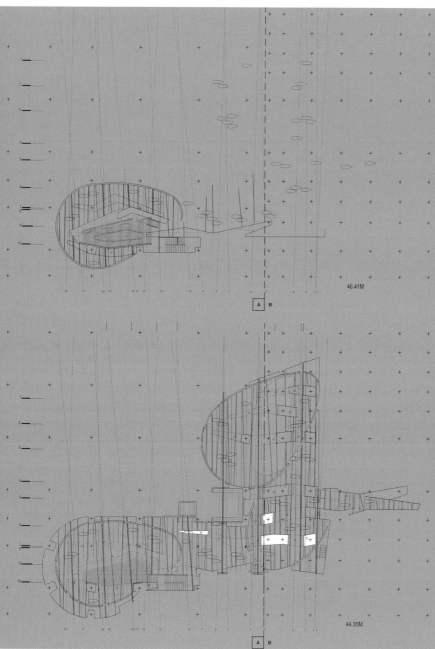

46.41M

A | B

44.35M

A | B

(the eggs) to assimilate an irregular structural field (the existing columns). "Taipei was a beginning for us, a transition to a new approach," says Mayne. "The computer has enabled the emergence of genuinely new morphologies for us, a new vocabulary ... virtual technology assisted us here, both in the translation of a complex series of forms into a logical construction sequence, and as a means of looking at the results of our inquiry ... understanding it in the terms of the program and assisting in its visualization ... Our client was able to see the physical environment as part of a creative, dynamic process of experimentation and inquisitiveness. They could participate vicariously in our investigations."

Morphosis is a transitional practice. Generationally situated somewhere between Frank Gehry's CATIA-driven studio and the younger, digitally savvy practices featured elsewhere in this book. Interestingly, Morphosis straddles both worlds comfortably, having moved confidently into large-scale commissions without abandoning the essentially experimental and radical nature of its early work. Morphosis has embraced the digital paradigm shift in architecture and, leaving behind any concerns about what might have been lost in the transformation, uncovered new freedoms and organizational principles.

ASE Design Center
opposite, above Axonometric view
opposite, below Plans
below Partially rendered wire-frame study
depicts the school of seed-pods or fish
"swimming" parallel to the blade-walls.

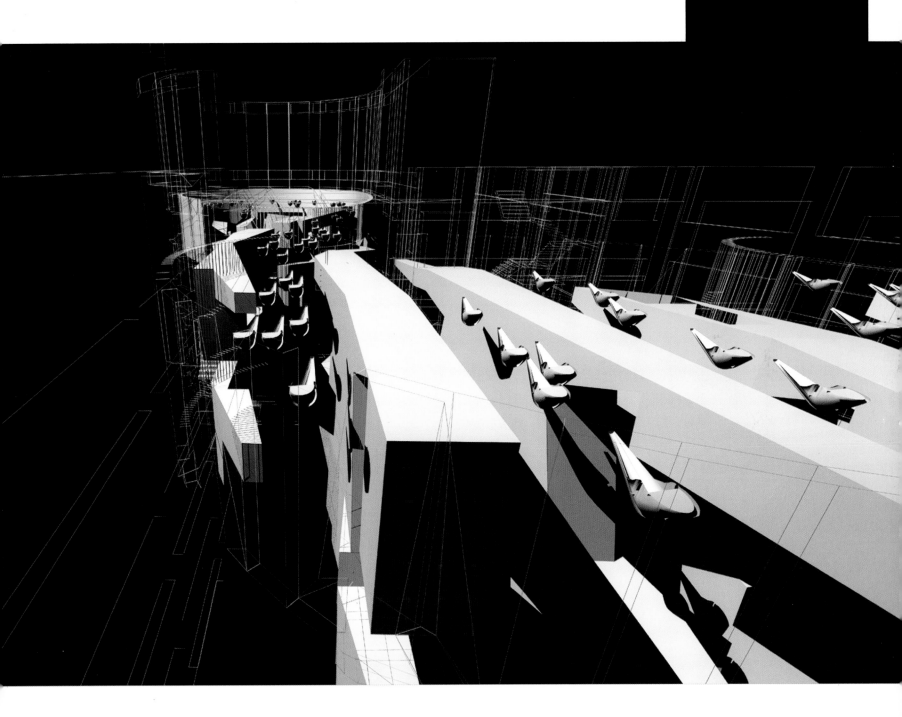

STEPHEN
PERR

ELLA
HYPERSURFACE SYSTEMS

Stephen Perrella's "hypersurfaces" intertwine surface and structure, image and object, to create emergent realms that dynamically rearrange the relationship betwen time and space, the real and the virtual.

Möbius House: Hypernurban Architecture

Designed in collaboration with Rebecca Carpenter, this theoretical project (1997–98) explores the hypersurface to consider how we dwell in relationship to communications media. A study-diagram for post-Cartesian dwelling, Möbius house is neither an interior space nor an exterior form, but a transversal membrane that reconceptualizes the conventional relationship between interior and exterior into a continuously deformed bridging surface. The structure's hypersurface was derived by initially atomizing the supporting geometry of a NURB (non-uniform rational B-spline) so that each of the five control points governing the NURB could be animated along the path of a Möbius surface.

above right Wire-frame plan view.
below right and **opposite** Eight rotational studies from the animated sequence. showing transformations of the interior and exterior surfaces.

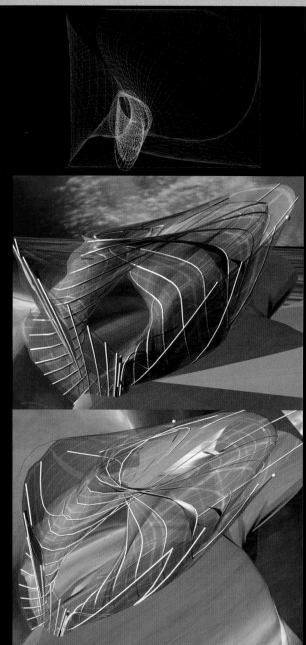

For architect and theorist Stephen Perrella, the entwining of the virtual and material strata of everyday life has produced an irresolvable, mutant culture, a schizophrenic yet fertile condition in which technology, consciousness, instrumentalities (forms and spaces), economy, representations (images) and identities have intermingled to produce a deep, dense swirling topology of real and mediated human affiliations. Perrella describes this condition as the "hypersurface," a problematic complex that emerges from the interaction of these commingling and competing constructs. Hypersurface theory explores this phenomenon to discover modes of cultural operation that might bypass schizophrenic or nihilistic responses that mirror the contemporary world. Neither matter nor media but something in-between, hypersurface architecture is Perrella's con-cept for informational and spatial structures that respond critically to these broader cultural transformations.

According to Perrella, virtual technologies produce new, heterogeneous interactive realms of human experience that bridge the real and the virtual, a relationship conventionally regarded as opposite and disconnected. Instead of simply adding another "dimension" to our three-dimensional world, Perrella contends that the virtual has folded itself *into* the world, contaminating our consciousness, physical experiences and colonizing our unconscious imagination. The virtual dimension, like a renegade mutagen, has insinuated itself into the physical systems of organization that define our most basic and traditional conception of space and time. If his description holds true, then the fundamental Cartesian notions of space-time that have long been the bedrock of architectural production can no longer stabilize our understanding of the world, let alone drive new building configurations and urban designs. The transitional boundaries between the real and the virtual can barely contain the proliferating interdependencies that exist between the two states. Emerging out of the middle of this hybrid, hypersurfaces propose a multidimensional spatial diagram for configuring the real and virtual realms that have now superseded our outmoded spatial paradigms.

Instead of the real and the ideal being separate realms, the divisions sustained by transcendental metaphysics, both divisions now impleat,

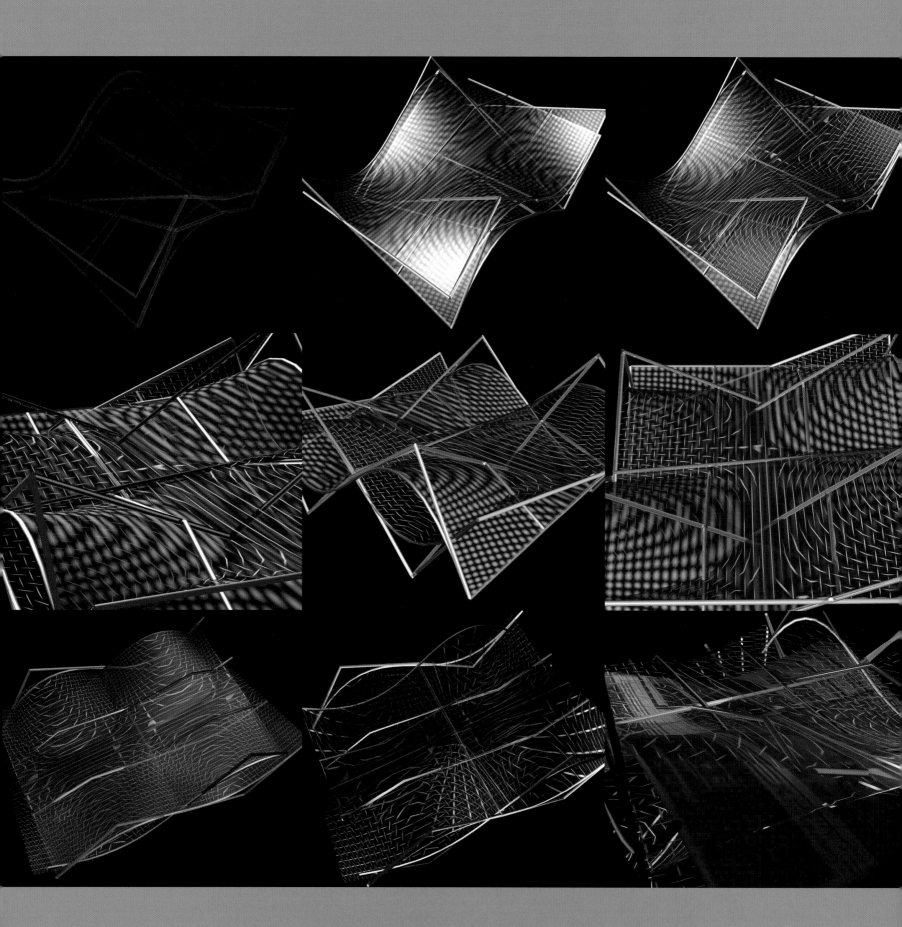

Hypersurface Panel Studies

Perrella developed this series of variant panel studies (1997–98) to present a viable physical model of a hypersurface. As structural diagrams the panels seamlessly enfold flexing lines of ribbed, interlocked steel or aluminium frames with a mesh-like metal fabric or synthetic membrane. The standard division between structure and skin usually upheld in architecture is totally enmeshed in these studies. Layers of fabric seem to implode, engendering fluid interpermutations with the intricate frames; panels demonstrate continuous and consistent inflections between structure and surface in which both intertwined systems interfold in unison to create transversal interrelations from specific distortions.

becoming interfused ... In our existing technologically saturated contexts there are horizons through which our lives are drawn ... The process and logic of pervasion stemming from tele-technology intermixes television with the Internet, the Internet impacts upon built infrastructure, and so forth, creating a convergent, enfolded, organization [of layered physical and electronic strata] ... These interpenetrating layers, fueled by consumer capitalism, will reconfigure the topology of human agency. Emergent forms of representation will unfold due to radical interweavings.

Engaging this multilayered condition by manifesting its complexity and by building fluid links between the layers, hypersurfaces address the disorder of the Information Age – the crisscrossing of communication lines, media types and technologies – and provide osmotic membranes for electronic transgressions that capture and draw energy from the proliferating multifariousness of our digital world.

Perrella's earliest hypersurface study, the Institute for Electronic Clothing experimented with texture-mapping and coordinate-manipulation using three-dimensional modeling softwares. First, a computer-generated wire-frame model was deconstructed. The model's coordinates were then stretched, pulled, tweaked, bent, enfolded and warped, resulting in a form that might be better described as a fabric than an object, over which or through which texture-mapping can be used. A seminal construct emerges: a radical relationship between form and image that rejects, rather than polarizes, the usual dichotomies of structure-ornament or substance-signification. Form is neither augmented nor enriched by image; rather, image and form are seamlessly integrated, flowing into each other as a unified topology – a hypersurface.

Recent projects explore hypersurfaces through software-based animation techniques. In contrast to some practices, Perrella considers animation software not as a means to inscribe time into static constructions but as a way to animate the temporal relationship between the architectural program (the event) and the user. Perrella is critical of architects who use the built-in temporality features of animation software but ultimately produce static sculptural forms that do not relate to the architec-

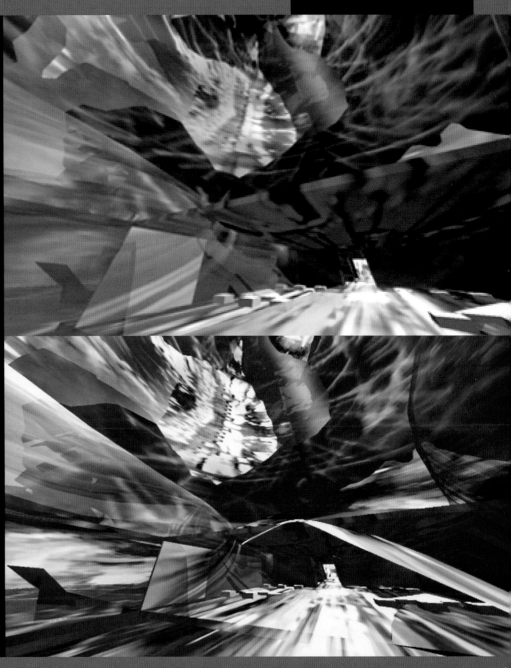

tural program. "Those form-makers," Perrella suggests, "are left-overs from the heroic-modern period where the promise of technology carried the hubris of ideality. Currently, however, technology is undermining ideality and the practices that chase it. As a result, reality is also debased (because reality and ideality are linked dialectically in Western metaphysics) so that middle-zone, or what I call the hypersurface, is where events are now unfolding; and for architecture that involves the event of the program, but not a sense for program that is merely immaterial."

The Möbius house, for example, designed in collaboration with Rebecca Carpenter, explores alternative uses of animation software and reconsiders domestic space with communications media. The concept of dwelling, Perrella reasons, has become problematic. In terms of Euclidean space, digital media have imploded distances and enfolded the inhabitant's perceptions into an endless barrage of electronic images, shifting the status of the home from an exclusively interior condition to one in which the public (the world) and the private (the home) are intermingled. The Möbius house is a study-diagram for post-Cartesian dwelling that accepts this hybrid condition as a starting point and proposes a domestic construction that neither encloses an interior space nor presents an exterior form. As a hypersurface, the house functions as a transversal membrane, reconfiguring the binary notions of interior and exterior into a continuously deformed intermediary surface.

The present phase of the study performs like a fluxing or phase-shifting diagram-membrane. Generated by animated inflections, the hypersurface contains temporal delays that were programmed into the form to avoid determinate, linear results, or "the stopping problem." This is the dilemma, defined by theorist-economist Akira Asada, that occurs when temporality is brought into architectural form through animation software but that must at some point in the design process be frozen and built as congealed. Perrella argues that hypersurfaces designed with animation software are not forms that are evolved over time and then "stopped" but media images intertwined with deconstructed forms, representing the collapse of two conventionally distinct realms, the domain of the media image and the territory of architecture. "Generally a hypersur-

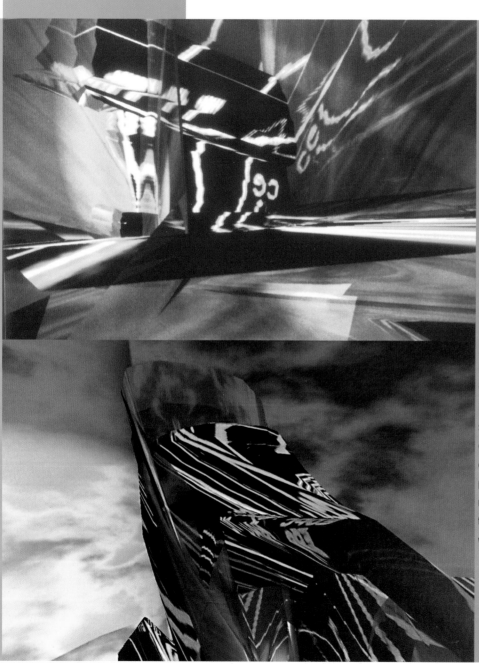

face has a range of effects, including and most significantly a surreality or hyperreality; a realism that is simultaneously uncanny, incomprehensible, and therefore a catalyst or a provocation ... A hypersurface is the informed topology of an interstitial terrain between the real and the unreal (or any other binary opposition) that then flows transversally into a stream of associations." Thus, the hypersurface, once animated, is irreducible and open to complex, temporal experiences.

Hypersurface architecture reconstructs spatial practice in terms that will allow such oppositions as structure-ornament, image-substance, inside-outside and ground-edifice to be released by a transformative dynamic. Most architectural engagements with virtual media until now have exploited and architecturalized the screen, the point of interface between the real and the virtual. This usually results in the back projection or front projection of fixed imagery onto architecture that retains the classic dualisms of subject-object and media surface–physical object; the superficial application of media images onto architecture is too often regarded as a sign of the future.

While the forms may be radical, image-to-form projection techniques were taken up long ago by the forces of global capitalism. Inventive form and image-making are part of the same process, an acceleration of innovation driven by the market's insatiable desire for the new. Radical research, Perrella therefore proposes, involves tapping into the vast realm of possibilities that technology offers. This implies understanding the computer's very nature and its extraordinary capacity for iterative production. Creation is no longer a matter of genius, but of the "genius" of media substances that emanate from the computer's strange logic into the world. Perrella's hypersurfaces attempt to reflect a social condition in which we have become an integral part of the media we have created.

The Institute for Electronic Clothing

The final figure, an involute form shrink-wrapped by an image, creates a smooth, reciprocal relation between a complex visual representation and a form that is further variegated by having an image run over its surface. The project proposes a radical relationship between form and image that transgresses the usual dichotomies of structure-ornament or substance-signification; form is neither augmented nor enriched by image. Rather, image and form are conjoined so that they flow together as a unified topology.

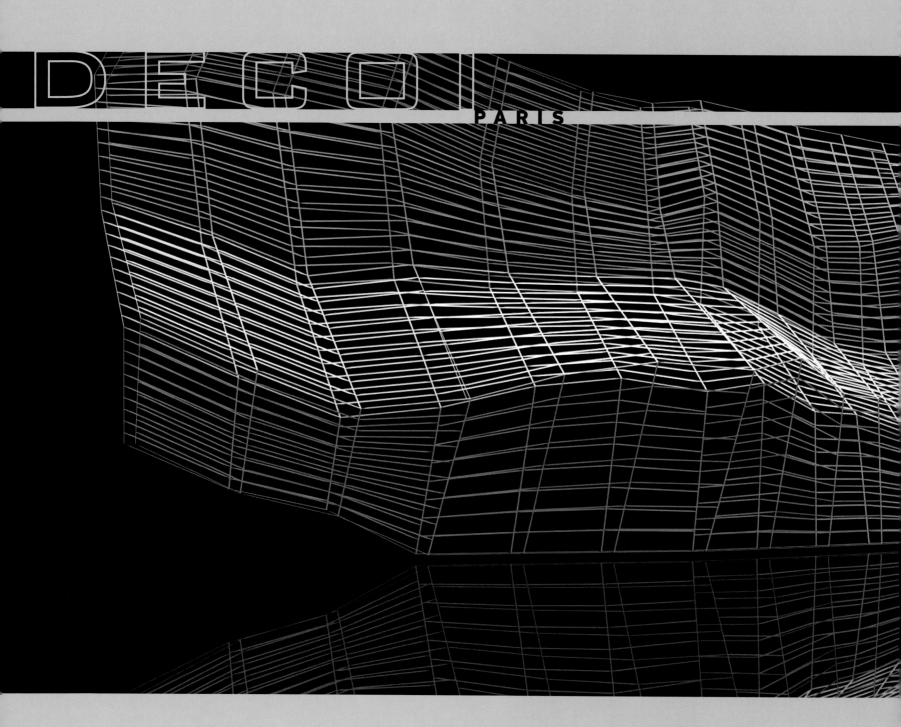

DECOI PARIS

Embracing collaboration, technical innovation and experimental manufacturing processes, dECOi creates architecture that is light, mobile and elegant.

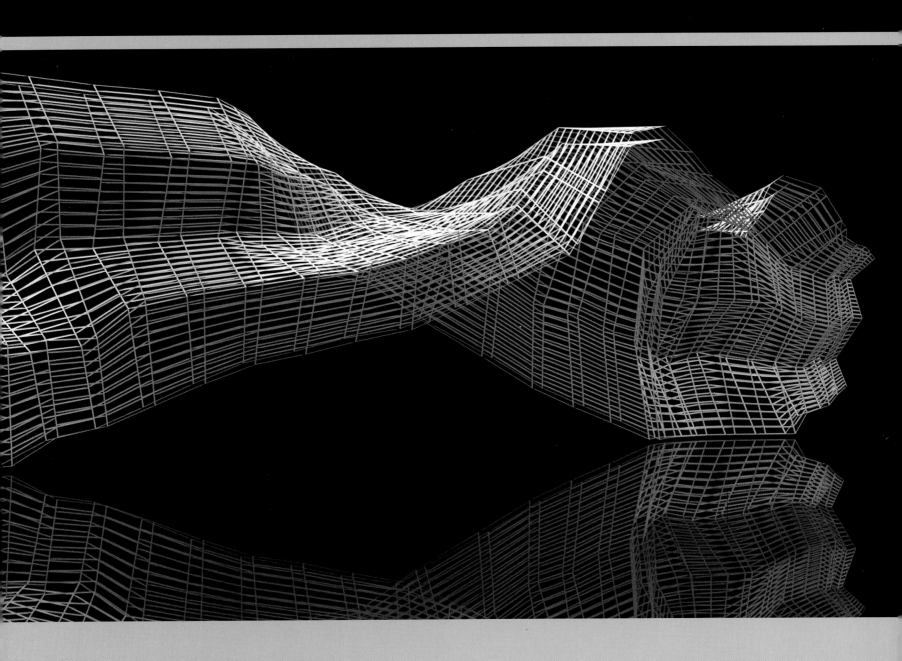

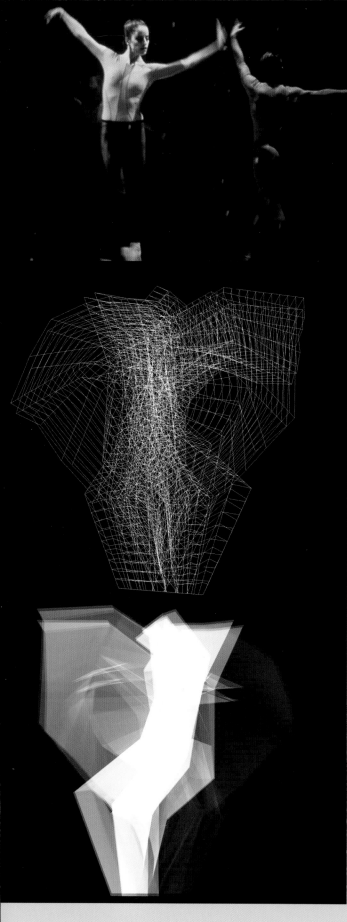

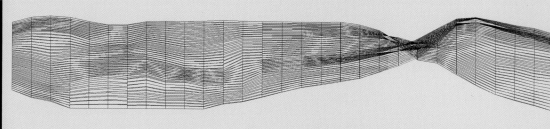

A light, mobile and nomadic architecture atelier, dECOi was established in Paris in 1991 as a research-based design collaborative. Initially focused on competitions, theory, experimental installations and teaching, dECOi has evolved into a "lubric" practice that now resides somewhere between Paris, Kuala Lumpur and London. Led by Mark Goulthorpe, the collective has not been merely influenced by the electronic revolution but rather radically formed by the conceptual, poetic and technical possibilities enabled by the speed, scalelessness and fluidity of the digital-technology paradigm shift.

dECOi's key operative aim has been to decipher how our desire for technical advancement – our drive to computerize practically every aspect of our lives – might be made manifest in architecture. The question of how to incorporate technology in our built forms is in counterpoint to the notion that efficiency or expression should drive technical researches. "Desire seems to be highly implicated in the question of such a change," Goulthorpe suggests, "emphasizing the extent to which technology as an 'extension of man' is never simply an external prosthesis but actively infiltrates the human organism, certainly in a cognitive sense ... It is interesting to speculate not just on what technology may 'put' (or better perhaps 'leave') in our objects, but on the extent to which its general impact on patterns of cognition may intersect with such 'process-objects.'" The concealed potential for architecture is therefore most effective when it is liberated through an examination of social capacities and ways of thinking that are induced by the latest technological advances.

It is precisely the accelerated change of technical states, logics and idioms – from the mechanical to the post-mechanical, linear to non-linear and solid to fluid – that informs and energizes dECOi's creative project. Goulthorpe has written,

We seem to find ourselves suspended in a sort of **smectic state** – that indeterminate liquid-crystal matrix at the point of imminent structuration or destructuration at which crystals begin to emerge from a generally depthless medium. We catch the sense of a shift in formal logics, an inscrutability or flotation within the generative matrix of the machine. But equally we sense the indeterminacy of current cultural operations

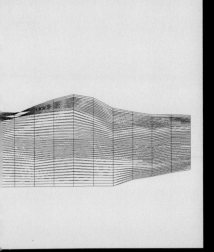

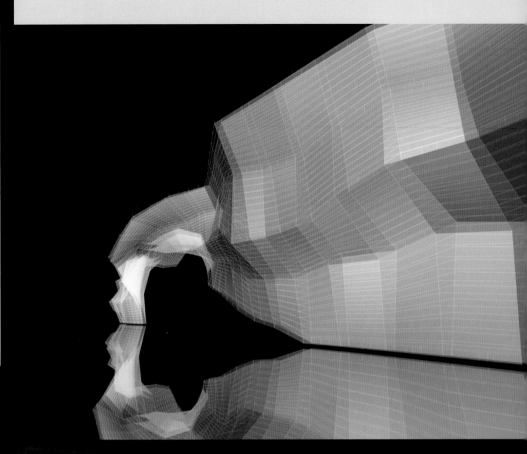

Ether/I

To commemorate the fiftieth anniversary of the United Nations in Geneva, dECOi was asked to create a work as part of a series of sculpted projects in which "image/surface/object [were] ambiguous in their status." Based on the negative trace of two dancers in space, captured on video - the trace that cannot be seen by the naked eye - Ether/I embodies a transitional phase from one state to another, "the trace of an absent presence," a surface that is also depth. The sculpture is a tribute to choreographer William Forsythe, who has broken down balletic movement into graphs. Developed like a rambling screen of aluminium reflections, Ether/I unfolds as if it were materializing the ghostlike traces of motion.

opposite, top Dancers in space
opposite, middle Video-captured trace of dancers depicted as a wire frame.
opposite, bottom Captured body traces as rendered form.
above left and **left** Plan and elevation
right and **below** Rendered studies

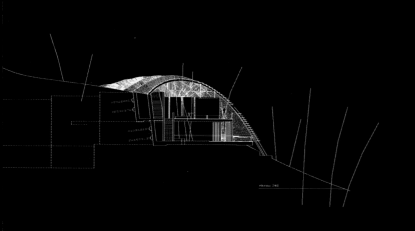

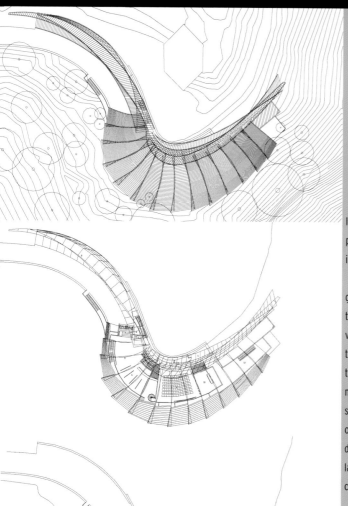

which fluctuate about their limit-case ... In both senses this smectic state seems to have been propagated by technical change: a physical dissolution corresponding to the evanescence of the screen; and a mental surfacing, an acceleration triggered by its proliferating imagery – an almost hallucinogenic tracery ... So the smectic, caught between the solid and the fluid, between visual and chemical logics, seems a suitably ambiguous or fluctuating term to capture our present change of state as we pass from a scriptual to electronic paradigm. And our sense is that such a suspension will carry to the manner in which we think of practice, to all manner of creative process, to a reversal of the process of our thought and to a drift or relative fluidity of consciousness.

It is a determined fascination with the unmeasured energy and unleashed potential of technical transubstantiation that drives dECOi's philosophical investigations, complex undertakings and even the atelier's functioning.

From its inception dECOi organized itself as a fundamentally loose global practice to capitalize on the collaborative possibilities offered by the speed of data transfer. Fixed neither by number nor location, dECOi validates the almost utopian multidisciplinary model of a flexible horizontal atelier, often working in close collaboration with technical experts and theorists in Europe and Asia via the Internet to exchange drawings, virtual models and instructions for computer-aided machining systems. To sustain this state of practice, dECOi has actively reinterpreted the very act of design – and in many instances the authorial role – of the architect. In dECOi's choreographed spatial productions the linear logic aligned with late-modern architecture has been abandoned in favor of lighter modes of collaborative thought, creativity and production.

In dECOi we thus discover a new model of architectural practice from the fracturing of the singular creator. The collective offers us the figure of the designer as the sampler or editor of state changes and shifts, of a greatly expanded range of formal, intellectual and technical possibilities. Goulthorpe describes this practice as a kind of "eclipse of IDEA, and of the conditional imperative it implies: an evaporation of the determinate, controlling architectural body, and a willingness to drift a bit." Encouraged by experimental speculation offered by computer generation, dECOi

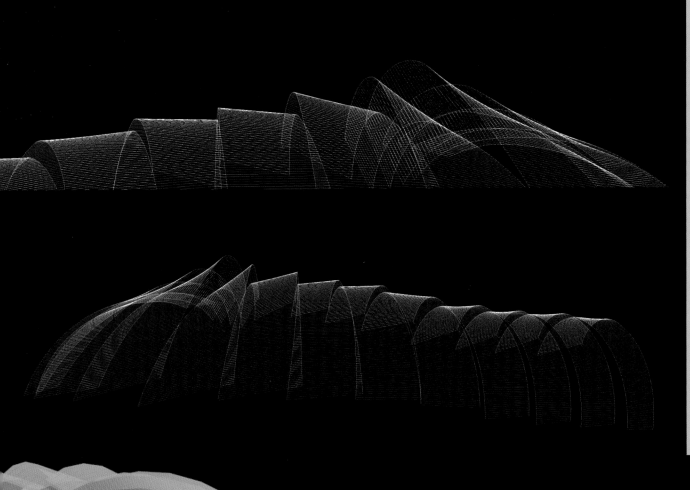

ECO Taal Environmental Center

The key building in a complex situated inside the crater of the Taal volcano in the Philippines, the project (1997) responds to the challenge of creating a form that would disappear into the site but remain clear in the mind, "a kind of psychological involution." Set on a steep, wooded slope, the structure appears like a carapace – a swelling of the surface that gives the slope back its form – which is interrupted by the program units (lecture and exhibition halls). The center's skin is composed of disjointed timber leaves, or slats, that are distorted, shifted and bent to the site's contour. The interior is streaked with rays of changing light, which introduces an unexpected and changeable sensation to curved space.

opposite, top Section showing the relationship of carapace to hillside.
opposite, middle to bottom Plans
this page Wire-frame and rendered studies reveal the sequencing and articulation of the carapace.

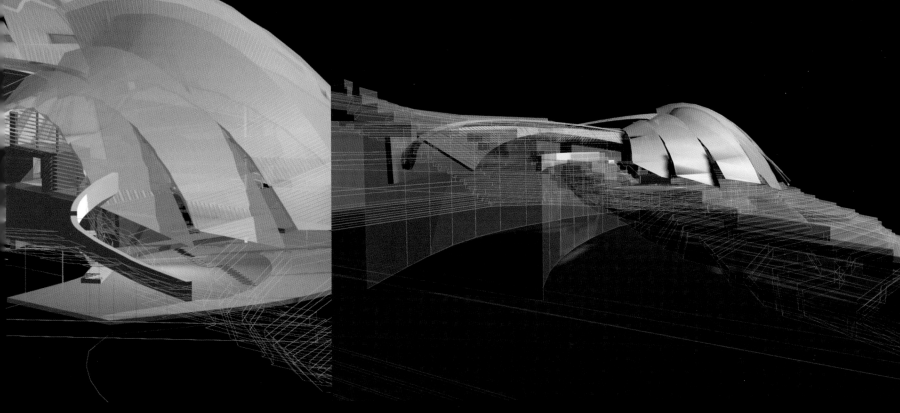

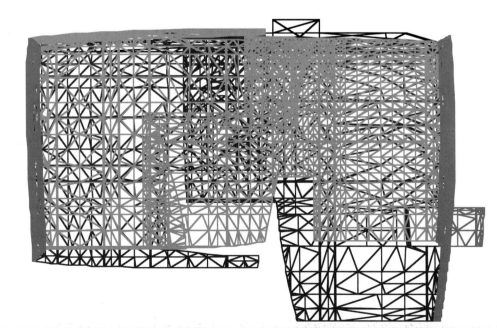

Pallas House

A collaboration with Objectile, the residence (1997) in Malaysia is articulated by two formal gestures: an excavation that carves out a "heavy negative" from the steep terrain and forms the entrance court, and the central form of the living spaces, poised on the site's bluff as a series of boxes wrapped in a curving and perforated shroud, a "light positive" cut from the air. Following the tropical precedent of layering filters against the sun and rain, the architects developed a "breathing skin" that liberates the house's formal expression and induces the house's monolithic form to appear suspended at the moment of transition from solid to fluid.

above Line study showing the composition of the exterior skin.
above left View from the interior through the perforated, "breathing" skin.
middle left X-ray perspective of the relationship between the house's box-like internal volumes and the perforated skin.
below left Study of the skin calculated as triangular polygon meshes.
above right Three exterior views showing how different skins – wooden slats, brise-soleils, the Hystera Protera glyphs [see page 69] – might be used.
right Rendered night study

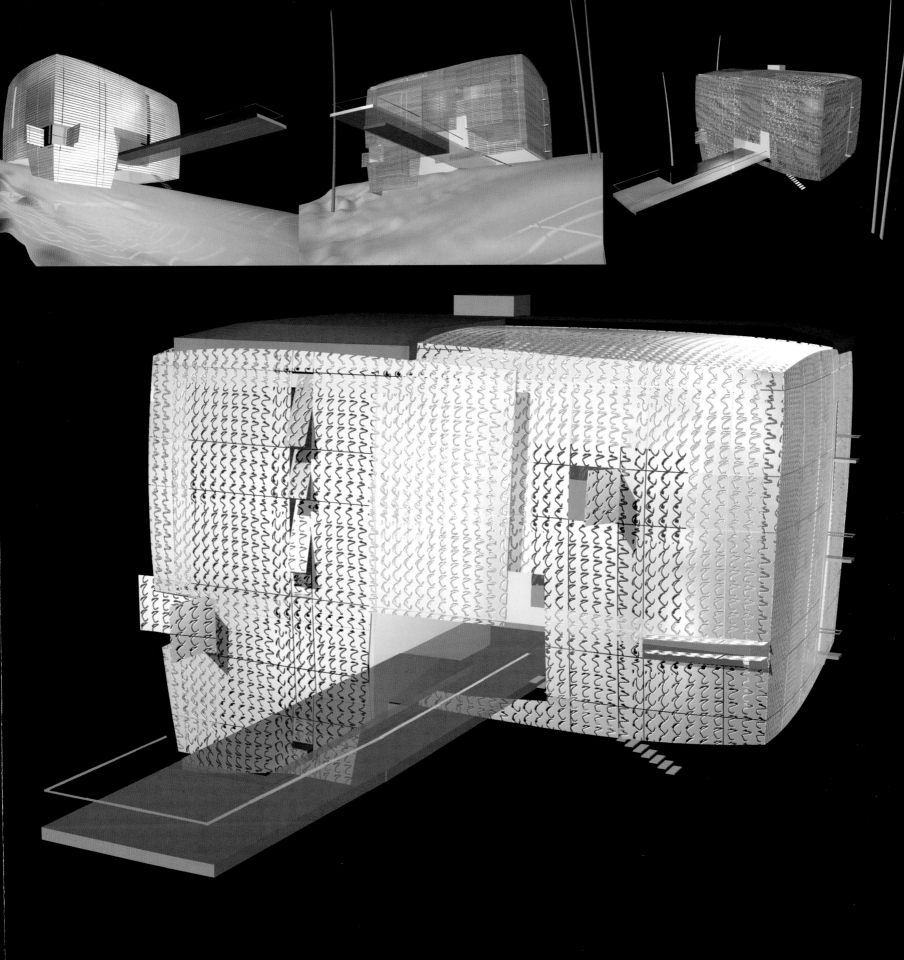

**Formal Research for a
Theater Complex**

Undertaken for Foster & Partners in
response to their design of a complex-
curved carapace that envelopes theaters of
three different sizes in Gateshead, England,
dECOi's study suggested a kind of "clam" in
which the volumetric swellings are
suggestive of the theaters below but
which overarches them to dramatically
enclose a larger public space.

**Formal Research for a
Theater Complex**

right Three aspects are noteworthy: First,
the form is complex in that its curvature is
constantly changing, suggesting the need
for nonstandard components. Second, with
the monolithic roof curving down to become
a wall – which required a view to the
adjacent river – there is a fusion of
elemental properties. Third, there is an
implicit merging of surface and structure,
the curvature of the carapace offering a
structural potential as a monoque shell.

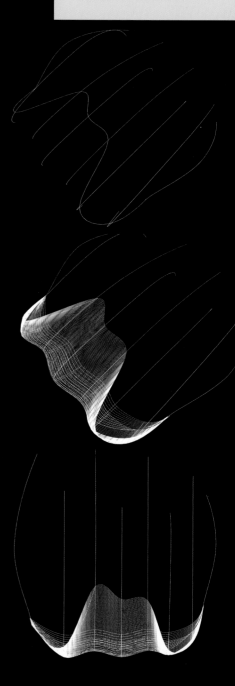

often works free from determinate ideas and with curiosity rather than dogma. This openmindedness allows the studio to discover new genres of form as they are coaxed from the seemingly infinite mind of the digital machine.

dECOi has embraced the process of working "blindly" with the computer, of allowing its algorithmic, animating dynamic to create unpredictable sequences and material excesses. The Hystera Protera project, for example, speculates on a generative approach to spatial patterning. The project was initiated by trapping lines derived from the rotation of animated amorphous forms. From these mappings, sequential traces called "glyphics" were generated. Operating in the interstices of the transition from industrial to post-industrial, the glyphics' beautiful traceries suggest that the computer's powers can create bizarre scriptual sequences that could be used as details, structures or decoration and capture the fluid and beguiling qualities of seriality, dynamic evolution and generative form.

The studio pursues "a whole variety of experimental processes," Goulthorpe says, "each being motivated by an internalized dynamic which develops over the course of a project, but which has rarely relied simply on the embedded capacity of a software system ... however, one concern that recurs throughout [dECOi's work] ... is a preoccupation with the transition to construction, which ... must develop from the investigative process itself, and be carried forward into production." Pallas House, a commission for a contemporary tropical house in Malaysia, was designed to test the formal limits of product design and computer-controlled manufacturing, construction and form-generation technology. The building's dominating feature is a "light positive" composition that constructs the house up from an arrangement of delicate boxes wrapped by a perforated steel shroud. The enveloping skin is composed of seven complex-curved metallic shells, each distinct but mathematically coherent, onto which a constantly shifting, numerically generated perforation is mapped (like the Hystera Protera). Developed with Objectile, a Paris-based design think-tank led by Bernard Cache, the perforations create optical movements, like a "shoal of fish" flickering on the house's surface. The lightweight

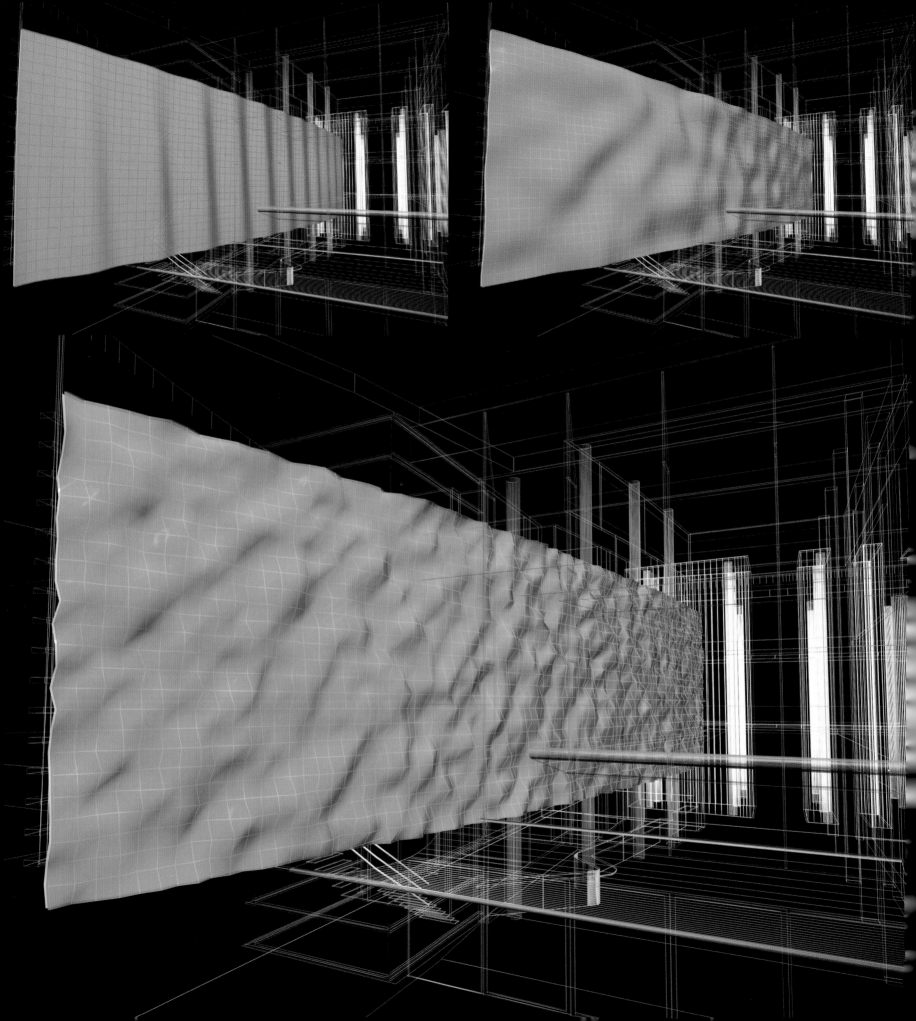

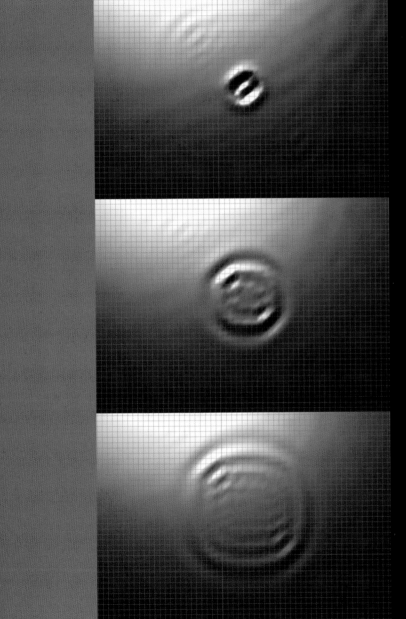

Aegis

The Birmingham Hippodrome theater requested an interactive art piece that would portray on the exterior what was happening on the interior. To explore the nature of translation, notation and the act of writing in contemporary culture, dECOi proposed a responsive surface, at an urban and architectural scale, that can respond to electronic or environmental stimuli. The surface is simple – metallic and faceted – penetrating from exterior to interior as a gently curving plane. At rest, the surface is a shimmering backdrop to events, but as a surface of potential, it can capture stimuli from the theater environment and dissolve them into movements, supple fluidity or complex patterning. The surface therefore embodies an act of translation, a sort of synaesthetic transfer device that crosswires the senses. Although the project involves an architect, parametric modeler, mathematician, programmer and a missile-deployment systems engineer (along with engineers, a piston manufacturer and a cladding company), the surface is ultimately not designed: it is generated by a random sampling and electronic sensory-input in which the designer's role becomes that of editor or sampler.

opposite The gently curving surface emerges prowlike from the theater's foyer to cantilever over the street. Composed of 94,000 metallic facets with a polarizing coating and powered by up to 10,000 pneumatic pistons, the surface is linked to the building's base electrical services. The sequence shows how one activated pattern might appear on the surface.
this page Two sets of surface details show the richness of possibility offered by the mathematical operations on which the surface's appearance are based.

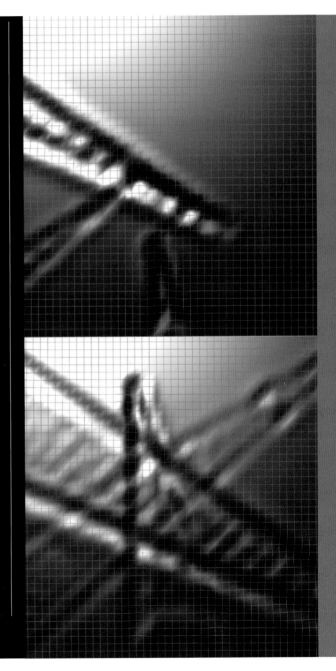

Missoni Boutique

This showroom (1996) in Paris's Rue du Faubourg was developed for the Italian knitwear company. Restricted space suggested opening up the basement and ramping the entire floor as a slowly descending spiral whose sweeping gesture is followed in the boutique's other surfaces – walls become floors become ceilings. The effect is an architectural equivalent of the company's hallucinogenic textiles.

below Rendered plan, isometric and semitransparent elevational studies.

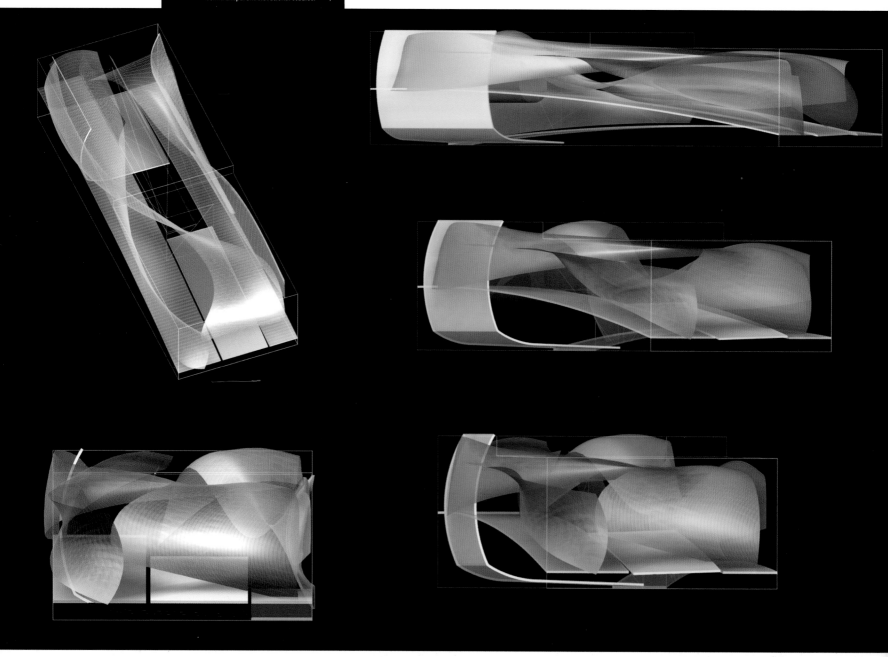

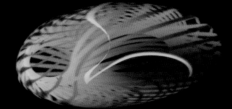
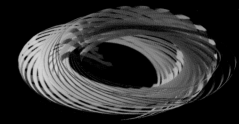

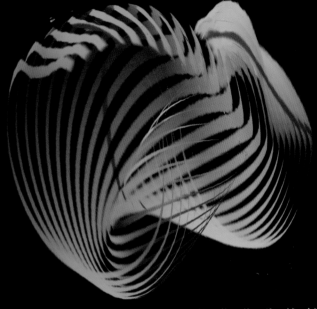

Hystera Protera Electroglyphs

Initiated by trapping lines derived from the rotation of animated amorphous forms, Hystera Protera – a term for which the masculine form means an inversion of natural or logical order – considers the possibilities of a generative approach to spatial patterning. Sequential traces called "glyphics" (as distinct from "graphics," which are determinate) were derived from the mappings. Floating enigmatically between abstract minimalism and decorative excess, the traceries suggest that the computer's powers can create bizarre scriptual sequences that can be used as details, structures or decoration.

above and **right** No longer simply a multiplying series of decorative devices, the glyphics capture the fluid and beguiling qualities of seriality, dynamic evolution and generative form. They trace an absent presence.
below right The decorative effect of the glyphics is fleeting – a light trace that moves beyond the minimal stricture of the Machine Age.

shell also functions as a climatic filter that uses the logic of contemporary cladding systems to overcome the tropical heat.

The Pallas House, dECOi stresses, is not simply a graphic exercise but the result of advanced research and production into the coordination of complex curve-generation software and numeric-command machining from computer models. While the design requires complex one-off components to be precisely produced, the manufacture of such a series of self-similar objects (mathematically coherent, but each different in its form) can be achieved relatively cheaply, thus giving rise to a radical transfer in the basic logic of construction and production.

"What seems highly suggestive," Goulthorpe explains, "is that we may begin to discern not only new generative techniques (everyone with his/her own electr(on)ic Wunderblock) but ... an entirely changed cognitive capacity ("desire"), and hints as to a radical realignment of cultural praxis." In order to assess dECOi's achievements so far it is imperative to appreciate the energetic divergence from traditional models of practice and design. Collectively, the studio has sought to eliminate the design ego in favor of an orchestration of concepts and possibilities. The powerful transformations registered in the work reveal not only a realignment in the manner in which buildings will be conceived and constructed but a reassessment of the way architects will think and construct in order to pursue the elastic possibilities of the emergent techno-cultural paradigm.

dECOi's investigations suggest a subtle and more fluid creative alternative to the linear, function-to-form trajectory of thinking that has doggedly hindered so much architectural reasoning in this century. What is ultimately so beguiling and perhaps curious about the studio's open processes of computational invention and technological improvisation is that the apparent mobility and lightness of its creative thinking and virtual technical application has resulted in architectural works of great physical complexity and intellectual rigor.

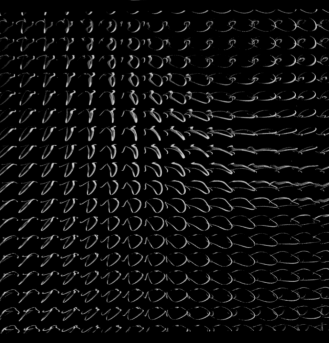

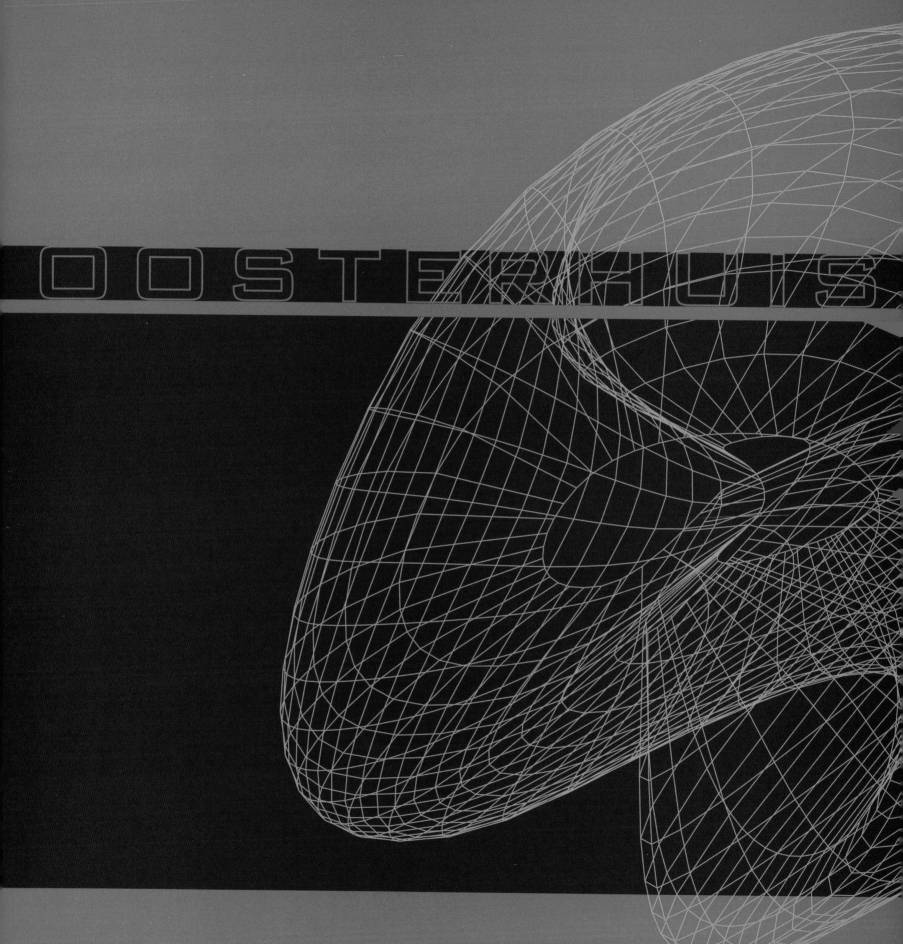

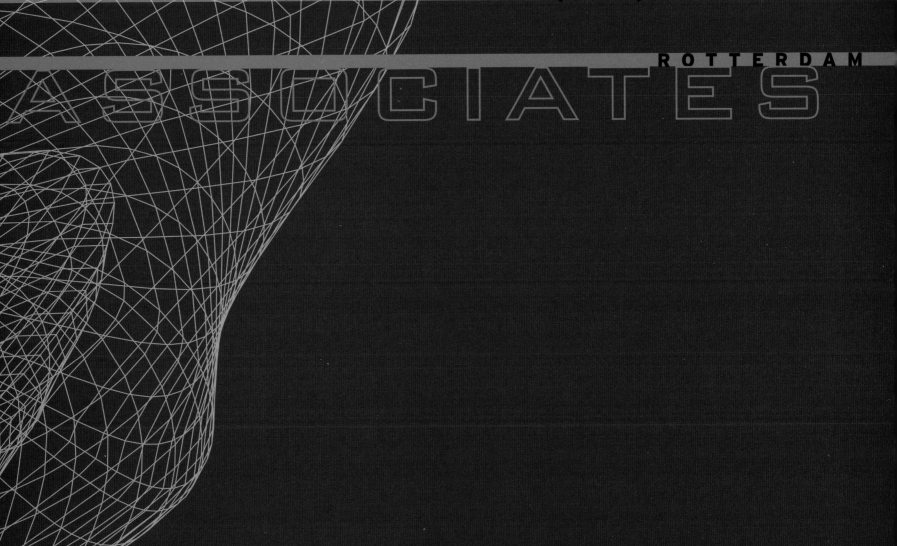

Synthesizing complex geometries, human action
and environmental data, Oosterhuisassociates
creates information-dense "body-buildings."

ROTTERDAM

ASSOCIATES

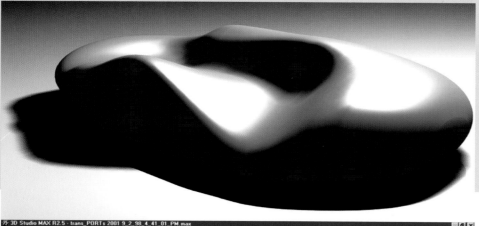

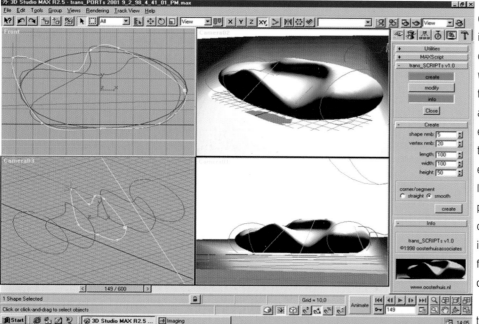

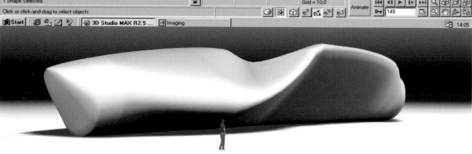

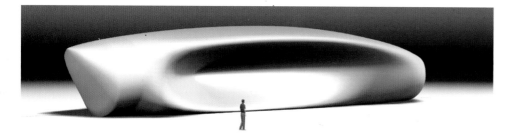

Kas Oosterhuis founded his multidisciplinary practice to look outside architectural circles and discourse for creative stimuli and procedural knowledge. As cofounder of the Attila Foundation, a research institution for the electronic fusion of art and architecture, he frequently collaborates with programmers and creative producers, most notably with the Rotterdam-based visual artist, Ilona Lénárd. Oosterhuis holds that the invention of building forms can no longer follow the paradigms and precepts of a classical discipline wedded to outdated codes and established traditions. Rejecting the dominance of Platonic geometries – the simple volumes of the cube, sphere, cylinder and cone as the basic elements of architecture – he views architecture as an evolving, technologically enhanced means of organizing sophisticated spatial data and programming information into structured mediums that synthesize complex geometries and aspects of human actions. These "body-buildings" embody behavioral rules that are derived from the integration of form and information and become environments that can develop their own intelligence.

Buildings are becoming data structures that we can no longer totally control and that can influence their immediate (and perhaps global) contexts according to unpredictable and unknowable behavior.

Building projects – architecture – is like placing an attractor into the future. All information will head towards that attractor from then on. This particular stream of information is thus energized and vectorized. All transport of materialized information, all immaterial concepts for assembling the product; all whimsical ideas are converging at that attractor at that place and at that particular time ... I suggest for the moment that I comfort myself with the notion that people are intermediate bodies – among many – absorbing data from a redundant stream of life (information) and excreting them again in modified form. When we make a film of, for example, a house and speed it up a thousand times (the *Koyaanisqatsi* effect) the house is acting like a living body. It absorbs all kinds of material, including a liquid stream of humans, pulsating in and out. It absorbs and it excretes them in a rhythmic pulsating manner. Who could tell the difference from biological life, seen at the speed we are living at? Since we are captured in our arbitrary speed of

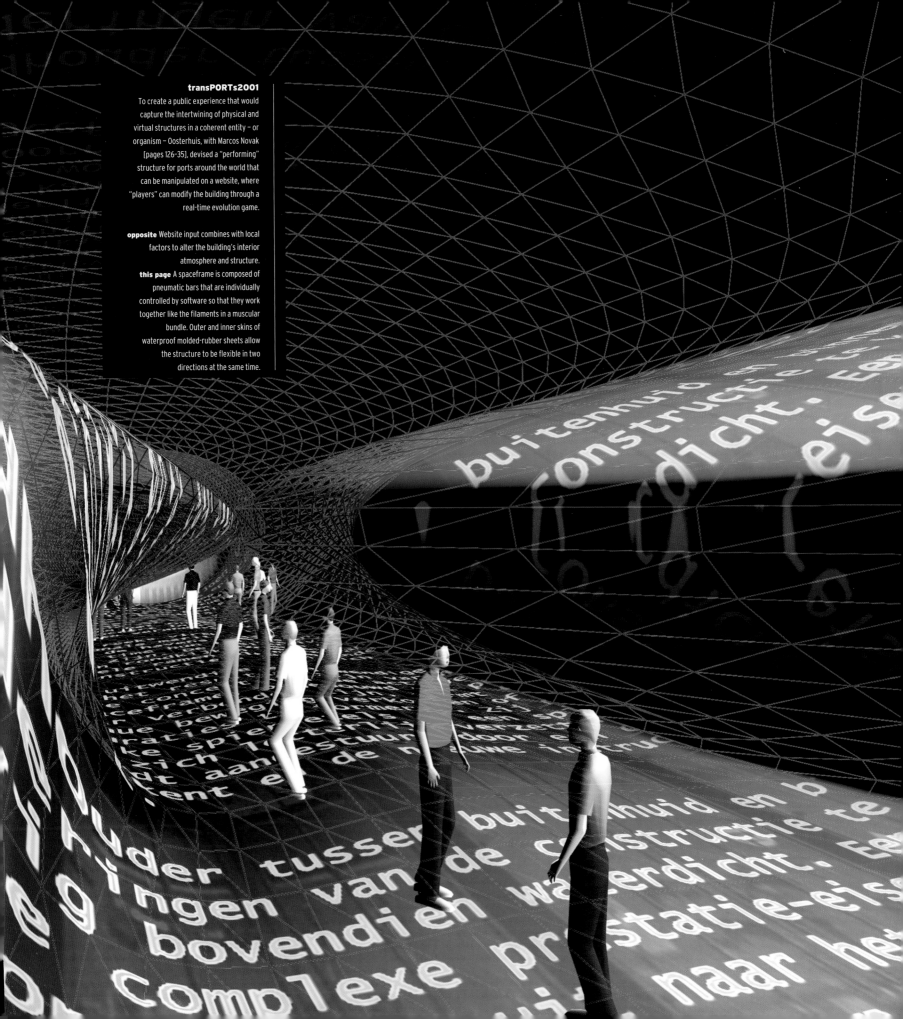

transPORTs2001

To create a public experience that would capture the intertwining of physical and virtual structures in a coherent entity – or organism – Oosterhuis, with Marcos Novak [pages 126–35], devised a "performing" structure for ports around the world that can be manipulated on a website, where "players" can modify the building through a real-time evolution game.

opposite Website input combines with local factors to alter the building's interior atmosphere and structure.

this page A spaceframe is composed of pneumatic bars that are individually controlled by software so that they work together like the filaments in a muscular bundle. Outer and inner skins of waterproof molded-rubber sheets allow the structure to be flexible in two directions at the same time.

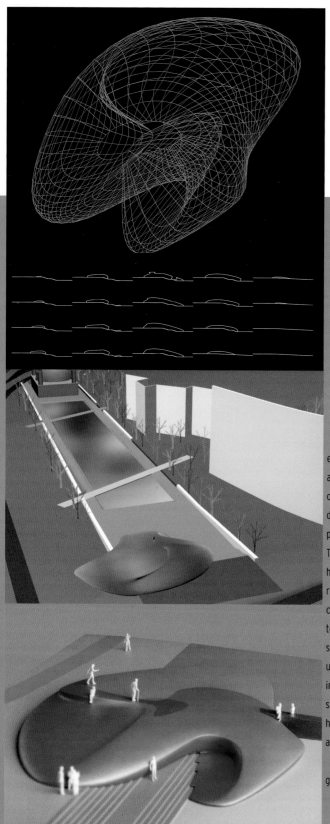

life we are unable to experience the consistency of other life forms which are living at a completely different pace ... this notion of life as a specific configuration of information is important ... in developing ideas about the realtime behavior of building bodies ... feeding the electrified body, fueling it with a wide variety of incoming matter and data. Ecological balance also includes people going in and out, and data being imported and exported, information flowing towards the body-building (feeding), through it (digesting) and away from it (excreting). In the end it does not matter if we call this evolution, proto-evolution, co-evolution or exo-evolution. The most important thing is how we make things work. How we will involve ourselves in building greater complexities of meaning and establishing an increased exchange of information between human body and building body?

Buildings, like other expressions of contemporary design – cars and electrical appliances, for example – contain ever-growing amounts of data and are increasingly semi-autonomous. Oosterhuis suggests that our overconstructed and encroaching artificial environments are, ironically, developing into virtual ecosystems of housing, electrical fields, transportation infrastructures, cars, communication devices and computers. These systems are evolving toward a form of independence in which humans will no longer be central to the ongoing correlation of data, materials and networks. In other words, once humans can no longer directly control the products and networks of communication that have given rise to the Information Age, a certain vitality will become resident in the systems and networks. Even now, the world's total data structure, products and processes are almost impossible to assess or to fully understand in the way the correlates of the mechanical regimes could be comprehensively explained and grasped. While we might have total access, we do not have complete control over even the smallest computers or networks, let alone large complex systems.

Oosterhuis therefore reasons that our complex systems are undergoing evolutions of their own, often without direct human intervention:

The central database is residing within the new species, not within their inventors. If we wait a hundred years or more, the systems will have

◄ **Parascape**
Conceived with visual artist Ilona Lénárd for the city of Rotterdam, Parascape is an active landscape and sculpture. A solid body with a supple skin that absorbs ambient sounds – like street sounds – then converts the samples into a soundscape, Parascape transforms a large lawn into a fluid field of sounds activated by visitors, an instrument the public can play. The genetic material for the growth of the Parascape's shape are two computer sketches that are projected toward each other following and scaled along a twisted path. A three-dimensional model is then placed in the environment and further adapted by skewing, bending and scaling the elastic mass.

top The wire-frame model depicts the final surface as the product of a computer-aided mathematical scaling operation; sections reveal the structure's organic form.
middle and **bottom** On the top is a "sweet spot," which is played by movement across its surface.

Polynuclear Landscape ➤
The large-scale housing-estate project (1998) develops a strategy of interactive growth, intuition and logic to evolve a loose grid of 320 housing plots folded into the earth in Almere East, The Netherlands. A database is used as a design instrument in which users (future home owners) manipulate the site's physical parameters to create the genetic material that will be used to set buildings into hills.

opposite, top The rippling effect on the grid.
opposite, middle Perpsective rendering
opposite, bottom Bird's-eye rendering shows how the structures are embedded in the site's contours.

more immersed intelligence [and] will make more decisions without humans ... The systems will definitely show behavior ... and we will experience a much more lively interaction with these products, machines and systems. A fundamental question here could be: 'Will the products keep evolving without human inventors, without human activation, without feeding upon humans?' ... [probably not] since new fresh data will always (or for a predictably long period of time) be carried by humans. In the end we might rather speak of co-evolution of humans and their product-extensions and of products evolving together than of human-centered exo-evolution.

The Saltwater Pavilion is perhaps the first work in Oosterhuis's oeuvre that convincingly explores the extraordinary prospects of his theories of co-evolving sentient technologies, techno-ecologies and body-buildings. Commissioned by the Dutch Ministry of Transport, Public Works and Water Management and located on the artificial island of Neeltje Jans (butting up to NOX's freshH$_2$O eXPO; see pages 118-19) southwest of Rotterdam, the pavilion is informationally rich architecture rooted in elementary "genetic" spatial rules, or scripts, that hold the key to the subsequent structural and material developments of the body-building. The particular "form-gene" underlying the structure's shape is an octagonalized ellipse that mutates into a quadrilateral along a three-dimensional curved path. The form was kneaded, stretched, bent, rescaled, morphed, styled and polished using computers that facilitated the mathematical description of its complex geometries. Oosterhuis created smooth interface and scripting tools that act like sliders in the seamless setting of parameters and that allow him to bypass typing in numbers and formulas, a process he calls "parametric modeling." Behind every line and surface in the pavilion's virtual model is an algorithm that sets the parameters and interactive dimensions for its geometry. The volume's curved path, for example, was pumped up then deflated to form the sharp nose that juts out 12 meters over the water. Accordingly, he refers to the pavilion as a "shaped container," a volume chased along a spline.

To optimize the building's construction Oosterhuis developed a three-dimensional database linked to a three-dimensional computer

Saltwater Pavilion
The building, on an island off Holland, is
informationally rich architecture developed
from "genetic" spatial rules.

below The data for the interior color
environment are generated from a bitmap
image whose values are processed
by a lighting computer.
middle Each spline represents one
fiber and its color shift (vertical
axis) in time (horizontal axis).
bottom The three-dimensional
landscape of all color shifts in time.

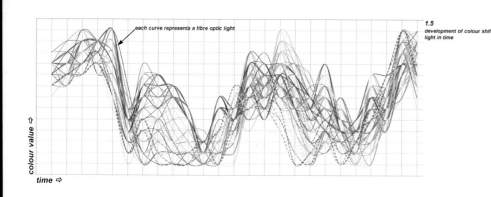

1.5
*development of colour shift per
light in time*

each curve represents a fibre optic light

colour value ⇩

time ⇨

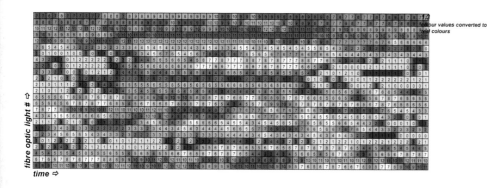

*colour values converted to their
real colours*

fibre optic light # ⇩

time ⇨

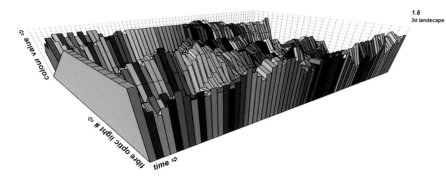

1.6
3d landscape of colour values

colour value ⇩

fibre optic light # ⇨

time ⇨

model. Using this database, he generated the relevant construction data for each building component – almost all of which are unique. The builder simply received a CD-ROM containing only a few principal details and tables of data related to the project's parametric values and scales. At the same time, the data were directly transferred to CNC-manufacturing machines that cut, warped and sized the steel structural elements. In the Saltwater Pavilion, as in Oosterhuis's most recent body-buildings, the building's basic genetic information (the particular physical and tectonic descriptions that explain its algorithmically shaped evolution) is written into the digital codes that are the germ of its final life-form.

"[The] idea of having to know it all," writes Oosterhuis, "is the legacy of the past, when humanity mistakenly had the feeling of being in control of its own inventions ... Today one should learn to swim in the deep data oceans and regard information as a vast and rich sea engulfing us ... a sea that is growing and is flooding the earth ... if we do not drown, we will learn to float freely in this excessive sea." Considering such a statement, we might ask if the production of conventional buildings, long regarded as icons of stability and discipline, can contribute critically to the enhancement of life, biological or otherwise, within the ever-expanding sea of global information. If the fruits of Oosterhuis's complex design orchestrations are any guide, it would seem that architecture would not be a mere buoy in a digital ocean but a vessel that surfs over data, enriching informational and biological life.

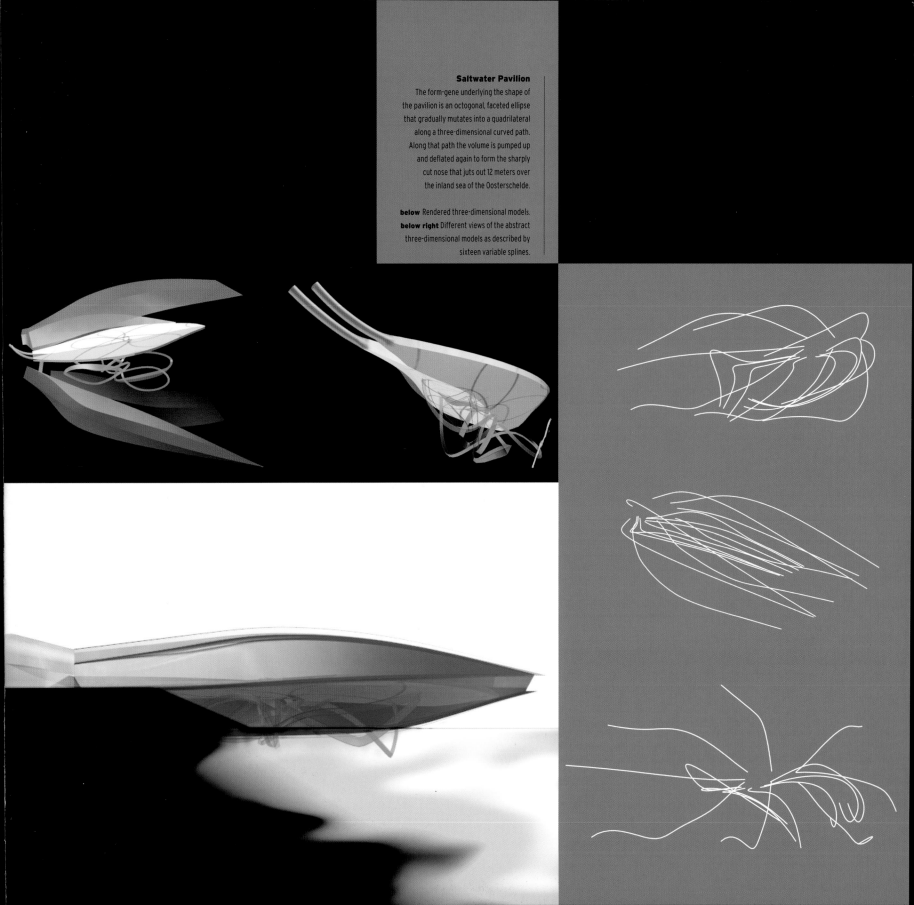

Saltwater Pavilion
The form-gene underlying the shape of
the pavilion is an octogonal, faceted ellipse
that gradually mutates into a quadrilateral
along a three-dimensional curved path.
Along that path the volume is pumped up
and deflated again to form the sharply
cut nose that juts out 12 meters over
the inland sea of the Oosterschelde.

below Rendered three-dimensional models.
below right Different views of the abstract
three-dimensional models as described by
sixteen variable splines.

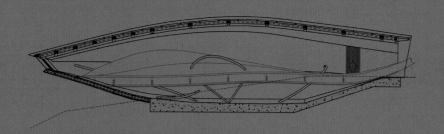

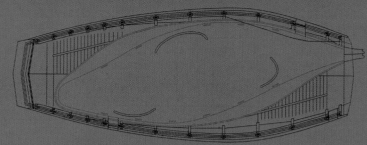

Saltwater Pavilion
above Two-dimensional horizontal and vertical sections cut out of a three-dimensional model and made for conventional communication with contractors.
left The hydra's trajectory in perspective reveals a secondary spline curling freely around the central steel spline. Flexible sheets of green latex are stretched between the two splines.
below left Exterior view from the north.
opposite, top The trajectory of the hydra, designed by Lénárd, is used partly as an information carrier and partly as a load-bearing structure (metal tubes stretched with latex).
below and **opposite, middle and below** Rendered views of the interior's interactive environment.

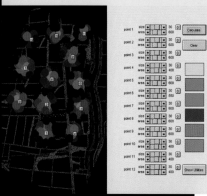
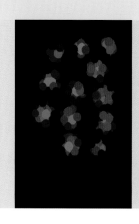
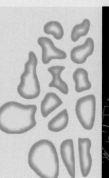

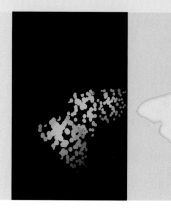

Attractor Game

The "park city" of Reitdiep in Groningen, Holland, is torn between being a city and an international ecological center for migrating birds. The plan (1996) proposed by Oosterhuis is conceived as a giant sponge that absorbs and discharges liquids, matter, people, impulses and information (data). The existing site forms the breeding ground for the park city that will be synonymous with the landscape. To give shape to the multiple processes that govern the city, a new design tool was developed: the Attractor Game, in which attractors and distractors are placed in the landscape. The interactive "game" allows non-experts, such as the future inhabitants, to play a part in a design process, which becomes transparent, flexible and fluid – like ecology itself.

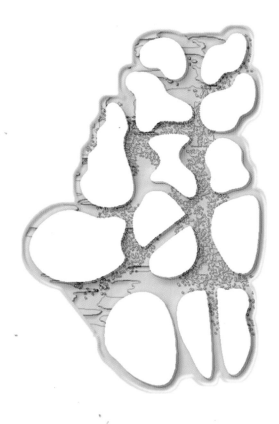
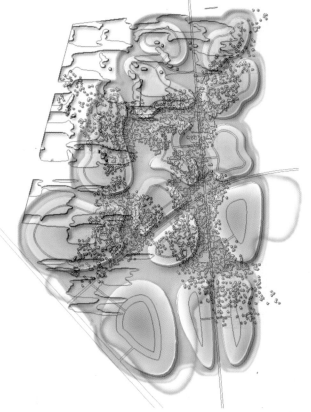

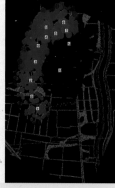
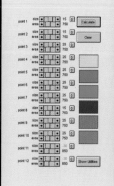

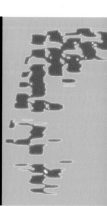

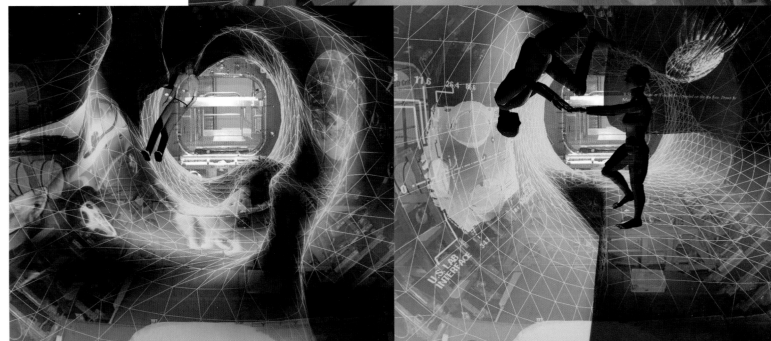

Active Innerskin

This project (1999) develops an active interior skin that could be used in a space-station module as a work and living environment. A high-resolution spaceframe in which each member is a pneumatic bar, the skin is a flexible membrane and a data-driven structure that works like a bundle of muscles. Bars are adjusted by a computer program that sends data to the pneumatic cylinders and can reset them within seconds, while the skin is covered in LED and LCD panels that can form images or texts. Astronauts, for example, could build text-environments or an image-cave according to their needs; a group of researchers could work together on a three-dimensional model or analyze data from experiments; or images of the exterior environments – nature or outer space – could be merged into the interior skin to create a real-time ambience.

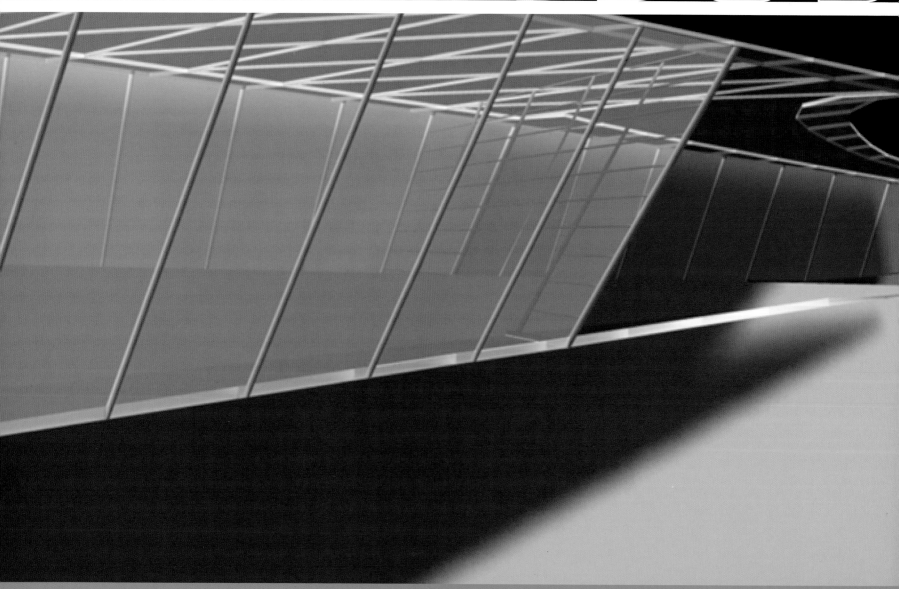

WINKA DUB

Winka Dubbeldam deploys the formal and organizational possibilities from the field of topology to create buildings that energetically combine competing spatial conditions with programmatic needs.

BELDAM
ARCHI·TECTONICS
NEW YORK

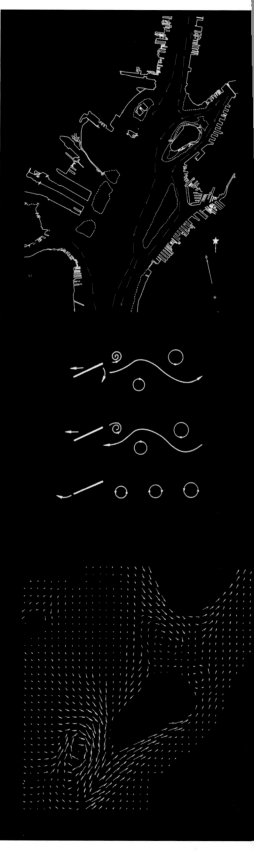

Trained in The Netherlands and based in New York, Winka Dubbeldam of Archi-tectonics argues that the architect should no longer be limited to the staid traditions of craftsmanship, proportionality and aesthetics in order to remain relevant in a developing culture driven by accelerating systems of knowledge and production. Confronted with a world made increasingly complex by the social, economic and technical materiali-zations of communications and media technologies, Dubbeldam contends that architects must study philosophy, mathematics and physics to understand the latest cultural transmutations, shifts that are moving us from mechanistic reasoning toward the process of oriented, organis-mic intelligence.

The most evident paradigms for Dubbeldam's architecture can be located in topology, the branch of mathematics concerned with the properties of geometric figures that are invariant under continuous trans-formations. She is fascinated by the formal, organizational and temporal possibilities that an investigation of topology has for architecture, and believes that it offers an exacting yet dynamic means of unifying dis-parate spaces, programs and conditions.

Topology is roughly divided into three overlapping branches of con-centration: point-set topology studies figures as sets of points that have such properties as being closed, open, compact or connected; algebraic topology makes extensive use of algebraic methods, particularly group theory; and combinatorial topology considers figures as combinations (complexes) of simple figures (simplexes) joined together in a regular, but endlessly variable manner. In a combinatorial transformation, for exam-ple, two figures are topologically commensurate if one can be deformed into the other through stretching, bending or twisting, but not through the actions of cutting, tearing or creasing. This distinction suggests that a topologically derived architecture differs from other recent tectonic and compositional methods of design that treat the surface in a crystalline or "fractalized" fashion, in particular "folding" and its related techniques. Rather than merely folding space, Dubbeldam exploits the agglomerative and "endless" possibilities – loopings, recurrent returns and circuits – in combinatorial topology.

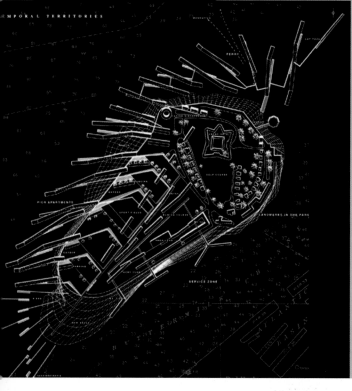

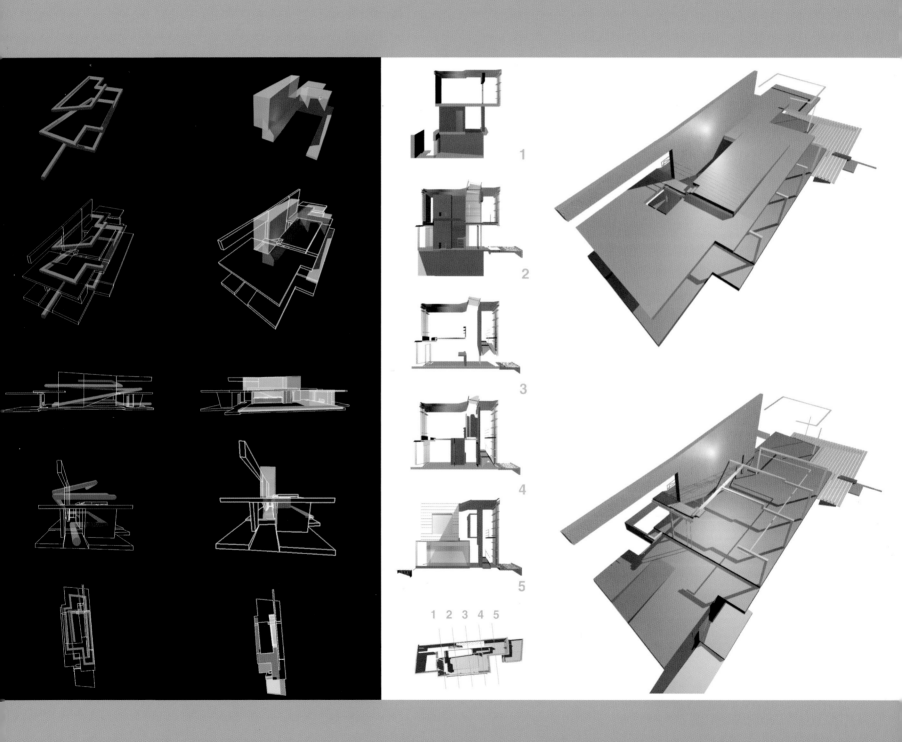

Millbrook House

Occupying the high point of its 34-acre site, the residence (1999) offers views over a valley and towards the Catskill Mountains. Traces of the site's former house are maintained, including a concrete-slab foundation and a stone fireplace and chimney. Recessed concrete plates extend to accommodate the elongated volume of the new house, while the structural system joins the heavy concrete elements to a light steel framework. A horizontal emphasis is continued across the long bands of the glass curtain wall and on the upper level across corrugated aluminum surfaces; sliding fiberglass panels mediate privacy and views along the bedroom wall. The roof is a folded plane of steel that follows the modulation of interior space. Set into the landscape of the hilly, wooded site are new concrete earthworks; one contains a lap pool, one forms a stairway down the hill, and others become linear flower beds that contain a single type of flower or herb.

opposite, left Rendered studies of the intertwined, wrapping circulation knot that winds through the house.
opposite, middle Elevations
opposite, right Exploded study of the floor-wall-roof configuration.
above right Preliminary perspective study
below right Plans

Specifically, she has deployed such elastic figures as the band, knot and loop as models for an architecture that reconciles buildings with disparate landscapes or urban situations. Topology thus affords Dubbeldam a "natural" link between the instability and fluidity of the world's urban constructs and the fluctuations and constant cyclings of the virtual world.

> The computer enables us to register complex mechanisms and simulate their spatial consequences. Here the matrix is the measuring device rather than the two-dimensional grid represented in Euclidean geometry. For example, in physics the matrix is used to define "phase space," a description of points notated in space over time, a dynamic system. In geology, the use of Land Sat and Sea Sat systems notate precise satellite measurements that are digitized by computer and visualized in three-dimensional imaging processes. They have enabled us to map parts of the world until then unknown. The new GPS systems can now measure not only static objects but also moving ones, constantly redefining their precise, but also relative, measurements. The computer enabled the scientist to visualize these complex phenomena and therefore to understand, investigate and develop them.

Dubbeldam's competition entry for an international cruise-ship terminal in Yokohama is arguably her most successful attempt to graft a topological architecture onto a large-scale, dense urban condition. Situated on artificially reclaimed land alongside the Yokohama harbor front, the project is part of a corporate identity-zone for the city. Formed as part of two conjoined urban zones – one close to the water and consisting primarily of business enterprises and high-rises, and the other a pier that is to be the future site for the international port terminal – Dubbeldam's building was designed to fuse the grainy density of the city with the expansive liquid surface of Yokohama Bay. With a central pier envisioned as a space of extreme movement, a zone of continuous arrival and departure, the project appropriates the notion of "time-delay," the stretching of time we often experience when we travel over continents and across time zones, as a starting point for a "twisted-rubber-band" topology that forms a dynamic yet "delayed" temporal space:

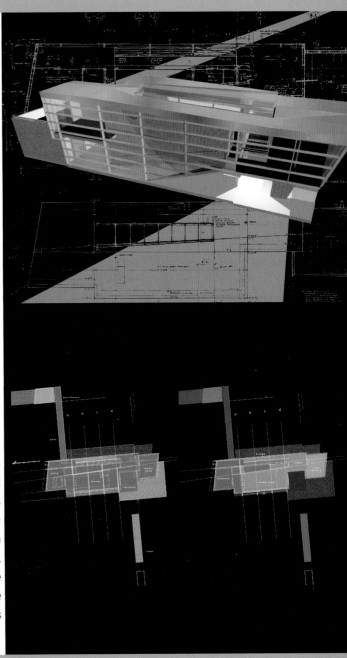

The notion of delay is introduced in a topological model where two twisted "rubber" bands distill this time-lapse in the space of the terminal. In the science of topology the "twisted rubber band" illustrates the unwinding and eventual disappearance of a texture. This notion of instability and temporality illustrates the fluidity of time and space. The two twisted surfaces are loaded with program, one the carrier of city functions, the other of terminal functions. The intersection of the two bands results in a series of functional modules which will simultaneously smoothly connect and slip past each other ... The fluidity of the terminal structure is determined by the interaction of the mathematical geometry of the computer and the material resistance factor of the physical model.

The terminal building is a long space located in the tightly intertwined figure formed by the intersection of two topological bands, which create a "bridge" structure that twists over the pier's surface. A continuous structural envelope/membrane – an aluminum frame wrapped with a translucent textured skin of aluminum ribbons and fiberglass – the terminal is enclosed by side walls of warping glass planes directed by the smooth curves of the bands' geometries. To articulate the complex structure's functions, the pier's surfaces are grained with variable materials that slip past each other with the movement of the winding volumes. For instance, the tip of the pier, used as the arrival area, is made of wooden slats, while the walkway ramp leading to the roof garden and restaurant and the pedestrian shopping areas are defined by rich stone surfaces.

The smooth transformation of the pier surface onto the terminal's twisting plane allows for access to the pier from the city to be drawn inconspicuously up a sloping "promenade" that is situated over the departure hall; an information center, exhibition hall and shops are located alongside. Shops have dual access for city pedestrians entering from the pier and for travelers entering through the departure and arrival hall. The space created by the complex crossing of the topological bands allows for a traffic plaza to be located under the main terminal, which in turn enables visitors to wander around the pier without having to cross the customs, immigration or quarantine areas. At the end of the pier, in the second of the loop sequences, a restaurant offers a view over the bay toward the

Millbrook House
left Interior and exterior perspectives reveal the residence's layered transparencies and reflections.
right Renderings show the residence's lean glass curtain walls suspended from a light steel framework.

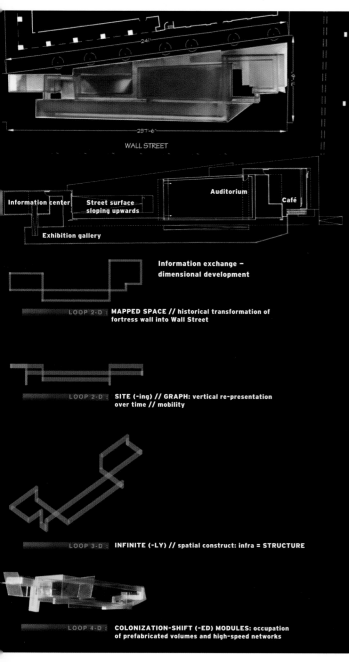

WALL STREET

Information center | Street surface sloping upwards | Auditorium | Café

Exhibition gallery

**Information exchange –
dimensional development**

LOOP 2-D : **MAPPED SPACE // historical transformation of
fortress wall into Wall Street**

LOOP 2-D : **SITE (-ing) // GRAPH: vertical re-presentation
over time // mobility**

LOOP 3-D : **INFINITE (-LY) // spatial construct: infra = STRUCTURE**

LOOP 4-D : **COLONIZATION-SHIFT (-ED) MODULES: occupation
of prefabricated volumes and high-speed networks**

city and over the bridge's powerful form that establishes a seamless conceptual, visual and metaphoric nexus between the oceanic and the urban.

A more recent project by Dubbeldam, the Millbrook House reconfigures an elastic-knot topology as a flexible domestic space set in an open natural environment. Weaving the house into, across and through its hillside setting in the Catskill Mountains, Dubbeldam created a spiraling "living-loop" that is open to dynamic interchange with its surroundings and reinforces the relationship between interior and exterior spaces. Like beads on a string, the house's knotted circulation figure draws out a series of spatial events along its length that are "gathered" together through movement, habitation and perception. Interleaving circulation, interior program and exterior spaces into a continuum, a single path winds up through the site and continues through the interior, which organizes living, working and sleeping spaces along upward and downward vectors. Dubbeldam envisions the house as an "anti-fortress ... Transparency between inside and out is maximized across lean glass walls, suspended like a curtain from a light steel framework. Visual connections multiply indoors as living areas fold around transparent, three-dimensional voids. These partly enclosed spaces, including an internal, open-air volume suspended above the entrance, wrap through the house to render it porous."

Architecture, Dubbeldam asserts, can be enriched and revitalized by a replacement of superannuated architectural orders – static temporality, linearity and idealism – with new, more pliant conceptual models and modalities of thought and practice. To overcome these conventions, architecture must be dedicated to decoding the emerging patterns of social organization through organizational models founded in complex geometries and new spatial types. Offering a precise yet energetic method of integrating manifold spatial programs and situations in the same way that the earth's surface unifies hill and valley, cliff and plain, topology is one such discipline. Finally, topology is also defined as the art of, or method for, assisting the memory – by associating a subject to be remembered with a place. Perhaps it is no coincidence, then, that Dubbeldam has deployed a topological architecture to recover the lost relationship between the natural and the urban.

**Wall Street Parallel
Network Center**

Formerly the location of the Dutch fortification, Wall Street is a global financial network that functions without a physical structure 24 hours a day. As a fragment of a changing urban environment a new center was proposed to provide access to this information and house an auditorium, café, information zone and gallery. Dubbeldam's concept (1998) was developed around the notion of simultaneity, in that the center's structures have dual functions (loop as structure and infrastructure), dual experiences (modular volumes can communicate on the inside and outside), dual connections (mental and physical) and interactivity (building reacts to occupation through sensors). In the proposal one can move through the structure, the loop, the site, in a single continuous movement. The center's spaces are therefore fluidly connected or completely separated by an active circulation loop that provides constant accessibility, an infrastructure, and a structure for modules to be attached onto a framework. Finally, the loop manifests the transformation of the original Dutch fortress into a new construct of shifting planes and tilting horizons, effectively a re-activation of historical development – or time itself.

left Roof plan and sequence of three-dimensional loop studies.
right Line and rendered elevations

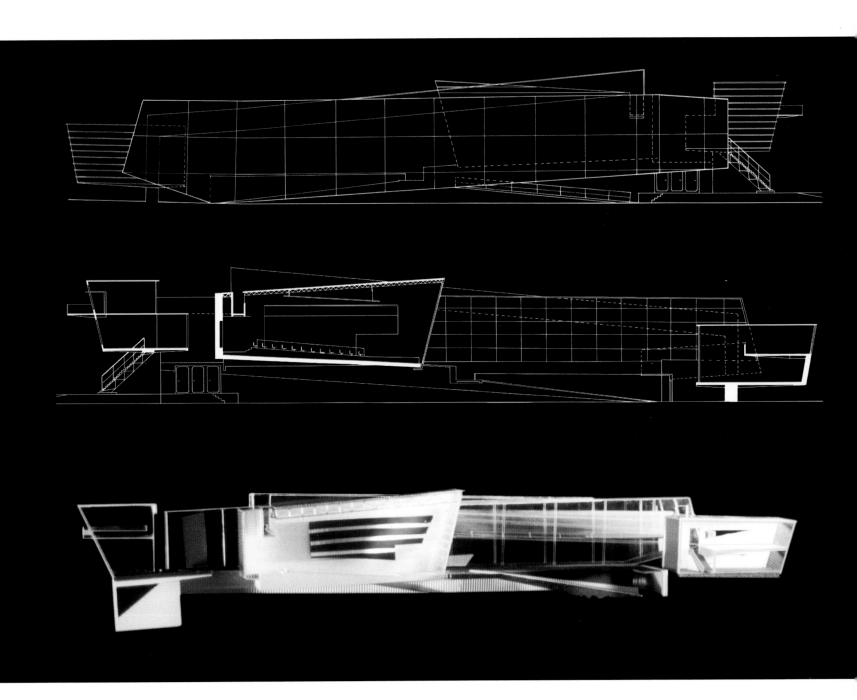

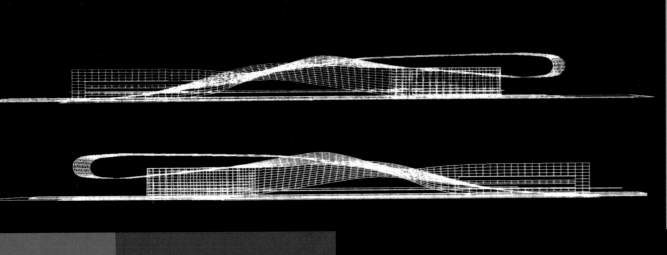

Yokohama Ocean Liner Terminal
A topological model in which two "twisted
rubber bands" distill a sense of delay
is the basis for the project (1994).

opposite, above, and below left Interior
and exterior studies show the aluminium
frame enclosed by the warping glass planes.
left Wire-frame elevations
below Sequential studies of the twisting
ribbons that unwind the program over the
pier's length • Sequential
cuts through the bands.

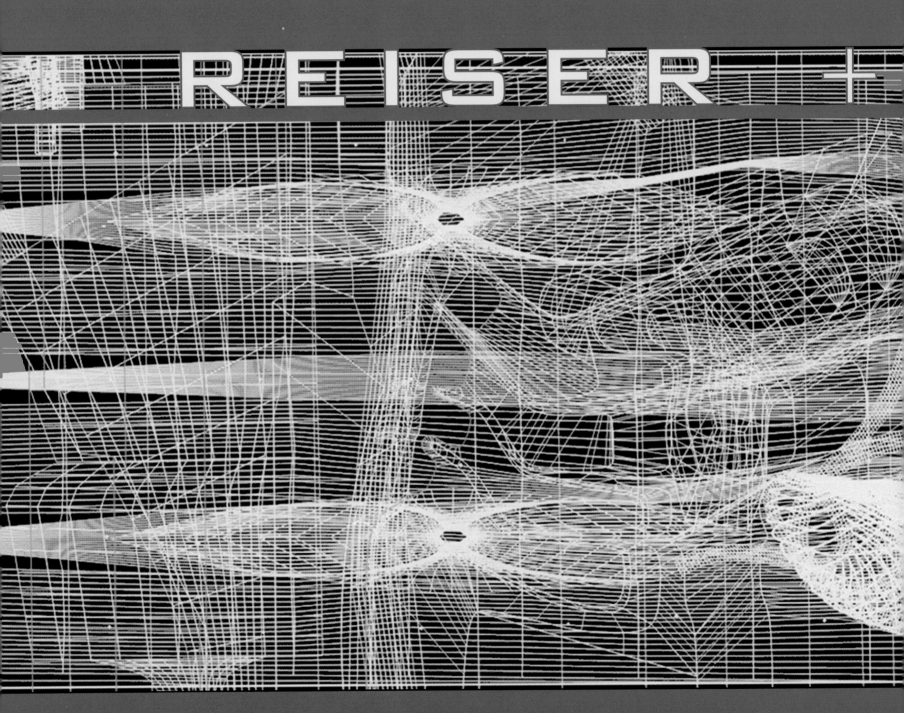

REISER +

Jesse Reiser and Nanako Umemoto draw from a wide base of scientific and theoretical thinking to infuse architecture with a rich mixture of novel concepts and working methods.

UMEMOTO

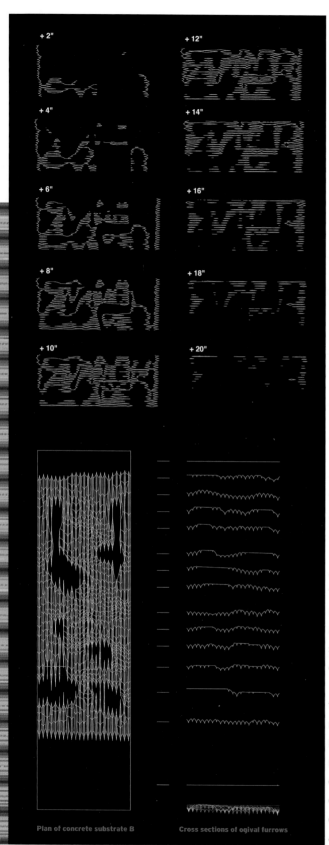

+2"

+4"

+6"

+8"

+10"

+12"

+14"

+16"

+18"

+20"

Plan of concrete substrate B

Cross sections of ogival furrows

Highly attuned to the cultural, technical and academic regressions that architecture has weathered since the collapse of modernism, Jesse Reiser and Nanako Umemoto have sought modes of thought and techniques to establish an inventive and critically reflective relationship between theory, technology and architectural practice. A scarcity of satisfactory models within architecture has driven Reiser and Umemoto to examine recent research in scientific and mathematical fields to break through the field's disastrous impasse. Incorporating such far-reaching concepts and theories as complexity and emergence into their work, they have initiated a dynamic and optimistic reappraisal of architecture's role in contemporary culture.

Complexity theory posits that evolution occurs most effectively through interaction between diverse agents or elements in a complex system – and not necessarily through competition along a linear trajectory. Complexity-based analysis studies the dynamic relationships among the components of a system, applying aspects of chaos theory, evolutionary theory and self-organization theory to understand the associations of individual elements that form a whole. Unlike the totalizing and undeviating systems of modernism, these analyses reveal unpredictable, intricate and often delicate properties or characteristics that can result directly from interactions within or among elaborate assemblages and environments. "Just as the sciences have experienced a sea change due to models of complexity first developed in the mathematics of dynamical systems," Reiser suggests, "we too benefit from their revolutionary em-ployment in architecture."

In their exploration of the chaotic or complex systems in architecture, Reiser and Umemoto have developed a fluid design process that can reveal innovative conceptual and productive territories. In an attempt to hurdle the traditional binary oppositions – between structure and ornament, programme and form, and typology and performance – that have dominated the discipline and resisted change, the architects seek loose couplings and productive codependencies. In contrast to the classical conception, for example, ornament for Reiser and Umemoto is not merely an image subservient to structure; rather, it can elaborate structural

Water Garden

A collaboration with David Ruy and Jeffrey Kipnis, this project (1997) in Columbus, Ohio, was conceived as a "primitive," inorganic, "dead" material geometry over which the flows of vital - organic - media will emerge. Extreme and unstable configurations in the garden's topology are built into the concrete substrate, defined by warps, dimples and folds, in order to express them in the vital media (water, soil, plant materials and chemical salts) of the "flow space" above. The fine-scale effects of surface tension and meniscus are amplified by the grooves' ogive geometry, so that every change in water level results in a new and changing constellation of puddles and walking surfaces. As the water rises, the effects are intensified because the ogive geometry tends toward the horizontal with every increase in the vertical dimension.

above left Water-level studies
below left Plan and section of ogival slab geometries.
opposite, top to bottom left
Topographic study • Computer-generated laminated object finished with zinc-alloy spray • Detail of computer model showing transition from interlaced earth berms to the ogival geometries of the flow space.
opposite, above and **below right**
Exploded computer study of the concrete tiling system • Projected isometric study of the ogival furrows.

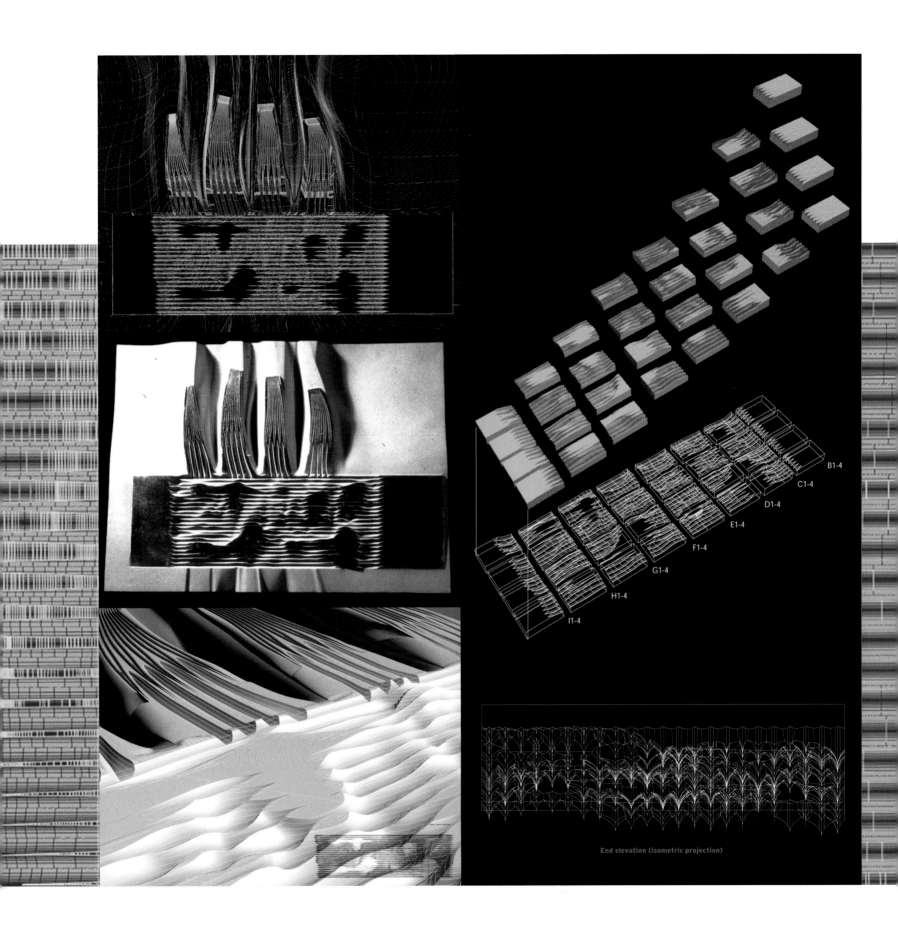

End elevation (isometric projection)

B1-4
C1-4
D1-4
E1-4
F1-4
G1-4
H1-4
I1-4

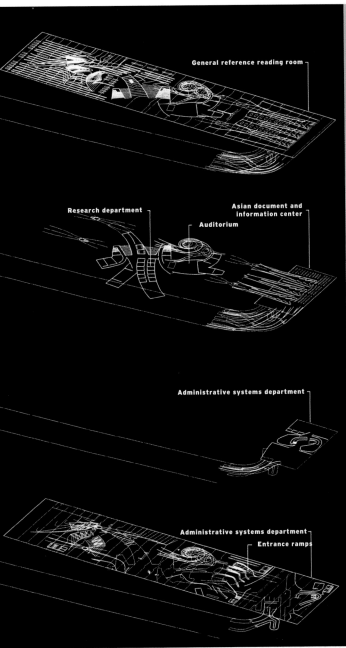

General reference reading room

Research department

Asian document and
information center

Auditorium

Administrative systems department

Administrative systems department

Entrance ramps

organization. As a consequence, ornament can serve as a powerful graphical tool capable of radically redistributing formal, material and program relationships. Similarly, program and form are not opposed or conjoined in a one-to-one fashion but instead share a "loose fit" – neither takes priority over the other.

Research into complexity theory has led the studio to rethink the constitution and manipulation of architectural classification and hierarchy in the design process. The architects have written, "material organization can be liberated from the meaning and intentionalities that might have prompted it; the work thus will always engender more meaning and find more unanticipated uses than is predicted within the confines of the concept. This is not to imply that all proposals are equivalent, rather that, except for the most limited and extreme circumstances, a necessarily loose fit exists between form, use-form, and meaning. Indeed such correspondences might foster a series of possible uses rather than define anyone." Complexity theory has also provided the architects with a compelling means of reformulating conventional architectural typologies. Instead of seeing building type as essential and static, modulated only by scale and material shifts, Reiser and Umemoto propose that typology be used to describe buildings as performative mediums, as confluences of various inflections and fluctuations of use.

Another fertile realm in which the studio operates is in the involved commingling of the virtual within the material. Whereas traditional architectural practice has accepted a fundamental division between the virtual and real, Reiser and Umemoto attempt to integrate the two territories. Rather than viewing the virtual as a force that drives architecture away from its material presence toward an immaterial space and idealized irreality, or constructing a "screened" interface between the digital realm and the physical world, they have developed the concept of a "solid-state" architecture to address the assumed chasm.

"Solid state" stresses the basic systems of physical organization that affect architecture – site, material, scale, structure – and permits an understanding of a building's internal economy as separate from its representation. Solid-state architecture conveys the notion that material

Kansai Library

Reiser+Umemoto's proposal (1996) attempts to address a paradox of the universal proliferation of data: the assumed placelessness of digital information and the necessity to find a definition for this condition in architecture. The building comprises three ramped slabs, which maximize continuity and interconnection among the public spaces and levels and which are suspended by cables from a steel roof carried on four steel piers. Topological deformations include cuts, mounds, ramps, ripples and stairs, which render the library a programmed landscape that permits the smooth functioning of the major programs and fosters unanticipated configurations of social space.

left Exploded isometric plan studies reveal the interweaving of the three slab levels.
opposite, top to bottom Aerial perspective showing topological geometries of the library building's roof form • Three rendered plans • Wire-frame view of the combined topological geometries of the library building's floor plates and roof.

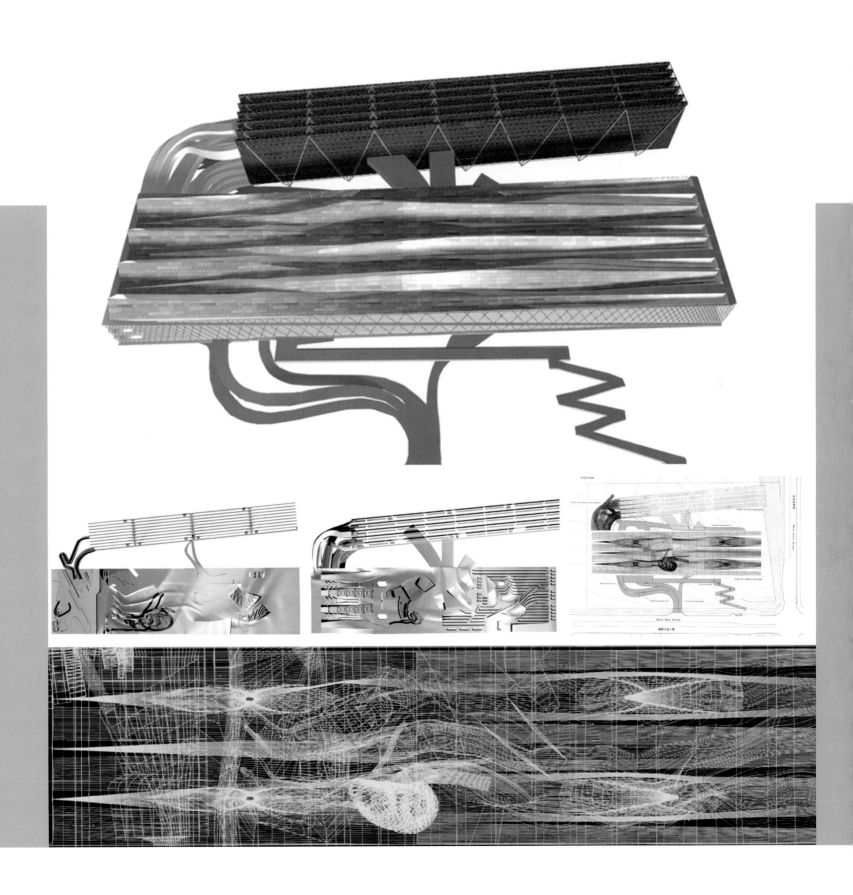

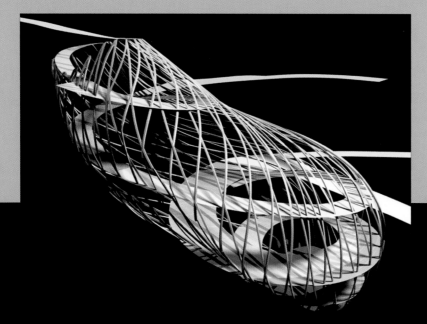

Kansai Library

opposite, above Detail of the Geodetic Store. A geodetic system carries loads along the shortest possible paths, which produces self-stabilizing members equalized by forces in the intersecting set of frames.

opposite, below and **right** Interior views of main library building showing topological deformations – cuts, mounds, ramps, ripples and stairs – in the hanging floor slabs.

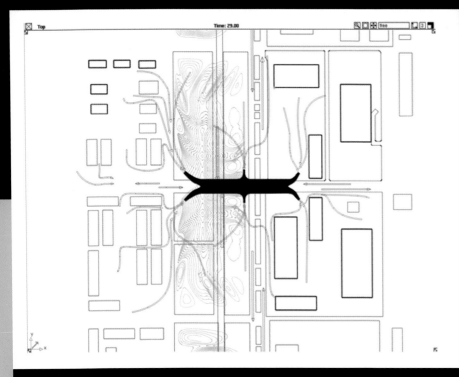

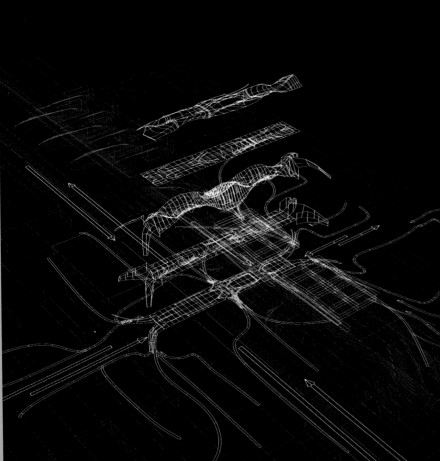

Illinois Institute of Technology Student Center

To address the Miesian order of the core campus, Reiser+Umemoto designed the building as an infrastructural system that enlists service, circulation and landscape. The flow-based organization, which interlaces "fingers" of space and new exterior and interior terrains, responds to the center's complex program and central position, while new construction kneads into adjacent streets and parking lots, feeding the existing pattern of the campus.

above left Site plan shows how the building's "fingers" accentuate the circulation flows into the rigid campus grid.
left Exploded axonometric wire-frame.
right Isostatic study of structural stresses.
opposite, above Exploded rendered study with landscape/roof, deeply punctured floor-plate stack, deformed skin and structural frame.
opposite, below Relational structure and skin study.

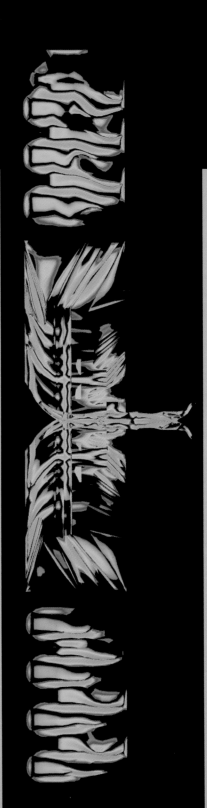

arrangements are inseparable from the mechanical axioms that are embedded in their fabric, rules that may be expressed and calculated in the form of computational information. Thus, the solid state in architecture furnishes models of physical material shifts, restructurings and realignments as forms of media and information manipulation. "Solid-state notions of architecture," Reiser writes, "engage media in two distinct but related ways: the first, media as a material substance (like a painter's medium); and, the second, media as a virtual or informing potential. These are not, in practice, separate notions, or procedures, but rather can be said to be folded together within the conception of the diagram."

The architects believe that diagramming typically either adheres to the standards of classical geometry and typology or is composed of succinct, organizational illustrations of utilitarian building relations and uses. Reiser and Umemoto believe that reconsidering these static models in favor of dynamic time-based diagramming of spatial relations would lead at least to a complexification of how buildings are conceived and executed, if not to an overturning of the traditional models. The "dynamical diagram," the architects contend, "has no essential original, and can be incarnated in multiple materials, scales and regimes. The use value of the diagram lies not in a capacity for representation, but rather in its latent potential for quantitative effects. Initially, the diagram carries a kind of proportionality which ultimately locks into a specific order, material and scale. This describes a particular and local necessity, a solid-state, wrought from an initially variable diagram. For architects, this means that the non-linear dynamics found, for example, in weather systems are already at the level of order, possible to instrumentalize architecturally with the stuff inherent in architecture; neither metaphor nor symbol, but a literal employment of the order itself."

Reiser and Umemoto's project for a water garden, with David Ruy and Jeffrey Kipnis, is an accomplished example of how the architects have orchestrated and instrumentalized complexity theory, dynamical diagrams and topological geometries to produce the full spectrum of implications for the solid state in architecture. Unlike the traditional

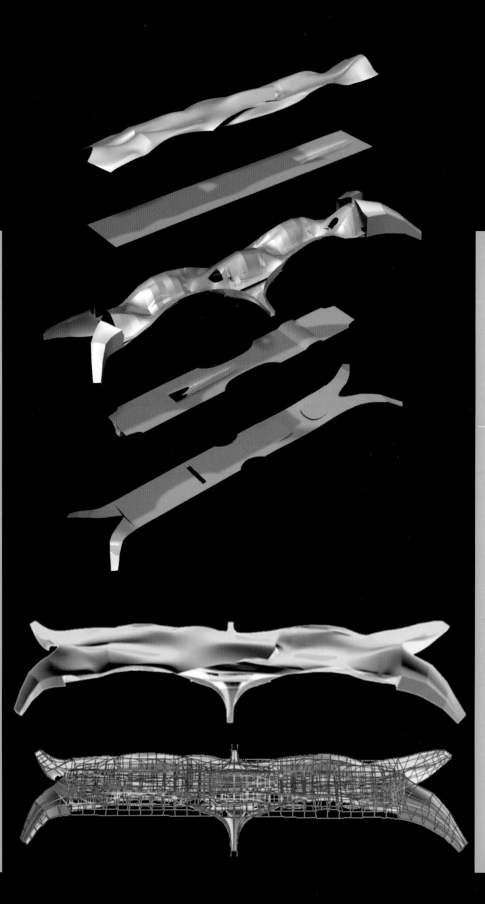

Graz Music Theater

The project (1998) for a theater complex in Austria is composed of two interlocking building forms: an educational/rehearsal section, housed in a translucent steel-and-glass plinth; and the main auditorium and its affiliated spaces, contained within and under a forthright concrete structure.

below Plans

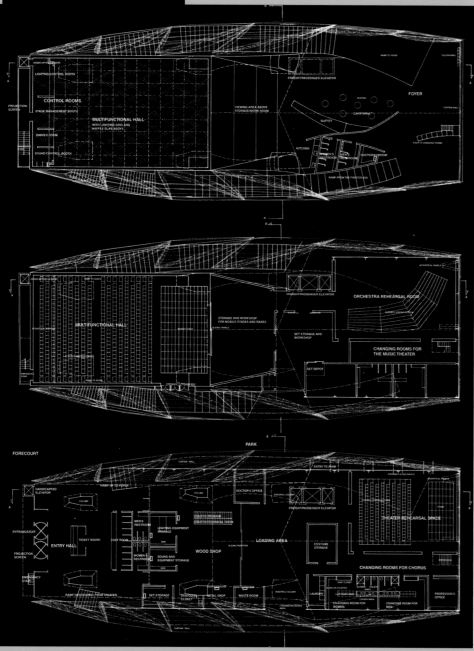

Western model of the garden – an eighteenth-century topiary, for instance – in which nature is forced into a static or fixed ideal, their water garden has been manipulated to possess the time-based or dynamical potential intrinsic to the organic world (nature) to produce or excite instability in the garden's "vital media" (water, soil, plant materials and chemical salts). In the architects' vision:

> Nature, then, is less a "creation" to be speculated on than an inventive and modifiable matrix of material becomings ... And here it will be necessary to set aside the nature/culture dialectics and focus instead on the processes that establish transverse developments across these regimes. The French philosopher Gilles Deleuze coined the concept of the "machinic phylum" to refer to the overall set of self-organizing processes in the universe. These include all processes in which a group of previously disconnected elements (organic and non-inorganic) suddenly reach a critical point at which they begin to "cooperate" to form a higher level entity. Recent advances in experimental mathematics have shown that the onset of these processes may be described by the same mathematical model. It is as if the principles that guide the self-assembly of these "machines are at some deep level essentially similar." The notion of a "machinic phylum" thus blurs the distinction between organic and non-organic life.

Allied with the notion of materiality mediated in architecture (as expressed by the solid state) is the function that topological geometries – in contrast to classical Euclidean geometries – can play in activating material flows in living and non-living systems. "It is important to clarify that this media-form relationship is not understood metaphorically or symbolically but as a direct instrumentality with the capability of producing architectural effects." The project was conceived as a topological surface, a "primitive," inorganic, "dead" material geometry over which the flows of vital media will emerge in new and unpredictable ways. Essentially a series of concrete tiles that form a surface defined by transformations like warps, dimples and folds, the artificial topology will express its logic in the "flow space" above it. As the architects see it:

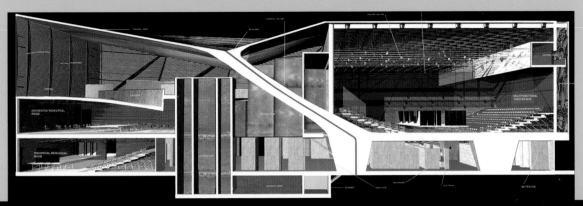

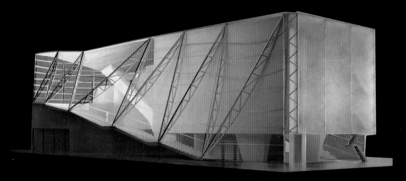

Roof drainage system

Diagonal column

Theater volume

Main entrance

Curtain wall

Bow-string truss structure

Foyer level

Kitchen

Ramp to theater

Cafeteria

Orchestra dressing room

Orchestra rehearsal room

Storage space

Music theater dressing rooms

Choir dressing room

Workshops

Theatrical rehearsal rooms

Elevator core

Mechanical level

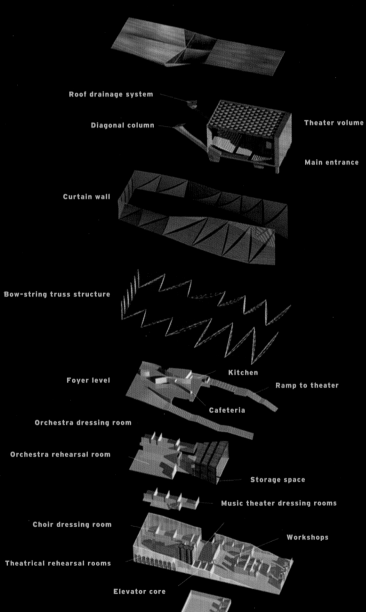

Graz Music Theater

The rehearsal spaces and main auditorium are conceived as open, flexible spaces that enjoy natural light and exposure to the adjacent park. A cantilevered double-conoidal roof partly covers the building and imparts a sculptural effect belied by the economy of its construction.

top Section showing relation between foyer, theater volume and conoidal roof.
above Perspective view with glazing system, theater volume and ramps.
left Exploded diagram
below Detailed computer study showing steel-truss members, glazing mullions and tightly knit aluminum rods.

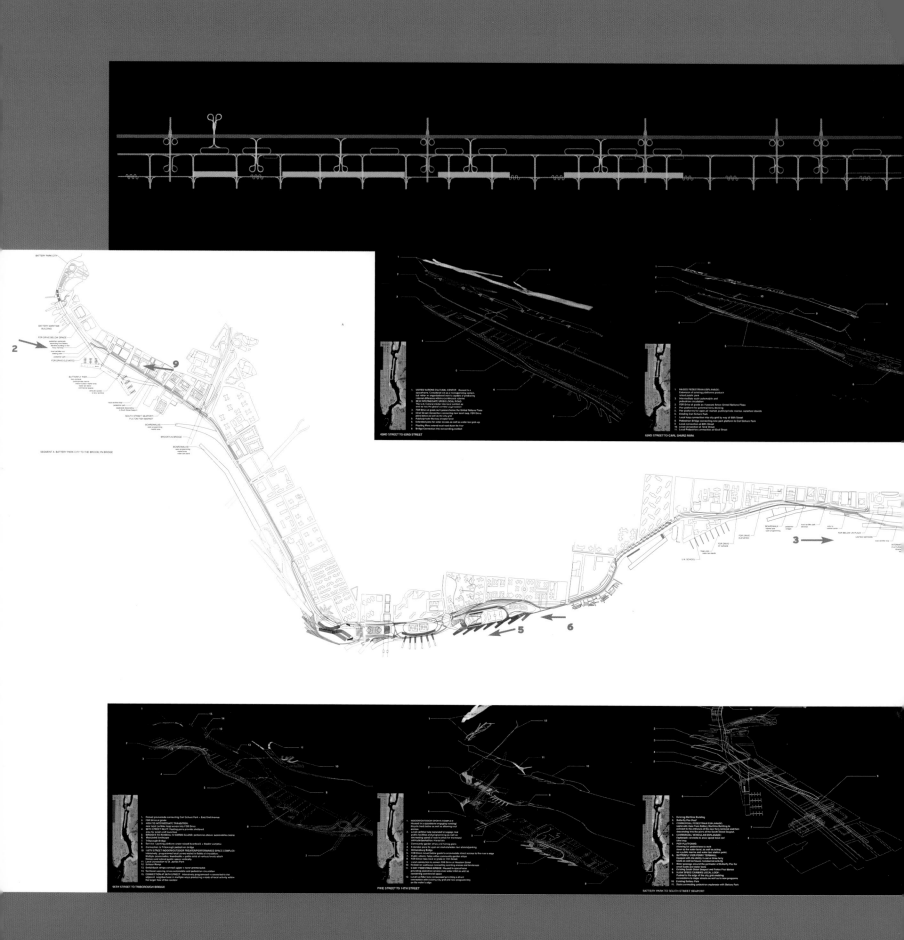

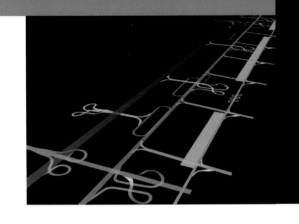

East River-front Development

For the development (1998) of a linear urban morphology that would integrate the diverse vehicular, pedestrian, commercial and cultural infrastructures on Manhattan's eastern shore, Reiser+Umemoto created a zone of transition between the high speed of passing cars and the comparatively low pace of the city grid. Rather than producing a collection of isolated objects, they established a twisting, interwoven system that organizes a series of moments within a continuum.

these pages Overall plan of development [numbers correspond to views at **right** and **bottom**].
opposite, top and **left** Roadway circulation studies along East River.
opposite, middle and bottom Diagram sequence
right and **bottom** Aerial views

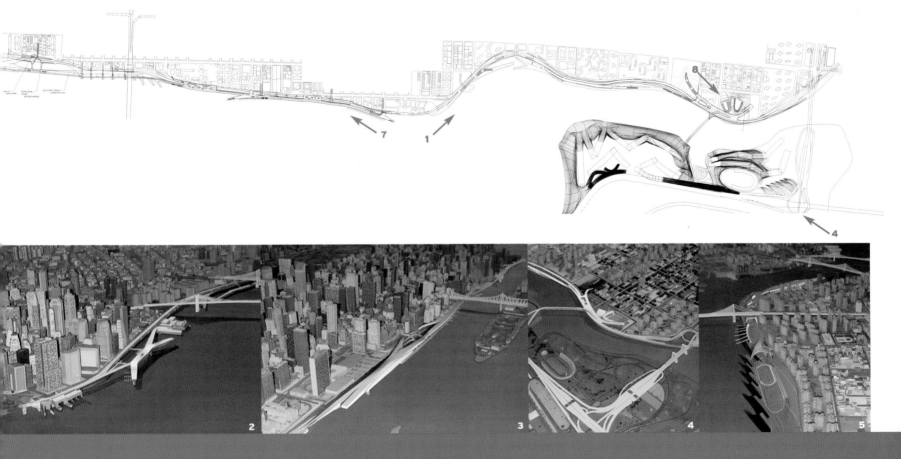

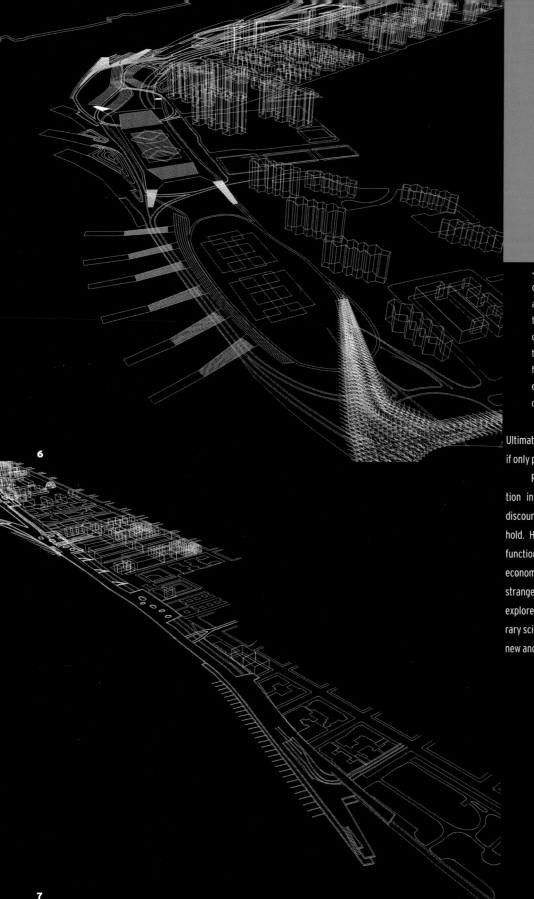

6

7

Just as the upward thrust of a mountain range breaks up the orderly flow of winds around the earth surface, creating local eddies which in turn affect temperature and rainfall, so the variegated peaks and valleys of the water garden exert a spectrum of turbulent effects on the flow space of the garden ... The dominant media (in this case the fixed geometry of the slab) exerts proportionally greater influence than the relatively weak forces of the flow space. This is not a simple dominance, however. Lateral efflorescences, mineral and vegetal, engender moving tracks that criss-cross the flow space, dissipating the determinations of the slab.

Ultimately, the dynamic organic/inorganic system will "yield a prodigious, if only partially manageable, field of blooms."

Recent discussions centered on issues of meaning or representation in architecture have inevitably failed to advance architectural discourse or keep pace with the broader cultural transformations taking hold. Hopelessly focused on issues of essential typologies, classical functions and geometries, debates have ignored the radical social, economic and cultural shifts that have placed those very categories into strange new relationships. Traversing these issues, Reiser and Umemoto explore the organization of the world and its effects through contemporary scientific thought, and in the process they have introduced genuinely new and vital architectural forms and techniques of invention.

8

9

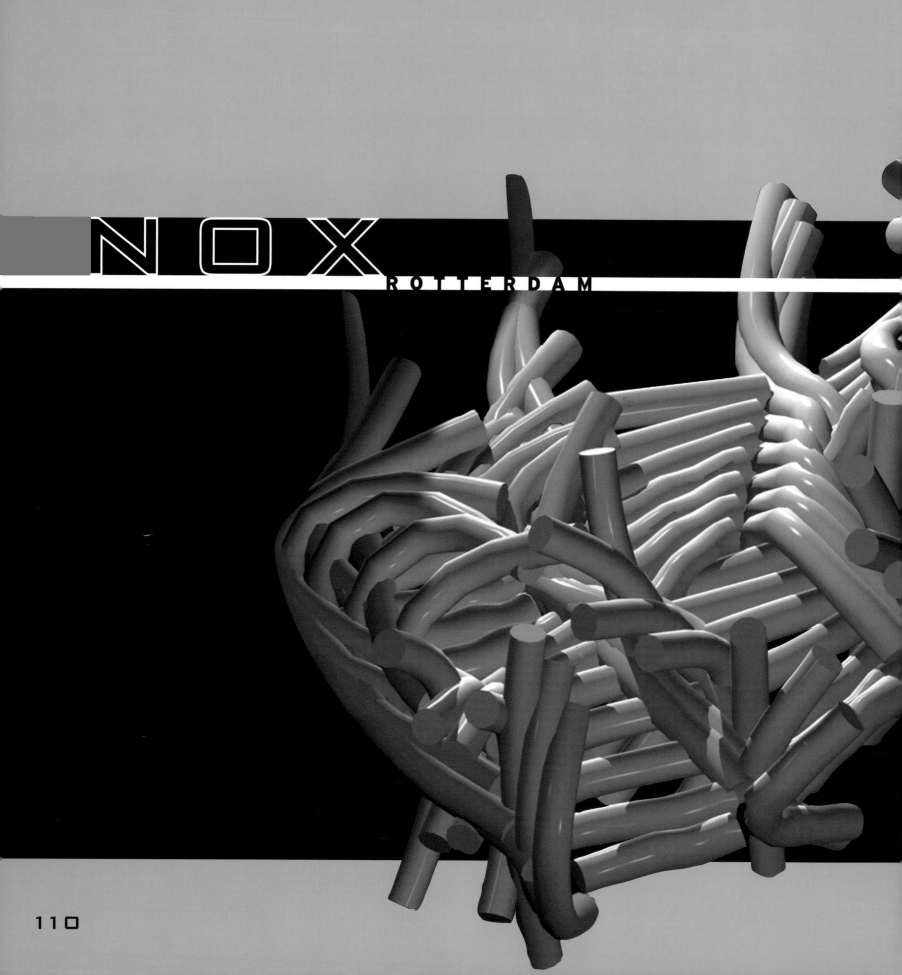

NOX
ROTTERDAM

110

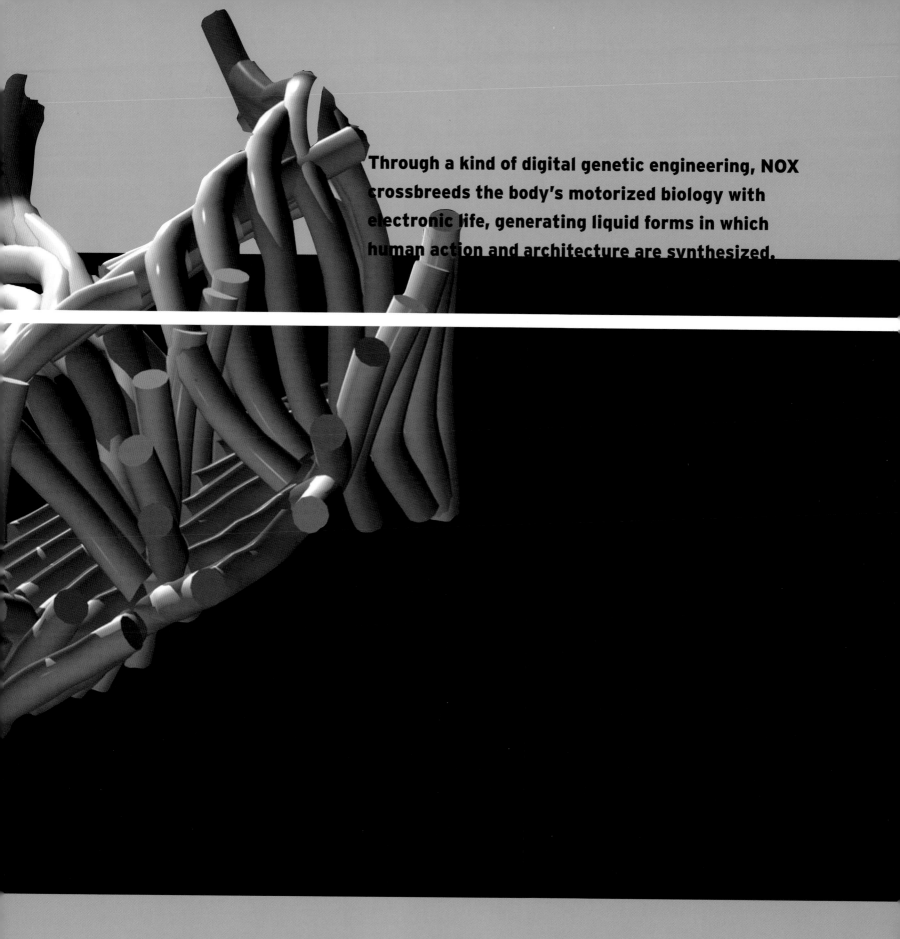

Through a kind of digital genetic engineering, NOX crossbreeds the body's motorized biology with electronic life, generating liquid forms in which human action and architecture are synthesized.

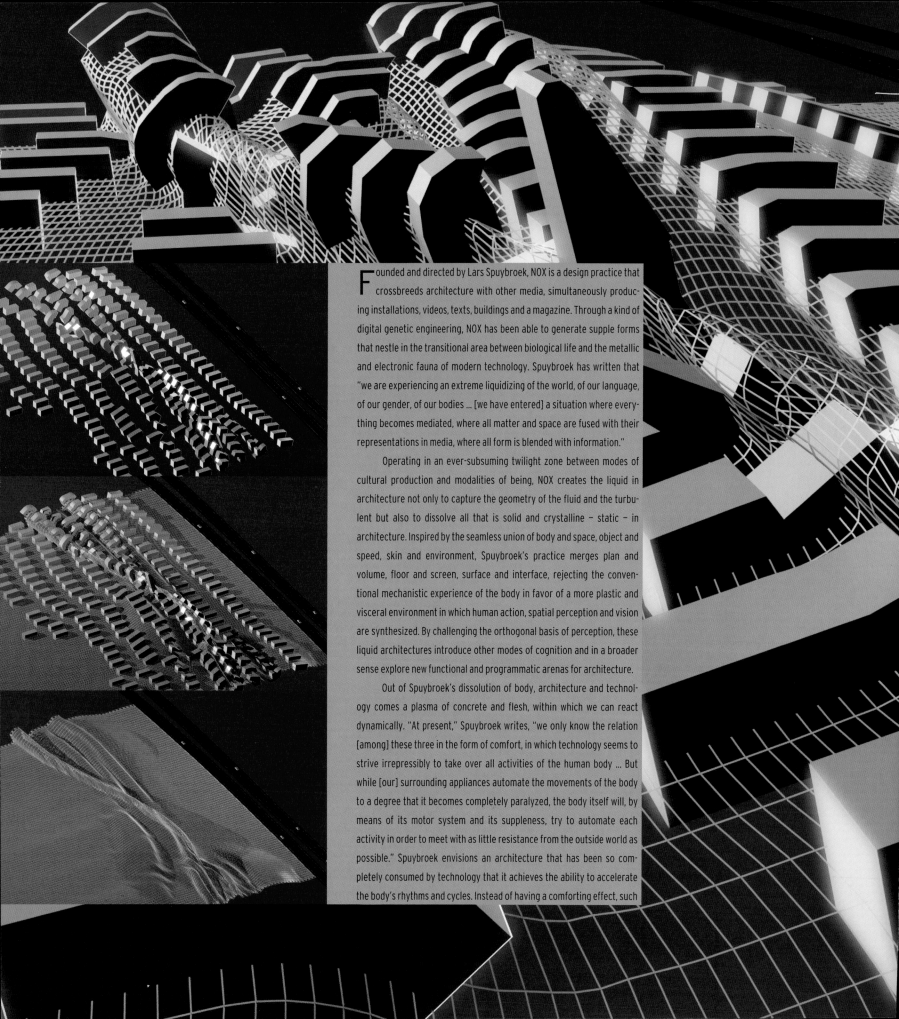

Founded and directed by Lars Spuybroek, NOX is a design practice that crossbreeds architecture with other media, simultaneously producing installations, videos, texts, buildings and a magazine. Through a kind of digital genetic engineering, NOX has been able to generate supple forms that nestle in the transitional area between biological life and the metallic and electronic fauna of modern technology. Spuybroek has written that "we are experiencing an extreme liquidizing of the world, of our language, of our gender, of our bodies ... [we have entered] a situation where everything becomes mediated, where all matter and space are fused with their representations in media, where all form is blended with information."

Operating in an ever-subsuming twilight zone between modes of cultural production and modalities of being, NOX creates the liquid in architecture not only to capture the geometry of the fluid and the turbulent but also to dissolve all that is solid and crystalline – static – in architecture. Inspired by the seamless union of body and space, object and speed, skin and environment, Spuybroek's practice merges plan and volume, floor and screen, surface and interface, rejecting the conventional mechanistic experience of the body in favor of a more plastic and visceral environment in which human action, spatial perception and vision are synthesized. By challenging the orthogonal basis of perception, these liquid architectures introduce other modes of cognition and in a broader sense explore new functional and programmatic arenas for architecture.

Out of Spuybroek's dissolution of body, architecture and technology comes a plasma of concrete and flesh, within which we can react dynamically. "At present," Spuybroek writes, "we only know the relation [among] these three in the form of comfort, in which technology seems to strive irrepressibly to take over all activities of the human body ... But while [our] surrounding appliances automate the movements of the body to a degree that it becomes completely paralyzed, the body itself will, by means of its motor system and its suppleness, try to automate each activity in order to meet with as little resistance from the outside world as possible." Spuybroek envisions an architecture that has been so completely consumed by technology that it achieves the ability to accelerate the body's rhythms and cycles. Instead of having a comforting effect, such

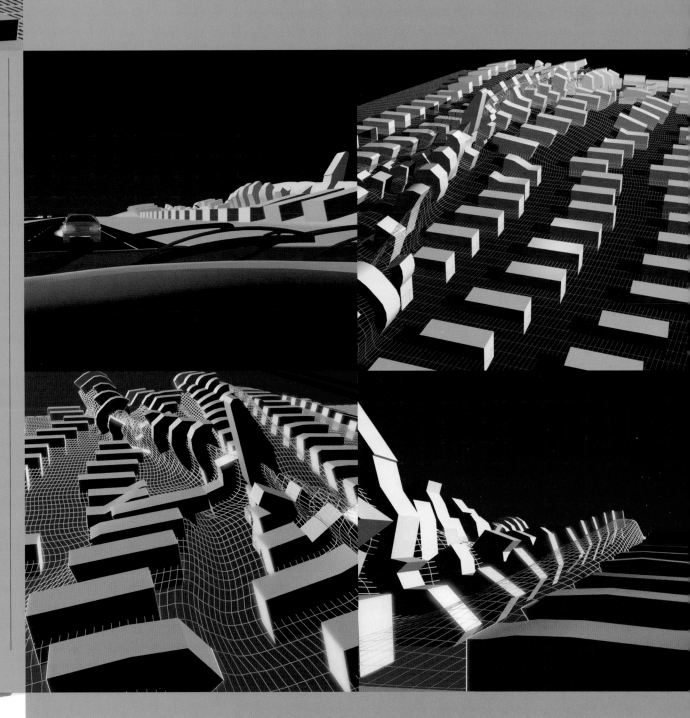

"Off-the-Road/103.8 Mhz" housing and noise barrier

Cities are being increasingly pervaded with the distinction between bits and bricks, between the technical and the visual, between media and architecture. Eindhoven in The Netherlands exemplifies this split: on the one side of the city lies a highway; on the other, the Blixembosch quarter. The two zones are separated by a noise barrier, Commissioned by the StadBeeld Committee, NOX attempted to reconcile this disjunction with a project that builds out a sound model of the site as it is experienced instead of creating a static no-man's land between road and housing. Using animation software, the existing noise barrier was translated into a system of "strings," which are given the properties of musical instruments and which "play" with the sound of passing traffic. Wave patterns triggered in the strings are recorded at intervals and "added up" into a sonic landscape. The form of each of the 208 houses is inflected by the sound patterns, flexing in all directions to differentiate program. Each house contains a sound system so that the collective noise – the tinkling of spoons, barking dogs, fighting spouses, the TV – of the residences is fed into a central computer that arranges the sounds in a real-time composition that is relayed to a transmitter and broadcast at 103.8 Mhz to car radios. Just as the residences' shapes interfere with the cars' sound, the cars' interiors resonate with the sound of the residences. Speed becomes inhabitable and the sound barrier is broken.

opposite, top to bottom The aerial sequence of traffic-generated wave diagram shows the site transformed into a sonic landscape.
above left View of proposed housing from the highway.
above right and **below left** Aerial views towards the highway.
below right Detail of the housing forms.

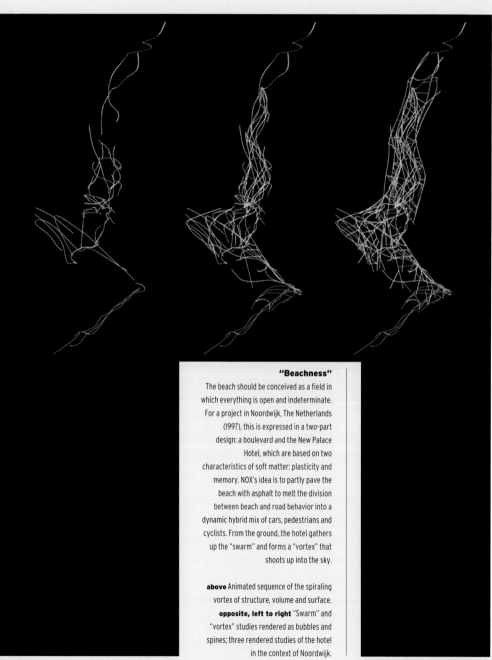

an architecture would absorb and enhance the plasticity and the suppleness of the human body in order to integrate it with an advancing technological environment. Thus the "soft" architecture that NOX creates is directly related to the mobility of the body, its speeds and movements. Eschewing static, classical architectural means, NOX's project implodes traditional categories of use and function to suggest an architecture of which we can never be certain.

Spuybroek argues that at present there are two dissimilar trends in architecture. The first mode of architectural production concentrates on a softness of form, a physical turbulence, that is literalized and understood as a phenomenological model of spatialization. The other architectural method develops a fluidity *within* the building program, so that softness is experienced indirectly through systems of circulation and internal spatial arrangement. While the former approach may neglect or willfully distort programmatic considerations, form in the latter can be subsumed and rendered completely neutral or generic. NOX's aims diverge from both: the architectural object is considered liquid and entirely capable of absorbing form (body) plus program (routine). NOX wants to connect the suppleness of the object to that of the body:

> The [other] possibilities that present themselves, exist because there are two other trends functioning within our culture as a whole; two trends which are diametrically opposed to the instrumentalist idea that technology mediates between the body and its environment without it essentially influencing or changing either the body or the world. First: the complete fusion of the architectural object with the technological object. And second: the fusion of the body with technology. In the first, architectural identity evaporates (which is not a tragedy). In the second the soul of the natural body has effortlessly moved into a bio-technical mutation (which is not a tragedy either). In both trends, technology seeks to calm the body down, to temper it and soothe it, to provide it with conditioned air, to keep it motionless while lifting it up a few floors, to make it fall asleep in the gentlest possible manner. But imagine the opposite: imagine a technology which is geared towards speeding the body up rather than calming it down. Imagine that architecture is swallowed up by technology so that it becomes completely capable of absorbing and

"Beachness"

The beach should be conceived as a field in which everything is open and indeterminate. For a project in Noordwijk, The Netherlands (1997), this is expressed in a two-part design: a boulevard and the New Palace Hotel, which are based on two characteristics of soft matter: plasticity and memory. NOX's idea is to partly pave the beach with asphalt to melt the division between beach and road behavior into a dynamic hybrid mix of cars, pedestrians and cyclists. From the ground, the hotel gathers up the "swarm" and forms a "vortex" that shoots up into the sky.

above Animated sequence of the spiraling vortex of structure, volume and surface.
opposite, left to right "Swarm" and "vortex" studies rendered as bubbles and spines; three rendered studies of the hotel in the context of Noordwijk.

enhancing the body's rhythm. That means that the body's rhythm will affect the form. And conversely it means that the form's rhythmicality will in turn activate the body.

The freshH$_2$O eXPO, a water pavilion and interactive installation designed and constructed by NOX for the Dutch Ministry of Transport, Public Works and Water Management, exemplifies the practice's inter-weaving of architectural design, interactive media orchestration and high technology. Located on an island southwest of the mainland, the pavilion was designed over three years as a building/exhibition/environment in which geometry, architecture and sensory-triggered multimedia installa-tions come together in an integrated spatial experience, and as such is one of the first built examples of NOX's soft architecture. A kind of "smart" building, the pavilion has an internal logic and sentience that responds to the activities and movements of the visitor. Inside the structure carefully positioned sets of sensors are connected to a 65-meter-long row of blue lamps attached to a sound system. Coordinated by banks of microproces-sors, this spine of light and sound pulses rhythmically with the visitors' movements. The freshH$_2$O eXPO's complex structure, developed and modeled on high-end workstations running advanced animation and simulation software, is a braid of sixteen splines that are shaped like a elongated worm of steel ellipses and semicircles. Within the 3D-modeling software used to design the pavilion, the splines are defined as active and reactive forms – unlike lines, which are traces of actions. When the splines are pulled in their virtual state, they deform in unison according to parameters determined by programming scripts and routines developed by NOX. In the built project this creates an environment in which floor blends into wall, wall into ceiling and where nothing is horizontal. At every moment the visitor is placed on a vector and must rely on his or her own motor system and haptic instinct to stay level – a firsthand experience of the NOX's challenge to Euclidean perception and space.

Spuybroek does not view media in competition or opposition to architecture. "Media come in waves, in tides, and it deals with space as a medium, as a field, that is a soft substance through which events are

"Beachness"

above and **right** As the swarm spirals upward enveloped in a structure of steel and translucent fabric, they shake off their relation with the ground and float into the sky. Inside, the space will feel like a flotation tank: a translucent capsule partly filled with very salty water in which one is immersed in diffuse light. All rooms will be equipped with such tanks, and many will have images projected on them. Visitors circulate on a spiraling steel staircase against the translucent fabric backdrop as wind and electronically amplified sounds howl past. Instead of having rooms that passively face the horizon, like any other Hotel Bellevue with rooms on the outer wall, the horizon is tilted over ninety degrees and wrapped with a landscape. The hotel therefore has No Rooms With A Sea View: People will find themselves on the horizon itself.

transported ... and become interrelated as a result of interference, amplification and decay ... Media are a way to inhabit time, as it were, a movement connected with our own movements, something far more sensitive and responsive than an architecture of frames, crystals and solids that is only capable of returning ... same answers to an experiential body." Ironically, NOX claims architecture as the first media machine, the first medium to connect behavior and action to time. Architecture, Spuybroek suggests, has evolved through a series of technical advancements closely correlated with advances in visual culture and representation.

Historically, this co-evolution began half a millennium ago with the introduction of static modes of representing space (the perspectival leap of the *quattrocento*), and over the past hundred years has derived kinetic energy from transportation, film and television.

> From perspective towards films ... and cars (all with their own architectural styles), moving eyes constructing spaces ... with computing – this step is not metaphorical anymore, now we not only incorporate and embody the conceptuality of a machine in design, we can now actually step into the screen and create reality from there ... design itself has become motorized, liquid, unstable, charged – the accelerating power of the computer is truly enormous ... But it is in the motor geometry, the geometry of the liquid that this machine becomes instrumental.

Motor geometry, or the computational instrumentalization of movement in space, creates a new type of three-dimensionality, deframing architecture so that a looping of perception and action (the optic and the haptic) induces a sense of endlessness. In this skewed field Spuybroek suggests that we react like skateboarders:

> We have a sense of direction, we have a sense of intentionality. We throw ourselves into time by movement. But then it is not a road or path we walk down. Our roads may be straight, but our tracks certainly are not. It is a vector with a point of action, and in that sense every act is an act of faith. Once underway we adapt, change our minds, engage other forces, but we do not just see these as resistance, no, they are like the curbs and obstacles for the skater. We use them as push-offs, as points of inflection

freshH$_2$O eXPO

On the island of Neeltje Jans in The Netherlands, NOX designed a turbulent alloy of the hard and the weak, of human flesh, concrete and metal, interactive electronics and water, a complete fusion of body, environment and technology. Based on the meta-stable aggregation of architecture and information, the building's form is shaped by the fluid deformation of fourteen ellipses spaced out over 65 meters. With no horizontal floors and no external relation to the horizon inside the pavilion, walking becomes akin to falling. The object's external deformation is mirrored in the interior environment, which registers the movements in seventeen sensors that respond interactively with the visitor. This building does not contain an exhibition per se, nor does it contain a program. In addition to events – ice, spraying mist, water on the floors, rain and an enormous well – three kinds of sensors create the "wave," the "ripple" and the "blob" in real-time projections and sound manipulation in a spline that holds 190 blue lamps.

opposite, top right, to lower right
Selected screen grabs from videos taken at the pavilion.
opposite, top to bottom View from roof and two interiors perspectives.
above left View from the side
left The project in relation to the adjacent Saltwater Pavilion by Oosterhuisassociates [pages 76-79].

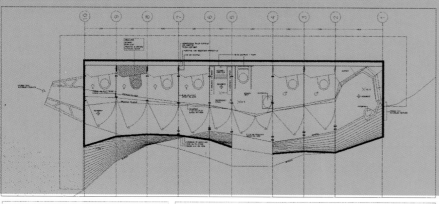

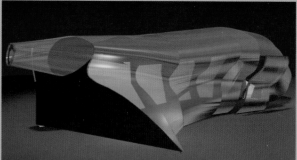

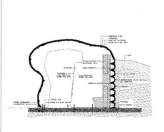

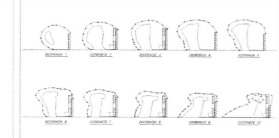

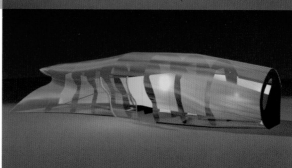

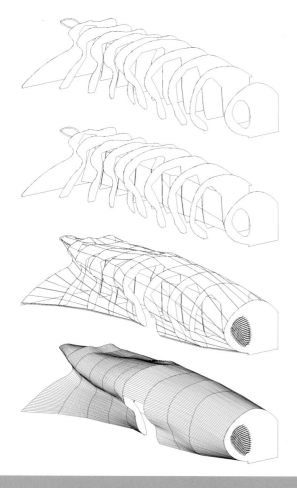

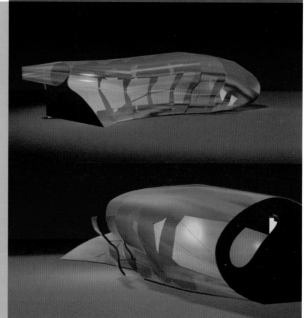

"Blow Out" toilet facility

This building (1997) in Zeeland, The Netherlands, establishes a dynamic equilibrium between internal pressure ("gotta go") and huge external forces. These forces are not just another natural element in the architecture: they represent media, a vector of the acts of other people and of the wind that carries the smell of others, their noises, their interiors. It is modeled in such a way that the wind blows through it at high speed (with a "grill" on the male side and an "exhaust pipe" on the female side). The liquid machine connects one interior with another, it shapes intimacy, builds it up, and releases it.

top Plans and sections
left Perspective studies of the evolution of structural plates, ribs, bracing and sprayed-on skin.
right, top to bottom Four semi-transparent computer studies and a view of the built project.

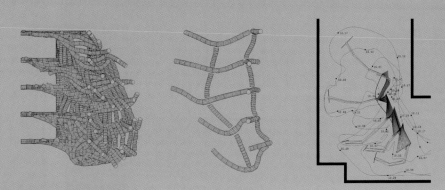

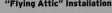

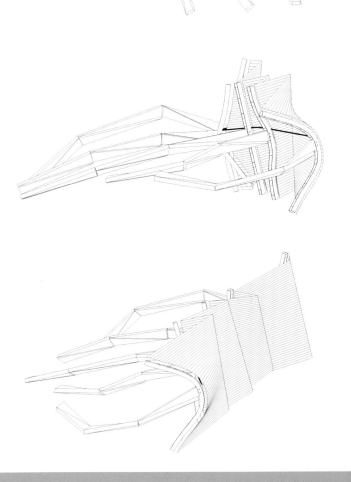

"Flying Attic" Installation

Structure is generally solid, static and tectonic, while images are variable, ephemeral and light. To blur this distinction "Flying Attic" (1998) takes an approach in which the constructive is conceived as textile. For each diagrammed stage of the process – vector, "rubber," materialization, transport, actualization – movement is structurally absorbed by the diagram and passed on to the next stage, ultimately capturing the interaction between human behavior and structure.

above Five stages of materialization
left and **right** Structural studies
below Final installation

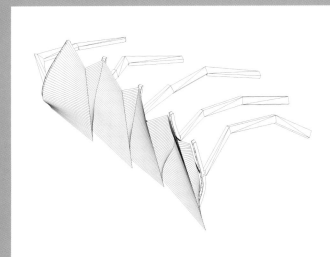

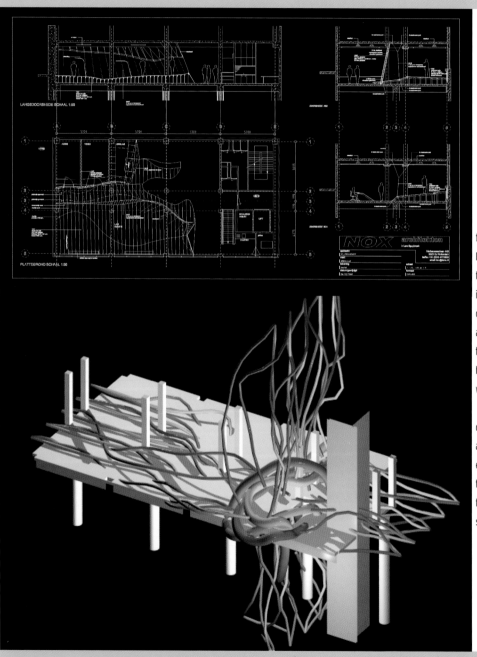

in the curve. That's it: a straight line goes from A to B, but while it leaves A it curves, trying to reach B. Architects have always misunderstood this position of B as something in space, instead of time. First of all the movement should be going from floor to wall and vice versa. That is: in the architecture itself ... it's about creating tension and suspense in the program ... on the one hand – with the desire to cool down behavior, to structure and separate actions, in short with the instrumentality of the program ... on the other hand, we vitalize action through animation, by replacing fixed points and fixed geometries by moving geometries, going from points to knots to springs, and we vitalize action through suspense, by shifting B from space to time, by multiplication of action.

It is not the fixation of the movement in the program, nor is it the fixation of motion in the form that drives NOX's extraordinary work. Rather, it is the desire to comprehensively relocate the body in architecture, to "motorize" reality and in so doing to radically transform how we inhabit and experience space. By fusing architecture with media technologies Spuybroek hopes that buildings will "soften," will become capable of acting and of reacting to human behavior and activity; at its most extreme, this softness could construct systems that could literally move with humans. In this scenario a "soft system" would refer to a condition in which events are mutual exchanges of the body and its environment.

Of course this image of a soft-moving architecture raises certain questions. Because an important aspect of architecture is its materiality, and most matter usually resists rapid transformation, one wonders exactly how a liquid architecture could be constructed. Given such a hypothetically liquid and soft state, might architecture change its focus from the production space into the instigation of soft media fields? Would we still call it architecture, or would we call it media?

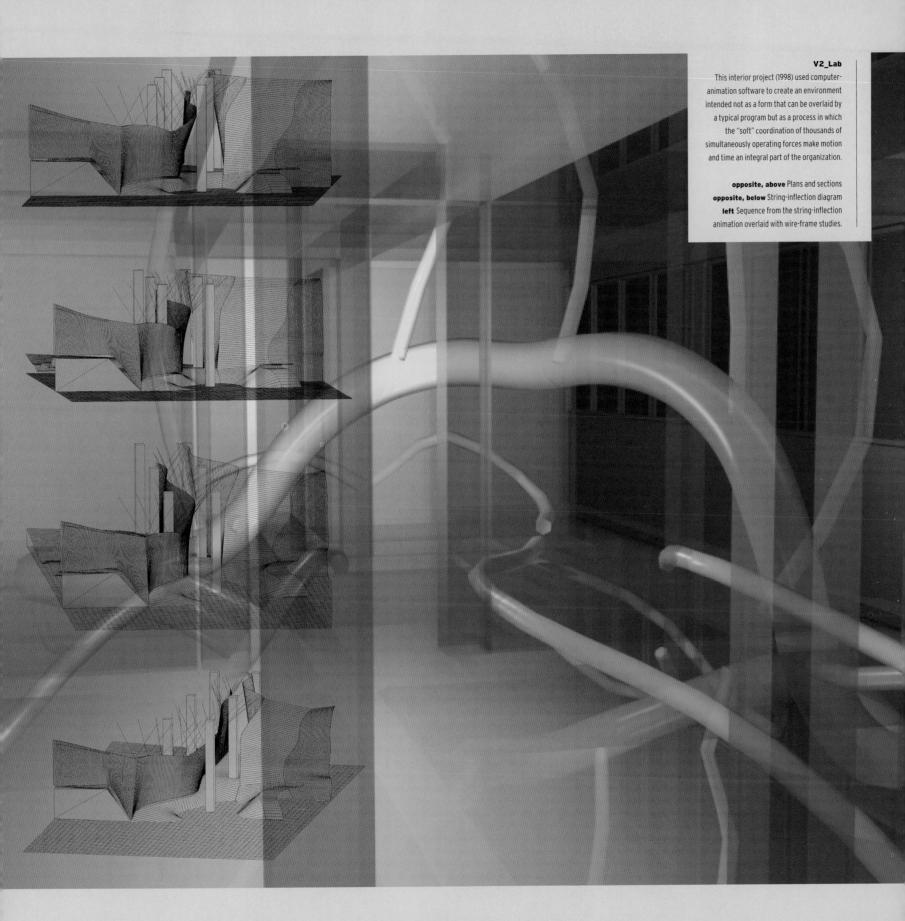

V2_Lab

This interior project (1998) used computer-animation software to create an environment intended not as a form that can be overlaid by a typical program but as a process in which the "soft" coordination of thousands of simultaneously operating forces make motion and time an integral part of the organization.

opposite, above Plans and sections
opposite, below String-inflection diagram
left Sequence from the string-inflection animation overlaid with wire-frame studies.

V2_Engine

left V2_Lab is part of V2_Engine, a future renovation project in Rotterdam, Holland. V2_Engine consists of a central void, a three-dimensional "window" where visitors enter the building, which will be partially finished with a translucent fabric and protrude from the façade. Borrowing the notion of a "spring," a scientific term for a non-static point capable of passing on force, a computer model was used to determine the location of the void. All forces within the spring are channeled towards the building's extremities by way of "strings." Sounds and images will be generated by a special software engine that will roam the Internet for images, which will be projected onto the exterior's fabric façade.

V2_Lab

right For the V2_Medialab, an international laboratory for unstable media, media are not perceived as a comfort-creating service but as a force that accelerates and destabilizes reality. The virtual is not a parallel world that exists on the other side of reality but something that continually charges up the present. Thus architecture here assumes the attitude of the furniture and textiles as catalytic agents that accelerate, vectorize, seduce and flex the space. One progresses seamlessly from a computer-generated process of forces, vectors and springs to inflections in plywood sheets and PVC pipes, to the undulations of the floor, to the chairs with adjustable spring legs, to tensions in the 4-millimeter-thick plastic wall (stretched with steel cables and springs), to the tensions within the body, to the arm and leg muscles that provide a constant neurological background to all human activity taking place there, indeed to all the human behaviors that may fall outside of the programmatic diagram.

MARCOS NO

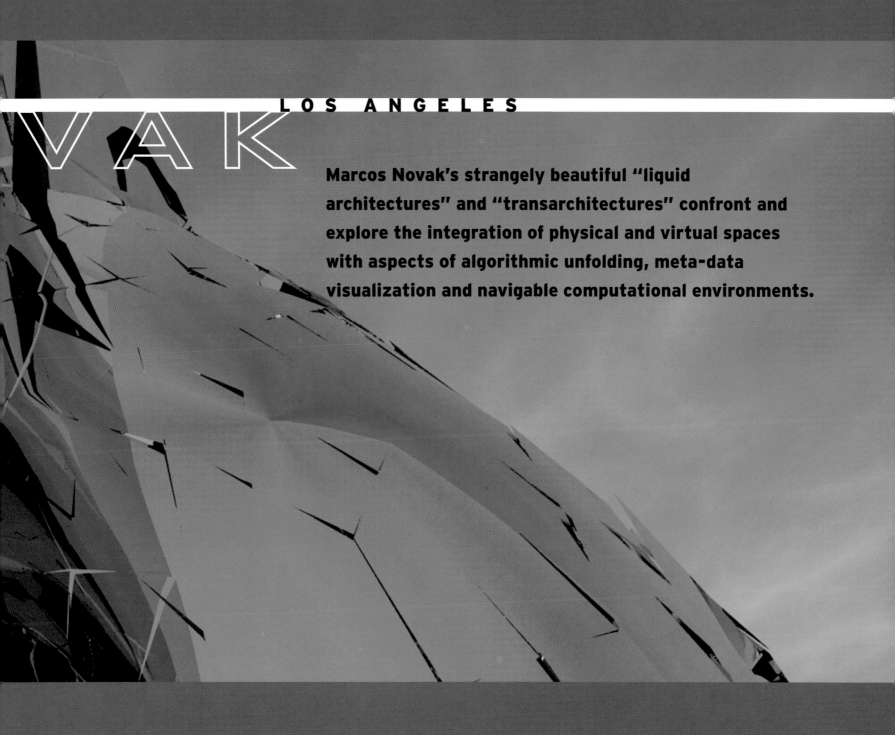

VAK

Marcos Novak's strangely beautiful "liquid architectures" and "transarchitectures" confront and explore the integration of physical and virtual spaces with aspects of algorithmic unfolding, meta-data visualization and navigable computational environments.

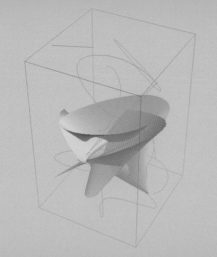

Marcos Novak is a visionary architect, theorist, intermedia artist who has actively developed strategies to address how physical space has been transformed by virtual space. Interweaving non-Euclidean spatial concepts with aspects of algorithmic unfolding, meta-data visualization (the mathematical modeling of data-space) and navigable computational and musical environments, Novak is widely regarded as the leading proponent of cyberspace as an autonomous architectural space.

To explore the integration of physical and virtual spaces, Novak has originated the concepts of "liquid architectures" and "transarchitectures," which outline the possibilities facing architecture and culture in the space-time vernacular of global communication and networked computation and which are part of a larger cultural phenomenon he terms "transmodernity." The concepts are grounded in the belief that today's constants in science and mathematics have been replaced by variables, a condition apparent not only in these fields but in the wider cultural arena. Throughout history we have accepted the reality that contradictory social entities or forces can often be combined without dissolution, which when successful can drive the cultural engine of an age, but we have rarely accepted such amalgams in the field of spatial production. Novak suggests a liquid architecture that combines opposites – the soft with the hard, the masculine with the feminine, the real and the virtual – to produce an alien, third condition. Although this notion was never intended merely to "architecturalize" cyberspace, liquid architectures have become inevitably associated with virtual reality.

Cyberspace itself was never limited to virtual reality ... but had more to do with our invention of a pervasive and inescapable information space completely enmeshed with culture ... I came across the idea of "transarchitectures" as a way of refocusing liquid architectures to broader issues, and to the idea of eversion, or turning-inside-out, of cyberspace. Transarchitecture is a superset that contains and extends liquid architectures. If liquid architectures have to do with the observation of liquid variability as a key part of our zeitgeist, transarchitectures are focused on the effect of that variability. We have moved from specialization of disciplines to multidisciplinarity, interdisciplinarity, and now transdisciplinarity. The prefix "trans-" implies not only the acceptance of the com-

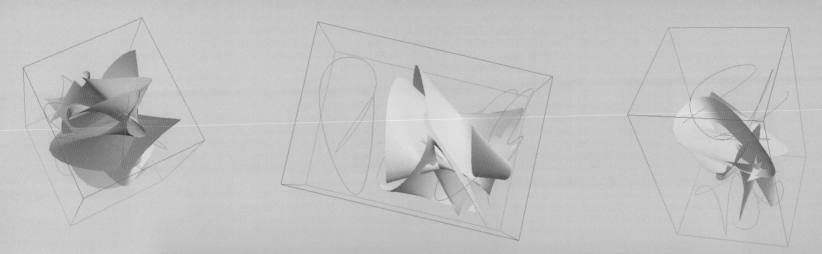

Data-Driven Forms

These images (1997–98) are the results of deriving forms from fields of found data. As spatial models, the forms explore two concepts: the delamination of passage (from one data set to another) and arbitrary cross-fade (between data sets). In the examples shown here, an algorithmic function extracted from linked Web pages as two sets of points in a three-dimensional matrix. Using spline-based interpolation, two sets of curves were generated. From further functions, the two sets of intertwined surfaces, or "lamina," were formed. A series of crossings or links (cross-fades) were then enframed between the conjoined surface-forms, producing a rich enmeshing of distorted frames and surface modulations.

above Data-driven surface evolutions **opposite** and **above right** Six images of cross-fades or linkages between the lamina calculated as distorted hyperframes. **right** and **far right** Variant studies of the conjoined lamina and cross-fade linkages.

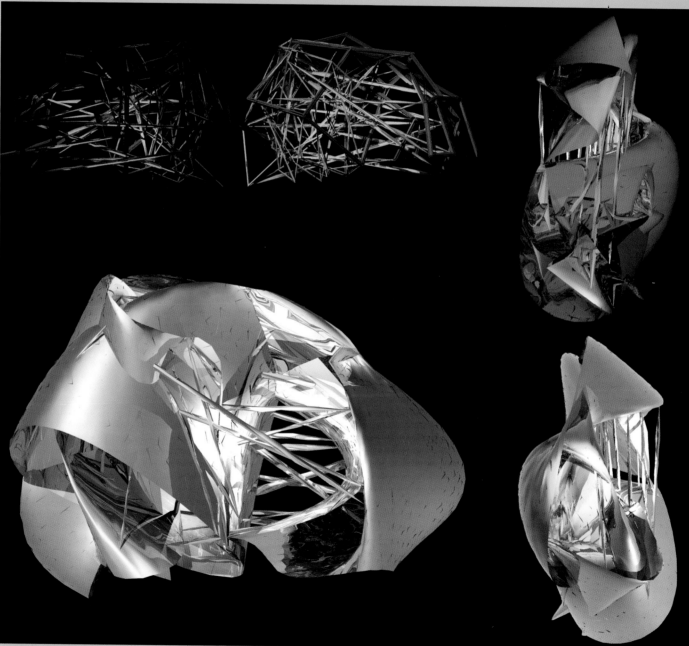

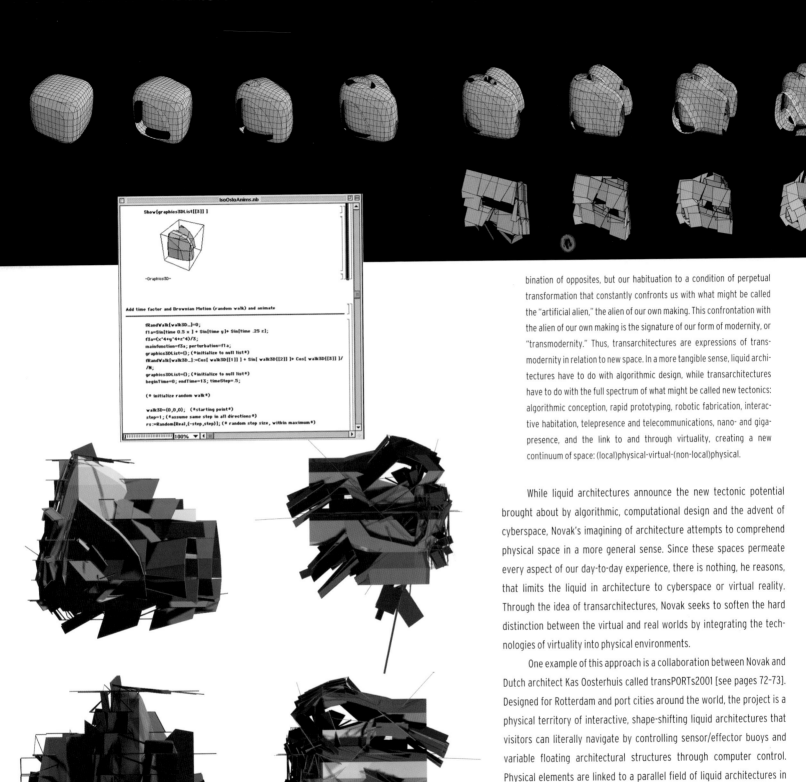

```
IsoOsloAnims.nb

Show[graphics3DList[[3]] ]

-Graphics3D-

Add time factor and Brownian Motion (random walk) and animate

fRandWalk[walk3D_]=0;
f1a=Sin[time 0.5 x ] + Sin[time y]+ Sin[time .25 z];
f3a=(x^4+y^4+z^4)/3;
mainfunction=f3a; perturbation=f1a;
graphics3DList={}; (*initialize to null list*)
fRandWalk[walk3D_]:=Cos[ walk3D[[1]] ] + Sin[ walk3D[[2]] ]+ Cos[ walk3D[[3]] ]/
/N;
graphics3DList={}; (*initialize to null list*)
beginTime=0; endTime=13; timeStep=.5;

(* initialize random walk*)

walk3D=(0,0,0);  (*starting point*)
step=1; (*assume same step in all directions*)
rs:=Random[Real,{-step,step}]; (* random step size, within maximum*)
```

bination of opposites, but our habituation to a condition of perpetual transformation that constantly confronts us with what might be called the "artificial alien," the alien of our own making. This confrontation with the alien of our own making is the signature of our form of modernity, or "transmodernity." Thus, transarchitectures are expressions of trans-modernity in relation to new space. In a more tangible sense, liquid archi-tectures have to do with algorithmic design, while transarchitectures have to do with the full spectrum of what might be called new tectonics: algorithmic conception, rapid prototyping, robotic fabrication, interac-tive habitation, telepresence and telecommunications, nano- and giga-presence, and the link to and through virtuality, creating a new continuum of space: (local)physical-virtual-(non-local)physical.

While liquid architectures announce the new tectonic potential brought about by algorithmic, computational design and the advent of cyberspace, Novak's imagining of architecture attempts to comprehend physical space in a more general sense. Since these spaces permeate every aspect of our day-to-day experience, there is nothing, he reasons, that limits the liquid in architecture to cyberspace or virtual reality. Through the idea of transarchitectures, Novak seeks to soften the hard distinction between the virtual and real worlds by integrating the tech-nologies of virtuality into physical environments.

One example of this approach is a collaboration between Novak and Dutch architect Kas Oosterhuis called transPORTs2001 [see pages 72-73]. Designed for Rotterdam and port cities around the world, the project is a physical territory of interactive, shape-shifting liquid architectures that visitors can literally navigate by controlling sensor/effector buoys and variable floating architectural structures through computer control. Physical elements are linked to a parallel field of liquid architectures in cyberspace, linking navigation in the harbor with virtual navigation on the Internet. This interleaving of the dissimilar and disconnected real and virtual strands of our existence involves not "merely collages or juxta-positions, as would characterize modernity, nor morphings, blendings, or folds, nor even formations of monstrosities that are still terrestrial, as would characterize recent postmodernity, but warpings into alien

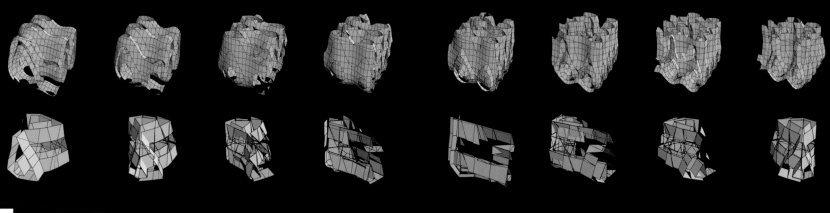

Data-Driven Forms

top The links between the lamina initially generated as plain lines, shown here in detail and on the previous pages at a later stage. Eventually they were parametricized and remodulated to vary along their lengths according to the contents of the original data sets. Since the algorithm for the forms was time-based as well as data-driven, the forms changed with the data sets. In cyberspace these forms are generated in real time at the point of transition from one information node to another.

opposite and **right** Detail studies of the intermingled surfaces and hyperframed linkages.

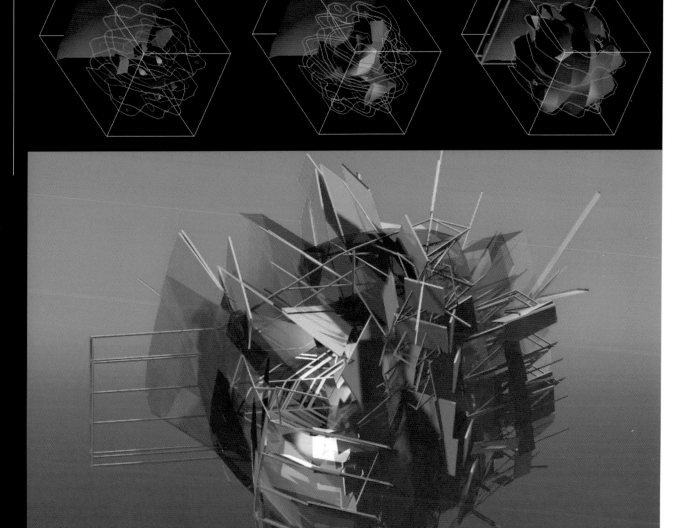

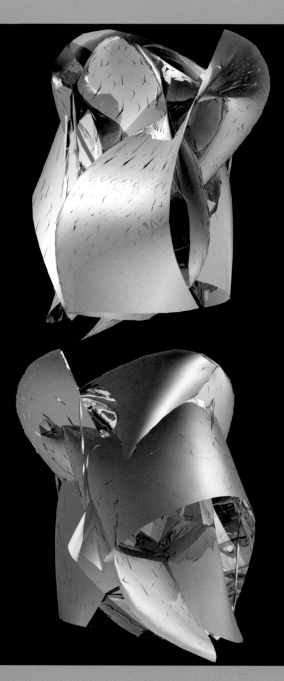

territory, true transmutations into unpredictable conceptual spaces, phase transitions into completely new states of being" – the transmodern state of being.

Novak describes this condition as "eversion," the turning inside-out of virtuality so that it is not just reliant on the technologies that support its existence but cast into the physical world. Historically, Novak defines eversion as the fifth stage of virtuality: the first used light and shadow projection, mirrors and shadow theaters, Plato's Cave in other words; the second stage encompassed zoetropes (early cinematic form), cinema, television, digital sound and processes of digital-to-analogue and analogue-to-digital conversion; the third virtuality is a form of inversion through computation, scientific visualization, simulation and special effects; the fourth is immersive, casting the world into cyberspace; eversion, the fifth, therefore, casts the virtual back into the material world, with the virtual-real and the actual-possible oppositions woven together. Novak writes,

> It is helpful to realize that we have already constructed pockets of intelligent space on our computer monitors. A computer screen can be seen as a two-dimensional prototype of a space whose extent is fixed but whose partitions and functions change as needed or desired. This space, once confined to the user-interfaces of our computers, has already escaped the monitor and has entered the three-dimensional world at large. The language of windows, menus, icons, tools and sundry controls to which we are already habituated is being extended to the third dimension.

Working across media and technologies, the full spatial spectrum of Novak's work includes algorithmically generated designs that explore extraordinary mathematical conceptions of space and music derived from four-dimensional computational operations. His work might be seen as total spatial performances configured in the interstices between the physical, the virtual and the interactive – all interwoven expressions of a larger project that transmutes models of inquiry into pure tectonics, pure form.

Variable Data Forms

This ongoing investigation seeks to create architectonic propositions that are liquid, algorithmic, transmissible and derived from the geometries of higher dimensionality. By "liquid," Novak intends a total but rigorous variability driven by data shifts in cyberspace that can be transformed into the physical world. By "algorithmic" Novak means that the forms are never manipulated through manual corrections: rather, the mathematical formula that generate them are adjusted to produce different results. By "transmissible" Novak means that his data-forms can be compressed into algorithmic codes for transmission to fabrication sites, machines or to virtual environments.

Novak begins by designing generative procedures and mathematical models that, while bound by diverse computational factors, are indifferent to practical concerns.

Form follows neither function nor form; rather, I am concerned with what the Situationists called the 'psychogeography' ["the study of specific effects of the geographical environment, consciously organized or not, on the emotions and behavior of individuals."] of these emergent spaces. Moving from algorithmic derivations to the Debordian 'derive,' from the 'naked city' to 'naked transarchitectures,' I am interested in drifting through these spaces in a condition at least momentarily uncontaminated by the spectacle. No purpose is needed; indeed, to the extent that the otherworldly spaces are fundamentally unfamiliar, it is possible to savor their inherent psychogeographic content in a moment of surprise just prior to the onslaught of references.

Still, external influences can be injected into Novak's uncontaminated forms to give them tangible, real-world expressions and responses. Each algorithmic variable or operation represents a "channel" into which an external rule or influence can be either statically or dynamically mapped. Once a physical model is constructed, a repeatable process may locate a set of factors extrapolated from the real world – motion captured from video, for instance – that can inflect the next iterative construction.

Marcos Novak's investigations offer the world a series of extreme provocations. Content neither to produce in the virtual realm nor particularly dedicated to the institutionalized art of building (as it is currently defined), his self-generated and undefinable projects reject the tired standards reserved for building and compel us to reflect on the architect's role as it is being reprogrammed by technical evolutions. Though striking in their virtuosity, Novak's spaces are hardly geared for popular consumption: they are uneasy and alien, resistant to convenience-driven late-capitalist production. The works are strangely beautiful, and because beauty is a form of equilibrium, Novak's project is ultimately balanced between radical progress, considered inquiry and poetic reflection.

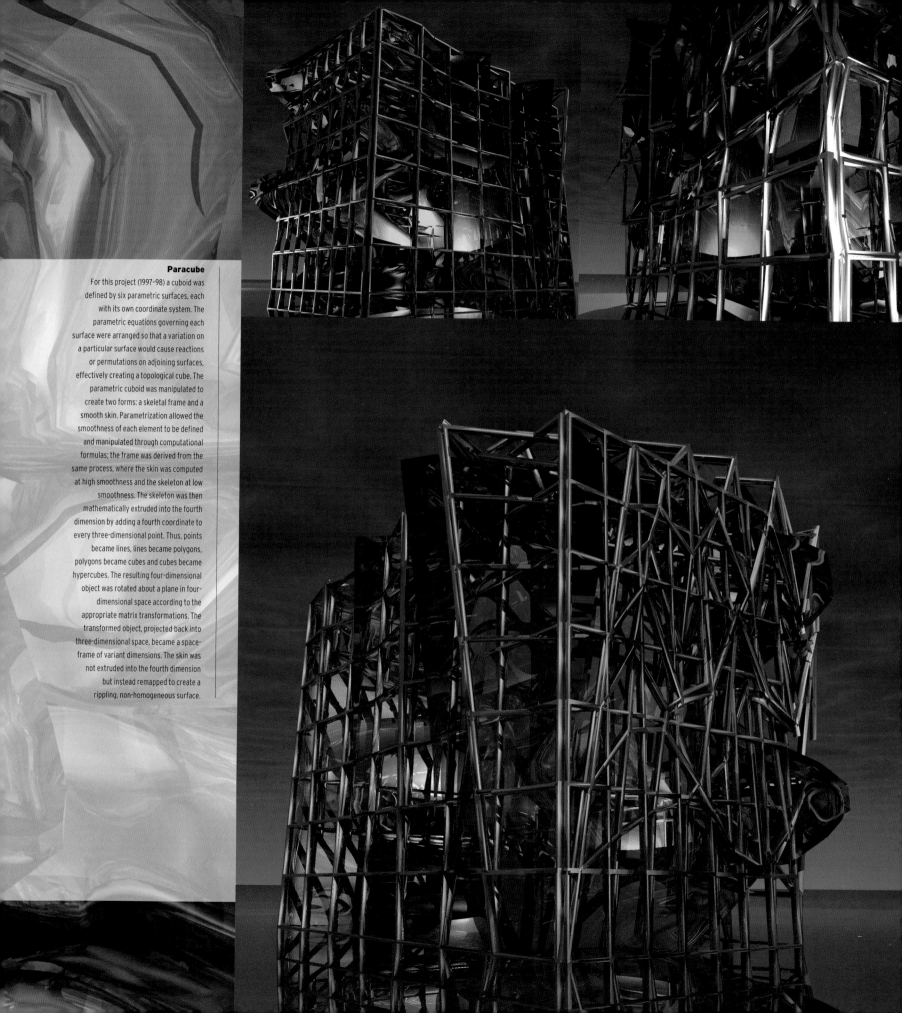

Paracube

For this project (1997–98) a cuboid was defined by six parametric surfaces, each with its own coordinate system. The parametric equations governing each surface were arranged so that a variation on a particular surface would cause reactions or permutations on adjoining surfaces, effectively creating a topological cube. The parametric cuboid was manipulated to create two forms: a skeletal frame and a smooth skin. Parametrization allowed the smoothness of each element to be defined and manipulated through computational formulas; the frame was derived from the same process, where the skin was computed at high smoothness and the skeleton at low smoothness. The skeleton was then mathematically extruded into the fourth dimension by adding a fourth coordinate to every three-dimensional point. Thus, points became lines, lines became polygons, polygons became cubes and cubes became hypercubes. The resulting four-dimensional object was rotated about a plane in four-dimensional space according to the appropriate matrix transformations. The transformed object, projected back into three-dimensional space, became a space-frame of variant dimensions. The skin was not extruded into the fourth dimension but instead remapped to create a rippling, non-homogeneous surface.

Constructed by and inside of flows, Greg Lynn's "animate forms" are designed within an unstable realm of variable, fluctuating dynamics and movements, leading away from an architecture of stasis to one of evolution.

GREG LYNN/

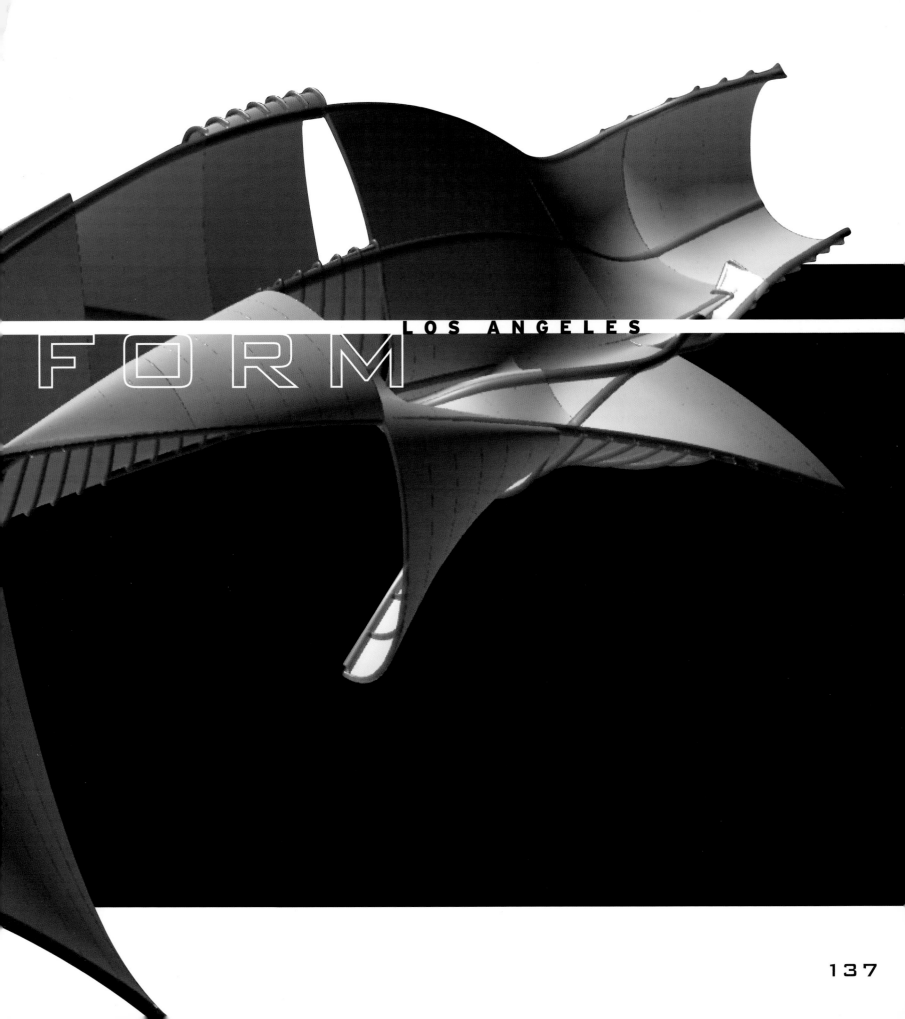

FORM

Animation is a term that differs from, but is often confused with, motion. While motion implies movement and action, animation implies the evolution of a form and its shaping forces; it suggests animalism, animism, growth, actuation, vitality and virtuality. In its manifold implications, animation touches on many of architecture's most deeply embedded assumptions about its structure. What makes animation so problematic for architects is that they have maintained an ethics of statics in their discipline. More than even its traditional role of providing shelter, architects are expected to provide culture with stasis. Because of its dedication to permanence, architecture is one of the last modes of thought based on the inert. Challenging these assumptions by introducing architecture to models of organization that are not inert will not threaten the essence of the discipline, but will advance it.

Greg Lynn, *Animate Form*

Over the last half decade Greg Lynn, principal of Form, has revolutionized architectural thinking about movement, permanence and structure by challenging and overturning long-held assumptions on statics and anti-statics. Using computational motion geometry and time-based dynamic-force simulations, Lynn's innovative approaches have made him a key contributor to contemporary architectural discourse. By exploding the dominate modes of architectural production – Cartesian geometry, verticality, orthogonal projection, typology and procession – Lynn has developed a model for translating the animate energies of our proliferating technologies into an ambitious and extraordinarily dynamic body of built and unbuilt projects.

Greg Lynn contends that the animate approach to architecture will reconstitute the inherited standard of stationary spatial description into a better expression of complex formulations and applications, to allow built form to be shaped with virtual movement and potential. "Animate design," Lynn maintains, "is defined by the co-presence of motion and force at the moment of formal conception. Force is an initial condition, the cause of both motion and the particular inflections of a form." Accepting that the term "virtual" has been so corrupted that it is usually misused to describe a computerized space, Lynn uses the word in its truer aspect to define a

Embryo House

This study (1998) into computer-controlled fabrication produced over 2048 panels, each of which is unique in shape and size. Individual panels were networked to one another so that a change in one panel could be transmitted throughout every other in the set. The variations to this surface are virtually endless, yet in each variation there is always a constant number of panels with a consistent relationship to the neighboring panels. The volume is defined as a soft flexible surface of curves rather than as a fixed set of rigid points, and instead of cutting window and door openings into this surface, an alternative strategy of torn, shredded and louvered openings was invented. Openings in the panels respect the curved envelopes' soft geometry and any dent or concavity is seamlessly integrated into the surface. The curved chips of the envelope are made of wood, polymers and steel, which were all fabricated with robotic computer-controlled milling and high-pressure water-jet cutting machinery.

opposite, top and **middle** Installation of resin-cast embryos.
opposite, bottom The water-jet cut rubber panels are unfolded into stars or fronds.
right Details of three-axis CNC-milled composite boards used as panel molds or form-works for casting.

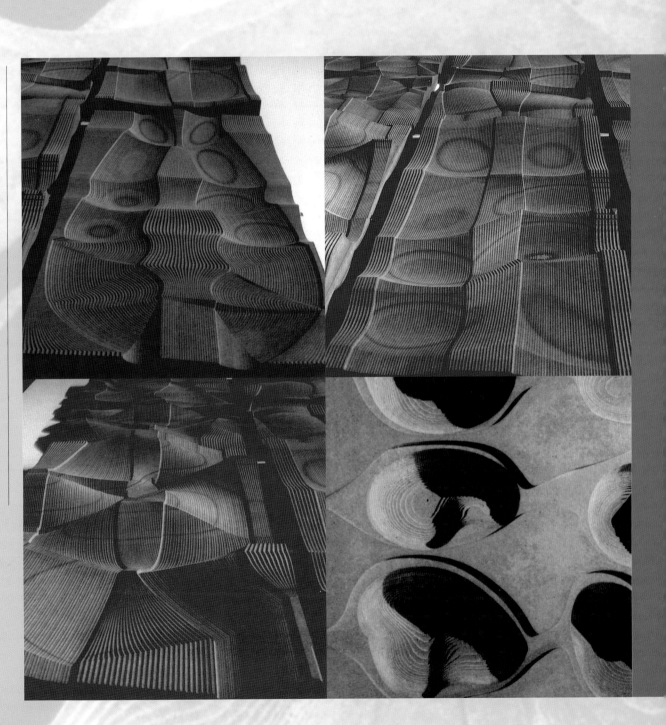

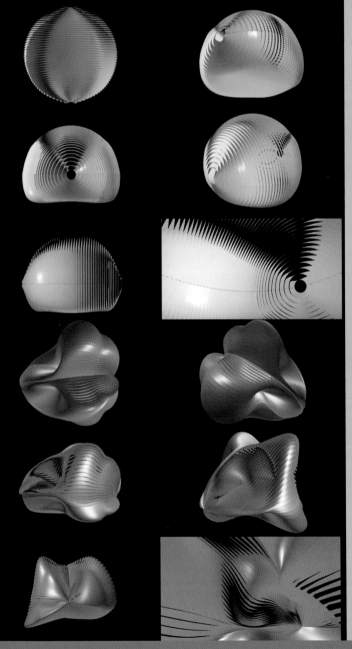

hypothetical state or condition that has the potential of coming into being. Architecture, Lynn argues, more than any other discipline, has been about the making of virtual representations and computations, but one function of virtuality that has been neglected is the conception of a virtual force and the divergent transformations it implies for architecture. Because buildings have conventionally been defined by Cartesian fixed-point coordinates and instituted through a state of inertia, Lynn argues that an object or building should also be determined as a vectorial motion defined or directed along a path.

This object-building can be positioned as a trajectory in relation to other virtual or real forces, scopes and fluxes, such as pedestrian or vehicular movement, natural or artificial climatic conditions. By allowing form and structure to be designed within an unstable realm of variable, fluctuating dynamics and movements, Lynn believes that architecture will move away from its inert logic of stationary relations into an operative rhythm of transmission and interaction. This in turn implies a fundamental shift from a classically figured architecture toward an architecture of context and particularity that could coexist with developments in other cultural arenas.

Lynn argues that because the primary manner in which motion in architecture has been understood in the late twentieth century is through the cinematic model of space, in which a succession of frames simulate movement, architecture has been assigned the role of the static frame through which motion progresses: "Force and motion are eliminated from form only to be reintroduced, after the fact of design, through concepts and techniques of optical procession." But with computer animation and special-effects softwares – unlike the cinematic model of spatial description – fluctuating dynamics and motions can be introduced at the moment of design conception. In addition to the special-effects and film industries, naval, aeronautical and automobile design employ similar approaches to modeling forms within spaces, which can be understood as mediums of movement and force. As computers are used increasingly as devices for design in architecture rather than for the mere task of rendering and imaging, Lynn believes that today's softwares and hardwares hold great promise for an animate architecture. One example is "inverse kinematic" animation, the potential action and contour of a form that is determined by accumulated vectors that refine themselves over time.

The traditional drawing tools that were used to design buildings – compasses and triangles – require little more than a working knowledge of basic algebra. Although a building's mechanical, acoustic and structural systems are usually resolved by engineers with calculus, architects rarely depend on advanced mathematics for design. But Lynn argues that despite architecture's fundamental reliance on geometrical descriptions of space, the lack of knowledge of higher mathematics has hindered the use of motion and flow in the design process, as these ideas require calculus-based operations. Today's, however, even the most rudimentary CAD applications make use of topology and time-and-force modeling, thus presenting architects with the novelty of sketching and drawing motion with calculus, courtesy the computer's powerful processing capability. "The challenge for contemporary architectural theory and design," Lynn writes, "is to try to understand the appearance of these tools in a more sophisticated way than as simply a new set of shapes. Issues of force, motion and time, which have perennially eluded architectural description due to their vague essence can now be experimented with by supplanting the traditional tools of exactitude and stasis with tools of gradients, flexible envelopes, temporal flows and forces."

Because his designs are determined by calculus, one of the first axioms of Lynn's topologically modeled surfaces or objects is that they may take a polyvalent form. In other words, structures are not composed of discrete points but of consistent streams of proportionate values that can be influenced (forced) by the field in which they are modeled. Rather than being objects or spaces shaped only by an internal definition or autonomous design logic, the surfaces or forms can be moved in space, and as they do so their shape changes according to a position within a gradient space. Thus, a topology or form duplicated but placed in a different context would behave differently depending on the influence and result in a unique and distinct configuration.

Lynn's Port Authority Gateway project in Manhattan provides an effective demonstration of how fluctuating dynamics can be incorporated

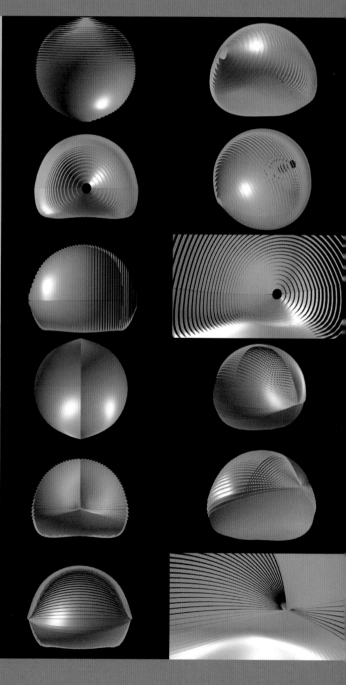

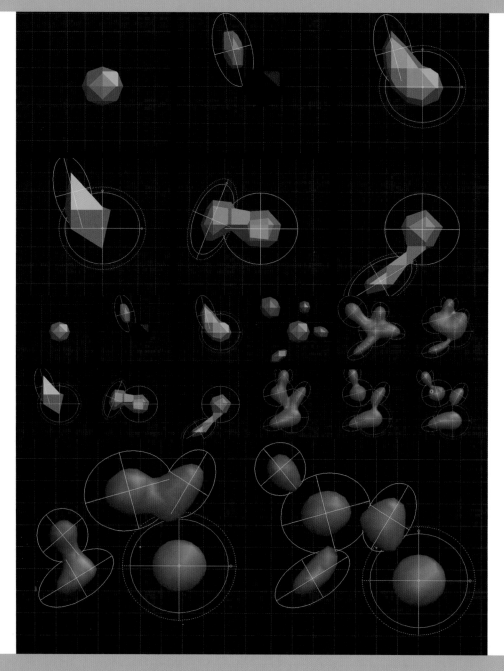

Korean Presbyterian Church

A collaboration with Michael McInturf and Douglas Garofalo, the project (1995-99) involved the adaptive re-use of and extension to a factory building to provide a center for cultural, social and religious activities.

left Meta-blob modeling studies. "Meta blobs" are specifically located nodes that interact according to their zones of assigned gravitational force. As the computer calculates the conditions of balance, the nodes grow and fuse into new nodes until they achieve a state of equilibrium.

opposite, clockwise from bottom left The "blob" strategy is used to generate a single volume that results from the growth and fusion process of separate programmatic nodes. What begins as a collection of single rooms (nodes), ends as a single surface that incorporates the entire program. · Phase portrait of the design's central object · Instead of radially arranged nodes, a parallel alignment is chosen and allows a different manipulation of the growth and fusion process. The shape of the single volume is then arrested after the separate nodes reach their equilibrium · A single shell is offset to generate a double-wall. The inner surface follows this offset, and each surface rib flares until its connects back to the outer skin, creating a stiff, three-dimensional structure needing no additional columns and an intricate interior volume of louvered-hung ceiling panels and faceted walls. · Phase portrait (in plan view) showing the generation of the entrance "tubes" for access and circulation into the factory building's existing column grid.

opposite, bottom right The exploded rendering reveals the existing factory, rib-like structural tubes and the interior and exterior envelope.

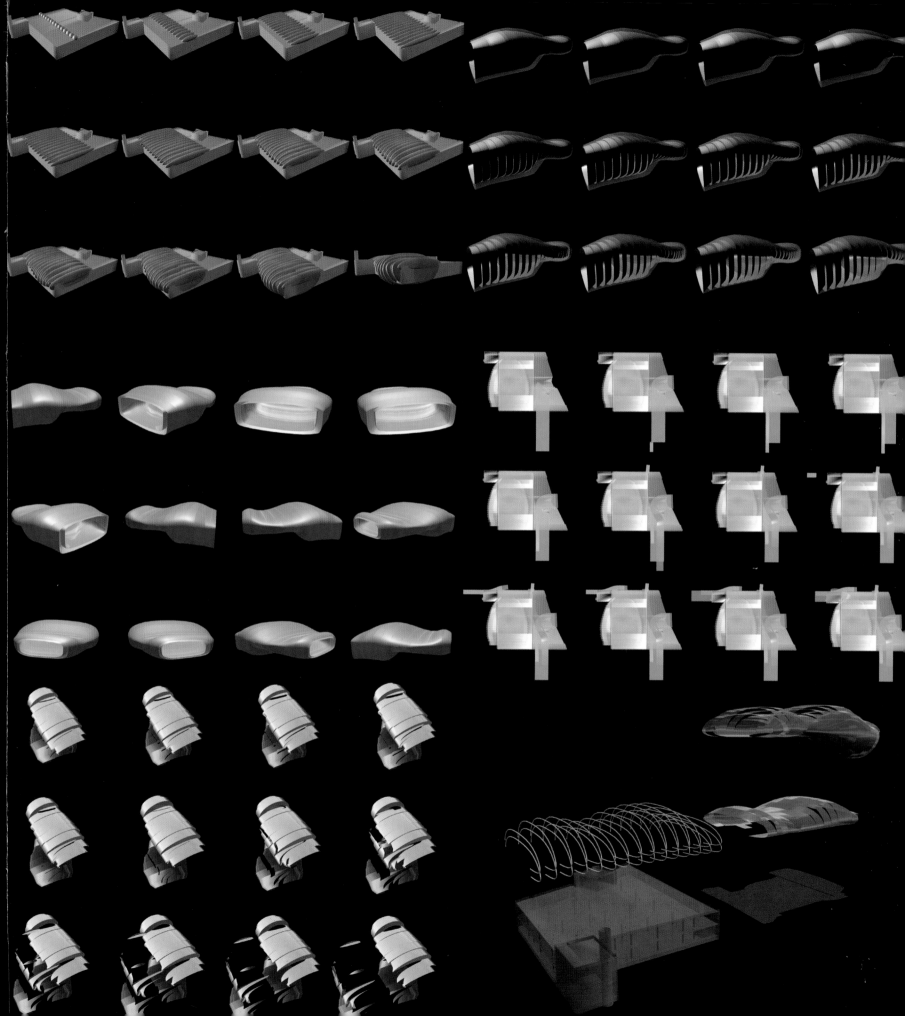

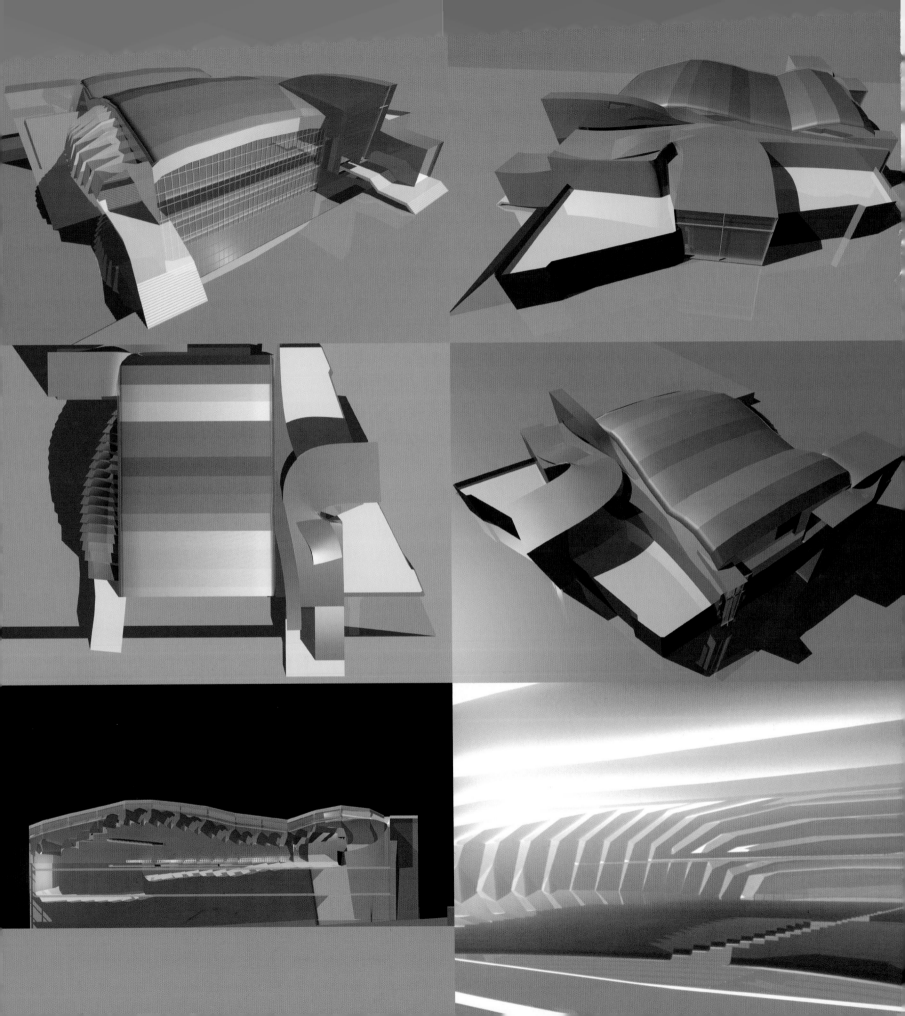

Korean Presbyterian Church

The building's multiple uses are reflected in the functions that support the central sanctuary space. The existing basement level includes church offices, five meeting halls, a day-care center, eight classrooms and a library; the first level contains a 600-seat wedding chapel, large cafeteria and exhibition hall. The design retained the factory building's industrial vocabulary and transformed its interior spaces and exterior massing into a new kind of religious building. The exterior was renovated to its original state, and the addition, which exploits the factory's eccentricities, is divided into two sections: a large shed structure with long span, open areas and a second structural system of varying steel elements. The addition relates to the existing areas with two types of construction: the shed is clad in metal and has an undulating shape that refers to the nearby rail lines; the second forms are stucco-clad entry tubes that snake vertically through the existing bays.

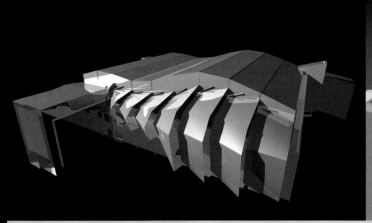

top Views of the church under construction.
above right Northwest aerial view, with the staircase's metal roof enclosure.
above far right Perspective of the first design proposal.
right The perspective view from the east shows the large elevation of the sanctuary space, entry tubes and elevator shaft.
opposite, clockwise from top left Aerial view from west showing the main elevation and the sanctuary space's curtain wall. • Aerial view from south with the entry "tube" turning towards the existing façade. • Southeast aerial view reveals the smaller entry tubes used as rehearsal space. • Interior view of wedding chapel. • The longitudinal section of the sanctuary space. • Bird's-eye view of the roof reveals the entire assemblage of building components: the sanctuary space's metal roof (center), exterior staircase (left), the entry tube (right) and the existing factory building.

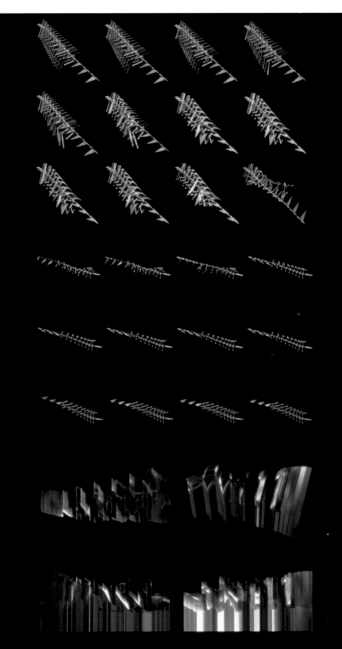

H2 House for the OMV

This multifunction visitor center for the display of new solar and low-energy technology was designed in 1996 for the Austrian Mineral Oil Processing Company in Schwechat, Austria. The design used computer simulation software to model the solar vault and the alignment and shape of the building's shading devices and photovoltaic cells. The north façade's form was shaped by simulating of the movement of the adjacent highway. Vehicles' motions were used to sweep a series of surfaces, which when viewed from the highway reveal the building's interior as a sequence. The forces were translated into the exterior form through a system of flexible surfaces that respond dynamically to the site.

left, top to bottom A series of snap-shot studies were executed to determine a deformable surface for the skeleton of the building envelope. In this case the test surface was defined as "stiff." · Other snap-shot series with the deformable frame-like surfaces. · Portrait of the surface as a continuous sweep. Using surface-manipulation software, the complex surface prototype was redesigned and abstracted to work out its expressive and architectural qualities.
right, top Studies of the building envelope prototype.
right, bottom View of model, with highway at lower right.
opposite Models of building envelope prototype. The interior is separated into two zones by a translucent fabric, on which computer animations, video sequences and still images can be projected. Mechanical systems are aligned so that when the lighting behind the screen is switched on, it becomes transparent, making the structure's experimental energy system visible to the visitor.

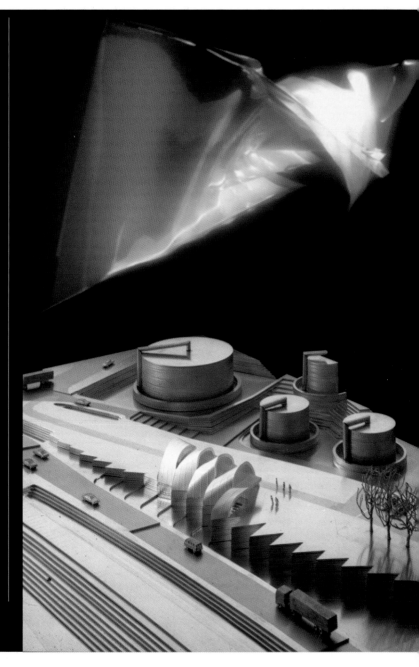

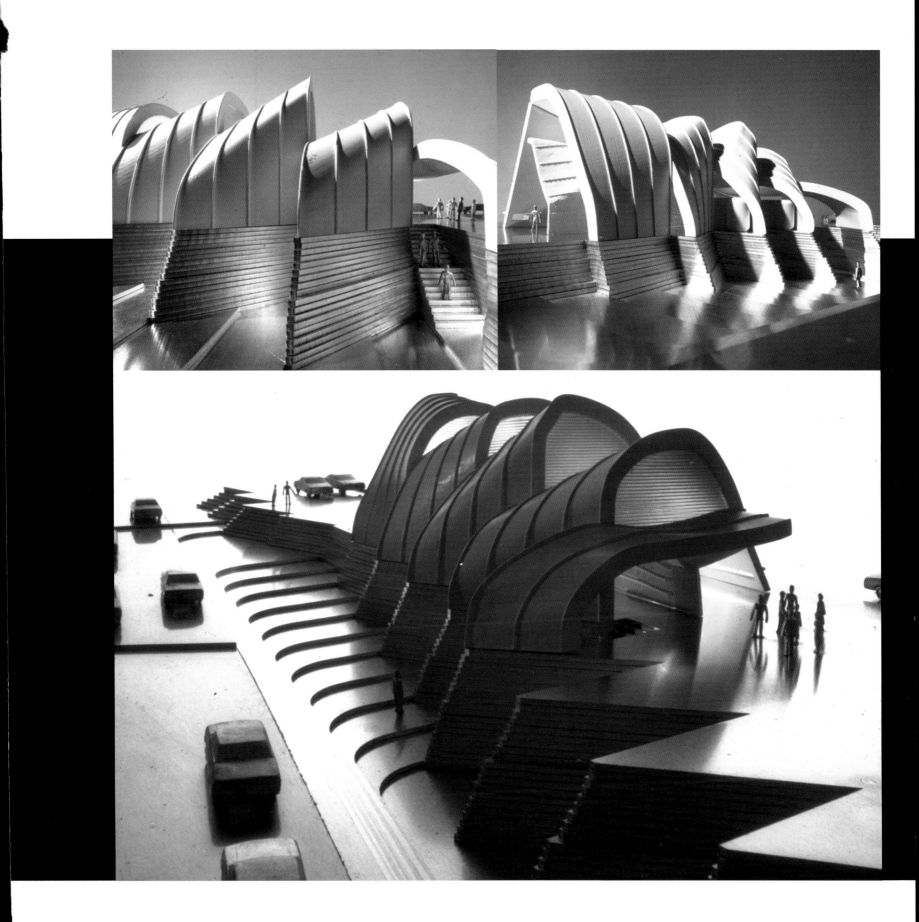

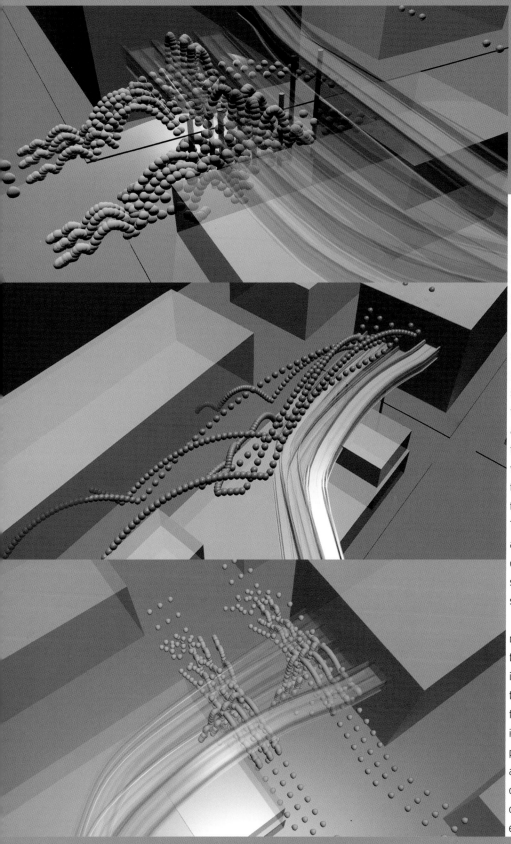

into architectural formations. Located near a series of massive steel bridges that connect the Lincoln Tunnel (running under the Hudson River between New Jersey with midtown Manhattan) to the upper decks of the Port Authority Bus Terminal along Ninth Avenue, 42nd and 43rd streets, Lynn's complex composition of steel tubular frames stretched over by translucent tensile surfaces that can act as screens for large media projections. The dynamic setting – cars and buses, pedestrians and vehicles, underground and overground, land and water – was initially simulated as a set of forces and geometric particle systems of varying physical weights and velocities. Using an animation and modeling software package, the simulations were generated to track a complex force-field of static and moving attractors around the site – car and bus and pedestrian flows, for example – each with different intensities, velocities and movements. At first the flows were modeled and animated as vaporous, light material emanating from the west façade of the bus terminal; gradually the weight of the simulated material was increased so that very heavy particle flows could be traced and frozen over the site. This particle material, translated visually into a collection of gravitationally weighted balls, was then amalgamated to create the building's tubular components. Because the dynamic movements were animated within a system of gravities, the tubes have ideal ballistic forms, critical for further static assumptions and calculations.

Rather than begin design conception within an idealized and neutral void, Lynn launches his architectural process in a charged space that instructs and informs architecture from and within a flow of mobile intensities. Spatial elements are embedded with vectorial effects before they are fired into a space regulated by gradients of motion and potential force. Rejecting the centuries-old tradition of perspectival projection in which architecture is frozen into an iconic semblance of itself, or exploiting the more recent artifice of the cinematic frame, Greg Lynn's animate architecture is constructed by and within dynamic flows. As a consequence, his endeavors point us towards the very radical possibility of an architecture no longer founded in stasis and permanence but in evolutionary growth.

Port Authority Gateway

This 1995 competition invited architects to design a protective roof and lighting scheme for the underside of the ramps leading into the Port Authority Bus Terminal in New York City. After simulating the flow of pedestrians, cars and buses across the site – each with differing speeds and intensities of movement along Ninth Avenue, 42nd and 43rd streets, as well as the four elevated bus ramps emerging from below the Hudson River – the forces of movement were used to establish a gradient field of attraction across the site. To find a form for this invisible field of attraction, Lynn introduced geometric particles that changed position and shape according to the influence of these forces. The particle studies were used to capture a series of phase portraits that showed cycles of movement over a given time period. The phase portraits were then swept with a secondary structure of tubular frames that linked the ramps, surrounding buildings and the terminal. Eleven tensile surfaces were stretched across the tubes as an enclosure and projection surface.

opposite, top to bottom The images show the result of "freezing" the system's complete movement, when the dynamic simulation is superimposed in a single image. · In one scenario the dynamic flight of a set of bouncing particle balls is traced after leaving the terminal along the bridge system. · In another situation, the particles run below the bridge system, behaving in a different manner, as the start location and distance to the force centers are different from the previous simulation.
top right Elevation
above right Roof plan
right Perspective view of the final design, showing the bridges, the bus terminal (left) and the translucent surface reserved for large-scale media projections like commercials and tourist and passenger information.

OCEAN

COLOGNE / HELSINKI /

An international network operating across six countries, OCEAN's
multidisciplinary and synergetic approach to architecture is
founded on a strong understanding of how digital technologies
are redistributing conventional models of cultural production.

LJUBLJANA / LONDON / OSLO / BOSTON

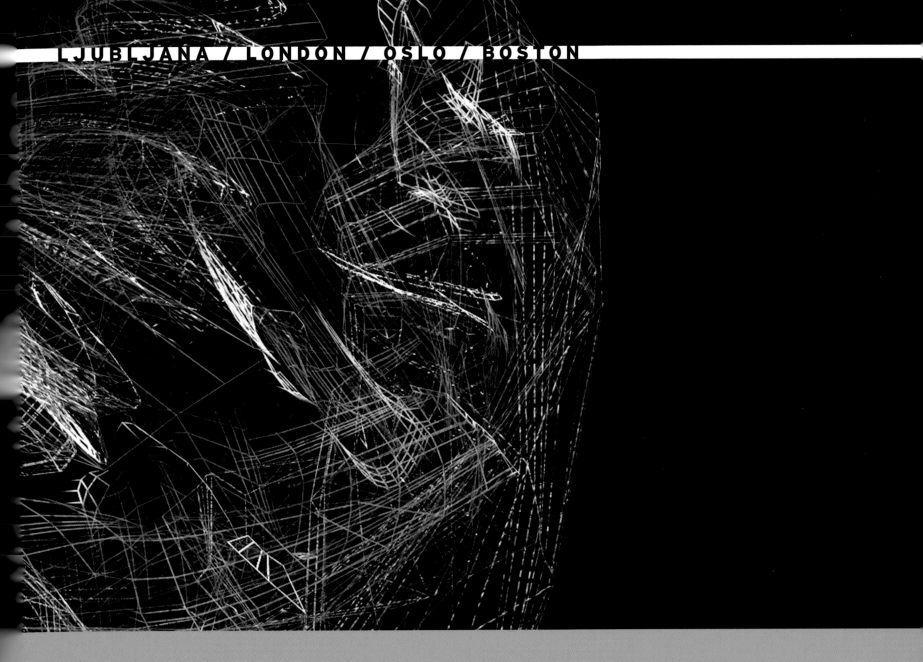

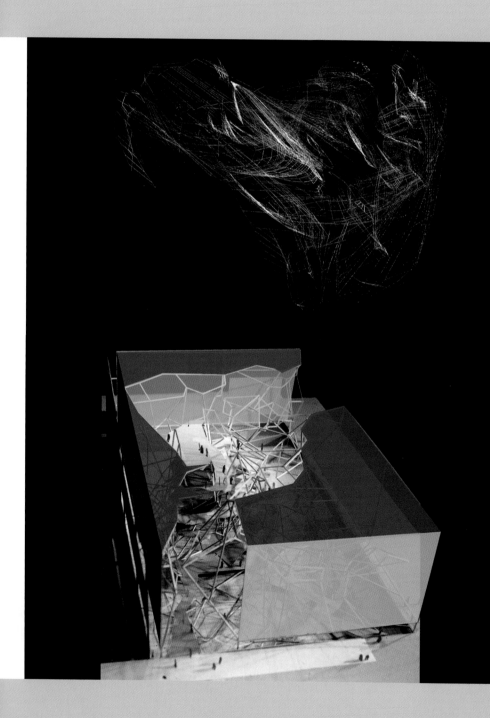

Originally configured in London in 1995, OCEAN is an international and interactive network of some twenty architects, designers and urbanists that operates across two continents, six countries and three time zones. Composed of six "local nodes" in Europe and North America, its organizational structure is collaborative, inclusive, multidisciplinary, synergetic and decentralized – an active system of architectural conception and production. OCEAN's dynamic and operative coherence is founded on a clear understanding of how digital technologies are dispersing and redistributing classically reinforced hierarchies, hegemonies and models of cultural production. Structured neither as a top-down practice nor as the utopic model of a collective, the network functions horizontally through dialogue, translation, interchange and transaction. Acknowledging a larger societal shift towards spatial paradigms that endorse the global condition of constant transition – the space of flows – and recombination, OCEAN works freely within the instability that defines our age. The studio's constructions and tasks vigorously resist binary oppositions (local-global, program-form, theory-practice) and reductive modes of production (form follows function, less is more) to seek a third condition that arises from processes of unfolding social diversity and accelerating technical connectivity.

Enabled by the widespread availability and cost-effectiveness of high-speed communication technologies, OCEAN's collaborative project is a synthetic merger of practice and research that exchanges singular authorship for mutual interests and engages a multitude of agendas. Each node (in Boston, Cologne, Helsinki, Ljubljana, London and Oslo) is simultaneously an independent local entity and an interdependent global agent in an energetic, evolving system of relations. Collaborations between and within the network's nodes are based on shifting reconfigurations of project-based teams for commissioned work, design competitions, research projects, education, exhibitions, publications, installations and public events. Working sessions are carried out in extended conference mode, as digital photographs, CAD models, drawings and texts for projects are electronically transferred between each node for editing or elaboration.

◄ Jyväskylä Music and Arts Center

OCEAN NO/UV's design for the Terra Cultura Music and Arts Center (1997) in Finland takes its cue from the surrounding topography, which is internalized inside the structure while the building's activities are exported to the outside in the form of a public performance space. Two categories were devised to achieve aspects of the inside-outside surface: "hard ground," natural rock, synthetic surfaces and wood platforms; and "soft ground," secondary data, such as trees, bushes and shrubs, and rods and free columns with synthetic nets. Because the ground condition comprises numerous kinds of hard and soft ground, the topological surface has varied materials and islands of secondary surfaces embedded in it.

above Virtual model constructed as a particle cloud. "Peels" taken from the cloud make up an "form-alphabet" from which the final building structure and interior forms were produced.
below Working model. View towards the tubular structure that configures the interior forms.

Töölö Football Stadium ►

Situated north of Helsinki city center, the site functioned as a wedge between the park surrounding the 1952 Olympic Stadium and a residential area. OCEAN NO/UV created a topological surface that would bridge these contexts and encompass seating, eating and movement, as well as forming as the stadium's roof structure. The surface is supported by two sets of columns, while the roof is held up by light-grey, twinned-steel reinforced concrete columns. Secondary support structures – like hollow-tube steel columns of varying diameter that look like a structural forest – are freely distributed throughout the space; some of the members double as lighting sources.

above The site plan shows the smoothing effect between the urban fabric and the park landscape.
below Aerial view of the final proposal.

Although each nodal point is immersed in a "total condition" produced by interacting with the other centers and is therefore continuously acting and reacting to other operations, each node maintains its autonomy by working on independent projects. The local points act as smaller-scale operations but can expand to suit larger project requirements. OCEAN's aim is to increase collaborative ventures within the network by sharing resources and expertise, but the design network also actively cultivates a logic of minor incoherence, so that the probability of contradiction has become a fundamental creative aspect to the multilinear organization. Indeed, discrepancies and disputes necessitate continual negotiations of basic operative and productive tasks, so that the work is in part the result of finding alternative ways of organizing its production and strategies.

The theoretical commitment of OCEAN is to search for new and differential modes in urban thinking and space-making, in particular toward fluid and topographical spatial organization and the invention of material effects. Made possible through the vectorial logics in digital softwares and hardwares, virtually all the group's work attempts to formulate a strategic urban architecture that validates and materializes contemporary culture and technology. The collaborative asserts that recent advances in digital space-modeling and scripting software will shift urban organi-zational principles from figure and field towards a vectorial logic of magnitude, trajectory and intensity. It follows that the emergence of digital procedures should catalyze city design. Mirroring developments in other creative fields (contemporary musical composition and notation, for example), these novel urban design procedures like smoothing and modulation will oppose conventional modes like delineation and figuration.

For instance, drawing from "sampling," the practice in popular music of "borrowing" fragments of songs, OCEAN has derived a kind of urban sampling to integrate spatial or differences by cataloguing the diverse or disjointed "bits" of the city. Stringing these bits together – in the manner that a D.J. "loops" pieces of musical tracks into a melodic and rhythmic synthesis – OCEAN produces a cohesive urban unity. Grafting

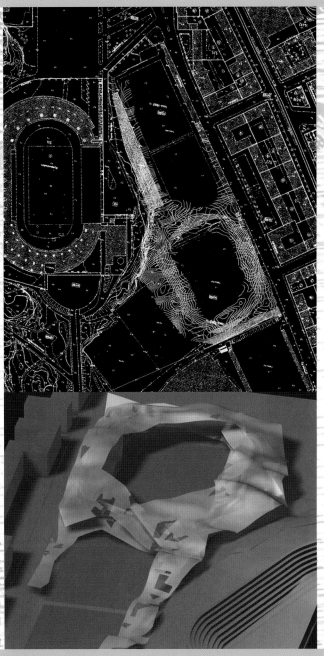

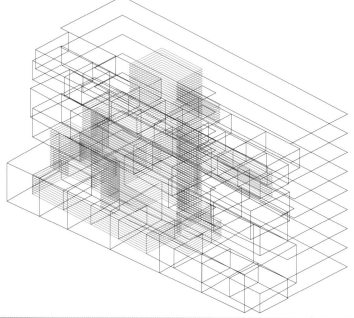

◀ Chamber of Commerce and Economy of Slovenia

OCEAN SI's design for an office building (1996-present) in Ljubljana, Slovenia, organizes the program and structure to reflect the deterritorialization of capital and the implications for its services – consulting, informing and educating. The architects sought to abolish conventional urban boundaries between public, semipublic and private activities. To optimize interaction, the semipublic areas are located not just on the ground floor (in the horizontal) but on all levels (in the vertical), achieving a smaller footprint on site and allowing an outdoor plaza to be created. A vertical hall functions as a "perforation point" in the structure's fabric, extending the plaza into the interior and forming a central core, around which the programs are arrayed.

above left Construction view and detail of the interleaved floors.
left Computer study of the deep Vierendel steel-beam structure.
below left Main façade as seen in the model. The exterior's horizontality conceals the vertical organization of the program inside.

Jeil's Hospital for Women ▶

OCEAN UK designed a 2000-square-meter hospital in Seoul that includes gynaecology, obstetrics, fertility departments and a 24-hour pharmacy. The double-skinned building encloses a space between two visually permeable surfaces that integrate the visual and public space of the city and the private territory of the building. The surfaces are mechanically adjustable to reveal or veil the interior spaces and to vary lighting conditions. The double envelope is composed of an orthogonal curtain wall and slab system, coupled with two systems of louvers, one tightly spaced for maximum visual buffering and a looser one for diffusing direct sunlight. The visual connection between the hospital and the urban topology is continued into the main entrance and folds up and into the building, creating a diagonal atrium that cuts through the seven levels; programs are distributed on both sides of the atrium.

oppposite above, left to right Axonometric studies showing floor and column structures, skin system and total envelope and structure.
opposite below, left and right Plans and model showing one effect of the double skin.

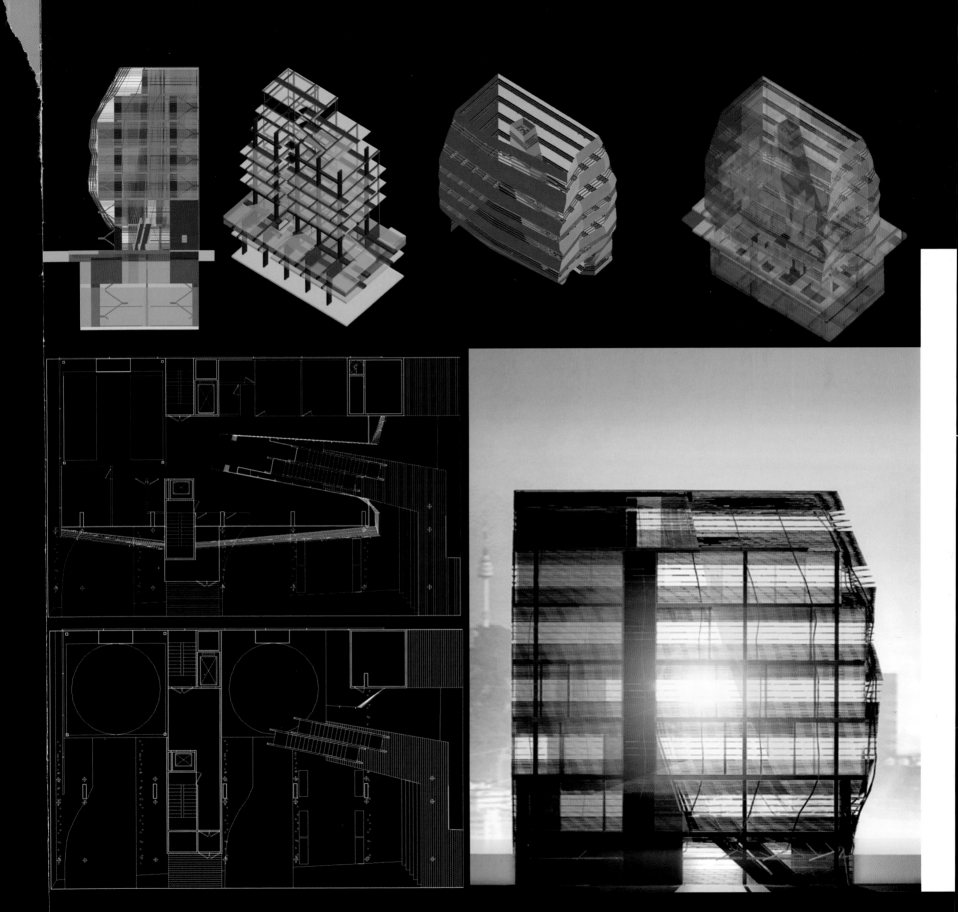

Synthetic Landscape
This research project (1995-98) is a
collaboration with the engineering firm,
Komtek AS, Devold AMT AS and OCEAN NO.
Anticipating an increased use of polymer
composites, the team was asked to explore how
complex geometries realized with new material
systems could be used in the heterogeneous
conditions of contemporary cities.

below and **opposite** The topological
renderings shows information mapped and
woven in three-dimensional colour that relates
to a given site, form, structure and program.

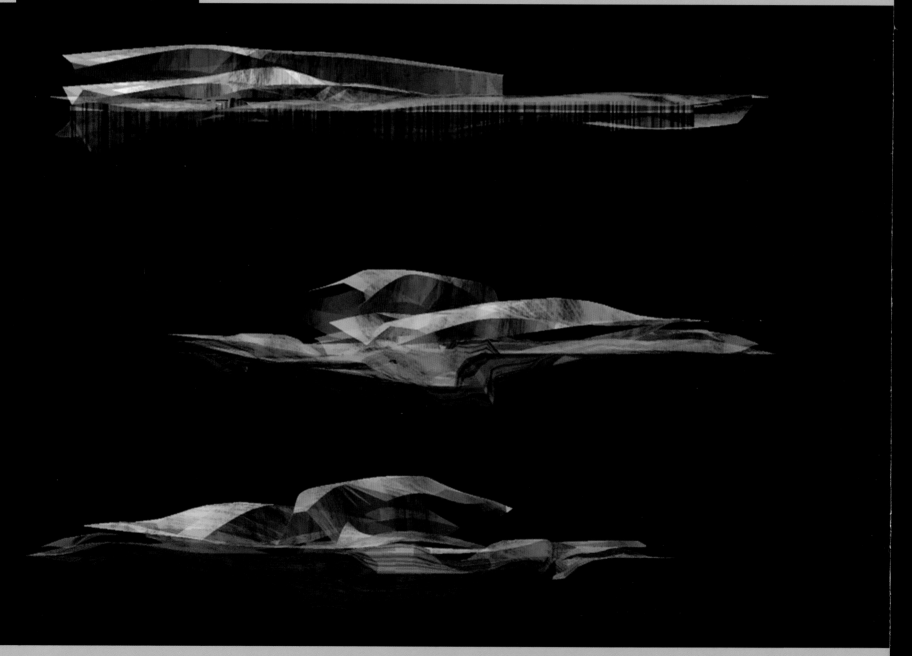

Europan 5

To promote Jeumont, France, as the region's commercial, cultural and infrastructural hub, stimulated by a projected TGV link for the town station, the project (1998) called for a multimodal transportation network that incorporated cars, trains, boats, bicycles and pedestrians, set within a preserved landscape ribbon along the River Sambre. Collaborating with engineers Ove Arup and Partners, OCEAN UK/US proposed an innovative form of urban connectivity. Exploiting disused land owned by the national railways, the architects configured the space as interlocking "strings" of distinct but mixable program zones in which three tectonic principles are injected. First, artificial ground surfaces negotiate the horizontal ground plane with pliant, parallel topographies. Second, looped skin structures (continuous helical bands in three dimensions) will occupy air-rights locations along and over the railway line; residential development will occupy the main massing of the helical core housing structures. Third, a transfer building will connect the motorway and retain the designated air-rights over the site; the mixed-use structure can adapt to variable economic conditions over the next twenty years. Overall, the organization creates cultural and recreational hubs that allow easy connections to the components of the multimodal transportation network.

right Study models of the looped, helical core housing structures.
opposite, left and middle Phase plan studies of the territorial zoning configured as a series of interlocking "strings."
opposite, right A sequence showing aerial views of the zoning as it develops over time.

urban bits together can act to smooth the fragmented roughness of the urban field, thus replacing the traditional figurations of figure-ground and object-background with notions of high and low resolution, definition and volume.

According to OCEAN, digital technology has shifted architecture from the production of singular, simplistic and fixed objects towards potentially dynamic and complex surfaces that are energized by their organizational syntax. At the same time, computerized production has created technical and material conditions that demand the development of new strategic operations in the design process. While the traditional mechanical techniques of reproduction translate simple graphical indexes (plans, sections, details) into basic material hierarchies, OCEAN has embraced digital production in order to advance complex material organizations that cannot be reduced to elementary notation.

For example, "Chamberworks," a project undertaken by OCEAN NO at the Ram Gallery in Oslo, was considered as a microcosm of the ever-changing intensities and flows of contemporary urban space, a space within which a changing set of animated interactions between the visitor, the gallery's spatial frame and the material, structural and ambient systems of the installation would come into play. The design itself resulted from a digital process in which "weighted particles" were shot into a virtual model of the gallery. The particles, which might describe lines of human movement within the gallery, were then animated. The resulting sequence of changing densities and formations was analyzed, and diagrams for the construction, lighting scheme and circulation patterns were deduced. From this underlay the design process went through stages of physical and virtual modeling, from which a final model was digitized, adjusted for structural and material soundness and construction drawings produced. The actual installation was built with steel rods and tubes welded together to form continuous linear arcs and spines.

The design for a stadium in Töölö, Finland, by OCEAN UV manifests the potential of digital production techniques at the urban scale. The designers started with the digital manipulation of a two-dimensional graphic based on local site conditions. This mapping was in turn trans-

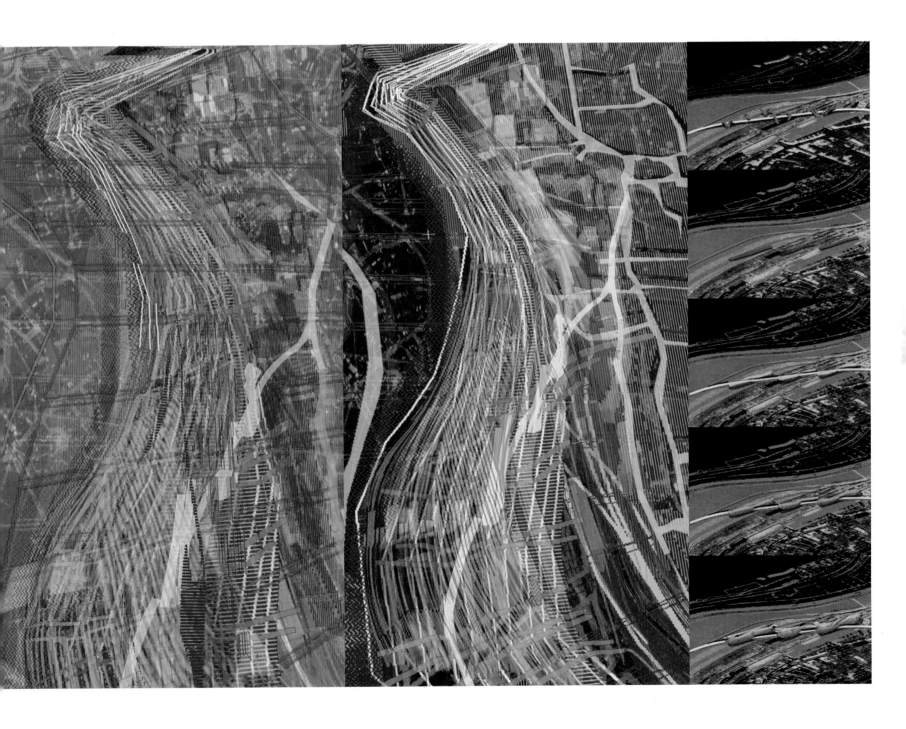

Europan 5

below Aerial study of the housing structures, transfer building and artificial ground topographies over the former railway yards.
opposite, above left Internal section perspective of the housing structures. The buildings' enclosures are defined by two skin systems: an exterior triangulated curtain wall follows the form of the pliant slab structure and a taut secondary surface relays high-loading stresses to the ground. The unbound skin is offset inward from the slab edges to create balcony spaces and outward to create controlled interior landscapes. The intersections of the slabs cause overlapping mezzanine spaces and double-height spaces.
opposite, above right and **below** Views of the helical core housing structures and transfer building.

posed into a three-dimensional construct, which through modeling was eventually used as the basis for the formal and structural design. The final result bridges the existing urban fabric (a natural park) and the stadium through the installation of a three-level continuous ringed surface.

By inventing adaptive techniques of site analysis and elaboration based on the vectorial logic of digitized surfaces, OCEAN have developed a self-organized urbanism of emergent properties and mobile influxes. Rather than resorting to the usual piecemeal systems of urban assemblage and bricolage, OCEAN mines the potential of a volatile and pliant urbanism and architecture for solutions. All projects, from furniture to landscape design, are understood as extensions of urban topologies and thus equally germane undertakings in relation to architecture. What springs from the group's multiple and inventive processes of design is therefore the creation of metropolitan ambiences and timbres that broaden the discussion on cities to include new technologies, greater complexities and other discourses.

Openly acknowledging the inclusiveness of a complex and heterogeneous organization that incorporates a wide spectrum of agents and intermediaries, OCEAN contends that the group's name is neither an acronym nor a metaphor. If the traditional notion of the "master architect" studio constructs hierarchies of working teams and orders to construct a unity of vision, OCEAN's methodology bravely seeks diversity and variability. Across the web of its nodes, this approach engenders and fosters a transgressive process of interaction and communication that is continually evolving according to an unanchored experimental logic. Given the social, political and moral complexities of our heterogeneous world, such an accomplishment is to be celebrated.

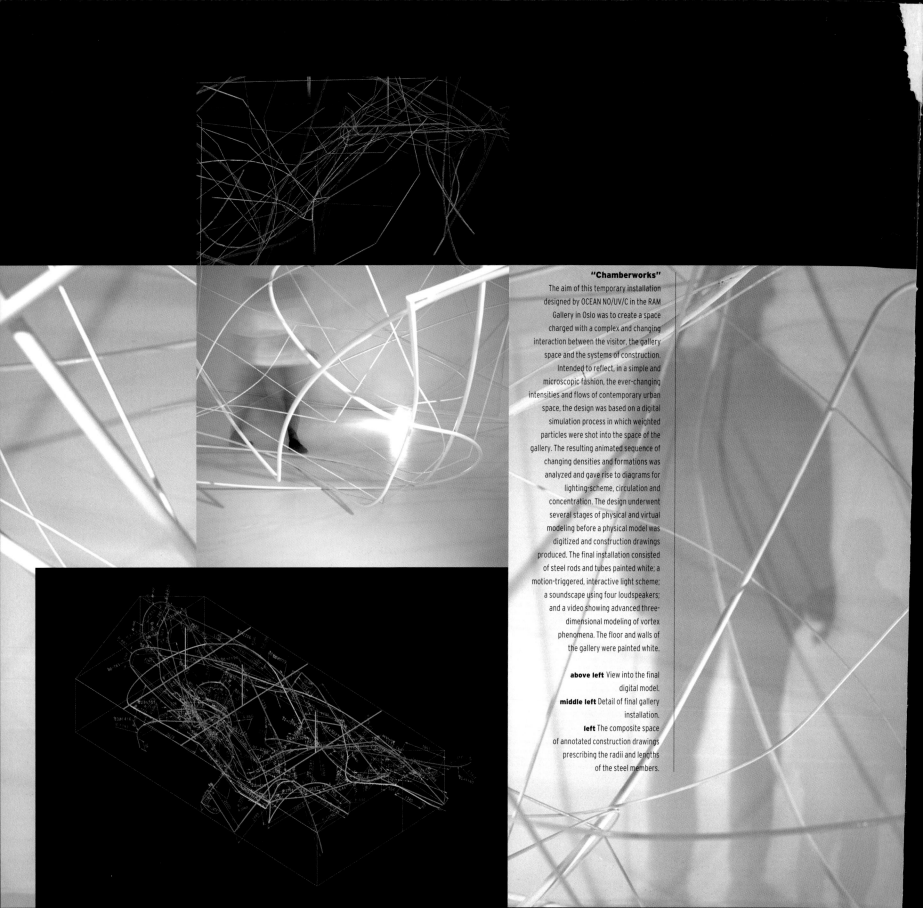

"Chamberworks"

The aim of this temporary installation designed by OCEAN NO/UV/C in the RAM Gallery in Oslo was to create a space charged with a complex and changing interaction between the visitor, the gallery space and the systems of construction. Intended to reflect, in a simple and microscopic fashion, the ever-changing intensities and flows of contemporary urban space, the design was based on a digital simulation process in which weighted particles were shot into the space of the gallery. The resulting animated sequence of changing densities and formations was analyzed and gave rise to diagrams for lighting-scheme, circulation and concentration. The design underwent several stages of physical and virtual modeling before a physical model was digitized and construction drawings produced. The final installation consisted of steel rods and tubes painted white; a motion-triggered, interactive light scheme; a soundscape using four loudspeakers; and a video showing advanced three-dimensional modeling of vortex phenomena. The floor and walls of the gallery were painted white.

above left View into the final digital model.
middle left Detail of final gallery installation.
left The composite space of annotated construction drawings prescribing the radii and lengths of the steel members.

```
        3.8
        3.6      7.2  5.2   2.8      3.0  4.7
        3.0   7.0 7.1  5.2  3.6      3.6  4.5   3.7
   0.5  2.8  6.2 6.2  5.1  5.3   2.5 4.8  4.4  3.9  2.0
   1.0  2.5  3.7 3.8  5.8  5.2  4.9 4.4  4.6  4.0  2.3  2.6  5.2  7.5
 1.6 1.5  2.3  2.6 2.7  5.0  5.1  4.8 4.2  4.7  3.6  2.1  2.5  5.2  7.5  9.0
2.4 1.5  2.2  2.1 2.3  2.4  4.8  4.9 4.7  4.1  4.2  3.1  1.6  2.6  5.2  7.4  9.0
3.8 4.0 3.8 2.6 1.8  2.1  2.5  3.6  4.3 4.6  4.0  3.5  2.4  1.2  2.6  5.0  7.2  8.5 10.0
4.8 4.8 4.3 3.5 2.0 1.8 2.1 2.7 5.0 5.6 6.0 5.4 2.7 0.0 0.7 2.6 4.5 6.8 8.2 10.5
4.2 4.7 3.9 2.4 2.0 2.0 2.4 2.8 6.8 7.9 8.3 5.8 1.6 0.6 0.8 2.2 4.0 6.5 7.5
4.0 4.3 3.0 1.6 1.5 1.9 2.5 3.3 7.5 8.0 10.1 7.5 2.2 1.3 1.1 2.1 3.2 6.3 6.5
3.9 2.5 2.6 1.0 0.0 1.3 2.4 3.6 8.3 9.1 12.0 7.8 3.1 2.4 1.4 1.7 1.9 5.9 5.6
    2.3 2.1 1.5 0.5 1.5 2.4 3.6 9.0 11.2    8.0 3.5 2.5 1.7 2.0 1.7 5.0 4.5
    2.2 2.0 4.2 0.8 1.7 2.3 2.4 11.0        7.7 4.1 2.5 2.0 2.3 1.9 2.6 3.4
        1.8 4.5 1.2 1.9 2.2 2.3             7.2 3.8 3.1 2.8 2.6 2.0 2.5
        1.7 4.7 1.5 2.0 2.2                 7.0 3.5 3.2 2.5 2.7 2.2 3.2
        5.6 4.9 1.8 2.2 2.3         5.0 4.0 3.3 2.5 2.4 2.8 2.6
        6.5 3.8 2.1                 2.5 1.0 0.5 0.0 0.7 1.5
        6.4
```

Extraterrains

Conceived by OCEAN UV as urban surfaces programmatically decoded into the scale of furniture, these furniture-objects cannot be autonomous entities, as they are integrated into a complex web of generic systems of several scales. Neither objects, idealized types nor static territories, Extraterrains do not inherit any prescriptive tradition of use, their momentary usefulness determined only by real-time occupation and by interactions between users. Made of vacuum-formed sheets of ABS-plastic, the objects are reinforced by a 5-millimeter layer of high-density polyurethane. Acrylics or fiberglass are also viable materials for lamination.

above Numerical datascape provides a map of the Extraterrain's surface deformations.
above right and **right** Detail of the integrated topographic surface-structure and an Extraterrain in situ.

UN STUDIO
BEN VAN BERKEL

Interpreting the patterns – "mobile forces" – woven
into the urban intensities and spaces of information flow,
UN Studio seeks to create an architecture in which the
deficiencies, differences and deformations of our cities are
desired, not rejected, for their transformative potential.

AND CAROLINE BOS   AMSTERDAM

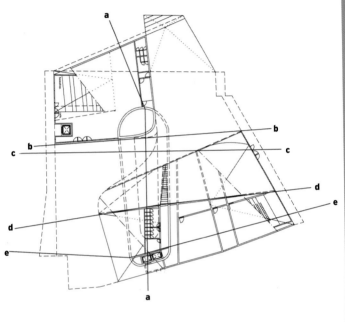

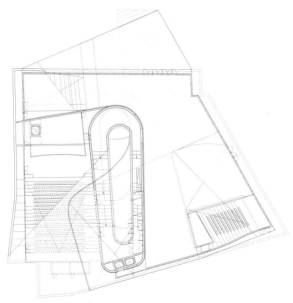

Ben van Berkel and Caroline Bos founded their practice in 1988. In the intervening years, they have designed and realized an exceptional number of projects, ranging from small houses to the Erasmus Bridge in Rotterdam. More recently, van Berkel and Bos have established a new firm, UN Studio, which will allow their practice to expand its interests, collaborations, connections and internal organization to undertake large-scale public projects. It is the studio's ongoing engagement with the question of public space – its forms and programmatic arrangement – in relation to the forces of globalization and cultural hybridization that drives their recent proposals. Their research has led them to detect new flows emerging in the urban domain, patterns woven into urban intensities and materializations and spaces of flow and transition that mirror the informational and commercial movement of international media. Working with these "mobile forces," van Berkel and Bos's term for these transformative energies, they seek to produce an architecture of inclusion and coherence without resorting to the totalizing phenomenon of the *Gesamtkunstwerk*.

In the context of the global continuum, van Berkel and Bos have conducted numerous researches into the technical, economic and political conditions that have resulted from the informational, technological and social reorganization of the last quarter-century. Attempting to instrumentalize and catalogue new social conditions, they have transformed abstract notions of time and movement into architectural and urban manifestations. Van Berkel and Bos write about "a policy of mobility":

> New computational techniques make it possible to lay bare the multiplicitous layering of experiences and activate this knowledge in new ways. When mapping movement patterns, the time-program relationship is not compartmentalized but is a reflection of synchronic continuous time. Infrastructural layers may be classified, calculated, and tested individually, then interwoven to achieve both effective flux and effective interaction. In this way temporal conditions are connected to programmatic themes in a simulation of the non-segmented manner in which time flows in a real situation.

Today urban space is subject to augmentations, inversions, evolutions and manipulations produced by the interactions of media communi-

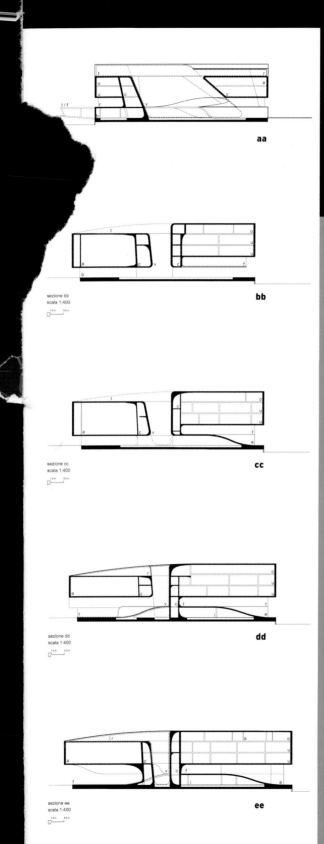

sezione bb
scala 1:400

sezione cc
scala 1:400

sezione dd
scala 1:400

sezione ee
scala 1:400

aa

bb

cc

dd

ee

University of Architecture Building, Venice (IUAV)

The university required an extension to its existing buildings, to be located on the Riva della Zattere and to house educational rooms, administrative offices, 500-seat auditorium, hall for conferences, exhibition space, library, restaurant and bar. One of the most important aspects of Venice's rich architectural history has been the radial orientation of Venetian palazzi toward the water, a principle that guided the placement of the new building toward the continuous quays surrounding the islands. Another feature borrowed from the organizational typology of the Venetian palazzo is the centrally oriented semipublic salon, which gives access to the private rooms. Diagrammatically, the linear form of the quayside is sucked in – rolling up like a long and gentle wave towards the ground-level entrance – then lifted straight up to form an elliptical pipe until it spills over into the roof volume at the top, thus becoming an indivisible part of the structural and infrastructural core. Arrayed around a diagonal open court that leads via a bridge to the university's headquarter on the next island, some public areas, like the café, exhibition hall and a bookshop, animate the public zone and make the IAUV accessible to passersby coming from the nearby cruise-ship terminal. As a result, there is a strong visual relationship between the existing and new buildings, and to the city that lies behind.

opposite, left Plans
opposite, right A sequence of abstracted diagrammatic plan formations shows the development of the elliptical pipe within the building envelope.
left Sections
right Transparent rotational series showing the elliptical pipe - or "mushroom" - in relation to the total structural mass.

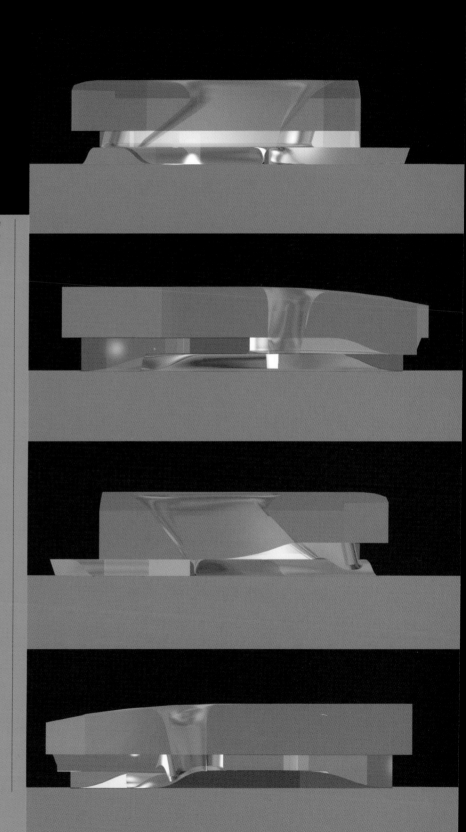

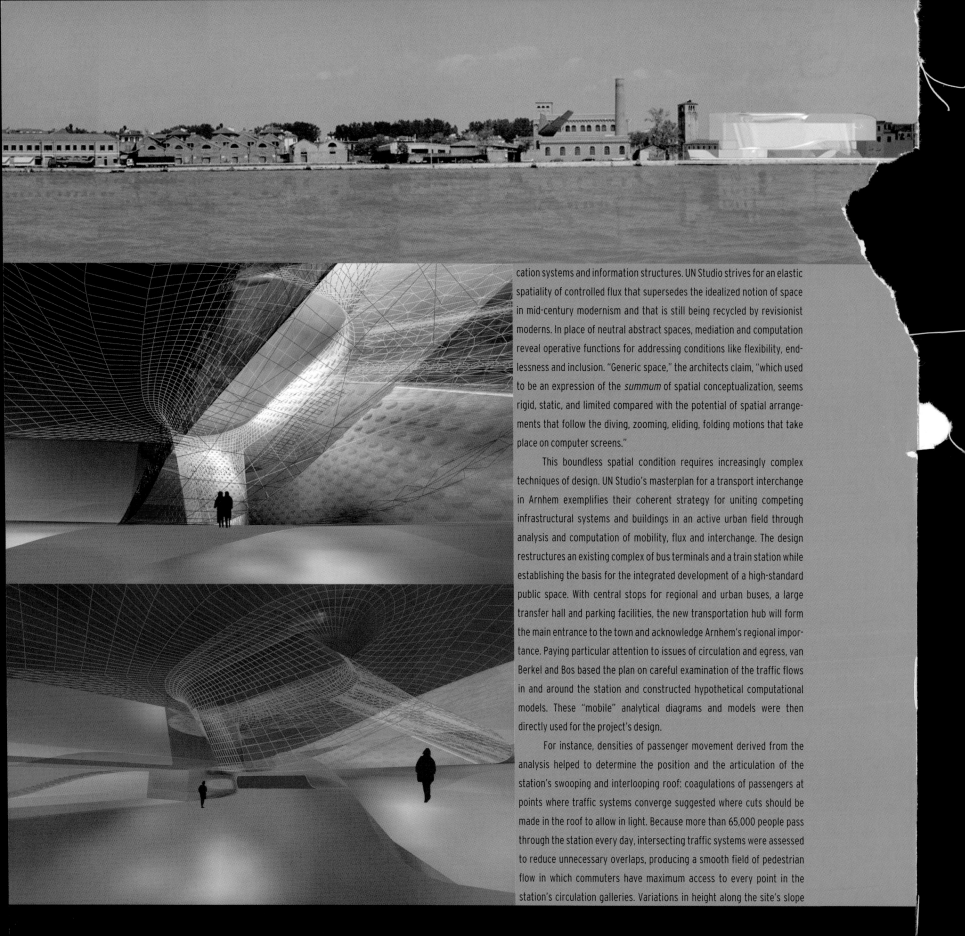

cation systems and information structures. UN Studio strives for an elastic spatiality of controlled flux that supersedes the idealized notion of space in mid-century modernism and that is still being recycled by revisionist moderns. In place of neutral abstract spaces, mediation and computation reveal operative functions for addressing conditions like flexibility, end-lessness and inclusion. "Generic space," the architects claim, "which used to be an expression of the *summum* of spatial conceptualization, seems rigid, static, and limited compared with the potential of spatial arrange-ments that follow the diving, zooming, eliding, folding motions that take place on computer screens."

This boundless spatial condition requires increasingly complex techniques of design. UN Studio's masterplan for a transport interchange in Arnhem exemplifies their coherent strategy for uniting competing infrastructural systems and buildings in an active urban field through analysis and computation of mobility, flux and interchange. The design restructures an existing complex of bus terminals and a train station while establishing the basis for the integrated development of a high-standard public space. With central stops for regional and urban buses, a large transfer hall and parking facilities, the new transportation hub will form the main entrance to the town and acknowledge Arnhem's regional impor-tance. Paying particular attention to issues of circulation and egress, van Berkel and Bos based the plan on careful examination of the traffic flows in and around the station and constructed hypothetical computational models. These "mobile" analytical diagrams and models were then directly used for the project's design.

For instance, densities of passenger movement derived from the analysis helped to determine the position and the articulation of the station's swooping and interlooping roof: coagulations of passengers at points where traffic systems converge suggested where cuts should be made in the roof to allow in light. Because more than 65,000 people pass through the station every day, intersecting traffic systems were assessed to reduce unnecessary overlaps, producing a smooth field of pedestrian flow in which commuters have maximum access to every point in the station's circulation galleries. Variations in height along the site's slope

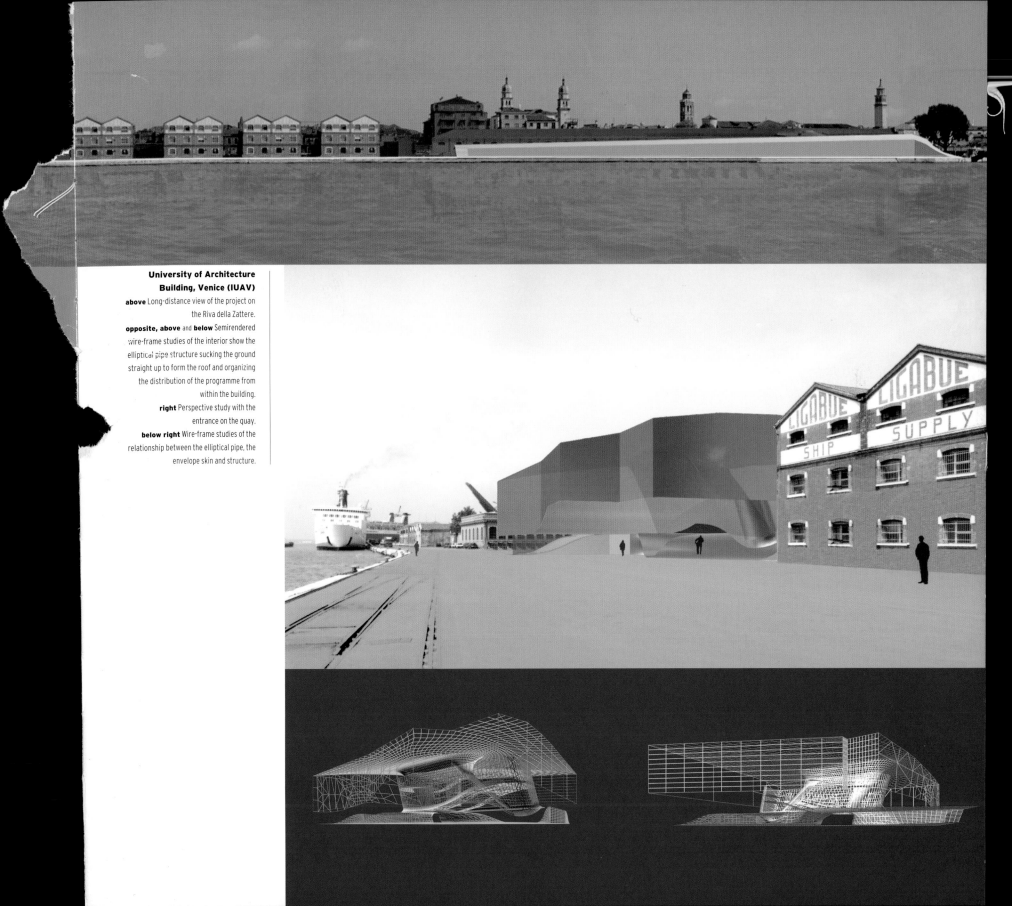

University of Architecture Building, Venice (IUAV)

above Long-distance view of the project on the Riva della Zattere.

opposite, above and **below** Semirendered wire-frame studies of the interior show the elliptical pipe structure sucking the ground straight up to form the roof and organizing the distribution of the programme from within the building.

right Perspective study with the entrance on the quay.

below right Wire-frame studies of the relationship between the elliptical pipe, the envelope skin and structure.

sezione aa
scala 1:400

Neutron Magnetic Resonance Facilities

Dedicated to research using Neutron Magnetic Resonance (NMR) for the Bijvoet University of Utrecht in The Netherlands, this laboratory building (1998) required facilities for eight spectrometers (high-frequency magnets), which all generate highly magnetic fields. Any irregularity close to the magnets can disturb tests results; equally, magnetic radiation can affect people, their computers, credit cards, pacemakers. The placement of magnets, therefore, determined the organization of the building's core, while the magnetic radii shaped the structure and surface, directed program and equipment and affected internal circulation.

opposite, above Conceptual and analytical diagrams.
opposite, below Sections
right Plans
below right Perspective view in the university context.

were exploited to create fluid walking routes (as opposed to distinct levels) and sight lines. Ease of orientation is enhanced by the penetration of natural light at key points, so the entrance area, for example, becomes an open interior court that optimizes access to trains, taxis, buses, bikes, parked cars, office spaces and the town center.

Diagrams, the architects contend, like computational analysis, can be used either as proliferating machines or as reductive mechanisms for the production of an architecture of inclusion. One such diagram, or ideogram, is the "manimal," the computer image of a lion, a snake and a human that have been morphed into something vaguely human, vaguely animal. This image

is the prime example of inclusiveness and intensive coherence as an effect ... in the manimal you don't see the seams anymore, nor can you retrieve information about the original component parts. There are not traces of the previous identities within the new image. In this respect the manimal relates to the non-existence of proper scale and proper identity, which is a structural effect already. Architecturally, the manimal could be read as an amalgamation of several different structures, which start to generate a new notion of scale, for which we don't yet even have a name. Your first impression of this diagram would be that it doesn't represent a building. We like diagrams which have organizational potential on several levels, making many readings possible.

The manimal is derived from a computational mediation technique (morphing) that differs significantly from other techniques of creative assembly, such as painting, collage or drawing. There is a clear difference in definition between diagram derived from morphing and a traditional graphic diagram used to clarify levels, and information type or organization. In the case of the manimal, the diagram is essentially a technical function (the morphing process), which has produced a visual image; that is, the traditional data diagram is inverted to become a form process. In this example, the diagram (image product) results from a process of computation (image process), and it is this seamless, decontextualizing recombination of discordant systems of information that makes the manimal such a seductive technique to van Berkel.

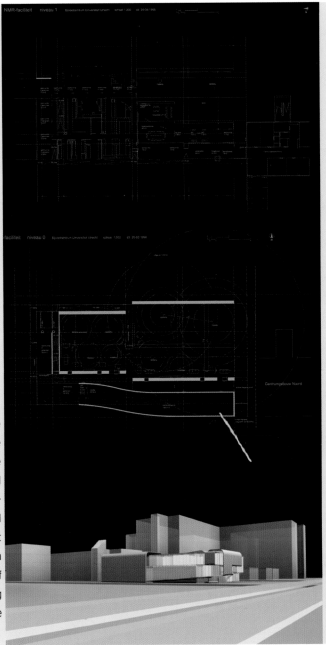

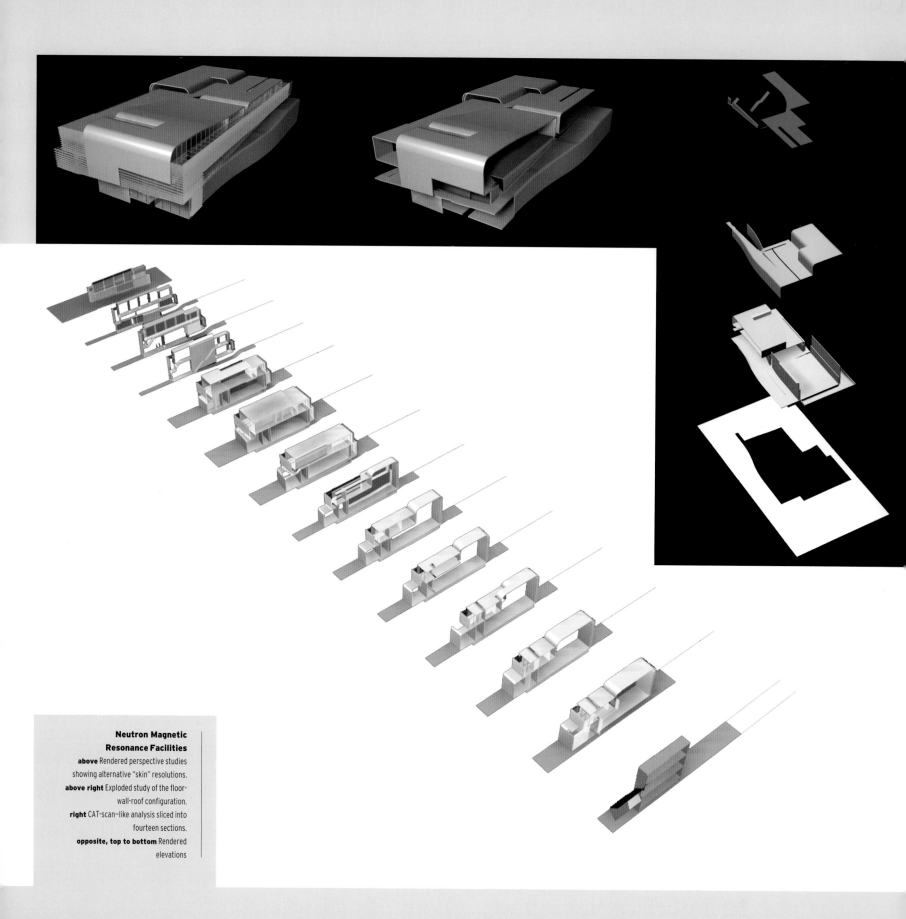

Neutron Magnetic Resonance Facilities
above Rendered perspective studies showing alternative "skin" resolutions.
above right Exploded study of the floor-wall-roof configuration.
right CAT-scan–like analysis sliced into fourteen sections.
opposite, top to bottom Rendered elevations

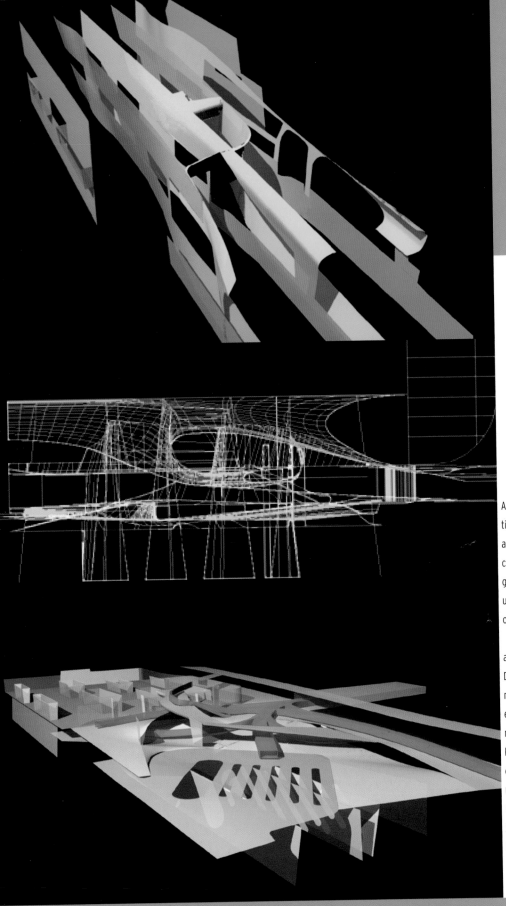

Because it sets off a process, it suggests a way of working that I can use. The manimal is important as an effect because it makes me wonder "How would something like this translate spatially?" and as a technique because it has been produced in a manner that is radically different from all pictorial technique that have been employed by artists before ... We talk about mediative space, for example, which in a sense represents a new type of public space. For me the manimal represents the social space. It would have been made by three or four designers, who all worked with different systems. I see an increasing tendency to investigate ways in which architecture might include more and more complexity in one comprehensive complexity. It looks more and more at organic systems; and the way in which, through the impact and effect of external forces, a generic model is differentiated. The model is perceived as a whole, but is at the same time very fragmented.

As an amalgamation of different organic compositions, systems of evolution and identity, the manimal could be a model for a new conception of architectural scale, structure and organization and suggests designing conditions of inclusion rather than singular building. Technical and programmatic organizations of competing and complex structures could be united through the integrating energy of computation and unfolded as open, "mobile" urban systems.

In the varied work of UN Studio we find a radical approach to architecture and the design of public structures and urban organizations. Dedicated to developing systems of analysis and production that will mobilize and materialize the dynamic economic, political and information energies that are reshaping our cities, they confidently reject the modernist obsession with stasis, monumentality and idealized urban life. Instead, they seek a "middle," a space in which the deficiencies, differences and deformations of our cities are desired, not rejected, for their mutational potential. The vitality of their work demonstrates a strong understanding of contemporary culture and technology, giving us hope for the invention of as-yet-unknown types of public spaces, conditions and humanity.

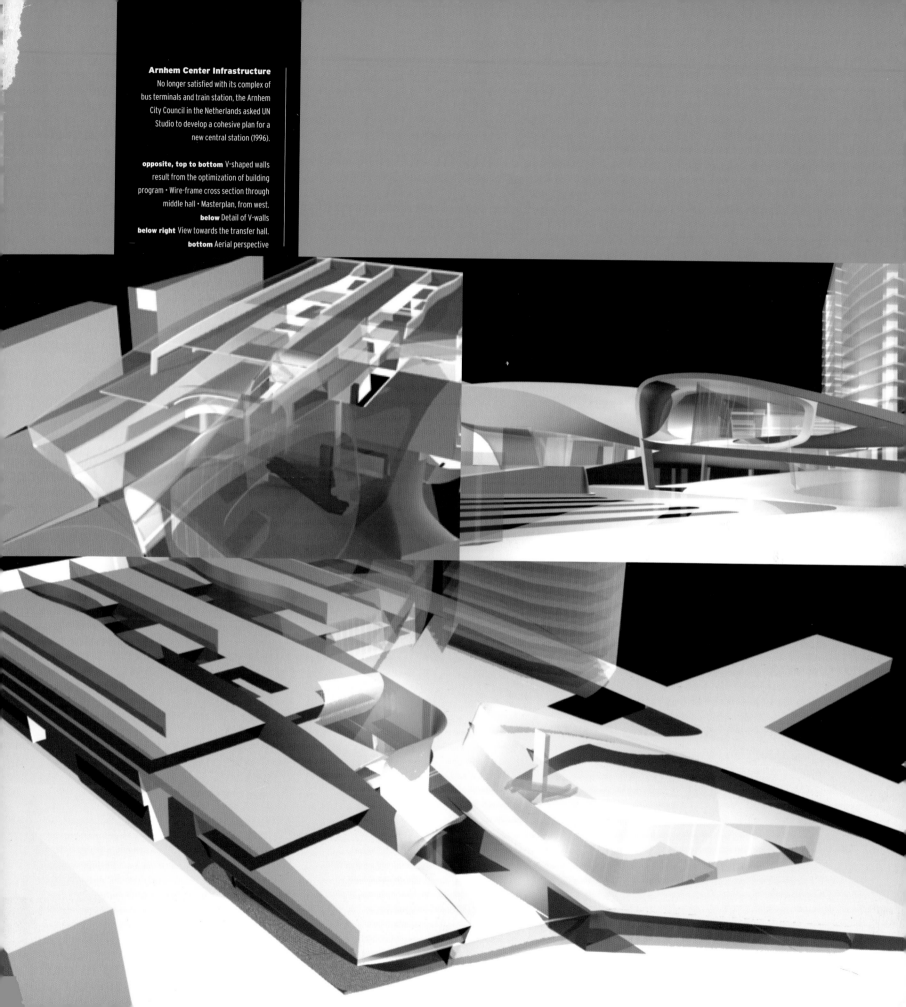

Arnhem Center Infrastructure
No longer satisfied with its complex of bus terminals and train station, the Arnhem City Council in the Netherlands asked UN Studio to develop a cohesive plan for a new central station (1996).

opposite, top to bottom V-shaped walls result from the optimization of building program · Wire-frame cross section through middle hall · Masterplan, from west.
below Detail of V-walls
below right View towards the transfer hall.
bottom Aerial perspective

Morphosis [pages 32-43]
2041 Colorado Avenue
Santa Monica, California 90404, U.S.A.
tel +1 310 453 2247/fax +1 310 829 3270
email morphosis@earthlink.net

Thom Mayne, AIA
education
Bachelor of Architecture, University of Southern California, Los Angeles
Master of Architecture, Graduate School of Design, Harvard University., Cambridge (MA)

recent projects

1996-99	Kartner Landes- und Hypothekenbank, Klagenfurt, Austria
1996-98	ASE Taipei Design Center, Taiwan
1998	Hippocampus (project)
1997-	Toronto Graduate Student Housing, Canada
1996-98	Diamond Ranch High School, Pomona, California
1996-98	Long Beach International Elementary School, Long Beach, California

select bibliography

1999   *GA International*, n. 99; "Progressive Architecture Awards" (Toronto), Apr;
        Architectural Record ("Record Houses"), Apr; "Sun Tower," C3 Publications,
        Feb; *Los Angeles Times*, Calendar Section, Jan 17;

1998   *Pacific Edge: Contemporary Architecture on the Pacific Rim*, P. Zellner (ed.)
        (Thames & Hudson, London; Rizzoli, New York); *Ediciones ARQ; Canadian
        Architect* (University of Toronto), Dec; *Projeto*, Oct; *Casabella*, Oct;
        Arredamento Mimarlik, Sep; *Interior Design*, Aug; *The New York Times*, Sect. B8,
        Aug 6; *GA Houses*, n. 55; *Dream Houses* (Arco Editorial, Spain); *Morphosis: The
        Crawford House* (Rizzoli, New York); *Diseño Interior*, Jan; *Architecture*, Jan; *GA
        Document Extra* (monograph), n. 9, Jan

1997   *Poar*, Dec; *American Houses Now: Contemporary Architectural Directions*,
        Boles, Daralice and Doubilet (Rizzoli, New York; Thames & Hudson, London);
        Interiors, Nov; *Metropolitan Home*, Nov; *Chinese Architect*, Oct; *Architecture*,
        Aug; *Dialogue*, Oct; *Architecture*, Aug; *Los Angeles Times*, Section M (Oct 5);
        Dialogue, May; *LA Architect*, Jun/Jul; *GA Houses*, n. 53, Jun; *Interior Design*, Mar;
        "Progressive Architecture Awards," *Architecture*, Jan; *Architektur and Bau
        Forum*, n. 86, Jan/Feb; *Architecture after Modernism*, D. Ghirardo (Thames &
        Hudson, London/New York)

project credits
Kartner Landes- und Hypothekenbank [pages 34-37]
Klagenfurt, Austria
Principal: Thom Mayne
Project architect: John Enright
Project team: Dave Grant, Martin Krammer, Fabian Kremkus, Ung Joo Scott Lee, Silvia
        Kuhle, David Plotkin, David Rindlaub, Robyn Sambo, Stephen Slaughter,
        Brandon Welling
Project assistants: Michael Folwell, Eugene Lee, Tomas Lenzen, Julianna Morais, Ulrike
        Nemeth, Brian Parish, Ivan Redi, Janice Shimizu, Bart Tucker, Ingo Waegner,
        Marion Wicher, Oliver Winkler

Hippocampus [pages 38-39]
Principal: Thom Mayne
Project assistants: Robyn Sambo, Ung Joo Scott Lee

A.S.E. Design Center [pages 40-43]
Taipei, Taiwan
Principal: Thom Mayne
Project designer: Patrick J. Tighe
Project assistants: Towan Kim, Richard Koschitz, Ung Joo Scott Lee, Michael O'Bryan,
        Robyn Sambo, Stephen Slaughter, Alan Tsaur, Bart Tucker, Oliver Winkler
Design consultant: CY Lee

text references
All quotations were taken from an interview with Thom Mayne in *GA Document
Extra*, n. 9 (Tokyo, Jan 1998).